FINGERPRINTS IN THE OPTICAL AND TRANSPORT PROPERTIES OF QUANTUM DOTS

Edited by **Ameenah Al-Ahmadi**

Fingerprints in the Optical and Transport Properties of Quantum Dots
Edited by Ameenah Al-Ahmadi

Published by InTech

Janeza Trdine 9, 51000 Rijeka, Croatia

Publishing Process Manager Molly Kaliman

Technical Editor Teodora Smiljanic

Cover Designer InTech Design Team

Additional hard copies can be obtained from orders@intechopen.com

Fingerprints in the Optical and Transport Properties of Quantum Dots,
Edited by Ameenah Al-Ahmadi
p. cm.
ISBN 978-953-51-0648-7

For EU product safety concerns:
IN TECH d.o.o., Prolaz Marije Krucifikse Kozulić 3, 51000 Rijeka, Croatia,
info@intechopen.com or visit our website at intechopen.com.

Meet the editor

Dr Ameenah N. Al-Ahmadi is currently a Professor of physics at Umm Al-Qura University in Makkah, Saudi Arabia. She received the M. Sc. degree in Physics in 2004 and awarded the Ph. D degree in Physics on June, 2006 from Ohio University at Athens, Ohio, USA. Her research interests is in the field of optical properties of nano scale system, including quantum dots, nanowires, nanotubes. She has contributed many papers in different International Journals in her area of research. She has two papers selected by Virtual Journal of Nanoscale Science & Technology in 2004 and 2006. She was the director of enrichment summer program for Giftedness & Creativity_ Sense the word_ 2010. She has been Vice-Dean of Applied science Faculty for the Academic Development and Community Services and graduate student advisor of Faculty of Applied science since 2009 and Vice-Dean of Quality and Academic Development at Umm Al-Qura University.

Contents

Preface

Quantum dots are one of the most promising types of nanoparticles, which are exceptionally useful for variety of new applications because of their unique properties.

This is a collaborative book sharing and providing the academic community with a base text that could serve as a reference in research by presenting up-to-date research work on the field of quantum dot systems. We are most grateful to all authors of the chapters for highlighting the important issue of the potential applications of quantum dot system with a high quality work of their research. We are especially thankful for the cooperation and support from InTech team who helped in publishing this book, in particular the publishing process manager of this book, Ms. Molly Kaliman for her hard effort and patience during the process of publishing the book.

"To my son Azuz"

Ameenah N. Al-Ahmadi, PhD
Associate Professor of Physics
Faculty of Apllied Science, Umm Al-Qura University,
KSA

Section 1

Optical Properties of Quantum Dot Systems

1

InAs Quantum Dots of Engineered Height for Fabrication of Broadband Superluminescent Diodes

S. Haffouz and P.J. Barrios
Institute for Microstructural Sciences, National Research Council of Canada, Ottawa, Ontario, Canada

1. Introduction

Superluminescent diodes (SLDs) are of great interest as optical sources for various field applications like fibre-optic gyroscopes (Culter et al, 1980), optical time-domain reflectometry (Takada et al, 1987), sensing systems (Burns et al, 1983) (such as Faraday-effect electric current sensors and distributed Bragg-grating sensor systems) and short and medium distance optical communication systems (Friebele & Kersey, 1994). One of the most attractive applications of SLDs has emerged after the successful demonstration of the optical coherence tomography (OCT) technique, and identification of its advantages compared to other imaging techniques in medical research and clinical practices. OCT is a real time and non-invasive imaging technique that uses low-coherence light to generate resolution down to the sub-micron-level, two- or three-dimensional cross-sectional images of materials and biological tissues. The earliest version of the OCT imaging technique was demonstrated in 1991 by Huang and co-workers (Huang et al, 1991), by probing the human retina *ex vivo.* Imaging was performed with 15μm axial resolution in tissue using a light source with a central wavelength of 830nm. Two years later, *in vivo* retinal images were reported independently by Fercher et al. (Fercher et al, 1993) and Swanson et al (Swanson et al, 1993). Although 800nm OCT systems can resolve all major microstructural layers of tissues, image quality can be severally degraded by light scattering phenomena. In low-coherence interferometry, the axial resolution is given by the width of the field autocorrelation function, which is inversely proportional to the bandwidth of the light source. In other words, light sources with broadband spectra are required to achieve high axial resolution. Although at longer wavelengths the bandwidth requirement increases, there is a significant advantage in using light sources of longer central wavelengths for which the light scattering is significantly reduced.

In recent few years, broadband light sources around 1μm have received considerable attention for their use in medical imaging technologies. It is due to the optimal compromise between water absorption and human tissue scattering that the 1000-1100 nm wavelength range has been proposed, and demonstrated, to be more suitable for OCT applications as compared to those that use a light source with a central wavelength of 800nm (Pavazay et al,

2003; Pavazay et al, 2007). There are a myriad of choices in selecting such OCT light sources i) femtosecond or fiber lasers that are dispersed to produce super-continuum light and swept source lasers (Hartl et al, 2001; Wang et al, 2003), ii) thermal sources, and iii) superluminescent diodes (Sun et al, 1999; Liu et al, 2005; Lv et al, 2008; Haffouz et al, 2010). Although the reported OCT tomograms with the highest axial resolution (1.8μm) were so far achieved in research laboratories with a photonic crystal fibre based source (Wang et al, 2003), superluminescent diodes are considerably lower in cost and complexity as well as being smaller in size, which makes them more attractive for mass production. Superluminescent diodes utilizing quantum-dots (QDs) in the active region are considered to be excellent candidates as light source for an OCT systems. The naturally wide dimensional fluctuations of the self-assembled quantum dots, grown by the Stranski-Krastanow mode, are very beneficial for broadening the gain spectra which enhances the spectral width of the SLDs. On the other hand, the three-dimensional carrier confinement provided by the dots' shape results in high radiative efficiency required for the OCT applications.

In this chapter the main governing factors to demonstrate ultrahigh-resolution OCT-based imaging tomographs will be reviewed in the second section. Research advances in the growth processes for engineering the gain spectrum of the quantum dots-based superluminescent diodes will be summarized in the third section of this chapter. Our approach for engineering the bandwidth of multiple stacks of InAs/GaAs QDs will be presented in the fourth section and demonstration of an ultra wide broadband InAs/GaAs quantum-dot superluminescent diodes (QD-SLDs) will be then reported in the last section of this chapter. Our approach is based on the use of SLDs where the broad spectrum is obtained by a combination of slightly shifted amplified spontaneous emission (ASE) spectra of few layers of dots of different heights. Spectral shaping and bandwidth optimization have been achieved and resulted in 3dB-bandwidth as high as ~190nm at central wavelength of 1020nm. An axial resolution of 2.4μm is calculated from our QD-SLDs.

2. Superluminescent diodes for ultrahigh-resolution optical coherence tomography (UHR-OCT)

Since its invention in the early 1990s (Huang et al, 1991), OCT enables non-invasive optical biopsy. OCT is a technique that provides *in-situ* imaging of biological tissue with a resolution approaching that of histology but without the need to excise and process specimens. OCT has had the most clinical impact in ophthalmology, where it provides structural and quantitative information that can not be obtained by any other modality. Cross-sectional images are generated by measuring the magnitude and echo time delay of backscattered light using the low-coherence interferometry technique. The earliest versions of OCT have provided images with an axial resolution of 10-15μm. OCT has then evolved very quickly, with two-dimensional (2D) and three-dimensional (3D) microstructural images of considerably improved axial resolution being reported (Drexler et al, 1999). These ultrahigh-resolution OCT systems (UHR-OCT) enable superior visualization of tissue microstructure, including all intraretinal layers in ophthalmic applications as well as cellular resolution OCT imaging in nontransparent tissues. The performance of an OCT system is mainly determined by its longitudinal (axial) resolution, transverse resolution, dynamic range (sensitivity) and data acquisition speed. Other decisive factors like depth penetration into the investigated tissue (governed by scattering, water absorption) and image contrast need to be carefully addressed. In addition, for field application, compactness, stability, and overall cost of the OCT system should be considered.

2.1 Factors governing OCT imaging performance

In this section we will review the key parameters that are directly or closely related to the light source used in the OCT technique. Other limiting factors, related to other optical, electronic and/or mechanical components can affect the resolution in OCT system when not properly addressed. For more details regarding OCT technology and applications, please refer to the book edited by Drexler and Fujimoto (Drexler & Fujimoto, 2008).

2.1.1 Transverse and axial resolution

As in conventional microscopy, the transverse resolution and the depth of focus are determined by the focused transverse sport size, defined as the $1/e^2$ beam waist of a Gaussian beam. Assuming Gaussian rays and only taking into account Gaussian optics, the transverse resolution can be defined by:

$$\Delta x = \frac{4\lambda}{\pi}\frac{f}{d} \tag{1}$$

where f is the focal length of the lens, d is the spot size of the objective lens and λ is the central wavelength of the light source. Finer transverse resolution can be achieved by increasing the numerical aperture that focuses the beam to a small spot size. At the same time, the transverse resolution is also related to the depth of the field or the confocal parameter b, which is $2z_R$, or two times the Rayleigh range:

$$b = 2z_R = \frac{\pi \Delta x^2}{\lambda} \tag{2}$$

Therefore, increasing the transverse resolution produces a decrease in the depth of the field, similar to that observed in conventional microscopy. Given the fact that the improvement of the transverse resolution involves a trade-off in depth of field, OCT imaging is typically performed with low numerical aperture focusing to have a large depth of field. To date, the majority of early studies have rather focused on improving the axial resolution.

Contrary to standard microscopy, the axial image resolution in OCT is independent of focusing conditions. In low-coherence interferometry, the axial resolution is given by the width of the field autocorrelation function, which is inversely proportional to the bandwidth of the light source. For a Gaussian spectrum, the axial (lateral) resolution is given by:

$$\Delta z = \frac{2Ln(2)}{\pi}\frac{\lambda^2}{\Delta\lambda} \tag{3}$$

where Δz is the full-width-at-half-maximum (FWHM) of the autocorrelation function, , and $\Delta\lambda$ is the FWHM of the power spectrum.

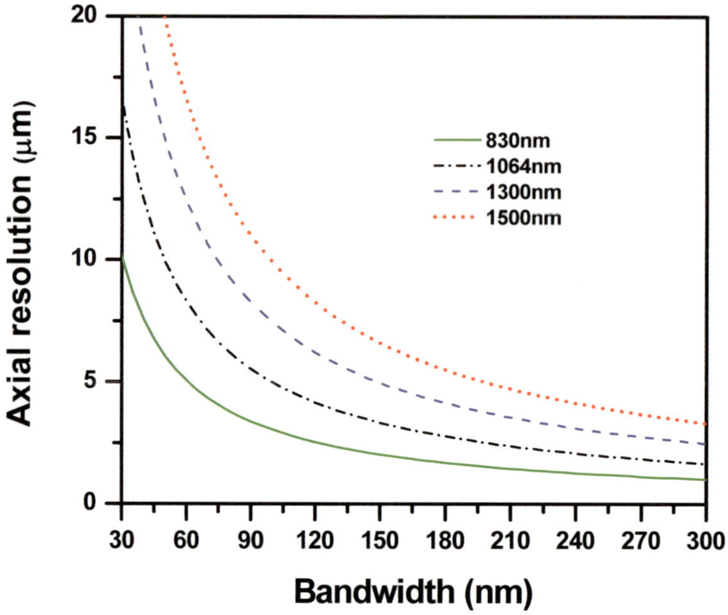

Fig. 1. Axial resolution versus bandwidth of light sources for central wavelengths of 830, 1064, 1300 and 1500nm.

Since the axial resolution is inversely proportional to the bandwidth of the light source, broadband light sources are required to achieve high axial resolution. For a given bandwidth, improving the axial OCT resolution can be also achieved by reducing the central wavelength of the light source (c.f. Figure 1). It should also be noticed that to achieve a given axial resolution the bandwidth requirement is increased at longer wavelengths. For example, to achieve an axial resolution of 5μm, the bandwidth required is only 50nm at central wavelength of 830nm, and three times higher when a light source of central wavelength of 1300nm is chosen.

2.1.2 Imaging speed-sensitivity in OCT

Detection sensitivity (detectable reflectivity) has a significant impact on the imaging speed capabilities of an OCT system. As the scan speed increases, the detection bandwidth should be increased proportionally, and therefore the sensitivity drops. The sensitivity of state-of-the-art time-domain OCT systems that operate at relatively low imaging speed (~2kHz A-line rate), ranges between -105 and -110dB. Increasing the optical power of the light source should in principle improve the sensitivity; however, the available sources and maximum permissible exposure levels of tissue represent significant practical limitations. The potential alternative technique for high-imaging speed is the use of Fourier/spectral domain detection (SD-OCT) or Fourier/swept source domain detection (SS-OCT) also known as optical frequency domain imaging (OFDI). The first approach, SD-OCT, uses an interferometer with a low-coherence light source (superluminescent diodes) and measures the interference spectrum using a spectrometer and a high-speed, line scan camera. The second approach, SS-OCT, uses an interferometer with a narrow-bandwidth, frequency-swept light source (swept laser sources) and detectors, which measure the interference output as a function of time. Fourier domain detection has a higher sensitivity as compared to time domain detection, since Fourier domain detection essentially measures all of the echoes of light simultaneously, improving sensitivity by a factor of 50-100 times (enabling a significant increase in the imaging speeds).

2.1.3 Image contrast and penetration depth in OCT

Tissue scattering and absorption are the main limiting factors for image contrast and penetration depth in OCT technology. Indeed, OCT penetration depth is significantly affected by light scattering within biological tissue, which scales as $1/\lambda^k$, where the coefficient k is dependent on the size, shape, and relative refractive index of the scattering particles. The difference in tissue scattering and absorption provides structural contrast for OCT. Since scattering depends strongly on wavelength and decreases for longer wavelengths, significantly larger image penetration depth can be achieved with light centered at 1300nm rather than 800nm. However, above 1300nm the water absorption becomes a problem. So far, the majority of clinical ophthalmic OCT studies have been performed in the 800 nm wavelength region. Excellent contrast, especially when sufficient axial resolution is accomplished, enables visualization of all major intraretinal layers, but only limited penetration beyond the retina. This limitation is mainly due to significant scattering and absorption phenomena.

Water is the most abundant chemical substance in the human body, accounting for up to 90% of most soft tissues. The most commonly used wavelength window of low water absorption (μ_a<0.1cm^{-1}) for OCT imaging is lying in the 200-900 nm range. Above 900 nm the absorption coefficient increases fairly rapidly to reach μ_a~ 0.5cm^{-1} at ~970 nm, drops back to ~0.13cm^{-1} at 1064nm, and then continues to increase at longer wavelengths into the mid-infrared. The region of low absorption around 1060nm acts as a 'window' of transparency, allowing near infrared spectroscopic measurements through several centimeters of tissue to be made. For this reason, OCT imaging at 1060 nm can achieve deeper tissue penetration into structures beneath the retinal pigment epithelium, as well as better delineation of choroidal structure.

2.2 Light source for ultrahigh resolution OCT

The light source is the key technological parameter of an OCT system. The performance characteristic of the light source, such as central wavelength, bandwidth, output power, spectral shape, and stability will directly affect the OCT image resolution. For this reason, a proper choice of the light source for optimized performance OCT system is imperative. In the recent years, there has been considerable interest in the use of broadband light sources around 1064nm for use in ophthalmic OCT applications. It is due to the optimal compromise between water absorption and human tissue scattering that the 1064nm wavelength '*window*' has been proposed, and demonstrated, to be more suitable for OCT applications as compared to those that use a light source with a central wavelength of 800nm (Povazay et al, 2007). There are a myriad of choices in selecting such OCT light sources i) femtosecond or fiber lasers that are dispersed to produce super-continuum light and swept source lasers, and ii) superluminescent diodes. Highly non-linear air-silica microstructure fibers and photonic crystal fibers (PCFs) can generate an extremely broadband continuous light spectrum from the visible to the near infrared by use of low-energy femtosecond pulses (Wang et al, 2003; Hartl et al, 2011). Spectral bandwidth up to 372nm was achieved at 1.1μm central wavelength. The super-continuum light source also has the advantage of achieving faster imaging speed with higher signal-to-noise ratio.

Although the reported OCT tomograms with the highest axial resolution (1.8μm) were so far achieved in research laboratories with a photonic crystal fibre based source (Wang et al, 2003), superluminescent diodes are considerably lower in cost and complexity as well as being smaller in size, which makes them more attractive for mass production. Superluminescent diodes utilizing quantum-dots in the active region are considered to be an excellent candidate as a light source for an OCT system. The naturally wide dimensional fluctuations of the self-assembled quantum dots, grown by the Stranski-Krastanow mode, are very beneficial for broadening the gain spectra which enhances the spectral width of the SLDs. On the other hand, the three-dimensional carrier confinement provided by the dots' shape results in high radiative efficiency required for the OCT applications.

3. Reported superluminescent diodes for bandwidth widening and their performance parameters

Since the first report in 1993 (Leonard et al, 1993), the formation of strained self-assembled quantum dots by heteroepitaxial growth in the Stranski–Krastanow mode has been studied extensively for their fundamental properties and applications in optoelectronics. Significant breakthroughs occurred over the last two decades with the fundamental understanding of the QDs systems and the demonstration of zero-dimensional novel devices. These achievements are directly related to the noticeable advances in the epitaxial materials deposition. With self-assembled QDs growth process, a certain size inhomogeneity is common and typically not less than 10%. It has been predicted (Sun & Ding, 1999) that the full width at half maximum of the SLDs output spectrum of the $In_{0.7}Ga_{0.3}As/GaAs$ quantum dot system, with a standard deviation in the average size of the QD ensemble of 10%, can be as high as 140nm. Increasing further the size variation of the dots to 30% should result in bandwidth as high as 160nm. The confinement potential between the dots and the barriers is another important factor for modifying the spectral width. With only 10% size variation increasing the potential confinement by using higher indium composition in the dots a spectral width of 230nm was predicted in the $In_{0.9}Ga_{0.1}As/GaAs$ quantum dot system (Sun & Ding, 1999). In general, such inhomogeneous size distribution of self-assembled QDs in the active region is disadvantageous for achieving lasing of QD-lasers. However, for the designed wide spectrum QD-SLDs it becomes an effective intrinsic advantage for broadening the emission spectrum. Experimentally, using five layers of InAs/GaAs QDs grown under identical growth conditions in a molecular beam epitaxy system (Liu et al., 2005), SLDs with full width at half maximum of ~110nm at a central wavelength of 1.1μm have been made. For high resolution optical coherence tomography applications around 1060nm an even wider broadband spectrum is required. Increasing further the bandwidth of the emission spectrum of the SLDs is a complicated process and requires more than just optimization of the growth conditions of the active region of the device. The precise control of the average size distribution of the dots within one layer is a very challenging process and is very difficult to reproduce. Very practical and successful ideas based on engineering the matrix surrounding the QDs have been also proposed and applied to the fabrication of broadband superluminescent diodes with central wavelength around 1060nm (Li et al, 2005; Ray et al., 2006; Yoo et al, 2007; Lv et al, 2008). Figure 2 shows examples of engineered energy band diagrams of the active region of QD-SLDs for increasing their spectral width.

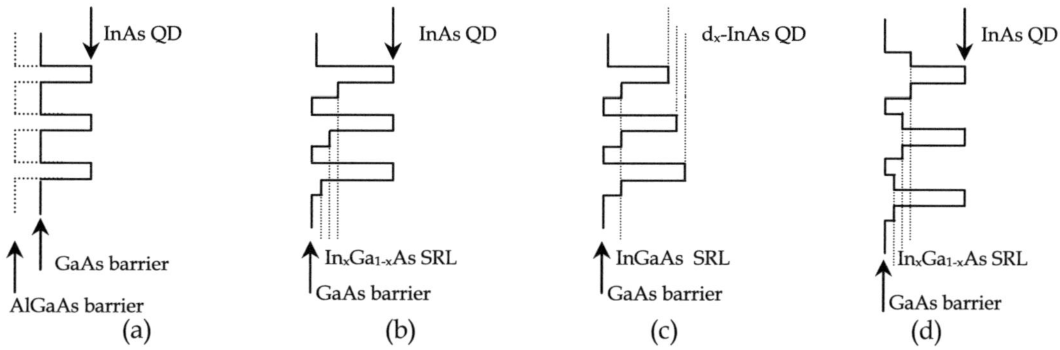

Fig. 2. Schematic band diagrams of some proposed schemes that have been reported in the literature: a) AlGaAs barrier instead of GaAs b) chirped QD structure with $In_xGa_{1-x}As$ strain-reducing layer (SRL), c) chirped QD structure with InGaAs SRL and InAs dots of different size by deposition of different InAs thicknesses, d) QD structure with dots in compositionally modulated quantum wells (DCMWELL).

The use of InAs QDs in $Al_{0.14}Ga_{0.86}As$ matrix instead of GaAs [fig.2 (a)] significantly affects the dot size and distribution and results in a light emitting diode with a spectral bandwidth of 142nm (Lv et al, 2008). The introduction of aluminum atoms reduces the migration length of the indium atoms on the AlGaAs surface. This results in an increase of the nucleation centers which favors the formation of smaller dots with higher density and of larger size fluctuation. For SLDs made using such approach, output power under pulsed conditions was 3mW at 4A driving current.

Another effective approach for changing the matrix surrounding the QDs was reported by Li and co-authors (Li et al, 2005). They have introduced a thin capping $In_xGa_{1-x}As$ strain-reducing layer (SRL) where the indium composition was increased from 9% to 15% by an interval of 1.5% for the five layers of InAs dots of the device [fig. 2 (b)]. QD-SLDs with 121nm bandwidth were demonstrated. The use of $In_xGa_{1-x}As$ SLR however red-shifted the central wavelength to 1165-1286nm range. The maximum achieved output power in these devices was limited to only 1.5mW in pulsed mode.

Introducing an $In_{0.15}Ga_{0.85}As$ SRL for all layers of dots, and changing the dots size from one layer to another by depositing different InAs thicknesses [fig. 2 (c)], is another approach that was proposed by Yoo *et al.* for broadening the gain spectrum of the QD-SLDs (Yoo et al, 2007). The resulted power spectrum was up to 98nm wide centered at ~1150nm. Output power of 32mW in continuous-wave operation mode was measured in these devices at 900mA injection current.

To control the bandwidth of the emission spectrum of QD-SLDs Ray and co-workers (Ray et al., 2006; Ray et al., 2007) proposed to use a dot in compositionally modulated well (DCMWELL) structure of different indium compositions within each well [fig. 2 (d)]. The indium compositions in this structure were chosen such that the separation of the peak wavelengths resulting from a dot-in-well (DWELL) of different compositions is equal to the linewidth of the individual DWELLs. Flat-topped spectral profile of 95nm full-width at half-maximum centered at 1270nm was demonstrated. The corresponding achieved output power in continuous-wave mode was 8mW at 900mA injection current.

Engineering the energy diagram of the surrounding matrix of the QDs is a precise and reproducible technique to manipulate the ground-state (GS) and the excited-states (ESs) peak positions for broadening the spectrum gain. Another powerful approach is to use *external* means to manipulate to peak positions of the GS and ESs of the dot. This was achieved by using multi-section ridge waveguide QD-SLDs. The multi-section SLDs consists of single ridge waveguide divided into three electrically isolated sections: the absorber (reverse-biased to eliminate back reflections) and the two gain sections that are independently biased at different current to favor either GS or ES emission from each section. In this configuration, adjusting the current densities and the lengths of the two SLD sections allows a control of the power output and bandwidth related to the GS and ES of the dots. Using such an approach, Xin and co-authors (Xin et al, 2007) were the first to use a multiple section QD-SLDs as a flexible device geometry that permits independent adjustment of the power and spectral bandwidth in the ground-state and the excited-states of the QDs. Emission spectrum with full width at half maximum of 164nm and 220nm were achieved with central wavelength of 1.15μm and 1.2μm, respectively. The maximum achievable output power in continuous-wave mode, at these wavelengths, was about 0.6mW and 0.15mW, respectively.

For fabrication of broadband SLDs around 1060nm, optimized postgrowth rapid thermal annealing at 750°C was also reported (Zhang et al, 2008). Compared to the as-grown structure, the bandwidth of the device was increased by a factor of two (to 146nm) with the central emission peak blueshift of 54nm (from 1038nm down to 984nm). However, this bandwidth increase was obtained at the expense of continuous-wave output power which decreased by a factor of six, down to 15mW.

A bipolar cascade SLD that uses tunneling junctions between distinct multiple quantum wells was also reported by Guol and co-authors (Guol et al, 2009) for bandwidth engineering. Emitting device with spectral bandwidth of 180nm at central wavelength of 1.04μm was demonstrated. The corresponding maximum continuous-wave output power was 0.65mW.

4. Spectral broadening using height engineered InAs/GaAs quantum dots

Tuning the emission properties of QDs assemblies by in-situ annealing after changing the growth kinetics during the capping (Garcia et al, 1998, Wang et al, 2006), or by post-growth annealing under a GaAs (Leon et al, 1996; Kosogov et al, 1996; Babinski et al, 2001) or SiO_2 proximity cap (Malik et al, 1997; Xu et al, 1998; Yang et al, 2007) have been extensively reported. At the National Research Council of Canada (NRC), we have previously reported (Wasilewski et al, 1999; Fafard et al, 1999) a growth technique, called *indium-flush,* to control the size and exciton levels of the self-assembled QDs. The indium-flush process consists in removing all surface resident indium at a certain position during the overgrowth of the GaAs cap layer. Using this process an additional degree of size and shape engineering, giving a much improved uniformity of the macroscopic ensemble of QDs with well-defined electron shells, was achieved. The process was also proven to be a very reproducible growth technique for improving the uniformity of the dots size distribution of QD ensembles in laser structures. In this chapter we will demonstrate that using the indium-flush process, to intentionally and precisely tune the GS peak position of dots from one layer to another in a superluminescent diode structure, is a controllable and effective approach to fabricate broadband emission spectra for ultrahigh resolution OCT applications (Haffouz et al, 2009; Haffouz et al, 2010; Haffouz et al, 2012).

4.1 The epitaxial growth procedure

The epitaxial growth of the InAs/GaAs QDs was carried out in a V80H VG molecular beam epitaxy (MBE) system using an As_2 molecular flux with arsenic pressure of ~1e-7 Torr. Solid source effusion cells were used for Ga and In elements. All the growths were done using a substrate rotation of 3s per turn to obtain uniformity throughout the wafers. The surface temperature was monitored by optical pyrometer. GaAs (100) substrate has been used as a template. Before introduction in the growth chamber, the GaAs substrates were outgassed under vacuum at 450°C for 2h. Oxide removal was carried out *in-situ* by either a thermal desorption process in the presence of As flux at high temperature or by first applying Ga pulses in the presence of As, partial removal of the oxide at lower temperatures via conversion of the stable Ga_2O_3 surface oxide into a volatile Ga_2O oxide, and then the high temperature standard oxide removal (Wasilewski et al, 2004). The later oxide removal technique was found to reduce the substrate surface roughness. The self-assembled InAs/GaAs QD layers were obtained using the spontaneous island formation at the initial stages of the Stranski-Krastanow growth mode during the epitaxy of highly strained InAs on GaAs. The growth rates of the GaAs and InAs used in these studies were 2Å/s and 0.23Å/s, respectively. The epitaxial growth procedure of the InAs QDs on GaAs buffer was performed as following: after growing the 200nm GaAs buffer layer at 600°C, the substrate

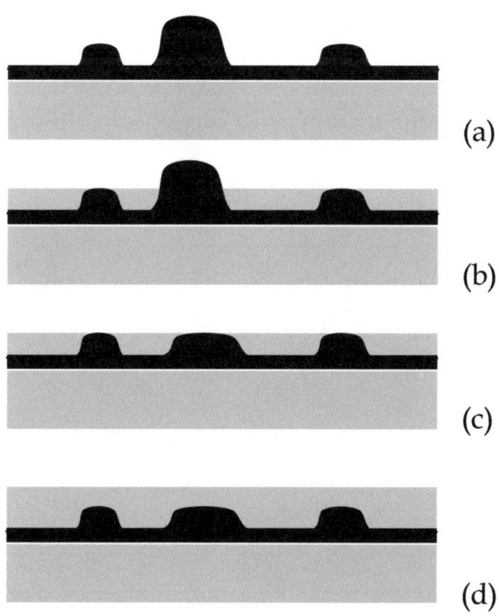

Fig. 3. Schematic drawing of the evolution of the dots during the overgrowth of the InAs with GaAs capping layer.

temperature was lowered to 480-505°C where an InAs layer of 1.95-ML thick was grown. Transition from streaky to spotty pattern measured by reflection high energy electron diffraction technique, which indicates the onset of the dot formation, was observed after approximately 26s of indium deposition [Fig. 3(a)]. A short anneal for 30 s at the same substrate temperature followed by a partial capping of the formed dots by a GaAs layer was applied [Fig. 3(b)]. The thickness of the GaAs layer in this case was varied in the range of 2.5- to 6.5-nm thick, thicknesses that are well below the typical average dots height (≥10nm). Right after the partial capping of the dots, the indium-flush was executed by interrupting the growth, raising rapidly the substrate temperature to 610°C and annealing for 70 s at that temperature. During this step, In/Ga interdiffusion was taking place and the non-protected resident indium desorbed [Fig. 3(c)]. The substrate temperature was then reduced to 600°C to complete the capping of the formed disk-like dots by growing a GaAs layer of total thickness of 100nm (Fig. 3(d)). For morphological analysis of the QDs, extra layer of dots (surface dots) was grown above the GaAs capping layer and left uncapped.

4.2 Tuning InAs quantum dots for high areal density

Epitaxial growth of InAs QD layers of high areal dot density and good optical quality is required to fabricate high optical gain devices like lasers, SLDs, SOAs, etc. Particularly, for broadband emission SLDs, high areal dot density should improve the optical properties of the QD-SLDs, since, unlike QD lasers, emission from QD-SLDs is contributed by QDs of all sizes. Size inhomogeneity in QD layers of low density is small compared to QD layers of high density. Therefore, the use of high areal dot density should introduce a wider emission energy range.

Fig.4 shows atomic force microscope (AFM) images of surface dots grown under identical growth conditions but different substrate temperature for the deposition of the InAs layer. In these QD layers, the indium-flush of the buried layers was executed after partial capping of the QDs with GaAs of 4.5nm thickness. When the InAs layer was deposited at substrate temperature of 505°C [fig. 4(a)] an areal dots density of $1.4 \times 10^{10} cm^{-2}$ was obtained. Decreasing the deposition temperature of the InAs layer to 480°C [fig. 4(b)] reduces the adatom migration length which led to the formation of new nucleation sites for the impinging adatoms, reducing the combination/coalescence with the existing dots. This resulted in the formation of denser dots with larger size inhomogeneity. The achieved areal dot density was about $1 \times 10^{11} cm^{-2}$.

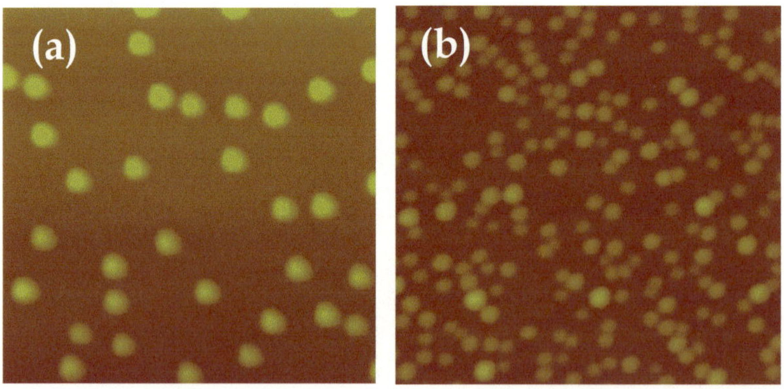

Fig. 4. Atomic Force Microscope (AFM) images of surface InAs QDs on GaAs buffer deposited at a substrate temperature of 505°C (a) and 480°C (b). The surface area is 500nm x 500nm.

Fig. 5 shows the optical properties as measured by photoluminescence (PL) at 77K of a single layer of InAs QDs grown at different substrate temperatures. The samples S_1, S_2, S_3 and S_4 correspond to growth temperatures of 505, 495, 485 and 480ºC, respectively. With identical monolayer coverage (1.95ML of InAs), the areal dot density can be directly controlled by the substrate temperature to achieve a dots density as high as ~1×10^{11}cm^{-2}. At high growth temperature (S_1), PL spectra with a ground-state (GS) peak position at 1.185eV and with well-resolved excited states peaks (n=1, 2, 3, 4) were obtained. Increasing further the dots density (S_2), the GS peak position remained unchanged (at 1.187eV), however the number of the excited-state transition peaks reduced (n=1,2,3). The measured intersublevel energy spacing was about 57meV for both samples. No noticeable change in the PL intensity was measured between S_1 and S_2. However, increasing the dot density to 6×10^{10}cm^{-2} significantly changed the PL spectrum which is now consisted in a single wideband centred at 1.222eV with a slightly reduced intensity. In sample S_4, where the dot density reached ~1×10^{11}cm^{-2}, the PL intensity was significantly reduced (by a factor of 100) and the central peak was blueshifted by 31meV. The spectra broadening in the case of S_3 and S_4 can be explained by the lateral coupling between the dots, the GS emission from the small dots overlapping with the emission from larger dots. However, the noticeable reduction in the PL intensity in S_4 was related to the formation of defective dots when their density was increased. For broadband SLDs fabrication, a compromise between high dot density and good optical properties had to be taken into account. For SLD fabrication an areal dot density of about 4-5×10^{10}cm^{-2} was chosen, using a growth temperature for the InAs layer of around 490°C.

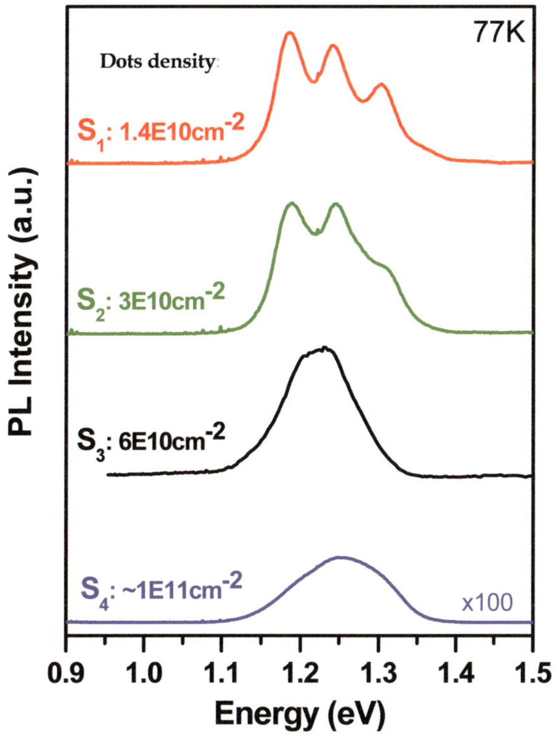

Fig. 5. Photoluminescence spectra measured at 77K of single layers of InAs QDs of different areal densities, capped with 100nm GaAs layers.

4.3 Height engineering of self-assembled InAs/GaAs QDs for wideband emission

The indium-flush process is a very reproducible and predictable process to engineer the QD height and is therefore a reliable tool for tuning the QD emission energy. By varying the thickness of the GaAs cap layer at which the indium-flush process is executed, the ground-state transition energy of the QDs can continuously be adjusted over a wide emission wavelength range. Combining selected layers of QDs with various dot heights offers the possibility to reliably broaden the emission bandwidth of the QD-SLD spectrum. With this motivation, we have carried out a study on tuning the dot height by growing a single layer of dots where the indium flux process was executed at different GaAs partial capping thickness. For all the samples a buffer layer of 200nm of GaAs was first deposited at 600°C on an un-doped GaAs (100) substrate before the growth of the InAs layer at a temperature of 490°C.

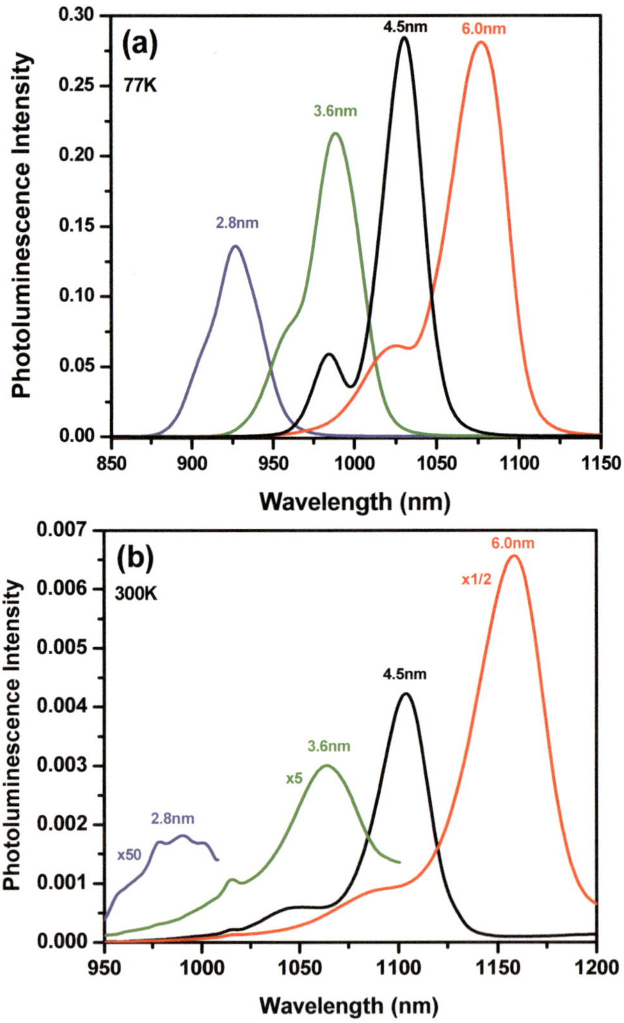

Fig. 6. Photoluminescence spectra at 77K (a) and at room-temperature (b) of single layers of InAs QDs grown with the indium-flush process that was executed at different thicknesses of GaAs capping layer.

Fig. 6(a) and (b) show the photoluminescence spectrum, measured respectively at 77K and at room-temperature, of a single layer of dots where the indium flux process was executed after the deposition of a thin GaAs cap layer between 2.8 and 6.0nm of thicknesses. The areal dot density in these layers was in the range of 3-4 $x10^{10}cm^{-2}$. Due to the high areal dot density, only the first intersublevel energy transition (s-shell) is observed. It should be noticed that decreasing the average dot height within one layer reduced the photoluminescence emission intensity very quickly at room-temperature whereas the decrease of the photoluminescence intensity was less pronounced at 77K. The photoluminescence intensity drop from one layer of dots to another at room-temperature can be explained by the reduction in the carrier confinement due to the reduced potential barrier for carriers in smaller dots. However, with suppressed non-radiative recombination at 77K, due to the reduced mobility of carriers at lower temperatures, the photoluminescence intensity drop from one sample to another was reduced. Nevertheless, PL intensity reduction at 77K by ~50% can still be observed in the layer of shorter dots as compared to longer ones.

Fig. 7 shows the variation of the GS and the first ES emission wavelength values at room-temperature and at 77K for the grown layer of dots as a function of the average dot height. From previous transmission electron microscopy studies (Haffouz et al, 2009), we found that the average dot height within one layer is approximately the thickness of the GaAs layer deposited at low temperature minus 2nm. With increasing dot height, by increasing the thickness of the deposited GaAs cap layer at low temperature before the indium-flush process, the GS peak wavelength of the emission spectrum shifted towards longer wavelength by about 150nm and 169nm at 77K and 300K, respectively. Combining these four layers of dots in the active region of a superluminescent diode could be very beneficial in generating a broadband emission spectrum.

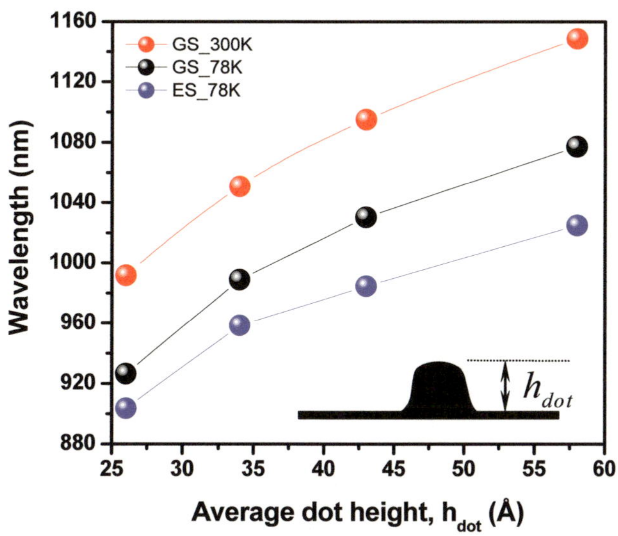

Fig. 7. Variation of the GS and ES peak wavelengths as extracted from the photoluminescence spectra as a function of average dot height.

5. Ultra wide bandwidth SLDs using InAs QDs of tuned heights: Epitaxial growth, fabrication and testing

Tuning the QD emission energy levels by varying the thickness of the GaAs cap layer (from 2.5 nm to 6.5 nm) at which the indium-flush process is executed, the ground-state emission wavelength peak position at 300K can be precisely adjusted from 990 nm to 1150 nm (Haffouz et al, 2009; Haffouz et al, 2012). Therefore, combining selected layers of QDs with various dots heights offers the possibility to reliably broaden the emission bandwidth of the QD-SLD spectrum. Demonstration of symmetric and regular shape emission spectra is required in order to avoid the presence of sidelobes in the OCT interferogram which could be a potential source of spurious structures in the OCT images. To maximize the bandwidth of the emission spectrum of the device, and in order to avoid the formation of dips in the power spectrum, the GS separation energy between adjacent layers of dots should be carefully tuned. We have chosen four different heights of dots in such a way that there was an overlap between the GS emission line from one layer of dots and the ES from the adjacent layer. The chosen average heights of the dots in each layer were about 2.6nm, 3.4nm, 4.8nm and 5.8nm. The dot layers were grown in such a way that the dot height was gradually increased from 2.6 nm up to 5.8 nm starting from the bottom to top in the epitaxial device structure (described in the paragraph hereafter).

The device structure was grown on an n+-GaAs (100) substrate in a solid-source V80H VG molecular beam epitaxy system. To achieve the 1μm emission line, InAs QDs inside a GaAs matrix have been grown. The InAs material growth temperature and growth rate were 490°C and 0.023nm/s, respectively. The obtained dots density was ~$5x10^{10}$dots/cm^2. The active region of the device consisted in two repeats of four layers of InAs quantum dots with GaAs barrier/cap layers, within a 300 nm thick waveguide. 200nm-thick graded index $Al_xGa_{1-x}As$ (x=0.1-0.33) layer was grown at 600°C around the QD core with a 1.5 μm $Al_{0.33}Ga_{0.67}As$:Be ($1x10^{18}$ cm^{-3}) upper cladding and a 1.5 μm $Al_{0.33}Ga_{0.67}As$:Si (2×10^{18} cm^{-3}) lower cladding layer. After each 97 nm of n-cladding growth, a 3 nm GaAs:Si layer was grown to smooth the surface. The top 100 nm GaAs:Be contact layer was doped to a level of 2×10^{19} cm^{-3}, while the bottom GaAs:Si buffer layer was doped to 2×10^{18} cm^{-3}.

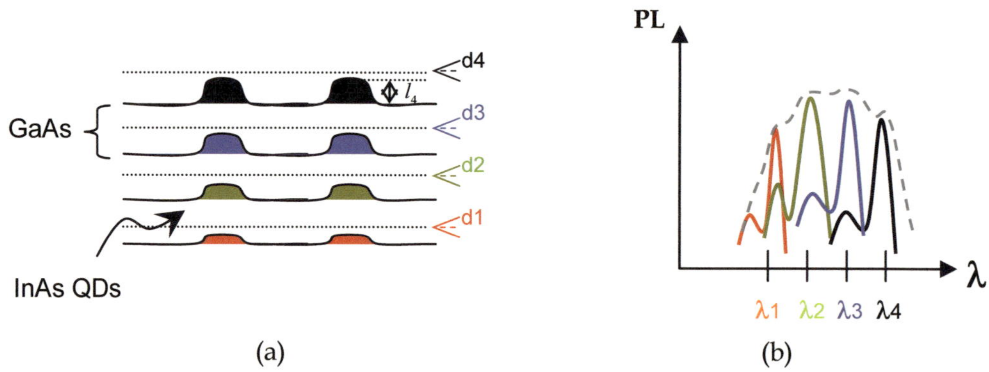

Fig. 8. (a) Schematic structure of the active region of the SLDs where the dot height was varied by controlling the thickness of the GaAs cap layers deposited at low temperature. The arrows indicate the position where the indium-flush was executed (d_x). l_4 is the height of the dots in layer 4. (b) Schematic diagram of the photoluminescence (PL) spectrum of such a stack of quantum dots.

Figure 8(a) shows a schematic drawing of the active region of the SLDs used in this study. The four layers of InAs QDs were grown using the same amount of InAs materials (1.95 ML), whereas the indium-flush was executed respectively at thicknesses equal to 2.8 nm, 3.6 nm, 4.5 nm and 6.5 nm from bottom to top layers. This resulted in the formation of truncated dots with their height typically about 2nm smaller than the corresponding indium-flush position. The predicted photoluminescence spectrum is an overlap of four spectra, each of which having a ground-state transition energy position inversely proportional to the average of the dots height in its corresponding layer [Fig. 8(b)].

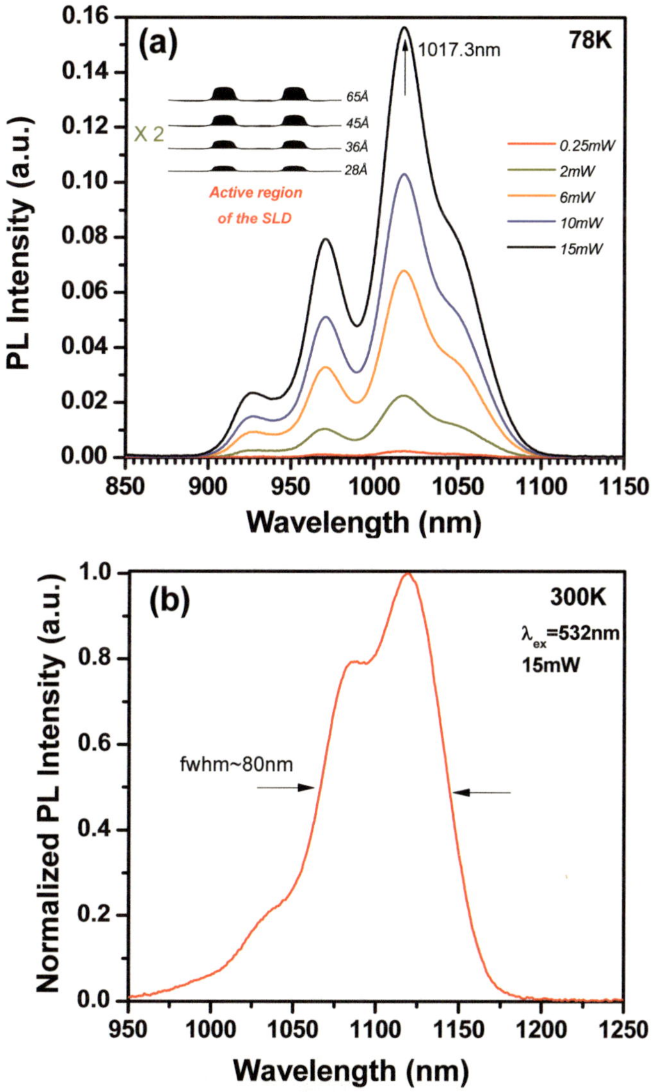

Fig. 9. Photoluminescence spectrum at a) 78K and b) room-temperature of the QD-SLD structure incorporating four layers of InAs QDs where the indium flushes were executed at different thicknesses (2.8 nm, 3.6 nm, 4.5 nm and 6.5 nm).

Photoluminescence measurements at low temperature (78K) with various excitation powers [Fig. 9(a)] and at room-temperature [Fig. 9(b)] have been carried out on this QD-SLD structure after removing the top contact layer and the AlGaAs claddings. As displayed in Fig. 9(a), the recorded photoluminescence spectra with varied excitation powers indicates that the four peaks at energy wavelengths of 926 nm, 970nm, 1017 nm and 1050 nm corresponding to the ground-state transitions wavelengths of the layers of dots with indium-flush positions at 2.8 nm, 3.6 nm, 4.5 nm and 6.5 nm, respectively. No state filling was observed with power excitation up to 15mW. Moreover, distinguishing the transitions related to the excited-states was render difficult by the fact that the combined layers of dots were designed in such a way that there was an overlapping between the GS emission peak from one layer of dots and the ES from the adjacent layer. Overlapping these four layers of QDs with deliberately varied heights has successfully resulted in a broad photoluminescence spectrum with a full width at half maximum of 82 nm and peak wavelength energy of 1.02 μm. Strong room-temperature photoluminescence emission [fig. 9(b)] was also observed around 1120nm with a full width at half maximum of 80nm. Further populating the GSs of the smaller dots, by increasing the power excitation, increased the photoluminescence emission intensity from the corresponding energy levels and considerably broadened the photoluminescence spectrum of the QDs' ensemble. A full-width at half-maximum of 125nm was measured with excitation power of 100mW (Haffouz et al, 2009).

Fig. 10 shows cross-section transmission electron microscopy images of a representative dot within each layer of the active region of the QD-SLDs. Truncated shape quantum dots were obtained from the indium-flush process. The variation in the dot height from one layer to another by varying the thickness of the GaAs layer during partial capping can be observed. This is consistent with the observation from the photoluminescence studies that the GS energy wavelength peak decreased when reducing the dot height.

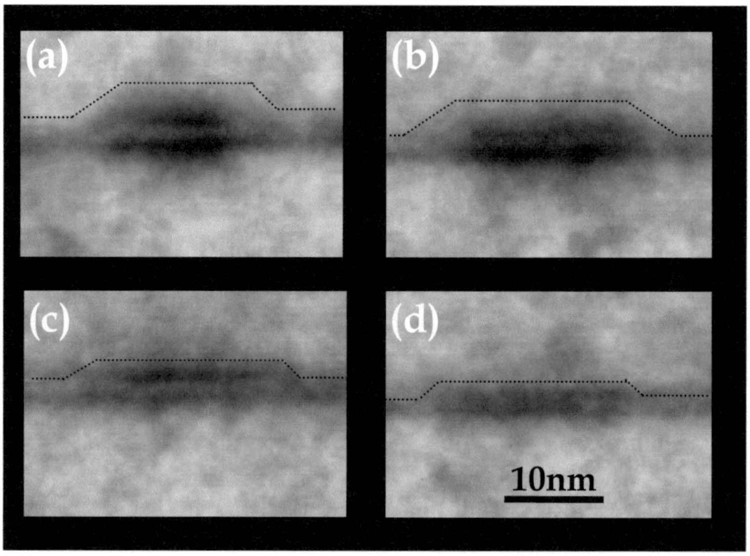

Fig. 10. Cross-section transmission electron microscopy images of representative dot from each layer where the dots height was deliberately tuned. The corresponding indium-flush positions were 6.5nm (a), 4.5nm (b), 3.6nm (c), and 2.8nm (d), respectively.

To fabricate a superluminescent diode, a ridge waveguide tilted by 5-8° off the normal facet is typically used to avoid lasing. It has been also reported (Koyama et al, 1993; Yamatoya et al, 1999) that to lower the effective reflectivity of the device facets, SLED fabricated into a tilted and tapered waveguide can be realized. Such a geometry improves the saturation output power and the quantum efficiency. A simplified schematic drawing, which we used in this study is shown in Fig.11. Devices with various tilt (α) and and tapered-angles (β) were fabricated and tested. For devices of 1-2mm long, the width of the emitting facets were in the range of 35.21 - 176.91 μm. The main difference between these waveguides of different sizes was reflected in the output power of the device. A deflector of 54.7° made in the shape of V-grooves (Middlemast et al, 1997) by wet etch was also implemented in the fabrication process in order to avoid the reflected light to re-enter the active region by deflecting it into the substrate. Finally, an unpumped absorption region of 300 μm wide was implemented at the narrow end of the tapered region just ahead the V-grooves. The devices were mounted p-side up on a Au-plated Cu-heatsink and the cooling was set to an operation temperature of 20°C. The devices were characterized under continuous-wave and pulsed operation conditions.

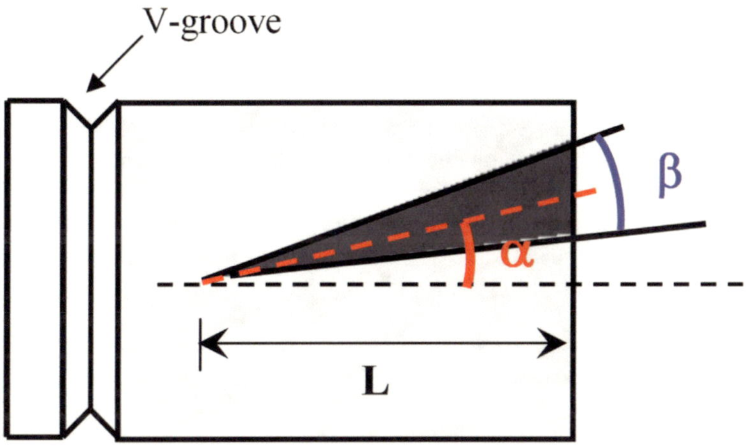

Fig. 11. Schematic drawing of the tilted and tapered waveguide design. L is the length of the tapered region.

Fig. 12 displays the recorded power spectrum of the QD-SLDs at various continuous-wave drive currents. The tapered region was 1-mm long whereas the tilt and tapered angles were 2° and 8°, respectively. The resulted width of the output facet was 35.49 μm. At very low drive current (2mA), a broadband emission centered at 1100nm with a bandwidth of 60nm was obtained. By increasing the injection current, the emission spectrum broadened further and the central wavelength blue-shifted by ~80nm. This can be explained by the carrier transfer effects between the different dots and the energy band filling from low to high energy levels.

Fig.13 summarizes the variation of the emission spectra bandwidth and its corresponding peak wavelength as a function of the injection current. The enhancement of the 3dB-bandwith and the continuous blue-shift of the corresponding peak wavelength with increasing injection current from 2mA to 700mA are attributed to the progressive increase of contribution of all dots to the emission mechanism. At a low injection current, the carriers excited in small dots, which have smaller exciton localization energies, may escape out of the dots and transfer to the large ones, and then radiate. When increasing the injection current, GS in larger dots become occupied, thus reducing the transfer of carriers between isolated dots. The energy states of small dots then begin to be filled and the shape of the emission spectrum slowly approaches that of the QDs' size distribution. With the further increase of injection current, the emission from GS of the QDs' ensemble saturates, and the carriers start to fill the ESs. Because of the ultrawide GS gain spectrum (as demonstrated in Fig. 12), the simultaneous contributions from multiple states do not change the spectral shape, but only broaden the emission bandwidth. Both effects are more beneficial for broadening the spectrum at shorter wavelengths, so that a blue-shift of emission spectra occurred simultaneously. Once all the energy levels (GS and ESs) of the different dots have been saturated, an emission spectrum fully represented by all dots' size distributions is achieved. A maximum 3dB-bandwidth of 190nm was measured at 700mA injection current, with a central wavelength of emission spectrum around 1020nm. Above 700mA drive current, the central wavelength of the spectrum did not change but we noticed a narrowing of the bandwidth of the spectrum down to 160nm. It should be mentioned that the modulation observed in the emission spectra are likely introduced during measurement due feedback effects. We have previously noticed that applying anti-reflection coating on the emitting facet significantly reduce such modulations.

It is believed that the engineered dot height was the reason for the ultra wide bandwidths observed even at relatively low injection current. The precise control of the dots height by controlling the thickness of the GaAs cap layer offers the possibility to precisely manipulate the ground-state peak position of each layer of dots and therefore engineer the bandwidth in a controlable manner. The calculated OCT axial resolution (Eq. 3) from the achieved 3dB-bandwidth of 190nm at central wavelength of 1020nm in our devices is about 2.4μm.

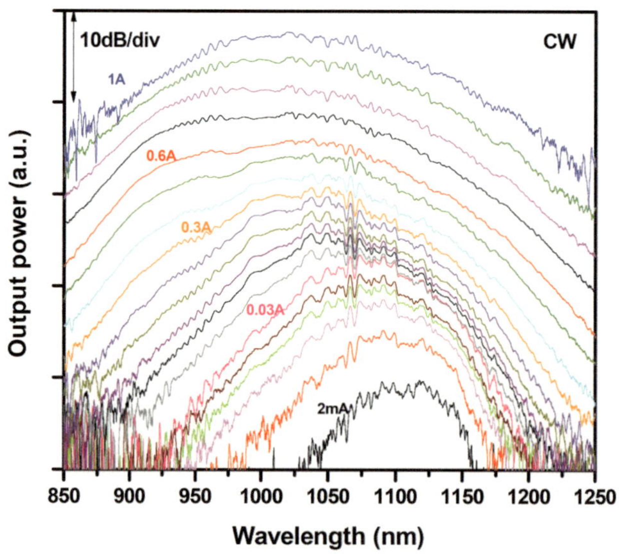

Fig. 12. Emission spectra of the fabricated device under various injection currents. The waveguide was 1mm long, with tilted (α) and tapered (β) angles of 8°and 2°, respectively.

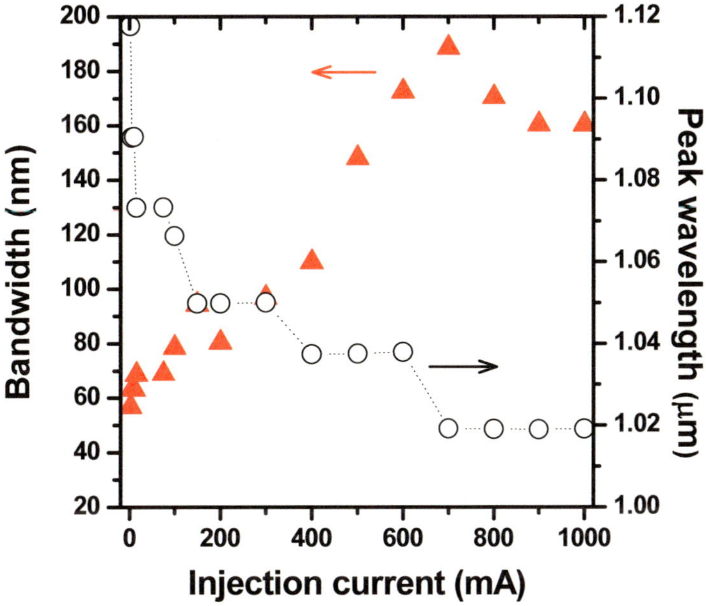

Fig. 13. Variation of the emission spectra bandwidth and its corresponding peak wavelength as a function of the injection current. The waveguide is 1mm long and with tilted (α) and tapered (β) angles of 8°and 2°, respectively.

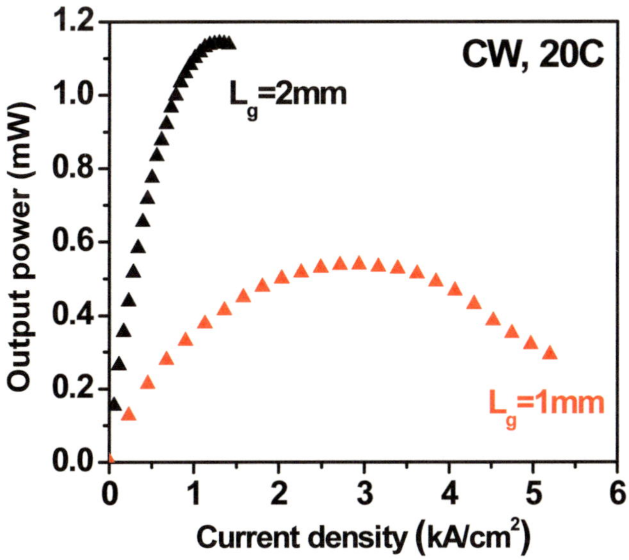

Fig. 14. Light output-injection current curves measured at 20°C under continuous-wave operation mode. The inset shows the schematic drawing of the tapered waveguide.

Fig. 14 shows the output power against the injection current density for 1mm and 2mm long devices measured at 20°C under continuous-wave operation conditions. A maximum output power of 0.54mW under an injection current of 3kA/cm^2 was achieved in 1mm long device.

Above 3kA/cm^2, the output power decreases due to the thermal effect caused by relatively high series resistance as measured in our devices. Increasing the device length resulted in an improved performance where an output power of 1.15mW was measured in 2mm-long device at injection current density of 1.3kA/cm^2. This may indicate that the single pass amplification of the spontaneous emission generated along the waveguide did take place in our devices under continuous wave operation conditions and it varies with the cavity length. However, from these L-I curves, there is no sign of the superluminescent phenomenon in our devices. The existence of the amplified spontaneous emission in our devices (that is varying with the cavity length) and the bandwidth narrowing are characteristics of an SLD but a light-emitting-diode like L-I curves were obtained under continuous wave operation mode. We believe that the super-linear behaviour in the L-I curve is "masked" by the heating problem which is causing L-I curve going down before reaching the injection current range where a super-linear behaviour would appear. To verify his, we have also tested the device of 1mm long under different pulse mode operations.

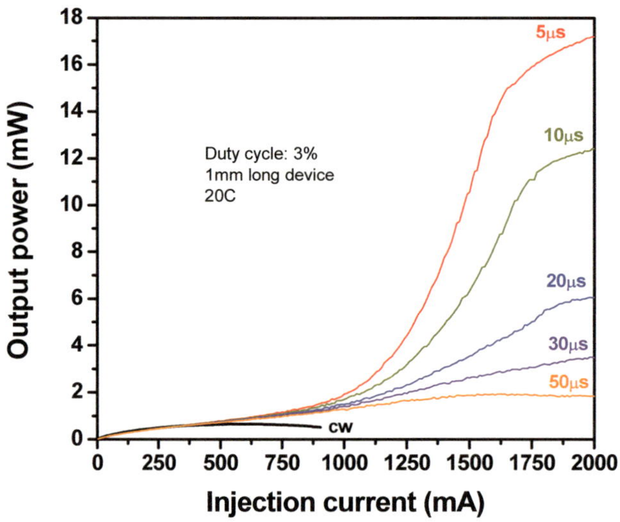

Fig. 15. Light output-injection current curve measured at 20°C under pulsed mode. The pulse width was varied in the range of 5-50μs whereas the duty cycle was 3%. The waveguide is 1mm long and with tilted (α) and tapered (β) angles of 8°and 2°, respectively.

Fig. 15 shows the L-I curves of 1mm long device measured under different pulsed operation conditions. A maximum achievable output power of 17mW under an injection current of 2A was measured with a pulse with of 5μm and a duty cycle of 3%. The super-linear increase of the power with the injection current was observed at threshold current of about 1A. However, increasing the pulse width reduces significantly the maximum achievable output power and the superlinear phenomenon gets less pronounced for pulse with higher than 50μs. As a comparison, we plotted the L-I curve under continuous wave mode together with the L-I curves measured under pulsed mode. As it can be seen, the device under continuous wave suffer from heating problems which causes the L-I curve to go-down (at ~ 500mA) well before reaching the injection current range when the super linear phenomenon kicks-in (at ~ 1000mA).

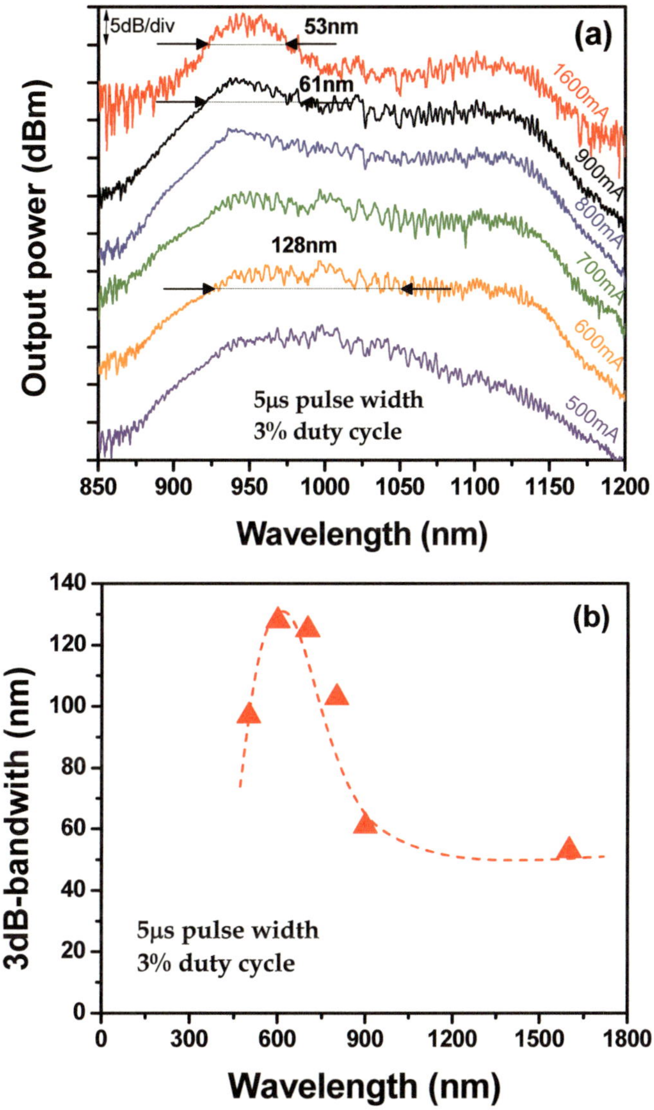

Fig. 16. a) Emission spectra of the fabricated device at different injection currents under pulse mode operation. b) Variation of the 3dB bandwidth of the emission spectrum as a function of injection current. The used pulse width and duty cycle is 5μs and 3%, respectively. The waveguide is 1mm long and with tilted (α) and tapered (β) angles of 8°and 2°, respectively.

Fig. 16 shows the emission spectra of the fabricated device at different injection currents under pulse mode operation [Fig. 16(a)] and its corresponding 3dB bandwidth [Fig. 16(b)]. Broadband spectrum with 3dB bandwidth of 128nm was measured at injection current of 600mA. Above this drive current, the bandwidth of the power spectrum decreases rapidly down to 53nm at injection current of 1600mA. As it can be seen from Fig. 16(b), the bandwidth narrowing above an injection current of 700mA corresponds well to the injection current range where the superlinear behavior was observed in L-I curves in Fig. 15. This bandwidth narrowing is caused by the disproportionate increase in gain for different wavelengths.

In summary, the largest 3dB-bandwidth measured in our devices was respectively 190nm and 128nm under continuous-wave and pulse-mode operation. The corresponding output power under continuous-wave and pulse-mode operation was 0.54mW and 1.2mW, respectively. Broadband QD-SLDs operating at central wavelength of 1.2-1.3μm with 3dB bandwidth not exceeding 100nm typically can deliver an output power of few tens of milliwatts (Yoo et al, 2007, Ray et al, 2007). However, for SLDs with much broader bandwidths (wider than 150nm), the amplified spontaneous emission level is reduced resulting in lower output power (Xin et al, 2007; Guol et al, 2009). For increasing the output power in extremely broadband SLDs, integrating an optical amplifier section has been demonstrated to be a very successful approach (Wang et al, 2011).

6. Conclusion

Broadband superluminescent diodes with central wavelengths around 1060nm have received considerable attention during the last few years due to their potential application in ultrahigh resolution OCT imaging systems for microstructural and biological tissues. Different approaches were proposed to engineer the bandwidth of the SLDs, using quantum wells and/or quantum dot structures. Ultrawide bandwidth emission spectra with output power of a few milliwatts were demonstrated. A calculated axial resolution in air of 2.4μm (Haffouz et al, 2012) is expected from the SLDs with widest bandwidth and central wavelength around 1060±40nm. This lateral (axial) resolution is approaching that observed when using the state of the art but bulky and expensive femtosecond laser sources (Wang et al, 2003). To achieve finer axial resolution at central wavelength around 1060nm with the use of superluminescent diodes, one needs to further improve the design of the quantum dot and/or quantum well active region of the device, for allowing a wider bandwidth and higher output power, thus opening possibilities for OCT imaging with 1-2μm axial resolution range.

7. References

Babinski A., Jasinski J., Bozek R., Szepielow A., & Baranowski J. M. (2001). Rapid thermal annealing of InAs/GaAs quantum dots under a GaAs proximity cap. *Applied Physics Letters*, Vol. 79, No. 16 (October 2001), pp. 2576-2578.

Burns W.K., Chen C.L., & Moeller R.P. (1983). Fiber optic gyroscope with broadband sources. *IEEE Journal of Lightwave Technology*, Vol. LT-1, No.1 (March 1983), pp.98-105.

Friebele E.J., & Kersey A.D. (1994). Fiberoptic sensors measure up for smart structures. *Laser Focus World*, Vol.30, No.5 (May 1994), pp.165-169.

Culter C.C., Newton S.A., Show H.J. (1980). Limitation of rotation sensing by scattering. *Optics Letters*, Vol.5, No.11 (November 1980), pp. 488-490.

Drexler W., Morgner U., Kartner F. X., Pitris C., Boppart S. A., Li X. D., Ippen E. P., & Fujimoto J. G. (1999). *In vivo* ultrahigh-resolution optical coherence tomography. *Optics Letters*, Vol.24, No.24 (September 1999), pp.1221-1223.

Drexler, W., & Fujimoto J. G. (2008). *Optical coherence tomography: Technology and applications.* Springer Berlin Heidelberg New York, 978-3-540-77549-2, Berlin, 2008.

Fafard S., Wasilewski Z. R., Allen C. Ni., Picard D., Spanner M., McCaffrey J. P., & Piva P. G., (1999). Manipulating the energy levels of semiconductor quantum dots. *Physics Review B*, Vol.59, No.23 (June 1999), pp. 15368-15373.

Fercher A.F., Hitzenberger C.K., Drexler W., Kamp G., & Sattmann H. (1993). In vivo optical coherence tomography. *American Journal of Ophthalmology*. Vol.116, No.1, (1993), pp.113-116.

Garcia J. M., Mankad T., Holtz P. O., Wellman P. J., & Pettrof P. M. (1997). Electronic states tuning of InAs self-assembled quantum dots. *Applied Physics Letters*, Vol.72, No.24 (June 1998), p. 3172-3174.

Guol S.-H., Wang Jr-H., Wu Y.-H., Lin W., Y.-J., Sun C.-K., Pan C.-L., & Shi J.-W. (2009). Bipolar cascade superluminescent diodes at the 1.04μm wavelength regime. *IEEE Photonics Technology Letters*, Vol.21, No.5 (March 2009), pp. 328-330.

Haffouz S., Raymond S., Lu Z.G., Barrios P.J., Roy-Guay D., Wu X., Liu J. R., & Poitras D. (2009). Growth and fabrication of quantum dots superluminescent diodes using the indium-flush technique: A new approach in controlling the bandwidth. *Journal of Crystal Growth*, Vol.311, No.7 (March 2009), pp. 1803-1806.

Haffouz S., Rodermans M., Barrios P.J., Lapointe J., Raymond S., Lu Z.G., & Poitras D. (2010). Broadband superluminescent diodes with height-engineered InAs-GaAs quantum dots. *Electronics Letters*, Vol.46, No.16 (August 2010), pp. 1144-1146.

Haffouz S., Barrios P.J., Normandin R., Poitras D., & Lu Z.G. (2012). Ultrawide bandwidth superluminescent light emitting diodes using InAs quantum dots of tuned height. *Optics Letters*, Vol. 37, No.6 (March 2012), pp. 1103-1105..

Hartl I., Li X. D., Chudoba C., Ghanta R. K., Ko T. H., Fujimoto J. G., J. K. Ranka and R. S. Windeler (2001). Ultrahigh resolution optical coherence tomography using continuum generation in an air–silica microstructure optical fiber. *Optics Letters*, Vol.26, N.9 (May 2001), pp.608-610.

Huang D., Swanson E.A., Lin C.P., Schuman J.S., Stinson W.G., Chang W., Hee M.R., Flotte T., Gregory K., Puliafito C.A., Fujimoto J.G. (1991). Optical coherence tomography. *Science*, Vol.254, (November 1991), pp.1178-1181.

Kosogov A. O., Werner P., Gosele U., Ledentsov N. N., Bimberg D., Ustinov V. M., Egorov A. Y., Zhukov A. E., Kop'ev P. S., Bert N. A., & Alferov Zh. I. (19996). Structural and optical peoperties of InAs-GaAs quantum dots subjected to high temperature annealing. *Applied Physics Letters*, Vol. 69, No. 20 (November 1996), pp. 3072-3074.

Koyama F., Liou K. -Y., Dentai A. G., Tanbun-ek T., & Burrus C. A. (1993). Multiple-quantum-well GaInAs/GaInAsP tapered broad-area amplifiers with monolithically integrated waveguide lens for high-power applications. *IEEE Photonics Technology Letters*, Vol. 5, No. 8 (August 1993), pp. 916-919.

Leonard D., Krishnamurthy M., Reaves C. M., Denbaars S. P., & Petroff P. M. (1993). Direct formation of quantum-sized dots from uniform coherent islands of InGaAs on GaAs surfaces. *Applied Physics Letters*, Vol. 63, No. 23 (December 1993), pp. 3203-3205,.

Leon R., Kim Y., Yagadish C., Gal M., Zou J., & Cockayne D. J. H. (1996). Effects of interdiffusion on the luminescence of InGaAs/GaAs quantum dots. *Applied Physics Letters*, Vol. 69, No. 13 (September 1996), pp. 1888-1890.

Li L. H., Rossetti M., Fiore A., Occhi L., & Velez C. (2005). Wide emission spectrum from superluminescent diodes with chirped quantum dot multilayers. *Electronics Letters*, Vol.41, No.1 (January 2005), pp. 41-43.

Liu N., Jin P., & Wang Z.-G. (2005). InAs/GaAs quantum-dot superluminescent diodes with 110nm bandwidth. *Electronics Letters*, Vol.41, No.25, (December 2005), pp. 1400-1402.

Lv X. Q., Liu N., & Wang Z. G. (2008). Broadband emitting superluminescent diodes with InAs quantum dots in AlGaAs matrix. *IEEE Photonics Technology Letters*, Vol.20, No.20 (October 2008), pp. 1742–1744.

Malik S., Roberts C., Murray R., & Pate M. (1997). Tuning self-assembled InAs quantum dots by rapid thermal annealing. *Applied Physics Letters*, Vol. 71, No. 14 (October 1997), pp. 1987-1989.

Middlemast I., Sarma J., & Ynus S. (1997). High power tapered superluminescent diodes using novel etched deflectors. *Electronics Letters*, Vol. 33, No. 10 (May 1997), pp. 903-904.

Pavazay B., Bizheva K., Hermann B., Unterhuber A., Sattmann H., Fercher A.F., Drexler W., Schubert C., Ahnelt P.K., Mei M., Holzwarth R., Wadsworth W. J., Knight J. C., Russel P. St. J. (2003). Enhanced visualization of choroidal vessels using ultrahigh resolution ophthalmic OCT at 1050 nm. *Optics Express*, Vol.11, No.17 (August 2003), pp.1980-1986.

Povazay B., Hermann B., Unterhuber A., Hofer B., Sattmann H., Zeiler F., Morgan J.E., Falkner-Radler C., Glittenberg C., Blinder S., & Drexler W. (2007). Three-dimensional optical coherence tomography at 1050nm versus 800nm in retinal pathologies: enhanced performance and choroidal penetration in cataract patients. *Journal of Biomedical Optics*, Vol.12, No.4 (July/August 2007), pp.04211-041211-7.

Ray S.K., Groom K.M., Beattie M.D., Liu H.Y., Hopkinson M., & Hogg R.A. (2006). Broad-band superluminescent light-emitting diodes incorporating quantum dots in compositionally modulated wells. *IEEE Photonics Technology Letters*, Vol.18, No.1 (January 2006), pp. 58-60.

Ray S.K., Choi T. L., Groom K. M., Stevens B. J., Liu H., Hopkinson M., & Hogg R. A. (2007). High-power and broadband quantum dot superluminescent diodes centered at 1250nm for optical coherence tomography. *IEEE Journal of Selected Topics in Quantum Electronics*, Vol.13, No.5 (September/October 2007), pp. 1267-1272.

Sun Z.-Z., Ding D., Gomg Q., Zhou W., Xu B., & Wang Z.-G. (1999). Quantum-dot superluminescent diode: A proposal for an ultra-wide output spectrum. *Optical and Quantum Electronics*, Vol.31, (1999), pp. 1235-1246.

Swanson E. A., Izatt J. A., Hee M. R., Huang D., Lin C. P., Schuman J. S., Puliafito C. A., & Fujimoto J. G. (1993). *In vivo* retinal imaging by optical coherence tomography. *Optics Letters*, Vol.18, No.21 (November 1993), pp.1864-1866.

Takada K., Yokohama I., Chida K., & Noda J. (1987). New measurement systems for fault location in optical waveguide devices based on an interferometric technique. *Applied Optics*, Vol.26, No.9 (May 1987) pp. 1603-1606.

Wang Y., Zhao Y., Nelson J. S., & Chen Z. (2003). Ultrahigh-resolution optical coherence tomography by broadband continuum generation from a photonic crystal fiber. *Optics Letters*, Vol.11, No.3 (February 2003), pp. 182-184.

Wang L., Rastelli A., & Schmidt O. G. (2006). Structural and optical properties of In(Ga)As/GaAs quantun dots treated by partial capping and annealing. *Journal of Applied Physics*, Vol.100, pp.064313_1-064313_4.

Wang Z.C., Jin P., Lv X.Q., Li X.K. & Wang Z.G. (2011). High-power quantum dot superluminescent diode with integrated optical amplifier section. *Electronics Letters*, Vol.47, No.21 (October 2011), pp. 1191-1193.

Wasilewski Z. R., Fafard S., & McCaffrey J. P., (1999). Size and shape engineering of vertically stacked self-assembled quantum dots. *Journal of Crystal growth*, Vol.201/202, (1999), pp. 1131-1135.

Wasilewski Z. R., Baribeau J.-M., Beaulieu M., Wu X., & Sproule G. I. (2004). Studies of oxide desorption from GaAs substrates via Ga2O3 to Ga2O conversion by exposure to Ga flux. *Journal of Vacuum Science and technology B*, Vol.22, No.3 (May/June 2004), pp. 1534-1538.

Xin Y,-C., Martinez A., Saiz T., Moscho A.J., Li Y., Nilsen T.A., Gray A.L., & Lester L.F. (2007). 1.3μm quantum-dot multisection superluminescent diodes with extremely broad bandwidth. *IEEE Photonics Technology Letters*, Vol.19, No.7 (April 2007), pp. 501–503.

Xu S. J., Wang X. C., Chua S. J., Wang C. H., Fan W. J., Jiang J., & Xie X. G. (1998). Effects of rapid thermal annealing on structure and luminescence of self-assembled InAs/GaAs quantum dots, *Applied Physics Letters*, Vol. 72, No. 25 (June 1998), pp. 3335-3337.

Yamatoya T., Mori S., Koyama F., & Iga K. (1999). High power GaInAsP/InP strained quantum well superluminescent diode with tapered active region. *Japanese Journal of Applied Physics*, Vol. 38, No. 9A (September 1999), pp. 5121-5122.

Yang T., Tatebayashi J., Aoki K., Nishioka M., & Arakawa Y. (2007). Effects of rapid thermal annealing on the emission properties of highly uniform self-assembled InAs/GaAs quantum dots emitting at 1.3μm, *Applied Physics Letters*, Vol. 90, pp. 111912_1-111912_3.

Yoo Y. C., Han I. K., & Lee J. I. (2007). High power broadband superluminescent diodes with chirped multiple quantum dots. *Electronics Letters*, Vol. 43, No.19 (September 2007), pp. 1045-1047.

Zhang Z. Y., Hogg R. A., Xu B., Jin P., & Wang Z. G. (2008). Realization of extremely broadband quantum-dot superluminescent light-emitting diodes by rapid thermal-annealing process. *Optics Letters*, Vol.33, No.11 (June 2008), pp.1210-1212.

2

Influence of Optical Phonons on Optical Transitions in Semiconductor Quantum Dots

Cheche Tiberius and Emil Barna
University of Bucharest/Faculty of Physics
Romania

1. Introduction

Accurate theoretical description of optical phenomena in semiconductor quantum dots (QDs) depends on the description accuracy of the energy structure of the QD. For the energy structure description the existent methods, such as, the tight-binding method, the effective bond-orbital model, the first-principles calculations or the multi-band approach within $\mathbf{k}\cdot\mathbf{p}$ theory, have some limitations either in the accuracy of the predicted electronic structures or the computation efficiency. In this context, the phonon influence on the optical properties makes the theoretical description of the optical phenomena in QDs more complex. In this chapter, we introduce several methods and techniques to describe the phonon influence on the emission and absorption spectra of semiconductor QDs. They are implemented on simplified models of QDs that can capture the main physics of the studied phenomena.

2. Phonons in optical transitions in nanocrystals. Theoretical background

The problem of the exciton-phonon interaction in zero-dimensional systems has a rich history. In principle, strong quantum confinement of the carriers or strong electron–phonon interaction induces increasing of the kinetic energy of the charge carriers involved in the optical transitions. In such cases, the optical transitions in nanocrystals are properly described by an *adiabatic* approach. On the other hand, if either the dynamic Jahn-Teller effect (in case the electronic levels are degenerate) or the pseudo-Jahn-Teller effect (in case the electronic inter-level energy is close to the optical phonon energy) is present, the electron-phonon system of the nanocrystal is properly described by a *non-adiabatic* approach. In this section, basic information regarding the optical transitions involving LO phonons, adapted to nano-crystals, is briefly introduced.

2.1 Adiabatic and non-adiabatic treatments of the optical transitions

Chemical compounds or solids, small or large molecules may be represented by an ensemble of interacting electrons and nuclei. Such complex systems are usually described by the Born-Oppenheimer approximation (Born & Oppenheimer, 1927), which separates the electronic and nuclear motions. This separation is made within the adiabatic approach, which means the electrons are much lighter and faster moving than the nuclei so they can follow the nuclei around and can adjust practically instantaneously their positions. The Hamiltonian of the global system is

$$H(r,Q) = T_Q + T_r + U(r,Q) + V(Q) \equiv T_Q + \mathrm{H}(r,Q) \tag{1}$$

where r, Q are the set of the generalised coordinates of electrons and nuclei, respectively, and $\mathrm{H}(r,Q)$ is the *electronic Hamiltonian.* $U(r,Q)$ represents the electron-electron plus the electron-nucleus interactions and $V(Q)$ represents nucleus-nucleus interactions. The kinetic energy operators are $T_r = -\sum_n \hbar^2 (2m)^{-1} \partial^2/\partial r_n^2$, $T_Q = -\sum_\alpha \hbar^2 (2M_\alpha)^{-1} \partial^2/\partial Q_\alpha{}^2$, where n and α are indices of individual electronic and nuclear coordinates, respectively; m and M_α are the electronic mass and mass of the α -th nucleus, respectively.

Next, following (Newton & Sutin, 1984), we introduce the *diabatic* and *adiabatic* description of the electronic system by expanding the vibronic wave functions $\psi\ (r,Q)$ in the basis set of the orthonormal electronic wave functions, $\{\Phi_n(r,Q)\}$, by $\psi\ (r,Q) = \sum_n \Phi_n(r,Q)\xi_n(Q)$, where $\xi_n(Q)$ are Q-dependent parameters. The orthonormal electronic wavefunctions are found by solving the electronic Schrödinger equation in the Born-Openheimer approximation taking Q as a parameter

$$\mathrm{H}(r,Q)\,\Phi(r,Q) = \mathrm{E}(Q)\Phi(r,Q)\,. \tag{2}$$

The solution $\mathrm{E}_\mathrm{n}(Q)$ of Eq. (2) corresponding to certain electronic wave function $\Phi_n(r,Q)$ are the so-called potential energy surfaces (PES). The expansion coefficients ξ can be found by solving the vibronic Schrödinger equation $H(r,Q)\psi(r,Q) = E\psi(r,Q)$, which leads to

$$\left[T_Q + (T_Q'')_{mm} + \mathrm{H}_{mm}(Q) - E\right]\xi_{m\alpha}(Q) = -\sum_{n\neq m}\left[\mathrm{H}_{mn}(Q) + (T_Q')_{mn} + (T_Q'')_{mn}\right]\xi_{n\alpha}(Q)\,. \tag{3}$$

In Eq. (3) ξ acquires a new index α which quantifies the nuclear states. The matrix elements are defined as $(T_Q'')_{mn} = \langle m|T_Q|n\rangle$, and $(T_Q')_{mn} = -\sum_k \hbar^2 (M_k)^{-1} \langle m|\partial/\partial Q_k|n\rangle \partial/\partial Q_k$ (Dirac notation is used). In Eq. (3) we can write

$$(T_Q')_{mn} + (T_Q'')_{mn} = -\int dq\Phi_m^* \sum_k \frac{\hbar^2}{2M_k}\left[\frac{\partial^2\Phi_n}{\partial Q_k{}^2} + 2\left(\frac{\partial\Phi_n}{\partial Q_k}\right)\frac{\partial}{\partial Q_k}\right] = L_{mn}$$

where the operator

$$L = -\sum_k \frac{\hbar^2}{2M_k}\left[\frac{\partial^2\Phi_n}{\partial Q_k{}^2} + 2\left(\frac{\partial\Phi_n}{\partial Q_k}\right)\frac{\partial}{\partial Q_k}\right] \tag{4}$$

is the so-called Born-Oppenheimer breakdown (nuclear coupling) or non-adiabaticity operator. $\mathrm{H}_{mn}(Q)$ from Eq. (3) is usually called electronic coupling term. In what follows for the clarity, we restrain discussion to only two electronic states. In studying the electron transition starting with Eq. (3) one frequently uses two basis sets:

i. The diabatic (non-stationary or localised) basis containing $\{\Phi_i, \Phi_f\}$ (see Fig. 1). They are chosen as set of eigenfunctions of the suitable zeroth-order electronic Hamiltonian, H , where the interaction between the two electronic states Φ_i and Φ_f is removed. The corresponding PESs are $\mathsf{H}_{ii}(Q) = \langle i|\mathsf{H}|i\rangle$ and $\mathsf{H}_{ff}(Q) = \langle f|\mathsf{H}|f\rangle$.
ii. The adiabatic (stationary or delocalised) basis containing $\{\Phi_1, \Phi_2\}$ (see Fig. 1). The corresponding PES are the *non-crossing* electronic terms, $\mathsf{H}_{11}(Q) = \langle 1|\mathsf{H}|1\rangle$ and $\mathsf{H}_{22}(Q) = \langle 2|\mathsf{H}|2\rangle$, and $\mathsf{H}_{mm} = \left[(\mathsf{H}_{ii} + \mathsf{H}_{ff}) \pm \left[(\mathsf{H}_{ii} - \mathsf{H}_{ff})^2 + 4|\mathsf{H}_{if}|^2\right]^{\frac{1}{2}}\right] / 2$ with $m = 1, 2$ is relation between the eigenvalues of the two basis sets. The smallest energy difference between the two non-intersecting adiabatic PESs is $2\mathsf{H}_{if}$. Transitions are classified as being adiabatic or non-adiabatic as function of the magnitude of the coupling matrix elements. The process is adiabatic if the matrix elements of T'_Q, T''_Q can be safely neglected, irrespective of basis used, either diabatic or adiabatic; if the adiabatic basis is chosen transition does not involve a tunneling between the two adiabatic states (surfaces). On the other hand, a reaction is non-adiabatic if there is no basis that permits the neglect of $(T'_Q)_{12}$, $(T''_Q)_{12}$; when the adiabatic basis is chosen transition involves a tunneling between the two adiabatic states (surfaces).

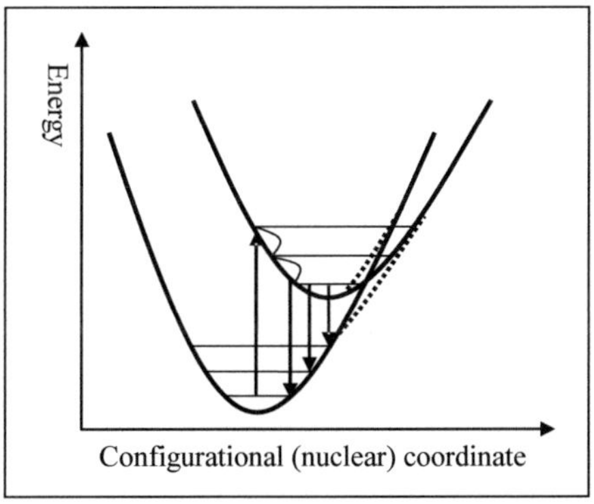

Fig. 1. Radiative adiabatic process in the diabatic/adiabatic picture.

Optical transitions can be produced by *tunnelling* or by *overcoming* the potential barrier. The PESs as function of a representative nuclear coordinate and the vibrational levels are sketched in Fig. 1 for the absorption/emission process by an adiabatic process. The vibrational levels represent the energy of longitudinal optical (LO) phonons, that have a fast relaxation. In this case, the possible tunneling induced by the nuclear coupling terms T' and T'' between the two adiabatic PESs has low probability, and the transition is radiative. The adiabatic PESs are drawn as non-intersecting PESs by the dotted line near the crossing point of the diabatic PESs (solid line). After photon absorption the nuclei in the material adjust position to their new equilibrium positions. The time of adjustment of order 10^{13}s is much faster than the spontaneous emission time of order 10^{8}s and the system relaxes to the lowest vibrational level of the excited state. Then radiative transitions to the vibrational states of the ground state are followed by subsequent relaxations to the lowest vibrational ground state. The mechanism explains the presence of LO phonon satellites in the photoluminescence (PL) spectra. In this scenario, one considers the transition from adiabatic upper PES to the lower adiabatic PES that is triggered by the nuclear coupling is of low probability. This low influence of the nuclear coupling on the process gives the adiabatic character of the transition. Otherwise, tunneling to the lowest adiabatic PES, which means a non-adiabatic process, followed by a non-radiative relaxation by muti-phonon emission is possible.

2.2 The Huang-Rhys factor

The Huang- Rhys factor is frequently used as roughly being a measure of the strength of the exciton-phonon coupling (Banyai & Koch, 1993; Woggon, 1997).

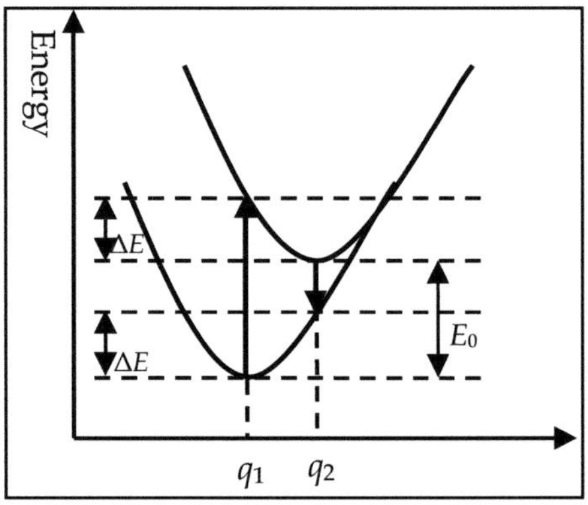

Fig. 2. PESs with the same force constant.

In a simple way, Huang-Rhys factor is introduced by using the configuration coordinate diagram, as sketched in Fig. 2. The PESs, E_g, E_e for the ground and excited states can be written for the model of a single frequency, ω, of the oscillators as $E_g = E_1 + \hbar\omega(q - q_1)^2/2$ and $E_e = E_2 + \hbar\omega(q - q_2)^2/2$, where q_1, q_2 are the equilibrium dimensionless coordinates. According to Fig. 2, we have

$$\Delta E = E_e(q_1) - E_2 = E_g(q_2) - E_1 = \frac{\hbar\omega}{2}(q_1 - q_2)^2 = g\hbar\omega\,, \tag{5}$$

and

$$\hbar\omega_{em} - \hbar\omega_{abs} = E_e(q_1) - E_g(q_1) - E_e(q_2) + E_g(q_2) = 2g\hbar\omega\,, \tag{6}$$

where g is the Huang-Rhys factor. In experiment, the energy difference between the maximum absorption peak and the emission peak is usually referred to as the Stokes shift. Eq. (6) is an approximate definition of the Stokes shift as well as of relation between the Stokes shift and the Huang-Rhys factor g (Ridley, 1988). Large values of g means a larger value between the minima of two PESs, $q_1 - q_2$.

Next, to make connection with the Hamiltonian of QD, we consider a system with the ground state $|g\rangle$, and two excited states $|i\rangle, |f\rangle$ in the diabatic representation. In this picture, the total Hamiltonian reads

$$H = |g\rangle H_g \langle g| + |i\rangle H_i \langle i| + |f\rangle H_f \langle f| + H_{if}\left(|i\rangle\langle f| + |f\rangle\langle i|\right) \quad (7)$$

where H_g is the ground state, H_i and H_f are the excited states, and H_{if} is the interaction between the excited states Hamiltonians. We consider for simplicity a single nuclear coordinate, the nuclear displacement relative to the equilibrium position, Q, with origin at the minimum of the ground state, and a single frequency of vibration of the phonon, ω. Assuming H_i, H_f, and H_{if} are linear dependent of Q, and a parabolic shape of the potential energy surfaces (PESs), we write $H_l = \varepsilon_l + P^2(2m)^{-1} + m\omega^2(Q+Q_l)^2/2$ (with $l = i, f$), $H_g = P^2(2m)^{-1} + m\omega^2 Q^2/2$, $H_{if} = \sqrt{2}(m\omega/\hbar)^{-1/2} C_{if} Q$, where ε_l is the zeroth-order energy separation between the l-th excited PES and the ground state PES, and C_{if} is a constant. Note that H_{if} is written in non-Condon approximation, that is, the nuclear coupling between the two excited states is Q dependent. Next, we introduce the dimensionless momentum $p = (\hbar m\omega)^{-1/2} P$, and dimensionless coordinate $q = (m\omega/\hbar)^{1/2} Q$, and obtain $H_g = \frac{\hbar\omega}{2}(p^2 + q^2)$, $H_i = \varepsilon_i + \frac{\hbar\omega}{2}\left[p^2 + (q+q_i)^2\right]$, $H_f = \varepsilon_f + \frac{\hbar\omega}{2}\left[p^2 + (q+q_f)^2\right]$, $H_{if} = \sqrt{2} C_{if} q$. Further progress is achieved by making the replacement $q = (a + a^+)/\sqrt{2}$, and $p = -i(a - a^+)/\sqrt{2}$, where a (a^+) are the usual annihilation (creation) boson operators. Thus, one obtains $H_g = \hbar\omega(a^+ a + 1/2)$, $H_{if} = C_{if}(a^+ + a)$, $H_i = \varepsilon_i + \hbar\omega\left[(a^+ a + 1/2) + (a^+ + a)q_i/\sqrt{2} + q_i^2\right]$, $H_f = \varepsilon_f + \hbar\omega\left[(a^+ a + 1/2) + (a^+ + a)q_f/\sqrt{2} + q_f^2\right]$.

With the closure relation $|g\rangle\langle g| + |i\rangle\langle i| + |f\rangle\langle f| = \mathbf{1}$ one obtains

$$H = \hbar\omega a^+ a + |i\rangle \varepsilon_i' \langle i| + |f\rangle \varepsilon_f' \langle f| + \left(M_i |i\rangle\langle i| + M_f |f\rangle\langle f|\right)\left(a^+ + a\right) + C_{if}\left(|i\rangle\langle f| + |f\rangle\langle i|\right)\left(a^+ + a\right) \quad (8)$$

where $\varepsilon_i' = \varepsilon_i + \hbar\omega\left(q_i^2 + 1\right)/2$, and $M_i = \hbar\omega q_i/\sqrt{2} = \hbar\omega\sqrt{g_i}$ in which g_i is the Huang-Rhys factor of the state $|i\rangle$ (similarly for the $|f\rangle$ state). In this single frequency model there is a simple dependence between the nuclear coupling and the Huang-Rhys factor, namely $M_i = \hbar\omega\sqrt{g_i}$. With introduction of the creation annihilation operators of electronic states by $|i\rangle\langle i| = C_i^+ C_i$ and $|i\rangle\langle f| = C_i^+ C_f$ (similarly for the $|f\rangle$ state), justified by $C_i^+ C_i |i\rangle = C_i^+ |g\rangle = |i\rangle$, $C_i^+ C_f |f\rangle = C_i^+ |g\rangle = |i\rangle = |i\rangle\langle f|f\rangle$, and $C_i^+ C_f |i\rangle = 0 = |i\rangle\langle f|i\rangle$, Eq. (8) reads

$$H = \hbar\omega a^+ a + \sum_{j=i,f} C_j^+ C_j \left[\varepsilon_j' + M_j\left(a^+ + a\right)\right] + C_{if}\left(C_i^+ C_f + C_f^+ C_i\right)\left(a^+ + a\right) \quad (9)$$

Eq. (9) describes interaction between an electronic system and phonons. It is of the form of the localized defect with several electronic states model, in the case the electronic states are mixed by phonons. The discrete structure of levels in this model is appropriate for description of the 'atomic' energy structure of QDs, and this Hamiltonian is often adopted in QDs problems. Regarding the type of approach, an adiabatic treatment of QD implies absence in the Hamiltonian of the nuclear coupling between PESs, that is C_{if}=0 or equivalently a non-mixing of the electronic states by phonons.

2.3 Absorption and emission spectra in nanocrystals

Often in experiment the Huang-Rhys factor for the LO phonons is calculated from the optical spectra as the ratio of 1LO and 0LO intensity lines, $I(1)/I(0)$. Justification is found within the adiabatic model of a localized impurity interacting with a set of mono-energetic phonons of frequency ω_0 (Einstein model, (Mahan, 2000)). Optical absorption spectrum is derived by evaluating the imaginary part of the one particle Green's function. One obtains that in limit of low temperatures the intensity ratio for the 1LO and 0LO spectral lines gives the Huang-Rhys factor, $g=\sum_{\mathbf{q}} M_{\mathbf{q}}^2/\left(\hbar^2\omega_0^2\right)$, with $M_{\mathbf{q}}$ the electron phonon coupling matrix element of the $\mathbf{q}$ mode phonon. At $T=0$ the phonon replicas follow a Poisson distribution, $I(n)\propto g^n e^{-g}/n!$, in which n is the number of phonons generated in the transition and $g=I(1)/I(0)$. Thus, calculation of the Huang Rhys factor from the optical spectra as the ratio $I(1)/I(0)$ should be cautiously considered as far as it is valid in the limit of an adiabatic approach that assumes absence of mixing of the electronic levels by phonons.

3. Longitudinal optical phonons in optical spectra of defect-free semiconductor quantum dots

The presence of the strong phonon replicas in PL spectra of QDs of weakly polar III-V compounds is a striking result since no such strong phonon replicas are usually observed in the luminescence of III-V compounds, and not always in the PL spectra of QDs of other semiconductor types. The exciton-phonon coupling is already accepted as being strongly enhanced in semiconductor QDs, see, e.g., (Fomin et. al., 1998; Verzelen et. al., 2002; Cheche & Chang, 2005), but there are few theoretical reports (Peter et. al., 2004; Axt et. al., 2005) on the optical spectra of multiexciton complexes which take into account the phonon coupling. For spherical QDs the one-band models by which conduction and valence states are computed from single-particle Schrödinger equations in the effective mass approximation are a good approximation for type I heterostructures (Sercel & Vahala, 1990). In what follows, in Section 3.1 two models built starting with such one-band single-particle states are introduced for spherical and cylindrical shapes of QDs. A short discussion about LO phonon confinement completes this section. In Sections 3.2 and 3.3 non-adiabatic, and adiabatic treatments are introduced to simulate the optical spectra of exciton and biexcton in interaction with LO phonons.

3.1 Quantum dots models

3.1.1 Spherical quantum dot

Within the effective mass approximation, following (Cheche, et. al., 2005) a spherical model is considered for the case of size-quantized energies of QD (or equivalently, QD with dimension smaller than its corresponding exciton Bohr radius, (Hanamura, 1988)). The confinement potential energy is $V_e(r_e)=0$ for $r_e \in [0,R_0]$, and $V_e(r_e)=V_{0e}$ for $r_e > R_0$ (similar equation is written for holes by replacing r_e by r_h); R_0 is the QD radius. The single-particle wave function is the product $\varphi_{nlm}(\mathbf{r}) = R_{nl}(r)Y_{lm}(\theta,\varphi)$, where $R_{nl}(r)$ is the radial function and $Y_{lm}(\theta,\varphi)$ is the spherical harmonics function. By using the second quantization language, and disregarding the spin dependence, the electron-hole pair (EHP) state may be written as (Takagahara, 1993) $|f\rangle \to |\varphi_{ab}\rangle = \int d\mathbf{r}_e\, d\mathbf{r}_h\, \varphi_a(\mathbf{r}_e)\,\varphi_b(\mathbf{r}_h) a_{\boldsymbol{e}}^{c+} a_{\boldsymbol{h}}^{v}|0\rangle$, where $a_{\boldsymbol{e}}^{c+}(a_{\boldsymbol{h}}^{v})$ are the creation (annihilation) fermionic operator of an electron in the conduction band at $\mathbf{r}_e$ (valence band at $\mathbf{r}_h$) and *a* (*b*) holds for the set of quantum numbers n_e, l_e, m_e (n_h, l_h, m_h) of electrons (holes). The single particle states composing the EHPs are obtained by optical excitation and we need to find the optical selection rules that dictate the allowed transitions. In the linear response theory and long wave approximation the particle-radiation Hamiltonian for a carrier of charge Q and mass M is given by $H_{Q-R} = -Q(Mc)^{-1}\mathbf{A}\cdot\mathbf{P}$, where *c* is the speed of light, $\mathbf{A}$ is the vector potential, and $\mathbf{P}$ is the carrier momentum. For monochromatic field of frequency ω, amplitude E_0, and direction of oscillation along the unit polarization vector $\boldsymbol{\varepsilon}$, the semi-classical EHP-radiation interaction form of H_{Q-R} reads

$$H_{EHP-R} = -\mathrm{e}E_0(m_0\omega)^{-1}\boldsymbol{\varepsilon}\cdot\sum_{f\neq 0}\left[\langle 0|\mathbf{P}|f\rangle B_f + \langle f|\mathbf{P}|0\rangle B_f^{+}\right]\sin\omega t \equiv W\sin\omega t \tag{10.a}$$

where $\mathbf{P} \equiv \sum_i \mathbf{p}_i$ the total electronic momentum (with $\mathbf{p}_i$ the electron momentum) and B_f^{+} (B_f) the creation (annihilation) exciton operators. The EHPs are considered as being bosons (EHP spin is an integer), a valid approximation in the dilute limit of excitons. Using an appropriate definition of the momentum (Takagahara, 1993), $\mathbf{P} = \mathbf{p}_{cv}^{0}\int d\mathbf{R}\, a_{\mathbf{R}}^{c+} a_{\mathbf{R}}^{v} + h.c.$, where $\mathbf{p}_{cv}^{0}$ is the momentum matrix element between the valence-band and the conduction-band at the Γ point and where $\mathbf{R}$ suggests integration over unit cell vectors, one obtains the optical matrix element

$$\langle\varphi_{ab}|\mathbf{P}|0\rangle = \mathbf{p}_{cv}^{0}\delta_{l_e l_h}\delta_{m_e m_h}\int_0^{\infty} dr\, r^2\, R_{n_e l_e}(r) R_{n_h l_h}(r) \equiv \mathbf{p}_{cv}^{0}\delta_{m_e m_h} A_{n_e n_h l}\,, \tag{10.b}$$

with $l_e = l_h = l$. Thus, one obtains that the optical selection rule requires $l_e = l_h$. The model takes into account the difference in the effective masses between the nano-sphere and its surroundings. Following (Chamberlain et. al., 1995), the expression of orthonormalized $R_{nl}(r)$ and the secular equation of energy are as follows

$$R_{nl}(r) = \sqrt{\frac{2}{R_0^3}}\left[j_l^2(x)k_{l-1}(y)k_{l+1}(y) - k_l^2(y)j_{l-1}(x)j_{l+1}(x)\right]\begin{cases} k_l(y)j_l(xr/R_0), & r < R_0 \\ j_l(x)k_l(yr/R_0), & r > R_0 \end{cases} \tag{11.a}$$

$$\mu_2 x k_l(y) j_l'(x) = \mu_1 y j_l(x) k_l'(y) \tag{11.b}$$

where $x = R_0\sqrt{(2\mu_1 E_{n,l})/\hbar^2}$, $y = R_0\sqrt{(2\mu_2(V_0^c - E_{n,l}))/\hbar^2}$, k_l is the modified spherical Bessel functions, μ_1 (μ_2) is the effective mass in the dot (surrounding medium), V_0^c the band offset of the carriers, and n, l stand for n_e, l_e (electrons) or n_h, l_h (holes).

For GaAs microcrystallites embedded in AlAs matrix, the compound discussed as an application, we use the parameters of material from (Menéndez et. al., 1997) : the GaAs energy gap $E_g = 1.5177$ eV, the GaAs (AlAs) electron effective mass $\mu_e/m_0 = 0.0665$ ($\mu_e/m_0 = 0.124$), the hole effective mass $\mu_h/m_0 = 0.45$ ($\mu_h/m_0 = 0.5$), the conduction band offset $V_0^e = 0.968$ eV, and the valence band offset $V_0^h = 0.6543$ eV; m_0 is the electron mass. The energy spectrum is obtained from Eqs. (11.a, b), and the EHP energy $E_{n_e,l_e;n_h,l_h} = E_g + E_{n_e,l_e} + E_{n_h,l_h}$ is computed as a function of the QD radius and shown in Fig. 3. Some particular levels are labeled by the set of quantum numbers, ($n_e, l_e, m_e; n_h, l_h, m_h$) as follows: $\mathsf{A}_0 \rightarrow (1,0,0;1,0,0)$, $\mathsf{B} \rightarrow (1,0,0;1,1,m_h)$ - dark level, $\mathsf{C} \rightarrow (1,0,0;1,2,m_h)$ - dark level, $\mathsf{D}_0 \rightarrow (1,0,0;2,0,0)$, $\mathsf{E} \rightarrow (1,1,m_e;1,0,0)$ -dark level, $\mathsf{F} \rightarrow (1,0,0;2,1,m_h)$ -dark level, $\mathsf{G}_0 \rightarrow (1,1,m_e;1,1,m_h)$. Based on the distribution of energy levels and taking into account the exciton Bohr radius (larger than 100Å), we consider $R_0 = 50$ Å as a reasonable upper-limit for neglecting the Coulombic interaction. On the other hand, possible phonon mixing effect could manifest starting with $R_0 \approx 23$ Å (see the ellipse mark at Fig. 3), between the optically active level G_0 and the dark level F. But, the phonon-assisted transition between G_0 and D_0 is improbable (at least in the low temperature limit) because for the intermediate transfer, $\mathsf{E} \rightarrow \mathsf{D}_0$, $(E_{\mathsf{E}} - E_{\mathsf{D}_0})/\hbar\omega_0 = 3.37$ (the LO phonon energy $\hbar\omega_0 = 36.2\text{meV}$). For the first two optically active levels, the adiabatic treatment is safe for $R_0 < 22$ Å and may be accepted as satisfactory for $R_0 < 32$ Å, beyond which the dark level C appears.

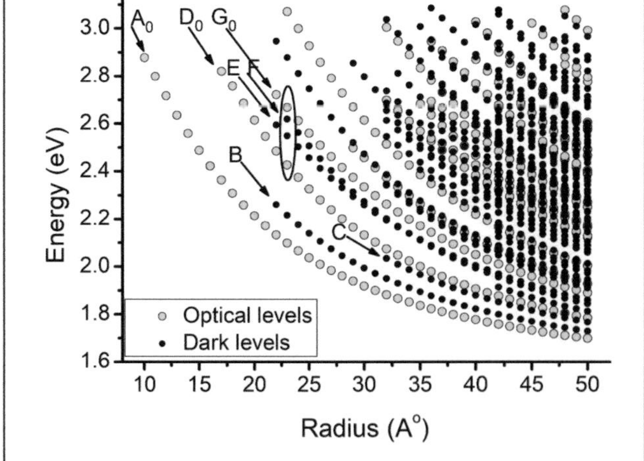

Fig. 3. The energy spectrum of small spherical GaAs/AlAs QDs.

3.1.2 Configurational interaction approach for cylindrical QDs

The energy levels of the exciton complexes can be obtained by the configurational interaction method (Hawrylak, 1999). Following (Cheche, 2009) we will describe a configurational interaction-based model for cylindrical semiconductor QDs. In the effective mass approximation the electron single particle wave function of QD can be approximated as the spin-orbital product (Haug & Koch, 1993) $\phi_{\alpha\sigma}(\mathbf{r}) = u_\sigma(\mathbf{r})\varphi_\alpha(\mathbf{r})$, where $\mathbf{r}$ is the carrier position vector. $\varphi_\alpha(\mathbf{r})$ is the envelope function, and $u_\sigma(\mathbf{r})$ is the periodical Bloch function at Γ point with spin dependence included. The same is valid for holes by replacing, notation wise, *e* by *h*, α by β, and σ by τ. σ and τ are the *z*-projections of the Bloch angular momentum, with $\sigma = \pm 1/2$ and $\tau = \pm 3/2,\ \pm 1/2$. By disregarding the band-mixing, we safely assume that the topmost states are formed from degenerate heavy-hole states, that is, $\tau = \pm 3/2$. With ρ, z, φ, cylindrical coordinates, we consider for the conduction electrons the confining potential made up of the in-plane parabolic potential $V_{\parallel}^{e}(\rho) = \mu_e \big/ \left(2\omega_e^2 \rho^2\right)$ and vertical potential, $V_{\perp}^{e}(z) = 0$ for $|z| \le L/2$ and $V_{\perp}^{e}(z) = V_b^e$ otherwise. The single-particle Hamiltonian, $H_e = H_{e\rho} + H_{ez}$, has the components

$$H_{e\rho} = -\frac{\hbar^2}{2\mu_e}\left[\frac{1}{\rho}\frac{\partial}{\partial\rho}\left(\rho\frac{\partial}{\partial\rho}\right)\right] + V_{\parallel}^{e}(\rho),\ H_{ez} = -\frac{\hbar^2}{2\mu_e}\frac{\partial^2}{\partial z^2} + V_{\perp}^{e}(z) \tag{12.a}$$

The corresponding Schrödinger equations read, $H_{e\rho}\psi(\rho,\varphi) = \varepsilon_{\rho\varphi}^{e}\psi(\rho,\varphi)$, and $H_{ez}\xi(z) = \varepsilon_z^e \xi(z)$. The electronic envelope wave functions $\varphi(\mathbf{r})$ is given by the product $\psi(\rho,\varphi)\xi(z)$, and has the concrete expression, $\varphi_\alpha(\mathbf{r}) = (2\pi)^{-1/2} e^{im_e\varphi} R_{n_e,m_e}(\rho)\xi_i^e(z)$, where α holds for the set of quantum numbers (n_e, m_e, i). For QD sufficiently narrow we may consider $i = 1$ level only, and take the approximate wave function of the first state in *z* direction as, $\xi_1^e(z) = (2/L_e)^{1/2}\cos(\pi z / L_e)$, where $L_e = L\left[1 + 2\hbar \big/ \left(L\sqrt{2\mu_e V_b^e}\right)\right]$ is the effective QD height including the band-offset, L is the QD height (Barker et. al., 1991). Thus, for the electron, the envelope wave function reads

$$\begin{aligned}\varphi_\alpha(\mathbf{r}) &= e^{im_e\varphi}(2\pi)^{-1/2}\left[(2n_e!)/(n_e + |m_e|)!\right]^{-1/2}(\rho/l_e)^{|m_e|} e^{-\frac{\rho^2}{2l_e^2}} l_e^{-1} L_{n_e}^{|m_e|}(\rho^2/l_e^2)\xi_1(z) \\ &\equiv (2\pi)^{-1/2} e^{im_e\varphi} R_\alpha(\rho)\xi_1(z)\end{aligned} \tag{12.b}$$

with $L_{n_e}^{|m_e|}$ denoting Laguerre polynomials, $n_e = 0,1,2,\ldots$, $m_e = 0, \pm 1, \pm 2, \ldots$, $l_e = \sqrt{\hbar/(\mu_e\omega_e)}$, and α re-denoting the set (n_e, m_e) for $i = 1$. The corresponding energy states are obtained as $\varepsilon_\alpha = \varepsilon_{n_e m_e} + \varepsilon_{1ez}$, where $\varepsilon_{n_e m_e}$ and ε_{1ez} are the quantized values of $\varepsilon_{\rho\varphi}^{e}$ and ε_z^e, respectively. The quantized energy for the in-plane motion is $\varepsilon_{n_e m_e} = \left(2n_e + |m_e| + 1\right)\hbar\omega_e$. The same considerations are valid for holes, by considering the effective mass in *z* direction, μ_{hz}, and the in-plane effective mass $\mu_{h\rho}$. An immediate analysis shows the spin-orbitals set

$\{\phi_{\alpha\sigma}(\mathbf{r}),\ \phi_{\beta\tau}(\mathbf{r})\}$ is orthonormal. The integrals involving spin-orbitals are solved by the usual decomposition in a product of two integrals, one over the space of the unit cells position vectors for slowly varying functions, and the other one over the unit cell space for rapidly varying functions. Thus, for example, one obtains, $\langle \phi_{\alpha\sigma} | \phi_{\beta\tau} \rangle = \langle \varphi_\alpha | \varphi_\beta \rangle_{all\,space} \langle u_\sigma | u_\tau \rangle_{\Omega_0} = 0$ from the orthonormality of the periodical Bloch functions (the indices show the volume of integration, with Ω_0 the unit cell volume). For such orthonormal basis set two equivalent ways, the language of the second quantization, and the technique of the determinantal states can be used to describe the energy structure of the system.

Next, we adopt the creation (annihilation) fermion operators, $c^+_{\alpha\sigma}(c_{\alpha\sigma})$ for electron in conduction band, and $h^+_{\beta\tau}(h_{\beta\tau})$ for hole in valence band; they create (annihilate) the carrier with spin projection σ for electrons and τ for holes. Considering negligible the piezoelectricity and the band-mixing effects, and disregarding the electron-hole exchange interaction, the QD Hamiltonian reads

$$H_D = \sum_{\alpha,\sigma} \varepsilon_\alpha c^+_{\alpha\sigma} c_{\alpha\sigma} + \sum_{\beta,\tau} \varepsilon_\beta h^+_{\beta\tau} h_{\beta\tau} + \frac{1}{2} \sum_{\substack{\alpha_1,\alpha_2,\alpha_3,\alpha_4 \\ \sigma_1,\sigma_2}} V^{ee}_{\substack{\alpha_1\sigma_1,\alpha_2\sigma_2 \\ \alpha_3\sigma_1,\alpha_4\sigma_2}} c^+_{\alpha_1\sigma_1} c^+_{\alpha_2\sigma_2} c_{\alpha_4\sigma_2} c_{\alpha_3\sigma_1}$$
$$+ \frac{1}{2} \sum_{\substack{\beta_1,\beta_2,\beta_3,\beta_4 \\ \tau_1,\tau_2}} V^{hh}_{\substack{\beta_1\tau_1,\beta_2\tau_2 \\ \beta_3\tau_1,\beta_4\tau_2}} h^+_{\beta_1\tau_1} h^+_{\beta_2\tau_2} h_{\beta_4\tau_2} h_{\beta_3\tau_1} + \sum_{\substack{\alpha_1,\beta_1,\alpha_2,\beta_2 \\ \sigma_1,\tau_1,\sigma_2,\tau_2}} V^{eh}_{\substack{\alpha_1\sigma_1,\beta_1\tau_1 \\ \alpha_2\sigma_2,\beta_2\tau_2}} c^+_{\alpha_1\sigma_1} h^+_{\beta_1\tau_1} h_{\beta_2\tau_2} c_{\alpha_2\sigma_2} \quad ,(13)$$

where the first, second, third, fourth, and fifth terms of right side stand for electrons, holes, electron-electron, hole-hole, and electron-hole Coulomb interactions, respectively.

Regarding the significance of terms in Eq. (13), we have (Takagahara, 1999)

$$V^{eh}_{\substack{\alpha_1\sigma_1,\beta_1\tau_1 \\ \alpha_2\sigma_2,\beta_2\tau_2}} = -\delta_{\sigma_1\sigma_2} \delta_{\tau_1\tau_2} \iint_V d\mathbf{R}_e d\mathbf{R}_h \varphi^*_{\alpha_1}(\mathbf{R}_e) \varphi^*_{\beta_1}(\mathbf{R}_h) \frac{\mathrm{e}^2}{4\pi\varepsilon |\boldsymbol{\rho}_e - \boldsymbol{\rho}_h|} \varphi_{\alpha_2}(\mathbf{R}_e) \varphi_{\beta_2}(\mathbf{R}_h), \quad (14)$$

where V is the volume of QD. Similar expressions hold for $V^{ee}_{\substack{\alpha_1\sigma_1,\alpha_2\sigma_2 \\ \alpha_3\sigma_1,\alpha_4\sigma_2}}$ and $V^{hh}_{\substack{\beta_1\tau_1,\beta_2\tau_2 \\ \beta_3\tau_1,\beta_4\tau_2}}$; the capital bold characters suggest integration over the 'coarse-grained' space of the unit cell position vectors. In Eq. (14), we considered an in-plane Coulombic interaction, with $\boldsymbol{\rho}$ the in-plane position vector. After integration over z, which gives unity, one obtains an integral over $\boldsymbol{\rho}$ only. Integral from Eq. (14) is solved as follows. The potential is written as a two-dimensional Fourier transform, $\mathrm{v}(|\boldsymbol{\rho}_e - \boldsymbol{\rho}_h|) \equiv 1/|\boldsymbol{\rho}_e - \boldsymbol{\rho}_h| = \int d\mathbf{q}\, \mathrm{v}(\mathbf{q}) e^{i\mathbf{q}\cdot(\boldsymbol{\rho}_e - \boldsymbol{\rho}_h)}$, and the inverse Fourier transform reads

$$\mathrm{v}(\mathbf{q}) = \frac{1}{4\pi^2} \int d\boldsymbol{\rho}\, \mathrm{v}(\rho) e^{-i\mathbf{q}\cdot\boldsymbol{\rho}} = \frac{1}{4\pi^2} \int_0^\infty d\rho\, \rho \int_0^{2\pi} d\varphi \frac{1}{\rho} e^{-iq\rho\cos\varphi}$$
$$= \frac{1}{4\pi^2} \int_0^\infty d\rho \int_0^{2\pi} d\varphi \sum_{m=-\infty}^{\infty} i^m e^{im\varphi} J_m(q\rho) = \int_0^\infty d\rho J_0(q\rho) = \frac{1}{2\pi q} \quad ,$$

where φ is the angle between $\mathbf{q}$ and $\boldsymbol{\rho}$. Using these expressions we write in Eq. (14)

$$\iint_V d\mathbf{R}_e d\mathbf{R}_h \varphi^*_{\alpha_1}(\mathbf{R}_e)\varphi^*_{\beta_1}(\mathbf{R}_h)\frac{1}{|\boldsymbol{\rho}_e-\boldsymbol{\rho}_h|}\varphi_{\alpha_2}(\mathbf{R}_e)\varphi_{\beta_2}(\mathbf{R}_h)$$

$$=\int_0^\infty dq\int_0^{2\pi} d\varphi\int_{S_0} d\boldsymbol{\rho}_e\, e^{i(m_{e2}-m_{e1})\varphi}R_{\alpha_1}(\rho_e)e^{i\mathbf{q}\cdot\boldsymbol{\rho}_e}R_{\alpha_2}(\rho_e)\int_{S_0} d\boldsymbol{\rho}_h\,(2\pi)^{-1}e^{-i(m_{h2}-m_{h1})\varphi}R_{\beta_1}(\rho_h)e^{-i\mathbf{q}\cdot\boldsymbol{\rho}_h}R_{\beta_2}(\rho_h)$$

where S_0 is the cylinder base surface. Next, we introduce

$$I^e_{\alpha_1\alpha_2}(\mathbf{q})=\frac{1}{2\pi}\int_{S_0} d\boldsymbol{\rho}\, e^{i(m_{e2}-m_{e1})\varphi}R_{\alpha_1}(\rho)e^{i\mathbf{q}\cdot\boldsymbol{\rho}}R_{\alpha_2}(\rho)=\frac{1}{2\pi}\int_0^\infty d\rho\,\rho R_{\alpha_1}(\rho)R_{\alpha_2}(\rho)\int_0^{2\pi} d\varphi e^{i(m_{e2}-m_{e1})\varphi}e^{iq\rho\cos\varphi}$$

and by using $\exp(iz\cos\varphi)=\sum_{p=-\infty}^{\infty} i^p J_p(z)\exp(ip\varphi)$, one obtains

$$I^e_{\alpha_1\alpha_2}(\mathbf{q})=\sum_{p=-\infty}^{\infty}\int_0^\infty d\rho\,\rho R_{\alpha_1}(\rho)R_{\alpha_2}(\rho)\delta_{p,m_{e2}-m_{e1}}i^p J_p(q\rho)=i^{m_{e2}-m_{e1}}\int_0^\infty d\rho\,\rho R_{\alpha_1}(\rho)R_{\alpha_2}(\rho)J_{m_{e2}-m_{e1}}(q\rho)$$

Similarly, for holes, $I^h_{\beta_1\beta_2}(\mathbf{q})=i^{-(m_{h2}-m_{h1})}\int_0^\infty d\rho\,\rho R_{\beta_1}(\rho)R_{\beta_2}(\rho)J_{m_{h1}-m_{h2}}(q\rho)$. Conservation of the angular momentum in z direction requires $m_{e1}=-m_{h1}$, and $m_{e2}=-m_{h2}$. For Eq. (14), after an integration over φ, we have $V^{eh}_{\substack{n,m,\sigma_1;n,-m,\tau_1\\ n',m',\sigma_2;n',-m'\tau_2}}=-\delta_{\sigma_1\sigma_2}\delta_{\tau_1\tau_2}\mathrm{e}^2(4\pi\varepsilon)^{-1}$ $\times\int_0^\infty dq I^e_{n,m;n',m'}(\mathbf{q})I^h_{n,-m;n',-m'}(\mathbf{q})$; such integrals have analytic solutions. General solutions of Coulombic integral for in-plane interaction can be found in (Jacak et. al., 1998).

The exciton state $|X^1_f\rangle$ is written as a linear combination of determinantal states,

$$|X^1_f\rangle=\sum_{\alpha\sigma,\beta\tau}C^f_{\alpha\sigma,\beta\tau}c^+_{\alpha\sigma}h^+_{\beta\tau}|0\rangle\equiv X^{1^+}_f|0\rangle, \tag{15.a}$$

with $|0\rangle$ standing for the exciton vacuum state (no excitons), the ground state (VS) of the sysytem. Similarly, the biexciton $|X^2_f\rangle$ state is written as linear combinations of determinantal states that differ of the VS by two of the spin-orbitals

$$|X^2_f\rangle=\sum_{\substack{\alpha_1\sigma_1,\alpha_2\sigma_2,\\ \beta_1\tau_1,\beta_2\tau_2}}C^f_{\substack{\alpha_1\sigma_1,\alpha_2\sigma_2,\\ \beta_1\tau_1,\beta_2\tau_2}}c^+_{\alpha_1\sigma_1}c^+_{\alpha_2\sigma_2}h^+_{\beta_1\tau_1}h^+_{\beta_2\tau_2}|0\rangle\equiv X^{2^+}_f|0\rangle. \tag{15.b}$$

The eigen-problem for exciton and biexcitons is solved through the equations $H_D|X^1_f\rangle=\varepsilon^{(1)}_f|X^1_f\rangle$, and $H_D|X^2_f\rangle=\varepsilon^{(2)}_f|X^2_f\rangle$. Their corresponding secular equations allow obtaining the eienvalues and eigenfunctions corresponding to the exciton and biexciton states. It is worth noting that the electron-electron and hole-hole Hamiltonians from Eq. (13) have no contribution to the secular equation associated to the exciton eigen-problem; the product of fermionic operators resulting from these Hamiltonians and from the exciton state forms sequence of operators which when acting on the VS gives zero.

Referring to the determinantal state technique, the VS is written as the ground-state Slater determinant $\Phi_0(\mathbf{r}_1,\tau_1,...,\mathbf{r}_\upsilon,\tau,...\mathbf{r}_N,\tau_N)=\mathsf{A}\left[\phi_{\beta_1\tau_1}(\mathbf{r}_1),...,\ \phi_{\beta\tau}(\mathbf{r}_\upsilon),...\phi_{\beta_N\tau_N}(\mathbf{r}_N)\right]$, where N is the total number of electrons in the system, and A is the antisymmetrizing operator. A single-substitution Slater determinant is written by promoting an electron from the occupied valence state $\phi_{\beta\tau}(\mathbf{r}_\upsilon)$ to the unoccupied conduction state $\phi_{\alpha\sigma}(\mathbf{r}_\upsilon)$

$$\Phi_{\beta\tau,\alpha\sigma}(\mathbf{r}_1,\tau_1,...,\ \mathbf{r}_\upsilon,\sigma,...\mathbf{r}_N,\tau_N)=\mathsf{A}\left[\phi_{\beta_1\tau_1}(\mathbf{r}_1),...,\ \phi_{\alpha\sigma}(\mathbf{r}_\upsilon),...\phi_{\beta_N\tau_N}(\mathbf{r}_N)\right].$$

The following equivalence between the single-substitution Slater determinant and configurations written in the language of the second quantization holds:

$$\Phi_{\beta\tau,\alpha\sigma}(\mathbf{r}_1,\tau_1,...,\ \mathbf{r}_\upsilon,\sigma,...\mathbf{r}_N,\tau_N)\leftrightarrow c^+_{\alpha\sigma}h^+_{\beta\tau}\left|0\right\rangle .$$

Taking the advantage of the determinantal states, we search for the optical selection rules that dictate the optically active pair states to be used in the linear combination from Eqs. (15.a, b). The radiation field is modeled as a single mode of polarized plane wave. In the limit of linear response theory and long-wave approximation, the semiclassical particle-field interaction Hamiltonian, for transitions $\left|X^m\right\rangle\leftrightarrow\left|X^{m-1}\right\rangle$ (with $\left|X^0\right\rangle\equiv\left|0\right\rangle$) is written as $H_{X^m-R}=\mathrm{e}E_0(m_0\omega)^{-1}\boldsymbol{\varepsilon}\cdot\mathbf{P}_{X^m}$, where the momentum operator is $\mathbf{P}_{X^m}=\sum_{f,i}\left|X^m_f\right\rangle\left\langle X^m_f\right|\mathbf{p}_i\left|X^{m-1}_f\right\rangle\left\langle X^{m-1}_f\right|+h.c.$, with $\mathbf{p}_i$ the momentum of the i electron and summation is done over all the electrons of the system and (multi)exciton states. Then, by using the algebra of determinantal states (Grosso & Parravicini, 2000), we have:

$$\left\langle X^1_f\right|\textstyle\sum_i\mathbf{p}_i\left|0\right\rangle=\sum_{\alpha\sigma,\beta\tau}C^{f*}_{\alpha\sigma,\beta\tau}\left\langle 0\right|h_{\beta\tau}c_{\alpha\sigma}\sum_i\mathbf{p}_i\left|0\right\rangle=\sum_{\alpha\sigma,\beta\tau}C^{f*}_{\alpha\sigma,\beta\tau}\left\langle\phi_{\alpha\sigma}\right|\mathbf{p}\left|\phi_{\beta\tau}\right\rangle, \tag{16.a}$$

$$\begin{aligned}\left\langle X^2_i\right|\textstyle\sum_i\mathbf{p}_i\left|X^1_f\right\rangle&=\sum_{\substack{\alpha_1\sigma_1,\alpha_2\sigma_2\\ \beta_1\tau_1,\beta_2\tau_2}}C^{i*}_{\substack{\alpha_1\sigma_1,\alpha_2\sigma_2\\ \beta_1\tau_1,\beta_2\tau_2}}\left\langle 0\right|h_{\beta_2\tau_2}h_{\beta_1\tau_1}c_{\alpha_2\sigma_2}c_{\alpha_1\sigma_1}\sum_i\mathbf{p}_i\sum_{\alpha\sigma,\beta\tau}C^f_{\alpha\sigma,\beta\tau}c^+_{\alpha\sigma}h^+_{\beta\tau}\left|0\right\rangle\\&=\sum_{\substack{\alpha\sigma,\alpha_1\sigma_1\\ \beta\tau,\beta_1\tau_1}}C^{i*}_{\substack{\alpha\sigma,\alpha_1\sigma_1\\ \beta\tau,\beta_1\tau_1}}C^f_{\alpha\sigma,\beta\tau}\left\langle\phi_{\alpha_1\sigma_1}\right|\mathbf{p}\left|\phi_{\beta_1\tau_1}\right\rangle\end{aligned} . \tag{16.b}$$

If we make use of the fact that the envelope functions vary relatively slowly over regions of the size of a unit cell, with $\mathbf{p}=-i\hbar\nabla$, we can write the integral

$$\begin{aligned}\left\langle\phi_{\alpha\sigma}\right|\mathbf{p}\left|\phi_{\beta\tau}\right\rangle&=\int_{\Omega_0}d\mathbf{r}u^{e*}_{\sigma}(\mathbf{r})\mathbf{p}u^h_{\tau}(\mathbf{r})\int_{all\ space}d\mathbf{r}\varphi^*_{\alpha}(\mathbf{r})\varphi_{\beta}(\mathbf{r})\\&+\int_{\Omega_0}d\mathbf{r}u^{e*}_{\sigma}(\mathbf{r})u^h_{\tau}(\mathbf{r})\int_{all\ space}d\mathbf{r}\varphi^*_{\alpha}(\mathbf{r})\mathbf{p}\varphi_{\beta}(\mathbf{r})\equiv\mathbf{p}^0_{cv}\int_{all\ space}d\mathbf{r}\varphi^*_{\alpha}(\mathbf{r})\varphi_{\beta}(\mathbf{r})\end{aligned} . \tag{16.c}$$

The second integral over unit cell of orthogonal Bloch periodical functions vanishes and Eq. (16.c) is in accordance with Eq. (10.b). Passing from the momentum matrix element to the dipole matrix element in Eqs. (16.a, b) we obtain the following (multi)exciton-field interaction Hamiltonians:

$$H_{X^1-R} = \mathrm{i}\mathrm{e}E_0\omega^{-1}\sum_f \sum_{\alpha\sigma,\beta\tau} C^{f*}_{\alpha\sigma,\beta\tau}\omega_{\alpha\beta}\langle\phi_{\alpha\sigma}|\boldsymbol{\varepsilon}\cdot\mathbf{r}|\phi_{\beta\tau}\rangle X_f^{1^+} - h.c. = \omega^{-1}\sum_f\left(C^f X_f^{1^+} + h.c.\right), \quad (17.a)$$

and

$$H_{X^2-R} = \mathrm{i}E_0\mathrm{e}(m_0\omega)^{-1}\left(\sum_{i,f}\sum_{\substack{\alpha\sigma,\alpha_1\sigma_1,\\ \beta\tau,\beta_1\tau_1}} C^{i*}_{\alpha\sigma,\alpha_1\sigma_1}C^{f}_{\alpha\sigma,\beta_1\tau_1}\omega_{\alpha_1\beta_1}\langle\phi_{\alpha_1\sigma_1}|\boldsymbol{\varepsilon}\cdot\mathbf{r}|\phi_{\beta_1\tau_1}\rangle X_i^{2^+}X_f^1 - h.c.\right). \quad (17.b)$$

$$\equiv \omega^{-1}\sum_{i,f}\left(\mathbf{C}^{if}X_i^{2^+}X_f^1 + h.c.\right)$$

We used the notations: $\omega_{\alpha\beta} = \left(\varepsilon_\alpha^e - \varepsilon_\beta^h\right)/\hbar$, and ω, E_0 for frequency, amplitude of the radiation field, respectively. We also introduced $X_f^{1^+} = |X_f^1\rangle\langle 0|$, $X_f^{2^+} = |X_f^2\rangle\langle 0|$, $X_i^{2^+}X_f^1 = |X_i^2\rangle\langle X_f^1|$. The optical selection rules for interband transitions are obtained from the $\boldsymbol{\varepsilon}\cdot\mathbf{r}$ matrix element. Thus

$$\langle\phi_{\alpha\sigma}|\boldsymbol{\varepsilon}\cdot\mathbf{r}|\phi_{\beta\tau}\rangle = \frac{1}{\Omega_0}\boldsymbol{\varepsilon}\cdot\int_{\Omega_0} d\mathbf{r}u_\sigma^{e*}(\mathbf{r})\,\mathbf{r}\,u_\tau^h(\mathbf{r})\int_{all\ space} d\mathbf{R}\varphi_\alpha^*(\mathbf{R})\varphi_\beta(\mathbf{R}), \quad (18)$$

By writing: i) the periodical Bloch functions at the Γ point, $u_{a^j}(\mathbf{r}) = \zeta_{a^j}(r)\chi_{a^j}(\theta,\varphi)$, where $j = e,h$, and $a^e = \pm 1/2$, $a^h = \pm 3/2$, as the following spinors (Merzbacher, 1988): $\chi^e_{1/2,1/2}(\theta,\varphi) = Y_0^0(\theta,\varphi)|\uparrow\rangle$, $\chi^e_{1/2,-1/2}(\theta,\varphi) = Y_0^0(\theta,\varphi)|\downarrow\rangle$, $\chi^h_{3/2,3/2}(\theta,\varphi) = Y_1^1(\theta,\varphi)|\uparrow\rangle$, $\chi^h_{3/2,-3/2}(\theta,\varphi) = Y_1^{-1}(\theta,\varphi)|\downarrow\rangle$, and ii) the position vector for light propagating in *z* direction, $\mathbf{r} = r\left(-Y_1^1\hat{\boldsymbol{\varepsilon}}_- + Y_1^{-1}\hat{\boldsymbol{\varepsilon}}_+\right)$ with $\hat{\boldsymbol{\varepsilon}}_-, \hat{\boldsymbol{\varepsilon}}_+$ the light helicity unit polarization vectors, we obtain the spin selection rules for the configurations. Thus, one finds that for *linearly* polarized light propagating in *z* direction, the only non-vanishing matrix elements involving the heavy-hole states correspond to the transitions $\sigma = 1/2 \leftrightarrow \tau = 3/2$ and $\sigma = -1/2 \leftrightarrow \tau = -3/2$. This result is guiding us in choosing the optically active configurations when using the configurational interaction method to obtain the energy structure of QD.

To obtain spin-polarized excitons, the linearly polarized light is used for photoexcitation. The nonequilibrium spin decays due to both carrier recombination and spin relaxation. Accordingly to (Paillard et al., 2001), and (Sénès et al., 2005), who studied polarization dynamics with linearly polarized light in InAs/GaAs self-assemled QD under (quasi)resonant excitation, following excitation the electron and hole spin states remain stable during the exciton lifetime for low temperatures. This is the case we assumed for the present discussion. Linearly polarized light is a linear combination of circularly polarized light with positive and negative helicity (Zutić, et. al, 2004), consequently, the configurations are obtained by respecting the optical selection rules for interband transitions for circularly polarized light with both positive and negative helicity.

Accordingly to our assumption that the electron and hole spins remain stable during the exciton lifetime the appearance of dark states (states with opposite spins of the electron and hole of a pair) is less probable, and we disregard them. Within the configurational interaction method we consider a limited number of states generated by the two lowest shells s and p configurations optically active, that is the pair states having $n_e = n_h = 0$, and $m_e = -m_h = 0, \pm 1$, as shown in Fig. 4. In Fig. 4, the filled (empty) triangles represent electrons (holes) of Bloch angular spin projection $\pm 1/2$ ($\pm 3/2$). The quantum numbers (n_e, m_e), (n_h, m_h) are shown for the single states.

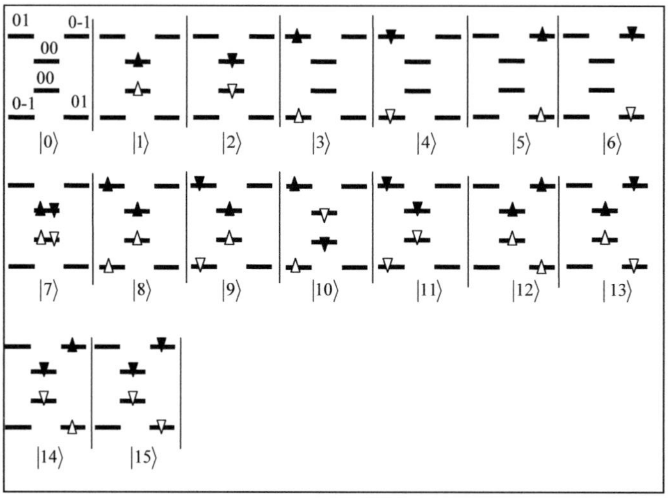

Fig. 4. Vacuum state, exciton and biexciton bright states with linearly polarized light.

Next, we apply the model to cylindrical InAs/AlAs QD. We use the following material parameters taking into account the presence of lattice mismatch strain: a) For InAs $\mu_e = 0.04m_0$, $\mu_{hz} = 0.41m_0$, $\mu_{h\rho} = 0.04m_0$, $\varepsilon_0 / \varepsilon_v = 11.74$, $\varepsilon_\infty / \varepsilon_v = 15.54$ (ε_v is the vacuum dielectric permittivity), $\hbar\omega_0 = 29.5\text{meV}$, and the energy gap, $E_g = 0.824\text{eV}$; b) For the InAs/AlAs the band-offsets are considered as $V_b^e = 1.5\text{eV}$, $V_b^h = 0.75\text{eV}$ (Vurgaftman et. al., 2001); c) For the value of QD height L=2.3nm which is considered, we find 1 electron and 3 hole levels in the quantum-well in the z direction. By setting $\hbar\omega_e = 0.065\text{eV}$ and $\omega_e / \omega_h = 3$, (according to the literature (Hawrylak, 1999; Shumway et. al., 2001) the exciton and biexciton eigenvalues obtained for this material parameters are as follows, $\varepsilon_1^{(1)} = 1.5792\text{eV}$, $\varepsilon_2^{(1)} = 1.6696\text{eV}$, $\varepsilon_3^{(1)} = 1.6736\text{eV}$ (all three two-fold degenerate), $\varepsilon_1^{(2)} = 3.1617\text{eV}$, $\varepsilon_2^{(2)} = 3.2429\text{eV}$ -three-fold degenerate, $\varepsilon_3^{(2)} = 3.24345\text{eV}$, and $\varepsilon_4^{(2)} = 3.24719\text{eV}$ -four-fold degenerate. Consequently, the inter-level bi/exciton energy is not close of the LO phonon energy and the mixing of the bi/exciton states by phonons is absent.

3.2 Confined optical phonons in semiconductor quantum dots

There are several theoretical models which investigates the optical phonon modes in semiconductors with low dimensionality. Generally, the LO phonons are considered as the main contributors to the electron-phonon coupling in polar semiconductors in the relaxation processes. Based on the continuum approach for long-wavelength optical phonons of (Born & Huang, 1998), macroscopic approaches, such as the dielectric continuum (DC) model (Fuchs & Kliewer, 1965; Klein et. al. 1990)), the multimode DC model (Klimin et. al., 1995), the mixed mechanical-electrostatic model (Roca et. al., 1994), and the hydrodynamic model (Ridley, 1989) have been developed. Microscopic approaches have also been proposed (Huang & Zhu, 1988; Rücker et. al., 1991).

The shape of QD plays a major role in setting the type of confined phonon modes and the strength of the exciton-phonon interaction. For spherical QD, the problem of the polaron was the most intensive studied case. One of the conclusions of the studies is that the inside QD, the electron-surface optical phonon interaction is absent (Melnikov & Fowler, 2001). Physically, this can be explained within the adiabatic picture: the electron is fast oscillating and in the ground state, which has a spherical symmetry of the charge distribution, the average surface ionic polarization charge is zero. For other shapes, the geometry itself brings additional complications in the study of the exciton-phonon interaction. Next, we extrapolate the above observation regarding the absence of electron-surface LO phonon interaction in spherical QD to the cylindrical shape case. The approximation is supported by the results obtained by (Cheche et. al., 2011), where calculus shows the exciton-bulk LO phonon interaction in such cylindrical QDs is dominant. Consequently, in the analysis of the optical spectra from the next sections, we consider the bulk LO phonons as the main contributors to the (multi)exciton-LO phonon interaction.

3.3 Optical spectra of spherical semiconductor quantum dots. A non-adiabatic treatment

Non-adiabatic treatments, necessary when the electron-hole pair (EHP) level spacing is comparable to the LO phonon energy, have been proposed (Cheche et. al., 2005; Fomin et. al., 1998; Takagahara, 1999; Vasilevskiy et. al., 2004; Verzelen et. al., 2002). Following (Cheche et. al. 2005; Cheche & Chang, 2005) in this section a non-adiabatic treatment of optical absorption in QDs is presented. The theoretical tool we develop: i) confirms existence of resonances accompanying the LO satellites in the optical spectra; ii) explains the temperature effect on the optical spectra. The Hamiltonian of the EHP-LO phonon reservoir we use is described by an extension of the Huang-Rhys model of F centers of the type described in Section 2.2,

$$H = H_{EHP} + H_{ph} + H_{EHP-ph}, \tag{19}$$

where $H_{EHP} \equiv \sum_f E_f B_f^+ B_f$, $H_{ph} \equiv \sum_{\mathbf{q}} \hbar\omega_{\mathbf{q}} b_{\mathbf{q}}^+ b_{\mathbf{q}}$, $H_{EHP-ph} \equiv \sum_{\mathbf{q},f,f'} M_{\mathbf{q}}^{ff'} B_f^+ B_{f'} (b_{\mathbf{q}} + b_{-\mathbf{q}}^+)$, B_f^+ (B_f) are the exciton operators already introduced in section 3.1.1, $b_{\mathbf{q}}^+$ ($b_{\mathbf{q}}$) are the bosonic creation (annihilation) operators of the phonons of mode $\mathbf{q}$, $M_{\mathbf{q}}^{ff'} \equiv \langle f | M_{\mathbf{q}} | f' \rangle$ is the coupling matrix element, $\omega_{\mathbf{q}}$ is the frequency of the phonon mode with wave vector $\mathbf{q}$, and E_f ($|f\rangle$) are the EHP eigenvalues (eigenstates) of the exciton system. The absorption coefficient for a single QD is given by (Mittin et al., 1999)

$$\alpha(\omega) = \frac{2\pi\hbar\omega}{ncE_0^2 V_0} R_{abs} \tag{20.a}$$

where ω is the frequency, E_0 is the amplitude of the monochromatic radiation field, n is the refractive index of the environment, V_0 is the absorptive volume, and R_{abs} is the radiation absorption rate. R_{abs} is calculated with the Fermi Golden Rule as follows.

$$R_{abs} = \frac{2\pi}{\hbar} \mathrm{Av} \sum_{\mathsf{GF}} |W_{\mathsf{GF}}|^2 \delta(\hbar\omega + E_{\mathsf{G}} - E_{\mathsf{F}}) \tag{20.b}$$

The average Av involved by Eq. (20.b) means a quantum average over the finite number of the exciton states in the QD and a statistical average over the phonon modes at thermal equilibrium. In Eq. (20.b), E_{G} is the energy of the system in the ground state (no exciton) $|\mathsf{G}\rangle = |0\rangle|\gamma\rangle$ ($|\gamma\rangle$ is the phonon state), E_{F} is the energy of the system in one of the exciton+phonons states $|\mathsf{F}\rangle = |f;\varphi\rangle$ ($|f\rangle, |\varphi\rangle$ is the exciton, phonon state, respectively), and $W_{\mathsf{GF}} = \langle \mathsf{G}|W|\mathsf{F}\rangle$ is the transition probability between the initial state $|\mathsf{G}\rangle$ and the final state $|\mathsf{F}\rangle$, with W from Eq. (10.a). Greek letters are used for phonon states, Latin letters for exciton states, and capital handwriting letters for all system. Eq. (20.b) can explicitly be written as follows

$$\begin{aligned} R_{abs} &= \frac{2\pi}{\hbar} Tr\left\{ \rho \sum_{\mathsf{GF}} |W_{\mathsf{GF}}|^2 \delta(\hbar\omega + E_{\mathsf{G}} - E_{\mathsf{F}}) \right\} \\ &= \frac{1}{\hbar^2} \int_{-\infty}^{\infty} dt e^{i\omega t} Tr\left\{ \rho \sum_{\mathsf{G}} \langle \mathsf{G}| e^{itH_{ph}/\hbar} W e^{-itH/\hbar} \left(\sum_{\mathsf{F}} |\mathsf{F}\rangle\langle \mathsf{F}| + |\mathsf{G}\rangle\langle \mathsf{G}| \right) W |\mathsf{G}\rangle \right\} \\ &= \frac{1}{\hbar^2} \int_{-\infty}^{\infty} dt e^{i\omega t} Tr\left\{ \rho \sum_{\mathsf{G}} \langle \mathsf{G}| e^{itH_{ph}/\hbar} W e^{-itH/\hbar} W |\mathsf{G}\rangle \right\} = \frac{1}{\hbar^2} \int_{-\infty}^{\infty} dt e^{i\omega t} \left\langle \langle 0|\widetilde{W}(t) W|0\rangle \right\rangle_0 \end{aligned} \tag{20.c}$$

where $\rho = \sum_{\nu} |\nu\rangle \rho_{\nu} \langle \nu|$ is the density matrix of the phonons, with $\rho_{\nu} = e^{-\beta E_{\nu}} / Tr\left(e^{-\beta H_{ph}}\right)$ the probability of the phonon state $|\nu\rangle$ in the equilibrium statistical ensemble of the phonons, and $Tr\{A\} = \sum_{\nu} \langle \nu|A|\nu\rangle = \sum_{\nu} A_{\nu\nu}$, $\langle A\rangle_0 \equiv Tr\{\rho A\} = \sum_{\nu} \rho_{\nu} \langle \nu|A|\nu\rangle = \sum_{\nu} \rho_{\nu} A_{\nu\nu}$. The closure relation $\sum_{\mathsf{F}} |\mathsf{F}\rangle\langle \mathsf{F}| + |\mathsf{G}\rangle\langle \mathsf{G}| = \mathbf{1}$ was used in the second equality of Eq. (20.c), where the operator $\sum_{\mathsf{G}} |\mathsf{G}\rangle\langle \mathsf{G}| = \sum_{\mu} |0\rangle|\mu\rangle\langle \mu|\langle 0|$, which has no effect on the matrix element was inserted. If using an adiabatic picture the state $|\mathsf{F}\rangle$ is written as a product of states, $|\mathsf{F}\rangle = |f;\varphi\rangle = |f\rangle|\varphi\rangle$, and the meaning of the closure relation is more transparent:

$$\sum_{\mathsf{F}} |\mathsf{F}\rangle\langle \mathsf{F}| + |\mathsf{G}\rangle\langle \mathsf{G}| = \sum_{f,\nu} |f\rangle|\nu\rangle\langle \nu|\langle f| + \sum_{\mu} |0\rangle|\mu\rangle\langle \mu|\langle 0| = \left(\sum_{f} |f\rangle\langle f| + |0\rangle\langle 0| \right) \sum_{\nu} |\nu\rangle\langle \nu| = \mathbf{1}.$$

In Eq. (20.c), $\widetilde{W}(t) = e^{itH_{ph}/\hbar} W e^{-itH/\hbar}$. Eqs. (20a-c) give

$$\alpha(\omega) = \frac{2\pi e^2}{ncm_0^2 \hbar\omega V_0} \int_{-\infty}^{\infty} dt e^{i\omega t} \left\langle \langle 0|\widetilde{W}(t) W|0\rangle \right\rangle_0 \tag{21.a}$$

By using the bosonic commutation rules for creation and annihilation of EHP and phonons, the operator relation $e^{A+B} = e^{A}e^{B}e^{-[A,B]/2}$, we write Eq. (21.a) as follows:

$$\alpha(\omega) = \frac{2\pi e^2}{ncm_0^2\hbar\omega V_0} \sum_{f,f'\neq 0} \left[P_{0f}P_{f'0} \int_{-\infty}^{\infty} dt \exp[i(\omega-\omega_f)t] \langle 0|B_f \left\langle \hat{T}\exp\left[-\frac{i}{\hbar}\int_0^t dt_1 \widetilde{V}(t_1)\right]\right\rangle_0 B_{f'}^{+}|0\rangle \right] \quad (21.b)$$

where $\hat{T}$ is the time-ordered operator, $\widetilde{V}(t) = \exp(itH_0/\hbar)H_{EHP-ph}\exp(-itH_0/\hbar)$, $\left\langle \hat{T}\exp\left[-\frac{i}{\hbar}\int_0^t dt_1 \widetilde{V}(t_1)\right]\right\rangle_0 \equiv \langle U(t)\rangle_0$, $H_0 \equiv H_{EHP} + H_{ph}$, $P_{0f} \equiv \langle 0|(\boldsymbol{\varepsilon}\cdot\mathbf{P})|f\rangle$, and $\mathbf{P} \equiv \sum_i \mathbf{p}_i$ is the total electronic momentum (with $\mathbf{p}_i$ the electron momentum). Further progress is achieved by using the cumulant expansion method in Eq. (21.b). For dispersionless LO phonons (Einstein model) of frequency ω_0, Eq. (21.b) can be approximated by the expression (Cheche and Chang, 2005)

$$\alpha(\omega) = \frac{2\pi e^2}{ncm_0^2\hbar\omega V_0} \sum_{p,s\neq 0} \left[P_{0p}P_{s0} \int_{-\infty}^{\infty} dt \exp[i(\omega-\omega_p)t] \exp(-\sum_{i\neq 0} G_{piis}) \right], \quad (22)$$

where $G_{piis} \equiv g_{piis}\omega_0^2 I(t,p,i,i,s)$, $g_{kk'pp'} \equiv \sum_{\mathbf{q}} \left[M_{\mathbf{q}}^{kk'} M_{-\mathbf{q}}^{pp'} / (\hbar\omega_0)^2 \right]$, ($g_{pppp} \equiv g_p$ is the Huang-Rhys factor), $I(t,k,k',k',p') = \int_0^t dt_1 \int_0^{t_1} dt_2 \exp(it_1\omega_{kk'})\exp(it_2\omega_{k'p'})D^0(t_1-t_2)$, $D^0(t_1-t_2) = \left[\overline{N}\exp(i\omega_0(t_1-t_2)) + (\overline{N}+1)\exp(-i\omega_0(t_1-t_2))\right]$, $\overline{N} = 1/\left(e^{\beta\hbar\omega_0}-1\right)$, and $\omega_{ij} = \left(E_i - E_j\right)/\hbar$.

If the off-diagonal coupling terms in Eq. (19) are disregarded then Eq. (22) is exact and it *recovers* the adiabatic limit (the Franck-Condon progression):

$$\alpha^{ad}(\omega) = \frac{4\pi^2 e^2}{ncm_0^2\hbar\omega V_0} \sum_{f\neq 0} \left\{ \left|P_{0f}\right|^2 \exp\left[-g_f(2\overline{N}+1)\right] \right. \\ \left. \times \sum_{n=-\infty}^{\infty} I_n\left(2g_f\sqrt{\overline{N}(\overline{N}+1)}\right) \exp(n\beta\hbar\omega_0/2) \right\} \delta\left(\omega-\omega_f+\Delta_{ad}^f - n\omega_0\right) \quad (23)$$

where I_n are the modified Bessel functions, and $\Delta_f^{ad} = \omega_0 \sum_{\mathbf{q}} \left(\left|M_{\mathbf{q}}^{ff}\right|^2 \hbar^{-2}\omega_0^{-2} \right) \equiv \omega_0 g_f$ is the self-energy. The relative intensity of absorption lines is given by the coefficients of the Dirac delta functions.

Next, we adopt the spherical model from section 3.1.1 for spherical GaAs microcrystallites embedded in AlAs matrix. The quantity $P_{0f}P_{f'0} = \left|\mathbf{p}_{cv}^0\right|^2 3^{-1} A_{n_e n_h l} A_{n_e' n_h' l'} \delta_{m_e m_h} \delta_{m_e' m_h'}$ in Eq. (21.b) is obtained by averaging over all space polarization directions. The Fröhlich coupling is written for dispersionless bulk LO phonons (for a spherical QD the interface modes do not couple with the exciton states (Melnikov & Fowler, 2002)). Within the pure-EHP approximation the EHP-phonon interaction reads (Voigt et. al., 1979; Nomura & Kobayashi ,1992)

$M_{\mathbf{q}}^{ff'} \to M_{\mathbf{q}}^{ab;a'b'} = V_0^{-1/2} f_0 q^{-1} \int d\mathbf{r}_e \, d\mathbf{r}_h \, \varphi_a^*(\mathbf{r}_e)\varphi_b^*(\mathbf{r}_h)\varphi_{a'}(\mathbf{r}_e)\varphi_{b'}(\mathbf{r}_h)[\exp(i\mathbf{q}\mathbf{r}_e) - \exp(i\mathbf{q}\mathbf{r}_h)]$, where f_0 is the Fröhlich coupling constant. Explicit expression of $M_{\mathbf{q}}^{ff'}$ for spherical QDs can be found in (Cheche and Chang, 2005).

For the only two optical levels which appear at $R_0 = 20$ (see Fig. 3), with an inter-level energy of approximately $11\hbar\omega_0$, the plot of absorption spectrum centred on the line A_0 obtained from Eq. (22) and that given by the adiabatic expression, Eq. (23) are, as expected, practically identical. Situation is different for $R_0 = 32$ Å, where the dark level D_1 is located between two optical levels A_0 and D_0 (see Fig. 3). Contribution of the optical and dark levels to the absorption centered on line A_0 is included in the following expression:

$$\alpha_1(\omega) = \frac{2\pi e^2 \left|\mathbf{p}_{cv}^0\right|^2}{3ncm_0^2\hbar\omega V_0} A_{110}^2 \exp(-\Lambda_1) \sum_{p=-\infty}^{\infty} \sum_{k,r=0}^{\infty} \sum_{s,t=0}^{\infty} \Bigg[I_p\left(2g_1\sqrt{\overline{N}(\overline{N}+1)}\right)$$
$$\times \left(\frac{\overline{N}g_{1221}\beta^2}{k!}\right)^k \left(\frac{(\overline{N}+1)g_{1221}\gamma^2}{r!}\right)^r \left(\frac{\overline{N}g_{1331}\overline{\beta}^2}{s!}\right)^s \left(\frac{(\overline{N}+1)g_{1331}\overline{\gamma}^2}{t!}\right)^t \exp\left(\frac{-p\hbar\omega_0}{2k_BT}\right) \quad (24)$$
$$\times \delta[\omega - \omega_1 + \Delta_1 + p\omega_0 - k(\omega_{21} - \omega_0) - r(\omega_{21} + \omega_0) - s(\omega_{31} - \omega_0) - t(\omega_{31} + \omega_0)]\Bigg]$$

with $\overline{\beta} \equiv \omega_0 / (\omega_{31} - \omega_0)$ and $\overline{\gamma} \equiv \omega_0 / (\omega_{31} + \omega_0)$. The non-adiabaticity effect expressed by Eq. (24) is shown in Fig. 5, where the absorption spectra at different temperatures are plotted (we dressed the lines by Lorentzians with a finite width of 15meV to simulate the EHP-acoustic phonons interaction). The adiabatic spectrum obtained with Eq. (23) has no temperature-induced shift and its maxima are not significantly changed with temperature. The following quantities obtained within the adopted QD model have been used: $E_1 = 1.8822\text{eV}$, $E_2 = 2.0738\text{eV}$, $\mathrm{E}_3 = 1.9496\text{eV}$, $g_1 = 0.039$, $g_{1221} = 0.234$, and $g_{1331} = 0.904$. The stronger accompanying resonances are marked by arrows. The energy of some resonances are indicated by factors which multiply the LO phonon energy; they are placed to the left of the lines or arrows. The temperature dependence of the spectra, weak in the case of adiabatic treatment, becomes important now. Thus, decrease of intensity (by 37%) and red shift (from 1.87eV to 1.85eV) of the 0PL lines are obtained when temperature increases from 10K to 300K. This agrees with the behavior observed experimentally for CdTe QDs (Besombes et. al., 2001). On the other hand, the simulated Huang-Rhys factors reach values larger by two orders of magnitude than those of the bulk phase (0.0079 obtained from (Nomura & Kobayashi, 1992)). A similar behavior is reported for small self-assembled InAs/GaAs QDs by (García-Cristobal et. al. 1999). Thus, by the non-adiabatic activated channel at +0.86LO, the simulated Huang-Rhys factor obtained as the ratio of the line intensities for this accompanying resonance increases from 0.084 at $T = 10\text{K}$ to 0.23 at $T = 200\text{K}$. On the other hand, the non-adiabaticity effect manifests by strong resonances at 2.9LO (see Fig. 5), close to the third LO phonon replica as reported by some experiments, see, e.g., (Heitz et. al., 1997). The usual Franck-Condon progression is obtained by the adiabatic treatment (see the dotted line in Fig. 5).

Concluding this section, the non-adiabatic treatment presented, in accordance with the experimental observation, predicts: (i) accompanying resonances to the LO phonon satellites in the optical spectra of QDs; (ii) red shift of the 0LO phonon lines and increased intensities of the accompanying resonances with temperature in the absorption spectra of QDs.

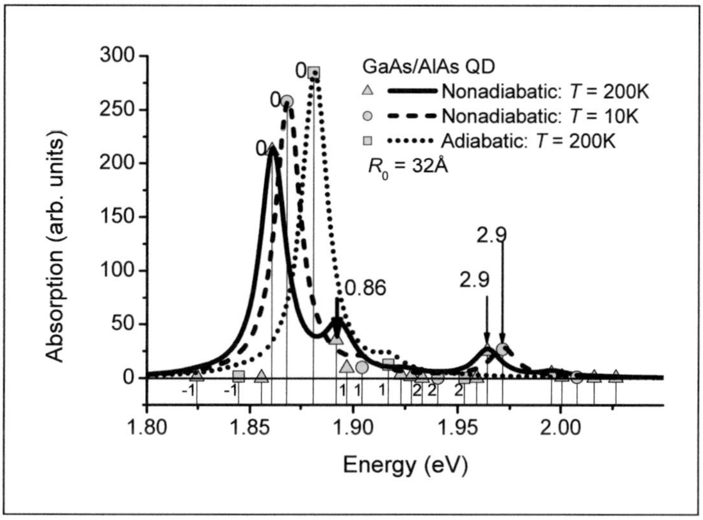

Fig. 5. Simulated absorption spectra of GaAs/AlAs nanocrystal QDs.

3.4 Phonon effect on the exciton and biexciton binding energy in cylindrical semiconductor quantum dots

In this section we discuss the exciton and biexciton emission spectra of polar semiconductor QDs within an adiabatic approach by using the configurational interaction method introduced in section 3.1.2. By taking into account the Fröhlich coupling between bi/exciton complexes and LO phonons, we simulate the *resonantly excited PL* spectrum (laser energy = detection energy + $n\cdot$LO energy, with n non-negative integer, (Sénès *et. al.* 2005)) with linearly polarized (LP) light of InAs/AlAs cylindrical QDs. The exciton and biexciton binding energy for such QDs is also evaluated. In accordance with Eq. (9), we consider the following (multi)exciton-phonon Hamiltonian:

$$H^{(m)}=\sum_f \varepsilon_f^{(m)} X_f^{m+} X_f^m+\sum_{\mathbf{q}} \hbar\omega_0 b_{\mathbf{q}}^+ b_{\mathbf{q}}+\sum_{\mathbf{q},f} M_{\mathbf{q}f}^{(m)} X_f^{m+} X_f^m (b_{\mathbf{q}}+b_{-\mathbf{q}}^+)=H_{QD}^{(m)}+H_{ph}+H_{QD-ph}^{(m)} \quad (25)$$

where $m=1$ for exciton, $m=2$ for biexciton, $b_{\mathbf{q}}^+$ ($b_{\mathbf{q}}$) are the bosonic creation (annihilation) operators of the phonons of mode $\mathbf{q}$, $M_{\mathbf{q}}^{(m)}$ is the Fröhlich coupling, $M_{\mathbf{q}f}^{(m)}=\langle X_f^m | M_{(m)\mathbf{q}} | X_f^m \rangle$ and $M_{\mathbf{q}f}^{(m)}=M_{-\mathbf{q}f}^{(m)*}$ (from Hermiticity of $H^{(m)}$), ω_0 is the frequency of the dispersionless LO phonons, and $\sum_f \varepsilon_f^{(m)} X_f^{m+} X_f^m$ is the (multi)exciton H_D from Eq. (13) written in the language of (multi)exciton complexes. According to discussion from section 3.1.3, the Fröhlich electron-bulk LO phonon coupling is an acceptable approach for QD with high geometrical symmetry, where the interface modes are usually weak. Thus, for the exciton-LO phonon coupling (Voigt et. al., 1979; Nomura & Kobayashi, 1992)

$$M_{\mathbf{q}f}^{(1)} \equiv \frac{f_0}{q\sqrt{V_0}}\left\langle X_f^1\left|e^{i\mathbf{q}\cdot\mathbf{R}_e}-e^{i\mathbf{q}\cdot\mathbf{R}_h}\right|X_f^1\right\rangle = \frac{f_0}{q\sqrt{V_0}}\sum\left|C_{\alpha\sigma,\beta\tau}^f\right|^2\left[\left\langle\varphi_\alpha\left|e^{i\mathbf{q}\cdot\mathbf{R}}\right|\varphi_\alpha\right\rangle-\left\langle\varphi_\beta\left|e^{i\mathbf{q}\cdot\mathbf{R}}\right|\varphi_\beta\right\rangle\right], \quad (26)$$

and for biexciton-LO phonon coupling (Peter et. al., 2004)

$$M_{\mathbf{q}f}^{(2)} = \frac{f_0}{q\sqrt{V_0}}\left\langle X_f^2\left|e^{i\mathbf{q}\cdot\mathbf{R}_{e1}}+e^{i\mathbf{q}\cdot\mathbf{R}_{e2}}-e^{i\mathbf{q}\cdot\mathbf{R}_{h1}}-e^{i\mathbf{q}\cdot\mathbf{R}_{h2}}\right|X_f^2\right\rangle = \frac{f_0}{q\sqrt{V_0}}\sum_{\substack{\alpha_1\sigma_1,\alpha_2\sigma_2,\\ \beta_1\tau_1,\beta_2\tau_2}}\left|C^f_{\substack{\alpha_1\sigma_1,\alpha_2\sigma_2,\\ \beta_1\tau_1,\beta_2\tau_2}}\right|^2$$
$$\times\left[\left\langle\varphi_{\alpha_1}\left|e^{i\mathbf{q}\cdot\mathbf{R}}\right|\varphi_{\alpha_1}\right\rangle+\left\langle\varphi_{\alpha_2}\left|e^{i\mathbf{q}\cdot\mathbf{R}}\right|\varphi_{\alpha_2}\right\rangle-\left\langle\varphi_{\beta_1}\left|e^{i\mathbf{q}\cdot\mathbf{R}}\right|\varphi_{\beta_1}\right\rangle-\left\langle\varphi_{\beta_2}\left|e^{i\mathbf{q}\cdot\mathbf{R}}\right|\varphi_{\beta_2}\right\rangle\right] \quad (27)$$

where $f_0 = \sqrt{2\pi\hbar\omega_0 e^2\left(\varepsilon_\infty^{-1}-\varepsilon_0^{-1}\right)}$ is the Fröhlich coupling constant, and V_0 is the QD volume.

The emission spectrum of *single* QD corresponding to exciton and biexciton-exciton recombinations is obtained with the Fermi Golden Rule, that should be adapted to the composed system, multi(exciton)+phonons. The statistical operator $e^{-\beta H^{(m)}}/Tr\left\{e^{-\beta H^{(m)}}\right\}$ is used for the statistical average in the Kubo formula of the optical conductivity. When applying the Fermi Golden Rule for the system multi(exciton)+phonons, we need to consider a *statistical* average for phonons and a *quantum* average for the *finite* number of multi(exciton) states in the QD. On the other hand, within the adiabatic approximation, the electronic potential energy surface is the potential for phonons in the QD. We imaginarily decompose temporally the absorption process and consider that before switching on the electron-phonon interaction, the electron-hole potential energy surface is raised vertically from the lowest potential energy surface of the exciton vacuum state to the excited potential energy surface (see dotted line parabola in Fig. 6). Then, we consider the electron-phonon interaction is switched on and as a result the potential energy surface is further modified to the new potential energy surface of the interacting multi(exciton)+phonon system, see upper solid line parabola in Fig. 6 and comments in (Odnoblyudov et. al., 1999).

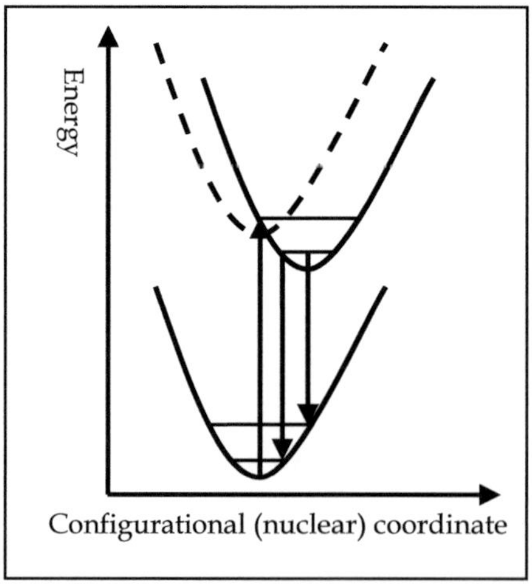

Fig. 6. Schematic exciton of the potential energy surface involved in transition.

Thus, according to its PES, each (multi)exciton state is characterized by its density matrix. To take into account the above considerations, we project the statistical operator of the phonon system interacting with the (multi)exciton state on the state $\left|X_i^m\right\rangle$ and write $\rho_i^{(m)} = \exp\left[-\beta\left(H_{ph} + h_i^{(m)}\right)\right]/Z_i^{(m)}$, with $Z_i^{(m)} = Tr\left\{\left\langle X_i^m \right| \exp\left[-\beta\left(H_{ph} + H_{QD-ph}^{(m)}\right)\right]\left|X_i^m\right\rangle\right\}$, and $h_i^{(m)} = \sum_{\mathbf{q}} M_i^{(m)}(b_{\mathbf{q}} + b_{-\mathbf{q}}^{+})$. The partition function $Z_i^{(m)} = \Pi_{\mathbf{q}} Z_{i\mathbf{q}}^{(m)}$ is the product of the partition functions for each mode having the wave vector $\mathbf{q}$. By dropping all $\mathbf{q}$ subscripts, the partition function for a single mode reads, $Z_i^{(m)} = Tr\left\{\exp\left[-\beta\left(\hbar\omega_0 b^+ b + M_i^{(m)}(b + b^+)\right)\right]\right\}$. It can be evaluated by using a canonical transformation

$$Z_i^{(m)} = Tr\left\{\exp\left(S_i^{(m)}\right)\exp\left[-\beta\left(\hbar\omega_0 b^+ b + M_i^{(m)}(b + b^+)\right)\right]\exp\left(-S_i^{(m)}\right)\right\} \tag{28.a}$$

where the anti-Hermitian operator is defined as $S_i^{(m)} = \left(M_i^{(m)*} b^+ - M_i^{(m)} b\right)/(\hbar\omega_0)$. With $\exp\left[S_i^{(m)}\right] b^+ \exp\left[-S_i^{(m)}\right] = b^+ - M_i^{(m)}/\hbar\omega_0$, $\exp\left[S_i^{(m)}\right] b \exp\left[-S_i^{(m)}\right] = b - M_i^{(m)}/\hbar\omega_0$ one obtains

$$Z_i^{(m)} = Tr\left\{\exp\left\{-\beta\left[\hbar\omega_0 b^+ b + (\hbar\omega_0)^{-1}\left|M_i^{(m)}\right|^2\right]\right\}\right\} = \left(1 - \exp(-\beta\hbar\omega_0)\right)^{-1}\exp\left[-\beta(\hbar\omega_0)^{-1}\left|M_i^{(m)}\right|^2\right],$$

and

$$Z_i^{(m)} = \exp\left(-\beta\hbar\omega_0 g_i^{(m)}\right)\prod_{\mathbf{q}}\left(1 - e^{-\beta\hbar\omega_0}\right)^{-1} \tag{28.b}$$

in which $g_i^{(m)} = \sum_{\mathbf{q}}\left|M_{\mathbf{q}i}^{(m)}\right|^2/\left(\hbar^2\omega_0^2\right)$ is the Huang-Rhys factor.

With the Fermi Golden Rule, the exciton emission spectrum is given by

$$\begin{aligned}
I_X(\omega) &= \frac{2\pi}{\hbar}\sum_{\mathrm{I}}\mathrm{Av}_{\mathrm{I}}^{(1)}\sum_{\mathrm{G}}\left|W_{\mathrm{GI}}\right|^2\delta\left(\hbar\omega + E_{\mathrm{G}} - E_{\mathrm{I}}\right) = \frac{1}{\hbar^2}\int_{-\infty}^{\infty} dt e^{\mathrm{i}\omega t} \\
&\times\sum_{\mathrm{I}}\mathrm{Av}_{\mathrm{I}}^{(1)}\left\langle \mathrm{X}_{i\nu}^1\right| e^{-\mathrm{i}tH^{(1)}/\hbar} H_{X^1-R}\left(\sum_{j,\mu}\left(\left|\mathrm{X}_{j\mu}^1\right\rangle\left\langle \mathrm{X}_{j\mu}^1\right|\right) + \left|\mathrm{G}\right\rangle\left\langle\mathrm{G}\right|\right) e^{\mathrm{i}tH_{ph}/\hbar} H_{X^1-R}\left|\mathrm{X}_{i\nu}^1\right\rangle \\
&= \frac{1}{\hbar^2}\int_{-\infty}^{\infty} dt e^{\mathrm{i}\omega t}\sum_{\nu,i}\left\langle \nu\right|\rho_i^{(1)}\left|\nu\right\rangle\left\langle\nu\right|\left\langle X_i^1\right| e^{-\mathrm{i}tH^{(1)}/\hbar} H_{X^1-R} e^{\mathrm{i}tH_{ph}/\hbar} H_{X^1-R}\left|X_i^1\right\rangle\left|\nu\right\rangle \\
&= \frac{1}{\hbar^2}\int_{-\infty}^{\infty} dt e^{\mathrm{i}\omega t}\sum_i Tr\left\{\left\langle X_i^1\right| e^{-\beta\left(H_{ph} + h_i^{(1)}\right)/\hbar} Z_i^{(1)-1} e^{-\mathrm{i}tH^{(1)}/\hbar} H_{X^1-R} e^{\mathrm{i}tH_{ph}/\hbar} H_{X^1-R}\left|X_i^1\right\rangle\right\}
\end{aligned} \tag{29}$$

where $\left|\mathrm{X}_{i\nu}^1\right\rangle \equiv \left|X_i^1\right\rangle\left|\nu\right\rangle$ is the initial state with energy E_{I} and $\left|\mathrm{G}\right\rangle$ is the ground state with energy E_{G}. H_{X^1-R} does not couple the exciton-phonon states, that is relation $\left\langle \mathrm{X}_{in}^1\right| H_{X^1-R}\left|\mathrm{X}_{jn}^1\right\rangle = 0$ holds, and in Eq. (29) we inserted $\sum_{j,\mu}\left(\left|\mathrm{X}_{j\mu}^1\right\rangle\left\langle\mathrm{X}_{j\mu}^1\right|\right)$ to make use of the closure relation $\sum_{j,\mu}\left(\left|X^1_{j\mu}\right\rangle\left\langle X^1_{j\mu}\right|\right)+|G\rangle\langle G|=\mathbf{1}$. Since $\left[H^{(1)}_{QD}, H_{ph}+H^{(1)}_{QD-ph}\right]=0$, by using the operator relation $e^{A+B}=e^{A}e^{B}e^{-[A,B]/2}$, we have

$$\left\langle X^1_i\right|e^{-itH^{(1)}/\hbar}=\left\langle X^1_i\right|e^{-itH^{(1)}_{QD}/\hbar}e^{-it(H_{ph}+H^{(1)}_{QD-ph})/\hbar}=e^{-it\varepsilon^{(1)}_i/\hbar}e^{-it(h^{(1)}_i+H_{ph})/\hbar}\left\langle X^1_i\right|, \tag{30.a}$$

and

$$H_{X^1-R}H_{X^1-R}\left|X^1_i\right\rangle=\omega^{-2}\sum_i\left|\mathbf{C}^i\right|^2\left|X^1_i\right\rangle, \tag{30.b}$$

With Eqs. (30.a, b), Eq. (29) reads

$$I_X(\omega)=\frac{1}{\hbar^2}\int_{-\infty}^{\infty}dte^{i\omega t}e^{-it\varepsilon^{(1)}_i/\hbar}\sum_i\left|\mathbf{C}^i\right|^2F^{(i)}_{1ph}(t) \tag{31}$$

where the correlation function is

$$F^{(i)}_{1ph}(t)=Tr\left\{e^{-\beta(h^{(1)}_i+H_{ph})/\hbar}Z^{(1)-1}_i e^{-it(h^{(1)}_i+H_{ph})/\hbar}e^{itH_{ph}/\hbar}\right\}. \tag{32.a}$$

Eq. (32.a) is transformed by using the canonical transformation,

$$\begin{aligned}F^{(i)}_{1ph}(t)&=Tr\left\{e^{S^{(1)}_i}e^{-\beta(h^{(1)}_i+H_{ph})/\hbar}Z^{(1)-1}_i e^{-S^{(1)}_i}e^{S^{(1)}_i}e^{-it(h^{(1)}_i+H_{ph})/\hbar}e^{-S^{(1)}_i}e^{S^{(1)}_i}e^{itH_{ph}/\hbar}e^{-S^{(1)}_i}\right\}\\&=Tr\left\{e^{-\beta(\overline{h}^{(1)}_i+\overline{H}_{ph})/\hbar}Z^{(1)-1}_i e^{-it(\overline{h}^{(1)}_i+\overline{H}_{ph})/\hbar}e^{it\overline{H}_{ph}/\hbar}\right\}\end{aligned} \tag{32.b}$$

where, generally ($m=1, 2$),

$$S^{(m)}_i=\frac{1}{\hbar\omega_0}\sum_{\mathbf{q}}M^{(m)}_{i\mathbf{q}}\left(b^+_{-\mathbf{q}}-b_{\mathbf{q}}\right), \tag{32.c}$$

and $\overline{H}_{ph}+\overline{h}^{(1)}_i=H_{ph}-\hbar g^{(1)}_i\omega_0$, $\overline{H}_{ph}=H_{ph}-h^{(1)}_i+\hbar g^{(1)}_i\omega_0$. With these two last equalities substituted in Eq. (32.b) we write

$$\begin{aligned}F^{(i)}_{ph}(t)&=e^{2ig^{(1)}_i\omega_0 t}Tr\left\{e^{-\beta H_{ph}/\hbar}Z^{-1}_{ph}e^{-itH_{ph}/\hbar}e^{it\left(H_{ph}-h^{(1)}_i\right)/\hbar}\right\}=e^{2ig^{(1)}_i\omega_0 t}Tr\left\{\rho e^{-itH_{ph}/\hbar}e^{it\left(H_{ph}-h^{(1)}_i\right)/\hbar}\right\}\\&=e^{2ig^{(1)}_i\omega_0 t}\left\langle e^{-itH_{ph}/\hbar}e^{it\left(H_{ph}-h^{(1)}_i\right)/\hbar}\right\rangle_0\equiv e^{2ig^{(1)}_i\omega_0 t}\left\langle U^{(i)}_{ph}(t)\right\rangle_0\end{aligned}, \tag{32.d}$$

where Z_{ph} is the partition function of the phonon system. By using the interaction representation, the correlation function reads

$$\left\langle U^{(i)}_{ph}(t)\right\rangle_0=T\left\langle\exp\left[-i\int_0^t dt_1\widetilde{h}^{(1)}_i(t_1)\right]\right\rangle_0 \tag{33.a}$$

where T is the time-ordering operator and

$$\widetilde{h}_i^{(1)}(t) = e^{-itH_{ph}/\hbar}\left(h_i^{(1)}/\hbar\right)e^{itH_{ph}/\hbar} = \sum_{\mathbf{q}} \frac{M_{\mathbf{q}i}^{(1)}}{\hbar}\left(b_{\mathbf{q}}e^{it\omega_0} + b_{-\mathbf{q}}^{+}e^{-it\omega_0}\right) \tag{33.b}$$

Next, to evaluate $\left\langle U_{ph}^{(i)}(t)\right\rangle_0$ we use the linked cluster expansion (Mahan, 2000)

$$\left\langle U_{ph}^{(i)}(t)\right\rangle_0 = \sum(-\mathrm{i})^n\left\langle U_n^{(i)}(t)\right\rangle_0, \quad \left\langle U_n^{(i)}(t)\right\rangle_0 = \frac{1}{n!}\int_0^t dt_1\int_0^t dt_2 \ldots \int_0^t dt_n\left\langle T\widetilde{h}_i^{(1)}(t_1)\ldots\widetilde{h}_i^{(1)}(t_n)\right\rangle_0 \tag{33.c}$$

and since $\widetilde{h}_i^{(1)}$ describes creation or annihilation of a phonon, they are grouped in pairs. Thus,

$$\left\langle U_0^{(i)}(t)\right\rangle_0 = 1, \quad \left\langle U_1^{(i)}(t)\right\rangle_0 = 0,$$

$$\begin{aligned}\left\langle U_2^{(i)}(t)\right\rangle_0 &= \frac{1}{2\hbar^2}\int_0^t dt_1\int_0^t dt_2\sum_{\mathbf{q}}\left|M_{\mathbf{q}i}^{(1)}\right|^2\left[\left(1+\overline{N}\right)e^{\mathrm{i}\omega_0|t_1-t_2|} + \overline{N}e^{-\mathrm{i}\omega_0|t_1-t_2|}\right] \\ &= \sum_{\mathbf{q}}\frac{2g_i^{(1)}}{2}\left[\left(1+\overline{N}\right)\left(1-e^{\mathrm{i}\omega_0 t}\right) + \overline{N}\left(1-e^{-\mathrm{i}\omega_0 t}\right) + \mathrm{i}\omega_0 t\right] = \sum_{\mathbf{q}}\frac{\phi_{\mathbf{q}i}^{(1)}(t)}{2}\end{aligned} \tag{33.d}$$

By using Wick's theorem to pair the boson operators for the terms of higher order one obtains (Mahan, 2000)

$$\left\langle U_{2m}^{(i)}(t)\right\rangle_0 = \frac{1}{m!}\left[\sum_{\mathbf{q}}\frac{\phi_{\mathbf{q}i}^{(1)}(t)}{2}\right]^m \tag{33.e}$$

and, consequently

$$\begin{aligned}\left\langle U_{ph}^{(i)}(t)\right\rangle_0 &= \sum_{m=0}^{\infty}(-1)^m\left\langle U_{2m}^{(i)}(t)\right\rangle_0 = \exp\left[-\sum_{\mathbf{q}}\frac{\phi_{\mathbf{q}i}^{(1)}(t)}{2}\right] \\ &= \exp\left\{-g_i^{(1)}\left[2\overline{N}+1-2\sqrt{\overline{N}(\overline{N}+1)}\cos[\omega_0(t+\mathrm{i}\beta\hbar/2] + \mathrm{i}\omega_0 t\right]\right\} \\ &= e^{-g_i^{(1)}(2\overline{N}+1)}\sum_{l=-\infty}^{\infty}I_l\left[2g_i^{(1)}\sqrt{\overline{N}(\overline{N}+1)}\right]e^{l\beta\hbar\omega_0/2}e^{\mathrm{i}l\omega_0 t}e^{-\mathrm{i}g_i^{(1)}\omega_0 t}\end{aligned} \tag{33.f}$$

With Eqs. (32.d) and (33.f), Eq. (29) that gives the exciton emission spectrum reads

$$I_X(\omega) = \frac{2\pi}{\hbar^2\omega^2}\sum_i\left|\mathbf{C}^i\right|^2 e^{-g_1^i(2\overline{N}+1)}\sum_{l=-\infty}^{\infty}I_l\left[2g_i^{(1)}\sqrt{\overline{N}(\overline{N}+1)}\right]e^{l\beta\hbar\omega_0/2}\delta\left[\omega - \varepsilon_i^{(1)}/\hbar + \left(g_i^{(1)}+l\right)\omega_0\right] \tag{34}$$

where $\mathbf{C}^i$ is defined in Eq. (17.a). I_l is the modified Bessel function obtained from expansion in Eq. (33.f), $\exp[z\cos\theta] = \sum_{l=-\infty}^{\infty}I_l(z)\exp(il\theta)$. Eq. (34) shows the usual phonon progression and comparatively to Eq. (23) in the argument of the Dirac delta function the sign of factor for the phonon progression is changed.

With the Fermi Golden Rule, the biexciton-exciton emission spectrum is given by

$$\begin{aligned} I_{XX}(\omega) &= \frac{2\pi}{\hbar}\sum_{\mathrm{I}} \mathrm{Av}_{\mathrm{I}}^{(2)} \sum_{\mathrm{F}} |W_{\mathrm{IF}}|^2 \delta(\hbar\omega + E_{\mathrm{F}} - E_{\mathrm{I}}) = \frac{1}{\hbar^2}\int_{-\infty}^{\infty} dt e^{\mathrm{i}\omega t} \sum_{\mathrm{I}} \mathrm{Av}_{\mathrm{I}}^{(2)} \langle \mathrm{X}_{i\nu}^{2} | e^{-itH^{(2)}/\hbar} H_{X^2-R} \\ &\times \left(\sum_{f,\mu} |\mathrm{X}_{f\mu}^{2}\rangle\langle \mathrm{X}_{f\mu}^{2}| + |\mathrm{G}\rangle\langle\mathrm{G}| + |\mathrm{X}_{f\mu}^{1}\rangle\langle\mathrm{X}_{f\mu}^{1}| \right) e^{itH^{(1)}/\hbar} H_{X^2-R} |\mathrm{X}_{i\nu}^{2}\rangle \\ &= \frac{1}{\hbar^2}\int_{-\infty}^{\infty} dt e^{\mathrm{i}\omega t} \sum_{\nu,i} \langle \nu | \rho_i^{(2)} | \nu \rangle \langle \nu | \langle X_i^2 | e^{-itH^{(2)}/\hbar} H_{X^2-R} e^{itH^{(1)}/\hbar} H_{X^2-R} | X_i^2 \rangle | \nu \rangle \\ &\equiv \frac{1}{\hbar^2 Z^{(2)}}\int_{-\infty}^{\infty} dt e^{\mathrm{i}\omega t} \sum_{i} Tr\left\{ \langle X_i^2 | e^{-\beta\left(H_{ph}+h_i^{(2)}\right)/\hbar} e^{-itH^{(2)}/\hbar} H_{X^2-R} e^{itH^{(1)}/\hbar} H_{X^2-R} | X_i^2 \rangle \right\} \end{aligned} \tag{35}$$

where $|\mathrm{X}_{i\nu}^{2}\rangle \equiv |X_i^2\rangle|\nu\rangle$ is the biexciton initial state with energy E_{I}, and $|\mathrm{X}_{f\nu}^{1}\rangle \equiv |X_f^1\rangle|\nu\rangle$ is the exciton final state with energy E_{F}. H_{X^2-R} does not couple the biexciton-phonon states and the ground state to the biexciton-phonon states, that is the relations $\langle \mathrm{X}_{i\nu}^{2} | H_{X^2-R} | \mathrm{X}_{j\mu}^{2}\rangle = 0$ and $\langle \mathrm{X}_{i\nu}^{2} | H_{X^2-R} |0\rangle|\mu\rangle = 0$ hold, and in Eq. (35) we inserted $|\mathrm{G}\rangle\langle\mathrm{G}| + \sum_{f,\mu} |\mathrm{X}_{f\mu}^{2}\rangle\langle\mathrm{X}_{f\mu}^{2}|$ to make use of the closure relation, $\sum_{f,\mu}\left(|\mathrm{X}_{f\mu}^{2}\rangle\langle\mathrm{X}_{f\mu}^{2}| + |\mathrm{X}_{f\mu}^{1}\rangle\langle\mathrm{X}_{f\mu}^{1}|\right) + |\mathrm{G}\rangle\langle\mathrm{G}| = \mathbf{1}$. Similarly to Eq. (30.a) we have $\langle X_i^2 | e^{-\mathrm{i}tH^{(2)}/\hbar} = e^{-\mathrm{i}t\varepsilon_i^{(2)}/\hbar} e^{-\mathrm{i}t(h_i^{(2)}+H_{ph})/\hbar} \langle X_i^1 |$. Next, Eq. (17.b) is inserted for H_{X^2-R} in Eq. (35). From the four terms containing four (multi)exciton operators involved by this substitution, only one has non-zero contribution and

$$\begin{aligned} &\langle X_i^2 | e^{-\beta\left(H_{ph}+h_i^{(2)}\right)/\hbar} e^{-itH^{(2)}/\hbar} H_{X^2-R} e^{itH^{(1)}/\hbar} H_{X^2-R} | X_i^2 \rangle \\ &= \omega^{-2} \sum_{a,b,c,d} \mathbf{C}^{ab}\mathbf{C}^{cd*} \langle X_i^2 | e^{-\beta\left(H_{ph}+h_i^{(2)}\right)/\hbar} e^{-itH^{(2)}/\hbar} X_a^{2+} X_b^1 e^{itH^{(1)}/\hbar} X_c^{1+} X_d^2 | X_i^2 \rangle \end{aligned}$$

With additional algebra and making use of $X_a^{2+}X_b^1 = |X_a^2\rangle\langle X_b^1|$, one obtains

$$\begin{aligned} &\langle X_i^2 | e^{-\beta\left(H_{ph}+h_i^{(2)}\right)/\hbar} e^{-itH^{(2)}/\hbar} H_{X^2-R} e^{itH^{(1)}/\hbar} H_{X^2-R} | X_i^2 \rangle \\ &= \omega^{-2} \sum_{f} |\mathbf{C}^{if}|^2 e^{-\mathrm{i}t\left(\varepsilon_i^{(2)}-\varepsilon_f^{(1)}\right)/\hbar} e^{-\beta\left(H_{ph}+h_i^{(2)}\right)/\hbar} e^{-\mathrm{i}t\left(H_{ph}+h_i^{(2)}\right)/\hbar} e^{\mathrm{i}t\left(H_{ph}+h_f^{(1)}\right)/\hbar} \end{aligned} \tag{36}$$

With Eqs. (36), Eq. (35) reads

$$I_{XX}(\omega) = \frac{1}{\hbar^2\omega^2}\int_{-\infty}^{\infty} dt e^{\mathrm{i}\omega t} \sum_{i,f} |\mathbf{C}^{if}|^2 e^{-\mathrm{i}t\left(\varepsilon_i^{(2)}-\varepsilon_f^{(1)}\right)/\hbar} F_{2ph}^{(if)}(t) \tag{37}$$

where the correlation function is

$$F_{2ph}^{(if)}(t) = Tr\left\{e^{-\beta H_{0i}/\hbar} Z_i^{(2)-1} e^{-\mathrm{i}tH_{0i}/\hbar} e^{\mathrm{i}t\left(H_{0i}+V_{if}\right)/\hbar}\right\} \tag{38.a}$$

with $H_{0i} = H_{ph} + h_i^{(2)}$, $H_{ph} + h_f^{(1)} = H_{0i} + h_f^{(1)} - h_i^{(2)} = H_{0i} + V_{if}$, and $Z_i^{(2)}$ is defined by Eq. (28.b). $F_{2ph}^{(if)}(t)$ is evaluated by the same procedure of the canonical transformation used for $F_{1ph}^{(i)}(t)$. Thus, in Eq. (38.a) one inserts the unitary operator $e^{-S_i^{(2)}} e^{S_i^{(2)}}$, with $S_i^{(2)}$ defined by Eq. (32.c) and one obtains

$$F_{2ph}^{(if)}(t) = Tr\left\{e^{-\beta \overline{H}_{0i}/\hbar} Z_i^{(2)-1} e^{-\mathrm{i}t\overline{H}_{0i}/\hbar} e^{\mathrm{i}t\left(\overline{H}_{0i}+\overline{V}_{if}\right)/\hbar}\right\} \tag{38.b}$$

with $\overline{H}_{0i} = H_{ph} - \hbar g_i^{(2)}\omega_0$, $\overline{H}_{0i} + \overline{V}_{if} = H_{ph} + h_{if} + \hbar\omega_0\left(g_{if} - g_b^{(1)}\right)$, $h_{if} = \sum_{\mathbf{q}}\left[\left(M_{\mathbf{q}f}^{(1)} - M_{\mathbf{q}i}^{(2)}\right)\right.$ $\left.\times(b_{\mathbf{q}} + b_{-\mathbf{q}}^{+})\right]$, and $g_{if} = (\hbar\omega_0)^{-2}\sum_{\mathbf{q}}\left|M_{\mathbf{q}f}^{(1)} - M_{\mathbf{q}i}^{(2)}\right|^2$. With these quantities, we rewrite Eq. (38.b) as follows

$$\begin{aligned} F_{2ph}^{(if)}(t) &= e^{\mathrm{i}t\left[g_i^{(2)} - g_f^{(1)} + g_{if}\right]\omega_0}\; Tr\left\{e^{-\beta H_{ph}/\hbar} Z_{ph}^{-1} e^{-\mathrm{i}tH_{ph}/\hbar} e^{\mathrm{i}t\left(H_{ph}+h_{if}\right)/\hbar}\right\} \\ &= e^{\mathrm{i}t\left[g_i^{(2)} - g_f^{(1)} + g_{if}\right]\omega_0}\; Tr\left\{\rho\, e^{-\mathrm{i}tH_{ph}/\hbar} e^{\mathrm{i}t\left(H_{ph}+h_{if}\right)/\hbar}\right\} \\ &\equiv e^{\mathrm{i}t\left[g_i^{(2)} - g_f^{(1)} + g_{if}\right]\omega_0} \left\langle e^{-\mathrm{i}tH_{ph}/\hbar} e^{\mathrm{i}t\left(H_{ph}+h_{if}\right)/\hbar}\right\rangle_0 \equiv e^{\mathrm{i}t\left[g_i^{(2)} - g_f^{(1)} + g_{if}\right]\omega_0} \left\langle U_{ph}^{(if)}(t)\right\rangle_0 \end{aligned} \quad . \tag{38.c}$$

Given the similarity between expressions of the correlation functions (see Eqs. (32.d) and (38.c)), we evaluate $\left\langle U_{ph}^{(if)}(t)\right\rangle_0$ by the same procedure used for evaluation of $\left\langle U_{ph}^{(i)}(t)\right\rangle_0$ and obtain

$$\begin{aligned} \left\langle U_{ph}^{(if)}(t)\right\rangle_0 &= \exp\left\{-\sum_{\mathbf{q}} g\left[\left(1+\overline{N}\right)\left(1-e^{\mathrm{i}\omega_0 t}\right) + \overline{N}\left(1-e^{-\mathrm{i}\omega_0 t}\right) + \mathrm{i}\omega_0 t\right]\right\} \\ &= e^{-g_{if}(2\overline{N}+1)} \sum_{l=-\infty}^{\infty} I_l\left[2g_{if}\sqrt{\overline{N}(\overline{N}+1)}\right] e^{l\beta\hbar\omega_0/2} e^{\mathrm{i}l\omega_0 t} e^{-\mathrm{i}g_{if}\omega_0 t} \end{aligned} \quad . \tag{38.d}$$

With Eqs. (38.c, d), Eq. (37) that gives the biexciton-exciton emission spectrum reads

$$\begin{aligned} I_{XX}(\omega) = \frac{2\pi}{\hbar^2\omega^2} \sum_{f,i} &\left\{\left(\left|\mathbf{C}^{if}\right|^2 \exp[-g_{if}(2\overline{N}+1)]\beta\hbar\omega_0]\right)\right. \\ &\left.\times \sum_{l=-\infty}^{\infty} I_l[2g_{if}\sqrt{\overline{N}(\overline{N}+1)}]\exp\left(l\beta\hbar\omega_0/2\right)\delta\left[\omega - \left(\varepsilon_i^{(2)} - \varepsilon_f^{(1)}\right)/\hbar + \left(g_i^{(2)} - g_f^{(1)} + l\right)\omega_0\right]\right\} \end{aligned} \quad . \tag{39}$$

with $\mathbf{C}^{if}$ defined by Eq. (17.b). g_{if} is function of the difference between the coupling of phonons to the initial biexciton $\left|X_i^2\right\rangle$ and the final exciton state, $\left|X_f^1\right\rangle$; it influences the intensity of the emission line. Note that g_{if} cancels out from the argument of the Dirac delta function from Eq. (39), instead and a difference of the Huang Rhys factors, $g_i^{(2)} - g_f^{(1)}$, is present. Eq. (39) has similarity with Eq. (34), and all characteristics of an emission spectrum are present. The spectra have ω^{-2} dependence. In Eqs. (34) and (39) the argument of the modified Bessel functions, I_l, plays major role in establishing the emission line intensity; a *larger* Huang-Rhys factor will result in more *intense* lines.

Next, we apply the theory to the resonantly excited photoluminescence for high barrier heterostructure of InAs/AlAs. According to the model from section 3.1.2, the mixing of the bi/exciton states by phonons is absent, and the formula (34) and (39) are valid. On the other hand, the bi/exciton degeneracy could make the dynamical Jahn-Teller effect (Jahn & Teller, 1937) to be effective. Accordingly to Eq. (39), the coupling Huang-Rhys factor g_{if} makes the degenerate lines to have different intensities. We approximate the intensity of emission lines by an average over the intensity of degenerate levels. The values of Huang-Rhys factors obtained, in accordance with (García-Cristóbal et. al., 1999; Cheche et. al., 2005) are large as follows: $g_1^{(1)} = 0.187$, $g_2^{(1)} = 0.103$, $g_3^{(1)} = 0.104$, $g_1^{(2)} = 0.747$, $g_2^{(2)} = 0.364$, $g_3^{(2)} = 0.365$, $g_4^{(2)} = 0.364$, and the g_{if} have values between 0.103 and 0.187, and larger values of 0.704 for g_{12}, and 0.706 for g_{13}. According to the presence of the Huang-Rhys factor in the argument of the modified Bessel functions, I_l, from Eqs. (34) and (39), a large Huang-Rhys factors obtained may be the sign of the appearance of strong phonon replicas in the optical spectra.

There is a variety of results regarding the biexciton binding energy, which reveal importance of shape, compounds, and size of QDs. In Fig. 7 the biexciton binding ground state (GS) energy, the difference of biexciton and exciton GS lines as given by Eqs. (34) and (39), i.e., $\tilde{\varepsilon}_b^{2X-X} = 2\varepsilon_1^{(1)} - \varepsilon_1^{(2)} - (2g_1^{(1)} - g_1^{(2)})\hbar\omega_0$, is obtained for different values of $\hbar\omega_e$ (with $\omega_e/\omega_h = 3$). Results from Fig. 7 show that the biexciton binding energy increases when the in-plane parabolic potential increases (QD radius decreases or exciton GS energy increases). This result is in agreement with the experimental data obtained for the same cylindrical shape of QD but with other compounds, InAs/InP (Chauvin et. al., 2006)), $In_{0.14}Ga_{0.86}As/GaAs$ (Bayer et. al., 1998) or with theoretical results obtained for GaAs.

QDs (Ikezawa et. al., 1998). An opposite behavior is reported for InAs/GaAs truncated pyramidal QDs by (Rodt, 2005). These facts might be related with the actual shape of the QDs. On the other hand, the binding character is obtained for smaller QDs ($\hbar\omega_e$ of order of tens of meV) and the antibinding character for larger QDs (for example, with $\hbar\omega_e = 0.001$ eV, we obtain $\tilde{\varepsilon}_b^{2X-X}$ = -0.0011 eV) in agreement with (Stier, 2001). Remarkable for the relevance of LO phonon influence on the spectra is the fact that without taking into account the self-energy (setting up $g_1^{(1)} = g_1^{(2)} = 0$ in Eq. (39)), $\varepsilon_b^{2X-X} = 2\varepsilon_1^{(1)} - \varepsilon_1^{(2)}$ is negative (increasing, e.g., from -0.0034 eV for $\hbar\omega_e = 0.065$ eV to -0.0008 eV for $\hbar\omega_e = 0.005$ eV) and $\tilde{\varepsilon}_b^{2X-X}$ becomes positive only by considering the phonon coupling.

These observations show that in addition to the shape, size, chemical composition, electron-hole exchange interaction, and piezoelectricity, the LO phonon coupling is an important factor which influences the anti/binding character of the biexciton. The extent to which the LO phonon coupling can not be neglected is a problem which can be addressed within a QD model of high enough accuracy. The confidence in the QD model we used is supported, in addition to the results obtained for biexciton, by those obtained for the exciton complex. As shown in Fig. 7, the magnitude of the exciton GS energy and decreasing of the exciton GS energy with QD size agree with other reports, see, e.g., (Ikezawa, 2006; Grundmann et. al., 1995). As the piezoelectricity in the case of cylindrical QD shape is expected to be less important (Miska, 2002) than for other QD shapes, the adopted QD model is suitable for describing the main physics of the bi/exciton-LO phonon coupling in cylindrical semiconductor QDs.

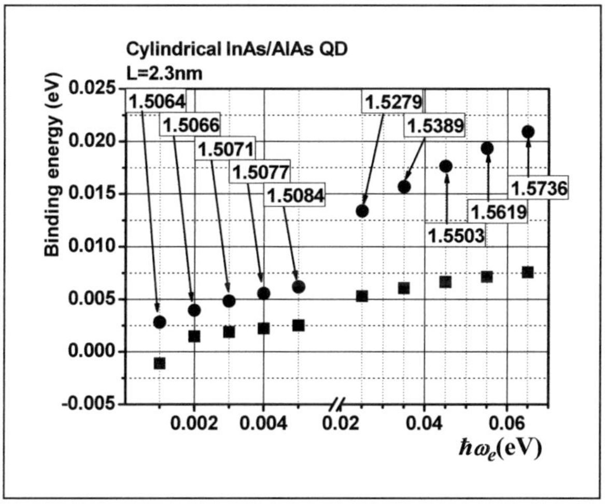

Fig. 7. The exciton (•) and biexciton (■) binding energies obtained for the simulated InAs/AlAs QD. The numbers show the energy of the exciton GS emission line.

The calculations show that value of exciton and biexciton binding energy is strongly influenced by diameter (in-plane confinement) and less by the height (perpendicular confinement) of cylindrical QDs. The binding character of the biexciton, with $\tilde{\varepsilon}_b^{2X-X} = 0.0076$ eV, and the exciton and biexciton GS emission lines of InAs/AlAs QD as reported by (Sarkar et. al., 2006) for $T = 9\text{K}$ are simulated in Fig. 8 by choosing $\hbar\omega_e = 0.065\text{eV}$ and $\hbar\omega_h = (0.065/3)\text{eV}$. Regarding the emission, the emission lines from Fig. 8 are labeled with three digits for transition from biexciton state (first digit) to exciton state (second digit), and with two digits for transition from exciton state (first digit) to the VS (reminding to the reader, VS means vacuum state, that is, the no excitons state); the last digit corresponds to the phonon replica. The open squares show the experimental results from

(Sarkar, 2006). The inset shows schematically the exciton resonant emission we simulated. Emission spectra of InAs/AlAs QDs are reported in a range of 1.5-1.9eV (Dawson et. al. 2005; Offermans et. al., 2005; Sarkar et. al., 2006). Our approach simulates the emission from exciton (1, 0) GS and biexciton (1, 1, 0) GS, in the range 1.56-1.68eV. In this interval the phonon replicas are predicted in accordance with the experimental data from (Sarkar et. al., 2005).

The literature regarding the presence of the excited states in emission spectra of QDs is rather scarce (Kamada, 1998; Khatsevich, 2005). The strong 0LO emission lines from excited states might explain the higher energy lines observed in the PL spectra reported by (Dawson et. al. 2005; Offermans et. al., 2005; Sarkar et. al., 2005). For small enough InAs/AlAs QDs the lowest energy state at Γ point in InAs moves above the AlAs X band edge, the electrons spread in the AlAs barrier, and appearance of high energy lines by this mechanism is forbidden. Instead, the exciton line (2, 0) and the biexciton-exciton emission lines (3, 1, 0), and (2, 1, 0) are candidates for explaining the high energy lines observed by (Offermans et. al., 2005). Accuracy of our QD model is not high enough to explain the fine-structure splitting reported by (Sarkar et. al., 2006) and shown in Fig. 8; the fine-structure is assigned to the electron-hole exchange interaction, which was neglected in our model. Prediction for higher temperatures is not reliable, as far as the possible dissociation of the biexciton with temperature had not been taken into by the present considerations. However, at larger, but still low temperatures, under 60K, the features of spectra predicted by our approach do not change significantly.

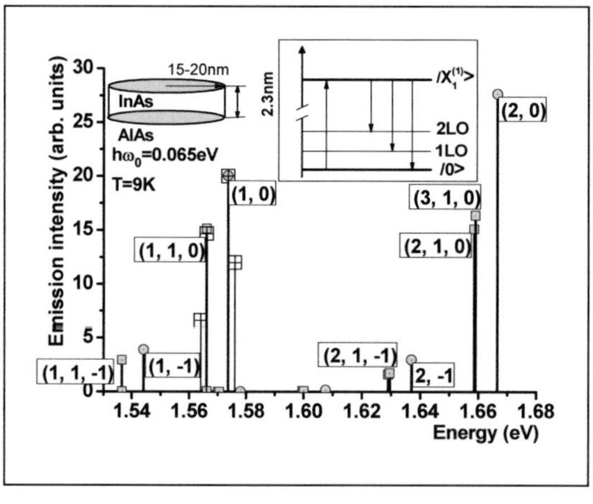

Fig. 8. The resonant emission spectrum of biexciton and exciton complexes.

Concluding this section, the theoretical approach we introduced is a useful tool for describing the influence of LO phonons on the resonant excitation emission at low temperatures. The high energy emission lines, that are obtained by configurational interaction calculations for cylindrical InAs/AlAs QDs, are associated to the emission from the excited states. One finds, in accordance with the experiment, that the biexciton binding energy has a binding character (positive value), which diminishes with decreasing the radius of QD, and becomes antibinding (negative value) for flat QDs. The simulated exciton and biexciton binding energies obtained, demonstrate that the phonon coupling is an essential factor, which should be integrated in the analyses for an accurate description of optical transitions in QDs. For the InAs/AlAs QDs, the presence of LO phonon replicas and emission from the excited states is explained as the consequence of large Huang-Rhys factors.

4. Outlook

To introduce the reader the problem of the electron-phonon interaction in QDs, three basic aspects are presented in the Sec. 2: i) the adiabatic and non-adiabatic transitions in the optical transitions; ii) the Huang-Rhys factor; iii) the Hamiltonian of localized defect with several electronic states mixed by phonons.

In Sec. 3.1, within the effective mass approximation two models describing the electronic energy structure of spherical GaAs/AlAs QDs and cylindrical InAs/AlAs QDs are introduced. For the optical transitions, the spherical QD model predicts the adiabatic treatment is appropriate for QD radius smaller than 32 Å, and a non-adiabatic is needed for larger radii. For the cylindrical QD both excitonic and biexcitonic complexes are considered by a configurational interaction method and for QD height of 2.3nm and parabolic confinement $\hbar\omega_e = 0.065\text{eV}$ and $\omega_e/\omega_h = 3$ the model predicts an adiabatic treatment is appropriate for describing optical transitions.

In Sec. 3.2 the Fermi Golden Rule and cumulant expansion method are used within a non-adiabatic treatment to spherical GaAs/AlAs QDs to obtain the absorption coefficient. In accordance with the experiment, we obtain: i) Large Huang-Rhys factors by two orders of magnitude than the bulk value with increasing values for smaller radii; ii) Accompanying resonances to the LO phonon satellites; iii) Red shift of the 0LO phonon lines and increased intensities of the accompanying resonances with temperature.

In Sec. 3.3 the Fermi Golden Rule and cumulant expansion method are used to describe the emission from the exciton and biexciton complexes of the cylindrical InAs/AlAs QDs. The presence of LO phonon replicas and emission from the excited states is explained as consequence of large Huang-Rhys factors. One finds, in accordance with the experiment, that the biexciton binding energy has a binding character (positive value), which diminishes with decreasing the radius of QD, and becomes antibinding (negative value) for flat QDs.

In conclusion, the present study emphasizes that the LO phonon coupling in the polar semiconductor QDs is an essential factor in understanding at a higher level of accuracy the optical transitions. The accordance between our results and experimental results show that the approaches we used, the Fermi Golden Rule and cumulant expansion method are useful tools in describing optical properties of semiconductor QDs. By the prediction of the Huang-Rhys factors and of the optical spectra shape, the present work is useful to people working in the field of semiconductor QDs optics, both theoreticians, in comparing different models, and experimentalists, in comparing theory and experiment.

5. Acknowledgements

The work was supported by the strategic grant POSDRU/89/1.5/S/58852, Project "Postdoctoral program for training scientific researcher" co-financed by the European Social Found within the Sectorial Operational Program Human Resources Development 2007-2013.

6. References

Axt V. M.; Kuhn T., Vagov A., & Peeters F. M. (September 2005). Phonon-induced pure dephasing in exciton-biexciton quantum dot systems driven by ultrafast laser pulse sequences, *Physical Review B,* Vol.72, No.12, pp. 125309-1-5, ISSN 1098-0121

Banyai, L. & Koch, S. W. (1993). *Semiconductor Quantum Dots,* World Scientific, ISBN 981-02-1390-5, Singapore

Barker B. I.; Rayborn G. H., Ioup J. W., & Ioup G. E. (November 1991). Approximating the finite square well with an infinite well: Energies and eigenfunctions, *American Journal of Physics,* Vol.59, No.11, pp. 1038-1042, ISSN 0295-5075

Bayer M.; Gutbrod T., Forchel A., Kulakovskii V. D., Gorbunov A., Michel M., Steffen R., & Wang K. H., Exciton complexes in $In_xGa_{1-x}As/GaAs$ quantum dots (August 1998). *Physical Review B,* Vol.58, No.8, pp. 4740–4753, ISSN 1098-0121

Besombes L.; Kheng K., Marsal L., & Mariette H. (March 2001). Acoustic phonon broadening mechanism in single quantum dot emission, *Physical Review B,* Vol.63, No.15, pp. 155307-1-5, ISSN 1098-0121

Born M. & Huang K. (1998). *Dynamical Theory of Crystal Lattices,* Clarendon, ISBN 0198503695, Oxford

Born, M. & Oppenheimer, M. (1927). Zur Quantentheorie der Molekeln. *Annalen der Physik,* Vol. 389, No. 20, pp. 457-484, ISSN 1521-3889

Chamberlain M. P.; Trallero-Giner C. & Cardona M. (January 1995). *Physical Review B,* Vol.51, No.3, pp. 1680–1693, ISSN 1098-0121

Chauvin N.; Salem B., Bremond G., Guillot G., Bru-Chevallier C., & Gendry. M (October 2006). Size and shape effects on excitons and biexcitons in single InAs/InP quantum dots, *Journal of Applied Physics,* Vol.100, No.7, pp. 073702-1-5, ISSN 0021-8979

Cheche T. O; Chang M. C., Lin S. H. (March 2005). Electron–phonon interaction in absorption and photoluminescence spectra of quantum dots, *Chemical Physics,* Vol.309, No.2-3, pp. 109-114, ISSN 0301-0104

Cheche T. O. & Chang M. C. (March 2005). Optical spectra of quantum dots: A non-adiabatic approach, *Chemical Physics Letters,* Vol. 406, pp. 479-482, ISSN 0009-2614

Cheche T. O. (June 2009). Phonon influence on emission spectra of biexciton and exciton complexes in semiconductor quantum dots, *EPL,* Vol.86, No.6, pp. 67011-1-6, ISSN 0295-5075

Cheche T. O., Barna E., Stamatin I. (to be published in Physica B)

Dawson P.; Göbel E. O., & Pierz K. (July 2005). *Journal of Applied Physics,* Vol. 98, No.1, pp. 013541-1-4, ISSN 0021-8979

Fomin V. M.; Gladilin V. N., Devreese J. T., Pokatilov E. P., Balaban S. N., & Klimin S. N. (January 1998). Photoluminescence of spherical quantum dots, *Physical Review B*, Vol.57, No.4, pp. 2415-2425, ISSN 1098-0121

Fuchs R. & Kliewer K.L. (December 1965). Optical Modes of Vibration in an Ionic Crystal Slab, *Physical Review*, Vol.140, No.6A, pp. A2076–A2088, ISSN 1943-2879

García-Cristóbal A.; A. W. E. Minnaert A. W. E., V. M. Fomin V. M., J. T. Devreese J. T., A. Yu. Silov A. Yu., J. E. M. HaverkortJ. E. M. , & J. H. Wolter J. H. (September 1999). Electronic Structure and Phonon-Assisted Luminescence in Self-Assembled Quantum Dots, *Physica Status Solidi* (*b*), Vol.215, No.1, pp. 331-336, ISSN 0370-1972

Grosso G. & P. Parravicini G. P. (2000). *Solid State Physics*, Academic Press, ISBN 0-12-304460-X, San Diego, San Francisco, New York, Boston, London, Sydney, Tokio, Toronto, Chapter 4

Grundmann M., Stier O.; & Bimberg D. (October 1995). InAs/GaAs pyramidal quantum dots: Strain distribution, optical phonons, and electronic structure, *Physical Review B*, Vol.52, No.16, pp. 11969–11981, ISSN 1098-0121

Hanamura E. (January 1988). Very large optical nonlinearity of semiconductor microcrystallites, *Physical Review B*, Vol.37, No.3, pp. 1273–1279, ISSN 1098-0121

Haug & Koch (1993). *Quantum theory of the optical and electronic properties of semiconductors*, World Scientific, ISBN 9812387560, Singapore

Hawrylak P. (August 1999). Exciton artificial atoms: Engineering optical properties of quantum dots, *Physical Review B*, Vol.60, No.8, pp. 5597–5608, ISSN 1098-0121;

Heitz R., Veit M.; Ledentsov N. N., Hoffmann A., Bimberg D., Ustinov V. M., Kop'ev P. S., & Alferov Zh. I. (October 1997). Energy relaxation by multiphonon processes in InAs/GaAs quantum dots, *Physical Review B*, Vol.56, No.16, pp. 10435–10445, ISSN 1098-0121

Huang K. & Zhu B. F. (December 1988). Dielectric continuum model and Fröhlich interaction in superlattices, *Physical Review B*, Vol.38, No.18, pp. 13377–13386, ISSN 1098-0121

Ikezawa M.; Nair S. V., Ren H.-W., Masumoto Y., & Ruda H., Biexciton binding energy in parabolic GaAs quantum dots (August 1998). *Physical Review B*, Vol.58, No.8, pp. 4740–4753, ISSN 1098-0121

Jacak L.; Hawrylak P., & Wójs A. (1998). *Quantum Dots*, Springer-Verlag Berlin and Heidelberg GmbH & Co. KG, ISBN 3540636536, Berlin

Jahn H. A. & Teller E. (July 1937). Stability of Polyatomic Molecules in Degenerate Electronic States. I. Orbital Degeneracy, *Proceedings of Royal Society A*, Vol.161, pp. 220-235, ISSN 1364-5021.

Kamada H.; Ando H., Temmyo J., & Tamamura T. (December 1998). *Physical Review B*, Vol.58, No.24, pp. 16243–16251, ISSN 1098-0121;

Khatsevich S.; Rich D. H., Kim E. –U., & Madhukar A. (June 2005). Journal of Applied Physics, Vol. 97, No.12, pp. 123520 -1-8, ISSN 0021-8979

Klein M. C.; Hache F., Ricard D., & Flytzanis C. (December 1990). Size dependence of electron-phonon coupling in semiconductor nanospheres: The case of CdSe, *Physical Review B*, Vol.42, No.17, pp. 11123–11132, ISSN 1098-0121

Klimin S. N.; Pokatilov E. P., & Fomin V. M. (August 1995). Polaronic Hamiltonian and Polar Optical Vibrations in Multilayer Structures, *physica status solidi (b)*, Vol.190, No.2, pp.441-453, ISSN 0370-1972

Mahan, G. D. (2000). *Many-Particle Physics*, Kluwer Academic/Plenum Publishers , ISBN 0-306-46338-5, New York, Boston, Dordrecht, London, Moscow, Chapter 4

Melnikov D. V. & Fowler W. B. (December 2001). Electron-phonon interaction in a spherical quantum dot with finite potential barriers: The Fröhlich Hamiltonian, *Physical Review B*, Vol.64, No.24, pp. 245320-1-9, ISSN 1098-0121

Menéndez E.; Trallero-Giner C. and Cardona M. (January 1997). Vibrational Resonant Raman Scattering in Spherical Quantum Dots: Exciton Effects, *Physica Status Solidi (b)*, Vol.199, No.1, pp. 81–94, ISSN 0370-1972

Merzbacher E. (1988). *Quantum Mechanics*, J. Wiley & Sons, ISBN 0-471-88702-1, USA

Miska P.; Paranthoen C., Even J., Bertru N., Le Corre A. & Dehaese O. (November 2002). *Journal of Physics: Condensed Matter*, Vol.14, No.47, pp. 12301, ISSN 0953-8984

Mittin V. V.; Kochelap V. A., & Stroscio M. A. (1999). *Quantum Heterostructures. Microelectronics and Optoelectronics*, Cambridge University Press, ISBN 0-521-63177-7, USA

Mori N. & Ando T. (September 1989). Electron–optical-phonon interaction in single and double heterostructures, *Physical Review B*, Vol.40, No.9, pp. 6175–6188, ISSN 1098-0121

Newton, M. D. & Sutin N. (1984). Electron Transfer Reactions in Condensed Phases. *Annual Reviews Physical Chemistry*, Vol. 35 (October 1984), pp. 437-480, ISSN 0066426X

Nomura S. & Kobayashi T. (January 1992). Exciton–LO-phonon couplings in spherical semiconductor microcrystallites, *Physical Review B*, Vol.45, No.3, pp. 1305–1316, ISSN 1098-0121

Odnoblyudov M. A.; Yassievich I. N., & Chao K. A. (December 1999). Polaron Effects in Quantum Dots, *Physical Review Letters*, Vol.83, No.23, pp. 4884–4887, ISSN 0031-9007

Offermans P.; Koenraad P. M., Wolter J. H., Pierz K., Roy M., & Maksym P. A. (October 2005). Atomic-scale structure and photoluminescence of InAs quantum dots in GaAs and AlAs, *Physical Review B*, Vol.72, No.16, pp. 165332-1-6, ISSN 1098-0121

Paillard M.; Marie X., Renucci P., Amand T., Jbeli A., & Gérard J. M. (February 2001). Spin Relaxation Quenching in Semiconductor Quantum Dots, *Physical Review Letters*, Vol.86, No.8, pp. 1634–1637, ISSN 0031-9007

Peter E.; Hours J., Senellart P., Vasanelli A., Cavanna A., Bloch J., & Gérard J. M. (January 2004). Phonon sidebands in exciton and biexciton emission from single GaAs quantum dots, *Physical Review B*, Vol.69, No.4, pp. 041307(R)-1-4, ISSN 1098-0121

Ridley B. K. (March 1989). Electron scattering by confined LO polar phonons in a quantum well, *Physical Review B*, Vol.39, No.8, pp. 5282–5286, ISSN 1098-0121

Ridley, B. K. (1988). *Quantum Processes in Semiconductors*, Clarendon Press, ISBN 019851171X, Oxford

Roca E.; Trallero-Giner C., & Cardona M. (May 1994). Polar optical vibrational modes in quantum dots, *Physical Review B*, Vol.49, No.19, pp. 13704–13711, ISSN 1098-0121

Rodt S.; Schliwa A., Pötschke K., Guffarth F., & Bimberg D. (April 2005), Correlation of structural and few-particle properties of self-organized InAs⁄GaAs quantum dots, *Physical Review B*, Vol.71, No.15, pp. 155325-1-7, ISSN 1098-0121

Rücker H.; Molinari E., & Lugli P. (August 1991). Electron-phonon interaction in quasi-two-dimensional systems, *Physical Review B*, Vol.44, No.7, pp. 3463–3466, ISSN 1098-0121

Sarkar D.; van der Meulen H. P., Calleja J. M., Becker J. M., Haug R. J., & K. Pierz K. (July 2006). *Journal of Applied Physics*, Vol. 100, No.2, pp. 023109-1-4, ISSN 0021-8979

Sarkar D.; van der Meulen H. P., Calleja J. M., Becker J. M., Haug R. J., & K. Pierz K. (February 2005). *Physical Review B*, Vol.71, No.8, pp. 081302(R)-1-4, ISSN 1098-0121

Sénès M.; Urbaszek B., Marie X., T., Tribollet J., Bernardot F., Testelin C., Chamarro M., & Gérard J.-M. (March 2005). *Physical Review B*, Vol.71, No.11, pp. 115334-1-6, ISSN 1098-0121

Sercel P. C. & Vahala K. J. (August 1990). Analytical formalism for determining quantum-wire and quantum-dot band structure in the multiband envelope-function approximation, *Physical Review B*, Vol.42, No.6, pp. 3690–3710, ISSN 1098-0121

Shumway J.; Franceschetti A., & Zunger A. (March 2001). Correlation versus mean-field contributions to excitons, multiexcitons, and charging energies in semiconductor quantum dots, Physical *Review B, Vol.63*, No.8, pp. 155316-1-13, ISSN 1098-0121

Stier O.; Schliwa A., Heitz R., Grundmann M., & Bimberg D. (March 2001). Stability of Biexcitons in Pyramidal InAs/GaAs Quantum Dots, *physica status solidi (b)*, Vol.215, No.1, pp.115-118, ISSN 0370-1972

Takagahara T. (February 1993). Effects of dielectric confinement and electron-hole exchange interaction on exciton states in semiconductor quantum dots, *Physical Review B*, Vol.47, No.8, pp. 4569–4584, ISSN 1098-0121

Takagahara T. (July 1999). Theory of exciton dephasing in semiconductor quantum dots, *Physical Review B*, Vol.60, No.4, pp. 2638–2652, ISSN 1098-0121

Vasilevskiy M. I.; Anda E. V., & Makle S.S. (July 2004). Electron-phonon interaction effects in semiconductor quantum dots: A nonperturabative approach, *Physical Review B*, Vol.72, No.12, pp. 125309-1-5, ISSN 1098-0121

Verzelen O.; Ferreira R., & Bastard G. (April 2002). Exciton Polarons in Semiconductor Quantum Dots, *Physical Review Letters*, Vol.88, No.14, pp. 146803-1-4, ISSN 0031-9007

Voigt J.; Spielgelberg F., & Senoner M., Band parameters of CdS and CdSe single crystals determined from optical exciton spectra (January, 1979). *Physica Status Solidi (b)*, Vol.91, No.1, pp. 189-199, ISSN 0370-1972

Vurgaftman I.; Meyer J. R., & Ram-Mohan L. R. (February 2001). Band parameters for III–V compound semiconductors and their alloys, *Journal of Applied Physics*, Vol.89, No.11, pp. 5815- 5875, ISSN 0021-8979

Woggon, U. (1997), *Optical Properties of Semiconductor Quantum Dots*, Springer, ISBN 3540609067, Berlin, Chapter 5

Zutić I.; Fabian J., & S. Das Sarma D. (2004), *Review of Modern Physics*, Vol.76, No.2, pp. 323–410, ISSN 0034-6861

3

Temperature-Dependent Optical Properties of Colloidal IV-VI Quantum Dots, Composed of Core/Shell Heterostructures with Alloy Components

Efrat Lifshitz, Georgy I. Maikov, Roman Vaxenburg, Diana Yanover, Anna Brusilovski, Jenya Tilchin and Aldona Sashchiuk
Schulich Faculty of Chemistry, Russell Berrie Nanotechnology Institute, Solid State Institute, Technion, Haifa, Israel

1. Introduction

Colloidal semiconductor nanocrystals attract worldwide scientific and technological interest due the ability to engineer their optical properties by the variation of size, shape, and surface properties.[1-3] Recent studies revealed new strategies related to composition control of the properties, including alloying,[4-7] doping,[8] and in particular the formation of core/shell heterostructures.[9-14] Whereas major effort has been devoted to the development of II-VI core/shell structures,[12-15] there are only a few reports concerning the heterostructures of IV-VI (PbSe, PbS) colloidal quantum dots (CQDs).[16-19] PbSe, PbS and $PbSe_xS_{1-x}$ alloyed CQDs are the focus of widespread interest due to their unique electronic and optical properties, with feasibility of applications in near infra-red (NIR) lasers, photovoltaic solar cells, Q-switches and nano-electronic devices.[20] These semiconductors have a simple cubic crystal structure with nearly identical lattice constants 5.93 Å and 6.12 Å at 300 K, respectively, which facilitates the formation of hetero-structures. Recently, high quality PbSe/PbS core/shell[16-19] and completely original $PbSe/PbSe_xS_{1-x}$ core/alloyed shell CQDs structures[19] were produced using a single injection process, offering the potential to tailor the crystallographic and dielectric mismatch between the core and the shell, forming a perfect crystalline hetero-structure. These structures present higher photoluminescence (PL) quantum yield (QY) with respect to those of core CQDs and tunability of the band-edge offset with variation of the shell thickness and composition, eventually controlling the electronic properties of the CQDs.

During the past few years, considerable interests have been focused on the thermally activated processes of the ground-state exciton emission of PbSe core CQDs.[21] The variation of the PL properties with temperature showed two thermal activation thresholds: the first in the temperature range 1.4–7 K connected with activation of acoustic phonon assisted dark exciton decay, and the second in the temperature range 100–200 K, connected with activation of bright excitons. This study also shows that the temperature coefficient of the energy gap and the optical phonon coupling were reduced with the decrease of the diameter, while the acoustic phonon coupling grew with the decrease of the diameter. Since the first report of experimentally prepared PbSe/PbS core-shell CQDs, some simple physical properties, such as electronic structure had been studied.[22] The previous theoretical work predicted a variation of the electronic structure of PbSe/PbS CQDs, pronounced in the variation of the carriers' radial distribution function, with the variation of the core-radius/shell-thickness ratio, showing a significant separation of the electron and hole wave functions only when the shell-thickness becomes equivalent or larger than the core radius. However, the electronic structure and optical properties of colloidal IV-VI quantum dots, composed of core/shell heterostructures with alloy components still lack systematic and in depth study.

Considering the significant potential of the IV-VI heterostructures the present work describes the structural and temperature-dependent optical characterization of PbSe/PbS core/shell (c/sh), PbSe/$PbSe_xS_{1-x}$ core/alloyed-shell (c/a-sh), and newly prepared $PbSe_yS_{1-y}$/ $PbSe_xS_{1-x}$ alloyed-core/alloyed shell (a-c/a-sh) CQDs,[23] with variable internal diameters and a radial gradient composition (when $0 < x < 1$, $0 < y < 1$) with respect to those of pure PbSe and PbS CQDs. The investigated CQDs were prepared by colloidal chemistry. The structure and composition of the CQDs were characterized by the use of high-resolution transmission electron microscopy (HR-TEM), selected area electron diffraction (SAED), energy-dispersive analysis of X-ray (EDAX). A thorough investigation of the optical properties was performed by following the variable temperature continuous-wave (cw) and transient (temporal and spectrally resolved) PL spectra, exploring energy shift, band edge temperature coefficient, alleviation of a dark-bright splitting (or exchange interaction), valley-valley interaction, emission QY, and radiative lifetime of the heterostructures, in comparison with the existing properties of the primary PbSe core CQDs[21] and PbS CQDs with equevalent size.[24]

This chapter is organized as follows. Section 2 presens the significant effect of thermally activated processes of the ground-state exciton emission of various $PbSe_xS_{1-x}$/$PbSe_yS_{1-y}$ a-c/a-sh CQDs structures, suggesting that cw-, temporal and spectrally resolved PL of PbSe/PbS (c/sh) , PbSe/$PbSe_xS_{1-x}$ (c/a-sh) and $PbSe_yS_{1-y}$/$PbSe_xS_{1-x}$ a-c/a-sh CQDs over a wide range of temperatures have distinguished properties in comparison with those of pure PbSe core CQDs with equivalent overall size (R_s) and identical core radius (R_c). Section 3 discusses the thermally activated processes of PL in PbS CQDs, while the theoretical insight into the electronic band structure of graded $PbSe_yS_{1-y}$/$PbSe_xS_{1-x}$ a-c/a-sh QDs structure with different composition and/or size using the multiband $\mathbf{k}\cdot\mathbf{p}$ envelope function method is given in Section 4. Section 5 presents the colloidal synthesis procedures and experimental techniques, used for CQDs structural and spectroscopic characterizations.

2. Temperature influence on composition-tunable optical properties of $PbSe_yS_{y-1}$/$PbSe_xS_{1-x}$ c/sh CQDs.

The investigated CQDs were prepared by colloidal chemistry, according to the short description given below in Section 5.1 and a detailed procedure reported in.[19] Figure 1 represents HR-TEM images of $PbSe_{0.5}S_{0.5}$/$PbSe_{0.27}S_{0.73}$ (a) and PbSe/PbS (b) CQDs.

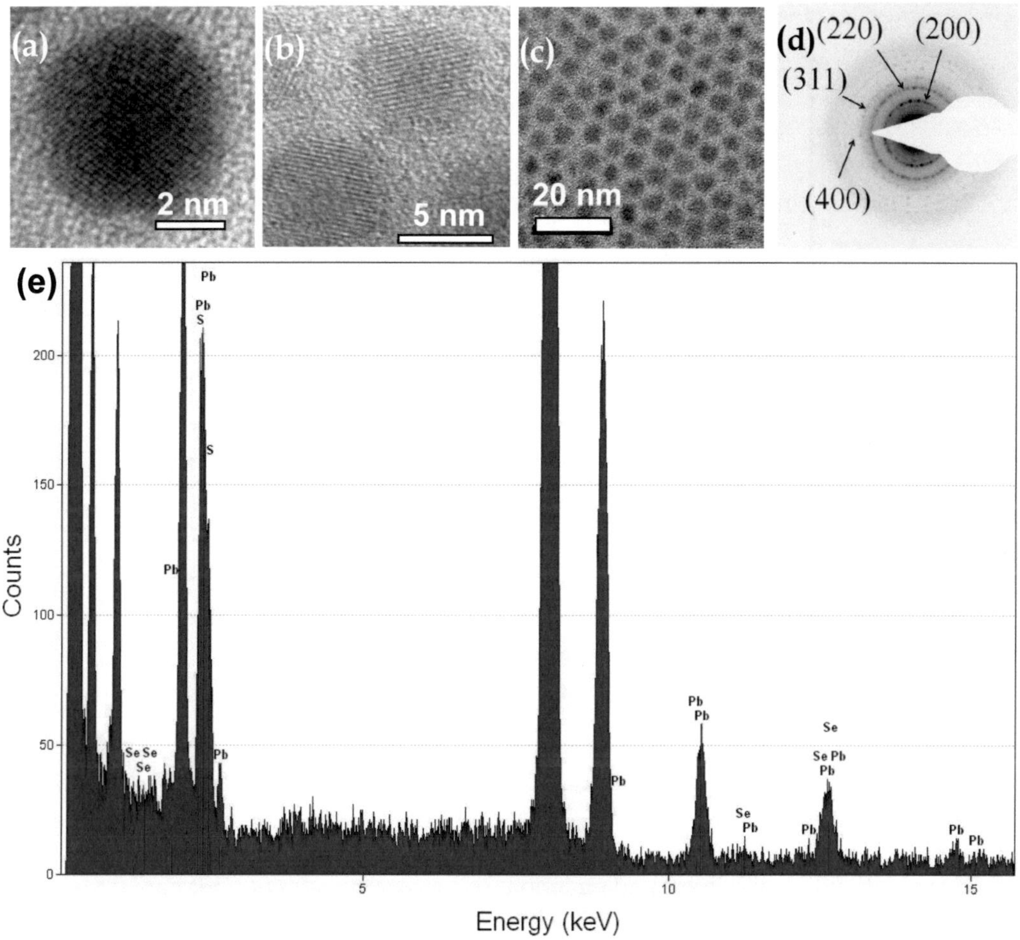

Fig. 1. Representative HR-TEM image of: single $PbSe_{0.5}S_{0.5}/PbSe_{0.27}S_{0.73}$ a-c/a-sh CQD with R_s of 2.0 nm (a), PbSe/PbS c/sh CQD with R_c of 1.3 nm and R_s of 2.5 nm (b), TEM image of an ensemble of CQDs shown in (a) (c), SAED image of the CQDs in (c) (d), EDAX spectrum of $PbSe_{0.5}S_{0.5}/PbSe_{0.27}S_{0.73}$ a-c/a-sh CQDs (e).

These images reveal distinguished crystal planes, supporting high crystallinity of the a-c/a-sh CQDs. In most cases the core/shell interface is indistinguishable in $PbSe_{0.5}S_{0.5}/PbSe_{0.27}S_{0.73}$ CQD (Panel (a)) due to the close proximity of the crystallographic components of PbSe and PbS semiconductors. However, a boundary is noted in PbSe/PbS CQD with a shell width > 3 nm (Panel (b)). A representative TEM image of CQDs shown in (a) is presented in Panel (c), exhibiting a size uniformity of ~ 5%. A representative SAED of CQDs shown in (c) is shown in Panel (d), confirming a rock-salt crystallographic structure ($Fm\bar{3}m$ space group). Similar rock-salt structures appeared in all the investigated samples. Representative EDAX spectra are presented in Panel (e). The Pb, Se, and S percentages of various samples are listed in Table 1.

Representative absorption (dashed lines) and cw-PL (solid lines) spectra of a few samples with various composition and size, measured at room temperature (RT) are shown in Figure 2.

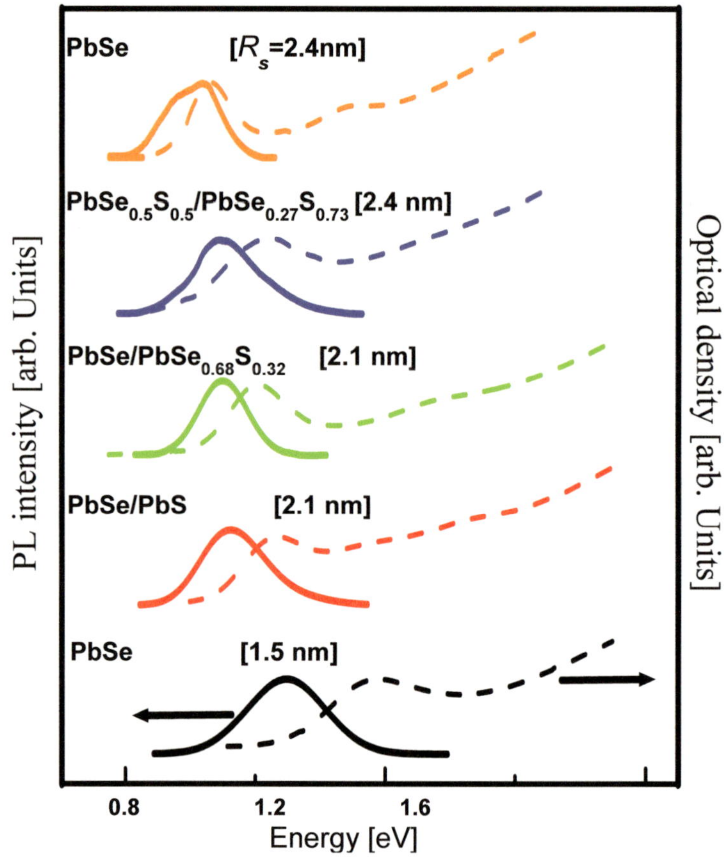

Fig. 2. Absorption (dashed lines) and emission (solid lines) spectra of PbSe core (bottom and top curves), PbSe/PbS c/sh, $PbSe/PbSe_{0.68}S_{0.32}$ c/a-sh and $PbSe_{0.5}S_{0.5}/PbSe_{0.27}S_{0.73}$ a-c/a-sh CQDs with overall radius, R_s, as labeled in the figure and measured at RT. The core radii of the heterostructures presented are in accordance with Table 1.

The bottom and top curves correspond to the spectra of PbSe core CQDs with average radius of R_S = 1.5 nm and 2.4 nm, respectively. The middle curves represent different heterostructures (c/sh, c/a-sh and a-c/a-sh), with composition and size as labeled in the Panel, when R_c = 1.5 nm and R_s up to 2.4 nm. These sets of curves suggest the occurrence of a red-shift of the absorption and emission spectra of the heterostructures with respect to those of the primary cores with R_S = 1.5 nm, but they are blue shifted with respect to that of pure PbSe cores with R_S = 2.4 nm. This midway shift is related to a quantum size effect combined with a composition tuning of the band edge energy.

The experimental band gap energy (E_g), estimated by the first excitonic absorption band, and the corresponding calculated values (discussed below) of the studied materials are listed in Table 1. The table also designates the PL quantum yield (η) of a few selected samples. The determination of the η is given in detail in the Section 5.3. Systematic improvement (up to 88%) of the η was found in c/sh, c/a-sh and a-c/a-sh CQDs *versus* those of the primary core CQDs. The relatively high η might be related to the improved quality of surfaces, *e.g.*, close crystallographic match between PbSe core and PbS or $PbSe_xS_{1-x}$ shell, as well as the increase of the sulfur precentege at the exterior surface, with a lower oxidation outcome. The cw-PL spectra shown in Figure 2 were pumped by a nonresonant excitation (1.54 eV), showing an asymmetric or a split band, Stokes shifted with respect to the first absorption band by an energy, E_s, as listed in Table 1.

Molecular formula of CQDs	Pb [%]	Se [%]	S [%]	R_c [nm]	R_s [nm]	E_g exp. [eV]	E_g calc. [eV]	η [%]	τ_{rad} [μs]	E_s [meV]
PbSe	51.5	48.5	0.0	1.5	1.5	1.17	1.03	48	2.99	112
PbSe	54.0	46.0	0.0	2.4	2.4	0.93	1.16	83	1.23	34
PbSe/PbS	55.2	17.1	27.7	1.5	2.1	1.03	1.10	65	2.62	55
PbSe/PbS	53.8	13.5	32.7	1.5	2.3	0.98	0.95	88	1.72	53
$PbSe/PbSe_{0.68}S_{0.32}$	50.0	40.0	10.0	1.5	2.1	1.00	1.14	68	2.21	75
$PbSe_{0.5}S_{0.5}$	50.8	25.1	24.2	1.6	1.6	1.18	1.30	27	5.50	103
$PbSe_{0.5}S_{0.5}/PbSe_{0.24}S_{0.76}$	49.2	20.3	30.5	1.6	1.9	1.09	1.10	46	2.37	73
$PbSe_{0.5}S_{0.5}/PbSe_{0.27}S_{0.73}$	48.8	17.5	33.7	1.6	2.4	1.02	0.97	65	1.96	45

Table 1. Relevant parameters of the investigated CQDs; Pb, Se, S percentages, core radius (R_c), overal radius (R_s), band gap energy (Eg), quantum yield (η), radiative lifetime (τrad) and Stokes shift (Es), all at RT.

Figure 3 compares the PL emission Stokes shift energy versus the experimental band gap energy, corresponding to the first excitonic absorption band of PbSe (black squares), PbSe/PbS c/sh (red circles) with primary R_c of 1.5 nm, $PbSe/PbSe_{0.8}S_{0.2}$ c/a-sh (green triangles) with primary R_c of 1.5 nm and $PbSe_{0.6}S_{0.4}/PbSe_{0.17}S_{0.83}$ a-c/a-sh (blue diamond's) with $PbSe_{0.6}S_{0.4}$ core radius of 1.5 nm CQDs at RT. The nonresonant Stokes shift has an interesting behavior: (i) a reduction of E_s in c/sh heterostructures with respect to their primary cores. The Stokes shift is related to a total growth of R_s with respect to R_c , as well as to the generation of an exciton fine-structure by valley-valley and electron-hole exchange interactions. These interactions may be reduced in c/sh structures, due to a *quasi*-type-II band alignment, as will be discussed below; (ii) E_s in c/a-sh and a-c/a-sh CQDs is smaller than in the corresponding cores, however, larger than that in core or c/sh CQDs of an equivalent size (see Table 1). Similar increase of E_s, in alloyed CQDs (in comparison with pure cores) was observed before in II-VI[5, 25] and III-V[26] quantum dots, and was associated with a nonlinear effect such as optical bowing, induced by a lattice constant deformation or different carriers' distribution in alloyed materials.[5]

Figure 4 illustrates the evolution of the cw-PL spectra of a few samples, excited at 1.54 eV and recorded at different temperatures from 1.4 K to 300 K as shown by the ruler in the figure. Panels (a) and (e) represent the spectra of reference PbSe cores, with an average radius, R_S = 1.5 nm and 2.4 nm, respectively. The CQDs were desperse in glass solution (GS). The cw-PL spectra correspond to a band edge exciton recombination emission at the L-point of the Brillouin zone of a PbSe semiconductor. Panels (b) and (c) show the spectra of

PbSe/PbS c/sh and PbSe/$PbSe_{0.68}S_{0.32}$ c/a-sh CQDs, respectively, both with R_c = 1.5 nm and R_s = 2.1 nm. Panel (d) exhibits the spectra of $PbSe_{0.5}S_{0.5}$/$PbSe_{0.27}S_{0.73}$ a-c/a-sh CQDs with R_s = 2.4 nm. Once again, at all temperatures, the emission energy of the heterostructures shows a midway shift between the emission energy of small and large reference cores (Panels (a) and (e)). The cw-PL spectra of the smallest PbSe cores CQDs are dominated by a single exciton band over the entire temperature range, similar to the observation found in Ref.[21]. However, the emission spectra of the larger PbSe cores occasionally exhibit a split band at elevated temperatures.

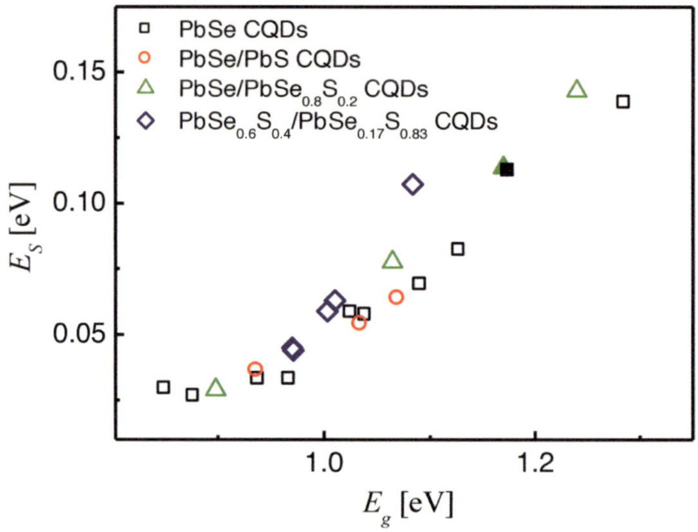

Fig. 3. Plot of the emission energy Stokes shift E_s versus E_g of the CQDs designated at the legend of the Figure, at RT. The full black square correspond to the primary core of the PbSe/PbS CQDs and the full green triangle corresponds to the primary core of the PbSe/$PbSe_8S_{0.2}$ CQDs.

Identicaly the cw-PL spectra of the various heterostructures show an asymmetric or a split shape at all temperatures, presumably consisting of two overlapping emission bands, where the energy split is of the order 30-55 meV, decreasing with the increase of the shell width and the S/Se ratio (see Figure 4). For convenience, we fitted the spectra to a sum of two Gaussian emission bands as demonstrated in Figure 5 for a $PbSe_{0.5}S_{0.5}$/$PbSe_{0.27}S_{0.73}$ a-c/a-sh sample at three representative temperatures (5 K, 150 K and 300 K). Interestingly, the split energy closely retains its value upon the increase in temperature, although the high energy component is gaining intensity with the increase in temperature. As a simple test of the possibility that the split band arises due to traps or defects on the surface of the CQDs, we checked the pumping intensity dependence of the spectra, as traps and defects should be saturated at high enough energies. We found that the spectrum not only maintained its shape but also the relative intensity between the two bands remained constant, while the pumping intensity varied over 10 orders of magnitude. The possibility of experimental artifacts was also rejected as the dual band still appears after intensity calibration on the experimental system and at different energies for different samples (core, c/sh, c/a-sh and a-c/a-sh). We take this evidence as proof that the dual band is an intrinsic property of the electronic structure of the samples.

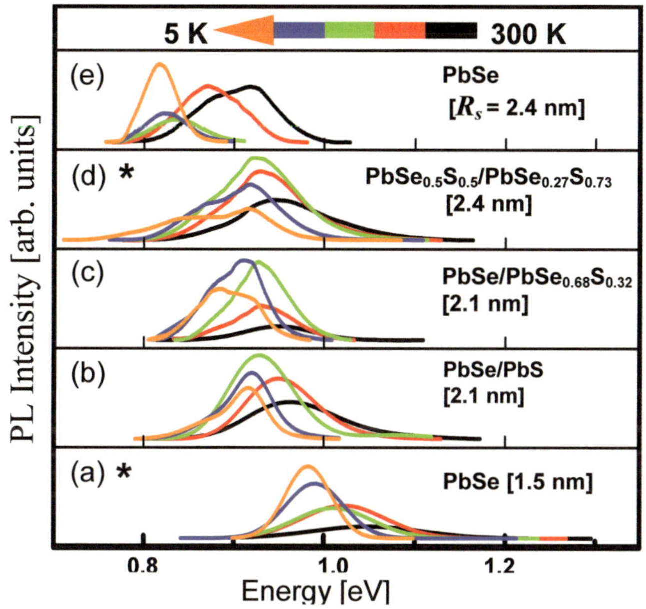

Fig. 4. cw-PL spectra of: PbSe core CQDs with core radius of R_S = 1.5 nm (a) and R_S = 2.4 nm (e); PbSe/PbS c/sh (b) and $PbSe/PbSe_{0.68}S_{0.32}$ c/a-sh CQDs (c), both with R_c = 1.5 nm and R_s = 2.1 nm; $PbSe_{0.5}S_{0.5}/PbSe_{0.27}S_{0.73}$ a-c/a-sh CQDs (d) with R_c = 1.6 nm and R_S = 2.4 nm. The CQDs were dispersed in GS. The data were recorded at various temperatures shown by the ruler. (* the PL intensity was multiplied by a factor of 3 at RT).

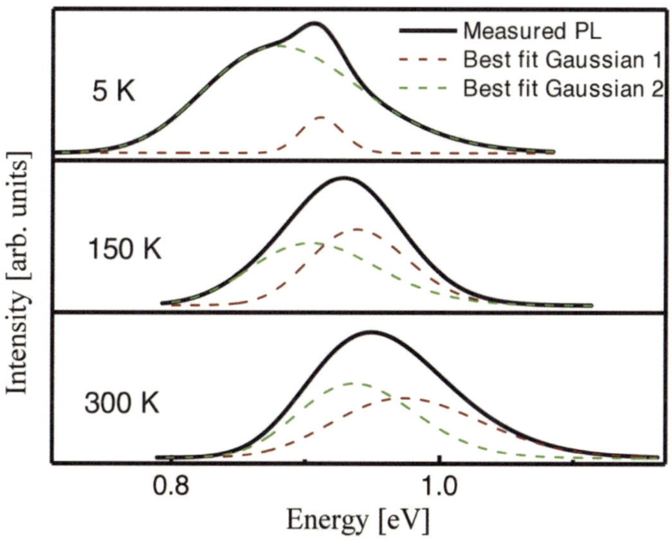

Fig. 5. PL spectra (black) and its best fit Gaussian curves (green and red) for $PbSe_{0.5}S_{0.5}/PbSe_{0.27}S_{0.73}$ a-c/a-sh CQDs, recorded at 5 K, 150 K and 300 K.

The splitting might be related to the occurrence of a recombination emission from two low lying excited states based on the following grounds: (i) a break of the four-fold degeneracy at the Brillouin L-point minima in IV-VI rock-salt structures, by confinement or valley-valley interaction. Indeed, valley-valley interaction, previously reported,[27-29] suggested a split of the band edge into two manifolds, each composed of bright and dark states induced by exchange interaction. The valley-valley split energy is in the range of 20-60 meV, increasing with the decrease of the CQDs' size, close to the experimental values attained in this work, which also increased with the reduction of R_s; (ii) simultaneous emission from both a dark and bright states, if a Boltzmann distribution at the cryogenic temperatures permits a heavy population of the higher energy bright state. This case is less probable, since the observed split of a few tenths of meV is substantially larger than a theoretical reported value for an exchange splitting between dark and bright states in pure PbSe cores; (iii) two emission processes can be related to parallel radiative recombination of a type-I and *quasi*-type II transitions, overlapping in an ensemble of CQDs. Presumably, such a case can be excluded, based on the observed uniformity, size and composition; (iv) occurrence of energy transfer between subgroups of small and large CQDs. The examined CQDs were dissolved in glass solutions, with a relatively low concentration, minimizing the energy transfer process. Thus, valley-valley interaction is the most probable mechanism for a split emission band.

The ΔE between the two bands for different samples are listed in Table 2. The table designates a general trend, a decrease of the ΔE with the increase of the R_s of the CQDs as well as the similarity in the value of ΔE of core CQDs and c/sh (or c/a-sh) CQDs of similar R_s (although their first exciton emission occurs at different energies).

Molecular formula	R_c [nm]	R_s [nm]	ΔE [meV]
PbSe	1.5	1.5	-
PbSe	2.0	2.0	53.1
PbSe	2.2	2.2	53
PbSe	2.4	2.4	52.6
PbSe/PbS	1.5	1.8	70.1
PbSe/PbS	1.5	2.1	59.1
PbSe/PbS	1.5	2.4	54.9
PbSe	1.5	1.5	67.3
$PbSe/PbSe_{0.68}S_{0.32}$	1.5	2.1	55.3
$PbSe/PbSe_{0.75}S_{0.25}$	1.5	2.5	49.2
$PbSe_{0.5}S_{0.5}$	1.6	1.6	63.4
$PbSe_{0.5}S_{0.5}/PbSe_{0.24}S_{0.76}$	1.6	1.9	43.2
$PbSe_{0.5}S_{0.5}/PbSe_{0.27}S_{0.73}$	1.6	2.4	40.1

Table 2. Splitting energy ΔE of the PL emission spectra in PbSe core, PbSe/PbS c/sh, $PbSe/PbSe_xS_{1-x}$ c/a-sh and $PbSe_yS_{1-y}/PbSe_xS_{1-x}$ a-c/a-sh CQDs at RT.

The nature of the two emitting states in PbSe CQDs have recently been investigated theoretically[27, 30] and exsperimentally[21, 23, 29, 31-33]. These studies have suggested that splitting within the exciton fine structure near the band gap may be responsible[23, 27, 29]. We proposed that the problem of two emitting bands need supplementary evidence, which can be gained from the temporal and spectrally resolved PL measurements.

The transient PL spectra were measured by exciting the sample with 1.17 eV and following the decay time (τ_0) of the emission intensity. Figure 6(a) displays decay curves of the samples given in the legend, measured at RT, monitoring the low energy PL component. Predominantly, the curves exhibit a single exponent behavior, where the value of τ_0 decreases upon the growth of a core radius from R_s = 1.5 nm to 2.4 nm. However, the τ_0 of the c/sh CQDs is longer than that of the primary PbSe cores. Spectrally resolved transient PL measurements provide more evidence of the two PL band nature. Plots of the values of τ_0 measured at various temperatures (as indicated by the arrow), monitored across the PL spectrum of PbSe cores and PbSe/PbS c/sh CQDs, are shown in the inset of Figure 6(a). The symbols are the experimental points and the solid lines are to guide the eye. It shows that the τ_0 of the core CQDs is approximately invariant across the PL spectrum. However, there is a pronounced decrease of the τ_0 when moving from the red to the blue side of a PL spectra of a c/sh sample, mainly pronounced at low *T*, but becoming insensitive to the detection energy at RT. The variation of τ_0 across the PL band supports the existence of emission of at least two manifold of states in the PL spectrum of the heterostructures, involving different radiative transitions, each of which having a distinct dependence on the temperature. τ_0 is correlated to the PL radiative lifetime (τ_{rad}) by the Eq. $\tau_{rad} = \tau_0 / \eta$ (η values are given in Table 1). Considering this relations, Figure 6(b) represents plots of τ_{rad} *versus* the measured *T* of the samples given in the legend of Panel (a), monitored only on the low energy PL component.

These plots reveal a drastic decrease of τ_{rad} with the increase in *T* in core and c/sh CQDs, related to a dark exciton emission,[21] however, only a minor change in c/a-sh and a-c/a-sh CQDs. The small variation of τ_{rad} in the later CQDs is also related to the diminished climax in the PL intensity versus *1/T* (see Figure 7(b)). Both effects suggests elevation of the dark exciton characteristic in alloyed CQDs. Figure 6(c) compares plots of τ_{rad} versus R_s of core and c/sh CQDs at three different *T* (5 K, 100 K and 300 K). These curves reflect a common behavior, characterized by a reduction which is turned into an extension of τ_{rad} with the increase in R_s (called bowing effect). The turning points are mainly pronounced when measured at low *T*. It is important to note that the decay processes in c/a-sh samples resemble those of the c/sh materials (not shown), but the turning points in c/sh and c/a-sh CQDs occur at a smaller R_s than that of cores of equivalent size. In fact, a turning point in the variation of τ_{rad} with size was already reported in PbSe core CQDs at RT for samples with R_S between 2 nm to 10 nm. Currently the mechanism of this behavior is not clearly understood, however it was previously suggested[24, 34] that the initial reduction of τ_{rad} with increasing size could be attributed to the size-dependence of the matrix element for spontaneous emission $\left|\langle\psi_f|\hat{\mathbf{p}}|\psi_i\rangle\right|^2$ which governs the recombination rate up to $R_s \approx 2$ nm.[35] For larger sizes, however, the matrix element is expected to become size-independent, so τ_{rad} becomes proportional to the wavelength of emitted radiation, which is consistent with the trend observed in Figure 6(c).

Figure 7(a) displays the temperature variation of the peak emission of the low energy cw-PL band of a few heterostructures, large core and small reference core, as given in the legend. The Figure demonstrates the peak energy at *T*, relative to that at 1.4 K (E_{PL}(T)-E_{PL}(1.4 K)), *versus T*. The plots disclose a blue-shift of the emission energy with increase in *T*, which is largest for PbSe cores; however, they illustrate a moderate change in the heterostructures. The symbols designate the experimental points, and the solid lines are best fitted curves. A tangent line to the fitted curve evaluates the slope, revealing the temperature coefficient,

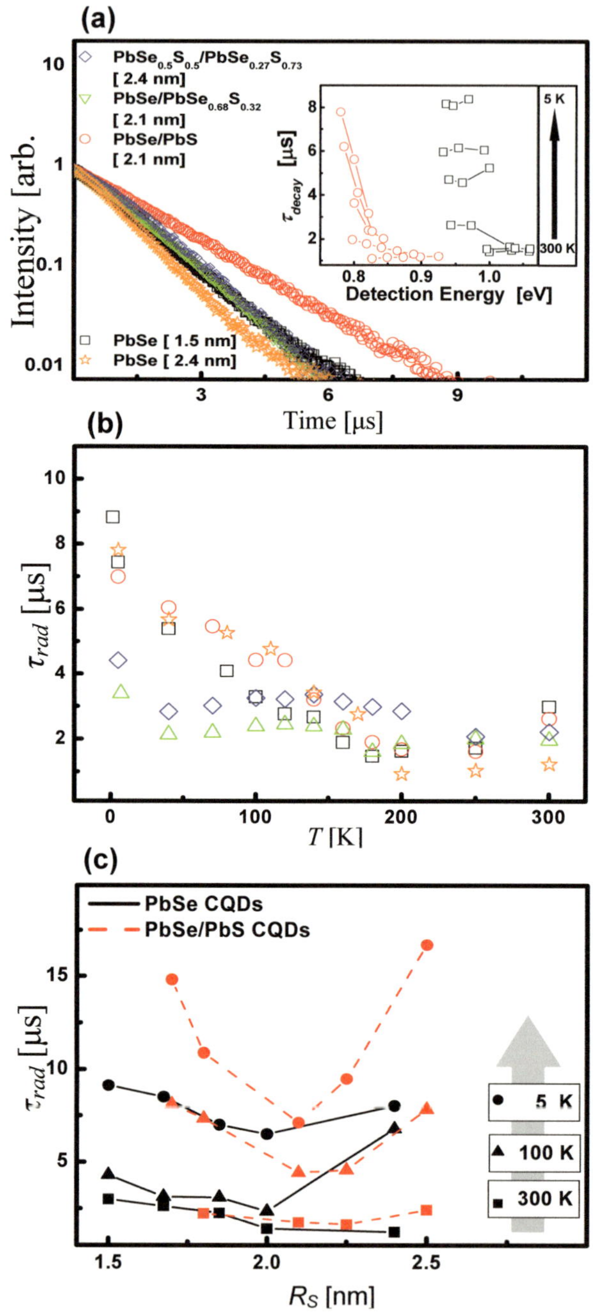

Fig. 6. Representative PL decay curves of the samples listed in the legend at RT (a); The inset in (a) presents plots of the measured τ_0 versus the detection energy across the PL spectrum, of core and c/sh CQDs, measured at different temperatures (5-300 K) as indicated by the arrow; Variation of the radiative lifetime versus the temperature of the samples listed in Panel (a) (b); Plots of the radiative lifetime versus the radius (R_s) of cores (solid line) and the corresponding c/sh (dashed line) CQDs, measured at the various temperatures as indicated in the figure (c).

dE/dT. This coefficient is most commonly derived from the temperature dependence of the first excitonic transition energy, using Varshni[36] relation, however, if the emission Stokes shift is invariant under the temperature, the coefficient derived in the present case should be relatively close to the band edge value, dE_g/dT. The best fitted coefficients of a few samples are listed in Table 3, indicating an increase of *dE/dT* with an increase in size[37] approaching the bulk value of dE_g/dT. Also, those coefficients of the heterostructures are reduced with respect to pure cores of equivalent R_s, which are mostly pronounced in a-c/a-sh CQDs. The temperature dependence of the coefficient dE_g/dT has dominant contributions from crystal dilation and electron-phonon interactions, as well as minor contributions from mechanical strain and thermal expansion of the wavefunction envelope.[37] Since the thermal expansion coefficients of bulk PbSe and PbS are almost identical,[38] it is expected that the electron-phonon coupling is the dominating effect responsible for the reduction of dE_g/dT in alloyed CQDs. A minor contribution might be also assigned to a reduction of a core/shell interface strain by a better crystallographic match. In any event, the low value of *dE/dT* in the alloyed CQDs suggests a thermal stability of the band edge properties, with a significant importance in various applications and in particular in solar energy panels.

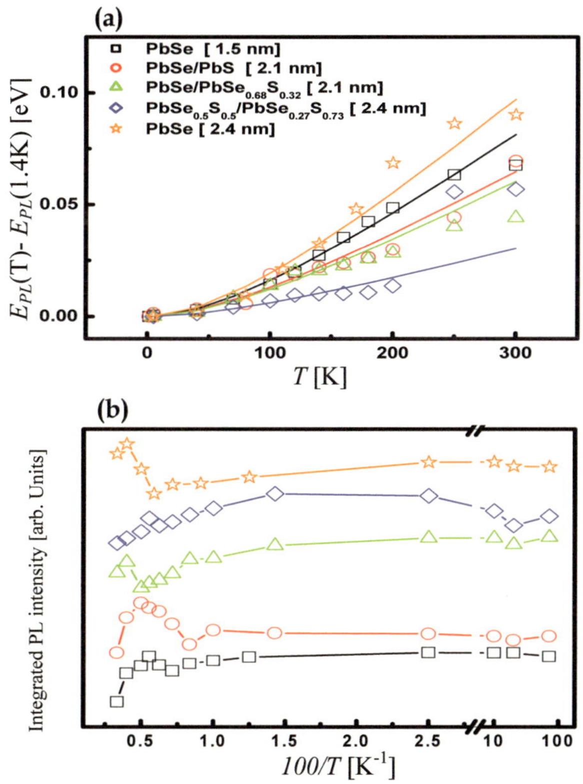

Fig. 7. Plot of the low cw-PL band emission peak energy relative to that at 1.4 K ($E_{PL}(T)$-$E_{PL}(1.4\ K)$) versus the temperature T of the samples mentioned in the legend (a). The symbols designate the experimental points, while the solid lines are the best fit curves; Plot of the integrated PL intensity versus *100/T* (b) of the samples presented in Panel (a). The solid lines here are drawn to guide the eye.

Figure 7 (b) demonstrates plots of the PL integrated intensity of the low energy cw-PL band *versus 100/T*. The plots show a similar tendency, including a plateau at the temperature range > 10 K to ~ 150 K, followed by low quenching of the intensity by an exciton-phonon interaction at higher temperatures, a typical behavior of a direct band gap semiconductor.[39] However, the trend is interrupted in two distinct points: (i) occurrence of unusual climax in the intensity profile at a temperature between 150 K to 250 K, appearing at higher temperature in c/sh and c/a-sh, compared with that of the primary cores, but with a very small effect in a-c/a-sh CQDs. This abnormal climax was previously explained[21] as a thermal activation between dark and bright states, with activation energy (ΔE_a) close to the LO phonon energy ($LO_{(PbSe)}$ = 16.8 meV, $LO_{(PbS)}$ = 26 meV). The values of ΔE_a of the investigated samples are listed in Table 3, spanning a range that is in close agreement to the suggested theoretical values of the dark-bright splitting in pure PbSe cores;[27] (ii) unexplained a minor decrease of the intensity <10 K with an activation energy ~ 0.4 meV, way below the acoustic phonon energy (LA, TA ~ 4-6 meV).[40] Worth to note, that the best fit shown in Panel (a) also show some deviation from perfection > 150 K, in correlation with the abrupt climax shown in Panel (b), due to a change in the emission mechanism from a dark to a bright state emission.

Molecular formula of CQDs	R_s [nm]	dE/dT [meV/K]	ΔE_a [meV]
PbSe	1.5	0.32	15.51
PbSe/PbS	2.1	0.17	17.23
$PbSe/PbSe_{0.68}S_{0.32}$	2.1	0.15	15.51
$PbSe_{0.5}S_{0.5}/PbSe_{0.27}S_{0.73}$	2.4	0.06	-
PbSe	2.4	0.48	21.54
Bulk material	dE_g/dT [meV/K]		
PbSe	0.51		
PbS	0.52		

Table 3. An energy band-gap temperature coefficient dE/dT and thermal activation energy ΔE_a of PbSe core, PbSe/PbS c/sh, $PbSe/PbSe_xS_{1-x}$ c/a-sh and $PbSe_yS_{1-y}/PbSe_xS_{1-x}$ a-c/a-sh CQDs.

3. Thermally activated photoluminescence processes in PbS CQDs

For deeper understanding of the thermally activated emission processes in $PbSe_yS_{y-1}/PbSe_xS_{1-x}$ hetero-structures we also examined pure PbS CQDs with equevalent sizes. In the recent years a progress towards achieving a detailed understanding of emission in PbS has been made.[41-46] However, important challenges still remain, thus providing a strong motivation to study the fundamental physics of fluorescence in this semiconductor. The investigated PbS CQDs were prepared by colloidal chemistry, according to the procedures given in Ref. [43] and will be described in Section 5.2. The room temperature absorption and cw -PL spectra of PbS CQDs, with diameters between 2.2 and 3.5 nm (all dispersed in GS), are shown in Figure 8, while a representative HR-TEM image of a single CQD is shown in the Inset. The crystallinity and composition of the materials were investigated by recording their HR-TEM images, revealing the formation of spherical CQDs with a high quality rock salt structure. The absorption spectra are comprised of the $1S_e$-$1S_h$ exciton transitions varying between 1.55 to 1.2 eV, corresponding to CQDs with a diameter between 2.2 to 3.5 nm and blue-shifted upon the decrease of diameter. Also, the $1S_e$-$1S_h$ exciton bands are blue-shifted with respect to that of the bulk exciton at 0.32 eV. The excitonic emission bands have a full width at half-maximum (FWHM) of about 200 meV, with E_s of 250 meV for the 2.2 nm CQDs, which reduces gradually with the increase of the CQDs size. For 3.5 nm CQDs E_s is 50 meV.

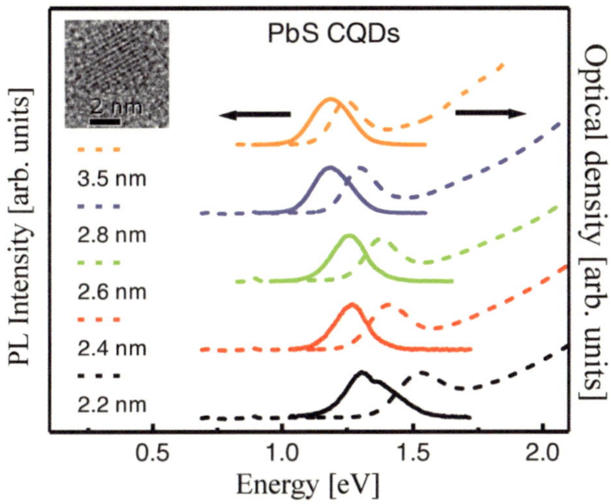

Fig. 8. Room temperature absorption (dash lines) and cw-PL spectra (solid lines) of PbS CQDs with an average diameter as indicated in the legend; Inset: HR-TEM image of a single PbS CQD; scale bar 2 nm.

This shift is the result of a split of the exciton manifold by the L-valley mixing and by the electron-hole exchange interaction, which further splits into dark and bright states. Figure 9(a) and 9(b) illustrate the cw-PL spectra of 2.6 and 3.5 nm PbS CQDs, respectively, dispersed in GS and recorded at various temperatures ranging from 4 K to 300 K.

At all temperatures the ground states exciton PL spectra of 2.6 nm CQDs are nealy symmetric, while PL spectra of 3.5 nm PbS CQDs have an asymmetric shape (like those of the equivalent size PbSe and $PbSe_yS_{1-y}/PbSe_xS_{1-x}$ a-c/a-sh hetrostructures), showing variations with the change in the temperature of the emission peak energy, FWHM and in integrated PL intensity.

The evolution of the PL peak energy with the increase in temperature of 2.6 and 3.5 nm PbS CQDs are shown in Figure 10(a) by the symbols (see legend in the figure) and the solid lines which are best fitted curves using Varshni relation.[36] The plots reveal a blue shift of the exciton emission energy with the increase in the temperature and can mainly attribute to the increase in the band gap energy with temperature, typical to the small band-gap IV-VI semiconductors. A tangent line to the fitted curve evaluates the slope, revealing the temperature coefficient, $\alpha = dE/dT$. The best fit α parameter for the PbS CQDs vary between - 0.17 meV/K (for the 2.6 nm) to 0.22 meV/K (for the 3.5 nm) and are smaller to similar sizes PbSe CQDs.[21]

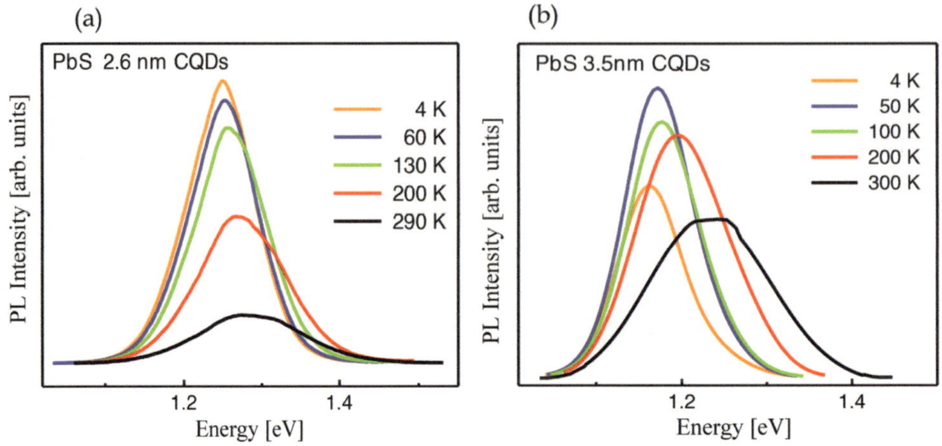

Fig. 9. Representative cw-PL spectra of PbS CQDs with diameter of 2.6 nm (a), and 3.5 nm (b) dispersed in GS and measured at various temperatures as indicated in the legend.

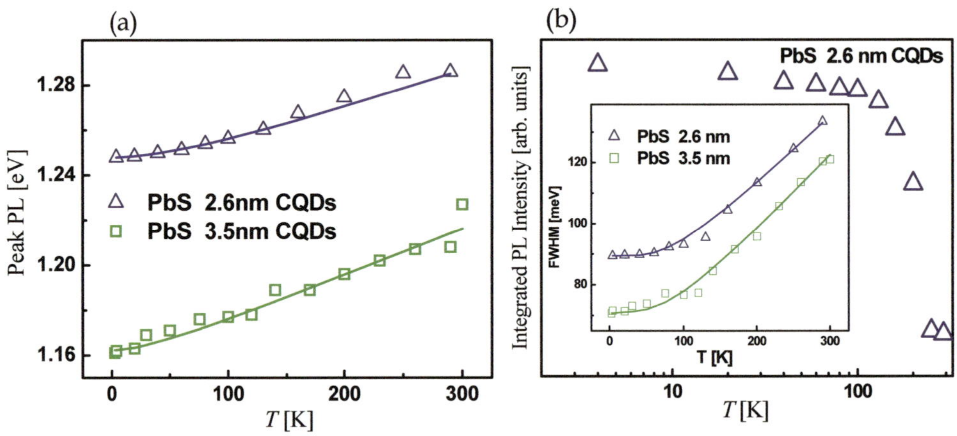

Fig. 10. Plots of the PL peak energy versus the temperature of PbS CQDs with dimensions as indicated in the legend (a); The solid lines represent a fit to a Varshni function[36] ;Integrated PL intensity of 2.6 nm PbS CQDs versus the temperature (b); inset: PL band's FWHM of PbS CQDs shown in Figure 9 versus the temperature. The solid lines represent a fit to the modified Bose-Einstein relation.[47]

Plot of the integrated PL intensity versus the temperature of 2.6 nm PbS CQDs is presented in Figure 10(b) and shows Arrhenius behavior. The activation energy extracted from the Arrhenius plot is 14 meV, and is smaller than the thermal activation energy ΔE_a obtained for PbSe CQDs'and $PbSe_yS_{1-y}$/$PbSe_xS_{1-x}$ a-c/a-sh CQDs heterostructures that are listed in Table 3.

The inset of Figure 10(b) represents a plot of the FWHM of PL spectra of the samples shown in Figure 10(a) as a function of temperature. The FWHM decreased with decreasing temperature and was best fitted to a modified Bose-Einstein relation given in Ref. [47]. The experimental data reveal a reduction of the ground-state exciton broadening in the 2.6 nm CQDs in comparison with that in 3.5 nm PbS CQDs. The best fit curves are shown by the solid line in the figure. The values that were obtained by the best fit parameters are: inhomogeneous broadening parameter Γ_{inh} (taken as $\Gamma_o(T)$ at 0 K) varies between 70-90 meV depending on the CQDs' size, acoustic phonon coupling σ=0.016 meV/K , optical phonon coupling Γ_{LO}=56 meV. The LO-phonon energy E_{LO} was extracted from Raman measurements (not shown here) and is about 23 meV. The values obtained from Bose-Einstein relation are with accordance to those presented in the literature.[41]

Representative transient PL curves of 3.5 nm PbS CQDs in GS, recorded at various temperatures, are shown in Figure 11(a). These curves were fitted to a single exponent decay curve, $I_{(t)}$= A exp(-t/ τ_0), showing an obvious change with variation of T.

The best-fit lifetime's τ_0 of 2.6 and 3.5 nm PbS CQDs versus T are plotted in Figure 11(b), revealing a moderate decrease of the lifetime from 5.2 to 4.0 μs for 2.6 nm and from 6.5 to 5 μs for 3.5 nm PbS CQDs when T increases from 2.4 to 100 K. A steeper decrease of the lifetime is observed when T increases from 120 to 290 K. At all temperature the τ_0 are longer than those of 2.6 nm PbS QDs.[44]

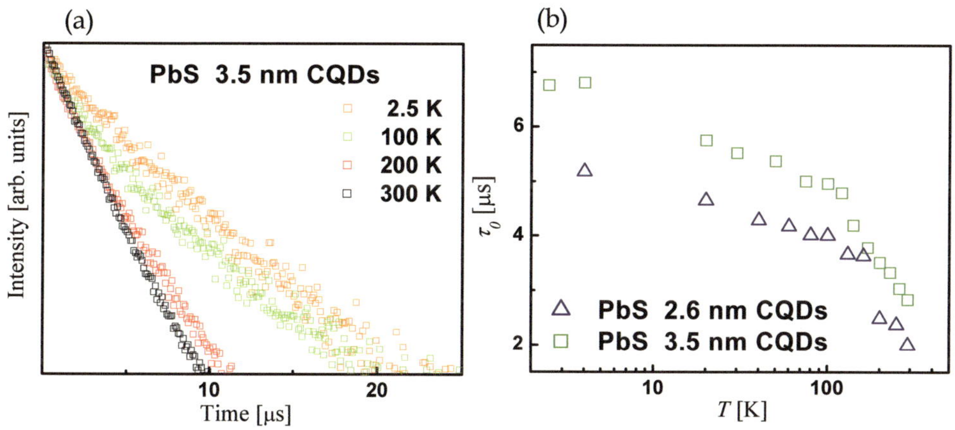

Fig. 11. Transient PL curves of 3.5 nm PbS CQDs dispersed in GS and recorded at various temperatures, as indicated in the legend (a); Plots of the lifetime, τ_0, versus T of PbS CQDs shown in Figure 9.

Spectrally resolved transient PL of 2.6 and 3.5 nm PbS CQDs measured at three different points of the PL band at various temperatures are shown in Figure 12(a) and 12(b). The arrows pointing to three different points of PL energy: red arrow- energy of the PL peak; black arrow - energy at the half PL intensity from the blue side of the PL, while the blue arrow is the energy at the half PL intensity from the red side of the PL. The obtained results reveal two emission bands with different τ_0 temperature dependent behaviour: the higher energy band τ_0 increase moderate with decrease of T from 290 to 4 K, while the low energy band show steeper τ_0 increase with decrease of T. This behavior suggests that the two band composed PL could correspond to bright and dark states induced by exchange interaction or to the two lowest valley-valley manifold exitonic states.

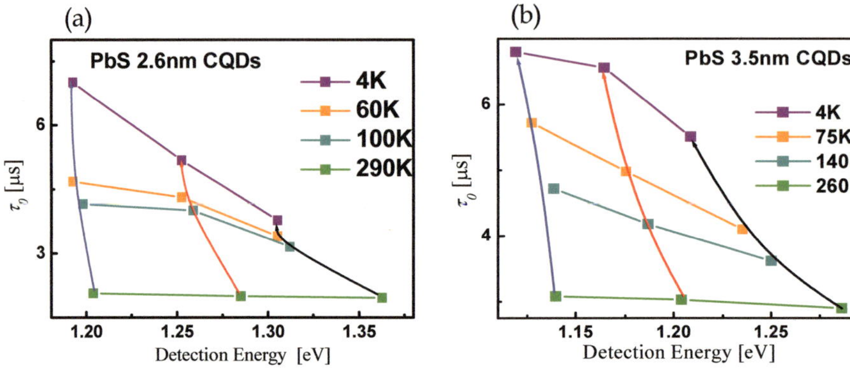

Fig. 12. Plots of the measured τ_0 versus the detection energy across the PL spectrum, of 2.6 (a) and 3.5 nm (b) PbS CQDs, measured at various temperatures (4-290 K) as indicated in the legends. The arrows pointing to three different points of PL energy: red arrow- energy of the PL peak; black arrow - energy at the half PL intensity from the blue side of the PL, while the blue arrow is the energy at the half PL intensity from the red side of the PL.

4. Theoretical prediction of the electronic properties of heterostructed QDs

The electronic band structure of the heterostructures (with/without alloy composition) quantum dots (QDs) was evaluated, using a k*p model, considering specific features, related to the discontinuity of the effective mass, crystal potential and dielectic constant at the core/shell, as well as at the shell/surrounding interface. The evaluation also considered anisotropy of effective masses, particularly in the IV-VI semiconductors. The evaluations explored interesting properties, associated with charge distribution between the core and the shell, effective electron-hole spatial separation, probability of transitions, coulomb interactions and tunability of band-edge and remote states' energy.[23, 48] Pre-engineering of the electronic band structure is done by theoretical consideration (of spherical particles alone for the moment), covering all cases, when, either the core or shell or both has alloy composition (see scheme in Figure 13). A few special points should be considered: anisotropy in effective mass (typical for IV-VI semiconductors), as well as the fact that each physical parameter dependents on its position (r) across the dot, and also may vary smoothly at the core/shell and the shell/surrounding interfaces, with a smoothing factor γ. The Hamiltonian was adjusted to the discontinuity at the PbSe/PbS interface by the appropriate choice of the kinetic energy term, ensuring probability current conservation and continuity of the envelope functions. In addition, the Hamiltonian potential energy term, included the heterostructure band offset, abrupt for the c/sh structure, but considered as a smooth function for the c/a-sh QDs.[49] Presumably, the smooth potential profile reflects the nature of the interface region in alloyed materials with gradient composition, being an extension of the standard treatment for semiconductor heterostructures. The overall band offset was chosen as that of the corresponding bulk PbSe and PbS materials (where the valence band maximum of bulk PbS lies 0.025 eV above that of PbSe, while the conduction band minimum lies 0.155 eV above that of PbSe). Diagonalization of the envelope function Hamiltonian yielded the electron and hole wavefunctions, as well as a good approximation of the energy values of the conduction and valence-band's states. The heterostructures investigated are ternary core or c/sh QDs, having a general formula $PbSe_xS_{1-x}/PbSe_yS_{1-y}$, covering the following cases: (a) x = y = 1 or x = y = 0 refers to a simple core PbSe or PbS, respectively; (b) 0 < x = y < 1 is a homogenous alloy core; (c) x = 1 and y = 0 is a simple PbSe/PbS c/sh; (d) x = 1 (y = 1) and y ≠ 0 (x ≠ 0) is a complex c/sh QD, when either the core or the shell has a homogenous alloyed composition. A schematic drawing of a ternary QD is shown in Figure 13. R_c and R_s designate the radius of a core and a c/sh QD, respectively.

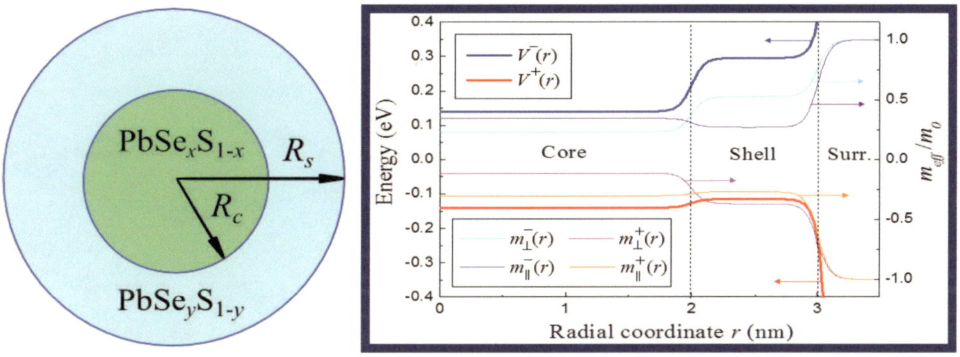

Fig. 13. Schematic drawing of a spherical c/sh $PbSe_xS_{1-x}/PbSe_yS_{1-y}$ QD. R_c and R_s are the core and the total radii, respectively. Radial variation of the bulk material parameters in spherical PbSe/PbS QD (R_c=2 nm, R_s=3 nm).

Three-dimensional plots of the electron and hole distribution functions on (111) cut-plane for a pure core (Panel (a)), c/sh QDs (Panels (b) and (c)) and c/a-sh (Panels (d) and (e)) of equivalent R_s, are shown in Figure 14. Panels (f) and (g) show the electron and the hole distribution for a c/a-sh with R_c=3 nm and R_s=5 nm. In the case of a pure core structure, the distribution of electron and hole is virtually identical, thus Panel (a) describes either one of the carriers. The choice of the (111) plane is made for the calculation convenience only, and is equivalent to choosing any other crystallographic plane for the distribution representation, since the ground state wavefunctions are spherically symmetric. These plots reveal a distinct trend, in which the lowest energy hole state, $|1/2, 1\rangle$ (1/2 denotes the total angular momentum j of the state, and ±1 corresponds to the parity π), is more delocalized with respect to its counter partner, the lowest energy electron state $|1/2, -1\rangle$, in c/sh and c/a-sh QDs, characteristic of *quasi*-type II configuration at the band edge. This electronic distribution explains the experimental observations shown in Figure 2 of the gradual red-shift of the absorption spectra with the increase of the shell width . The electron and hole spatial distribution functions in c/sh CQDs shown in Figure 14(b) and 14(c) suggest that the distribution of both carriers is similar to that of a c/a-sh structure (Figure 14(d) and 14(e)); however, the distribution differs from the case of a simple core QD of comparable total size (Figure 14(a)). The calculations show that for heterostructured particles with R_c< 3 nm the energies of the lowest lying electron levels exceed the potential barrier height (either abrupt or gradual), located at the interface between the core and the shell, thus significantly reducing the effect of quantum confinement induced by the shell layer. On the other hand, in the case of larger c/sh or c/a-sh particles the electron energy is lower than the barrier height, thus enforcing the confinement (and subsequently the localization) of the electron in the core region (*cf.* Figure 14(b) and 14(f)). However, the energy difference between the valence band edges of PbSe and PbS is almost an order of magnitude smaller than that between the conduction band edges, hence the hole distribution is influenced by the shell to a much lesser extent (*cf.* Figure 14(c) and 14(g)).

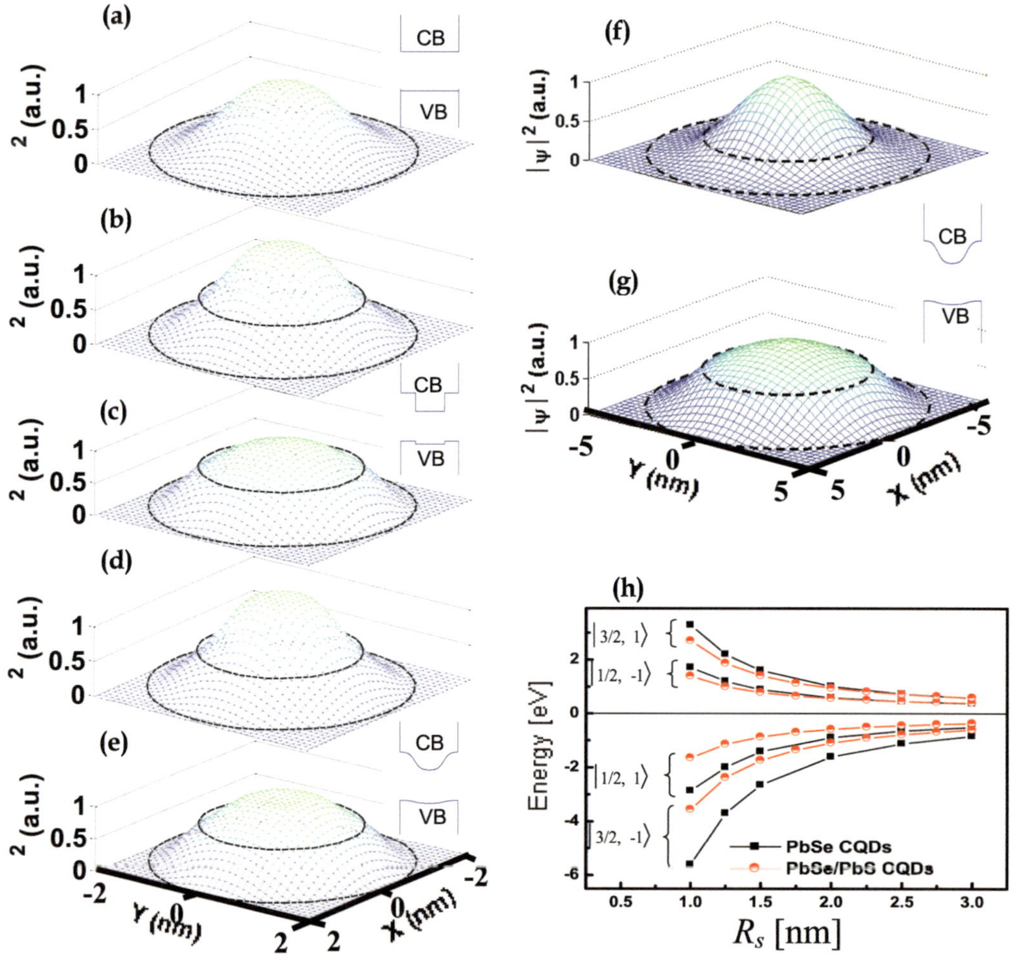

Fig. 14. Probability density on the (111) cut-plane for electron and hole in (a) PbSe QDs, R_s=2.4 nm. (b) Electron and (c) hole in PbSe/PbS CQDs, R_c=1.5 nm, R_s=2.4 nm. (d) Electron and (e) hole in c/a-sh QDs, R_c =1.6 nm, R_s =2.4 nm. (f) Electron and (g) hole in c/a-sh QDs, R_c =3 nm, R_s =5 nm (Both dashed circles, outer and inner, represent the external particle boundary and the core/shell interface location, respectively). The insets marked VB (valence band) and CB (conduction band) schematically represent the radial potential energy profile used in the calculation in each case. (h) Energy as a function of R_s of two lowest states of electron and hole in PbSe core and PbSe/PbS c/sh QDs with R_s / R_c =3/2.

Figure 14(h) displays the calculated energy of two lowest energy states versus the R_s of core and c/sh QDs with R_s / R_c =3/2. This Figure reveals a pronounced influence of the shell on the energy levels of the carriers.

In the case of a c/sh structure, both the electron and hole levels are lowered in energy relative to a core structure of the same size, with a larger influence on the hole levels. In the framework of this model the energy levels of c/a-sh structures are almost identical to those of c/sh, hence are not shown here. This finding is consistent with the experimental observation of the red-shift in the emission energy of the c/sh and c/a-sh heterostructures, relative to the cores of corresponding size. The theoretical $|1/2,1\rangle \rightarrow |1/2,-1\rangle$ transition energies (which is the first excitonic transition) are listed in Table 1 and are compared with the experimental absorption band edge energies, with a close agreement for QDs with R_s> 1.5 nm (Apparently, the accuracy of the model is not satisfactory for very small sizes due to the breakdown of the major assumption that the envelope function is slowly varying on the scale of the unit cell). The model reproduced the band edge energies of the QDs with relatively close agreement with the experiment, as well as predicted varying delocalization extent of the electrons in the lowest conduction band. The explanation of the reported variation of various physical properties of c/sh and c/a-sh heterostructures would demand further theoretical considerations (e.g., mass anisotropy, exchange interactions), which are beyond the scope of the discussed model, however, will be done in the future.

Evolution of the ten lowest conduction and valence band levels as a function of structure and composition of QD with R_s are shown in Figure 15.

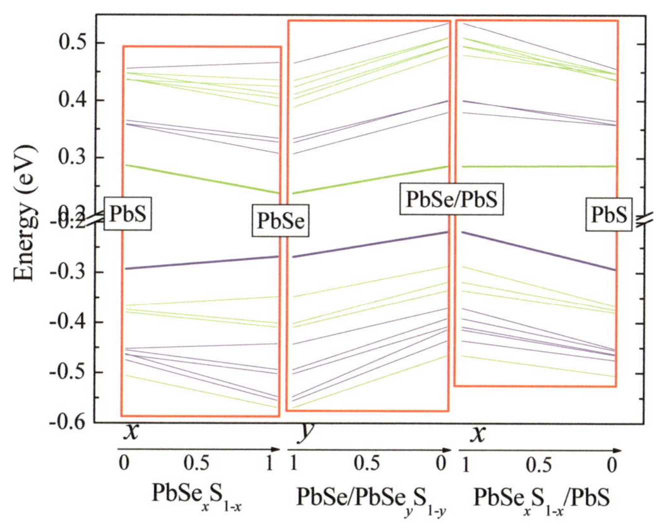

Fig. 15. Evolution of the energy of the conduction and valence band energy levels through a series of composition and structural changes, while maintaining a constant QD radius of 2.5 nm.

First, the QD structure evolves from PbS core to PbSe core via the intermediate alloyed $PbSe_xS_{1-x}$ structures (Left Panel). Next, the QD is divided into a 3 nm PbSe core and a 2 nm thick $PbSe_yS_{1-y}$ shell (*i.e.*, R_c=3 nm, R_s=5 nm). The shell composition then varies from y=1 to y=0, corresponding to a transition from pure PbSe to PbSe/PbS c/sh via intermediate PbSe/$PbSe_yS_{1-y}$ c/sh structures (middle panel). Finally, the composition of the core constituent evolves from x=1 to x=0, corresponding to a transition from PbSe/PbS c/sh to a pure PbS core, which completes the cycle (Right Panel). States of even (+) and odd (-) parity are marked by purple and green lines, respectively. CB and VB correspond to the conduction and valence bands, respectively.

Figure 16 shows the energy levels (blue) and density of states (DOS) (green) of PbS (left panel), $PbSe_{0.5}S_{0.5}$ (middle panel) and PbSe (right panel) having R_s=2.5 nm. The density of states calculated by broadening each energy level by 25 meV Gaussian. The composition is found to have a significant impact on the energy spectrum of spherical QDs (apart from the band gap energy). For instance, when looking at the all three cases of $PbSe_xS_{1-x}$ core structures of the same size, in $PbSe_{0.5}S_{0.5}$ the levels are arranged into more dense discrete groups, while in PbS and PbSe QDs they are more evenly distributed, as can be seen in Figure 16.

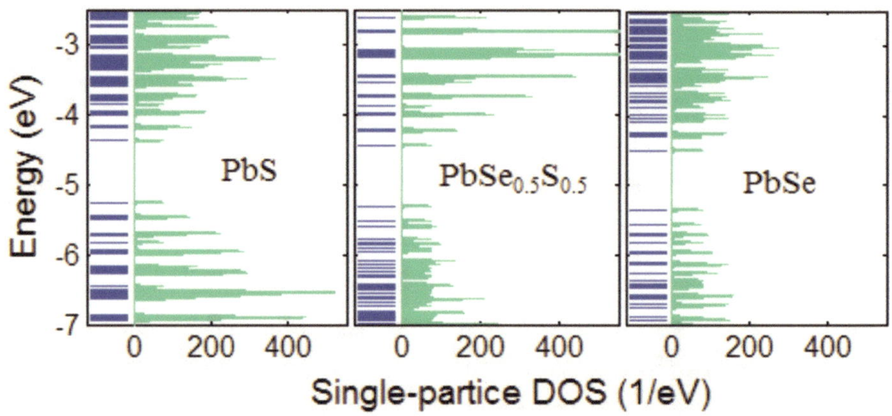

Fig. 16. Single-particle energy levels (blue) and density of states (green) of PbS (left panel), $PbSe_{0.5}S_{0.5}$ (middle panel) and PbSe (right panel) QDs having R_s=2.5 nm. The density of states calculated by broadening each energy level by 25 meV Gaussian.

It is intuitively clear that the origin of this composition-dependent variance of the DOS (that even reaches degeneracy in some cases) should be linked to some symmetry property of the system. In our case this property is the shape of the energy isosurface (EI) of $PbSe_xS_{1-x}$ bulk materials in the reciprocal space. As mentioned above, in the case of lead chalcogenides the general shape of the EI resembles a spheroid, having its principle axis in the L-direction, and it varies as a function of energy $E(\mathbf{k})$.

Calculated values of the fundamental gap energies E_g evaluated for various core and c/sh QDs, compared with experimental data are presented in Figure 17 and shows a qualitativly agreement between the theoretical and experimental results.

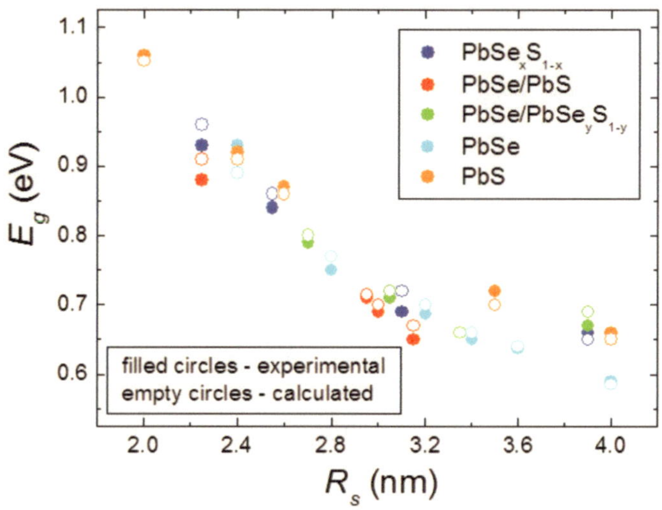

Fig. 17. Calculated (empty circles) and the corresponding experimental (filled circles) values of the band gap energies evaluated for several core and c/sh QDs.

5. Synthesis procedures and experimental techniques used for core/shell heterostructures with alloy components

5.1 Synthesis of PbSe, PbSe/PbS c/sh, PbSe/ $PbSe_xS_{1-x}$ c/a-sh and $PbSe_yS_{1-y}/PbSe_xS_{1-x}$ a-c/a-sh CQDs

The synthesis of core PbSe CQDs followed a modified procedure to that given by Murray et al.[50] , following the procedure given in[19] and including the preceding stages: (1) 0.71 gr of lead(II) acetate trihydrate [Pb-ac] ($Pb[CH_3COO]_2 \cdot 3H_2O$, GR, Merck) were dissolved in a solution composed of 2 mL diphenyl ether [PhEt] ($C_6H_5OC_6H_5$, 99%, Aldrich), 1.5 mL oleic acid (OA) ($CH_3(CH_2)_7$ $CHCH(CH_2)_7COOH$, 99.8%, Aldrich) and 8 mL TOP ($(C_8H_{17})_3P$, Tech, Aldrich), under standard inert conditions in the glove box, and were inserted into a three-neck flask (flask I); (2) 10 mL of PhEt were inserted into a three-neck flask (flask II) under the inert conditions of a glove box; (3) both flasks were taken out of the glove box, placed on a Schlenk line and heated under a vacuum to 100 - 120°C for an hour; (4) flask I was cooled to 45°C, while flask II was heated to 180 -210°C, both under a fledging of an argon gas; (5) 0.155 gr of selenium powder (Se, 99.995%, Aldrich) was dissolved in 2.0 mL TOP, forming a TOP:Se solution, under standard inert conditions of a glove box. Then, 1.7 mL of this solution was injected into flask I on the Schlenk line; (6) the content of flask I, containing the reaction precursors, was injected rapidly into the PhEt solution in flask II, reducing its temperature to 100 - 130°C, leading to the formation of PbSe CQDs within the first 15 min of the reaction. The described procedure produced nearly mono-dispersed CQDs with < 8% size distribution, with average size between 3 and 9 nm, controlled by the temperature and by the time duration of the reaction.

The preparation of PbSe/PbS c/sh CQDs by a two-injection process[19] begins with formation of core PbSe CQDs and their isolation from the initial reaction solution, according to the procedure described previolsy. Those core CQDs were re-dissolved in chloroform solution, forming a solution of 50 mg/mL weight concentration. The quality of 1.4 mL of TOP was then added to the CQDs solution, while the chloroform molecules were removed by distillation under vacuum and heating at 60°C. In parallel, 0.2 gr of a Pb precursor, Pb-ac, was dissolved in a mixture of 2 mL PhEt, 1.5 mL of OA, and 8 mL of TOP, heated to 120°C for an hour, and then cooled to 45°C. Also, 0.03–0.10 gr of sulphur (S, 99.99 + %, Aldrich) was dissolved in 0.3 mL of TOP and was premixed with a PbSe core CQDs in a TOP solution. This mixture was injected into the Pb-ac solution. All reagents were then injected into a PhEt mother solution and kept on a Schlenk line at 180°C, causing a reduction in temperature of the mother solution to 120°C. The indicated chemical portions caused the precipitation of 1–3 monolayers of PbS shell over the PbSe core surface within the first 15 min of the reaction.

The preparation of $PbSe/PbSe_xS_{1-x}$ c/a-sh and $PbSe_xS_{1-x}/PbSe_yS_{1-y}$ a-c/a-sh structures[19] is nearly identical to that of the core PbSe CQDs, described previosly using a single injection of the precursors into a single round flask. However, step (5) was altered by the use of an alternative chalcogen precursor stock solution. A stock solution of Se and S was prepared by mixing 0.15 gr Se dissolved in 1.4 mL TOP, with 0.03–0.10 gr S dissolved in 0.3 mL TOP. The amount of S in the new stock solution corresponded to a stoichiometric amount of 1–2 ML of the PbS compound. Thus, the mole ratio of the precursors Pb:Se:S ranged from 1:1:0.5 to 1:1:1.3. Aliquots were drawn periodically from the mother solutions while quenching process to RT terminated the CQDs' growth. They were isolated from the aliquots solution by the addition of methanol, and by centrifugation. The isolated CQDs were further purified by dissolving them in chloroform, followed by filtering several times through a 0.02 micron membrane. A preliminary injection of Pb/Se/S ions ratio of 1/1/0.5 led to the nucleation of a pure PbSe core, due to the faster reactivity of the Se precursors at the nucleation stage. However, the increase of the S/Se ratio (S/Se> 1.5/1) enabled an immediate integration of both elements with the nuclei (monitored already in the first aliquot). Further aliquots revealed a gradient increase of the S/Se ratio when moving from the interface toward the exterior surface. For simplicity, the samples were labeled as $PbSe_yS_{1-y}/PbSe_xS_{1-x}$.

5.2 Synthesis of PbS CQDs

PbS CQDs were prepared according to literature procedure.[43] The lead oleate precursor was prepared by heating 0.09 g PbO in 4 ml OA under N_2 at 120°C for 1h. A solution of 42 µl bis(trimethylsilyl)sulfide (TMS) in 2 ml octadecene (ODE) was injected into the vigorously stirring lead oleate solution. The final particle size was controlled by injection temperature (120-150 °C), where lower temperature leads to smaller sizes, and by varying the Pb:OA molar ratios (2:32 to 2:4, where the total volume of lead oleate solution was kept at 4 ml by dilution with ODE). The reaction was quenched by cooling it to RT and the CQDs were precipitated with acetone, then subsequently redispersed in chloroform and precipitated again.

5.3 Experimenthal methods

The morphology and crystallography of the colloidal CQDs were examined by X-ray diffraction, TEM, HR-TEM and SAED. The TEM specimens were prepared by injecting small liquid droplets of the solution on a copper grid (300 mesh) coated with amorphous carbon film and then dried at room temperature. The elemental analysis were examined by EDAX, and inductively coupled plasma atomic emission spectroscopy (ICP-AES).

The absorption spectra of the samples were recorded on a JASCO V-570 UV-VIS-NIR spectrometer. The cw-PL spectra were obtained by exciting the samples with a tunable Ti:Sapphire laser, (E_{exc} = 1.48-1.80 eV). The PL spectra of the materials studied were recorded at a temperature range of 1.4 K to 300 K, while immersing the samples in a variable temperature Janis cryostat, and detecting the emission with an Acton Spectrapro 2300i monochromator, which was equipped with a cooled InGaAs CCD or cooled Ge photo detector. The transient PL curves were recorded by exciting the samples with a Nd:YAG laser, (E_{exc} = 1.17 eV). The measurements utilized a laser flux <0.1 mJ/cm^2, corresponding to a photon fluence of j_p~10^{11} photons/cm^2 per pulse. Considering the absorption cross-section of σ_0~10^{-15} cm^2, measured in reference,[14] the number of photo-generated excitons is given by $<N_0>= j_p \cdot \sigma_0$, and estimated to be $10^{-4}<<1$, ensuring the generation of single excitons. The transient-PL curves were monitored by a photon multiplier tube, Hamamatsu NIR-PMT H10330-75. The measurements were carried out at temperature range from 1.4 K to 300 K. The PL quantum yield was measured utilizing integrating sphere technique described by Friend.[51] A solution of CQDs was placed inside an integrating sphere and excited by a monochromatic light. Luminescence spectra were detected by a fiber-coupled spectrometer equipped with a liquid nitrogen-cooled Ge photo detector. The entire system response was normalized against a calibrated detector, and care was taken to ensure that the sample absorption was more than 20%.

5.4 Storage conditions

The $PbSe_yS_{1-y}/PbSe_xS_{1-x}$ a-c/a-sh ($0<x<1;0<y<1$) CQDs were stored either in a various solutions (chloroform, hexane) or were embedded in a polymer film or dissolved in a GS (2,2,4,4,6,8,8,-heptamethyl-nonane) for the optical measurements. The polymer-CQDs solution was prepared by mixing PbSe CQDs in chloroform solution with poly-methyl-methacrylate (PMMA) ($[-CH_2C(CH_3)(CO_2CH_3)-]_n$, analytical grade, Aldrich) polymer solution. The resultant mixture was spread on a quartz substrate and dried over 24 h to a uniform film. The stability of these CQDs was examined over a period of time by recording the absorption spectra and following the consistency of the low exciton energy. Plots of this exciton energy versus time suggest that the exciton energy in the core samples is blue-shifted by ~ 400 meV over a period of days for the CQDs kept in a chloroform solution. Such a blue-shift, however, occurs over months for the samples kept as dry powders. On the other hand, the energy shift is smaller for the PbSe/PbS c/sh samples, and even nearly disappeared for the CQDs coated with three PbS shell monolayers. It is presumed that the exciton energy blue-shift is due to oxidation of the surface, and a decrease of the effective size of the core. Obviously, the penetration of oxygen through the PbS shell is reduced with the growth of the shell width. Furthermore, storage of the CQDs in a nitrogen environment nearly eliminates any spectral drift over a period of months, even extending to two or three years. It should be noted that most of the optical measurements were carried out at cryogenic temperatures, inducing a He gas environment around the samples. But unpublished results determined degradation of the samples when exposed to intense pulsed UV radiation, which is avoided completely in the current study.

6. Conclusion and future directions

Unique alloyed c/sh heterostructures, such as $PbSe_yS_{1-y}$ /$PbSe_xS_{1-x}$, ($0<x\leq1$; $0<y\leq1$) were developed, offering good crystallographic and dielectric match at the c/sh interface,

regulating carriers' delocalization and/or charge separation by tunability of the band off-set, showing an exceptionally high emission quantum yield, chemical stability, and an option to stabilize the emission intensity (blinking free behavior), as well as sustain the biexciton lifetime over a nanosecond. The last can be of a valuable benefit in the use of CQDs in gain devices and photovoltaic cells.

A smooth potential at the c/sh interface was applied here for the determination of the electronic structure of IV-VI CQDs, using a $\mathbf{k}\cdot\mathbf{p}$ model, covered a wide physical aspects, including an effective mass anisotropy, dielectric variation between the constituents, showing a ground for tailoring heterostructures with the desired composition and optical properties.

A thorough investigation of the optical properties was performed by following variable temperature cw- and transient spectrally resolved PL spectra, exploring energy shift, band edge temperature coefficient, alleviation of a dark-bright splitting (or exchange interaction), valley-valley interaction, emission quantum yield, and radiative lifetime of the $PbSe_yS_{1-y}$ /$PbSe_xS_{1-x}$ heterostructures, in comparison with the existing properties of the primary PbSe core CQDs. Temporally and spectrally resolved PL spectra provide more-systematic evidence of the two emissive centers nature. The results reflect the uniqueness of the electronic properties of the heterostructures, controlled by shell width and alloyed composition.

The discussed heterostructures could be of significant importance in applications where the CQDs' size is restricted, *e.g.*, biological markers or self-assembled CQDs in opto-electronic devices, while at the same time, there are stringent demands regarding the optical tunability. We showed that the restriction can be overcome by the discussed new strategies gaining property control using: (a) alloyed ternary or quaternary compounds, when all elements can be either distributed homogeneously or exhibit a graded composition along a selective direction; (b) c/sh heterostructures, comprised of a semiconductor core, covered by a shell, of another semiconductor, when the band-edge off-set at the core/shell interface, can be tuned from a type-I (when shell band-edge is rapping that of the core), through *quasi*-type-II, to a type-II (when, band-edge of the constituents are staggered) alignment. Moreover, one of the constituents (core or shell) may have alloyed composition.

7. Acknowledgment

The authors thank G. Kventsel for helpful discussions, providing many insightful comments, A. Bartnik and F. Wise for helpful discussions and guidance on the theoretical model, A. Efros for the useful scientific discussions, and O. Solomeshch for assistance in the quantum yield measurements. The authors acknowledge support from the Israel Science Foundation (Projects No. 1009/07 and No. 1425/04), the USA-Israel Binational Science Foundation (No. 2006-225), and the Ministry of Science (No. 3-896).

8. References

[1] Steigerwald, M. L.; Brus, L. E., Semiconductor crystallites: a class of large molecules. *Accounts of Chemical Research* 1990, *23* (6), 183-188.

[2] Peng, X.; Manna, L.; Yang, W.; Wickham, J.; Scher, E.; Kadavanich, A.; Alivisatos, A. P., Shape control of CdSe nanocrystals. *Nature* 2000, *404* (6773), 59-61.

[3] Akamatsu, K.; Tsuruoka, T.; Nawafune, H., Band Gap Engineering of CdTe Nanocrystals through Chemical Surface Modification. *Journal of the American Chemical Society* 2005, *127* (6), 1634-1635.

[4] Ma, W.; Luther, J. M.; Zheng, H.; Wu, Y.; Alivisatos, A. P., Photovoltaic Devices Employing Ternary PbS_xSe_{1-x} Nanocrystals. *Nano Letters* 2009, *9* (4), 1699-1703.

[5] Bailey, R. E.; Nie, S., Alloyed Semiconductor Quantum Dots: Tuning the Optical Properties without Changing the Particle Size. *Journal of the American Chemical Society* 2003, *125* (23), 7100-7106.

[6] Piven, N.; Susha, A. S.; Döblinger, M.; Rogach, A. L., Aqueous Synthesis of Alloyed CdSexTe1-x Nanocrystals. *The Journal of Physical Chemistry C* 2008, *112* (39), 15253-15259.

[7] Erwin, S. C.; Zu, L.; Haftel, M. I.; Efros, A. L.; Kennedy, T. A.; Norris, D. J., Doping semiconductor nanocrystals. *Nature* 2005, *436* (7047), 91-94.

[8] Danek, M.; Jensen, K. F.; Murray, C. B.; Bawendi, M. G., Synthesis of Luminescent Thin-Film CdSe/ZnSe Quantum Dot Composites Using CdSe Quantum Dots Passivated with an Overlayer of ZnSe. *Chemistry of Materials* 1996, *8* (1), 173-180.

[9] Ivanov, S. A.; Piryatinski, A.; Nanda, J.; Tretiak, S.; Zavadil, K. R.; Wallace, W. O.; Werder, D.; Klimov, V. I., Type-II Core/Shell CdS/ZnSe Nanocrystals: Synthesis, Electronic Structures, and Spectroscopic Properties. *Journal of the American Chemical Society* 2007, *129* (38), 11708-11719.

[10] Piryatinski, A.; Ivanov, S. A.; Tretiak, S.; Klimov, V. I., Effect of Quantum and Dielectric Confinement on the Exciton-Exciton Interaction Energy in Type II Core/Shell Semiconductor Nanocrystals. *Nano Letters* 2006, *7* (1), 108-115.

[11] Oron, D.; Kazes, M.; Banin, U., Multiexcitons in type-II colloidal semiconductor quantum dots. *Physical Review B* 2007, *75* (3), 035330.

[12] Mamutin, V. V.; et al., Molecular beam epitaxy growth methods of wavelength control for InAs/(In)GaAsN/GaAs heterostructures. *Nanotechnology* 2008, *19* (44), 445715.

[13] Lauhon, L. J.; Gudiksen, M. S.; Wang, D.; Lieber, C. M., Epitaxial core-shell and core-multishell nanowire heterostructures. *Nature* 2002, *420* (6911), 57-61.

[14] Dorfs, D.; Franzl, T.; Osovsky, R.; Brumer, M.; Lifshitz, E.; Klar, T. A.; Eychmuller, A., Type-I and Type-II Nanoscale Heterostructures Based on CdTe Nanocrystals: A Comparative Study. *Small* 2008, *4* (8), 1148-1152.

[15] Nonoguchi, Y.; Nakashima, T.; Kawai, T., Tuning Band Offsets of Core/Shell CdS/CdTe Nanocrystals. *Small* 2009, *5* (21), 2403-2406.

[16] Sashchiuk, A.; Langof, L.; Chaim, R.; Lifshitz, E., Synthesis and characterization of PbSe and PbSe/PbS core-shell colloidal nanocrystals. *Journal of Crystal Growth* 2002, *240* (3-4), 431-438.

[17] Stouwdam, J. W.; Shan, J.; van Veggel, F. C. J. M.; Pattantyus-Abraham, A. G.; Young, J. F.; Raudsepp, M., Photostability of Colloidal PbSe and PbSe/PbS Core/Shell Nanocrystals in Solution and in the Solid State. *The Journal of Physical Chemistry C* 2006, *111* (3), 1086-1092.

[18] Xu, J.; Cui, D.; Zhu, T.; Paradee, G.; Liang, Z.; Wang, Q.; Xu, S.; Wang, A. Y., Synthesis and surface modification of PbSe/PbS core–shell nanocrystals for potential device applications. *Nanotechnology* 2006, *17* (5428).

[19] Brumer, M.; Kigel, A.; Amirav, L.; Sashchiuk, A.; Solomesch, O.; Tessler, N.; Lifshitz, E., PbSe/PbS and $PbSe/PbSe_xS_{1-x}$ Core/Shell Nanocrystals. *Advanced Functional Materials* 2005, *15* (7), 1111-1116.
[20] Lifshitz, E.; Brumer, M.; Kigel, A.; Sashchiuk, A.; Bashouti, M.; Sirota, M.; Galun, E.; Burshtein, Z.; Le Quang, A. Q.; Ledoux-Rak, I.; Zyss, J., Air-Stable PbSe/PbS and $PbSe/PbSe_xS_{1-x}$ Core/Shell Nanocrystal Quantum Dots and Their Applications. *The Journal of Physical Chemistry B* 2006, *110* (50), 25356-25365.
[21] Kigel, A.; Brumer, M.; Maikov, G. I.; Sashchiuk, A.; Lifshitz, E., Thermally Activated Photoluminescence in Lead Selenide Colloidal Quantum Dots. *Small* 2009, *5* (14), 1675-1681.
[22] Bartnik, A. C.; Wise, F. W.; Kigel, A.; Lifshitz, E., Electronic structure of PbSe/PbS core-shell quantum dots. *Physical Review B* 2007, *75* (24), 245424.
[23] Maikov, G. I.; Vaxenburg, R.; Sashchiuk, A.; Lifshitz, E., Composition-Tunable Optical Properties of Colloidal IV−VI Quantum Dots, Composed of Core/Shell Heterostructures with Alloy Components. *ACS Nano* 2010, *4* (11), 6547-6556.
[24] Moreels, I.; Lambert, K.; Smeets, D.; De Muynck, D.; Nollet, T.; Martins, J. C.; Vanhaecke, F.; Vantomme, A.; Delerue, C.; Allan, G.; Hens, Z., Size-Dependent Optical Properties of Colloidal PbS Quantum Dots. *ACS Nano* 2009, *3* (10), 3023-3030.
[25] Garrett, M. D.; Dukes Iii, A. D.; McBride, J. R.; Smith, N. J.; Pennycook, S. J.; Rosenthal, S. J., Band Edge Recombination in CdSe, CdS and CdSxSe1−x Alloy Nanocrystals Observed by Ultrafast Fluorescence Upconversion: The Effect of Surface Trap States. *The Journal of Physical Chemistry C* 2008, *112* (33), 12736-12746.
[26] Huang, Y. H.; Cheng, C. L.; Chen, T. T.; Chen, Y. F.; Tsen, K. T., Studies of Stokes shift in InxGa1?xN alloys. *Journal of Applied Physics* 2007, *101.*
[27] An, J. M.; Franceschetti, A.; Zunger, A., The Excitonic Exchange Splitting and Radiative Lifetime in PbSe Quantum Dots. *Nano Letters* 2007, *7* (7), 2129-2135.
[28] Allan, G.; Delerue, C., Confinement effects in PbSe quantum wells and nanocrystals. *Physical Review B* 2004, *70* (24), 245321.
[29] Harbold, J. M.; Wise, F. W., Photoluminescence spectroscopy of PbSe nanocrystals. *Physical Review B* 2007, *76* (12), 125304.
[30] Tischler, J. G.; Kennedy, T. A.; Glaser, E. R.; Efros, A. L.; Foos, E. E.; Boercker, J. E.; Zega, T. J.; Stroud, R. M.; Erwin, S. C., Band-edge excitons in PbSe nanocrystals and nanorods. *Physical Review B* 2010, *82* (24), 245303.
[31] Schaller, R. D.; Crooker, S. A.; Bussian, D. A.; Pietryga, J. M.; Joo, J.; Klimov, V. I., Revealing the Exciton Fine Structure of PbSe Nanocrystal Quantum Dots Using Optical Spectroscopy in High Magnetic Fields. *Physical Review Letters* 2010, *105* (6), 067403.
[32] Chappell, H. E.; Hughes, B. K.; Beard, M. C.; Nozik, A. J.; Johnson, J. C., Emission Quenching in PbSe Quantum Dot Arrays by Short-Term Air Exposure. *The Journal of Physical Chemistry Letters* 2011, *2* (8), 889-893.
[33] Abel, K. A.; Qiao, H.; Young, J. F.; van Veggel, F. C. J. M., Four-Fold Enhancement of the Activation Energy for Nonradiative Decay of Excitons in PbSe/CdSe Core/Shell versus PbSe Colloidal Quantum Dots. *The Journal of Physical Chemistry Letters* 2010, *1* (15), 2334-2338.

[34] Liu, H.; Guyot-Sionnest, P., Photoluminescence Lifetime of Lead Selenide Colloidal Quantum Dots. *The Journal of Physical Chemistry C* 2010, *114* (35), 14860-14863.

[35] van Driel, A. F.; Allan, G.; Delerue, C.; Lodahl, P.; Vos, W. L.; Vanmaekelbergh, D., Frequency-Dependent Spontaneous Emission Rate from CdSe and CdTe Nanocrystals: Influence of Dark States. *Physical Review Letters* 2005, *95* (23), 236804.

[36] Varshni, Y. P., Temperature dependence of the energy gap in semiconductors. *Physica* 1967, *34* (1), 149-154.

[37] Olkhovets, A.; Hsu, R. C.; Lipovskii, A.; Wise, F. W., Size-Dependent Temperature Variation of the Energy Gap in Lead-Salt Quantum Dots. *Physical Review Letters* 1998, *81* (16), 3539.

[38] Madelung, O., *Semiconductors: Data Handbook*. 3rd ed.; Springer: 2004.

[39] Morello, G.; De Giorgi, M.; Kudera, S.; Manna, L.; Cingolani, R.; Anni, M., Temperature and Size Dependence of Nonradiative Relaxation and Exciton−Phonon Coupling in Colloidal CdTe Quantum Dots. *The Journal of Physical Chemistry C* 2007, *111* (16), 5846-5849.

[40] Upadhyaya, K. S.; Yadav, M.; Upadhyaya, G. K., Lattice Dynamics of IV–VI Ionic Semiconductors: An Application to Lead Chalcogenides. *physica status solidi (b)* 2002, *229* (3), 1129-1138.

[41] Turyanska, L.; Patane, A.; Henini, M.; Hennequin, B.; Thomas, N. R., Temperature dependence of the photoluminescence emission from thiol-capped PbS quantum dots. *Applied Physics Letters* 2007, *90* (10), 101913-3.

[42] Sargent, E. H., Infrared photovoltaics made by solution processing. *Nat Photon* 2009, *3* (6), 325-331.

[43] Hines, M. A.; Scholes, G. D., Colloidal PbS nanocrystals with size-tunable near-infrared emission: Observation of post-synthesis self-narrowing of the particle size distribution. *Advanced Materials* 2003, *15* (21), 1844-1849.

[44] Moreels, I.; Justo, Y.; De Geyter, B.; Haustraete, K.; Martins, J. C.; Hens, Z., Size-Tunable, Bright, and Stable PbS Quantum Dots: A Surface Chemistry Study. *Acs Nano* 2011, *5* (3), 2004-2012.

[45] Fernée, M. J.; Jensen, P.; Rubinsztein-Dunlop, H., Bistable Switching between Low and High Absorbance States in Oleate-Capped PbS Quantum Dots. *ACS Nano* 2009, *3* (9), 2731-2739.

[46] Fernée, M. J.; Jensen, P.; Rubinsztein-Dunlop, H., Origin of the Large Homogeneous Line Widths Obtained from Strongly Quantum Confined PbS Nanocrystals at Room Temperature. *The Journal of Physical Chemistry C* 2007, *111* (13), 4984-4989.

[47] O'Donnell, K., Temperature dependence of semiconductor band gaps. *Appl. Phys. Lett.* 1991, *58* (25), 2924.

[48] Trinh, M. T.; Polak, L.; Schins, J. M.; Houtepen, A. J.; Vaxenburg, R.; Maikov, G. I.; Grinbom, G.; Midgett, A. G.; Luther, J. M.; Beard, M. C.; Nozik, A. J.; Bonn, M.; Lifshitz, E.; Siebbeles, L. D. A., Anomalous Independence of Multiple Exciton Generation on Different Group IV−VI Quantum Dot Architectures. *Nano Letters* 2011, *11* (4), 1623-1629.

[49] Cragg, G. E.; Efros, A. L., Suppression of Auger Processes in Confined Structures. *Nano Letters* 2009, *10* (1), 313-317.

[50] Murray, C. B.; Shouheng, S.; Gaschler, W.; Doyle, H.; Betley, T. A.; Kagan, C. R., Colloidal synthesis of nanocrystals and nanocrystal superlattices. *IBM J. Res. & Dev.* 2001, *45*, 47.
[51] de Mello, J. C.; Wittmann, H. F.; Friend, R. H., An improved experimental determination of external photoluminescence quantum efficiency. *Advanced Materials* 1997, *9* (3), 230-232.

4

Optical Properties of Spherical Colloidal Nanocrystals

Giovanni Morello
[1]*Nanoscience Institute of CNR, National Nanotechnology Laboratory (NNL),*
[2]*Center for Biomolecular Nanotechnologies @UNILE, IIT, Arnesano (LE),*
Italy

1. Introduction

The desire to fabricate materials with novel or improved properties is a powerful stimulus for the development of materials science. Thermal and electrical conduction, optical response, energy conversion and storage are just a few of the large number of properties underwent a very fast evolution in the last decade thank to the birth of a new branch of materials science and technology defined "Nanotechnology". Nanotechnology includes the totality of the physical, chemical, biological and engineering knowledge involving artificial structures whose properties are controlled at the nanometer level. Among the multitude of nanomaterials created, a particular class of them is becoming very popular and represents nowadays the most fascinating and potentially revolutionary inorganic semiconductor structure, which is the family of the colloidal quantum dots. They are often referred to as "nanocrystals" and the colloidal definition reveal their chemical origin. Actually, the chemical synthesis currently represents the most effective way to obtain high quality (in terms of size control, narrow size distribution, good crystalline structure and high optical performances) nano-objects on a gram scale which can be handled as ordinary chemical substances and implemented in several opto-electronic devices as well as biological ambient. Today, colloidal nanocrystals are successfully used as active media in lasers (Chan, et al., 2004; Klimov et al., 2000, 2007), LEDs (Anikeeva et al., 2009; Caruge et al., 2008), photovoltaic (Gur et al., 2005; Huynh et al., 2002; Kim et al., 2003), sensors (Oertel et al., 2005), biological labelling (Deka et al., 2009; Michalet et al., 2001), photo catalysis (Hewa-Kasakarage et al., 2010). Chemical syntheses allow for the fabrication of nanocrystals (NCs) with nearly atomic precision. They are currently prepared in a variety of compositions as nearly spherical particles (Peng et al., 2000), elongated nanorods (Krahne et al., 2011), and other more complex structures like tetrapods (Fiore et al., 2009) and octapods (Miszta et al., 2011; Zhang et al., 2011). Moreover, the sophistication with which inorganic nanoparticles can be prepared has inspired the creation of multi-combined nano-systems having different compositions ranging from the all-semiconductor to hybrid semiconductor-metal nanoparticles possessing on demand properties in terms of electronic levels energy. As a consequence, the in-depth knowledge of the electronic structure of nanocrystals became of fundamental importance and a huge effort on the theoretical and experimental point of view has been made in the two last decades dedicated to their study and comprehension.

This chapter wants to provide an overview of the main studies carried out on spherical nanocrystals on what concerns their electronic structure and optical properties. Particular attention will be devoted to the review of the role played by defect states (especially surface states) on the final optical performances by means of steady-state and time-resolved spectroscopy. For the sake of clarity, the reported discussions will concern two main kinds of nanocrystals, namely CdTe and CdSe. They represent the ideal cases study, since these two materials have historically been the most studied in the field of semiconductor nanostructures. The reason lies on the fact that they cover great part of the optical properties possessed by a number of different other nanocrystals and moreover they present the two main crystallographic symmetries, namely cubic (CdTe) and wurtzite (CdSe).

The chapter is organized as follows: Section 2 will be dedicated to the basic concepts of the low-dimensional systems. The idea is to provide a comprehensive overview of the physics at the basis of the systems studied. We will start from the most common problem consisting in solving the 1-D Schrödinger equation for an electron in a box, extending the discussion to the types of confinement (1-D, 2-D, 3-D) a nano-object undergoes. The size dependence will be reviewed, by distinguishing the different degrees of confinement. A final part will be devoted to the study of the electronic structure of spherical nanocrystals. Section 3 will provide an extensive discussion about the most important recombination processes occurring in spherical nanocrystals. These processes will be basically separated in two big families: radiative and non radiative processes. In the first case we will deal with relaxation processes involving the emission of one or more photons due to electron-hole annihilation. In the second case the excess energy is released as heat inside the materials and/or by excitation of new electron-hole pairs by Auger-like processes. Since great part of the potential applications of colloidal nanocrystals concerns with light emitting devices, particular emphasis is dedicated to the non radiative pathways limiting their optical performances. A review of the main studies reported in literature will be presented, with particular attention to the parameters affecting them, which are the size, the temperature, the excitation density and the surface quality. About the latter, the impact of surface states on the optical properties of CdSe NCs will be treated by means of a four-level model. A last section is dedicated to the Conclusions.

2. Fundamentals of nano-physics

This first section wants to stress the point that any development in nanoscience necessarily requires an understanding of the physical laws governing the matter at the nanoscale and of how the interplay of the various physical properties of a nanoscopic system translates into some novel behaviour or into a new physical property. In this sense, the section will be an overview of the basic physical laws that govern the nanomaterials, with particular emphasis on quantum dots, being the subject of the chapter.

2.1 Quantum confinement on low dimensional systems

The phenomenon of the quantum confinement (i. e. the size quantization) can be observed in systems where the motion of electrons or other particles (holes, excitons, etc.) is restricted at least in one dimension by some potential energy profile. Such a system is usually referred to as a "low dimensional" system. The energy spectra and the wavefunctions localization depend on the type of restriction in one, two or three dimensions, as well as on the size of the nanostructure. In particular, the quantum phenomena start to be noticeable when the lateral extension of the potential well becomes comparable to the particle wavelength. In order to better understand this concept, let us consider the case of the electrons. As elementary particles they exhibit the wave-particle duality of the matter following the "de Broglie" relation (de Broglie, 1924,1925). When immersed in a solid the electron is treated as a particle having an effective mass m* accounting for the periodicity of the crystal potential. Its linear momentum p can be written in terms of its wave-like nature, $p = \hbar k$, where $\hbar$ is the Dirac's constant (the Planck's constant divided by 2π) and k represents its wavenumber, associated to the de Broglie wavelength $\lambda = 2\pi / k$. Electrons in a bulk solid are treated as particles not feeling the borders by imposing the periodic boundary conditions, so that the wavefunctions and the energies are not affected by the real spatial extension of the solid. When the solid dimensions approach the electron wavelength, the permitted wavefunctions and energies undergo a series of restrictions in terms of continuity and absolute values. In few words, the system starts to be considered as "quantized". The simplest example of quantum confinement is an electron enclosed in a one-dimensional quantum box having lateral size a and infinitely high walls. In figure 1 we can see the situation in terms of energy and wavefunctions.

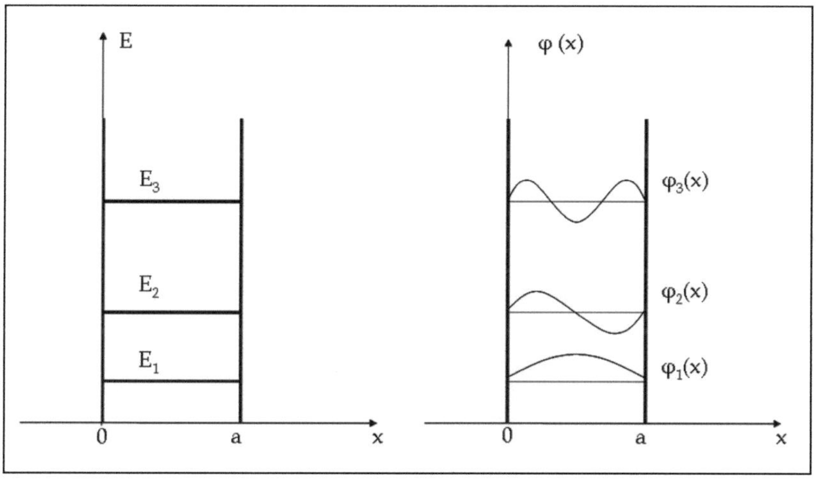

Fig. 1. Electron in a potential well. Energy levels and wavefunctions.

The problem constitutes a classical textbook case that can be approached by solving the Schrödinger equation in one dimension (Yoffe, 2001):

$$\frac{d^2\varphi}{dx^2} + \frac{2me}{b^2} = V \tag{1}$$

with V = 0, 0<x<a and V = ∞, x≤0, x≥a. The solutions of this second-order linear differential equation are in the form:

$$\varphi(x) = A \cdot e^{ikx} + B \cdot e^{-ikx}. \tag{2}$$

For an infinitely high potential barrier, the electron wavefunction must vanish at the borders of the potential well, such as $\varphi(0)=0$ and $\varphi(a)=0$. The condition $\varphi(0)=0$ produces B=-A, such that:

$$\varphi(x) = A(e^{ikx} - e^{-ikx}) = 2iA\sin(kx). \tag{3}$$

Substituting φ(x) in the Schrödinger equation one can obtain the following expression for the electron energy:

$$E = \frac{\hbar^2 k^2}{2m}. \tag{4}$$

The condition φ(a) =0 produces the following identity:

$$\varphi(a) = 2iA\sin(ka) = 0. \tag{5}$$

The identity is verified if ka = nπ, with n=1, 2, 3, ... The parameter k, therefore, results quantized, and the separation between two consecutive allowed values is Δk=π/a. In terms of energy:

$$E = \frac{\hbar^2 k^2}{2m} = \frac{\hbar^2 \pi^2 n^2}{2ma^2}. \tag{6}$$

The lowest energy for the electron, called Ground State energy, is obtained for n = 1,

$$E = \frac{\hbar^2 \pi^2}{2ma^2}. \tag{7}$$

The expression above defines the minimum energy possessed by a quantized system and is termed "point-zero energy" and constitutes a fascinating manifestation of the Heisenberg Uncertainty Principle. In fact, from

$$\Delta x \cdot \Delta p \geq \frac{\hbar}{2} \tag{8}$$

follows

$$\Delta p \geq \frac{\hbar}{2\Delta x} = \frac{\hbar}{a} \tag{9}$$

since a represents the potential well dimension and Δx ≤ a/2. In terms of averaged kinetic energy:

$$\langle E_k \rangle = \left\langle \frac{p^2}{2m} \right\rangle \geq \frac{(\Delta p)^2}{2m} \Rightarrow \langle E_k \rangle \geq \frac{\hbar^2}{2ma^2}. \tag{10}$$

This expression gives the theoretical lower limit of the possible kinetic energy value for a quantum particle confined in a box (one-dimensional, in this case) having size a. In the reality we find an actual value about $\pi^2 \approx 10$ times larger.

The role of the system size on the confinement energy is played by the $1/a^2$ factor according to which larger the size a, smaller both the point-zero energy and the spacing in the k-space. This leads to smaller absolute energy spacing, whereas the relative spacing is expressed by

$$E_{n+1} - E_n = \frac{\hbar^2 \pi^2}{2ma^2}\left[(n+1)^2 - n^2\right] = \frac{\hbar^2 \pi^2}{2ma^2}(2n+1), \tag{11}$$

and increases with increasing n.

The wavefunctions look like the standing waves on a string. Figure 1 plots the energies of the allowed states and the relative wavefunctions. They have increasing number of nodes with increasing energy according to the fact that more nodes mean shorter wavelength and higher momentum (i.e., energy).

2.2 Classification of quantum confined systems

In general, all the quantum confined systems can be classified on the basis of the number of dimensions along which the motion of electrons is coerced. The usual terminology refers to as 1-, 2-, 3-D confinement, in which the more evident effect consists in the modulation of the density of states function and a restriction of the allowed energies. In order to better understand what happens to a system when its dimensions start to shrink, one needs to begin from a non-confined structure, such as a bulk material. Here, one can assume that N electrons are not bound to individual atoms such that they can be considered "free" to move in three directions. If we suppose that, as a first approximation, the electron-electron interactions and the crystal potential are negligible (free electron gas model) (Pines, 1963) we can write the kinetic energy of an electron moving in the solid with velocity $\vec{v}=(v_x,v_y,v_z)$:

$$E = \frac{1}{2}m\vec{v}^2 = \frac{1}{2}m({v_x}^2 + {v_y}^2 + {v_z}^2). \tag{12}$$

The corresponding wavevector is derived from the relation

$$\vec{p} = m\vec{v} = \hbar\vec{k} \tag{13}$$

and the corresponding wavelength $\lambda=2\pi/|k|$.

On the wavefunction point of view the condition of infinite solid is expressed by imposing the so called boundary conditions. They consist in the continuity of the wavefunction $\varphi(x,y,z)$ at the border of the real finite solid of dimensions d_x, d_y, d_z:

$$\begin{cases} \varphi(x,y,z) = \varphi(x+d_x,y,z) \\ \varphi(x,y,z) = \varphi(x,y+d_y,z) \cdot \\ \varphi(x,y,z) = \varphi(x,y,z+d_z) \end{cases} \tag{14}$$

The solution of the 3-D Schrödinger equation gives a factored function, written as the product of three independent functions:

$$\varphi(x,y,z) = \varphi(x)\varphi(y)\varphi(z) = A\exp(ik_x x)\exp(ik_y y)\exp(ik_z z) \tag{15}$$

In the argument of the exponential functions, $k_{x,y,z}$ is such that $\pm\Delta k_{x,y,z}=\pm n2\pi/d_{x,y,z}$ with n integer. In a bulk (having d much larger than the electronic wavefunction), this condition tells us that all the values of allowed k are contained in a sphere in the k-space with a quasi-continuous distribution of states. At this point, it is useful to introduce the concept of Density of States (DOS) function $D_{3d}(k)$, intended as the number of states for unitary interval of wavenumbers. The electrons in a solid having a wavenumber k included in the interval between k and k+Δk belong to the function $D_{3d}(k)\Delta k$. The total number of electrons contained in the sphere (having a maximum wavenumber k_{max}) is

$$N = \int_{0}^{k_{max}} D_{3d}(k)dk \tag{16}$$

Since the volume of the sphere is proportional to k^3 the number ΔN (k) of electrons inside the interval k+Δk is proportional to $k^2\Delta k$. Therefore

$$D_{3d}(k) = \frac{dN(k)}{dk} \propto k^2 \tag{17}$$

Now, it is possible to give the expression of the number of states in a unitary interval of energy D_{3d} (E). Since $E(k) \propto k^2$, thus $k \propto \sqrt{E}$, $dk/dE \propto 1/\sqrt{E}$. It follows

$$D_{3d}(E) = \frac{dN(E)}{dE} = \frac{dN(k)}{dk}\frac{dk}{dE} \propto E/\sqrt{E} = \sqrt{E} \tag{18}$$

In figure 2 we can see the situation in terms of Energy and Density of States.

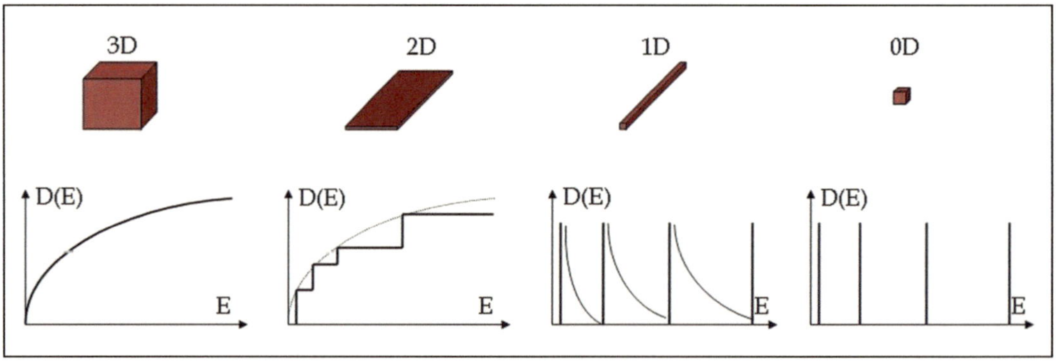

Fig. 2. Evolution of the "Density of States" (DOS) function by varying the degree of confinement. Reprinted from Physics Reports, 501, Krahne, R.; Morello, G.; Figuerola, A.; George, C.; Deka, S. & Manna, L., Physical properties of elongated inorganic nanoparticles, 75-221, Copyright (2012), with permission from Elsevier.

The next step is to consider a system in which just a spatial dimension is reduced down to few nanometres, so that the electrons can freely move on a plan giving rise to the so called "2-D electron gas" (Davies, 1998). The quantization acts in the shrunk dimension (let us say the direction $\vec{z}$) along which the wavenumber assumes discrete values, $k_z=n_z\Delta k_z$ where

$\Delta k_z = \pi / d_z$ (d_z being the size in the $\vec{z}$ direction). On the Density of States point of view, we can consider a continuous distribution in the x-y plane and an ensemble of infinite states equidistant one to each other. The number of states with k in the interval $k \div k + \Delta k$ is now proportional to $k \cdot \Delta k$ and the consequent DOS function

$$D_{2d}(k) = \frac{dN(k)}{dk} \propto k \tag{19}$$

In terms of energy:

$$D_{2d}(E) = \frac{dN(E)}{dE} = \frac{dN(k)}{dk}\frac{dk}{dE} \propto \sqrt{E} / \sqrt{E} \propto C . \tag{20}$$

This expression means that in the motion plane the density of the states is constant. Obviously, this holds for a fixed value of k_z. By changing k_z results in an increasing of the density D_{2d} (E) which assumes now a step-like function.

In a one-dimensional system another direction of confinement is introduced making the structure resembling to a long wire. In the long dimension the DOS function becomes

$$D_{1d}(k) = \frac{dN(k)}{dk} \propto C \tag{21}$$

from which

$$D_{1d}(E) = \frac{dN(E)}{dE} = \frac{dN(k)}{dk}\frac{dk}{dE} \propto 1 / \sqrt{E} . \tag{22}$$

As in the 2-D case, an energy quantization is present and the function D_{1d} (E) follows a saw-tooth-like function. In particular, it is characterized by a quasi-continuous distribution of states (the hyperbolas in figure 2), but delimited by a series of singularities.

When the carriers are confined in all the three spatial dimensions the system undergoes a triple quantization. As a consequence the wavenumbers of the allowed states (k_x, k_y, k_z) are discrete and can be represented by a point in the k-space. In terms of energy, the Density of States collapses into a series of discrete values assuming a Dirac's Delta function, as shown in figure 2, resembling to an atomic-like distribution.

On the optical point of view, the main quantity considered in studying the properties of semiconductor materials (either bulk or confined systems) is their energy gap. It represents the energy spacing between the last occupied and the first unoccupied electronic state at the ideal temperature of 0 K and is defined as

$$E_g = E_0 + E_q + E_C . \tag{23}$$

Here, E_0 represents the energy gap at 0 K of the bulk semiconductor, E_q is the contribution of the quantization effect and E_C accounts for the Coulomb interaction between the e-h pairs present in the semiconductor. The relative contribution of the last two terms determines the degree of confinement that a system undergoes. To this purpose, it is useful to introduce the fundamental concept of the "exciton". An exciton is a bound system composed of an electron and a hole experiencing a mutual Coulomb interaction. An important parameter of the exciton is the so called Bohr radius. It is a characteristic quantity of each bulk material and represents the natural extension of the e-h pair when the carriers are free to orbit one around to each other. A parameter linked to the exciton and to its Bohr radius a_B is the binding energy of the exciton E_b:

$$E_b = \frac{\hbar^2}{2\mu a_B{}^2} \cdot \tag{24}$$

where μ is the reduced mass of the exciton, defined as $\frac{1}{\mu} = \frac{1}{m_e} + \frac{1}{m_h}$, m_e and m_h being the effective masses of electron and hole, respectively. This energy assumes always negative values (being a binding energy) and represents the energy needed to ionize the exciton, that is to sustain the electron and the hole as separate entities. The Bohr radius can be expressed in terms of some physical parameters, such as the dielectric constant of the semiconductor ε, the reduced mass of the exciton μ and the electron charge e:

$$a_B = \frac{\hbar^2 \varepsilon}{\mu e^2} \cdot \tag{25}$$

Such quantity is different for each material and particle (i.e. electrons, holes, excitons, etc.) and ranges from a few nm to some tens of nm. Since the confinement acts as a constrainer for the e-h wavefunction extension, we can say that each material undergoes proper quantization effects, depending on its size and Bohr radius. In few words, the degree of confinement of a system expresses a measure of how much the real dimensions of the material affect the motion and the lifetime of an exciton in such material. Indeed, comparing the Bohr radius and the real size of a nanoparticle makes possible to define three different regimes of quantization: weak, intermediate and strong confinement. Let a_e, a_h and a_{ex} be the Bohr radii of electron, hole and exciton, respectively; if a is the size of the system considered (for instance the radius of a spherical nanoparticle), when a is smaller than a_e, a_h and a_{ex} the system is in the strong confinement regime since both the electron and the hole strongly feel the boundary of the nanocrystal. In the weak confinement regime, the nanocrystal size is larger than the electron and hole Bohr radius but smaller than the exciton Bohr radius. If a falls in between a_e and a_h, then the nanoparticle experiences the intermediate confinement effect. It is important to note, this point, that the degree of confinement featuring a semiconductor nanocrystal depends on the particular material constituting itself, since all the playing quantities are characteristic of each component. Therefore, nanocrystals of different materials but having the same size undergo a different degree of confinement. As a few examples we can consider three materials: CuCl, CdSe, InAs with radius of about 6 nm; their Bohr radii are 0.68 nm (Ohmura & Nakamura, 1999), 5.7 nm (Millo et el., 2004) and 34 nm (Kong et al., 2006), respectively. As a consequence, the effect of the quantum confinement effect is different in each nanocrystal: CuCl experiences a weak confinement, CdSe an intermediate one and InAs undergoes a strong confinement regime.

2.3 Exciton fine structure of wurtzite CdSe and cubic CdTe nanocrystals

CdSe and CdTe nanocrystals probably represent the most investigated nanostructures, for what concerns both their optical and their electronic properties. About the electronic structure of the lowest energy state, we will highlight some of the pioneering works which have represented important milestones in the field, namely the theoretical and experimental studies of Efros et al. (Efros et al., 1996). The concepts discussed here can be considered as of general validity for nanocrystals having hexagonal or cubic crystallographic structure and are applicable to nanocrystals of a wide range of dimensions.

In our discussion, we consider a nanocrystal with dimensions that are much larger than its lattice constant, such that the effective mass approximation (Efros et al., 1996) is applicable. This condition is practically fulfilled in all cases, since the nanocrystal diameter is hardly smaller than 2-3 nm. The notation used to name the quantum states of a nanocrystal closely follows that of an atomic system, consistently to the energy expression previously determined. Therefore, we define the total angular momentum J=(L+S) as the sum of the total orbital angular momentum L and the multiplicity term S, i.e. the electron spin, and the relative momentum projections: j, l, and s. The electron ground state has s-symmetry and presents a double degeneracy, which is due exclusively to the spin momentum. Thus J=0+1/2, its projections are j=±1/2, and the state is conventionally indicated as $1S_e$. On the other hand, the first hole level, having a p-symmetry, is fourfold degenerate, having J=1+1/2=3/2 (j=3/2, 1/2, -1/2, -3/2), and is named $1S_{3/2}$. The composition of the two ground states yields the eightfold degenerate exciton ground state $1S_{3/2}1S_e$.

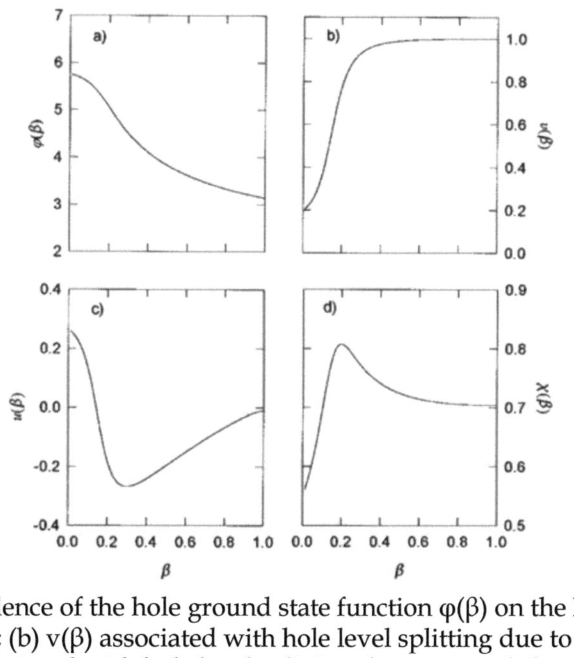

Fig. 3. (a) The dependence of the hole ground state function φ(β) on the light to heavy hole effective mass ratio β; (b) v(β) associated with hole level splitting due to hexagonal lattice structure; (c) u(β) associated with hole level splitting due to crystal shape asymmetry; (d) χ(β) associated with exciton splitting due to the electron-hole exchange interaction. Reprinted figure with permission from Efros, Al. L.; Rosen, M.; Kuno, M.; Nirmal, M.; Norris, D. J.& Bawendi, M. Physical Review B, 54, 4843-4856, 1996. Copyright (2012) by the American Physical Society.

The degeneracy of the ground state is lifted by various effects, such the internal crystal field, shape effects and the electron–hole exchange interaction. The first effect arises from an intrinsic property of semiconductors having hexagonal lattice structure (like CdSe) and therefore manifests itself in both bulk and nanoscale materials. The second effect takes into account the deviation from the ideal spherical shape of nanocrystals, while the third effect accounts for mixing of electron and hole spins. The first two effects can be grouped together, as they arise from the intrinsic asymmetry of the material/nanocrystal. The intrinsic crystal field produces a first splitting of the valence band, i.e. the lowest hole state, in the so called Kramers doublet, which consists of two doubly degenerate states with $j=\pm1/2$ and $j=\pm3/2$ (Efros et al., 1996). Let us define a parameter β as the ratio between the mass of the light hole m_{lh} and the mass of the heavy hole m_{hh} (hence $\beta = m_{lh}/m_{hh}$). The energetic splitting due to the intrinsic crystal field is then expressed as (Efros et al., 1996):

$$\Delta_1 = \Delta_{CF}\nu(\beta) \tag{26}$$

where Δ_{CF} is a parameter related to the crystal field (CF) splitting in a crystal with hexagonal structure, contributing to determine the hole ground state with $|j|=3/2$, while v (β) is a function which is unique for each material (see (Efros et al., 1996) for details). It is worth to stress that Δ_1 does not depend on the size of the nanocrystal. Moreover, since v (β) is always positive (see figure 3 b), the lowest hole level is fixed with the heavy hole state with $|j|=3/2$. In order to take the shape anisotropy into account, we can model a nanocrystal as an axially symmetric ellipsoidal particle (i.e. an ellipsoid with principal axes $a = b < c$), and define the ratio of the major to minor axes as $c/b = 1+\varepsilon$, ε being the ellipticity. The induced splitting in this case is:

$$\Delta_2 = 2u(\beta)E_{3/2}(\beta)\varepsilon \tag{27}$$

Here, u (β) is a dimensionless function associated with the hole level splitting due to the crystal shape (for details see (Efros et al., 1996), figure 3 c) and $E_{3/2}$ (β) is the hole ground state energy which can be written as:

$$E_{3/2}(\beta) = \frac{\hbar^2\varphi^2(\beta)}{2m_{hh}a^2} \tag{28}$$

where $\varphi^2(\beta)$ is a term related to the spherical Bessel functions and a is related to the nanocrystal size in the sense that for quasi-spherical nanocrystals $a = (b^2c)^{1/3}$.

Concerning Δ_2, an important point is the trend in the function u (β), in particular for what concerns its sign. As shown in Figure 3 c, u (β) reverses its sign past a certain value of β, meaning that for some materials the shape anisotropy induces a negative splitting resulting in a possible inversion of the hole ground state between $|j|=3/2$ and $|j|=1/2$, since the global energy splitting is the sum of the single asymmetry contributions ($\Delta_t = \Delta_1 + \Delta_2$). A negative Δ_2 is found, for example, in elongated CdSe nanorods, for which $\beta = 0.28$, where a possible inversion would depend on the hole ground state energy and on the radius of rods (Krahne et al., 2011).

The exciton ground state results until now split in two fourfold degenerate excitonic states, having total angular momentum F=1 and F=2. The exchange interaction further contributes to an increase of the splitting of the remaining states, defining the fine structure for nanocrystals having a series of possible shapes, as depicted in figure 4 (Efros et al., 1996). The final configuration of the fine structure is based on the definition of the projection f of the total angular momentum F. It assumes different values: one state with $f=\pm2$, two states with $f=\pm1$ (named Upper and Lower, depending on the branch they originate) and two others with $f=0$ (Upper and Lower). Three of them are optically active, namely the states 0^U, $\pm1^U$ and $\pm1^L$, and the remaining ±2 and 0^L states are passive. The ±2 state is optically forbidden because of the restrictions about the angular momentum conservation for photons (which cannot have an angular momentum ±2, for example). The 0^L has zero optical transition probability because of an interference phenomenon between the two indistinguishable states with zero angular momentum (Efros & Rosen, 2000), due to the influence of the electron-hole exchange interaction. The shape of the nanocrystal plays the significant role on defining which of the above states represents the exciton ground state. For perfectly spherical nanocrystals the ±2 is the ground state, whereas in prolate nanocrystals an inversion of the ±2 with the 0^L state can occur, because the state ±2 originates from the hole state with $|j|=3/2$, whilst 0^L arises from the state $|j|=1/2$. When the conditions for the sign change of Δ_t are met, the ground state is inverted. A natural generalization of this concept can be found in the electronic structure of elongated nanocrystals (also named Nanorods (Krahne et al., 2011)) which can be approximated by axially symmetric prolate ellipsoids with ellipticity ε defined as $\varepsilon = (2a_B/b)-1$ (here, the long axis c can be replaced by the Bohr radius), with b being the ellipsoid diameter and a_B the Bohr radius. In the case of strong lateral confinement ($b<2a_B$), the ellipsoids are subject to a possible inversion of the ground state between ±2 and 0^L. This can happen because Δ_2 becomes increasingly important in the strong confinement regime, and at some point it would cause the light hole state with $j=\pm1/2$ to become the hole ground state. The coupling with the electron state $1S_e$ yields a fourfold degenerate state, with angular momentum 0 (two states) and ±1. The hole state with $j=\pm3/2$ yields the second doubly degenerate state with momentum ±1 and ±2. In practice, the new lowest exciton level would be the state 0^L and the exciton fine structure resembles that of figure 4 c. For what concerns nanocrystals having a cubic structure, we will refer to another case study represented by CdTe nanocrystals. Efros et al. (Efros et al., 1996) calculated the size dependence of the band-edge splitting, showed in figure 5 for different shapes. In spherical nanocrystals the e-h exchange interaction split the exciton ground state into two states. The state at lowest energy is fivefold degenerate, presents total angular momentum F=2 and results dark. The higher energy state is threefold degenerate, has a total angular momentum F=1 and is bright. Contrary to the wurtzite case, the crystal field does not act as splitter, and just the shape anisotropy contributes to determine the order of the excitonic states in the real system (Efros et al., 1996).

On the experimental point of view, the fine structure of CdSe nanocrystals has been investigated by a number of experiments, more than CdTe one which presents a lot of degeneracy. The main difficulty in studying the fine structure is given by the size distribution of a sample of colloidal nanocrystals leading to an inhomogeneous broadening of the optical spectra which hides the fine distribution of the states. On the other hand, single nanoparticle experiments can provide information only ideally, since several effects contribute to destroy the advantages of spectral narrowing, namely intermittent emission (blinking) (Schlegel et al., 2002), spectral diffusion (Empedocles & Bawendi, 1999) and possible photodegradation (Wang et al., 2003). Fluorescence line narrowing and photoluminescence excitation (Norris et al., 1996) experiments are the most useful steady-state methods to access a subset of the nanocrystals ensemble, whereas the cross polarized,

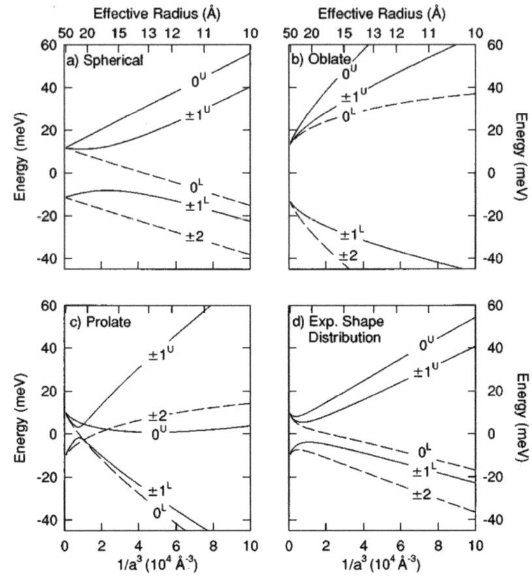

Fig. 4. Size dependence of the exciton ground state of CdSe NCs. (a) spherical NCs; (b) oblate NCs; (c) prolate NCs; (d) NCs having a size dependent ellipticity as determined from Efros et al., 1996. Solid/dashed lines indicate optically active/passive levels. Reprinted figure with permission from Efros, Al. L.; Rosen, M.; Kuno, M.; Nirmal, M.; Norris, D. J.& Bawendi, M. Physical Review B, 54, 4843-4856, 1996. Copyright (2012) by the American Physical Society.

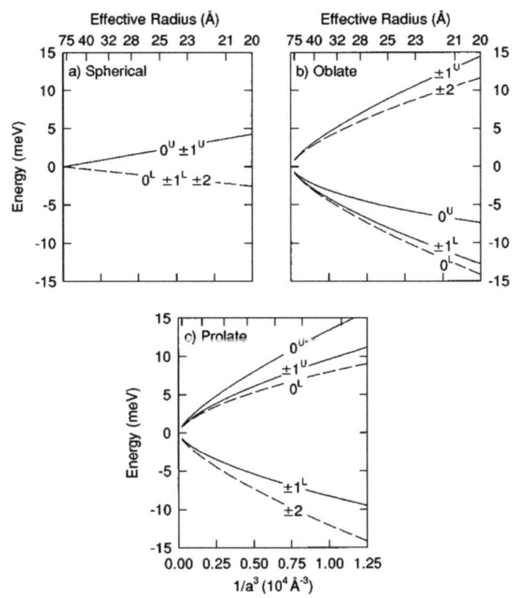

Fig. 5. The exciton fine structure of cubic CdTe NCs. (a) spherical NCs; (b) oblate NCs; (c) prolate dots NCs; Solid/dashed lines indicate optically active/passive levels. Reprinted figure with permission from Efros, Al. L.; Rosen, M.; Kuno, M.; Nirmal, M.; Norris, D. J.& Bawendi, M. Physical Review B, 54, 4843-4856, 1996. Copyright (2012) by the American Physical Society.

heterodyne detected third-order transient grating (CPH-3TG) method (Kim et al., 2009) has been employed to probe ultrafast transient dynamics in the fine structure distribution. On the luminescence point of view, the main difficulty is to be able to detect the very fast signal arising from the higher energy state of the fine structure. On this regard, a few time resolved PL experiments have been reported in the literature, showing emission from higher states on the picoseconds time scale (Morello et al., 2007a; Moreels et al., 2011) and only by using a streak camera. Also in these experiments the inhomogeneous broadening contributes to hide some spectral features and the situation is further complicated by both the possible exchange among the states due to the unavoidable asymmetric shape (Moreels et al., 2011) and the intermixing between intrinsic and surface states (Califano et al., 2005; Morello et al., 2007). All these effects participate to the definition of the actual ground state inside the fine structure. Indeed, the ideal case of perfectly spherical and pure nanocrystals is often replaced by the real situation in which slightly asymmetric and defected nanocrystals are investigated. Therefore, it is not rare to assist to the inversion between bright and dark states (it is the case of most nanorods systems (Krahne et al., 2011) and some not perfectly spherical NCs) or to the mixing between dark and surface states leading to the partial brightening of the same (Califano et al., 2005). The role of surface states in terms of quantum yield and lifetime will be treated in the last part of the chapter.

3. Excitation and relaxation processes in colloidal nanocrystals

After excitation of charge carriers, the system tends to restore the initial equilibrium state losing the excess energy. All the processes involved in this action constitute the whole of the relaxation processes of the system. Restoring of the initial equilibrium state can occur in several ways, and dealing with nanoparticles instead of bulk semiconductor introduces some complications that will be analysed in the following paragraphs. The most common energetic relaxation process is the heating of the material. It consists in the trigger of an emission of phonons (lattice vibrations), mediated by the electron-phonon interaction. This kind of interaction does not affect only the decay toward lower energy states (carrier cooling (Nozik, 2001)), but it can influence the promotion of charge carriers to higher states through the absorption of phonons inducing, in particular cases, the ejection of charges away from the material. This particular process, holding for both electrons and holes, is strongly dependent on the temperature and is often referred to as "thermal escape" (Valerini et al., 2005; Morello et al., 2007b). In the following we will give a complete description of thermal escape occurring in CdSe (Valerini et al., 2005) and CdTe nanocrystals (Morello et al., 2007b). Another phenomenon of electron-hole annihilation is due to carrier-carrier scattering, via the so called "Auger effect". It manifests when more than one e-h pair is present into the semiconductor and consists in the transfer of the energy owned by an e-h couple to a third particle (another electron, hole, or exciton). This relaxation pathway is strongly dependent on both the morphology of the semiconductor (shape and dimensions) and the level of charge carrier injection. The alternative way for exciton relaxation is the emission of photons that, in general, is less efficient than heating. The reason lies on the fact that the emission (likewise absorption) of photons is hindered by the selection rules of the optical transitions: the photons can involve states with angular momentum equal to 0, ±1, whereas the branches of phonons are distributed on the whole first Brillouin's region, i.e. they possess all the possible electron momenta, thus allowing for connection of states having different angular momentum. This condition is particularly favourable in taking place of intraband relaxation by means of phonon emission instead of radiative emission.

In a semiconductor, the absorption of a photon with energy larger than the energy gap, creates an e-h pair. Often, the energy of the exciting photon does not match perfectly the energy gap of the material. In particular, this is true in bulk for which the continuum of states allows for a continuous distribution of excited electrons in the conduction band. Such a distribution, at the thermal equilibrium, follows a Boltzmann's statistics depending on the temperature of the system, with electrons being within k_BT from the bottom of the conduction band and holes within k_BT from the top of the valence band (Nozik, 2001). Quantum confinement strongly affects carrier cooling efficiency by phonon scattering. In bulk semiconductors, exciton energy relaxation results mainly from cascade LO-phonon emissions, whereas interaction with acoustic phonons is less important. In nanocrystals smaller than 10 nm, or in general when the discreteness of electron-hole states becomes essential, optical phonons cannot provide an efficient relaxation channel. The dispersion curve of optical phonons is nearly wavenumber-independent, and LO-phonon energies can only be weakly deviated from the bulk ones (Trallero-Giner et al., 1998). Therefore, the relaxation via multiple LO-phonon emission requires $\Delta E_{1,2} \approx nE_{LO}$, where n is an integer number. Unlike in bulk crystals where there exists a continuous spectrum for quasi-particles, in nanocrystals there are not intermediate states between the two lowest excited ones. Therefore, the probability of the above process is very low because multiple phonon emission should occur via virtual intermediate states. The corresponding relaxation rate is lower than the typical electron-hole recombination rates. Thus, the relaxation between the excited states in nanocrystals is essentially inhibited. This effect is the so called "phonon bottleneck"(Gaponenko, 1998; Nozik, 2001). Phonon bottleneck in nanoparticles has been studied in the past years, and the results are controversial (Guyot-Sionnest et al., 1999; Klimov & McBranch, 1998; Woggon et al., 1996). The excited carriers should have an infinite relaxation lifetime for the extreme, limiting condition of a phonon bottleneck; thus, the carrier lifetime would be determined by non radiative processes, and PL would be absent (Gaponenko, 1998). In the case that the relaxation times are not excessively long and PL is observed, the results are not indicative of a phonon bottleneck, although relatively long hot electron relaxation times (tens of picoseconds) compared with what is observed in bulk semiconductors are observed. Some researchers (Klimov, 2000) also found very fast electron relaxation dynamics on the order of 300 fs from the first-excited 1P to the ground 1S state on II-VI CdSe colloidal NCs using interband pump-probe spectroscopy, which was attributed to an Auger process for electron relaxation that bypassed the phonon bottleneck. In this specific Auger process, the excess electron energy is rapidly transferred to the hole, which relaxes through its dense spectrum of states. The final equilibration step results in complete relaxation of the system. The electrons and holes can recombine, either radiatively or non radiatively, to produce the population densities that existed before photoexcitation.

A detailed study of the NCs photophysics with a particular attention to non radiative processes is not only interesting for fundamental physics, but it is also relevant to the exploitation of nanocrystals in practical applications. In this context we will deal with relaxation of carriers after they reach the lowest, potentially radiative, excited state. Again, the following discussion is based on the properties of CdSe and CdTe nanocrystals. Several relaxation processes have been proposed to explain the photophysics of CdSe QDs, including the thermally activated exciton transition from dark to bright states (Crooker et al., 2003) and carriers surface localization in trap states (Lee et al., 2005). Moreover, it has been shown that at room temperature the main non radiative process in CdSe/ZnS core/shell QDs is thermal escape, assisted by multiple longitudinal optical (LO) phonons absorption (Valerini et al., 2005), while at low temperature evidence for carrier trapping at surface defects was found. Despite these results, the role and the chemical origin (Wang et al., 2003b) of the surface defect states in the radiative and non radiative relaxation in nanocrystals has not been clarified completely. The existence of surface states due to unpassivated dangling bonds has been invoked to explain anomalous red-shifted emission bands in colloidal nanocrystals (Lee et al., 2005). On the contrary, it has been shown that above-gap trap states (Bawendi et al., 1992) affect the ultrafast relaxation dynamics (Cretì et al., 2005) and the single nanoparticle PL spectra of CdSe quantum rods (Rothenberg et al., 2005) due to charge trapping and local electric field fluctuations. These effects are expected to be dependent on the chemical composition of the QDs, on the density of surface defects and on the nanocrystals size. In what follows we will review the main processes affecting the light emission properties of nanocrystals, namely exciton-phonon coupling, Auger interactions and surface quality.

3.1 Exciton-phonon interaction

The main features that distinguish nanostructures from bulk materials originate from the localized character of the electron and hole wavefunction and the discrete nature of their optical transitions. In terms of carrier-phonon coupling, it essentially affects and defines the properties of homogeneous broadening of the PL signal. In the bulk, the broadening is mainly determined by the polar coupling of both electrons and holes to optical phonons (Liebler et al., 1985). The piezoelectric and deformation potential coupling of both carriers to acoustic phonons is usually not very important (Takagahara, 1996). This situation is different in nanostructures, where the local charge neutrality character of the exciton becomes predominant, producing an ideal null polar coupling of the exciton to optical phonons (Schmitt-Rink et al., 1987). This holds true for infinite barriers, such that the electron and hole wavefunctions are practically identical. In general, in real systems (finite barrier) just a decrease of the polar coupling with increasing barrier is expected (Muljarov & Zimmerman, 2007; Nomura & Kobayashi, 1992; Schmitt-Rink et al., 1987). On the other hand, since the deformation potential coupling is proportional to $1/R^2$ (R being the radius of a spherical dot), the coupling strength to acoustic phonons is increased as the dimensions are reduced below the Bohr radius (Gindele et al., 2000; Takagahara, 1996). Also, the temperature affects each of these contributions. The temperature dependence of the spectral line width can be expressed as (Rudin et al., 1990):

$$\Gamma(T) = \Gamma_0 + \sigma T + \gamma N_{LO}(T) \tag{29}$$

where Γ_0 is the inhomogeneous broadening, σ is the exciton-acoustic phonons coupling coefficient, γ is the coefficient accounting for the exciton-optical (LO) phonon coupling and N_{LO} represents the Bose function for LO phonon occupation:

$$N_{LO} = \frac{1}{e^{E_{LO}/k_B T} - 1}. \tag{30}$$

In the expression above E_{LO} is the energy of the longitudinal optical phonon with momentum k = 0 (i.e. the phonon that preferably couples to the lowest exciton state) and k_B

is the Boltzmann's constant. Due to the different energetic dispersion curves of acoustic and optical phonons, the two couplings dominate at different temperature ranges. The acoustic phonons, having smaller energies (a few meV) heavily contribute to the broadening at low temperature (until 50-70 K), whereas the optical phonons (with energies of a few tens of meV) dominate at higher temperature (Morello et al., 2007b). In a general fit procedure, the experimental data of the broadening are extracted by fitting the PL spectra to a convolution of a number of Gaussian peaks (usually not more than three) from which the broadening of the PL is extracted. In order to discern the inhomogeneous and homogeneous components the extracted values are fitted to eq. 29. In figure 6, it is shown the case of a set of three samples of CdTe NCs having different size whose temperature dependent broadenings have been analyzed by eq. 29 (Morello et al., 2007b).

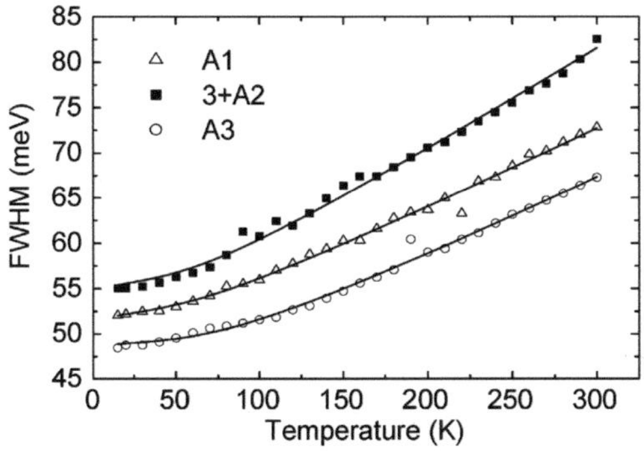

Fig. 6. PL broadening as a function of temperature (symbols) and relative best fit to eq. 29 (continuous lines) for three different size of CdTe NCs. Reprinted with permission from Morello, G.; De Giorgi, M.; Kudera, S.; Manna, L.; Cingolani, R. & Anni, M., Temperature and size dependence of nonradiative relaxation and exciton-phonon coupling in colloidal CdTe quantum dots. J. Phys. Chem. C, 111, 5846-5849. Copyright (2012) American Chemical Society.

The three samples were called A1 (average diameter of 4.2 nm), A2 (4.9 nm) and A3 (5.9 nm) and the relative PL was well fitted by one Gaussian peak. As expected from eqs. 29-30 the broadening increases with the temperature by virtue of an increasing probability of existence of (optical) phonons. The above mentioned analysis produced the expected size-dependent behavior of the coupling coefficients between excitons and acoustic/optical phonons. In particular, it was found a coupling to acoustic phonons about three orders of magnitude larger than in the bulk system and increasing σ with decreasing the NC size, consistently with the theoretical prediction of a strong increase of the coupling to acoustic phonons in zero-dimensional systems (Valerini et al., 2005). The exciton-LO phonon coupling, on the contrary, showed a smaller value respect to the bulk counterpart and moreover it was reduced in the smallest NCs, accordingly to an expected ideally null polar coupling in zero-dimensional systems.

An important contribution of the exciton-phonon interaction to the emission properties consists in the influence on the temperature dependence of the PL quantum yield. As usually observed in colloidal nanocrystals, the PL intensity exhibits a considerable decrease with increasing the temperature. Starting from 10 K, the decrease is moderate in the low temperature regime (let us say until 100 K) becoming heavier at temperature higher than about 150 K. In general, the temperature-induced quenching includes a number of processes that could be identified by analyzing the trend of the PL integrated area as recorded at the different temperatures. The plot of the intensity as a function of $1/k_BT$ (where k_B is the Boltzmann constant and T is the temperature) on a semi-logarithmic scale allows for the determination of the activation energy of the thermal processes triggered at high temperature, by fitting the experimental data with an Arrhenius law, using the expression

$$I(T) = \frac{I_0}{1 + \sum_{i=1}^{n} a_i e^{(-E_i/k_B T)}}. \quad (31)$$

Here, n is the number of thermal processes, E_i represents the activation energy of the i-th process and a_i is a fitting parameter accounting for the relative weight of each exponential term.

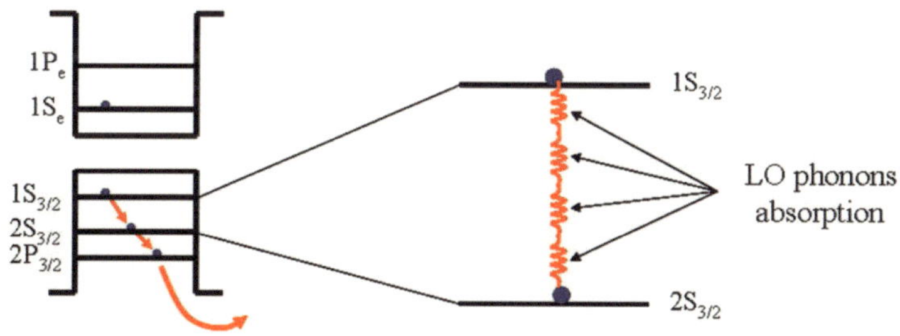

Fig. 7. Schematic description of thermal escape in CdTe NCs.

Among the thermal processes affecting the PL quantum efficiency of nanocrystals and involving exciton-phonon coupling we recall the "thermal escape". This process involves the carriers occupying the emitting state and consists in the absorption of a number of optical phonons such as to cover the energetic spacing between the emitting and the first, high energy, allowed state. Since the protagonist physical phenomenon is the absorption of phonons, the essential requirement to occur is the nonzero probability of existence of a certain number of optical phonons in the material (eq. 30). Figure 7 schematizes the mechanism for a CdTe NC: after excitation, electrons and holes undergo ultrafast intraband relaxation, until they occupy the lowest permitted energetic level before annihilation. This point, the e-h pair can recombine in several ways, among which we find the radiative emission, trapping, Auger effects and exciton-phonon coupling. If we consider a high degree of purity (neglecting trapping) and a low excitation density level (avoiding Auger processes) the photon emission must compete only with carrier-phonon interaction. When the latter has a high rate (at high temperature) the absorption of a number of optical phonons leads one or both the carriers (the hole, in the case of CdTe) to occupy the higher energy level. From such state the further absorption of phonons is facilitated by the reduced energetic spacing required to jump into the successive hole level, since a smaller number of phonons is needed. The process is iterated until the carrier is ejected from the nanocrystal. Therefore, the involved physical quantities are the temperature (according to eqs. 29-30) and the energetic spacing between the first two excited states (defining the maximum number of absorbed phonons) being the most energetically spaced. A good treatment of the thermal escape on the theoretical and experimental point of view has been provided by Valerini et al. (Valerini et al., 2005) and Morello et al. (Morello et al., 2007b).

We can consider the following rate equation for the carrier density n (Valerini et al., 2005):

$$\frac{dn}{dt} = g(t) - \frac{n}{\tau_{rad}} - \frac{n}{\tau_{act}} - \frac{n}{\tau_{esc}} \tag{32}$$

where g(t) is the generation term, $1/\tau_{rad}$ is the radiative recombination rate, and $1/\tau_{esc}$ is the thermal escape rate given by

$$\frac{1}{\tau_{esc}} = \frac{1}{\tau_0}(e^{E_{LO}/k_B T} - 1)^{-m} \tag{33}$$

where $1/\tau_0$ is a fitting parameter acting as a weight for the probability of carrier–LO-phonon scattering, and m is the number of LO phonons involved in the process. The rate of a generic thermally activated process is given by

$$\frac{1}{\tau_{act}} = \frac{1}{\tau_a} e^{-E_a/k_B T}, \tag{34}$$

where E_a is the activation energy and $1/\tau_a$ is a fitting parameter acting as a weight for the probability of this process. The intensity of the PL emitted per unit time is given by

$$I_{PL}(t) = \frac{n(t)}{\tau} = \frac{n_0}{\tau} e^{-t/\tau}, \tag{35}$$

where n_0 is the initial carriers population and τ is the temperature-dependent PL decay time given by

$$\frac{1}{\tau} = \frac{1}{\tau_{rad}} + \frac{1}{\tau_{act}} + \frac{1}{\tau_{esc}}. \tag{36}$$

The integrated PL intensity is instead given by

$$I_{PL}(T) = \int_0^\infty I_{PL}(t)dt = \frac{n_0}{1 + \tau_{rad}/\tau_{act} + \tau_{rad}/\tau_{esc}}. \tag{37}$$

Considering only the thermal escape the expression above becomes:

$$I(T) = \int_0^\infty I(t) \frac{n_0}{1 + \tau_{rad}/\tau_{esc}} \Leftrightarrow I(T) = \frac{n_0}{1 + \frac{\tau_{rad}}{\tau_{esc}}(e^{E_{LO}/k_B} - 1)^{-m}} \tag{38}$$

As a case study, we can consider a set of three samples of CdTe nanocrystals having different diameter, namely 4.2 nm, 4.9 nm and 5.9 nm (Morello et al., 2007b). From the absorption spectra it is possible to firstly deduce the energetic spacing among the lowest quantized states. In CdTe the two lowest absorption peaks arise from absorption of the lowest degenerate electron state $1S_e$ and the two lowest hole states $1S_{3/2}$ and $2S_{3/2}$, as shown in figure 8. Therefore, their energetic separation determines the jump to be executed by the charge carriers (in this case the holes) in order to escape from the nanocrystals. Due to the quantization effect, the first two states observed in absorption have the largest energy separation, thus they represent the major obstacle to the process. In the specific case of figure 8, the separations are 124.5, 96.5 and 82.2 meV for samples A1, A2 and A3, respectively. By analysing the PL intensity vs. temperature with an Arrhenius function like eq. 38, it is possible to extract a number m of optical phonons involved in the quenching process, namely 5.6, 4.9 and 4. Such number is not integer due to both the size dispersion of the samples and the statistic character of the physical quantity considered. By multiplying this number to the energy separation arising from the absorption spectra one obtains an energy value comparable to the separation between the two lowest excited states giving evidence for the occurred thermal escape process (see Morello et al., 2007b for details).

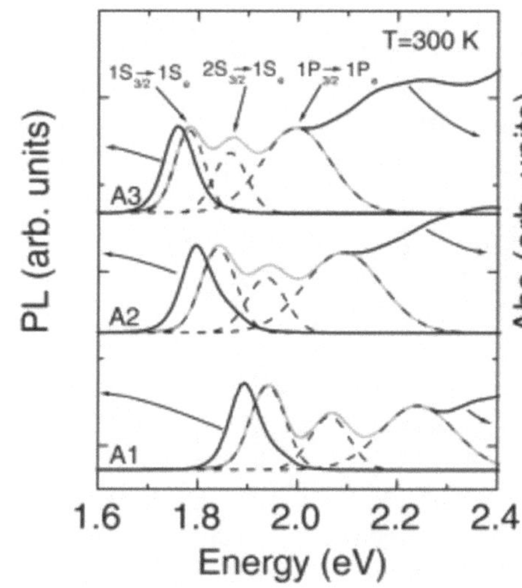

Fig. 8. Absorption and emission spectra at room temperature of the three samples of CdTe NCs studied by Morello et al, 2007b. Gray lines are the best fit to the convolution of three Gaussian curves for the first three absorption peaks. Reprinted with permission from Morello, G.; De Giorgi, M.; Kudera, S.; Manna, L.; Cingolani, R. & Anni, M., Temperature and size dependence of nonradiative relaxation and exciton-phonon coupling in colloidal CdTe quantum dots. J. Phys. Chem. C, 111, 5846-5849. Copyright (2012) American Chemical Society.

Thermal escape is not the only non radiative process featuring colloidal nanocrystals. We have to mention al least two other phenomena contributing to the global lowering of the PL, namely trapping at defect surface states and Auger processes. They are just partially dependent on the temperature and for this reason they are shifty and difficult to analyse.

3.2 Auger-like interactions

While the intrinsic decay of singly excited NCs is generally due to the e-h recombination via photon emission, the deactivation of two e-h pairs contemporarily living into the dot is dominated by non radiative Auger recombination. In such a case, the excess energy is not released as a photon but is transferred to a third particle (an electron, a hole or an exciton) that is re-excited to higher energy states. The efficiency of Auger processes, which are mediated by Coulomb electron-electron interactions, differs greatly between the atomic and the bulk semiconductor case. In atomic systems (the extreme case of a nanocrystal), for which the electron-electron coupling is much stronger than the electron-photon coupling, the rates of Auger transitions are significantly greater than the rates of the radiative transitions. As a result, the decay of the multi-electron states is dominated by Auger processes. Their efficiency is greatly reduced in bulk materials because of the reduced Coulomb e-e coupling and kinematic restrictions imposed by energy and momentum conservation. As the carriers (electrons and/or holes) occupying higher energy states have larger momentum respect to the ones lying at the band edge, the probability of Auger recombination is near to zero. In order to be allowed, Auger recombination must involve electrons and holes having a high momentum in their lowest energy state, meaning that a rapid spatial variation of the wavefunctions is required. Such situation is absent in bulk semiconductors which are characterized by negligible Auger recombination rates. The situation changes, however, in nanoparticles in which the abrupt truncation of the wavefunctions at the borders of the NCs makes possible high momenta also for the lowest wavefunctions without nodes. The collapse of the restrictions about the momentum conservation makes nanocrystals the ideal candidates for the exploitation of a physical phenomenon sought for a long time, namely the "direct carrier multiplication" (DCM). It consists in the generation of multiple excitons by using the excess energy possessed by a single electron-hole pair excited at higher energy levels (Califano et al., 2004a, 2004b; Velizhanin & Piryatinski, 2011). Following the concept of Califano et al. (Califano et al., 2004a) such process can be considered an inverse Auger recombination for which a highly excited carrier decaying into its ground state is able to excite a valence electron, thus producing a second e-h pair (see figure 9). In principle, DCM could happen every time the excess energy ΔE exceeds the energy gap E_g; in the reality it must compete with phonon scattering, radiative recombination and Auger cooling (this latter process foresees the transfer of the excess energy of an electron to a hole by Coulomb scattering leading to the jump of the hole to deeper valence states as depicted in figure 9) and in general an energetic threshold is associated. In the bulk, due to the restrictions of momentum and energy conservation such threshold could reach very high values (up to 1 eV (Harrison et al., 1999; Wolf et al., 1998)). In nanocrystals, on the contrary, the overcoming of the momentum restrictions makes DCM process possible with threshold energy close to E_g. In the last four/five years numerous researchers have claimed the reached conditions for carrier multiplication in a number of different semiconductor nanomaterials. Since the first studies the most promising systems, on this regard, seemed to be PbSe (Velizhanin & Piryatinski, 2011) followed by CdSe (Califano et al., 2004a, 2004b; Lin et al., 2011), InAs (Schaller et al., 2007), InP (Stubbs et al., 2010), Si (Gali et al., 2011), Sn (Allan & Delerue, 2011). It should be noted, however, that the publication of some works reporting several discrepancies among them about the exact determination of the efficiency of carrier multiplication, indicated that the experimental conditions and managements rather than the actual material type were responsible of some higher and/or lower measured thresholds (Nair et al., 2011; Trinh et al., 2011; McGuire et al., 2010; Rabani & Baer, 2010). Nowadays, the controversies seem to be far away to be solved although the publication of recent works which attempt to unify the main results. Nevertheless the global significance of the discovery can not be undermined, and the great potential of DCM in technological exploitation well justifies the current exited debate.

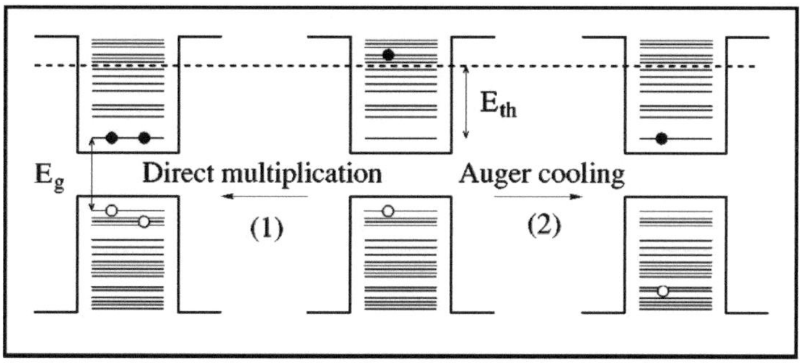

Fig. 9. Schematics of the main hot electron relaxation pathways: Direct Carrier Multiplication (1) and Auger Cooling (2). Reprinted with permission from Califano M.; Zunger A.& Franceschetti A., Direct carrier multiplication due to inverse Auger scattering in CdSe quantum dots. Appl. Phys. Lett., 84, 2409-2411. Copyright (2012), American Institute of Physics.

3.3 Surface-related properties

The surface plays a nontrivial role in defining the optical properties of NCs, especially for what concerns the existence of surface states and their influence on the carrier dynamics. The efficiency of the Auger processes, for instance, may be affected by means of trapping at such states (Cretì et al., 2007). This has important implications on the carrier cooling from higher excited states to the band-edge, which depends on the Auger efficiency, as demonstrated by Klimov and co-workers (Achermann et al., 2006). In their study, for instance, they showed that in nanorods multicarrier generation can cause an Auger heating, consisting in the increase of the energy of an exciton, as a consequence of an energy transfer event deriving from a second exciton recombination. The consequent increase in carrier temperature competes with the more classic phonon emission as a relaxation channel, and decreases the carrier cooling in the high excitation density regime. On the other hand, the surface of NCs constitutes the most probable place to be affected by defect and/or charge accumulation. The reason lies on the fact that in nanostructures the surface to volume ratio is greater with respect to other bulky systems, so that in the strong confinement regime the number of atoms on the surface could reach values larger than the internal ones. Moreover, these atoms present dangling bonds leading to an unavoidable charge imbalance and then to a very high degree of reactivity with the surrounding ambient. As a consequence, the surface atoms constitute the real weak point of each nanoparticle on what concerns the exciton recombination, and a high quality surface reconstruction (passivation) is mandatory in order to obtain good optical performances. The most traditional and easy way to produce a good passivation is to cover the bare surface of nanocrystals with an organic capping layer with the aim to remove potential charge accumulation sites (Bertoni et al., 2007; Cao et al., 2007; Kairdolf et al., 2008; Kalyuzhny & Murray, 2007; Puzde et al., 2004; Sharma et al., 2009;). A major advance in the improving of optical performances has been reached by surface passivation with an outer solid state material deposited onto the active one and acting as a protective layer (shell). This way, the good choice of the shell could allow for an almost complete reconstruction of the superficial atoms thanks to a good lattice constant matching between the two materials (Isnaeni et al., 2011; Pandey & Guyot-Sionnest, 2008; Zhang et al., 2010; Zheng et al., 2010).

It has been shown that the presence of hole surface states can affect the actual order of the single states in the fine structure distribution of the ground state (Califano et al., 2005). Depending on the energetic range covered we can refer to two kinds surface defect states. First, the existence of dangling (unpassivated) bonds on the surface introduces trap states, lying in the forbidden band-gap energy, that can be radiative or not. If they radiate, a typical red shifted (of some hundreds of meV) shoulder on the main intrinsic PL spectrum appears (Landes et al., 2001; Underwood et al., 2001; Lim et al., 2008); if they act as non radiative centres, the reduction of the PL quantum yield is the typically observed effect (Burda et al., 2001; Baker & Kamat, 2010). The growth of a shell usually overcomes the formation of such deep trap states. The second class of traps consists of surface states inducing smaller red shift, typically of the order of a few tens of meV (for this reason they are referred to as shallow trap states), and it may be correlated to the different atomic arrangement of the surface with respect the inside of the dots. All these superficial effects also depend on the size of the nanocrystals, since the fraction of superficial atoms increases as the size decreases and than they define a lower limit attainable for the size of high quality nanocrystals.

In general, surface states are prevalently studied by means of transient absorption measurements (Burda et al., 2001), where the tuning of the probe wavelengths offers the possibility to investigate energies over a very broad range (from UV to NIR on the same sample). If on one hand pump-probe technique allows for discovering numerous properties on what concerns the distribution of trap and intrinsic states, as well as their filling and depopulation dynamics, on the other hand it provides few information about their actual weight in influencing the optical performances of potential devices having nanocrystals as active media. An efficient way to carry out such investigation is to study the dynamics of the radiative relaxation of nanocrystals as a function of the surface passivation. In principle, such a study requires the analysis of the PL decay on a time scale spanning a broad range from picoseconds to microseconds. Great part of the literature regards the analysis of long living radiative emission from deep trap states located at energies well below the lowest optically allowed one. These states present lifetimes on the order of micro-milliseconds depending on their origin and the nature of the host material (Lim et al., 2008). Few works report on the role of shallow surface states on the luminescence properties of NCs and most of them do not discern the nature of the multiple emissive states characteristic, for instance, of nanocrystals having wurtzite structure. As explained in Section 2 the fine structure of CdSe spherical nanocrystals presents a manifold of bright states lying at energies higher than the lowest dark state, complicating the exact determination of the dynamics of the single states. It is obvious that the presence of surface trap states further contributes to complicate the situation. Several theoretical studies report the lifetimes expected for the dark and the bright states, together with the expected effects resulting from mixing of dark and surface states on these lifetimes (Califano et al., 2005). A potentially successful method for discerning the different emitting states (either intrinsic or not) would consists in the optical characterization at single nanoparticle level. However, some intrinsic limitations makes such measurements inadequate. Dynamical intermittent emission and spectral diffusion heavily contribute to cancel information about the actual emission energy and the decay time.

As for the steady-state optical properties, the most complete lifetime studies on colloidal nanocrystals regard CdSe. In what follows, we recall a treatment on the impact of shallow surface states on the optical properties of CdSe NCs, especially about their radiative relaxation properties, as reported by Morello et al. (Morello et al., 2007a). In general, radiative relaxation channels in CdSe NCs have been investigated by time resolved photoluminescence (TRPL), typically on the nanosecond time scale. These experiments usually exhibit bi-exponential decay traces (Javier et al., 2003; Wang et al., 2003). The shortest lifetime is of the order of several nanoseconds, whereas the longest one is on the time scale of tens of nanoseconds. The origin of these processes (and other longer up to microseconds) is still matter of debate. The longest lifetime is usually attributed to surface states emission (Wang et al., 2003), whereas non-exponential traces, in the same temporal range (Schöps et al., 2006), have been explained in terms of superposition of bright and dark states, and of incomplete surface passivation. Relaxation processes on the microsecond time scale have also been observed and associated to dark state emission (Crooker et al., 2003). Califano et al. (Califano et al., 2005) have shown that the microsecond decay time is actually due to dark-bright state emission induced by the presence of surface states, while the dark state has been predicted to have a millisecond lifetime. At the single dot level, TRPL on CdSe QDs has revealed nanosecond lifetimes (Labeau et al., 2003) (probably, emission from dark-bright states) as well as nonexponential decays arising from fluctuating non radiative relaxation channels (Fisher et al., 2004; Schlegel et al., 2002).

An interesting issue concerns the role of the bright states on the temporal dynamics of the PL, and their interplay with dark and/or surface states. These properties have been investigated for the first time by Wang et al. (Wang et al., 2006) who showed the carrier relaxation from bright to dark and surface defect states. Since such kind of relaxation is predicted to be faster than the natural radiative emission lifetime (less than 100 ps) the role of $\pm 1^U$ and $\pm 1^L$ bright intrinsic states (see Section 2 and Efros et al., 1996 for details) in presence of emitting surface states has been only postulated (Bawendi et al., 1992; Jungnickel & Henneberger, 1996). As representative examples we can cite Bawendi et al. (Bawendi et al., 1992) who have showed transient emission from CdSe QDs involving surface states. These authors distinguished both a short lifetime of the order of their time resolution (about 100 ps) probably arising from intrinsic emission, and a temperature-dependent interplay between the band-edge and surface states. Jungnickel and Henneberger (Jungnickel & Henneberger, 1996) have found similar transient behaviour on the same time scale, stressing the long decay time of radiative surface state emission. de Mello Donegá et al. (de Mello Donegà et al., 2006) have investigated the temperature dependence of the exciton lifetime in CdSe QDs, finding evidence for a fast component in the time trace in the low temperature "radiative regime", again within the temporal resolution of their system (700 ps). Such contribution was ascribed to rapid carrier thermalization from bright to dark states. A reliable study able to resolve fast emission from bright states (relaxing into the surface ones) can be performed by means of a streak camera (temporal resolution below 10 ps), with the possibility to find evidence for the thermal evolution of the population of the single states (Morello et al. 2007a). As a case study, we report here a comparative investigation between core CdSe and core/shell CdSe/ZnS NCs having the same core dimension by TRPL measurements in the temperature range of 15-300 K (Morello et al., 2007a).

In figure 10 (left panel) the temporal evolution of CdSe/ZnS QDs PL spectra at 20 K is shown. After 1.7 ns a small red shift is observed, completed after 12 ns. By analysing the spectrum at 0 ps delay (figure 11) it is possible to discern the single emitting entities responsible for the observed PL. In this particular case the PL results as a convolution of 3 emitting states lying at different energies, as shown in figure 11. The dynamical red shift, therefore, reveals the different dynamics the 3 states undergo. In the study of Morello et al. (Morello et al., 2007a) two samples of NCs were investigated, namely CdSe and CdSe/ZnS (both samples having the same core dimensions). They found the following energetic separations among the states: $E_{1,2}$=21 meV and $E_{2,3}$=16 meV for core NCs; $E_{1,2}$=21 meV and $E_{2,3}$=13 meV for core/shell NCs.

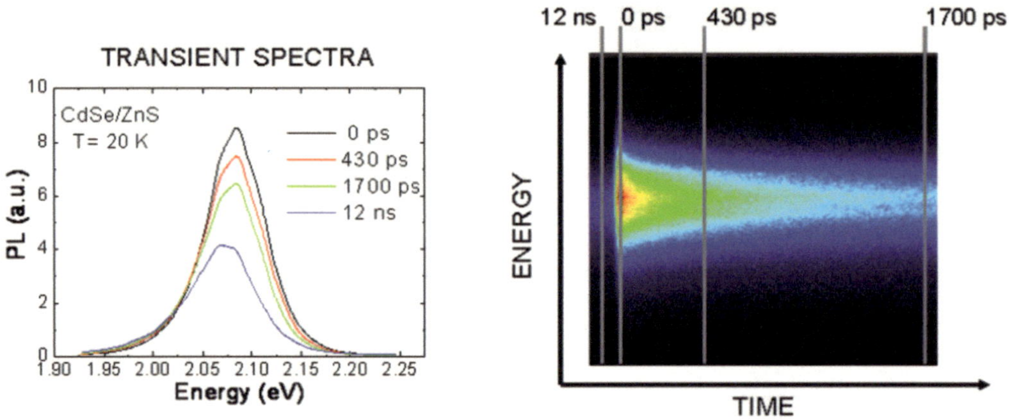

Fig. 10. Left panel: transient PL spectra of CdSe/ZnS NCs at 20 K from 0 ps to 12 ns after the pump pulse. Right panel: PL decay trace showing the temporal slices reported in the left panel.

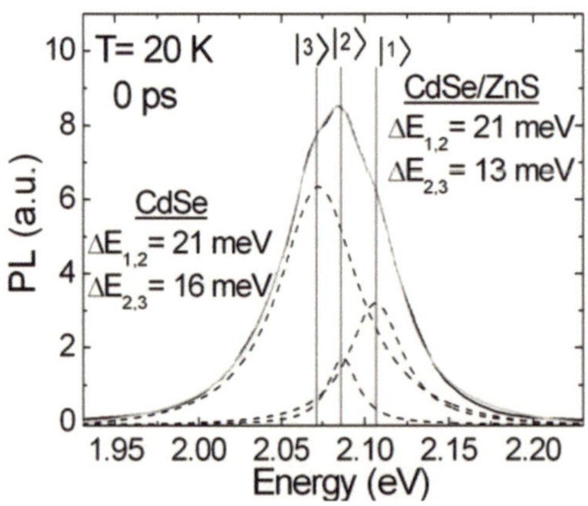

Fig. 11. PL spectrum at 20 K of CdSe/ZnS NCs taken at 0 ps delay looked as the result of three superposed transitions.

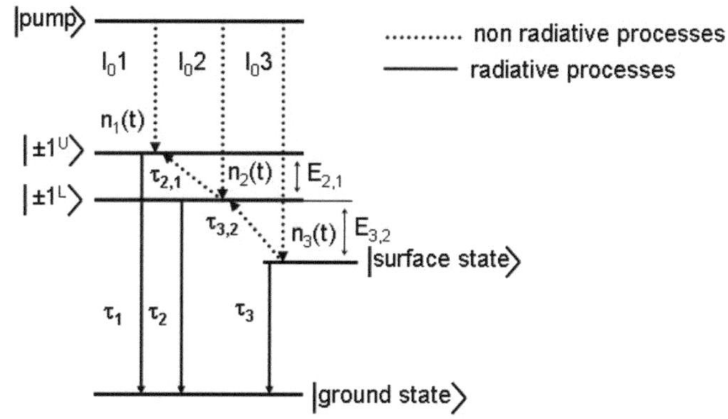

Fig. 12. Schematics of the radiative and non radiative processes occurring among two intrinsic and a surface state in CdSe nanocrystals in the low temperature range (10-70 K). Reprinted with permission from Morello, G.; Anni, M.; Cozzoli, P. D.; Manna, L.; Cingolani, R. & De Giorgi, M., Picosecond Photoluminescence Decay Time in Colloidal Nanocrystals: The Role of Intrinsic and Surface States. J. Phys. Chem. C , 111, 10541- 10545. Copyright (2012) American Chemical Society.

The PL time decay for core and core/shell samples (figure 10, right panel, shows an image of the time trace as recorded by the Streak Camera) was well reproduced by a triexponential decay function for both the samples studied in a temperature range of 15-300 K:

$$I(t) = A_1 \cdot e^{-(t-t_0)/t_1} + A_2 \cdot e^{-(t-t_0)/t_2} + A_3 \cdot e^{-(t-t_0)/t_3} \tag{39}$$

where t_0 is the delay at which I(t) is maximum, t_1, t_2, t_3 are the lifetimes and A_1,A_2,A_3 are the weights of each process, respectively. The time constants t_1 and t_2 were the typical carrier relaxation times from intrinsic bright states of the fine structure of spherical CdSe NCs into the surface defect states (Wang et al., 2006), and t_3 was comparable with typical lifetime of surface-related emission in CdSe NCs (Wang et al., 2006). Moreover, the extracted energy splitting $E_{1,2}$ was the same in core and core/shell sample, similar to the splitting between the lowest bright states $\pm1^U$ and $\pm1^L$ in CdSe QDs (20 meV) (Efros et al., 1996), whereas $E_{2,3}$ was different in the two studied samples, revealing an extrinsic nature of the reddest transition. A credible origin was ascribed to surface states, also considering that the dark state lifetime should be of the order of micro-milliseconds. From temperature dependent measurements it was possible to analyse the contribution of the single emitting states by integration of the single exponential terms of eq. 39:

$$\begin{aligned} I_1(T) &= \int_0^\infty A_1 \cdot e^{-(t-t_0)/t_1} dt, \\ I_2(T) &= \int_0^\infty A_2 \cdot e^{-(t-t_0)/t_2} dt, \\ I_3(T) &= \int_0^\infty A_3 \cdot e^{-(t-t_0)/t_3} dt. \end{aligned} \tag{40}$$

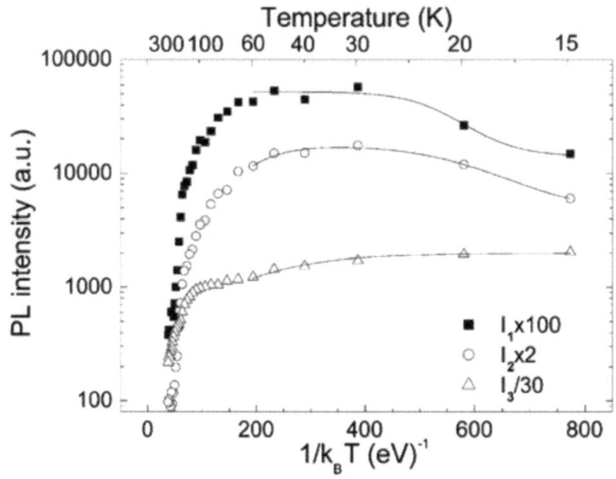

Fig. 13. PL intensity of the three states as a function of temperature for CdSe/ZnS QDs. The continuous lines are the fit curves to eqs 42. Reprinted with permission from Morello, G.; Anni, M.; Cozzoli, P. D.; Manna, L.; Cingolani, R. & De Giorgi, M., Picosecond Photoluminescence Decay Time in Colloidal Nanocrystals: The Role of Intrinsic and Surface States. J. Phys. Chem. C , 111, 10541– 10545. Copyright (2012) American Chemical Society.

Figure 13 shows the temperature dependence of the PL intensities I_1, I_2 and I_3 for the three states from which it is clear the existence of several thermal regimes of population and depopulation processes. The PL thermal increase for I_1 and I_2 in the range of 15-70 K suggests that thermally induced population of the high energy states occurs, fed by the lowest energy state. At higher temperatures, on the other hand, the overall PL intensity strongly decreases, indicating the occurrence of thermal activation of non radiative relaxation channels, such as thermal escape induced by multiple LO phonons absorption. The dynamics occurring at low temperature (up to 70 K) was well explained by considering a four-level system, as depicted in figure 12. After laser excitation, carriers relax non radiatively into states $\pm 1^U$, $\pm 1^L$ and into surface states, from which they relax radiatively. The lifetime of states $\pm 1^U$ and $\pm 1^L$ is shorter than the intrinsic radiative decay time, because of the fast carrier relaxation into the surface state, which instead can emits with its intrinsic radiative decay time. According to this picture, the two shortest lifetimes were attributed to carriers relaxing from $\pm 1^U$ and $\pm 1^L$ bright states into the surface one. The effectiveness of the model could be verified by solving the equation system associated to thermal population and depopulation of the states with non radiative lifetime:

$$\begin{cases} \dfrac{dn_3(t)}{dt} = n_{03}\delta(t) - \dfrac{n_3(t)}{\tau_{3,2}} e^{-E_{3,2}/k_BT} - \dfrac{n_3(t)}{\tau_3} \\ \dfrac{dn_2(t)}{dt} = n_{02}\delta(t) - \dfrac{n_2(t)}{\tau_{2,1}} e^{-E_{2,1}/k_BT} + \dfrac{n_3(t)}{\tau_{3,2}} e^{-E_{3,2}/k_BT} - \dfrac{n_2(t)}{\tau_2} \\ \dfrac{dn_1(t)}{dt} = n_{01}\delta(t) + \dfrac{n_2(t)}{\tau_{2,1}} e^{-E_{2,1}/k_BT} - \dfrac{n_1(t)}{\tau_1} \end{cases} \tag{41}$$

where n_{01}, n_{02}, and n_{03} represent the respective initial population of the three states as generated by the laser pulse (the latter is treated as a δ function because the pump pulse is much shorter than all the processes considered here); τ_1, τ_2, and τ_3 are the intrinsic radiative decay times of the states; $1/\tau_{2,1}$ and $1/\tau_{3,2}$ are the depletion rates of the state $\pm1^L$ and the surface state, respectively; and $E_{2,1}$ and $E_{3,2}$ are the energy separations among the three states as indicated in figure 12. By solving the system, we obtain the expressions for the time-dependent populations $n_1(t)$, $n_2(t)$, and $n_3(t)$ of the states involved in the emission process. Thus, the PL integrated areas $I_1(T)$, $I_2(T)$, and $I_3(T)$ are as follows:

$$\begin{aligned} I_1(T) &= \int_0^\infty \frac{n_1(t)}{\tau_1}dt = I_{01} + \frac{\tau_2}{\tau_{2,1}} e^{-E_{2,1}/k_B T}(I_{02} + I_{03}\rho), \\ I_2(T) &= \int_0^\infty \frac{n_2(t)}{\tau_2}dt = \frac{I_{02} + I_{03}\rho}{1 + \frac{\tau_2}{\tau_{2,1}} e^{-E_{2,1}/k_B T}}, \\ I_3(T) &= \int_0^\infty \frac{n_3(t)}{\tau_3}dt = \frac{I_{03}}{1 + \frac{\tau_3}{\tau_{3,2}} e^{-E_{3,2}/k_B T}}, \end{aligned} \tag{42}$$

where

$$\rho = \frac{\frac{\tau_3}{\tau_{3,2}} e^{-E_{3,2}/k_B T}}{1 + \frac{\tau_3}{\tau_{3,2}} e^{-E_{3,2}/k_B T}}. \tag{43}$$

The three expressions represent the theoretical counterpart of eqs. 40 and could be, thus, employed to fit the experimental data of figure 13 in the low temperature range with the aim to extract the actual energetic separation among the states and the relative lifetimes.

4. Conclusion

The importance of spherical nanocrystals relies of the fact that they represent the perfect model of a quantum confined system. This leads to consider these nanoparticles ideal for the study of the fundamental physics at the basis of the quantum mechanics and several applications in technological field. In the latter context, chemically synthesized nanoparticles are the most promising candidates to be used as active media in opto-electronic devices, due to the relatively easy methods of production and to the versatility of the colloidal solution form in which they appear. As pioneering nano-objects, spherical nanocrystals have attracted remarkable attention by researchers and constitute the most studied systems in terms of opto-electric properties. In this chapter we covered several aspects on what concerns their optical properties. We started with an overview of the main consequences resulting from the progressive shrinking of the spatial extension of a solid system, and of the implications on the modulation of the DOS function and the quantization of the energy. Particular attention has been dedicated to the textbook case of the Schrödinger equation in one dimension. A brief excursus has been dedicated to the electronic configuration of the fine structure of the ground state in wurtzite and cubic crystal structure. As examples of interest we have tackled the cases of CdSe (wurtzite) and CdTe (cubic) nanocrystals since they are, in absolute, the most studied systems, following the pioneering works of Efros. Regarding the optical properties, we have concentrated our attention to the effects deteriorating the optical performances in terms of quantum efficiency (especially acting at room temperature), which depend on both intrinsic and extrinsic factors. The firsts include all the effects which can not be avoided by any synthesis improvement, because they are typical of the physics of the material, such as exciton-phonon coupling and Auger effects. The seconds are related to the global purity of the nanocrystals in terms of crystalline structure and surface quality. The latter, in particular, constitutes the major contributor to the observation of poor emission performances or, sometimes, spurious effects. A review of the main studies about this topic has been provided with a particular attention to the methods for studying and discerning the role of surface states. The reported examples wanted to stress the importance of a critical point in the definition of the optical performances of colloidal nanocrystals which is the presence of surface states. Their study is constantly evolving as well as the attempts to reduce their negative effects by improving the chemical methods. The first is preparatory for the second and often the second contributes to complicate the first, such that nowadays the duel seems to be far away to end.

5. Acknowledgments

The author acknowledges M. De Giorgi, L. Manna, M. Anni, R. Cingolani, P. D. Cozzoli, for fundamental contributions and P. Cazzato for precious technical assistance. Finally, he wants to thank the CBN centre of IIT for the complete funding of this work

6. References

Achermann, M.; Bartko, A. P.; Hollingsworth, J. A. & Klimov, V. I., (2006). The effect of Auger heating on intraband carrier relaxation in semiconductor quantum rods. *Nat. Phys.*, Vol. 2, No. 8, pp. 557-561.

Allan, G. & Delerue, C., (2011). Optimization of Carrier Multiplication for More Effcient Solar Cells: The Case of Sn Quantum Dots. *ACS Nano*, Vol. 5, No. 9, pp. 7318–7323.

Anikeeva, P. O.; Halpert, J. E.; Bawendi, M. G. & Bulović, V. (2009). Quantum Dot Light-Emitting Devices with Electroluminescence Tunable over the Entire Visible Spectrum. *Nano Letters*, Vol. 9, No. 7, (September 2009), pp. 2532-2536.

Baker, D. R. & Kamat, P. V., (2010). Tuning the Emission of CdSe Quantum Dots by Controlled Trap Enhancement. *Langmuir*, Vol. 26, No. 13, pp. 11272–11276.

Bawendi, M. G.; Carroll, P. J.; Wilson, W. L.& Brus, L. E., (1992). Luminescence properties of CdSe quantum crystallites: Resonance between interior and surface localized states. *J. Chem. Phys.*, Vol. 96, No. 2, pp. 946-954.

Bertoni, C.; Gallardo, D.; Dunn, S.; Gaponik, N. & Eychmüller, A., (2007). Fabrication and characterization of red-emitting electroluminescent devices based on thiol-stabilized semiconductor nanocrystals. *Appl. Phys. Lett.*, Vol. 90, 034107-034109.

Burda, C.; Link, S.; Mohamed, M. & El-Sayed, M., (2001). The relaxation pathways of CdSe nanoparticles monitored with femtosecond time-resolution from the visible to the

IR: Assignment of the transient features by carrier quenching. *J. Phys. Chem. B*, Vol. 105, No. 49. pp. 12286-12292.

Califano M; Zunger A & Franceschetti A., (2004a). Direct carrier multiplication due to inverse Auger scattering in CdSe quantum dots. *Appl. Phys. Lett.*, Vol. 84, No. 13, pp. 2409-2411.

Califano M; Zunger A & Franceschetti A., (2004b). Efficient inverse Auger recombination at threshold in CdSe nanocrystals. *Nano Lett.*, Vol. 4, No. 3, pp. 525-531.

Califano, M.; Franceschetti, A. & Zunger, A., (2005). Temperature Dependence of Excitonic Radiative Decay in CdSe Quantum Dots: The Role of Surface Hole Traps. *Nano Lett.*, Vol. 5, No. 12, pp. 2360–2364.

Cao, X.; Li, C. M.; Bao, H.; Bao, Q. & Dong, H., (2007). Fabrication of Strongly Fluorescent Quantum Dot-Polymer Composite in Aqueous Solution. *Chem. Mater.*, Vol. 19, pp. 3773-3779.

Caruge, J. M.; Halpert, J. E.; Wood, V.; Bulović, V. & Bawendi, M. G. (2008). Colloidal quantum-dot light-emitting diodes with metal-oxide charge transport layers. *Nature Photonics*, Vol. 2, pp. 247-250.

Chan, Y.; Caruge, J. M.; Snee, P. T. & Bawendi, M. G. (2004). Multiexcitonic two-state lasing in a CdSe nanocrystal laser. *Applied Physics Letters*, Vol. 85, No. 13, (December 2005), pp. 2460-2462.

Cretì, A.; Anni, M.; Zavelani-Rossi, M.; Lanzani, G.; Leo, G.; Della Sala, F.; Manna, L. & Lomascolo, M., (2005). Ultrafast carrier dynamics in core and core/shell CdSe quantum rods: Role of the surface and interface defects. *Phys. Rev. B*, Vol. 72, No. 12, pp. 125346-125355.

Creti, A.; Anni, M.; Zavelani-Rossi, M.; Lanzani, G.; Manna, L. & Lomascolo, M., (2007). Role of defect states on Auger processes in resonantly pumped CdSe nanorods. *Appl. Phys. Lett.*, Vol. 91, No. 9, pp. 093106-093108.

Crooker, S. A.; Barrick, T.; Hollingsworth, J. A. & Klimov, V. I., (2003). Multiple temperature regimes of radiative decay in CdSe nanocrystal quantum dots: Intrinsic limits to the dark-exciton lifetime. *Appl. Phys. Lett.*, Vol. 82, No. 17, pp. 2793-2795.

de Broglie, L. Recherches sur la théorie des quanta (Researches on the quantum theory), *Thesis*, Paris.

de Broglie, L. (1925), *Ann. Phys.* (Paris), Vol. 3, pp. 22.

de Mello Donega, C.; Bode, M. & Meijerink, A., (2006). Size- and temperature-dependence of exciton lifetimes in CdSe quantum dots. *Phys. Rev. B*, Vol. 74, No. 8, pp. 085320-085328.

Davies, J. H. (1998).*Physics of Low Dimensional Semiconductors*, Cambridge University Press, ISBN 0-521-48148-1, Cambridge.

Deka, S.; Quarta, A.; Lupo, M. G.; Falqui, A.; Boninelli, S.; Giannini, C.; Morello, G.; De Giorgi, M.; Lanzani, G.; Spinella, C.; Cingolani, R.; Pellegrino, T. & Manna, L. (2009). CdSe/CdS/ZnS Double Shell Nanorods with High Photoluminescence Efficiency and Their Exploitation As Biolabeling Probes. *J. Am. Che. Soc.*, Vol. 131, pp. 2948-2958.

Efros, Al. L.; Rosen, M.; Kuno, M.; Nirmal, M.; Norris, D. J.& Bawendi, M. (1996) Band-edge exciton in quantum dots of semiconductors with a degenerate valence band: Dark and bright exciton states. *Phys. Rev. B*, Vol. 54, No. 7, pp. 4843-4856.

Efros, Al. L. & Rosen, M. (2000). The electronics structure of semiconductor nanocrystals. *Annu. Rev. Mater. Sci.*, Vol. 30, pp. 475–521.

Empedocles, S. A. & Bawendi, M. G. (1999). Influence of Spectral Diffusion on the Line Shapes of Single CdSe Nanocrystallite Quantum Dots. *J. Phys. Chem. B*, Vol. *103*, No. 11, pp. 1826–1830.

Fiore, A.; Mastria, R.; Lupo, M. G.; Lanzani, G.; Giannini, C.; Carlino, E.; Morello, G.; De Giorgi, M.; Li, Y.; Cingolani, R. & Manna, L. (2009). Tetrapod-Shaped Colloidal Nanocrystals of II-VI Semiconductors Prepared by Seeded Growth. *J. Am. Chem. Soc.*, Vol. 131, No. 6, pp. 2274-2282.

Fisher, B. R.; Eisler, H. J.; Stott, N. E. & Bawendi, M. G., (2004). Emission intensity dependence and single-exponential behavior in single colloidal quantum dot fluorescence lifetimes. *J. Phys. Chem. B*, Vol. 108, No. 1, pp. 143-148.

Gali, A.; Kaxiras, E.; Zimanyi, G. T. & Meng, S., (2011). Effect of symmetry breaking on the optical absorption of semiconductor nanoparticles. *Phys. Rev. B*, Vol. 84, 035325-035330.

Gaponenko, S. V., (1998). *Optical Properties of Semiconductor Nanocrystals* (2005). Cambridge University Press, 9780521019231, Cambridge, U.K..

Gindele, F.; Hild, K.; Langbein, W. & Woggon, U., (2000). Temperature-dependent line widths of single excitons and biexcitons. *J. Lumin.*, Vol. 87-89, pp. 381-383.

Gur, I.; Fromer, N. A.; Geier, M. L. & Alivisatos, A. P. (2005). Air-Stable All-Inorganic Nanocrystal Solar Cells Processed from Solution. *Science*, Vol. 310, No. 5747, pp. 462–465.

Guyot-Sionnest, P.; Shim, M.; Matranga, C. & Hines, M., (1999). Intraband relaxation in CdSe quantum dots. *Phys. Rev. B*, Vol. 60, No. 4, pp. R2181-R2184.

Harrison, D.; Abram, R. A. & Brand, S., (1999). Characteristics of impact ionization rates in direct and indirect gap semiconductors. *J. Appl. Phys.*, Vol. 85, No. 12, pp. 8186-8192.

Hewa-Kasakarage, N. N.; El-Khoury, P. Z.; Tarnovsky, A. N.; Kirsanova, M.; Nemitz, I.; Nemchinov, A. & Zamkov, M. (2010). Ultrafast carrier dynamics in type II ZnSe/CdS/ZnSe nanobarbells. *ACS Nano*, Vol. 4, pp. 1837–1844.

Huynh, W. U.; Dittmer, J.J. & Alivisatos, A. P. (2002). Hybrid Nanorod-Polymer Solar Cells. *Science*, Vol. 295 (October 2011), pp. 2425–2427.

Isnaeni; Kim, K. H.; Nguyen, D. L.; Lim, H.; Pham, T. N. & Cho, Y.-H., (2011). Shell layer dependence of photoblinking in CdSe/ZnSe/ZnS quantum dots. *Appl. Phys. Lett.*, Vol. 98, No. 1, pp. 12109-12111.

Javier, A.; Magana, D.; Jennings, T. & Strouse, G. F., (2003). Nanosecond exciton recombination dynamics in colloidal CdSe quantum dots under ambient conditions. *Appl. Phys. Lett.*, Vol. 83, No. 7, pp. 1423-1425.

Jungnickel, V. & Henneberger, F., (1996). Luminescence related processes in semiconductor nanocrystals - The strong confinement regime. *J. Lumin.*, Vol. 70, pp. 238-252.

Kairdolf, B. A.; Smith, A. M. & Nie, S, (2007). One-Pot Synthesis, Encapsulation, and Solubilization of Size-Tuned Quantum Dots with Amphiphilic Multidentate Ligands. *J. Am. Chem. Soc.*, Vol. 130, pp. 12866–12867.

Kalyuzhny, G. & Murray, R. W., (2005). Ligand Effects on Optical Properties of CdSe Nanocrystals. *J. Phys. Chem. B*, Vol. 109, pp. 7012-7021.

Kim, J.; Wong, C. Y. & Scholes, G. D. (2009). Exciton fine structure and spin relaxation in semiconductor colloidal quantum dots. *Acc. Chem. Res.*, Vol. 42, No. 8, pp. 1037-1046.

Kim, S.; Fisher, B.; Eisler, H.-J. & Bawendi, M. G. (2003). Type-II Quantum Dots: CdTe/CdSe(Core/Shell) and CdSe/ZnTe(Core/Shell) Heterostructures. *Journal of the American Chemical Society*, Vol. 125, (September 2003), pp. 11466–11467.

Klimov, V. I. & McBranch, D. W., (1998). Femtosecond 1P-to-1S electron relaxation in strongly confined semiconductor nanocrystals. *Phys. Rev. Lett.*, Vol. 80, No. 18, pp. 4028-4031.

Klimov, V. I., (2000). Optical nonlinearities and ultrafast carrier dynamics in semiconductor nanocrystals. *J. Phys. Chem. B*, Vol. 104, No. 26, pp. 6112-6123.

Klimov, V. I.; Mikhailovsky, A. A.; Xu, S.; Malko, A.; Hollingsworth, J. A.; Leatherdale, C. A.; Eisler,H.-J. & Bawendi, M. G. (2000). Optical Gain and Stimulated Emission in Nanocrystal Quantum Dots. *Science*, Vol. 290, (October 2000), pp. 314-317.

Klimov, V. I.; Ivanov, S. A.; Nanda, J.; Achermann, M.; Bezel, I.; McGuire, J. A. & Piryatinski, A. (2007). Single-exciton optical gain in semiconductor nanocrystals. *Nature*, Vol. 447, (May 2007), pp. 441-446.

Kong, L. M.; Cai, J. F.; Wu, Z. W.; Gong, Z.; Niu, Z. C. & Feng, Z. C. (2006). Time-resolved photoluminescence spectra of self-assembled InAs/GaAs quantum dots. *Thin Solid Films*, Vol. 498, No. 1-2, pp. 188-192.

Krahne, R.; Morello, G.; Figuerola, A.; George, C.; Deka, S. & Manna, L. (2011). Physical properties of elongated inorganic nanoparticles. *Phys. Rep.*, Vol. 501, No. 3-5, pp. 75-221.

Labeau, O.; Tamarat, P. & Lounis, B., (2003). Temperature dependence of the luminescence lifetime of single CdSe/ZnS quantum dots. *Phys. Rev. Lett.*, Vol. 90, No. 25, pp. 257404-257408.

Landes, C. F.; Braun, M. & El-Sayed, M. A., (2001). On the nanoparticle to molecular size transition: Fluorescence quenching studies. *J. Phys. Chem. B*, Vol. 105, No. 43, pp. 10554-10558.

Lee, W. Z.; Shu, G. W.; Wang, J. S.; Shen, J. L.; Lin, C. A.; Chang, W. H.; Ruaan, R. C.; Chou, W. C.; Lu, C. H. & Lee, Y. C., (2005). Recombination dynamics of luminescence in colloidal CdSe/ZrS quantum dots. *Nanotechnology*, Vol. 16, No. 9, pp. 1517-1521.

Liebler, J. G.; Schmitt-Rink, S. & Haug, H., (1985). Theory of the absorption tail of wannier excitons in polar semiconductors. *J. Lumin.*, Vol. 34, No. 1-2, pp. 1-7.

Lim, S. J.; Chon, B.; Joo, T. & Shin, S. K., (2008). Synthesis and Characterization of Zinc-Blende CdSe-Based Core/Shell Nanocrystals and Their Luminescence in Water. *J. Phys. Chem C*, Vol. 112, pp. 1744-1747.

Lin, Z.; Franceschetti, A. & Lusk, M. T., (2011). Size Dependence of the Multiple Exciton Generation Rate in CdSe Quantum Dots. *ACS Nano*, Vol. 5, No. 4, pp. 2503–2511.

McGuire, J. A.; Sykora, M.; Joo, J.; Pietryga, J. M. & Klimov, V. I., (2010). Apparent Versus True Carrier Multiplication Yields in Semiconductor Nanocrystals. *Nano Lett.*, Vol. 10, pp. 2049–2057.

Michalet, X. ; Pinaud, F. ; Lacoste, T. D.; Dahan, M.; Bruchez, M. P. ; Alivisatos, A. P. & Weiss, S. (2001). Properties of Fluorescent Semiconductor Nanocrystals and their Application to Biological Labeling. *Single Mol.*, Vol. 2, pp. 261-276.

Millo, O.; Katz, D.; Steiner, D.; Rothenberg, E.; Mokari, T.; Kazes, M. & Banin, U. (2004). Charging and quantum size effects in tunnelling and optical spectroscopy of CdSe nanorods. *Nanotechnology*, Vol. 15, pp. R1-R6.

Miszta, K.; Dorfs, D.; Genovese, A.; Kim, M. R. & Manna, L. (2011). Cation Exchange Reactions in Colloidal Branched Nanocrystals. *ACS Nano*, Vol. 5, No. 9, pp. 7176-7183.

Moreels, I.; Rainò, G.; Gomes, R.; Hens, Z.; Stöferle, T. & Mahrt, R. F., (2011). Band-Edge Exciton Fine Structure of Small, Nearly Spherical Colloidal CdSe/ZnS Quantum Dots. *ACS Nano*, Vol. 5, No. 10, pp. 8033-8039.

Morello, G.; Anni, M.; Cozzoli, P. D.; Manna, L.; Cingolani, R. & De Giorgi, M. (2007a). Picosecond Photoluminescence Decay Time in Colloidal Nanocrystals: The Role of Intrinsic and Surface States. *J. Phys. Chem. C* , Vol. 111, No. 28, pp. 10541–10545.

Morello, G.; De Giorgi, M.; Kudera, S.; Manna, L.; Cingolani, R. & Anni, M., (2007b). Temperature and size dependence of nonradiative relaxation and exciton-phonon coupling in colloidal CdTe quantum dots. *J. Phys. Chem. C*, Vol. 111, No. 16, pp. 5846-5849.

Muljarov, E. A. & Zimmermann, R., (2007). Exciton Dephasing in Quantum Dots due to LO-Phonon Coupling: An Exactly Solvable Model. *Phys. Rev. Lett.*, Vol. 98, No. 18, pp. 187401-187404.

Nair, G.; Chang, L.-Y.; Geyer, S. M. & Bawendi, M. G., (2011). Perspective on the Prospects of a Carrier Multiplication Nanocrystal Solar Cell. *Nano Lett.*, Vol. 11, pp. 2145–2151.

Nomura, S. & Kobayashi, T., (1992). Exciton–LO-phonon couplings in spherical semiconductor microcrystallites. *Phys. Rev. B*, Vol. 45, No. 3, pp. 1305–1316.

Norris, D. J.; Efros, Al. L.; Rosen, M. & Bawendi, M. G. (1996). Size dependence of exciton fine structure in CdSe quantum dots. *Phys. Rev. B*, Vol. 53, No. 24, pp. 16347-16354.

Nozik, A. J., (2001). Spectroscopy and hot electron relaxation dynamics in semiconductor quantum wells and quantum dots. *Annu. Rev. Phys. Chem.*, Vol. 52, pp.193–231.

Oertel, D. C.; Bawendi, M. G.; Arango, A. C. & Bulović, V. (2005). Photodetectors based on treated CdSe quantum-dot films. *Appl. Phys. Lett.*, Vol. 87, pp. 213505-213507.

Ohmura, H. & Nakamura, A. (1999). Quantum beats of confined exciton-LO phonon complexes in CuCl nanocrystals. *Phys. Rev. B*, Vol. 59, No, 19, pp. 12216-12219.

Pandey, A. & Guyot-Sionnest, P., (2008). Slow Electron Cooling in Colloidal Quantum Dots. *Science*, Vol. 322, No. 5903, pp. 929-932.

Peng, X. G.; Manna, L.; Yang, W. D.; Wickham, J.; Scher, E.; Kadavanich, A. & Alivisatos, A. P. (2000). Shape control of CdSe nanocrystals. *Nature*, Vol. 404, No. 6773, pp. 59-61.

Pines, D. (1963), *Elementary excitations in solids*, W. A. Benjamin Inc., New York.

Puzder, A.; Williamson, A. J.; Zaitseva, N. & Galli, G., (2004). The Effect of Organic Ligand Binding on the Growth of CdSe Nanoparticles Probed by Ab Initio Calculations. *Nano Lett.*, Vol. 4, No. 12, pp. 2361-2365.

Rabani, E. & Baer, R., (2010). Theory of multiexciton generation in semiconductor nanocrystals. *Chem. Phys. Lett.*, Vol. 496, pp. 227–235.

Rothenberg, E.; Kazes, M.; Shaviv, E. & Banin, U., (2005). Electric field induced switching of the fluorescence of single semiconductor quantum rods. *Nano Lett.*, Vol. 5, No. 8, pp. 1581-1586.

Rudin, S.; Reinecke, T. L. & Segall, B., (1990). Temperature-dependent exciton linewidths in semiconductors. *Phys. Rev. B*, Vol. 42, No. 17, pp. 11218–11231.

Schaller, R. D.; Pietryga, J. M. & Klimov, V. I., (2007). Carrier Multiplication in InAs Nanocrystal Quantum Dots with an Onset Defined by the Energy Conservation Limit. *Nano Lett.*, Vol. 7, No. 11, pp. 3469-3476.

Sharma, H.; Sharma, S. N.; Kumar, U.; Singh, V. N.; Mehta, B. R.; Singh, G.; Shivaprasad, S. M. & Kakkar, R., (2009). Formation of water-soluble and biocompatible TOPO-capped CdSe quantum dots with efficient photoluminescence. *J. Mater. Sci.: Mater. Med.*, Vol. 20, pp. S123–S130.

Schlegel, G.; Bohnenberger, J.; Potapova, I. & Mews, A. (2002). Fluorescence Decay Time of Single Semiconductor Nanocrystals. *Phys. Rev. Lett.*, Vol. 88, pp. 137401-197404.

Schmitt-Rink, S.; Miller, D.A.B. & Chemla, D.S., (1987). Theory of the linear and nonlinear optical properties of semiconductor microcrystallites. *Phys. Rev. B*, Vol. 35, No. 15, pp. 8113–8125.

Schöps, O.; Le Thomas, N.; Woggon, U. & Artemyev, M. V., (2006). Recombination dynamics of CdTe/CdS core-shell nanocrystals. *J. Phys. Chem. B*, Vol. 110, No. 5, pp. 2074-2079.

Stubbs, S. K.; Hardman, S. J. O.; Graham, D. M.; Spencer, B. F,; Flavell, W. R.; Glarvey, P.; Masala, O.; Pickett, N. L. & Binks, D. J., (2010). Efficient carrier multiplication in InP nanoparticles. *Phys. Rev. B*, Vol. 81, pp. 081303-081306 (R).

Takagahara, T., (1996). Electron—phonon interactions in semiconductor nanocrystals. *J. Lumin.*, Vol. 70, No. 1-6, pp. 129-143.

Trallero-Giner, C.; Debernardi, A.; Cardona, M.; Menendez-Proupin, E. & Ekimov, A. I., (1998). Optical vibrons in CdSe dots and dispersion relation of the bulk material. *Phys. Rev. B*, Vol. 57, No. 8, pp. 4664–4669.

Trinh, M. T.; Polak, L.; Schins, J. M.; Houtepen, A. J.; Vaxenburg, R.; Maikov, G. I.; Grinbom, G.; Midgett, A. G.; Luther, J. M.; Beard, M. C.; Nozik, A, J.; Bonn, M.; Lifshitz, E. & Siebbeles, L. D. A., (2011). Anomalous Independence of Multiple Exciton Generation on Different Group IV-VI Quantum Dot Architectures. *Nano Lett.*, Vol. 11, pp. 1623–1629.

Underwood, D. F.; Kippeny, T. & Rosenthal, S. J., (2001). Ultrafast carrier dynamics in CdSe nanocrystals determined by femtosecond fluorescence upconversion spectroscopy. *J. Phys. Chem. B*, Vol. 105, No. 2, pp. 436-443.

Valerini, D.; Cretì, A.; Lomascolo, M.; Manna, L.; Cingolani, R. & Anni, M. Temperature dependence of the photoluminescence properties of colloidal CdSe/ZnS core/shell quantum dots embedded in a polystyrene matrix. *Phys. Rev. B*, Vol. 71, No. 23, pp. 235409-235414.

Velizhanin, K. A. & Piryatinski, A., (2011). Numerical Study of Carrier Multiplication Pathways in Photoexcited Nanocrystal and Bulk Forms of PbSe. *Phys. Rev. Lett.*, Vol. 106, pp. 207401-207404.

Wang, H. Y. ; de Mello Donegà, C.; Meijerink, A. & Glasbeek, M., (2006). Ultrafast exciton dynamics in CdSe quantum dots studied from bleaching recovery and fluorescence transients. *J. Phys. Chem. B*, Vol. 110, No. 2, pp. 733-737.

Wang, X.; Qu, L.; Zhang, J.; Peng, X. & Xiao, M. (2003a). Surface-Related Emission in Highly Luminescent CdSe Quantum Dots. *Nano Lett.*, Vol. 3, No. 8, pp. 1103-1106.

Wang, X. Y.; Yu, W. W.; Zhang, J. Y.; Aldana, J.; Peng, X. & Xiao, M., (2003b). Photoluminescence upconversion in colloidal CdTe quantum dots. *Phys. Rev. B*, Vol. 68, No. 12, pp. 125318-125323.

Woggon, U.; Giessen, H.; Gindele, F.; Wind, O.; Fluegel, B. & Peyghambarian, N., (1996). Ultrafast energy relaxation in quantum dots. *Phys. Rev. B*, Vol. 54, No. 24, pp. 17681-17690.

Wolf, M.; Brendel, R.; Werner, J. H. & Queisser, H. J., (1998). Solar cell efficiency and carrier multiplication in $Si_{1-x}Ge_x$ alloys. *J. Appl. Phys.*, Vol. 83, No. 8, pp. 4213-4222.

Yoffe, A. D. (2001). Semiconductor quantum dots and related systems: electronic, optical, luminescence and related properties of low dimensional systems. *Adv. Phys.*, Vol. 50, pp. 1-208.

Zhang, H.; Chen, D.; Zhang, J.; Wang, Z.; Cui, Y. & Shen, L., (2010). Effect of Shell Layers on Luminescence of Colloidal CdSe/$Zn_{0.5}Cd_{0.5}Se$/ZnSe/ZnS Core/Multishell Quantum Dots. *Journal of nanoscience and nanotechnology*, Vol. 10, No: 11, pp. 7587-7591.

Zhang, Y.; Miszta, K.; Kudera, S.; Manna, L.; Di Fabrizio, E. & Krahne, R. (2011). Spatially resolved photoconductivity of thin films formed by colloidal octapod-shaped CdSe/CdS nanocrystals. *Nanoscale*, Vol. 3, No. 7, pp. 2964-2970.

Zheng, J. J.; Ji, W. Y.; Wang, X. Y.; Ikezawa, M.; Jing, P.; Liu, X.; Li, H.; Zhao, J. & Masumoto, Y., (2010). Improved Photoluminescence of MnS/ZnS Core/Shell Nanocrystals by Controlling Diffusion of Mn Ions into the ZnS Shell. *J. Phys. Chem. C*, Vol. 114, No. 36, pp. 15331-15336.

5

Molecular States of Electrons: Emission of Single Molecules in Self-Organized InP/GaInP Quantum Dots

Alexander M. Mintairov[1], James L. Merz[1] and Steven A. Blundell[2]

[1]*University of Notre Dame*

[2]*INAC/SPSMS, CEA/UJF-Grenoble*

[1]*USA*

[2]*France*

1. Introduction

Correlation between particles in finite quantum systems leads to a complex behavior and novel states of matter. One remarkable example of such a correlated system is expected to occur in an electron gas confined in a quantum dot (QD), where at vanishing electron density the Coulomb interaction between electrons rigidly fixes their relative positions like those of the nuclei in a molecule. Unlike real molecules, however, which have sizes and properties fixed by their chemical constituents, the size, shape and electronic density of such confined electronic structures, referred to as Wigner molecules (WM), can be varied experimentally using various combinations of semiconductor materials, types of nanostructures, numbers of electrons, electrostatic potentials and magnetic fields. Thus these WMs present a novel and compelling field for fundamental and applied research. So far, however, the properties of WMs and their underlying fundamental physics have been studied primarily by theory; what little experimental evidence there is for their existence consists only of measurements of charging energies and light-scattering spectra of GaAs/AlGaAs quantum dots created from modulation doped 2D-electron gas heterostructures.

Here we present the results of an experimental study of correlated states of electrons in a WM in self-organized InP/GaInP quantum dots. The unique properties of these QDs are their relatively large lateral size (~80-200 nm) and their ability to accommodate up to 20 electrons, providing electron density up to 2×10^{11} cm^{-2}. The dots have strong emission intensity which allows us using photoluminescence spectroscopy for their study. We used a high-spatial-resolution low-temperature near-field scanning optical microscopy (NSOM) having spatial resolution up to 30 nm in combination with a high magnetic field to resolve emission spectra of single QDs. Using emission spectra of single dots we observed crossover from a Fermi liquid to WM behavior at a critical density of 5×10^{10} cm^{-2}. A magnetic-field-induced molecular-droplet transition has been observed in the Fermi liquid regime. In the Wigner molecule regime we observed a rich vibrational structure of the emission spectra, which opens way to identify the electron arrangement in the WM. These results are discussed in detail and compared with existing literature data.

We also present theoretical calculations of electron correlation in quantum dots using an accurate configuration-interaction method employing a numerical mean-field basis set and analysis of vibrational modes in WMs using the classical limit.

2. Wigner localization in semiconductor quantum dots

A Wigner phase is a strongly correlated state of an electron system, in which electrons occupy separate sites forming a regular lattice. The possibility of crystallization of an electron gas at densities below a certain critical value (n_s) was predicted by Wigner in 1934 (Wigner, 1934). Experimentally such crystallization has been observed in two dimensional electron systems on the surface of liquid He (Grimes & Adams, 1979), in a GaAs/GaAlAs heterojunction (Andrei et al., 1988) and in Si (Pudalov et al., 1993) using detection of the metal-insulator transition.

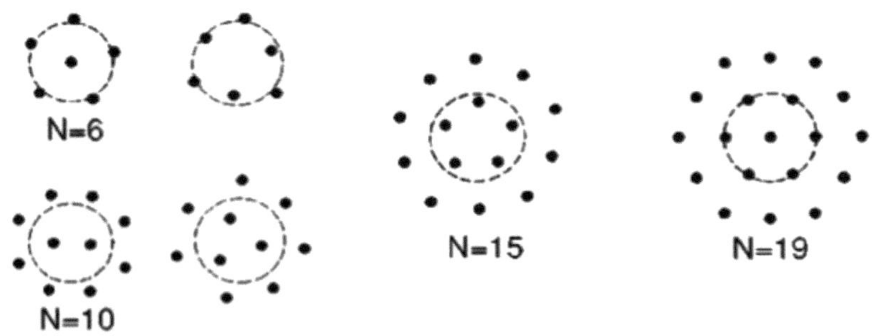

Fig. 1. Schematic view of the classical electron configurations in a parabolic potential for N=5, 6, 10, 15 and 19 (Bolton & Rössler, 1993).

The electrons confined in traps having volume >$1/n_s$ form Wigner Molecules (WMs). Wigner localization of electrons in such traps formed by interface fluctuations is responsible for the quantum Hall effect in high mobility semiconductor heterostructures (Ilani et al., 2004; Laughlin, 1983). The Wigner localization regime can be realized in single electron transistors (SETs) using GaInAs/AlGaAs quantum dots (QDs) nano-fabricated from modulation doped quantum well structures (Kastner, 1993; Ashoori, 1996; Tarucha, 1996). Coulomb blockade measurements and theoretical analysis (Maksym et al., 2000; Reimann& Manninen, 2002; Szafran et al., 2003) have shown that a WM in such SETs can reveal a rich set of electron arrangements and spin states, which are controlled by the number of electrons present, and by an applied magnetic field. While electron arrangement in WMs will depend on the shape of the confinement potential and the number of electrons (N), theoretical analysis for the ideal 2D parabolic potential shows that in the classical approximation for systems having N<6, simple polygons are formed. The first nontrivial configurations are found for N=6 (see Fig.1). In addition to the ground state with five electrons surrounding a single electron at the center, metastable states and isomers at very similar energies exist, such as the 6-electron hexagon. Further, for N=10, the ground state is a "dimer" in the center with eight surrounding electrons. The hexagonal lattice of a Wigner crystal is not restored until N~200. In addition to the existence of metastable states, theory also predicts that the application of a magnetic field re-arranges the electron distribution in a QD and leads to specific phase transitions, which are called molecular-droplet transitions.

The effect of Wigner localization on electronic states is characterized by the dimensionless density parameter (Wigner-Seitz radius) $r_s=1/[a^*_B(\pi \cdot n)^{0.5}]$, where a^*_B is the effective Bohr radius and n is the average electron density in the plane of the dot. This parameter is approximately equal to the ratio of Coulomb-to-kinetic energy. It can also be expressed approximately via the parabolic (harmonic) potential frequency ω_0 via $r_s^3=1/[\omega_0^2 N^{0.5}]$], where ω_0 is expressed in units of effective Hartrees Ha*, and N is the number of electrons in the dot. For $r_s<1$, i.e. in the Fermi liquid regime (strong confinement potential), the electrons in a QD behave similarly to the electrons in an atom and their energy spectrum for a 2D parabolic potential is $E_{K,L}= \hbar\omega_0(2K+|L|+1)$, with the K quantum number corresponding to the number of radial nodes in the electron wave function and L is the azimuthal quantum number. Note that this formula neglects the electron-electron interaction and is modified somewhat by the electronic mean-field potential. In the classical limit $r_s>>10$, and the electrons behave as point charges (Fig.1). According to the theoretical calculations for small numbers of confined electrons (say, up to 10), the onset of electron localization occurs gradually as r_s increases with partial Wigner localization in distinct WM geometries occurring already for r_s~4 (Egger et al., 1999).

The classical regime, resembling the point charge arrangements presented in Fig.1, was realized in the macroscopic experiment in which the electrons were represented by negatively charged (up to 10^9 electrons) metallic balls having 0.8 mm diameter and the trap (QD) was a positively charged cylindrical electrode having 10 mm in diameter (Saint Jean et al., 2001). For III-V semiconductor materials and in particular GaAs, the only semiconductor used so far for SETs, the effective Bohr radius having value a^*_B ~10 nm is relatively large and the classical regime requires potentials $\hbar\omega_0<0.2$ meV and large QD sizes >500 nm, which seems to be hardly achievable experimentally.

So far experimentally the signatures of formation of WMs were observed using a GaAs/AlGaAs heterostructure system with a two-dimensional electron gas. Using electrostatically-defined two-electron QDs having ω_0~5 meV and r_s~1.55, the existence of electron correlation was detected by identification of the singlet nature of the lowest excited state at finite magnetic field (Ellenberger et al., 2006). Light scattering spectra were used to observe spin and charge modes in nanofabricated QDs having two (Singha et al, 2010) and four (Kalliakos et al., 2008) electrons. These two and four electron dots have $\hbar\omega_0$~1.6 meV/r_s~3.4 and $\hbar\omega_0$~3.8 meV/r_s~1.7, respectively. The effect of formation of electronic molecules in these investigations was revealed from a fit of the observed energies using a configuration interaction approach.

In the present contribution we introduce InP/GaInP QDs as a natural WM system providing a great variety of electron occupation and sizes and, using high-spatial resolution nano-optical methods, we report the observation of emission of different types of WMs.

3. Excited states in Wigner molecules: Example for N=2 and N=6

3.1 Rovibrational states for N=2

In the Wigner localization regime the correlated state is characterized by the separation of the center-of-mass (c.m.) motion, having frequency ω_0, and a relative (rovibrational-spin) electron motion. According to the Kohn theorem this separation is an exact result for a circular parabolic confining potential at any electron density (Jacak et al., 1998). For the simplest case of a two-electron molecule (2e-WM) the equations of motions allow an exact solution (Yannouleas&Landman, 2000). These solutions have shown that for r_s=200 the energy spectrum of 2e-WM has a well developed and separable rovibrational contribution exhibiting collective rotations, as well as stretching and bending vibrations:

$$E_{KL,kl}=Cl^2+(k+1/2)\hbar\omega_s+(2K+|L|+1)\,\hbar\omega_b, \quad (1)$$

where the rotational constant C≈0.037, the phonon for the stretching vibration has energy $1.75\omega_0$ and the phonon for the bending vibration coincides with that of the c.m. motion, i.e. $\omega_b=\omega_0$. Note that the bending vibration can itself carry an angular momentum $\hbar L$ and thus rotational angular momentum $\hbar l$ does not necessary coincide with the total angular momentum $\hbar(L+l)$. The calculations have shown that the molecule preserves its structure at r_s =3 ($\hbar\omega_0$~1 meV for GaAs), i.e. below the "theoretical" Fermi liquid to WM transition. Here the rotational sequence shows nearly equal spacing having value $\omega_0/2$ but the stretching vibration does not change.

3.2 Excited states of six-electron Wigner molecules

3.2.1 Configuration-interaction calculations of spin states

We used an accurate configuration-interaction (CI) method employing a numerical mean-field basis set to study the excitation spectrum of a six-electron WM (Blundell & Chacko, 2011). The CI method (Szabo&Ostlund, 1996; Blundell&Joshi, 2010) is more suitable for the systematic study of excited states than other methods such as Hartree-Fock-based methods and variational Quantum Monte-Carlo (which can treat only the lowest-energy state of a given symmetry). Dots with N ≥ 6 electrons in general have more than one classical isomer (see, for example, the [1,5] and [0,6] isomers in Fig. 1) and therefore isomeric states should form an important part of the phenomenology of excited states. Now, for a circularly

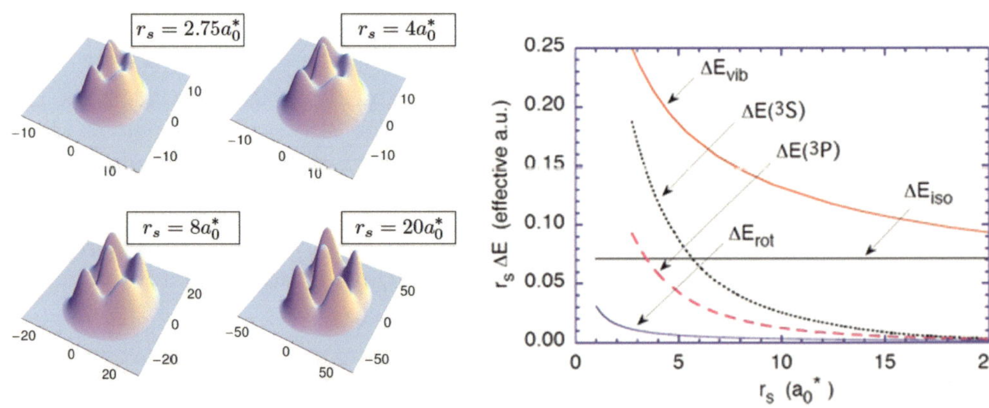

Fig. 2. Electron spatial pair correlation functions for r_s=2.75, 4, 8 and 20 (left) and excitation energies of the six-electron dot versus Wigner-Seitz radius r_s (right). The quantities ΔE_{rot}, ΔE_{vib}, and ΔE_{iso} are approximate rotational, vibrational, and isomeric excitation energies, respectively, inferred from a classical model (see text); $\Delta E(^3S)$ and $\Delta E(^3P)$ are excitation energies to the lowest ^3S and ^3P states calculated by CI. All excitation energies are scaled by r_s.

symmetric external potential and a state of definite L_z, the electronic density (in the "laboratory" frame) in 2D must also be circularly symmetric (Hirose&Wingreen, 1999) and in the Wigner limit the density therefore becomes a series of concentric rings (see, for example, Ghosal, et al., 2006). To reveal the Wigner localization, we therefore consider the internal many-body correlations by means of the electronic (charge-charge) pair-correlation functions (PCFs) $g(\mathbf{r};\ \mathbf{r}_0)$ (Maksym, 1996; Reimann et al., 2000; Yannouleas&Landman, 2000). The quantity $g(\mathbf{r};\ \mathbf{r}_0)$ is proportional to the conditional probability of finding an electron at the position $\mathbf{r}$ given that another (reference) electron is present at $\mathbf{r}_0$. Calculated $g(\mathbf{r};\ \mathbf{r}_0)$ functions are presented for r_s = 2.75, 4, 8, and 20 in Fig. 2, from which it is seen that a partially correlated state is observed even at r_s = 2.75. Recall that r_s is expressed in units of the effective Bohr radius a_B*, where $a_B^* \approx 8.7$ nm for the InP/GaInP dots in our experiments.

Our calculations have shown that the evolution of the excitation energy of the lowest ^3S, ^5S, ^7S, and ^3P states relative to the ^1S ground state versus r_s yields approximately parallel straight lines on a logarithmic plot $6 \le r_s \le 10$, the excitation energy ΔE of these states being well fit by an expression of the form $\Delta E = c\ \exp(-m\ r_s)$, with $c(^3P) = 0.020$ Ha*, $c(^5S) = 0.028$ Ha*, $c(^3S) = 0.048$ Ha*, $c(^7S) = 0.054$ Ha*, and $m \approx 0.30\ (a_B^*)^{-1}$. The energy units Ha* here are effective Hartrees, with 1 Ha* ≈ 13.2 meV for the InP/GaInP dots used in our experiments.

3.2.2 Low-lying excitations in the classical limit

At large r_s the quantum excitation energy of a quasi-2D Wigner molecule may be written approximately in a way analogous to that for a planar molecule

$$E(\mathrm{P}) = E_{\mathrm{cl}}(\mathrm{P}) + L_z^2/(2I_P) + \Sigma_a\,[\Omega_a(\mathrm{P})(n_a+1/2)] + E_{\mathrm{spin}}, \qquad (2)$$

where $E_{\mathrm{cl}}(\mathrm{P})$ is the classical electrostatic energy of isomer P, I_P is its moment of inertia, and n_a is the number of vibrational quanta in a normal mode with frequency $\Omega_a(\mathrm{P})$. The energy E_{spin} is the spin-spin interaction energy of the spins of the localized electrons. As an example of a typical rotational excitation energy, we note that the ground-state [1,5] isomer has a moment of inertia $I_P = 8.9\ r_s^2 N^{1/3}$, and it then follows using Eq. (2) that the *S*- to *P*-wave excitation energy is given in the classical limit by (in effective a.u.) $\Delta E_{\mathrm{rot}} = 0.0309\ r_s^{-2}$. Similarly, noting that the frequency of the first classical normal mode of the [1,5] isomer is $\Omega_1 = 0.650\omega_0$ (see below), we find that the excitation energy of one vibrational quantum in this mode is (in effective a.u.) $\Delta E_{\mathrm{vib}} = 0.415\ r_s^{-3/2}$. For isomeric excitations we found (Blundell &Chacko, 2011) $\Delta E_{\mathrm{iso}} = E_{\mathrm{cl}}\,(0, 6) - E_{\mathrm{cl}}\,(1, 5) = 0.0714\ r_s^{-1}$. To clarify the role of rotational excitations, we show in Fig. 2 the classical estimate of the *S*- to *P*-wave excitation energy ΔE_{rot} as a function of r_s, together with the excitation energy calculated by CI (from the ^1S ground state to the lowest ^3P state, which is the lowest-lying *P*-wave state).

We also show a typical spin excitation energy $\Delta E(^3S)$, defined as the energy of the lowest ^3S state relative to the ^1S ground state, as calculated by CI. From Fig. 2, one sees that the smallest rotational excitation for r_s <10 is in fact somewhat larger than the classical estimate ΔE_{rot}. This is simply because a spin excitation is also involved. One also sees in Fig. 2 that for $r_s < 6$, the "spin" excitation energy (due to atomic-like exchange and correlation effects) is nominally comparable to the isomeric and vibrational energies. Thus, although at these values of r_s it is possible to find partial Wigner localization in a recognizable geometry, it is not generally possible to discuss isomeric and vibrational excitations separately from spin excitations at these densities.

3.2.3 Vibrational modes

The classical normal modes for the [1,5] ground-state isomer are shown in Fig. 3. There are 2N - 1 = 11 normal modes grouped into five doubly degenerate modes and one nondegenerate mode, which is a breathing mode at high frequency. The lowest frequency mode can be thought of as a dipolar oscillation of the central electron accompanied by a distortion of the outer ring. The third and fourth modes, at Ω=1.223ω_0 and Ω=1.314ω_0, correspond to quadrupole and octupole distortions, respectively, of the outer ring, with the central electron remaining fixed. The second mode is a collective dipolar oscillation of the c.m. of the system at frequency Ω=ω_0 (exactly), in which the whole structure remains undistorted during the oscillation. The existence of such a classical mode can be shown to be a general result for a system in a harmonic confining potential having an interaction depending only on the relative coordinates of the particles. The quantum mechanical analog of this result is the Kohn theorem, according to which under the same circumstances the c.m. motion decouples exactly from the "relative coordinates," and one can describe the system by a wave function in relative coordinates combined with oscillations of the c.m. in the harmonic confining potential. Note that the breathing mode of the six electron molecule has frequency Ω=1.732ω_0, which is nearly the same as the frequency of the breathing mode of the two electron molecule considered above.

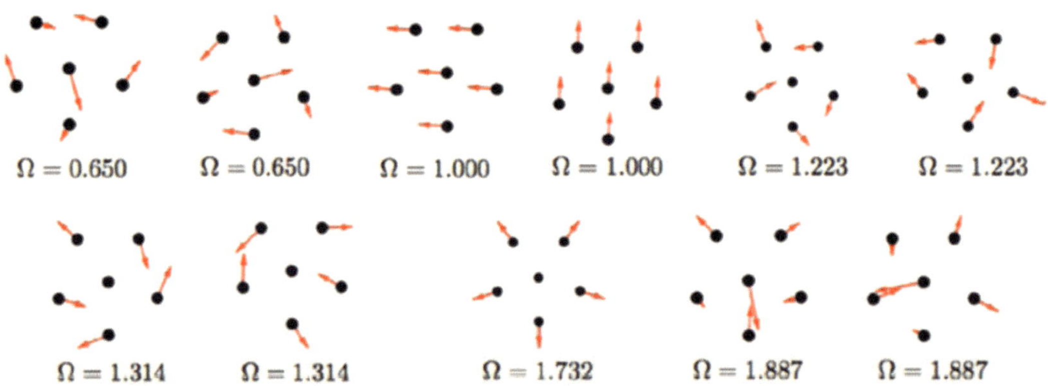

Fig. 3. Normal modes of the ground-state isomer of the classical six-electron dot. The normal mode frequency Ω is indicated as a multiple of the frequency of the parabolic potential ω_0.

3.3 Generalization to larger size N

We can use the classical arguments that work well for $N = 6$ to show that aspects of the same excited-state phenomenology apply to larger N as well. The classical model yields more than one isomer for $N = 6$ and for $N \geq 9$. We have used the "basin hopping" algorithm to generate and study the classical isomers in the size range up to $N = 20$. As N increases, the energy separation of isomers tends to become smaller. Thus the excitation energy $\Delta E_{iso}(N)$ for N electrons satisfies $\Delta E_{iso}(6) = 0.10\omega_0^{2/3}$, $\Delta E_{iso}(9) = 0.044\omega_0^{2/3}$ and $\Delta E_{iso}(19) = 0.013\omega_0^{2/3}$ (in effective a.u.). It is then generally the case that the first excited level (spin multiplet) for fixed L_z

at intermediate r_s is an isomer rather than a vibrational excitation of the ground state. We thus expect low-lying isomeric states to be a generic feature of the excitation spectrum of dots with $N = 6$ and $N \geq 9$ electrons at intermediate r_s values. Also, the rotational excitation energies ΔE_{rot} are generally found to be small compared to ΔE_{iso} and ΔE_{vib}, similar to Fig. 2 for $N = 6$.

4. InP/GaInP quantum dots as natural electronic molecules

4.1 Structural properties

Our InP/GaInP QD samples were grown by Metal-Organic Chemical Vapor Deposition (MOCVD) in a horizontal AIX200/4 reactor under pressure of 100 mbar. Trimethylgallium (TMGa), trimethylaluminium (TMAl) and trimethylindium (TMIn) metalorganic compounds were used as the group III element sources. Arsine (AsH_3) and phosphine (PH_3) were used as the group V element sources. (100) GaAs substrates misoriented by 2° towards the [110] direction were used. Initially a 50 nm-thick GaAs layer was deposited on the wafer. Then 50 nm thick $Ga_{0.52}In_{0.48}P$ (GaInP) lattice-matched with GaAs layer was grown. The QDs were grown at 725° C by depositing 7 monolayers (ML) of InP (Vinokurov et al,. 1999; Chu et al., 2009,). We studied the structures with uncapped dots and structures having GaInP cap layer thickness 5, 20, 40 and 60 nm.

The dot density (~$2*10^9$ cm^{-2}) and their sizes (base ~10-200 nm, height ~5-60 nm) were measured using atomic force microscopy (AFM) for the uncapped samples and transmission electron microscopy (TEM) for the capped samples. These data are presented in Fig.4a-d. A clear bimodal size distribution is seen from AFM images in Fig.4a consisting of large dots having size 100-200 nm and density 0.6×10^9 cm^{-2} and the small dots having sizes 10-70 nm and density 1.2×10^9 cm^{-2}. From TEM measurements a lens shape of the large dots was revealed (see Fig.4c and d). Due to a residual n-type doping of the GaInP layer in the MOCVD growth process (n~10^{16} cm^{-3}) the modulation doping of InP QDs occurs and they can contain up to 20 electrons (Hessman et al., 2001). Thus these QDs can represent natural WM and for dot sized 150-200 nm the r_s ~4 can be achieved for few electron dots. The electron density can be varied from 10^{10} up to 5×10^{11} cm^{-2} and thus these InP QDs offer much more flexibility in varying of WM parameters then "conventional" GaAs/AlGaAs QDs used in SETs (Singha et al, 2010).

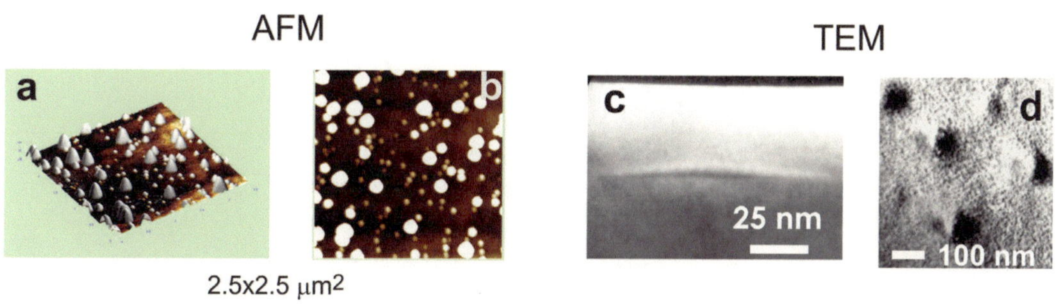

Fig. 4. Atomic force (a and b, size 2.5x2.5 μm^2) and transmission electron (c - plan-view, d - cross section) microscopy images of InP/GaInP QDs structures.

We estimated that our InP QDs can contain up to 10% Ga. Pure InP QDs can be grown by depositing nominally 0.5 ML of InP at growth temperatures 580 C and using in-situ annealing/growth interruption. Such growth conditions suppress Ga and In intermixing and result in the pyramidal dot shape having base ~60 nm and height ~15 nm (Georgsson et al., 1995). Even smaller InP QDs having base 30 nm and height 7 nm were grown using 4ML InP deposition at 550 C (Ren et al., 1999).

In Fig.5a we present a cartoon showing the formation of a WM in our InP/GaInP QDs. Nine electrons come from the adjacent donor atoms located in the GaInP. The classical arrangement of a 9e-WM consists of eight electrons surrounding one electron at the center. The corresponding tentative band diagram of this QD is shown in Fig.5b. The barrier is formed by GaInP having band gap energy 1.97 eV (Janssens et al., 2003; Pryor et al., 1996). The "bandgap" of the QD material includes vertical confinement energy (~150 meV) and bandgap increase (~50 mev) due to Ga/In intermixing. We estimated the QD material bandgap to be ~1.7 eV, which is nearly 200 meV higher than the bandgap of InP. Due to strain effects and Ga/In intermixing we expect type-I band alignment between the QD and barrier material (Janssens et al., 2003; Pryor et al., 1997).

4.2 Photoluminescence of Wigner molecules

Under photexcitation an electron-hole pair is generated in the dot and it forms a trion with the central electron (see in Fig.5a). The formation of trions in a dilute electron gas is well established (Finkelstein et al., 1995) and thus we can assume that a radiative recombination of the trion also forms the photoluminescence (PL) spectra of the WM. The selection rules for radiative transitions of the trion involving the electron having spin ± 1/2 and the hole having spin ± 3/2 do not change the electron spin and thus only charge excitations are expected to dominate the emission spectra of the WM.

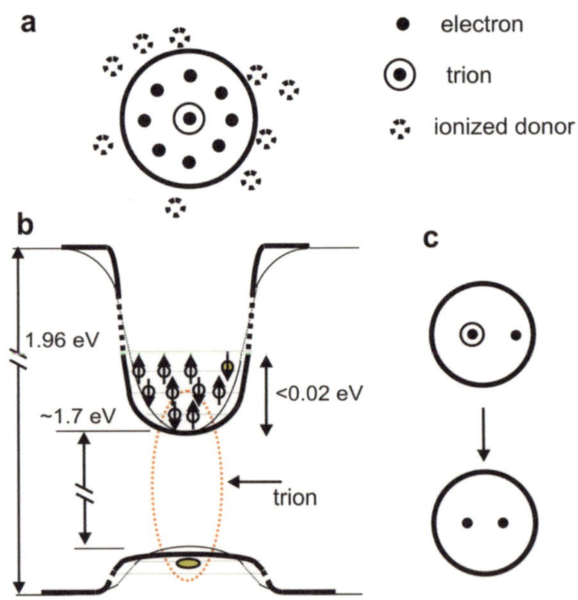

Fig. 5. Arrangement of electrons (classical positions) (a) and corresponding band diagram (b) of photo-excited nine-electron InP QD (base ~150 nm); classical positions of the trion and electrons in two electron WM before (upper) and after (lower) radiative recombination (c). Trion and adjacent ionized donors are also shown in a. Thick(thin) curves in b are the band diagrams for filled(empty) dot.

In Fig.5c we show classical positions of the trion and the electrons in the 2e-WM before (upper cartoon) and after (lower cartoon) the radiative recombination. It can be shown (Govorov et al., 2006) using the minimization of the classical energy of the electrons that the classical distance d^i_e of the electron to the dot center is a factor $\beta = m_{tr\omega0tr}{}^2/m_{e\omega0}{}^2$ larger than d^i_{tr}, the corresponding distance for the center of mass of the trion. This is because the trion confinement potential and corresponding harmonic frequency ω_{0tr}, is stronger than the electron confinement potential and corresponding harmonic frequency ω_0. In a similar way, we can determine the classical configuration of the two electrons left in the final state after photon emission. Their classical coordinate d^f_e in the ground state obeys relation $d^i_e > d^f_e > d^i_{tr}$ as seen in Fig.5c. The wave functions of the initial and final states are peaked at the classical coordinates and decay exponentially along the radial axis away from their peak positions (see for example Fig.2). Since $\beta \neq 1$, the classical coordinates for the initial and final states are different. This gives a coupling of the radiative recombination transition with vibrational modes of the WM which is the origin of the shake-up process in the photoluminescence of WM, to be discussed in the experimental part. Such "electron-phonon" coupling can be expressed via Huang-Rhys factors (Huang&Rhys, 1950).

From symmetry considerations one can see from Fig.5c that the annihilation of the trion induces stretching and center-of-mass distortion along the 2e-WM axis, thus generating the stretching and translation modes discussed above. Thus the emission spectra of a 2e-WM are expected to have contributions from the "zero-phonon" line (ZPL) and two sets of Stokes phonon replicas having energies $n\hbar\omega_0$ and $n\hbar 1.7\omega_0$, where n=1, 2, … . This is similar to the vibronic structure of conventional molecules.

Similar symmetry considerations based on classical electron arrangements shown in Fig.1 predict weak phonon Stokes emission for WM having a central electron, i.e. for N=6-9, and strong phonon Stokes emission for the two electrons at the center, i.e. N=2 and N=10-14.

5. The near-field scanning optical microscopy (NSOM) technique

5.1 Optical spectroscopy of quantum dots

For study of the effects of the electron localization in single InP/GaInP QDs we used the near-field scanning optical microscopy (NSOM) technique (Betzig&Trautman, 1993) in combination with magneto-PL spectroscopy. The basic principle of NSOM (Synge, 1928), providing a way to overcome the diffraction limit of light of conventional optics, is to use the coupling of the evanescent electromagnetic field and the radiative electromagnetic waves in the vicinity of a nano-probe placed near the boundary between two media. This principle was realized nearly three decades ago using nano-apertures (Lewis et al., 1984, Pohl et al., 1984), metallic (Fischer and Pohl ,1989) and dielectric (Coutjon et al 1989, Reddick at al., 1989) nano-tips, allowing nanometer-scale spatial resolution in optical experiments with a wide range of applications including experiments with single semiconductor QDs (Flack et al., 1996, Toda et al., 1996). The high spatial resolution, scanning ability and non-destructive character of the experiment in combination with a high magnetic field and time-resolved techniques, allows the use of NSOM to study structural parameters (Mintairov et al., 2001), the spin structure of exciton states (Ortner et al., 2003, Toda et al., 1998), the temporal coherence of the wave functions (Toda et al., 2000), and the mechanisms of carrier migration (K. Matsuda et al., 2000) and relaxation (Toda et al., 1999) in semiconductor QDs.

Achieving spatial resolution as high as 30 nm allows optical mapping of exciton wave functions in a single QD (Matsuda et al., 2003). Such spatial resolution is of the order of the electron separation in the Wigner localization regime and thus the NSOM technique opens the possibility to probe the position of the individual electrons in the electronic molecules. However, present GaAs/AlGaAs SET structures require relatively thick AlGaAs cap layers (~70 nm), which provides vertical modulation doping but does not allow achievement of spatial resolution below 200 nm (Mintairov et al., 2003). In contrast InP/GaInP QDs, having lateral modulation doping (see Fig.5b), can form a WM for cap layer thickness down to zero nm (see below). Thus the tip-QD distance separation can be zero and spatial resolution as high as 10 nm can be possible, as was demonstrated for single molecules (Hosaka& Saiki, 2001).

5.2 Experimental details

5.2.1 NSOM set up

For our low-temperature magneto-PL measurements we used an Oxford Instruments CryoSXP cryogenic scanner together with a liquid helium cryostat with a superconducting magnet providing magnetic fields up to 12 T. Shear-force tip-sample distance control and scanning are governed by a Vecco AFM controller from Digital Instruments, providing a scan range over an area of 3x3 μm^2 at 4.2K. For characterization of InP/GaInP structures at room temperature we used an NSOM-2000 system from NANONICs Inc. having a scan range 90x90 μm^2. The near-field photoluminescence (PL) spectra were taken in collection-illumination mode, i.e. laser excitation and PL emission collection using the same NSOM fiber probe. Emission was excited by the 488 nm line of an Ar ion laser and dispersed using a 270 mm focal length monochromator. The spectra were measured using a nitrogen-cooled CCD detector and 1200 gr/mm grating. Monochromatic NSOM images were measured using a 300 gr/mm grating and a GaAs photomultiplier working in the photon-counting regime with an accumulation time of 20 ms per pixel. Excitation power density was ~20-100 and 0.1-1 W/cm-2 for 300 and 10 K, respectively. The spectral resolution of the system is 0.2-0.4 meV. We measured spectra in a magnetic field up to 10 T. A quarter wave-plate and linear polarizer were used for the separation of right- and left-hand circular polarization.

5.2.2 NSOM probes

As near-field probes we used tapered single mode fiber tips. We used both metal coated and uncoated tips. Coated tips were prepared by electron beam deposition of metal (Al or Au in combination with Ti) having thickness ~50-200 nm. The physical structure of the aperture and its optical quality were controlled using scanning electron microscopy (SEM) imaging, visual observation of the light coming from the tip under a microscope with x100 objective, and by measuring its far-field transmission for 632.8 nm wavelength (HeNe laser). SEM images of typical near-field optical fiber probes are presented in Fig. 6a-f. The taper was prepared by using a pulling technique ((Lewis et al., 1984, Pohl et al., 1984). We used two pulling regimes in which the taper angle increases towards the tip end (see Fig.6a and b). Both regimes give tips (apertures) with diameters 200 nm and aperture angles 30-40° (see Fig. 6c) but within a few micrometers from the aperture they have different taper angles. To make smaller apertures we used further chemical etching in a hydrofluoric acid solution (Otsu, 1998). We also use a focused ion beam milling technique to flatten the apertures (Fig.6f). The transmission of our metal coated fiber probes was 10-

^{4}-10^{-2}. Uncoated fiber tips provide an order of magnitude higher PL signals but in general they have poorer spatial resolution. Below, however, we will demonstrate an apertureless mode of uncoated tip in which spatial resolution is determined just by the apex diameter and thus can be as high as coated ones.

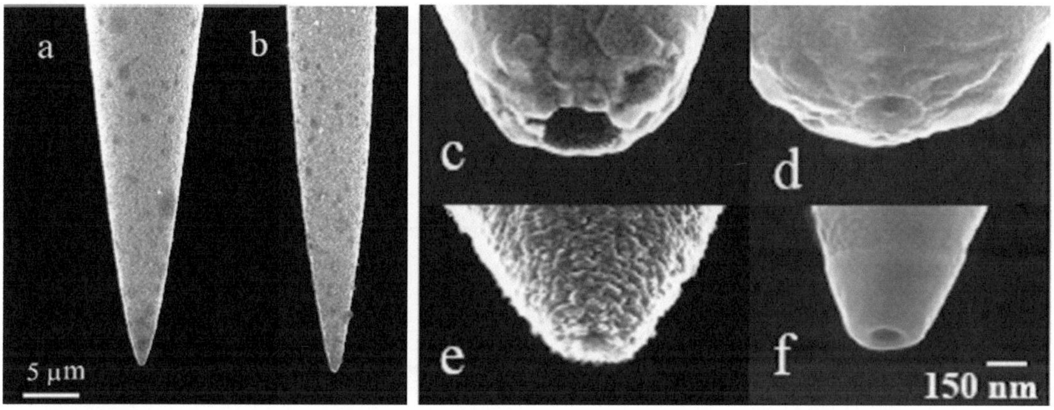

Fig. 6. SEM images of the near-field optical fiber probes: side view of the two types of tapers used (a, b); 45° tilt images of apertures having diameters (nm) and coating: c - 280 and Au, d - 270 and Al, e - 60 and Al/Ti and f- 120 and Al. The apertures were prepared by pulling (c), etching (d-f) and FIB trimming (f).

5.3 NSOM imaging

Fig.7a-d presents our results of NSOM imaging of ~ 2 μm long section of a CdSe nanowire (NW) having diameter 40 nm taken with an uncoated fiber tip (Mintairov et al., 2010). It shows topographic (a) and monochromatic NSOM (b) images together with a set of twelve spectra measured along a 400 nm size central section of the wire (c) and a false-color wavelength-position plot of these spectra (d). Note that the topographic image is taken with the same fiber as the NSOM image. For further discussion we will denote the plot in Fig.7d as linear scan spectra (LSS) plot (image). Since we use accumulation time up to few seconds for each point of LSS image it can be considered as "static" image. The monochromatic NSOM image (see Fig.7b), taken at 20 ms per pixel, thus can be considered as a "dynamic" image.

In the topographic image (Fig.7a) the NW is seen as a vertical stripe having width ~100 nm. It has some distortions arising from the noise of the tuning-fork feedback control. We can estimate the width (W) and height (H) of the wire in the topography image to be W~100 and H~50 nm, respectively, and accounting for the fact that topographical height is equal to the wire diameter we can get value of the tip apex size to be $TA=W-H$~50 nm.

In the NSOM image (Fig.7b) the wire has a much larger width (~250 nm). It also reveals ~500 and ~100 nm intensity modulation along the wire. The fine scale modulation ~100 nm has the same scale as the topography. The nearly three times larger width of the NW in the NSOM image compared to the topography indicates an optical coupling between the wire and the tip before their physical contact, and reflects the "apertureless" nature of imaging using an uncoated tip. In such an apertureless regime the spatial resolution along the NW depends on the tip-wire distance and it is equal to the tip apex size (~50 nm for our tip) when the tip and the wire are in contact. Thus the 100 nm size modulation of the emission intensity seen in the contact part of the NSOM image corresponds to an intrinsic uniformity of the NW emission of ~50 nm.

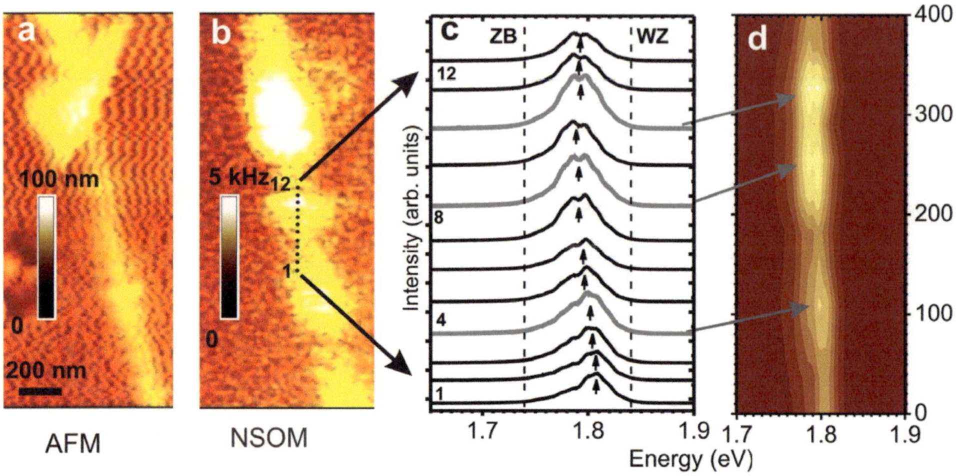

Fig. 7. Topography (a) and NSOM (b) images of NW1 at 50 K together with the set of twelve spectra taken during linear scan having length 400 nm (c) and their false-color wavelength-tip-position plot (d). Fiber tip positions for linear scan are marked by dots in (b). Image size in (a) and (b) is 0.8x2 μm^2. Detection energy in (b) is 1.8 eV

Comparison of Fig.7b and Fig.7c shows that variations of the intensity of the spectra follow the intensity variations in the monochromatic NSOM image. For example, the strong (weak) NSOM intensity in Fig.7b at points 9-11 (1-6) corresponds to strong (weak) spectra in Fig.7c taken at corresponding points. Some intensity variations seen in the spectra (like the intensity increase at point 8 and 4) are not observed in the NSOM image due to photon counting noise arising from the "dynamic" character of the image. Analysis of Fig.7c and d shows that intensity fluctuations are accompanied by changes of the spectral position and the width of the NW emission peak. For neighboring points separated by only 36 nm the changes of peak position and width by a few meV can be detected.

6. NSOM characterization and emission spectra of InP/GaInP structures

6.1 Room temperature NSOM imaging

Fig.8a-d shows the results of room temperature NSOM imaging of InP/GaInP QD structures having cap layer thickness d=0, 5, 20 and 60 nm. The images were taken at detection wavelength 750 nm. NSOM images of our structures (see inserts in Fig.8a and Fig.8b-d) show a set of bright spots having density ~20 μm^{-2} and size ~150-250 nm related to single QDs. Clearly-resolved single QD images are observed for 5 and 20 nm capped QDs as well as for the uncapped QDs. It is important to note the strong emission intensity for the uncapped QDs at room temperature. We believe that this is the first observation of such strong emission from uncapped QDs allowing ultra-high spatial-resolution. Accounting for the ~100 nm base of the QDs we estimated the tip apex size (spatial resolution) to be 25-75 nm. For a 60 nm cap the image contrast strongly decreases, which demonstrates the expected reduction of the spatial resolution (down to 150 nm) due to increased tip-dot separation. By positioning the tip on the bright spots we measured the spectra of individual InP/GaInP QDs at room temperature. Three such spectra for the uncapped sample presented in Fig.8a show that the spectra of single InP QDs at room temperature consists of a single band having wavelength in the range of 710-790 nm (1.75-1.56 eV) and halfwidth γ ~60 nm (~150 meV).

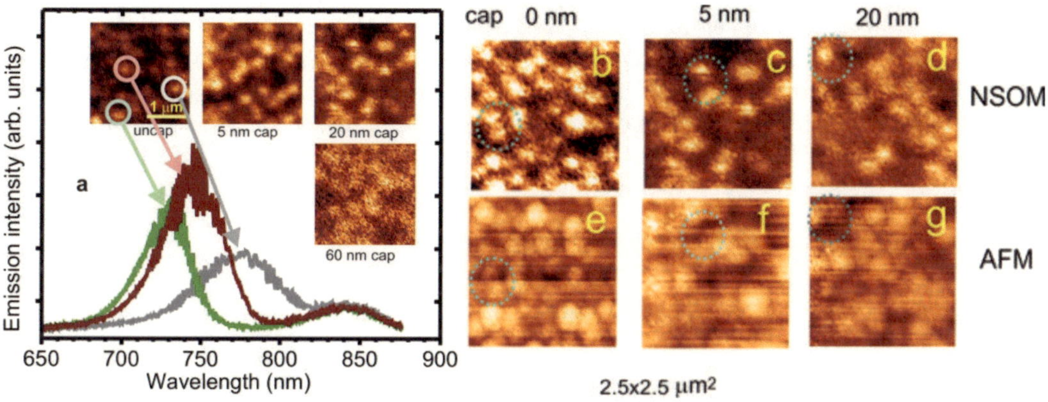

Fig. 8. Room temperature NSOM spectra (a) together with NSOM (b-d) and AFM (e-g) images (size 2.5x2.5 μm^2) of InP/GaInP QD structures having cap layer thickness (nm): 0 (b and e), 5 (c and f) and 20 (d and g). The ~500 nm diameter dashed circles in b-g outline the same area in the NSOM and AFM images. Inserts in 8a show NSOM images of the structures having cap layer thickness 0, 5, 20 and 60 nm.

In Fig.8b-g we present simultaneously measured NSOM (b, c and d) and AFM (e, f and g) images for d=0, 5 and 20 nm. The InP QDs having height 20-40 nm are clearly seen in the AFM image of the uncapped sample (Fig.8e). However the QD base observed is slightly (~50 nm) larger for the AFM image than in the NSOM image (see the encircled 1 dots in Fig.8b and e), which demonstrates the effect of electron localization. In the capped samples the QD location corresponds to the "valley" of the AFM images; this is seen by comparing the AFM and NSOM images (see the encircled dots for both the 5 and 20 nm cap) and the QD height decreases to 20-10 nm. Thus we observed the growth of the GaInP matrix material at the edges of the QDs. This together with the height reduction demonstrates additional growth control of the dot shape and thus confinement potential in InP/GaInP QDs.

We should also note that the use of the topographic images for the uncapped samples (see Fig.8b and e) allows the experimental determination of the size of the specific QD, which is one of the key parameters determining the electron correlation regime.

6.2 Low-temperature emission spectra

Fig.9 shows low-temperature (T=10 K) NSOM emission spectra of an InP/GaInP structure taken with spatial resolution ~200 nm at two fiber tip positions centred at two different QDs ($QD2_e$ and $QD3_m$) separated by 400 nm. (This notation will be clarified below). In the range

1.67-2.0 eV (740-620 nm) the spectra are dominated by the bands of the "central" dots, i.e. by $QD2_e$ in position 1 and by $QD3_m$ in position 2, having emission energy ~1.70 eV. Weaker bands of neighbouring dots located close to the tip edge are also seen at ~20 meV higher energy. The band shape of the large QDs reveals multipeak (manifold) structures, which will be discussed in detail below. The spectra also contain sharp lines observed in the range 1.85-1.95 eV, which are related to small InP QDs, and a broader band at 1.97 eV which is related to the GaInP matrix. The emission energy of our large InP QD is ~50 meV higher than the emission energy of the pyramidal InP QDs having base 60 nm, observed by other authors (Blome et al., 2000; Hessman et al., 2001), which indicates Ga/In intermixing. Using this energy difference we can estimate the value of the Ga composition of our InP QDs to be ~10%. Such intermixing resulting in significant increase of the dot size is favourable for the WM formation. Using low-temperature NSOM we measured the emission spectra of ~50 single InP QDs, allowing us to observe the effects of Wigner localization, which will be analysed below.

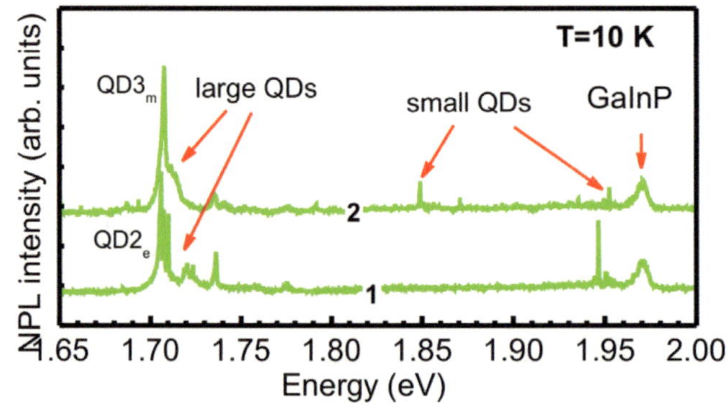

Fig. 9. 10K-NSOM spectra of InP/GaInP QD structures taken with spatial resolution 200 nm

7. Experimental study of Wigner molecules

7.1 Fermi liquid to Wigner molecule transition

Fig.10a shows an LSS plot, taken over a linear scan of 1.6 μm, for a 60 nm capped sample measured at T=10 K using a coated fiber having a 100 nm aperture. Four QDs denoted by $QD1_m$, $QD2_e$, $QD3_m$ and $QD4_e$ are observed in Fig.10a in the spectral range 1.72-1.68 eV. The spectra at tip positions 1 and 2 centered at $QD2_e$ and $QD3_m$ were shown in Fig.9 over a wider spectra range. Fig.10b-e compares the spectra of three large dots $QD5_m$, $QD1_i$, $QD2_e$ and a small lone dot $QD1_s$. The LSS image in Fig.10a demonstrates a drastic difference in the fine structure of the emission manifold of the dots having subscript m and e. This difference is also clear from the spectra of dots $QD5_m$ and $QD2_e$ in Fig.10c and e, respectively. In the m-type dots (further referred to as metallic) up to three components (s, p and d) of the emission manifold are observed and they have an energy splitting (ΔE) of ~4 meV and halfwidth (γ) of 3-5 meV. For the e-type dots (further referred to as excitonic) a fine structure of the s and p components consisting of several ultranarrow lines is clearly seen in Fig.10a, having γ<0.2 meV and ΔE as small as 1-2 meV. In Fig.10d we present the spectrum of a dot having a mixed structure consisting of few sharp lines and wider peaks. In contrast, for a small dot (Fig.10b) having ΔE >15 meV a single line related to a neutral exciton is observed (Sugisaki et al., 2002). The manifold structure was also observed for ΔE values as large as 10 meV (Blome et al., 2000; Hessman et al., 2001) for pyramidal InP/GaInP QDs having base 60 nm.

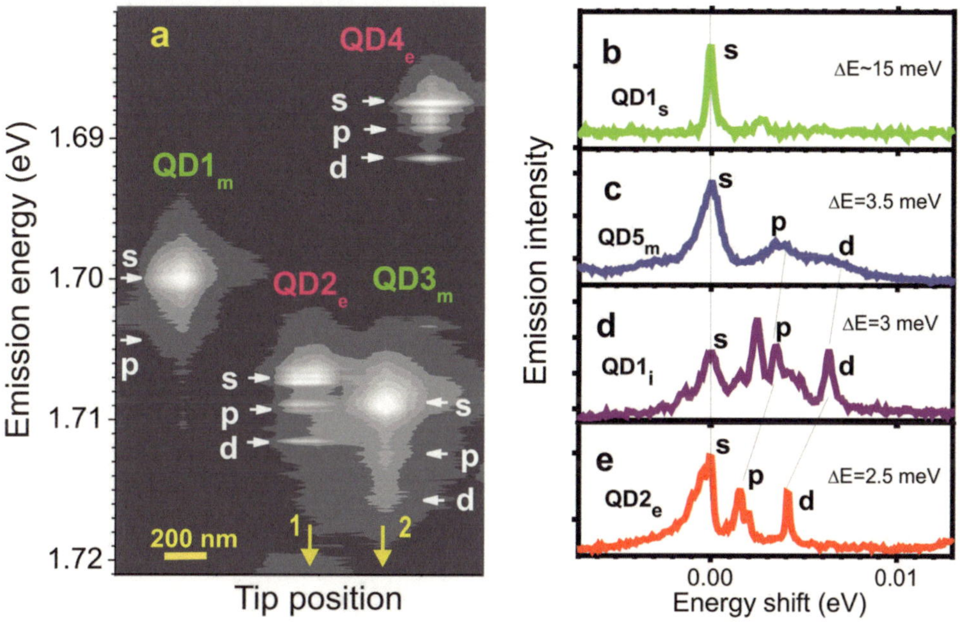

Fig. 10. Low-temperature (10K) NSOM LSS plot of InP/GaInP QD structure (a) and spectra of individual dots (b-e) The s-peak energy (in eV) in (b-e) is 1.9241 for $QD1_s$, 1.6994 for $QD5_m$, 1.7662 for $QD1_i$, and 1.7073 for $QD2_e$.

Our measurements of metallic QDs using a magnetic field discussed below have shown that the observed multi-peak structure of the emission spectra of InP/GaInP in Fig.10, c-e results from the filling of several electron shells. Here the spectral line shape is created by the radiative recombination of a photo-excited hole localized inside QD with electrons resulting from the "metallic character" of the dot; this is similar to the acceptor-related emission of a two-dimensional electron gas (2DEG) (Hawrylak, 1992). However in our QDs, i.e. a confined 2DEG, the spectral shape is determined by zero-dimensional confinement selection rules, and has maxima at the lowest s-state energy due to the s-state hole involved in the recombination process (see also the band diagram in Fig.5b), while in the 2DEG the spectral shape is determined by the density of the electronic states and is dominated by the high energy Fermi edge state (Hawrylak, 1992; Kukushkin et al., 1989).

The observed number of shells and the shell spacing allows us to estimate the number of electrons and the "effective" dot size, assuming a parabolic confinement potential (Jacak et al., 1998). The dots $QD5_m$ and $QD3_e$ in Fig.10c and e have spacing 3.5 and 2.5 meV, giving dot sizes ~90 and ~100 nm, respectively. They have three occupied s, p and d shells giving the possible number of electrons to be 7-12. Thus the electron density in these QD is ~5-10x10^{10} cm^{-2}, which is similar to the density used in 2DEG GaAs/AlGaAs structures. Using the values of the dot size and the number of electrons we estimate r_s values of the InP QDs to be 1.5-2.5. Our observation of the narrowing of shell peaks and their fine structure for excitonic dots indicates a formation of the excitonic/trion emission at a critical electron density at or below ~5x10^{10} cm^{-2}. Such a transition (from a broad Fermi sea emission to exciton and trion narrow lines) was observed in a 2DEG at higher densities (Finkelstein et al., 1995). Since the formation of the excitonic/trion emission in a many-electron system implies exciton/trion *localization* we can conclude that the multi-shell structure in these QDs reflects the formation of a Wigner molecule (WM). We believe this to be the first observation of a WM in a semiconductor QD.

7.2 Electron shell filling in the Fermi liquid regime

In Fig.11a and b we show the spectra of five metallic dots having varying ΔE and shell fillings. The QD7, QD8 and QD9 in Fig.11a have emission from three shells (~10 electrons) and show a progressive decrease of ΔE from 5.9 to 4.6 and to 3.5 meV, corresponding to sizes of the confining potential changing from ~65 to ~80 and to ~90 nm. On the other hand, QD10, QD7, and QD11 in Fig.11b show a progressive increase of the number of the shells from two for QD10 to three for QD7 and to four for QD11 (see spectra in a magnetic field below), thus demonstrating changes of the number of electrons from ~6 to ~20. The increase of electron numbers is accompanied by a decrease of ΔE, as expected.

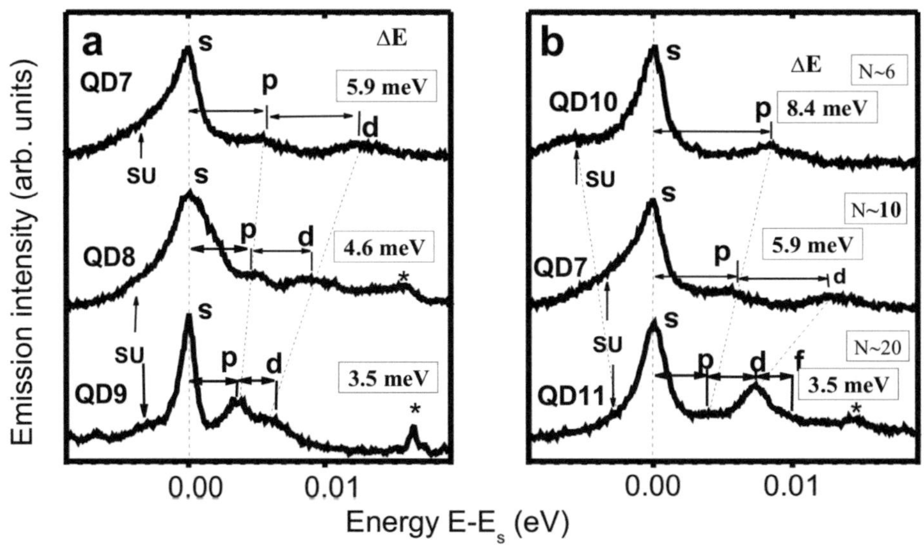

Fig. 11. Low-temperature (10K) NSOM spectra of five metallic single InP/GaInP QDs . The s-peak energy E_S is 1.7064, 1.6992, 1.6994 and 1.7041, 1.6981 eV for QD7, QD8, QD9 and QD10, QD11, respectively. Peaks * are contributions of neighboring QDs.

In addition to the shell peaks the spectra of single InP QDs display emission features related to so-called shake up or Auger processes denoted in Fig11a and b as SU. These features appear as low energy tails of the s-shell peak and for some dots they resolve into a separate band shifted by ~ΔE/2 (see QD9 and QD10). The scaling of the SU emission energy with ΔE is seen in Fig.11b. The SU emission occurs in many-electron systems when a recombining electron-hole pair excites surrounding electrons via the Coulomb interaction. In a 2DEG the SU emission appears at high magnetic fields via excitation of electrons into higher Landau levels and related magneto-plasmons (Butov et al., 1992; Hawrylak&Potemski, 1997; Nash et al., 1993). SU emission from the excited states was also observed in an InAs QD ensemble (Paskov et al., 2000). In the Wigner localization regime one can expect a vibronic structure of the SU emission, as was discussed in section 4.2.

To probe Wigner localization in the metallic dots we used a magnetic field which effectively increases r_s by squeezing the electron motion. With increase of the magnetic field the WM is formed above the molecular-droplet transition at magnetic field >4T ((Maksym et al., 2000; Reimann&Manninen, 2002; Szafran et al., 2003). Details of this phenomenon follow.

7.3 Emission spectra of metallic dots in a magnetic field

7.3.1 Optically induced intra-dot magnetic field

Fig. 12a and b presents circular polarized emission spectra of QD11 and QD8 measured at magnetic field B=0, 1, 2 ... 10T.

The s-shell band in Fig. 12a (QD11) at zero field is not polarized. With increasing magnetic field it becomes circularly polarized. The dominant emission of this band is σ^+-polarized for B=1-3 and 9-10T and σ^--polarized for B=5-7T. Zeeman splitting varies from +0.3 meV for 4T to -0.3 meV for 8T. The band shifts to higher energies with magnetic field.

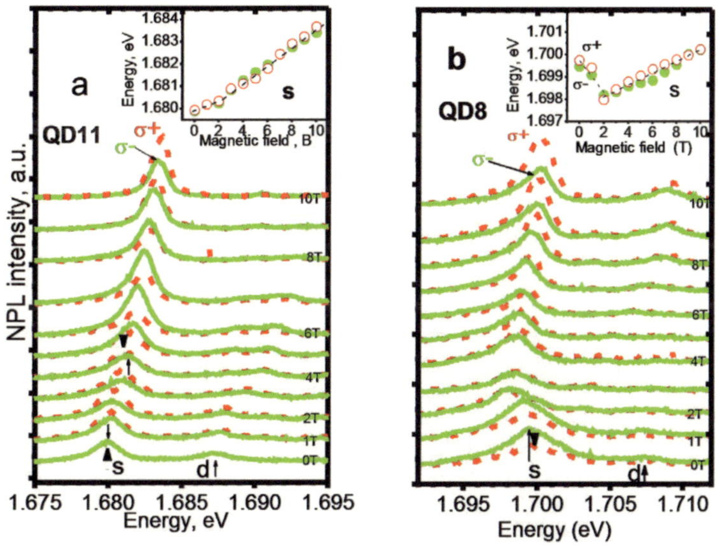

Fig. 12. Circularly polarized components (σ- - solid, σ+ - dotted) of emission spectra of QD11 and QD8 in magnetic field (0, 1, 2, ...10 T). Inserts show position of the energy of the s-peak versus magnetic field. Dashed lines are drawn for clarity.

The s-shell band in Fig. 12b (QD8) shows different magnetic field behavior. First, unlike QD11, it has strong circular polarization and Zeeman splitting at zero magnetic field. Both circular polarization and Zeeman splitting disappear at 2T. Second, the band has a strong *low energy shift* (slope~0.8 meV/T) for magnetic fields 0-2T. For higher fields it has a high energy shift with a slope of 0.25 meV/T. Our observations of the circular polarization at zero magnetic field and negative magnetic field shift at B=0-2 T for QD8 reflect a strong internal magnetic field. It arises from optical pumping of the nuclear spins (Brown et al., 1996; Maleinsky et al., 2007; Tratakovskii et al., 2007), inducing an internal magnetic field (B_{int}) of 2T which is anti-parallel to the external field. Such a field can induce the Wigner localization even at zero external magnetic field and thus using optical pumping gives additional control of WM formation.

7.3.2 Observation of the molecular-droplet transition

The measurements of the behavior of the multi-shell peaks in a magnetic field allow observation of a magnetic-field-induced phase transition of a WM using magneto-NSOM spectroscopy which is similar to that observed in Coulomb blockade measurements using nano-lithographically-defined GaInAs/AlGaAs QDs in a SET (Ashoori, 1996; Kastner, 1993; Tarucha, 1996). These measurements also confirm a shell type nature of an emission manifold of InP/GaInP QDs and estimate "valence" shell filling. In Fig.13a and b we show the unpolarized emission spectra versus magnetic field of QD8 and QD11 having ΔE=4.5 and 3.5 meV and N~10 and 20, respectively (see Fig.11a and b). In Fig.13c and d we compare the magnetic-field-induced shifts of the shell peaks for these QDs with the energy levels of the Fock-Darwin (FD) Hamiltonian. From Fig.13b and d one can see that for QD11 having larger size (i.e. smaller ΔE) than QD8, the f-shell peak is observed. Here the FD levels follow the experimentally observed shifts only for field up to 3T indicating filling of all f shell states, which gives electron number 19-20. At higher fields peak positions shift slightly to a lower energy and the d peak transforms to an x peak. This can indicate a transition of the electrons from the third (υ=2) to the second (υ=1) Landau levels. However, FD levels follow experiment only approximately and no distinct assignments of the shell peaks can be made at B>3T. We should point out that at 10T only a s-shell peak dominates the spectra and all other shells strongly suppressed.

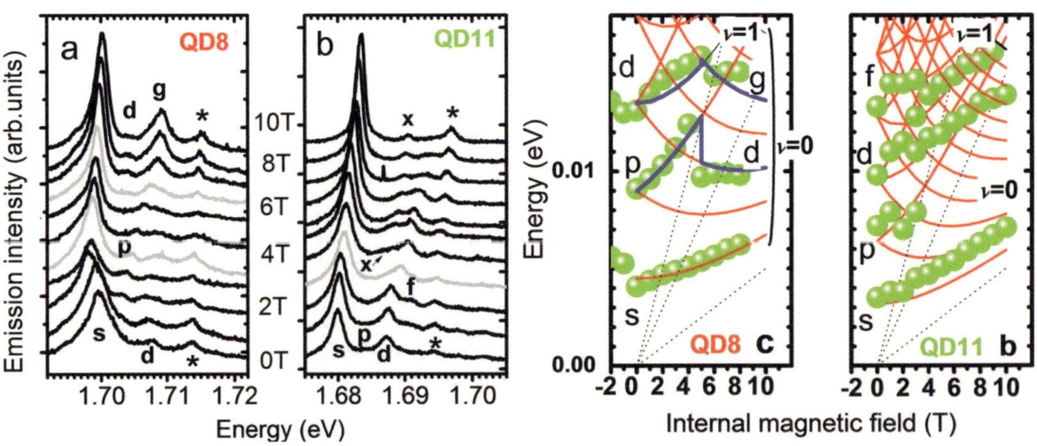

Fig. 13. Near-field spectra at 10 K representing shell structure (peaks s-g) of QD8 (a) and QD11 (b) at magnetic fields 0, 1,..,10T. Peaks * are contributions of neighboring QDs. Energy shift of emission peaks of QD8 (a) and QD11 (b) versus magnetic field (solid circles) are shown in (c) and (d), together with calculated Fock-Darwin energy levels (solid curves) and Landau levels υ=0 and 1 (dashed lines). In (c) and (d) the abscissa is the net *internal* magnetic field.

For QD8 (Fig.13a and c) the observed magnetic-field-induced shell shifts follow the FD energy levels remarkably well. Here two magnetic-field induced transitions are observed. The first transition occurs at B=2T arising from an internal magnetic field (B_{int}) of 0T which was discussed above (see Fig12.b). In the unpolarized spectra presented in Fig.13a, B_{int} is responsible for the shift to lower energy of the s- and d-shell peaks at B<2. For B>2 (B_{int}>0) the peak positions shift to increasing energy (see Fig.13c). In the range B=2-7 T (B_{int}=0-5 T) the p-shell peak appears. At B>7 T (second transition; B_{int}=5 T) the p-shell peak transforms into a d-shell, and d into g-shell. As can be seen from Fig.13c the second transition (at 7T) corresponds to transition of electrons from the second (υ=1, p/d-shells) to the first Landau level (υ=0, d/g-shells). Such a transition was previously observed in the magnetic field dependence of Coulomb blockade and can be described as a molecular-droplet transition (Oosterkamp et al., 1999). It arises from formation of a maximum-density-droplet at B=2, and its decomposition is accompanied by a WM formation at B>4 T (Maksym et al., 2000; Reimann& Manninen, 2002; Szafran et al., 2003). In the spectra the WM formation at a high magnetic field corresponds to a strong increase of the intensity of the g-shell peak (in contrast to QD11).

7.4 High-spatial resolution NSOM imaging

We used the uncapped structure to perform ultra-high-spatial resolution imaging and spectroscopy of InP/GaInP QDs using uncoated fiber probes working in an apertureless regime. In Fig.14a and b we present monochromatic 5K-NSOM images (detection energy 1.713 eV) taken for the same area (size 500x500 nm^2) in two separate scans. Fig.14c and d shows an LSS plot for the dashed line along the dot center shown in Fig.14a, and seven selected spectra from this scan (denoted 1-7), respectively. Fig.14e shows classical positions of the electrons and the trion for nine (cartoons I and II) and ten (cartoon III) electron WMs. In the NSOM image in Fig.14a and b in addition to the bright resonant QD denoted Qa (emission energy 1.713 eV), two weaker non-resonant QDs denoted Qb and Qc (emission energy 1.701 and 1.695 eV, respectively) appear in the images. Resonant Qa dot has ΔE=5.5 meV and three shells are filled, as can be seen from the spectra in Fig.14d. The image of Qa has size ~120 nm and it reveals a strong (up to 50%) intensity fluctuation on a length scale ~30 nm, which is of the order of the single electron separation in a WM in the classical limit. We determined, using LSS in Fig.14c and d, that the photon counting detection noise only partly contributes to the spatial intensity fluctuation in Fig.14a and b (note that photons are counted only for 20 ms in monochromatic imaging). In the image in Fig.14c plotted from spectra measured by a CCD with accumulation time 1 s we observed similar spatial fluctuations of the emission intensity. From Fig.14c and d it is seen that intensity fluctuations are accompanied by changes of the SU-part of the emission spectra and by the spectral diffusion of a few meV. Spectral diffusion indicates the effects of rearrangement of the charge distribution inside and/or around the QDs under near- field excitation. It is seen from Fig.14c and d that for positions 1, 2, 4 and 7 the SU-emission consists of two peaks SU1 and SU2, having energy shift from the s-peak of ~3 and 5 meV, respectively. For positions 3 and 6 the SU1 peak reveals a splitting of ~2 meV, and for position 5 a SU0 peak having energy shift ~2 meV appears. For position 5 the intensity of the SU0 emission peak is very strong: it is nearly the same intensity as the intensity of the s-peak. Using the analysis of the "electron-phonon" interaction discussed in section 4.2 and WM excited states discussed in section 3 we suppose that the SU1 and SU2 peaks can be related to the translational and breathing vibrational modes and that the splitting of the SU1 peak is related to a other normal mode of a 9e-WM. We assign the SU0 peak to isomeric excitation of a 10e-WM. The appearance of the different (from translational and breathing) modes for positions 3 and 6 can be explained by photoexcitation at the edge of the dot, in which recombination of an "edge" trion generates the WM rotational motion (cartoon II in Fig.14e). For photoexcitation at the center of the dot (position 4), no rotations are generated by symmetry (see cartoon I in Fig.14e). Positions 1 and 7 are non-contact positions for which the photexcitation can be considered uniform. We can expect that for such excitation the emission is generated by the trion bound to the central electron. We also suppose that the enhanced coupling to the vibrational mode arises in a 10e-WM due to the formation of a two-electron isomeric arrangement at the center see Fig.14e, cartoon III). We obtained further evidence of vibronic structure of the SU-emission from analysis of spectra of the excitonic dots.

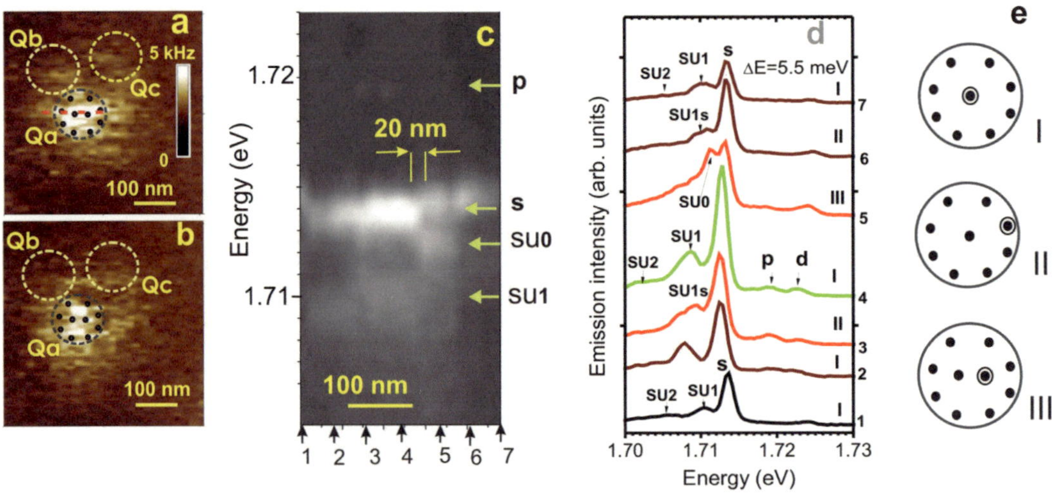

Fig. 14. High-spatial resolution NSOM measurements of InP QDs at 5 K: (a and b) monochromatic NSOM image, (c) false-color LSS-plot for a scan along the QD (see dashed horizontal line in a); (d) seven selected spectra from LSS (see vertical arrows at the bottom of c) from the linear scan; (e) classical arrangements of electrons and trion in photoexcited state of nine (upper and central) and ten electron WMs. In (a) and (b) Qa, Qb and Qc dashed large circles outline different QDs, while the small circles in dot Qa represent possible WM electron positions.

7.5 Emission spectra of Wigner molecules

Fig. 15a-d show spectra of four excitonic dots ($QD2_e$, $QD3_e$ $QD4_e$ and $QD5_e$) having different intensity distributions of the shell and SU peaks. The ΔE for these dots have values between 1 and 1.9 meV. Three shell peaks are observed for $QD2_e$, $QD3_e$, and $QD4_e$ and two for $QD5_e$. The p-shell peaks for $QD2_e$ and $QD4_e$ reveal two components having splitting 0.4-0.5 meV whereas no components are observed for $QD3_e$ and $QD5_e$. The p- and d-shell peaks are strong for $QD2_e$, $QD4_e$ and $QD5_e$ and weak for $QD3_e$. The intensity of the SU emission is strong for $QD2_e$, $QD3_e$ and $QD5_e$ and very weak for $QD4_e$. The energy shift of the SU peaks in units of ΔE is equal to 0.2, 0.4 and 1.5 for $QD2_e$, $QD3_e$ and $QD5_e$. The right inserts in Fig. 15 shows our suggestions for the WM structure, which could explain the observed differences in the spectra.

We suppose that $QD2_e$, $QD3_e$, $QD4_e$, and $QD5_e$ have ten, ten, nine, and two electrons, respectively. Three of the electrons in $QD2_e$ are centrally located, whereas only two are centrally located in $QD3_e$. For $QD2_e$ the optical transition interacts equally with all three electrons, giving the high intensity of the p- and d-shells. The SU structure arises from the interaction with the other WM arrangement ($QD3_e$) having two central electrons, and the energy 0.2ΔE is the splitting between these arrangements, which is the energy difference of the isomeric excitations. The intensity distribution for $QD3_e$ is similar to the metallic dot Qa in Fig.14 (see also Fig16b below).

The $QD4_e$ dot must have electron number between 7 and 9 since the d shell is populated and the existence of the central electron suppresses the SU emission. The SU structure of $QD5_e$ allows us to assume that this is a two electron QD. This is supported from our analysis of the SU peak structure presented in Fig.16a.

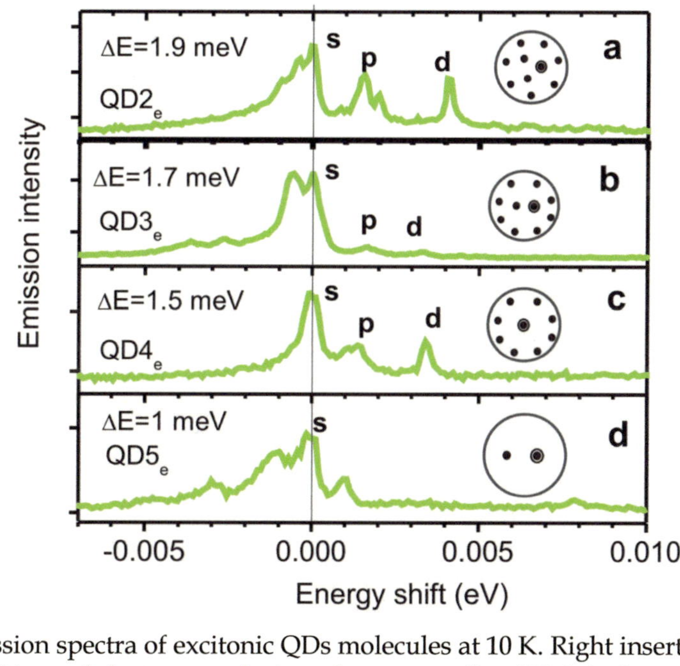

Fig. 15. Emission spectra of excitonic QDs molecules at 10 K. Right inserts show possible classical positions of electrons and trion of corresponding WMs in the excited state. E_s energy (in eV) is 1.7073 for QD_{2e}, 1.766 for $QD3_e$, 1.687 for $QD4_e$ and 1.7438 for $QD5_e$

7.5.1 Vibronic structure of a two-electron Wigner molecule

We use six Gaussian peaks to model the SU structure of this $QD5_e$ (see Fig.16a). Our analysis has shown that the energy of these peaks can be grouped in two sequences – one is n*ΔE and the other is n*1.5ΔE, or $n\omega_0$ and $n*1.5\omega_0=n*\omega_S$, where n is an integer. These energies are naturally assigned to bending and stretching vibrations of a 2e-WM (see Eq. 1), which according to the theoretical calculations have values ω_0 and $1.7\omega_0$. The smaller value of the frequency of the stretching vibration observed experimentally ($1.5\omega_0$ instead of $1.7\omega_0$) can result from deviations of the real confinement potential from the ideal 2D potential. The anti-Stokes peak ω_0 arises from the thermal activation and its relative intensity is well described by the thermal factor $\exp(-\Delta E/kT)$, which for T=10K is equal to 0.3. From the intensities of the peaks we can estimate Huang-Rhys factors for bending and stretching vibrations to be ~0.2 and ~0.1.

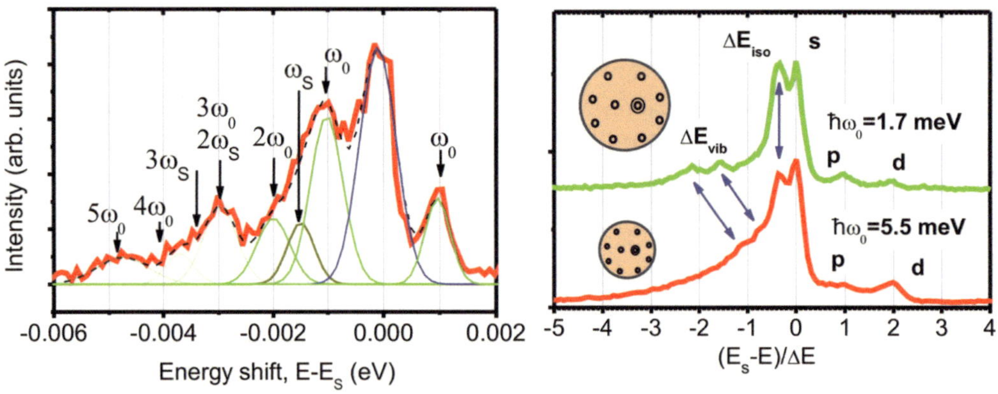

Fig. 16. Gaussian contour decomposition of emission spectra of $QD5_e$ (a) and comparison of emission spectra of two QDs having shell splitting ΔE=1.7 and 5.5 meV (b). The horizontal axis in b has reduced units $(E_s\text{-}E)/\Delta E$.

7.5.2 Size-dependent structure

Fig. 16b shows the comparison of the emission spectra of the metallic and the excitonic dots having similar intensity distributions of the SU peaks. These are the Qa dot (lower spectrum in Fig. 16b), the spectrum of which is also shown in Fig.14d for tip position 5, and the QD3e dot (upper spectrum), the spectrum of which is shown in Fig.15b. The shell splitting of these two dots is 1.7 and 5.5 meV, which corresponds to sizes 120 and ~70 nm, respectively. We suppose that these two dots have the same electron occupation equal to ten, which corresponds to r_s~2 and ~1.2. The horizontal axis in Fig.16b is in reduced units $n=(E_S\text{-}E)/\Delta E$ or $n=(E_s\text{-}E)/\hbar\omega_0$ and from Fig.16b a very good coincidence of the spectral shape is seen. This coincidence is remarkable, accounting for a nearly three times difference in the absolute spectral range, which is equal 15.3 and 45.5 meV for QD3e and Qa, respectively, for n values from -5 to 4 in Fig.16b. These dots have a strong SU peak having the same intensity as the intensity of the "zero-phonon" line, i.e. the s-peak. The weaker structure also observed near n=2 for QD3e and n=1 for Qa. The SU peak can be assigned to the isomeric excitation between the ground isomer having [2, 8] classical electron geometry and excited isomer having [3,7] geometry. This follows from the fact that this is the only excitation which linearly scales with r_s (see Fig.2). The weaker structure does not scale with r_s and it shows shrinkage of the energy. Such shrinkage is expected for a transition from partial Wigner localization to Fermi liquid discussed in section 3.1 for 2e-WM. For the excitonic dot the energy of the weak structure is $\sim 2\hbar\omega_0$, which fits well to the high frequency vibrational modes. These modes include the breathing mode, having frequency $1.7\omega_0$ (see Fig.3). Reduction of this energy down to $\hbar\omega_0$ for a metallic dot reflects the loosing of the molecular character of the electron distribution for $r_s \sim 1.2$ and failure of the classical approximation. In this Fermi liquid regime the electron motion is described only by translation vibration having frequency ω_0.

We should also point out the stronger contribution of p- and d-shell peaks in the smaller Qa dot, which reflect stronger overlap of hole and electrons wave functions.

8. Conclusions

We presented the results of an experimental (photoluminescence) study of correlated states of electrons in a WM in self-organized InP/GaInP quantum dots. The unique properties of these QDs are their relatively large lateral size (~80-200 nm) and their ability to accommodate up to 20 electrons, providing electron density up to 2×10^{11} cm^{-2} and strong emission intensity. Using high-spatial-resolution low-temperature near-field scanning optical microscopy (NSOM) having spatial resolution up to 30 nm in combination with a high magnetic field, we were able to resolve emission spectra of single QDs, and to observe crossover from a Fermi liquid to WM behavior at a critical density of 5×10^{10} cm^{-2}. A magnetic-field-induced molecular-droplet transition has been observed in the Fermi liquid regime. In the Wigner molecule regime we observed a rich vibrational structure, which opens the way to identify electron arrangement in WMs. Further experiments are in progress.

9. Acknowledgment

Development of the NSOM technique was supported by a National Science Foundation NIRT grant No. ECS-0609249. The authors wish to thank Nikolay A. Kalugnyy, Vladimir. M. Lantratov, Sergei A. Mintairov and Dmitrii A. Vinokurov from the Ioffe Physical-Technical Institute (St. Petersburg, Russia) for growth InP/GaInP structures. The authors wish to thank Alexei Vlasov and Tatiana Prutskih for making fiber tips and for help in NSOM measurements.

10. References

Andrei, E.Y.; Deville, G.; Williams, F.; Paris, I.B. & Etienne, B. (1988). Observation of Magnetically Induced Wigner Solids, *Physical Review Letters*,Vol.60, No.26 ,(27 June), pp.2765-2768

Ashoori, R.C. (1996). Electrons in artificial atoms, *Nature*, Vol.379, (1 February), pp.413-419

Betzig, E. & Trautman, J. K. (1992). Near-Field Optics: Microscopy, Spectroscopy, and Surface Modification Beyond Diffraction Limit, *Science* Vol.257, (10 July), pp. 189-195

Blome, P. G.; Wenderoth, M.; Hubaner, M. & Ulbrich, R. G. (2000). Temperature-dependent linewidth of single InP/$Ga_xIn_{1-x}P$ quantum dots: Interaction with surrounding charge configuration, *Physical Review B*, Vol.61, No.12, (15 March) pp.8382-8387

Blundell, S. A. & Chacko, S. (2011). Excited states of incipient Wigner molecules *Physical Review B,* Vol.83, No.19, (31 May) pp.195444-11

Blundell, S. A. & Joshi, K. (2010). Precise correlation energies in small parabolic quantum dots from configuration interaction, *Physical Review B,* Vol.81, No.11, pp.115232-1-11

Bolton, F. & Rössler, U. (1993). Classical model of a Wigner crystal in a quantum dot, *Superlattices and Miucrostructures,* Vol.13, No.2 (March) pp.139-145

Brown, S. W.; Kennedy, T. A.; Gammon, D. & Snow, E. S. (1996). Spectrally resolved Overhauser shifts in single GaAs/$Al_xGa_{1-x}As$ quantum dots, *Physical Review B, Vol.*54, No.24 (15 December) pp.R17339-R17342

Butov, L. V.; Grinev, V. I. ; Kulakovskii, V. D. & T. G. Anderson (1992) Direct observation of magnetoplasmon-phonon coupled modes in the magnetophotoluminescence spectra of the two-dimensional electron gas in $In_xGa_{1-x}As$/GaAs quantum wells, *Physical Review B,* Vol.46, No.20 (15 November) pp. 13627-13630

Chu, Y.; Mintairov, A. M. ; He, Y.; Merz, J. L.; Kalyuzhnyy, N. A.; Lantratov, V. M.& Mintairov S. A. (2009). Lasing of whispering-gallery modes in asymmetric waveguide GaInP micro-disks with InP quantum dots, *Physics Letters A,* Vol.373, 1185-1188

Coutjon, D.; Sarayeddine, K.& Spajer, M. (1989). Scanning tunneling optical microscopy, *Optics Communication,* Vol.71, No.1,2 (1 May) pp.23-28

Egger, R.; Häusler, W.; Mak, C. H. & Grabert, H. (1999). Cossover from Fermi Liquid to Wigner Molecule Behavior in Quantum Dots *Physical Review Letters* 82 3320

Ellenberger, C. Ihn, T. Yannouleas, C Landman, U. Ensslin, K. Driscoll, D. & Gossard A. C. (2006) Excitation Spectrum of Two Correlated Electrons in a Lateral Quantum Dot with Negligible Zeeman Splitting, *Physical Review Letters* 96, 126806

Finkelstein, G.; Shtrikman, H. & Bar-Joseph, I. (1995). Optical Spectroscopy of Two-Dimensional Electron Gas near the Metal-Insulator Transition, *Physical Review Letters* Vol.74, No.6 (6 February) pp.976-979

Fischer, U. C. & Pohl, D. W. (1989). Observation of Single-Particle Plasmons by Near-Field Optical Microscopy, *Physical Review Letters,* Vol.62, No.4 (23 January), pp.458-461

Flack, F. ; Samarth, N.; Nikitin, V.; Crowell, P. A.; Shi, J.; Levy, J.& Awschalom, D. D. (1996). Near-field optical spectroscopy of localized excitons in strained CdSe quantum dots, *Physical Review B,*Vol.54, No.24 (15 december) pp.R17312-R17315

Georgsson, K.; Carlsson, N.; Samuelson, L.; Seifert, W.& Wallenberg L. R. (1995) Transmission electron microscopy of the morphology of InP Stranski-Krastanov islands grown by metalorganic chemical deposition *Appl. Phys. Lett.* Vol.67, No.20 (13 November) pp. 2981-2982

Govorov, A. O.; Schulhauser, C.; Haft, D.; Kalameitsev, A. V.; Chaplik, A.; Warburton, R. J.; Karrai, K.; Schoenfeld, W.; Garcia, J. M.& Petroff P. M. (2002). Magneto-optical properties of charged excitons in quantum dots, (26 February, 2002) Available from arXiv:cond-mat/0202480 v1

Ghosal, A.;. G¨uc¸l¨u, A. D.; Umrigar, C. J.; Ullmo, D. & Baranger H. U. (2006). Correlation-induced inhomogeneity in circular quantumdots, *Naure Physics.* Vol.2, No.5 (23 April) pp. 336-340

Grimes, C.C.& Adams, G. (1979) Evidence of Liquid-to-crystal Phase Transition in a Classical Two-Dimensional Sheet of Electrons *Physical Review Letters Vol.*42, No.12, (19 March) pp.795-798

Hawrylak, P. (1992). Many electron effects in acceptor-related radiative recombination of quasi-two-dimensional electrons, *Physical Review B,* Vol.45, No.8 (15 February), pp. 4237-42400

Hawrylak, P. & Potemski, M. (1997). Theory of photoluminescence from an interacting two-dimensional electron gas in strong magnetic fields, *Physical Review B* , Vol.65, No.19 (15 November) pp.12386-12384

Hirose, K. & Wingreen, N. S. (1999). Spin-density-functional theory of circular and elliptical quantum dots, *Physical Review B,* Vol.59, No. 7, (15 February) pp. 4604

Jacak, L.; Hawrylak, P. & Wojs, A. (1998)Quantum dots (Springer, Berlin, p. 176)

Hessman, D.; Persson, J.; Pistol, M-E.; Pryor, C.& Samuelson, L. (2001). Electron accumulation in single InP quantum dots observed by photoluminescence *Physical Review B* Vol.64, pp. 233308-1-11

Hosaka, N. & Saiki, T. (2001) Near-field fluorescence imaging of single molecules with spatial resolution in a range of 10 nm, *Journal of Microscopy,* Vol.202 No.2 (May), pp.362-364

Huang, K.& Rhys, A. (1950). Theory of light absorption and non-radiative transitions in F-centers, *Proceedings of the Royal Society of London. Series A, Mathematical and Physical Sciences,* Vol.204, No. 1078 (22 December), pp. 406-423

Ilani, S.; Martin, J.; Teitelbaum, E.; Smet, J.H.; Mahalu, D.; Umansky, V. & Yacoby, A. (2004). The microscopic nature of localization in the quantum Hall effect, *Nature,* Vol.427, (22 December) pp. 328-332

Janssens, K. L.; Partoens, B. & Peeters F. M. (2003). Effect of strain on the magnetoexciton ground state in InP/GaxIn1ÀxP quantum disks *Phisical Review B* Vol.67, pp.235325-8

Kalliakos, S. Rontani, M. Pellegrini, V. GARC´IA, C. P. Pinczuk, A. Goldoni, G. Molinari, E. N. Pfeiffer, L. West K. W. (2008) A molecular state of correlated electrons in a quantum dot *Nature Physics* 4 469

Kastner, M.A. (1993). Artificial atoms, *Physics Today,* (January) pp.24-31

Kukushkin, I. V.; Klitzing, K. von; Ploog, K. & V. B. Timofeev (1989). Radiative recombination of two-dimensional electrons in acceptor δ-doped GaAs-$Al_xGa_{1-x}As$ single heterojunctions, *Physical Review B,* Vol.40, No.11, (15 October) pp.7788-7792

Laughlin, R.B. (1983). Quantized motion of three two-dimensional electrons in a strong magnetic field, *Physical Review,* Vol.27, No.6, (15 March) pp. 3383-3389.

Lewis, A.; Isaacson, M.; Harootunian, A. & Muray, A. (1984). Development of a 500 Å spatial resolution light microscope: I. light is efficiently transmitted through λ/16 diameter apertures, *Ultramicroscopy,* Vol. 13, pp.227-232

Maksym, P.A.; Imamura, H.; Mallon, G.P. & Aoki, H. (2000). Molecular aspects of electron correlation in quantum dots, *Journal of Physics: Condensed. Matter,* Vol.12, R299-334

Maksym, P. A. (1996). Eckardt frame theory of interacting electrons in quantum dots, *Physical Review B,* Vol. 53, No. 16 (15 April) pp.10871-10886

Maleinsky P, Lai C W, Bodalato A, and Imamoglu, A. (2007). Nonlinear dynamics of quantum dot nuclear spins, *Physical Review B,* Vol.75, 035409-7

Matsuda, K.; Matsumoto, T.; Saito, H.; Nishi, K.& Saiki, T. (2000). Direct observation of variations of optical properties in single quantum dots by using time-resolved near-field scanning optical microscope *Physica E,* Vol.7, No.3-4, (May), pp.377-382

Matsuda, K.; Saiki, T.; Nomura S., Mihara, M.; Aoyagi, Y.; Nair S. & Takagahara, T. (2003) "Near-Field Optical Mapping of Exciton Wave Functions in a GaAs Quantum Dot" *Phys. Rev. Lett.* Vol.91, No.17 (24 october), pp.177401-4

Mintairov, A. M.; Kosel, T. H.; Merz, J. L.; Blagnov, P. A.; Vlasov, A. S.; Ustinov, V. M. &Cook, R. E. (2001). Near-Field Magnetophotoluminescence Spectroscopy of Composition Fluctuations in InGaAsN, *Physical Review Letters,* Vol.87, No.27 (31 December), pp.277401-14

Mintairov, A. M.; Sun, K.; Merz, J. L.; Li, C.; Vlasov, A. S.; Vinokurov, D. A.; Kovalenkov, O. V.;Tokranov, V. & Oktyabrsky, S. (2004). Nanoindentation and near-field spectroscopy of single semiconductor quantum dots, *Physical Review B,* Vol.69, No.15 (15 April), pp.155306-12

Mintairov, A. M.; Herzog, J.; Kuno, M. & Merz, J. L. (2010). Near-field scanning optical microscopy of colloidal CdSe nanowires, *Phys. Status Solidi B* Vol.247, No. 6, pp.1416–1419

Nash, K. J.; Scolnick, M. S.; Saker, M. K. & Bass, S. J. (1993). Many Body Shakeup in Quantum Well Luminescence Spectra, *Phys. Rev. Lett.* Vol.70, No.20 (17 May), pp.3115-3118

Oosterkamp, T. H.; Janssen, J. W.; Kouwenhoven, L. P.; Austing, D. G.; Honda, T.& Tarucha, S. (1999). Maximum-Density Droplet and Charge Redistributions in Quantum Dots at High Magnetic Fields, *Physical Review Letters,* Vol. 82, No.14 (5 April) pp.2931-2934

Ortner, G.; Bayer, M.; Larionov, A.; Timofeev, V. B.; Forchel, A.; Lyanda-Geller, Y. B.; Reinecke, T. L.; Hawrylak, P.; Fafard, S. & Wasilewski, Z. (2003). Fine Structure of Excitons in InAs/GaAs Coupled Quantum Dots: A Sensitive Test of Electronic Coupling, *Physical Review Letters,* Vol.90, No.8, (28 February) pp. 086404-4

Otsu, M. (1998) Near-Field Nano/Atom Optics and Technology, *Springer-Verlag, Tokyo.*

Paskov, P.P.; Holtz, P.O.; Wongmanerod, S.; Monemar, B.; Garcia, J.M.; Schoenfeld, W.V.& Petroff, P.M. (2000). Auger processes in InAs self-assembled quantum dots, *Physica E,* Vol.6, 440-443

Pohl, D. W.; Denk, W. & Lanz, M. (1984). Optical stethoscpy: Image recording with resolution $\lambda/20$, *Applied Physics Letters,* Vol.44, No.&, (1 April), pp.651-653

Pryor, C.; Pistol, M-E. & Samuelson, L. (1997) Elecronic structure of strained InP/$Ga_{0.51}In_{0.49}P$ quantun dots. *Physical Review B,* Vol.56, No.16, (October 15) pp. 10404-10411

Pudalov, V.M.; D'Iotio, M.; Kravchenko, S.V.& Campbell J.W. (1993) Zero-Magnetic-Field Collective Insulator Phase in a Dilute 2D Electron System, *Physical Review Letters,* Vol.70, No.12, (22 March) pp. 1866-1869

Reddick, R. C.; Warmack, R. J. & Ferrell, T. L. (1989). New form of scanning optical microscopy, *Physical Review B,* Vol.39, No.1, (1 January), pp. 767-770

Reimann, S. M.; Koskinen, M. &. Manninen, M. (2000). Formation of Wigner molecules in small quantum dots, *Physical Review B* Vol.62, No.12 (15 September) pp.8108-8113

Reimann, S.M. & Manninen, M. (2002). Electronic structure of quantum dots, *Review of Modern Physics*, Vol.74, No.4, (October), pp.1283-1342

Ren, H.-W.; Sugizaki, M.; Lee, J.-S.; Sugou, S.&Masumoto, Y.,Highly Uniform and Small InP/GaInP Self-Assembled Quantum Dots Grown by Metal-Organic Vapor Phase Epitaxy (1999) *Japanese Journal of Applied Physics* Vol.38, No.1B, Part 1 (January) pp. 507-510

Saint Jean, M; Even, C. & Guthmann, C. (2001). Macroscopic 2D Wigner islands, *Europhysics Letters,* Vol.55, No.1 (1 July), pp. 45–51

Singha A, V. Pellegrini, A. Pinczuk, L. N. Pfeiffer, K. W. West, and M. Rontani, (2010)Correlated Electrons in Optically Tunable Quantum Dots: Building an Electron Dimer Molecule *Phys. Rev. Lett.* 104, 246802

Sugizaki, M.; Ren, H.-R.; Nair, S. V.; Nishi, K&Masumoto, Y. (2002) External-field effects on the optical spectra of self-assembled InP quantum dots. *Physical Review B,*Vol.66, pp.235309-1-10

Synge, E. H. (1928). A suggested method for extending microscopic resolution into the ultra-microscopic region, *Phyloofical Maazine Series 7,* Vol.6, No.35, pp.356-362

Szabo, A. & Ostlund, N. (1996) Modern Quantum Chemistry, *Dover, New York*

Szafran, B.; Bednarek, S. & Adamowski, J. (2003). Magnetic-field-induced phase transitions in Wigner molecules, *Journal of Physics: Condensed Matter*, Vol.15, pp.4189-4205.

Tratakovskii, A. I.; Wright, T.; Russell, A.; Fal'ko, V. I.; Van'kov, A. B.; Skiba-Szymanska, J.; Drouzas, I.; Kolodka, R. S.; Skolnik, M. S.; Fry, P. W.; Taharaoui, A.; Lui, H-Y. & Hopkinson, M. (2007). Nuclear Spin Switch in Semiconductor Quantum Dots, *Physical Review Letters,* Vol.98, No.2 (12 January) pp. 026806-4

Tarucha, S.; Austing, D.G.; Honda, T.; Van der Hage, R.J.; Kouwenhoven, L.P. (1996). Shell Filling and Spin Effects in a Few Electron Quantum Dots, *Physical Review Letters,* Vol.77, No.17 (21 October), pp.3613-3616

Toda, Y.; Kourogi, M.; Otsu, M.; Nagamune, Y.& Arakawa Y. (1996). Spatially and spectrally resolved imaging of GaAs quantum-dot structures using near-field optical technique, *Applied Physics Letters* Vol.69, No.6, (5 August) pp.827-829

Toda, Y.; Shinomori, S.; Suzuki, K. & Arakawa, Y. (1998). Polarized photoluminescence spectroscopy of single self-assembled InAs quantum dots, *Physical Review B*, Vol.58, No.16 (15 October) pp.R10147-10150

Toda, Y.; Sugimoto, T.; Nishioka, M. & Arakawa, Y. (2000). Near-field coherent excitation spectroscopy of InGaAs/GaAs self-assembled quantum dots, *Applied Physics Letters* Vol.76, No.26 (26 June) pp.3887-3889

Toda, Y.; Moriwaki, O.; Nishioka, M. & Arakawa, Y. (1999). Efficient Carrier Relaxation Mechanism in InGaAs/GaAs Self-Assembled Quantum Dots Based on the Existence of Continuum States, *Phys. Rev. Lett.* Vol.82, No.20 (17 May) pp.4114-4117

Vinokurov, D. A.; Kapitonov, V. A.; Nikolaev, D. N.; Sokolova, Z. N. & Tarasov I. S. (1999). Self-organized nanoscale InP islands in an InGaP/GaAs host and InAs islands in an InGaAs/InP, *Semiconductors*, Vol.35, No.7, (July) pp.788-791

Yannouleas, C. & Landman, U. (2000). Collective and Independent-Particle Motion in Two-Electron Artificial Atoms *Physical Review Letters*, Vol.85, No.8, (21 August) pp.1726-1729

Wigner, E.P. (1934) On the Interaction of Electrons in Metals, *Physical Review*, Vol.46, (1 December), pp.1002-1011.

6

InAs Quantum Dots in Symmetric InGaAs/GaAs Quantum Wells

Tetyana V. Torchynska
ESFM-National Polytechnic Institute, Mexico D.F., Mexico

1. Introduction

The self-assembled InAs quantum dots (QDs) are the subject of substantial interest during last fifteen years due to both fundamental scientific and application reasons. In these systems, the strong localization of an electronic wave function leads to an atomic-like electronic density of states and permits to realize the novel and improved photonic and electronic devices. Microlectronic and optoelectronic devices based on quantum wells (QWs) with InAs QDs have been the subject of investigation for the applications in semiconductor lasers for the optical fiber communication [1-3], infrared photo-detectors [4-6], electronic memory devices [7,8], as well as single electron devices and single photon light sources on the base of single-QD structures for the quantum information applications [9-12]. QDs are especially attractive for the applications in semiconductor lasers. For laser or photodiode applications the surface density of QDs has to be high, but for single-QD devices the QD density has to be low. As a result, there is an extensive effort to manipulate and control the position, size, shape and density of QDs [13-19].

Self-assembled InAs/GaAs QDs in lattice-mismatched systems can be achieved by using the Stranski–Krastanow (S–K) growth mode [20-22]. In the process of InAs/GaAs QDs growth using the S–K growth mode, the InAs mismatched layer growths two dimensionally on the GaAs substrate during the initial stage; then, above a critical thickness, the strain increased and the dislocation-free QDs with a three-dimensional shape are formed on a residual two-dimensional wetting layer (WL) [23,24]. InAs/GaAs QD structures grown using the S–K mode have inherent several problems: the density random distribution, the large temperature-dependent variation of the photoluminescence (PL) intensity and line width resulting from the nonuniform size and density of InAs QDs [25] etc. Precise control of the QD size and the homogeneity of InAs QD distribution is necessary for achieving the high-performance devices.

2. Advances of InAs QDs in InGaAs/GaAs quantum wells

InAs/GaAs QDs are especially attractive for the applications in lasers because the QD based lasers have a higher differential gain, lower threshold current density, and improved temperature performance in comparison with QW lasers [1-3]. The band structures of InAs/GaAs QDs are well suited for covering the 1.30 and 1.55 μm spectral ranges, important in optical fiber communications. These wavelengths are impossible to achieve in quantum well (QW) InAs/GaAs structures due to the strain-limited thicknesses. It is shown that the QD density in laser structures can be enlarged significantly by growing the dots within $In_xGa_{1-x}As$/GaAs QWs, in so called dot–in-a-well (DWELL) structures [3, 4]. In these structures photoluminescence has been enhanced due to better crystal quality of the layer surrounding QDs and more effective exciton capture into the QW and into QDs.

A crucial aspect for the realization of efficient light-emitting devices operated at room temperature is the understanding of a temperature dependence of QD photoluminescence. PL intensity decay in InAs QDs as a rule attributed to thermal escape of carriers from QDs into the wetting layer or into the GaAs barrier [5-8], as well as to thermally activated capture of excitons by nonradiative defects in the GaAs barrier or at the GaAs/InAs interface [5, 9, 10]. The unusual variations of emission energy and the full width at half maximum (FWHM) of PL bands in the InAs/GaAs self-assembled QDs have been investigated earlier as well [11, 12]. A decrease in the FWHM, together with a red shift of the emission wavelength were explained by the re-localization of carriers between dots caused by their inhomogeneous sizes. A set of theoretical works has been devoted to these questions, which described the carrier dynamics in QD systems using a rate equation model [13-19].

However, it is not still clear the details of exciton capture and thermal escape in high quality QD structures. Two main ways are discussed: a) thermal escape (capture) of carriers (electron and hole) in/from QDs takes place as an exciton or correlated electron-hole pairs [17, 18]; or b) the excitons dissociate and single electrons (holes) thermally escape or capture independently [19-21].

The first mentioned mechanism is the subsequence of the fact that the activation energy of ground state PL thermal decay measured in the regime of strong quenching matches the difference between the GaAs band gap and QD ground state (GS) energy (the sum of barrier heights for electrons and holes) [17-19]. The same effect was earlier revealed in strained InGaAs/GaAs QWs, where the activation energy of PL thermal decay corresponded to the difference between the GS energies of QW and the GaAs barrier, defined as total confinement energy, i.e. the sum of electron and hole potential depths. The main conclusion follows from this fact: excitons or electron-hole pairs are emitted from QWs into the barrier [22].

The motivation for the second mechanism was presented in [19-21]. It was shown in [19] at the InAs/GaAs QD PL investigation under low excitation and high temperature the ground state PL intensity varies quadratically with excitation density. This fact was explained as an evidence of independent occupation of QDs by electrons and holes with the probability which is proportional to the multiplication of their concentrations ($n_e n_p$).

Other evidences for supporting the second approach were shown in [20] at the investigation of cathodoluminescence thermal quenching in InAs/GaAs self-assembled QDs at high-excitation conditions. The significant reductions in the thermal activation energies in the 230–300°K temperature range for the ground (GS) and excited states (ES) are found. It was suggested that excitons dissociate and the thermal reemission of single holes from QD states into the GaAs is responsible for the observed temperature dependence. The same explanation was supposed in [21] at the joint investigation of InAs QD and InGaAs QW PL thermal decays.

In DWELL structures the introduction of surrounding $In_xGa_{1-x}As/GaAs$ QWs changes the QD density, the elastic strain in structures, the height of potential barriers for exciton capture and thermal escape in/from QDs as well as can increase in some cases the density of nonradiative (NR) centres. In these structures the mechanism of PL intensity thermal decay could depend strongly on parameters of DWELL. Improved understanding of the operation and design of InAs/InGaAs QD-based devices could be emanated from studies involving the excitation and temperature dependences of PL in such structures.

Note the highest emission intensity of QDs was achieved in the symmetric wells with the composition of quantum well (QW) capping/buffer layers of $In_{0.15}Ga_{0.85}As/GaAs$ [3]. One of the best methods of manipulating the InAs QD density and sizes inside of QWs related to controlling the QD growth temperatures [3,12-15]. But even for the optimal QD growth parameters and capping/buffer layer compositions, the InAs QD structures are characterized by photoluminescence (PL) inhomogeneity along the wafers [16-19].

The inhomogeneity of InAs QD parameters across the wafer due to the size, chemical composition and stress variations leads to a broadening of emission spectra. This type of problems in QD structures was investigated using scanning PL spectroscopy [23, 26-29], high resolution transmission electron microscopy [30], scanning tunneling microscopy [31], as well as spatially resolved scanning tunneling luminescence [32]. However, in dot-in-a well structures, where the InAs QDs coupled with InGaAs/GaAs QWs, the physical reasons of emission inhomogeneity still have to be discussed. The technology of growth of InAs QD structures has become more reliable recently enabling the systematic studies their physical properties and emission inhomogeneity of QD ensembles.

3. PL spectra of symmetric $GaAs/In_{0.15}Ga_{0.85}As/GaAs$ QWs with the different density of InAs QDs

The solid-source molecular beam epitaxy (MBE) in V80H reactor was used to grow the waveguide structures consisting of the layer of InAs self-organized QDs inserted into 12 nm $In_{0.15}Ga_{0.85}As/GaAs$ QWs. The thickness of the buffer $In_{0.15}Ga_{0.85}As$ layer was 2nm, which was grown on the 200.nm GaAs buffer layer and the 2 inch (100) GaAs SI substrate (Fig.1). Then an equivalent coverage of 2.4 monolayers of InAs QDs were confined by the capping (10nm) $In_{0.15}Ga_{0.85}As$ layer and by the 100nm GaAs barrier. Investigated structures are grown under As-stabilized conditions at five different temperatures: 470°C (#1), 490°C (#2), 510°C (#3), 525°C (#4) and 535°C (#5), during the deposition of the InAs active region and InGaAs wells, and at 590-610°C for the rest of layers. All layers were grown with the growth rate of 0.30 ML/s, but for the QD formation the process provides deposition of 2.4 ML with the growth rate of 0.053 ML/s.

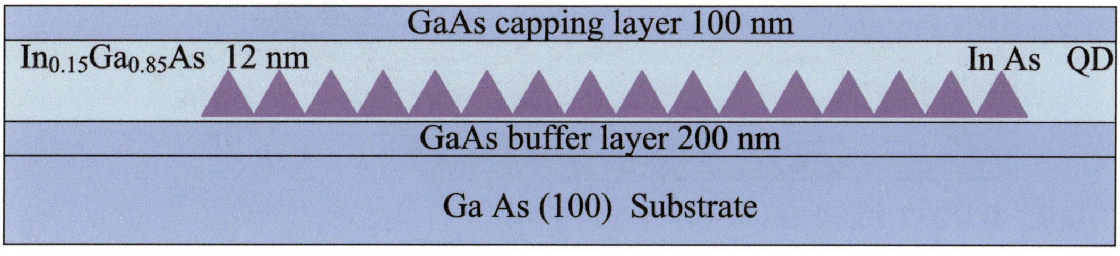

Fig. 1. The schematic design of studied QD structures

For AFM measurement the process of growth was stopped for satelite samples after the formation of QDs. The in-plane density of QDs changed from $1.1x10^{11}$ to $1.3x10^{10}cm^{-2}$ (Table 1) when the temperature increases from 470 up to 535 °C [4, 33]. The samples were mounted in a closed-cycle He cryostat where the temperature is varied in the range of 10 - 300 K. PL spectra were measured under the excitation of the 514.5 nm line of a cw Ar+-laser at an excitation power density form the range of 10-1000 W/cm^2. The setups used for PL and PL excitation spectroscopies were presented earlier in [5] and [17], respectively.

T_{gr} (°C)	N_{QD} cm^{-2}	D_{QD} nm	h nm	h/D	S_{QD}^{single} x 10^{14}, cm^2	Surface Area of QDs, cm^2
470	1.1×10^{11}	12	6	0.50	113.0	0.124
490	7.0×10^{10}	14	8	0.57	153.9	0.108
510	3.4×10^{10}	18	13	0.72	254.3	0.086
525	1.8×10^{10}	24	11	0.46	452.2	0.081
535	1.3×10^{10}	28	10	0.36	615.4	0.080

Table 1. Average parameters of InAs QDs estimated by AFM [33]

Typical PL spectra of the DWELL structures #2, #3 and #4 measured at different temperatures at the excitation light density of 500 W/cm^2 are shown in Fig.2. The structures #1 and #2 grown at low temperatures 470 and 490°C are characterized by one PL band only (Fig.2a) with a higher value (Fig.3b) of the full width at half maximum (FWHM). In structures #3, #4 and #5 two PL bands appear due to the recombination of excitons localized at a ground state (GS) and at first excited state (1ES) in QDs (Fig.2b,c).

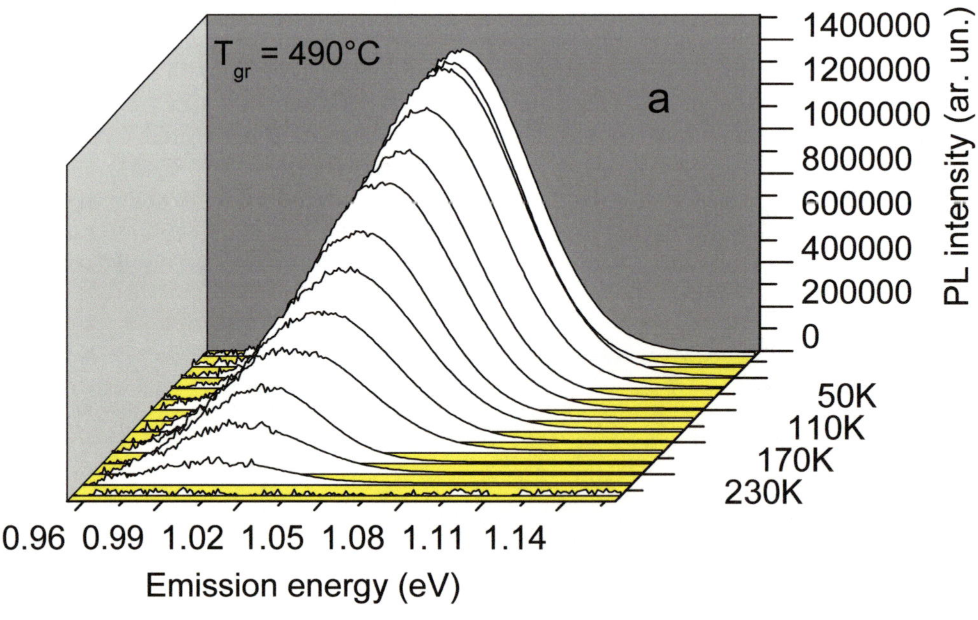

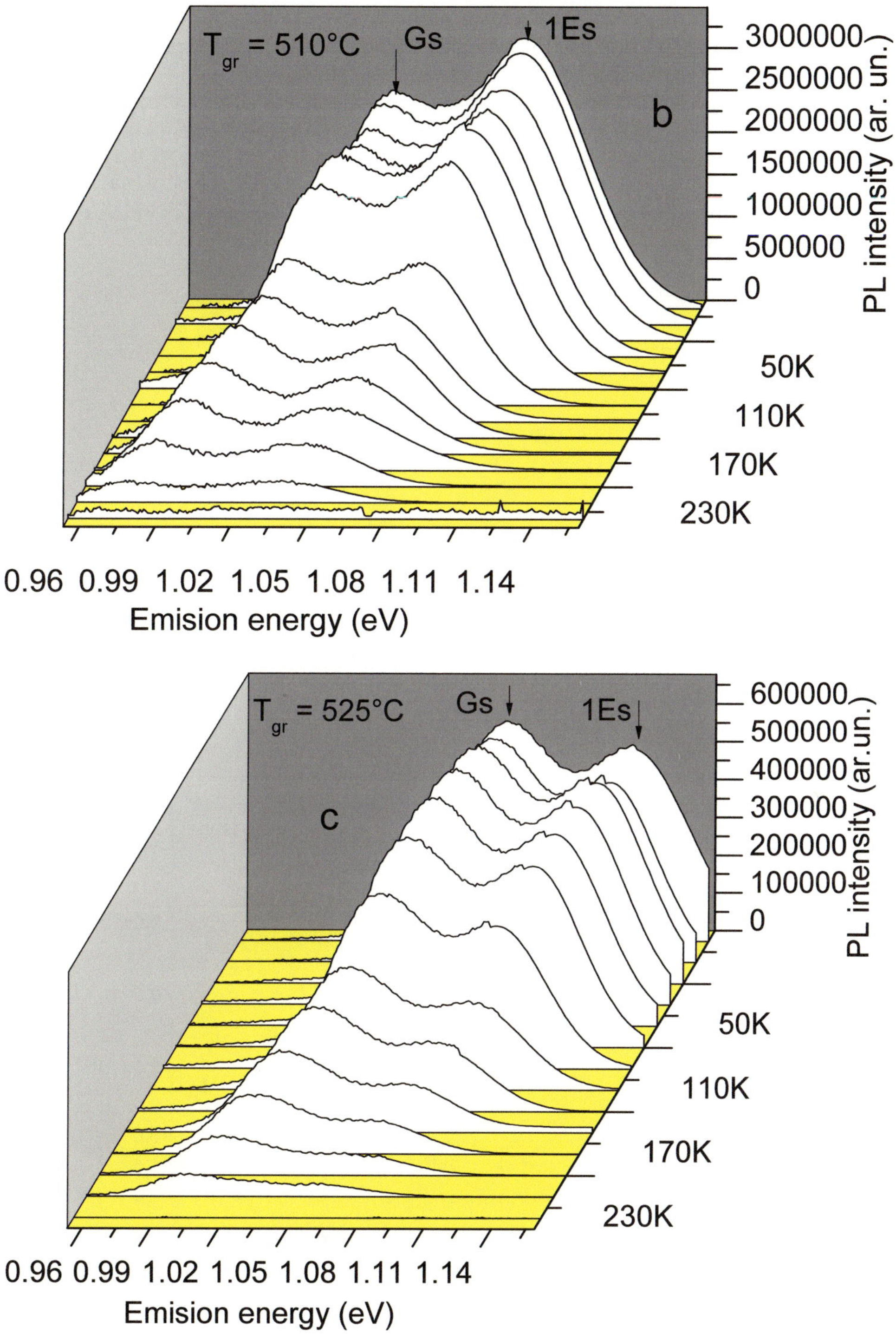

Fig. 2. Typical PL spectra measured at different temperatures for the DWELL structure #2(a), #3(b) and #4(c) at the excitation light density of 500 W/cm^2.

The density of QDs decreases monotonically with the rise of QD growth temperatures from 470 up to 535 °C (Table 1). Thus the PL intensity diminishing can be expected with reducing a QD density. The QD diameters in studied structures increase monotonically from 12 up to 24-28 nm with the rise of QD growth temperatures from 470 up to 535 °C. Thus it is possible to expect that the PL peak position in QDs has to shift monotonically to low energy.

Fig.3 presents the variation of in plane QD densities, estimated by AFM on satelite samples, as well as the average GS integrated PL intensities and the GS peak FWHM measured at 300K in studied structures.

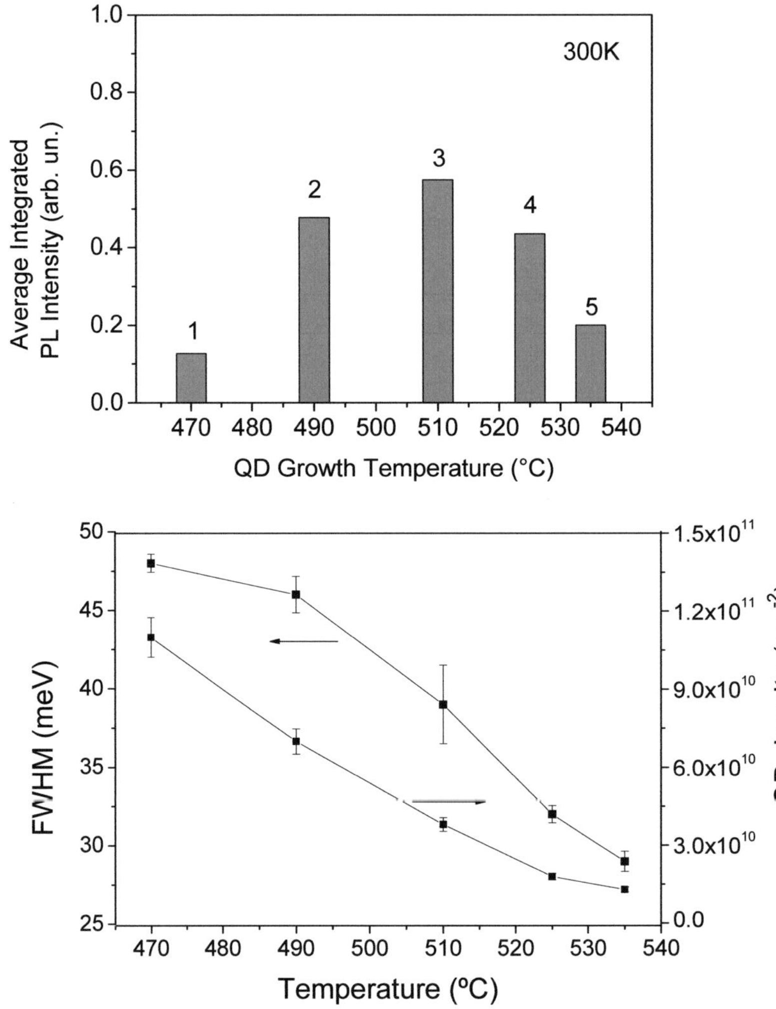

Fig. 3. The average integrated PL intensities measured in DWELLs with QDs grown at different temperatures (a) and the FWHM and QD density in DWELLs with QDs grown at different temperatures (b).

However the PL intensity increases (Fig.3a) and the GS peak shifts to low energy in structures with QDs grown at 490 and 510°C (Fig.2a,b). On the contrary the QD structures grown at 525 and 535°C are characterized by lower intensities of GS emission (Fig.3a), by smallest FWHM (Fig.3b), and the peak position of GS PL bands shifts into higher energy (Fig.2b,c). Note that lower PL peak energy corresponds to higher PL intensity (Fig.2b and 3a). Thus the variation non monotonous the PL intensity ans peak positions versus QD density (QD growth temperature) has been revealed in studied QD structures.

4. PL excitation spectra of InAs QD structure and PL excitation power dependences

The typical PL excitation spectrum measured at 80K is shown in Fig.4.

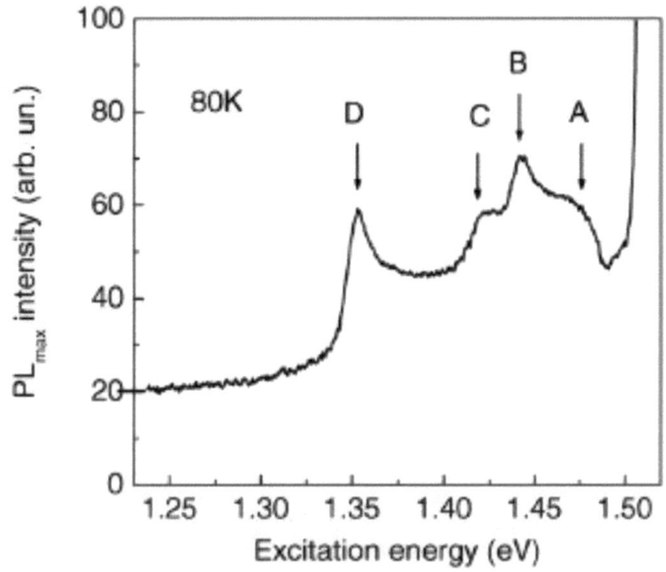

Fig. 4. PL excitation spectrum for the structure #3.

In high excitation energy range the spectrum presents a sharp PL intensity increase due to the fundamental light absorption (at 1.51 eV) in the GaAs barriesr. In the low energy region the PL excitation spectrum can be considered as a superposition of the four absorption bands: A, B, C and D (Fig.4, Table 2).

Optical transitions	GaAs Band gap	peak A	peak B	peak C	peak D	QD-PL	QW-PL
E, eV	1.51	1.46-1.47	1.44	1.42	1.35-1.36	1.06-1.11	1.32-1.33

Table 2. Optical transitions in DWELL structures at 80K

The peak A spectral position (1.46-1.47eV) is close to the GS resonant absorption related to the WL in InAs/GaAs QD structures [17, 24, 25]. Note the authors [25] registrated at 100K in PL excitation spectrum two overlapping maxima (1.45-1.47eV) which were attributed to the photon absorption in the 2D wetting layer between the heavy hole and light hole GS subbands to the GS electronic subband. In studied DWELL structures the GS resonant absorption in the 2 nm buffer $In_{0.15}Ga_{0.85}As$ QW can contribute in other peaks B and C. Thus both layers (the $In_{0.15}Ga_{0.85}As$ buffer QW and WL) are the same in all studied DWELL structures and can be responsible for the peak A, B and C in the PL excitation spectrum. The spectral position of the lowest energy absorption band (1.35 eV, peak D) is close to one of the PL band (1.31-1.32eV) caused by GS exciton recombination in the capping InGaAs QW (Table 2) [34]. Thus the peak D can be attributed to the GS resonant excitation in the capping $In_{0.15}Ga_{0.85}As$ /GaAs QW (Fig.4).

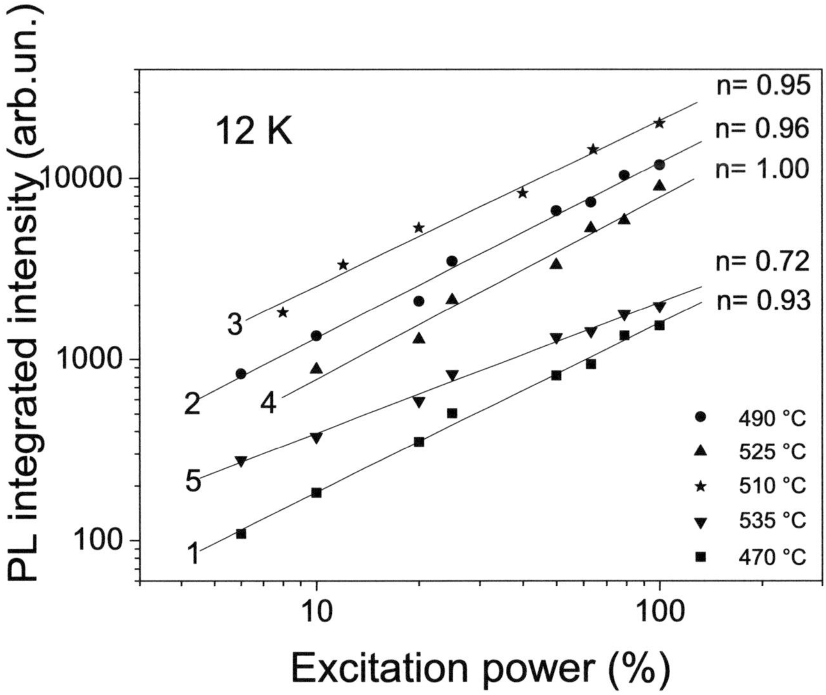

Fig. 5. PL integrated intensity versus excitation power measured for all studied structures

The dependences of integrated PL intensity versus excitation light power, measured at 12K with the aim to avoid the thermal decay process, are presented in Fig.5 for all studied structures. The QD structures #2, #3 and #4 with highest emission intensity are characterized by the linear dependence of the integrated PL intensity (I) versus excitation power ($I \approx P^n$), n=0.95-1.00 (Fig.5). In QD structures with low emission (#1, #5) the integrated PL intensity changes sublineary (n=0.72-0.93) with excitation light power (Fig.5).

5. Emission inhomogeneity along the QD structures

5.1 Scanning PL spectroscopy at 300K

Photoluminescence inhomogeneity of QD ensembles is studied along the scanning line crossed the wafers from the periphery of QD structures to the center. In this case PL spectra were measured at 300 K in a set of points on the QD structures under the excitation by the 804 nm line of a solid state IR laser at an excitation power density of 100 W/cm^2. PL spectra were dispersed by a SPEX 500M spectrometer and recorded by a liquid-nitrogen-cooled Ge detector with a standard lock-in technique.

The inhomogeneity of QD ensemble emission along the line crossed the structures #1 and #2 related mainly to the variation of PL intensity (Fig.6). The integrated PL intensity in the structure #1 is low in comparison with #2, and the variations of PL peak positions are small for both DWELLs (Fig.6): $\Delta h\nu_m$ ~0.003 eV (#1) and $\Delta h\nu_m$ ~0.010 eV (#2). Thus in QD structures with InAs QDs grown at low temperatures (470-490°C) the less integrated PL intensity (Fig.3a), the high dispersion of QD sizes and, as a result, the larger value of FWHM (Fig.3b) of GS PL bands, but the low dispersion of QD ensemble parameters along the line crossed the structures #1 and #2 have been detected (Fig.6a,b).

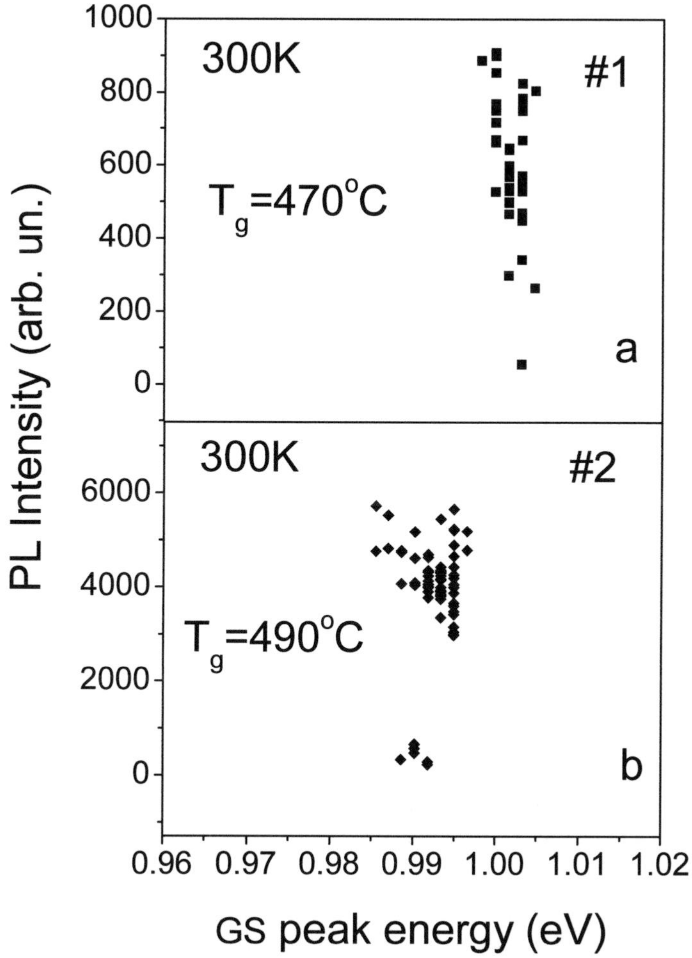

Fig. 6. The PL intensity and GS PL peak positions measured along the line scan from the center of QD structures #1 (a) and #2 (b) to their periphery.

As it follows from fig.7 the decrease of integrated PL intensity, measured along the line from the center to periphery in QD structures #3, #4 and #5, is accompanied by the "blue" energy shift of the PL maximum. The FWHM of GS PL bands in the structures #3, #4 and #5 was equal to 35-40 meV (Fig.3b) and it is smaller than those in mentioned above structures #1 and #2. The inhomogeneity of QD ensemble emission along the lines crossed the structures #3, #4 and #5 is characterized by the essential variations of GS peak positions (Fig.7): $\Delta h\nu_m$

~ 0.02 eV(#3), 0.03 eV(#4) and $\Delta h\nu_m$ ~0.04 eV(#5), respectively. The integrated PL intensity of QD ensembles in the structure #3 in two- or five-fold higher than in #4 and #5, respectively. Thus in the structures #3, #4 and #5 the low dispersion of QD sizes and, as a result, less values of FWHM, but the essential PL energy shift and the dispersion of QD ensemble parameters along the line crossed the structures have been detected (Fig.7).

The emission inhomogeneity in the QD structures #1 and #2 is related mainly to the change of PL intensity owing the density variations of QDs and/or of nonradiative centers (NR) in QD structures. At the same time the PL peak positions vary negligibly (3-10 meV) testifying that the difference in QD sizes is small. In the structures #3, #4 and #5, as it follows from fig.7, the decrease of PL intensity along the line scan is accompanied by the "blue" energy shift of PL maximum. This effect leads at the wafer periphery to shallower QD localized states (i.e. smallest electron and hole binding energy), poorer carrier localization and a higher probability of carrier thermal escape, which reduces the integrated QD PL intensity at 300K.

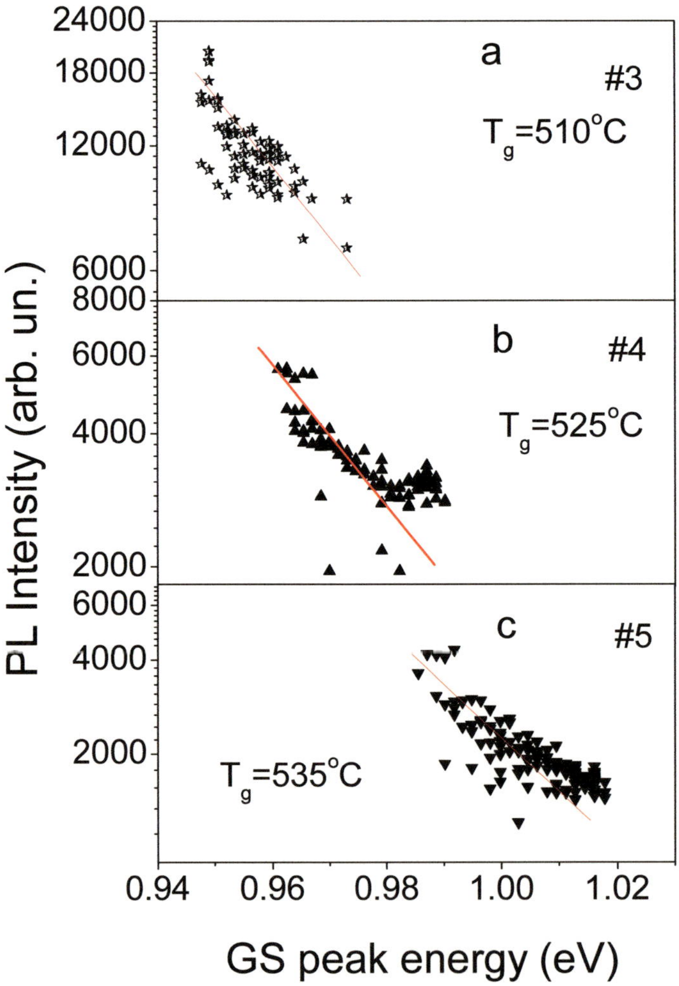

Fig. 7. The PL intensity and GS PL peak positions measured along the line scan from the center of QD structures #3 (a), #4 (b) and #5 (c) to their periphery.

5.2 Analyses of PL inhomogenouty reasons in studied QD structures at 300K

The ground state PL intensity (I_{PL}) of InAs/InGaAs QDs is directly proportional to internal quantum efficiency η which can be presented as: $\eta = \frac{\varpi_R}{\varpi_R + \varpi_{NR}}$. For QD emission the radiative recombination rate is [35]: $\varpi_R = \sum_{i=1}^{N_D} \frac{f_i^e f_i^p}{\tau_R}$, where f_i^e and f_i^p are the occupation probabilities for electrons and holes at ground state levels given by the Fermi-Dirac distribution functions, f^e, f^p=(exp[($E_{n,p}$ −$\mu_{n,p}$)/kT]+1]$^{-1}$, where $\mu_{n,p}$ are the quasi-Fermi-levels for the conduction and valence bands, respectively, measured from the QD band edges, $E_{n,p}$ are the quantized energy levels of an electron and a hole in th conduction and valence bands of a QD, measured from the QD band edges, N_D is QD density and τ_R is electron–hole radiative recombination time in the QD [36]. At low excitation light intensity (100 W/cm^2), used during GS PL scanning, which is well below the GS saturation intensity, we can present the occupation probabilities using Maxwell -Boltzmann distribution functions f^e, f^p=exp(−$E_{n,p}$+ $\mu_{n,p}$ /kT). Taking into account that excitation light power is not changed during GS PL scanning experiment, we can assume that $\mu_{n,p}$ are constant along the scanning line and the PL intensity variation occurs due to parameters $E_{n,p}$ only [35]. Thus $f^e \approx \exp \frac{E_{loc}^e}{kT}$ and $f^p \approx \exp \frac{E_{loc}^p}{kT}$, where E_{loc}^p and E_{loc}^e are the binding energy of electron and hole on GS levels in a QD. In this case for GS emission intensity at room temperature is possible to write:

$$I_{PL} \approx \tau_R^{-1} \approx \sum_1^N \frac{1}{\tau_{RQD}} \exp(\frac{E_{loc}^e + E_{loc}^p}{kT}) \approx \sum_1^N \frac{1}{\tau_{RQD}} \exp(\frac{E_{GS}^{InGaAs} - E_{GS}^{QD}}{kT}) \approx \exp \frac{-E_{GS}^{QD}}{kT} \qquad (1)$$

where E_{GS}^{InGaAs} is energy gap between electron and hole quantized levels in narrow-gap $In_{0.15}Ga_{0.85}As$ layer and E_{GS}^{QD} is energy gap between electron and hole quantized levels in QD. Finally the GS PL optical transition energy ($h\upsilon_{\max}^{GS}$) is the difference between E_{GS}^{QD} and exciton binding energy E_{bin}^{ex} in QDs. Exciton binding energies were computed as the function of QD size using 8-band **k·p** approach and are estimated as 22 meV for QDs with the base size 14 nm [37]. Thus the E_{bin}^{ex} value is small in comparison with GS PL optical transition energy ($h\upsilon_{\max}^{GS}$) equal to 1.06-1.11 eV. If the InGaAs composition is the same in any points of the capping/buffer layers, the linear dependence of GS PL intensity versus PL peak positions in QD ensembles in semi-logarithmic plot follows from Eq.1 (Fig.7). At the same time the slope of this linear dependence is proportional to kT. The fitting procedure was applied to the analysis of PL intensity data presented in Fig.7 and the slopes of these three lines were estimated as ~25 meV that is exactly the value of kT at 300K. Thus the analysis of linear dependence in Fig.7 testifies that the variation of GS PL intensity in the QD ensemble along the line scan in the structures #3, #4 and #5 is owing to the QD size variation – its decreasing from the wafer center to the periphery [38]. The last effect can be connected with temperature inhomogeneity along the wafer at the QD growth process.

6. Temperature dependences of PL integrated intensities and peak positions

Two reasons can explain the variation non monotonically of PL peak positions and the PL intensity in studied QD structures (Fig.2 and 3a): (i) the change of QD composition due to the Ga/In inter diffusion between the InAs QDs and capping/buffer $In_{0.15}Ga_{0.85}As$ QW layers or (ii) the different levels of elastic strains in QD structures due to the difference in QD densities and sizes. To distinguish these two reasons PL spectra at different temperatures in the range 12-300 K have been studied. The temperature dependences of integrated PL intensities and peak positions of GS PL bands for the structures #1, #3, #5 are shown in Fig.8.

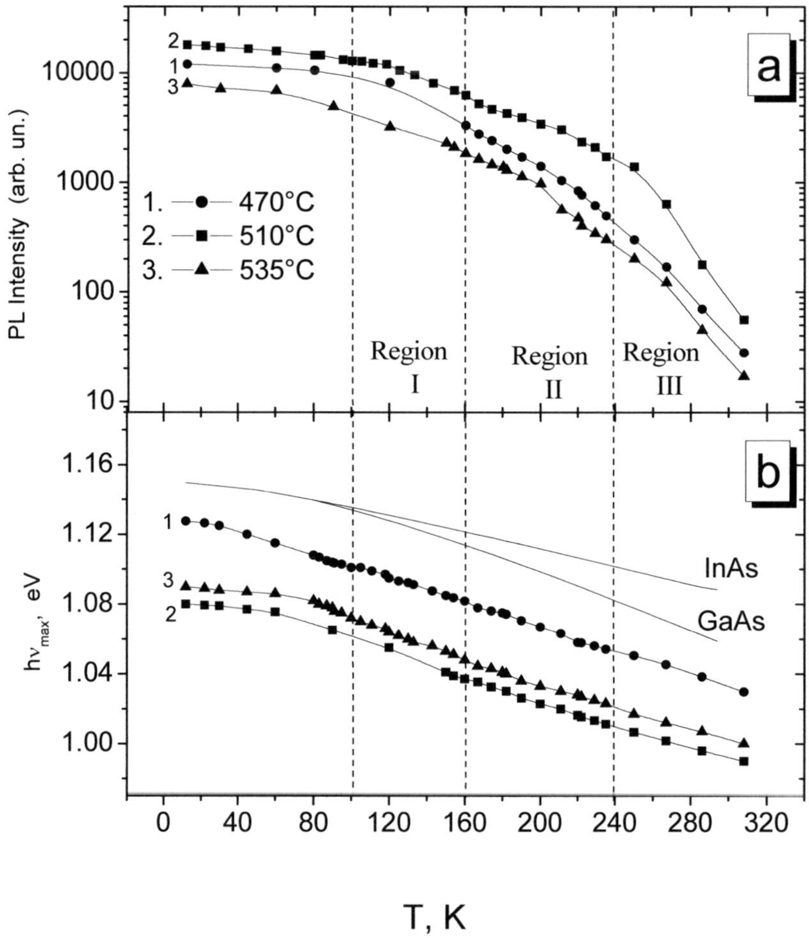

Fig. 8. Temperature dependencies of GS integrated PL intensities (a) and GS peak positions (b) measured for the structures #1, #3 and #5. Lines present the fitting results.

The process of PL thermal decay starts at 80-120 K and it is characterized by different rates in the ranges T=100-180 K, T=180-250 K and T=250-300 K. At low temperatures up to 100 K the integrated PL intensity does not change, but the energy of GS peak decreases. The process of PL intensity decay starts at 100-120K in high quility structures #2, #3 and #4, but in structures of low quality (#1, #5) this process appears at 80-90K.

7. The analysis of Ga/In interdiffusion in QD structures with different density of InAs QDs

Fig. 9 presents the variation of PL peak positions versus temperature in studied structures. PL peaks shift to low energy with increasing temperature due to the optical gap shrinkage. The lines in Fig.9 are the fitting results analyzed on the base of Varshni relation that presents the energy gap variation with temperature as [39]:

$$E(T)=E_o - \frac{aT^2}{T+b} \tag{2}$$

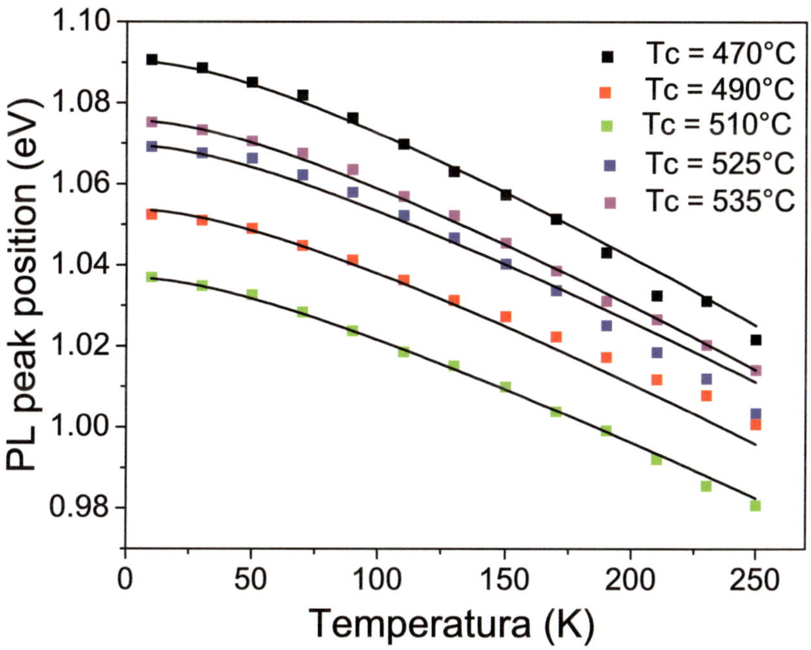

Fig. 9. The variation of PL peak positions versus temperature. The lines present the Varshni fitting results: 1- #1, 2- #2, 3- #3, 4-#4 and 5 - #5.

The comparison of fitting parameters with the variation of energy band gap versus temperature in the bulk InAs and GaAs crystals (Table 3) has revealed that in studied QD structures the fitting parameter "a" and 'b' in the temperature range 12-250 K are very close to their values for the bulk InAs crystal only in the structures #2 and #3. But in other QD structures the fitting parameter "a" and 'b' are different a little bit from the values in the bulk InAs crystal (Table 3). The last fact testifies that the process of Ga/In inter diffusion takes place in these QD structures [33,42-44]. Note that the process of Ga/In inter diffusion in studied structures passes non monotonically versus QD growth temperatures. It means that not only temperature but some other factors discussed below are essential as well.

	E_0	A	b
Structure numbers	eV	meV/ °K	°K
#1	1.082	0.355	110
#2	1.089	0.346	98
#3	1.010	0.300	95
#4	1.049	0.330	110
#5	1.079	0.335	130
InAs [40]	0.415	0.276	93
GaAs [41]	1.519	0.540	204

Table 3. The Varshni fitting parameters

8. Model for the dependence of PL integrated intensity versus temperature

An analysis of the thermal behavior of QD luminescence indicates that excitons are the dominant electronic particles. Thus to modeling the dependence of the ground state PL intensity versus temperature the simple assumption is applied: the carriers are considered to behave as excitons or correlated electron-hole pairs. This assumption is a common feature of most exiting models [15-18]. The motivation of the choosing this approach in this paper will be clear from presented esperimental results as well.

For InAs/InGaAs structures, the two-stage processes of exciton capture and thermal escape in/from QDs have been considered [17, 33]. These processes in QD structures can occur not only through the wetting layer (WL) states, as was proposed earlier [16, 18], but also through the capping/buffer InGaAs QW layers [17, 33]. However in [17] the localization of nonradiative defects was considered in GaAs barrier mainly. In present model two types of nonradiative defects: in the GaAs barrier (NR1) and in the capping/buffer $In_{0.15}Ga_{0.85}As$ QWs (NR2) are taken into account [33]. To simplify the rate equations, one intermediate level, referred as QW, will be considered in the model.

It is supposed that photo-generated excitons are created in GaAs barrier and in $In_{0.15}Ga_{0.85}As$ QWs with generation rates G_{GaAs} and G_{InGaAs}, respectively. The exciton recombination takes place in the GaAs barrier, InGaAs QWs and in InAs QDs. In this model all QDs in an ensamble are assumed to be identical with the same properties and the process of carrier thermal redistribution between QDs is inessential. This is supported by the experimental fact that the FWHM of the ground state PL band is measured to be constant in the temperature range of 10-300K. The system of rate equations for exciton concentrations in the GaAs barrier ($C_{o,}$), QWs (C_1) and QDs (C_2) can be written as:

$$\frac{dC_0}{dt} = G_{GaAs} - \omega_{QW} C_0 N_{QW} + \omega_{QW} C_1 N_{GaAs} \exp\left(-\frac{\Delta E^{GaAs-QW}}{kT}\right) - \frac{C_0}{\tau_{NR1}} \tag{3}$$

$$\begin{aligned} \frac{dC_1}{dt} = G_{InGaAs} + \omega_{QW} C_0 N_{QW} - \omega_{QW} C_1 N_{GaAs} \exp\left(-\frac{\Delta E^{GaAs-QW}}{kT}\right) - \omega_{QD} C_1 N_{QD} + \\ + \omega_{QD} C_2 N_{QW} \exp\left(-\frac{\Delta E^{QW-QD}}{kT}\right) - \frac{C_1}{\tau_{RQW}} - \frac{C_1}{\tau_{NR2}} \end{aligned} \tag{4}$$

$$\frac{dC_2}{dt} = \omega_{QD} C_1 N_{QD} - \omega_{QD} C_2 N_{QW} \exp\left(-\frac{\Delta E^{QW-QD}}{kT}\right) - \frac{C_2}{\tau_{RQD}} \tag{5}$$

Here $\varpi_{QW} N_{QW} = \tau_{QW}^{-1}$ and $\varpi_{QD} N_{QD} = \tau_{QD}^{-1}$ are exciton thermalization (capture) rates from the GaAs barrier into QWs and from QWs into QDs, respectively, N_{QD}, N_{QW}, N_{GaAs} is the density of states in QDs, QWs and in the GaAs barrier, ϖ_{QW}, ϖ_{QD} are the exciton capture coefficients into QWs and QDs, τ_{RQW}^{-1}, τ_{RQD}^{-1}, τ_{NR1}^{-1} and τ_{NR2}^{-1} are the radiative exciton recombination rates in QWs and QDs as well as the nonradiative recombination rates in the GaAs barrier and in QWs, respectively. The values $E^{GaAs\text{-}QW}$, $E^{QW\text{-}QD}$, $E^{GaAs\text{-}QD}$ are the energy differences between: (i) the GaAs band gap and the GS energy in QWs, (ii) the GS energy in QWs and in QDs, (iii) the GaAs band gap and the GS energy in QDs, respectively.

The temperature dependence of the exciton escape rates from QWs and QDs is taking into account, but temperature dependences of other parameters (trapping and recombination coefficients, density of states, exciton thermalization rates) are neglected. This is motivated by simplifying the calculation process and by minimization of the number of parameters samulteniously with significant progress in understanding the experimental results. Several comments deal with the experimental base for such simplifications are discussed below.

In presented model the variation of the GS PL intensity (I(T)) in QDs versus temperature in the stationary state can be described by the formula:

$$I(T) = \frac{G_{InGaAs} + G_{GaAs}\left(1 - \frac{\tau_{NR1}^{-1}}{\tau_{QW}^{-1} + \tau_{NR1}^{-1}}\right)}{1 + \frac{\tau_{RQW}^{-1} + \tau_{NR2}^{-1}}{\tau_{QD}^{-1}} + \frac{\tau_{NR1}^{-1}\tau_{QW}^{-1} N_{GaAs}}{\tau_{QD}^{-1}\left(\tau_{QW}^{-1} + \tau_{NR1}^{-1}\right) N_{QW}} \exp\left(-\frac{\Delta E^{GaAs-QW}}{kT}\right) + \frac{\left(\tau_{RQW}^{-1} + \tau_{NR2}^{-1}\right) N_{QW}}{\tau_{RQD}^{-1} N_{QD}} \exp\left(-\frac{\Delta E^{QW-QD}}{kT}\right) + \frac{\tau_{QW}^{-1}\tau_{NR1}^{-1} N_{GaAs}}{\tau_{RQD}^{-1}\left(\tau_{QW}^{-1} + \tau_{NR1}^{-1}\right) N_{QD}} \exp\left(-\frac{\Delta E^{GaAs-QD}}{kT}\right)} \tag{6}$$

Here the radiative lifetime in QDs (τ_{RQD}) is assumed to be constant with T as expected theoretically for strong confinement in three dimensions [15-18]. The radiative recombination rate in InGaAS/GaAs QWs (τ_{RQW}), as was shown in [22], is controled mainly by nonradiative recombination processes at the T≥100K. The dependence of the nonradiative rate in InGaAs/GaAs QWs was investigated as well in [22]. Its value in QWs, as a rule, increases at low temperatures and saturates at 120-150K at a constant value depending on the quality of the structure. As result it is possible to neglect by temperature dependences of the parameters: τ_{RQW}^{-1}, τ_{NR1}^{-1} and τ_{NR2}^{-1} at the T≥120-150K.

The exciton thermalization (capture) rates from the GaAs barrier into a QW τ_{QW}^{-1} and from a QW into QDs, τ_{QD}^{-1}, are the multiphonon-assisted processes deal with the scattering via multiple longitudinal optical phonon (LO) emission [41-43]. The values τ_{QW}^{-1} and τ_{QD}^{-1} have to increase versus temperature due to the enlargement of the phonon number. However, it was shown in [42] that the PL rise time in a QW (dependent on the exciton capture rate into QWs, τ_{QW}^{-1}) for all temperatures from the range 50-300K is the same about 1-2 ps.

The dominant multiphonon capture mechanism and the InGaAs/GaAs QD capture time τ_{QD}^{-1} decreasing with temperature were confirmed experimentally in time-resolved experiments [45-47]. It was shown using the investigation of the PL rise time for the GS in QDs that its value decreases: from 6.5ps to 3.5ps upon increathing the temperature from 4 to 300K [45], or from 15ps to 7ps at the temperature rise from 50 to 300K [46]. Thus nearly two-fold decreasing of the exciton capture time into QDs is revealed experimentally versus temperature. It is essential that the temperature dependence of the exciton escape rates from QWs and from QDs obviously much stronger and it is taking into account in the model.

9. Discussion of PL integrated intensity dependences versus temperature

It is clear from the formula (6), that in high quality QD structures (#2, #3 and #4) with low concentrations of the NR1 and NR2 centres the PL intensity I(T) is linearly dependent on excitation power (or generation rates, G_{InGaAs}, G_{GaAs}), as it is demonstrated in Fig.5. In low quality QD structures (#1 and #5) photo-generated excitons recombine partially via NR1 and NR2 centres and W (T) changes Sublineary versus power (Fig.5).

The analysis of PL temperature dependences is resonable to provide for the temeparture ranges I - III separately (Fig.8). At low temperature (< 100K) the processes of exciton thermal escape from a QW into the GaAs barrier and from QDs into a QW are not essential, and the PL intensity I (T) does not change (Fig. 8).

In the temperature ranges I00-300K following the model the activation energy of PL thermal decay has correspond to the values $E^{GaAs-QW}, E^{QW-QD}, E^{GaAs-QD}$. To quantitative determination of activation energies the $\{I_{max}/I(T) - 1\}$ dependences for the GS and ES PL bands were plotted versus temperature in Arrhenius plots (Fig.10). Three distinct linear regions (I-III) with the corresponding activation energies Ea (I), Ea (II) and E_a(III) are observed for different temperatures (Fig.10).

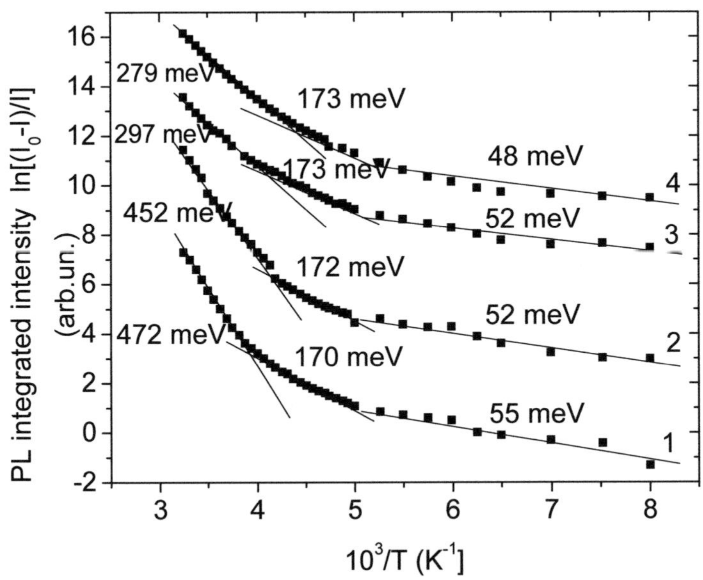

Fig. 10. Arrhenius plots for the thermal decay of the GS integrated PL intensity measured at the power density of 500W/cm² for the structures #3 (1), #2 (2), #1 (3) and #5 (4).

In the range I (100-180K) the activation energy (Ea (I) = 48-55 meV) of PL decay measured at low excitation power for GS PL bands (Fig.10) is very close to the energy difference between the GaAs band gap and the GS energy of the WL and/or the buffer InGaAs QW layer (peak A, B, C in Fig.4, Table 2). It can be supposed that thermal quenching of the GS PL intensity in the temperature range I is due to decreasing an exciton flow to QDs caused by thermal escape of excitons from the WL (or buffer InGaAs QW) into the GaAs barrier where they are lost through subsequent nonradiative recombination. If the above mentioned mechanism takes place in QD structures the same activation energy ($E^{GaAs-QW}$) has to be detected in the range I for thermal decay of PL bands connected with the GS and ES of QDs, as well as for the QW PL band. Actually the results presented in Fig.11 testify that in the range I thermal decay of PL bands deal with the GS, four ES in QDs, as well as with a QW is characterized by the same activation energy from the range 48-53meV.

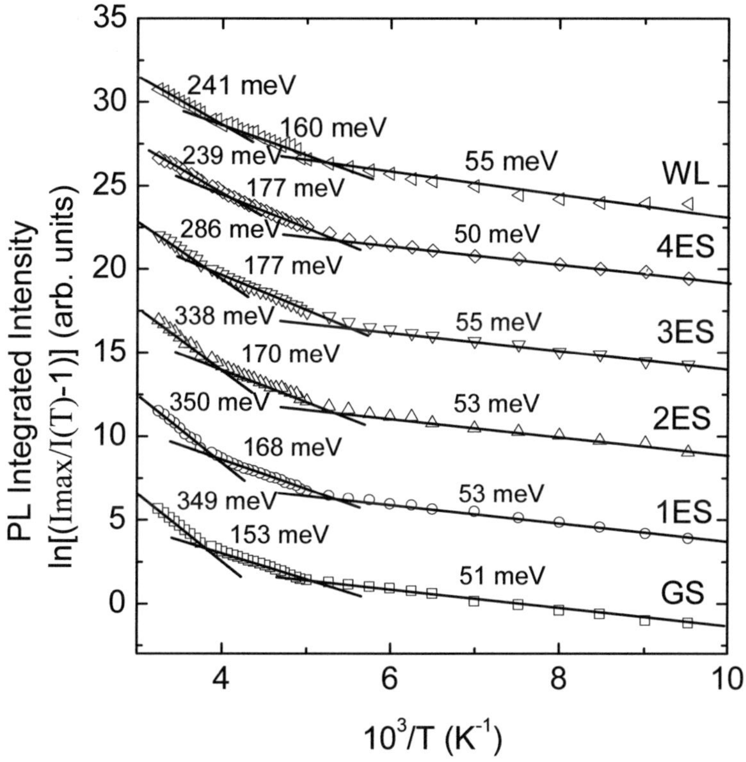

Fig. 11. Arrhenius plots for the thermal decay of the integrated PL intensity measured at the power density of 1000W/cm² for the QD structures #3

In the temperature range II (180-250K) the activation energy (Ea (II) = 170-173 meV) of PL decay for GS bands (Fig.10 and 11) is very close to the energy difference between the GaAs band gap and the GS energy for the capping $In_{0.15}Ga_{0.85}As$ QW (peak D, Fig.4, Table 2). Thus the process of thermal decay of the QD PL intensity in the range II can be attributed to diminishing of an exciton flow to QDs caused by thermal escape of excitons from the capping $In_{0.15}Ga_{0.85}As$ QW into the GaAs barrier. This explanation is confirmed as well by the thermal quenching of PL bands deal with the GS and 1ES-4ES in QDs, and with the capping QW. All PL bands in the range II demonstrate the activation energy from the range of 153-177 meV (Fig.11).

At high temperatures, stage III (250-300K), the activation energy of PL thermal decay depends on the quality of DWELL structures (Fig.10). As one follows from the formula (6) the activation energy of PL thermal decay will be close to the value $\Delta E^{GaAs-QD}$ if the follows two relations are correct:

$$\frac{\left(\tau_{RQW}^{-1}+\tau_{NR2}^{-1}\right)N_{QW}}{\tau_{RQD}^{-1}N_{QD}}\exp\left(-\frac{\Delta E^{QW-QD}}{kT}\right) << \frac{\tau_{QW}^{-1}\tau_{NR1}^{-1}N_{GaAs}}{\tau_{RQD}^{-1}\left(\tau_{QW}^{-1}+\tau_{NR1}^{-1}\right)N_{QD}}\exp\left(-\frac{\Delta E^{GaAs-QD}}{kT}\right) \quad (7)$$

or after the simplification of the (7):

$$(\tau_{RQW}^{-1}+\tau_{NR2}^{-1})N_{QW}\exp\left(-\frac{E_{GS}{}^{QW}}{kT}\right) << \frac{\tau_{QW}^{-1}\tau_{NR1}^{-1}N_{GaAs}}{\left(\tau_{QW}^{-1}+\tau_{NR1}^{-1}\right)}\exp\left(-\frac{E_g{}^{GaAs}}{kT}\right) \quad (8)$$

It means that if the concentration of nonradiative defects in QWs (NR2) is low the activation energy of PL thermal decay for the GS PL band in QDs will approach to the value $\Delta E^{GaAs-QD}$. Actually for the high quality structures (#2, #3) the activation energies (452 and 472 meV) in the range III (Fig.10) are close to the $\Delta E^{GaAs-QD}$. Note this value is the sum of the barrier energy for electrons and holes in QDs. It testifies that just excitons thermally escape in the range III from QDs into the GaAs barrier with subsequent NR recombination. In the case of measurement at the high excittaion power (Fig.11) the decrease of activation energy for GS thermal decay deals with exciton re-localization from the ES to the GS of other QDs which slow down the process of GS thermal quenching.

In low efficient QD structures (#1, #5) the process of GS PL thermal decay is characterized by smaller activation energies, 279 and 297 meV (Fig.10). For these structures it is natural to supose the high concentration of the NR2 centers in InGaAs QWs. It means the conditions (7) and (8) do not satisfy. As one follows from the formula (6) in this case the activation energy of GS PL thermal decay will approach to the value ΔE^{QW-QD}.

Note that obtained experimental results do not present an evidence for the exciton thermal dissociation in QWs or in QDs at high temperatures with subsiquent re-localization, thermal escape or tunneling of separated electrons or holes. If the process of exciton thermal dissociation is realized the activation energy of GS PL thermal decay should be smaller and comparative with the value of barriers for electrons or holes in the QD structures that was not observed.

10. Fitting of the data of PL integrated intensity thermal decay

The fitting procedure was applied to the experimental curves for thermal decay of integrated PL intensities presented in Fig.8. As it follows from Eq.6 the integrated PL intensity thermal decay can be simulated as:

$$I = \frac{I_0}{1 + K_1 \exp(-E_1/kT) + K_2 \exp(-E_2/kT) + K_3 \exp(-E_3/kT)} \quad (9)$$

The values I_o and the activation energies E_1, E_2 and E_3 for different temperature ranges (I-III) were taken from the experimental results presented in Fig.10 (Table 4). Parameters K_1, K_2 and K_3 have been obtained from the numerical simulation procedure (Table 4) [48,49]. As it follows from Eq.6 the coefficients K_1 and K_3 related to the exciton nonradiative recombination rate (τ_{NR1}^{-1}) in the GaAs barrier, but the coefficient K_2 depends on the exciton nonradiative recombination rate (τ_{NR2}^{-1}) in InGaAs QWs. Numerical simulation results presented in Table 4 have shown that the coefficients K_1 and K_3 related to the exciton nonradiative recombination rate (τ_{NR1}^{-1}) in the GaAs barrier of the structure #3 are less in comparison with those in #1 and #5 [48-50]. Simultaneously the coefficient K_2 related to the exciton nonradiative recombination rate (τ_{NR2}^{-1}) in InGaAs QWs for the structure #3 is one order smaller than in #1 and #5. These facts are the reason of fast PL thermal decay and lower integrated PL intensities in structures #1 and #5 in comparison with the structure #3.

Structure	I_o	E_1 meV	E_2 meV	E_3 meV	K_1	K_2	K_3
#1	$1.1\ 10^4$	53	174	451	100	$2.0\ 10^5$	$6.7\ 10^9$
#3	$1.7\ 10^4$	52	175	452	85	$2.3\ 10^4$	$6.0\ 10^9$
#5	$8.9\ 10^3$	54	173	450	100	$2.3\ 10^5$	10^9

Table 4. Fitting parameters estimated from the PL thermal decay

Thus, it is shown that the excitonic nature of carriers is important for the processes of capture and thermal escape in/from QDs in DWELL structures. At low temperatures the GS and ES PL thermal decays are attributed to the reduction of exciton flow into the QDs due to the exciton thermal escape from the WL or InGaAs buffer layers (100–180 K) or from the GS of capping $In_{0.15}$ $Ga_{0.85}$ As layers (180–250 K) into the GaAs barrier with subsequent NR recombination. At high temperatures (250–300 K) the activation energy of PL thermal decay depends on the QD density and the quality of DWELL structures. In DWELL structures with high emission (#2,# 3, and #4) the activation energy matches the energy difference between the GaAs band gap and the GS level of QDs. In structures with weak emission (#1 and #5) the activation energy is close to the energy difference between the GS level of QDs and the GS energy level in InGaAs/GaAs QWs. The reasons of DWELL quality variation versus QD density are discussed below.

11. X ray diffraction study in InAs QD structures with the different densities of QDs

It is important to discuss the reasons of the quality change in studied QD structures with different InAs QD densities. The application of the capping/buffer In_xGa_{1-x} As layers in the QD structure has been demonstrated as an effective means to the QD density increase

[3], to tune the GS PL transition to the 1.3 μm spectral region [3,17, 51] and to narrow a PL line width [52]. However, the structural and electronic properties of InAs QDs coupled with In_xGa_{1-x} As/GaAs QWs are still understood partially. It is generally supposed that this type of growth provides the potential to strain engineering of structural and electronic properties due to efficient strain relaxation altering the electronic potential of capped QDs [51, 53]. Additionally the In_xGa_{1-x} As layer reduces the inhomogeneous surface stress enhancing the ability to grow a multitude of identical uncorrelated QD layers [51]. The majority of publications considered the stress variation in the vicinity of QDs or ordered arrays of QDs [52-58].

To investigate the strain levels in QD structures the X ray diffraction (XRD) has been studied. The XRD experiments were made using the XRD equipment model of D-8 advanced (Bruker Co.) with $K_{\alpha 1}$ line from the Cu source (λ=1.5406Å). Figure 12 present the superposition of XRD peaks related to the diffraction of $K_{\alpha 1}$ and $K_{\alpha 2}$ lines of Cu source from the (400) crystal planes of the cubic GaAs substrate and GaAs layers in studied InGaAs/GaAs QWs [59.60].

	(400)				(200)		
Material	$2\theta_1$ (degree) $K_{\alpha 1}$	$2\theta_1$ (degree) K_β	d_1, A	$2\theta_2$ (degree) $K_{\alpha 2}$	$2\theta_1$ (degree) $K_{\alpha 1}$	d_2, A	$2\theta_3$ (degree) $K_{\alpha 2}$
GaAs Bulk	66.044	59.0165	1.414	66.225	31.63	2.828	31.71

Table 5. Values of 2θ angles for the diffraction of $K_{\alpha 1}$, $K_{\alpha 2}$ and K_β X-ray beams of Cu source from the (400) and (200) cubic GaAs crystal planes [61].

As one can see the peaks (66.05° and 66.24°) related to the diffraction of $K_{\alpha 1}$ and $K_{\alpha 2}$ lines from the (400) crystal planes in GaAs QW structures #2, #3 and #4 with QDs grown at 490-525 °C locate very close to corresponding XRD peaks (66.044° and 66.225° [61]) of the bulk cubic GaAs (Fig.12, Table 5). The last fact testifies that the level of elastic strain in InGaAs/GaAs QWs of #2, #3, #4 is minimum. In contrary in the structures with QDs grown at 470 and 535 °C the corresponding XRD peaks shift to 66.10° and 66.29° (Fig.12) testifying the higher levels of compressive strains in InGaAs/GaAs QWs of these structures.

Figure 13 presents the additional confirmation of conclusions mentioned above. It shows the superposition of XRD peaks related to the diffraction of K_β line from the same (400) crystal planes of cubic GaAs substrate and of GaAs layers in InGaAs/GaAs QWs. As one can see the peaks related to the K_β line diffraction from the (400) crystal plan of GaAs layers in DWELLs with QDs grown at 490-525 °C (59.036-59.042°) locate close to the corresponding XRD peak in the bulk GaAs (Table 5, Fig.12) for the same (400) crystal plane. Thus the level of elastic strain in these DWELLs is minimum. In DWELLs with QDs grown at 470 and 535 °C the corresponding XRD peaks shift to 59.08° testifing the higher level of compresive strain. Note that the partial relaxation of elastic strain can stimulate the creation of nonradiative recombination centers and due to this decreasing the PL intensity in the structures with QDs grown at 470 and 535 °C.

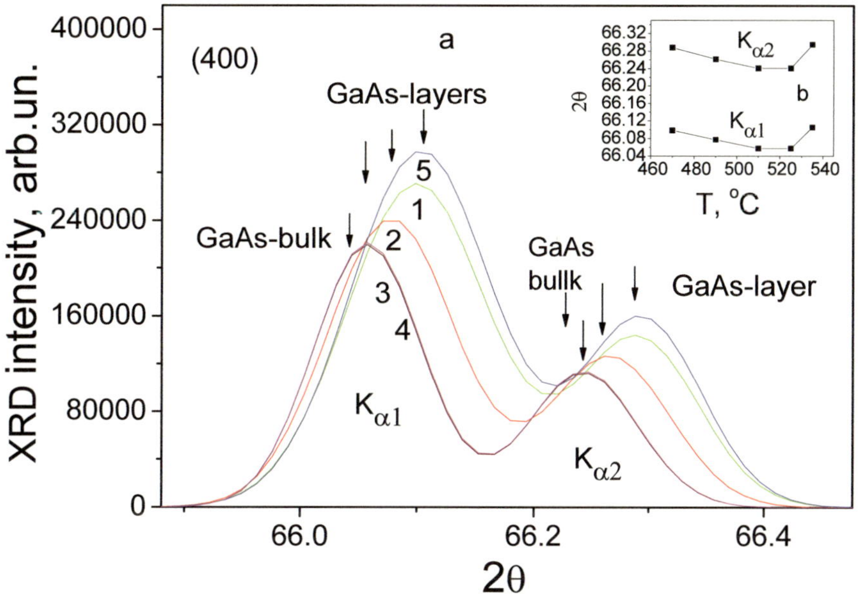

Fig. 12. XRD peaks related to the diffraction of $K_{\alpha 1}$ and $K_{\alpha 2}$ lines on (400) cubic GaAs crystal plane for the structures with QDs grown at 470 (1), 490 (2), 510 (3), 525(4) and 535 °C (5).

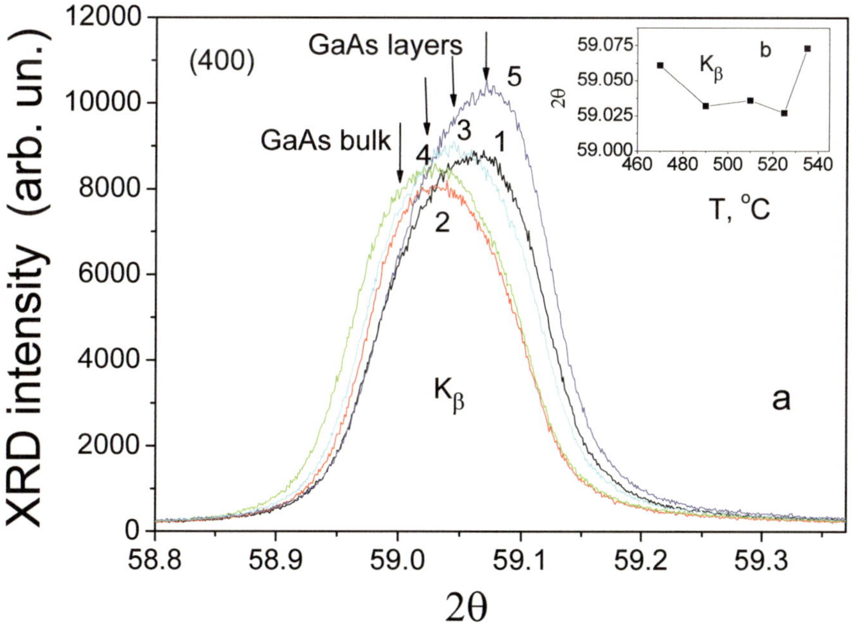

Fig. 13. XRD peaks related to the diffraction of K_{β} line on (400) cubic GaAs crystal plane in DWELLs with QDs grown at 470 (1), 490 (2), 510 (3), 525(4) and 535 °C (5).

Figures 14 presents the superposition of XRD peaks related to the diffraction of $K_{\alpha 1}$ and $K_{\alpha 2}$ lines of the X-ray Cu source from the (200) crystal planes of cubic GaAs substrate and GaAs layers in studied $In_{0.15}$ $Ga_{0.85}As$ /GaAs QWs. As one can see the peaks (31.69-31.70° and 31.77-31.78°) related to the diffraction of $K_{\alpha1}$ and $K_{\alpha2}$ lines from the (200) crystal planes in GaAs QW layers with QDs grown at 490-525 °C locate more close to the corresponding XRD peaks (31.63° and 31.71° [22]) of the bulk cubic GaAs (Fig.14). The last fact indicates that the level of elastic strain in $In_{0.15}$ $Ga_{0.85}As$/GaAs QWs of #2, #3, #4 is smaller than in the structures #1 and #5.

In the QD structures with QDs grown at 470 and 535 °C the corresponding XRD peaks shift to higher angles (31.72° for $K_{\alpha1}$ and 31.80° for $K_{\alpha2}$) testifying the higher levels of compressive strain in the QWs of structures #1 and #5 (Fig.14). The lowest integrated PL intensities have been detected in the QD structures #1 and #5, apparently, due to the high concentration of nonradiative (NR) defects. The high level of elastic strain enhances, apparently, partial stress relaxation in the structures #1 and #5 that accompanies by the appearance of NR defects.

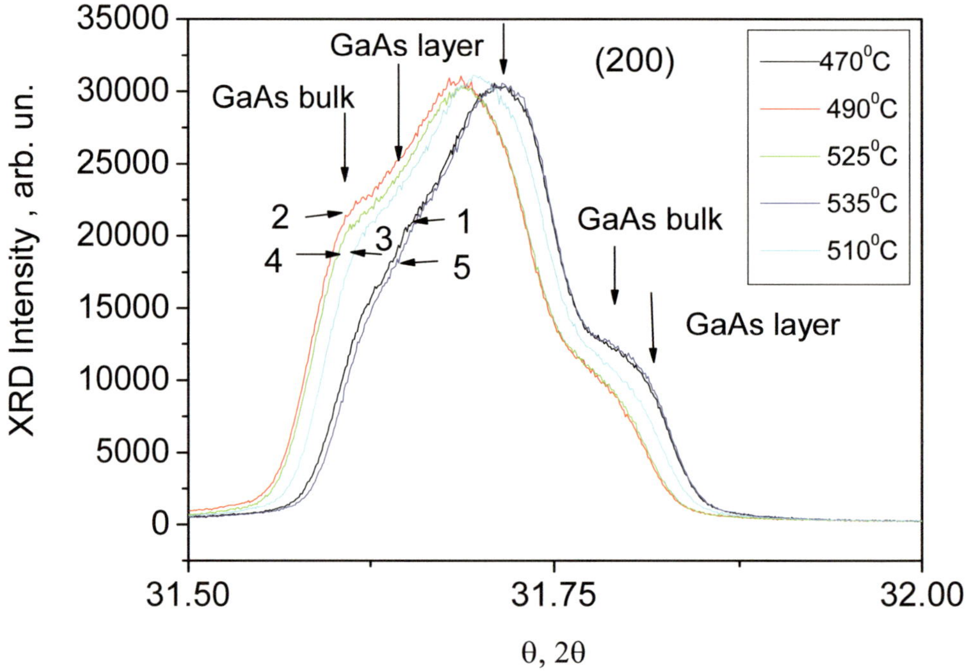

Fig. 14. XRD peaks related to the diffraction of $K_{\alpha1}$ and $K_{\alpha2}$ lines of the X-ray Cu source from the (200) crystal planes in the GaAs substrate and GaAs QW layers of studied structures: 1-#1, 2- #2, 3- #3, 4-#4 and 5 - #5.

12. Elastic strain in symmetric InAs QD structures with different QD densities

Let us to discuss the reason of quality changes in studied DWELL structures. It is essential that the application of the buffer and capping $In_xGa_{1-x}As$ layers coupled with InAs QDs stimulates the lattice mismatch and stress decreasing in the vicinity of QDs, but simultaneously to enhance the lattice mismatch and stress increasing at the InGaAs /GaAs interface for the surface area between QDs. With reducing the QD density versus growth temperature the surface area of QDs reduces (Table 1) and the InGaAs/GaAs interface area between the QDs enlarges. The variation non-monotonically of the integrated PL intensity versus QD density in studied structures, apparently, is connected with the competition of mentioned above two effects.

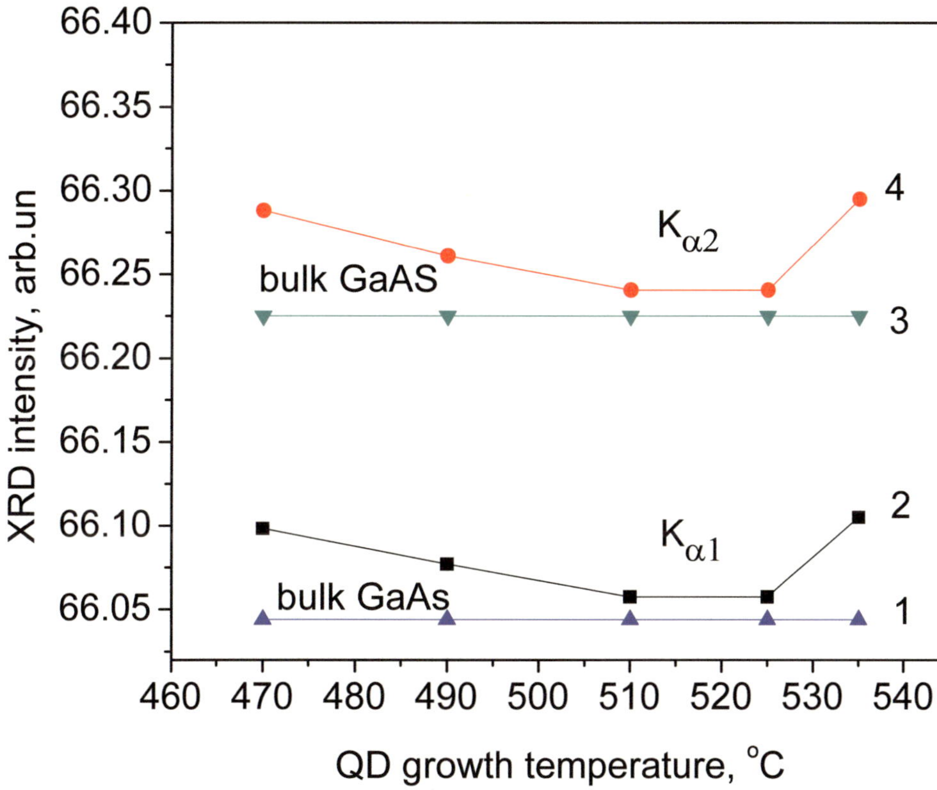

Fig. 15. The 2Θ angles for the diffraction of $K_{\alpha 1}$ and $K_{\alpha 2}$ lines from the (400) crystal planes of the GaAs substrate and GaAs QW layers in DWELLs with QD grown at different temperatures (or with different QD densities).

The X-ray difraction results confirm the mentioned explication. It is knom that the value of elastic deformation can be estimated by following [62]:

$$\varepsilon = -(\theta - \theta_o)\cot\theta_o \,, \tag{10}$$

where Θ and Θ_o are the diffraction angles: in the strained layer (Θ) and in the reference layer without strain (Θ_o). The Θ_o value in present cases has been choosen as the diffraction angle meaning in the bulk cubic GaAs [61]. Thus in studied structures the elastic deformation (or elastic strain) is proportional to the difference between the diffraction angles measured for the GaAs QW layers and the bulk GaAs (Fig.15). Actually as it follows from Fig.15 the variation of elastic deformation in studied structures versus QD growth temperatures (QD densities) has the non monotonous behavior: decreasing in the structures #2, #3 and #4 and increasing in the structures #1 and #5 [38].

Note, that in #3, #4 and #5 the shift of the dependence of PL intensity versus PL peak position into the high energy range (Fig.7) can be explained as well by elastic strain increasing with the enlargement of QD growth temperatures. The high level of elastic strain enhances, apparently, stress relaxation partially in the low quality structures #1 and #5 with the appearance of NR defects (NR1 and NR2) in the GaAs barriers and InGaAs QWs respectively. Simultaneously the shift of GS PL peak to high energy in #5 with larger QD lateral sizes (Table 1) is the subsequence of essential compressive strain in this structure. Note that the compresive strain can stimulate the Ga/In inter-diffusion process in #1 and #5 that is accompained by the PL peak shift into high energy side as well (Fig.7) [63,64].

The low PL intensity in #1 can related to coupling and/or the coalescence of QDs (revealed by AFM for highest concentration of QDs) which can stimulate the NR defect generation, as well as the activation of exciton (or electron/hole) tunneling between the QDs [38]. Finally, the structures with less levels of elastic strain (#2, #3 and #4), have been characterized by the high PL intensity (Fig.3a) and by the shift of GS PL peak emission in low energy spectral range (Fig.2).

13. Conclusions

The photoluminescence, its temperature and power dependences as well as PL inhomogeneity and X ray diffraction has been studied in the symmetric $In_{0.15}Ga_{1-0.15}As/GaAs$ quantum wells coupled with InAs quantum dots. The different QD densities in DWELLs were achieved by the variation of QD growth temperatures between 470 and 535 ^oC. It is shown four reasons for the variation of emission intensities, PL peak positions and PL inhomogeneity in studied QD structures: i) the high concentration of nonradiative recombination centers in the capping $In_{0.15}Ga_{1-0.15}As$ layer at low QD growth temperatures (470^oC), ii) the QD density and size distributions along the wafer for DWELLs with QD grown at 510-535^oC, ii) the high concentration of nonradiative recombination centers in the GaAs barrier at the QD growth temperature of 535^oC and iv) the non monotonous behavior of elastic strain in DWELLs versus QD density. XRD testifies that with decreasing the density of QDs from 1.1 10^{11} cm^{-2} down to 1.3 10^{10} cm^{-2} the level of compressive strain in DWELLs varies not monotonously. The DWELLs with minimum of elastic strain (#2, #3 and #4) are characterized by the higher PL intensity and by the shift of PL peak to 1.3μm (300K) that is important for the application in optical fiber lasers.

14. Acknowledgements

The work was supported by CONACYT Mexico (project 130387) and by SIP-IPN, Mexico. The author thanks Dr. A. Stintz from Center of High Technology Materials at University of New Mexico, Albuquerque, USA, for growing the studied QD structures and Dr. G. Gómez Gasga for XRD measurements.

15. References

[1] D. Bimberg, M. Grundman, N. N. Ledentsov, Quantum Dot Heterostructures, Ed. Wiley & Sons (2001) 328.

[2] V. M. Ustinov, N. A. Maleev, A. E. Shukov, A. R. Kovsh, A. Yu .Egorov, A. V. Lunev, B. V. Volovik, I. L. Krestnikov, Yu. G. Musikhin, N. A. Bert, P. S. Kopev, Zh .I. Alferov, N. N. Ledentsov, D. Bimberg, Appl.Phys.Lett. 74, 2815 (1999).

[3] G. T. Liu, A. Stintz, H. Li, K. J. Malloy and L. F. Lester, Electron Lett, 35, 1163 (1999).

[4] Stintz, G. T. Liu, L. Gray, R. Spillers, S. M. Delgado, K. J. Malloy, J. Vac. Sci.Technol. B. 18(3), 1496 (2000).

[5] T. V. Torchynska, J. L. Casas Espínola, E. Velazquez Losada, P. G. Eliseev, A. Stintz, K. J. Malloy, R. Peña Sierra, Surface Science 532, 848 (2003).

[6] Y. T. Dai, J. C. Fan, Y. F. Chen, R. M. Lin, S. C. Lee, H. H. Lin, J. Appl. Phys. 82, 4489 (1997).

[7] M. A. Kapteyn, M. Lion, R. Heitz, and D. Bimberg, P. N. Brunkov, B. V. Volovik, S. G. Konnikov, A. R. Kovsh, and V. M. Ustinov, Appl. Phys. Lett. 76, 1573 (2000)

[8] A. Duarte, E. C. F. da Silva, A. A. Quivy, M. J. da Silva, S. Martini, J. R. Leite E. A. Meneses and E. Lauretto, J. Appl. Phys., 93, 6279 (2003).

[9] X.Q. Meng, B. Xu, P. Jin, X.L. Ye, Z.Y. Zhang, C.M. Li, Z.G. Wang, Journal of Crystal Growth 243, 432 (2002).

[10] L. Seravalli, P. Frigeri, M. Minelli, P. Allegri, V. Avanzini, S. Franchi, Appl. Phys. Lett. 87, 063101 (2005).

[11] I. Lubyshev, P. P. Gonzalez-Borrero, E. Marega, Jr. E. Petitprez, N. La Scala, Jr. and P. Basmaji, Appl. Phys. Lett. 68, 205 (1996).

[12] C. Lobo, R. Leon, S. Marcinkevičius, W. Yang, P. C. Sercel, X. Z. Liao, J. Zou, and D. J. H. Cockayne, Phys. Rev. B 60, 16647 (1999).

[13] M. Grundmann and D. Bimberg, Phys. Rev. B 55, 9740 (1997).

[14] W. H. Chang, T. M. Hsu, N. T. Yeh, and J. I. Chyi, Phys. Rev. B 62, 13040 (2000).

[15] J. W. Tomm, T. Elsaesser, Y. I. Mazur, H. Kissel, G. G. Tarasov, Z. Y. Zhuchenko, and W. T. Masselink, Phys. Rev. B 67, 045326 (2003).

[16] T. E. Nee, Y. F. Wu, Ch. Ch. Cheng, and H. T. Shen, J. Appl. Phys. 99, 013506 (2006).

[17] T. V. Torchynska, J. L. Casas Espinola, L. V. Borkovska, S. Ostapenko, M. Dybic, O. Polupan, N. O. Korsunska, A. Stintz, P. G. Eliseev, K. J. Malloy, J. Appl. Phys., 101, 024323 (2007).
[18] S. Sanguinetti, M. Henini, M. Grassi Alessi, M. Capizzi, P. Frigeri, S. Franchi, Phys. Rev. B 60, 8276 (1999)
[19] C. Le Ru, J. Fack and R. Murray, Rhys. Rev. B. 67, 245318 (2003).
[20] S. Khatsevich, D. H. Rich, E.T. Kim and A. Madhukar, J. Appl. Phys. 97, 123520 (2005).
[21] Yu. I. Mazur, B. L. Liang, Zh. M. Wang, D. Guzun, G. J. Salamo, Z. Ya. Zhuchenko and G. G. Tarasov, Appl. Phys. Lett., 89,151914 (2006).
[22] Bacher, C. Hartmann, H. Schweizer, T. Held, G. Mahler, H. Nickel, Phys. Rev. B. 47, 9545 (1993).
[23] T. V. Torchynska, M. Dybiec, S. Ostapenko, Phys. Rev. B. 72, 195341 (2005).
[24] Guffarth, R. Heitz, A. Schliwa, O. Stier, A. R. Kovsh, V. Ustinov, N. N. Ledentsov, D. Bimberg, phys. stat. sol. (b) 224, 61 (2001).
[25] S. Sauvage, P. Boucaud, J. M. Gerard, V. Thierry-Mieg, J. Appl. Phys. 84, 4356 (1998).
[26] S.-K. Eah, W. Jhe, Y. Arakawa, Appl. Phys. Lett. 80, 2779 (2002).
[27] M. Dybiec, L. Borkovska, S. Ostapenko, T. V. Torchynska, J. L. Casas Espinola, A. Stintz and K. J. Malloy, Photoluminescence scanning on InAs/InGaAs Quantum Dot Structures, Applid. Surface Science, 252, 5542 (2006).
[28] M. Dybiec, S. Ostapenko, T. V. Torchynska, E.Velasquez Lozada, P. G. Eliseev, A. Stintz, K. J. Malloy, Photoluminescence mapping on InAs/InGaAs dot structures, phys. stat. sol. (c), 2, n.8, 2951-2954 (2005).
[29] M. Dybiec, S. Ostapenko, T. V. Torchynska, E.Velasquez Losada, Scanning Photoluminescence Spectroscopy in InAs/InGaAs Quantum Dot Structures, Appl. Phys. Lett. 84, 5165-5167 (2004).
[30] M. Catalano, A. Taurino, M. Lomescolo, L. Vasanelli, M. De Giorgi, A. Passasco, R. Rinaldi, R. Cingolani, J. Appl. Phys. 87, 2261 (2000).
[31] Y. Temko, T. Siziki, K. Jacobi, Appl. Phys. Lett. 82, 2142 (2003).
[32] S.E.J. Jacobs, M. Kemerink, P.M. Koenroad, M. Hopkinson, H.W.M. Salimink, J.H. Walter, Appl. Phys. Lett. 83, 290 (2003).
[33] T.V. Torchynska, J. Appl. Phys., 104, 074315 , n.7 (2008).
[34] T.V. Torchynskaa,, A. Vivas Hernandezb, G. Polupanb, E. Velazquez Lozada, Photoluminescence study and parameter evaluation in InAs quantum dot-in-a-well structures, Material Science and Engineering B. 176 , 331–333 (2011).
[35] L.V. Asryan, R.A. Suris, Semicond. Sci. Technol. 11, 554 (1996).
[36] T.V. Torchynska, J.L. Casas Espinola, P.G. Eliseev, A. Stintz, K.J. Malloy, R. Pena Sierra, Phys. Status Solidi (a) 195 , 209 (2003).
[37] C. Pryor, Phys. Rev. B 57, 7190 (1998).
[38] T.V.Torchynska, A. Stintz, Some aspects of emission variation in InAs quantum dots coupled with symmetric quantum wells, J. Applied Physics, 108, 2, 024316 (2010).
[39] Y.P. Varshni, Physica 34 , 149 (1967).

[40] P.G. Eliseev, H. Li, G.T. Iu, A. Stintz, T.C. Nevell, L.F. Lester, J. Mally, IEEE J. Select. 35 Topics Quant. Electron. 7 (3) (2001) 135.
[41] Y.S. Huang, H. Qiang, F.H. Pollack, G.D. Pettit, P.D. Kirtchner, J.M. Woodall, H. Stagier, L.B. Soresen, J. Appl. Pys. 70 (12) (1991) 7537
[42] Torchynska, T.V., Casas Espinola, J.L., Lopez, H.M.A., Eliseev, P.G., Stintz, A., Malloy, K.J., Pena Sierra, R. *Institute of Physics Conference Series*, 174, 69-71 (2003).
[43] J. L. Casas Espinola, M. Dybiec, S. Ostapenko, T. V. Torchynska and G. Polupan, J. of Physics: Conference Series, 61, 180-184 (2007).
[44] L. Casas Espinola, 1 T. V. Torchynska, G. Polupan, R. Pena Sierra, phys. stat. sol. (c), 4, n.2, 379-381 (2007).
[45] M. De Giorgi, C. Lingk, G. von Plessen, J. Feldmann, S. De Rinaldis, A. Passaseo, M. De Vittorio, R. Cingolani, M. Lomascolo, Appl. Phys. Lett. 79, 3968 (2001).
[46] S. Marcinkevicius, R. Leon, Phys. Rev. B., 59, 4630 (1999).
[47] B. Ohnesorge, M. Albrecht, J. Oshinowo, A. Forchel, Y. Arakawa, Phys. Rev. B., 54, 11532 (1996).
[48] T.V. Torchynska, E. Velazquez Lozada, J.L. Casas Espinola, J. Vac. Scien. & Tech. B 27(2), 919-922, (2009)
[49] T.V. Torchynska, Superlattice and Microstructure, 45, 349-355 (2009)
[50] T. V. Torchynska, J. L. Casas Espinola, E. Velasquez Lozada, L. V. Shcherbyna, A. Stintz, R. Pena Sierra, Localization of defects in InAs QD symmetric InGaAs-GaAs DWELL structures, Physica B: Condensed Matter, 401-402, 584-586 (2007), ISSN 0921-4526.
[51] E. Zhukov, A. R. Kovsh, N. A. Maleev, S. S. Mikhrin, V. M. Ustinov, A. F. Tsatsulnikov, M. V. Maximov, B. V. Volovik, D. A. Bedarev, Y. M. Shernyakov, P. S. Kopev, Z. I. Alferov, N. N. Ledentsov, D. Bimberg, Appl. Phys. Lett. 75, 1926 (1999).
[52] K. Nishi, H. Saito, S. Sugou, J. S. Lee, Appl. Phys. Lett. 74, 1111 (1999).
[53] N. T. Yeh, T. E. Nee, J. I. Chyi, T. M. Hsu, C. C. Huang, Appl. Phy. Lett. 76, 1567 (2000).
[54] F. Guffarth, R. Heitz, A. Schliwa, O. Stier, N. N. Ledentsov, A. R. Kovsh, V. M. Ustinov, D. Bimberg, Phys. Rev. B, 64, 085305 (2001).
[55] E. Romanov, G. E. Beltz, W. T. Fischer, P. M. Petroff, J. S. Speck, J. Appl. Phys., 89, 4523 (2001).
[56] Y. Y. Lin, J. Singh, J. Appl. Phys. 92, 6205 (2002).
[57] B. Yang, J. Appl. Phys. 92, 3704 (2002).
[58] Y. Nabetani, T. Matsumoto, G. Sasikala, I. Suemune, J. Appl. Phys. 98, 063502 (2005).
[59] T.V.Torchynska, J. Palacios Gomez, G. Gomez Gasga, A. Vivas Hernandez, E. Velazquez Lozada, G. Polupan, Ye. S. Shcherbyna, J. of Physics, Conference Ser. 245, 012060 (2010).
[60] J.L. Casas Espinola, T V Torchynska, J Palacios Gomez, G Gómez Gasga, A Vivas Hernandez and R. Cisneros Tamayo, phys.stat.solid. (c). 8, No. 4, 1391-1393 (2011)
[61] Yeh CY, Lu ZW, Froyen S and Zunger A, *Phys. Rev.* B, 46, 10086 (1992).
[62] S. Ejiri, T. Sasaki, and Y. Hirose, Thin Solid Films 307, 178 (1997)

[63] J. L. Casas Espínola, T. V. Torchynska, G. Polupan, and M. Ojeda Martínez, phys.stat.solid.(c) 8, No. 4, 1388–1390 (2011)
[64] J. Casas Espinola, V. Torchynska, G 1 . Polupan, E. Velazquez Lozada, Material Science and Engineering, B., 165, 115- 117 (2009)

7

Photoionization Cross Sections of Atomic Impurities in Spherical Quantum Dots

C. Y. Lin and Y. K. Ho
Institute of Atomic and Molecular Sciences, Academia Sinica
Taiwan

1. Introduction

With the advances of experimental techniques in fabrication and investigation of nano-scale structures, confined atomic systems become practical and useful models for the illustration of interesting phenomena arising from a system in dimensions comparable to the electronic de Broglie wavelength. The confined atomic models are widely used to study a variety of physical systems, such as impurities in quantum dots (Lin & Ho (2011)), atoms encapsulated in fullerenes (Connerade et al. (1999); Dolmatov et al. (2004)), and atoms under high pressure (de Groot & ten Seldam (1946); Michels et al. (1937)). In this article, we focus on the quantum confinement occurring in quantum dots. The emphasis is placed on the variation of electronic structures and photoionization properties of atomic impurities under the spatial confinement effect of quantum dots.

As a quantum confinement system, the quantum dot has attracted considerable attention due to not only its theoretical but also practical significance. In addition to the analogies of discrete structure in their optical and electrical features between a quantum dot and an atom, the coupled quantum dots provide a model to mimic molecules with tunable bonds (Alivisatos (1996); Schedelbeck et al. (1997)). On the other hand, the quantum dots also serve as contrast agents in bioimaging for biotechnological applications (Michalet et al. (2005)). It is well known that the quantum dot with atomic impurities is a suitable model for studying the semiconductor heterostructures. Recently, the enhancement of semiconductor nano-crystal performance due to the impurities has been reported in the literature (Cao (2011)), which indicates, for instance, that magnetic impurities can be doped to tune optical and magnetic properties.

The physical properties of confined atomic systems are greatly influenced by confinement potentials, which are unable to be determined through the direct measurement of experiment. Although *ab initio* calculations can comprehensively deal with the interaction in confined atomic systems, they may not provide a direct and simple physical interpretation. The usage of semi-empirical model potentials to mimic the interaction of confined atom and surrounding environment provides an efficient way to study the complex systems. The appropriate models, which might not treat the system comprehensively but take the important interaction into account, give a clear physical insight into complex problems. The confinement potentials associated with the structures of quantum dots are often modelled by the rectangular potential well

$$V_{\mathrm{RECT}}(r) = \begin{cases} -V_0 & r \leq R; \\ 0 & r > R, \end{cases} \quad (1)$$

or the harmonic oscillator (parabolic) potential.

$$V_{\mathrm{HO}}(r) = -V_0 + \frac{V_0}{R^2} r^2, \quad (2)$$

where R determines the size of quantum dot, and V_0 gives the strength of confinement. The rectangular potential well has a simple but unrealistic form due to the non-parabolic shape at the center of quantum dots. Although the harmonic oscillator potential fulfils the parabolic property, the infinite depth and range of potential restrict the calculation of continuum states and fail to describe the charging of quantum dots with the finite number of electrons.

The Woods-Saxon potential given as

$$V_{\mathrm{WS}}(r) = \frac{V_0}{1 + \exp\left[(R - r)/\gamma\right]}, \quad (3)$$

where γ controls the slope of confinement potential, also has been used in the study of confined quantum system (Costa et al. (1999); Xie (2009)). It should be noted that the Woods-Saxon potential turns to be the rectangular potential well as $\gamma \to 0$. Another confinement potential flexible to model the different type of quantum dots is the so called power-exponential potential (Ciurla et al. (2002)),

$$V_{\mathrm{EP}}(r) = -V_0 \exp\left[-(r/R)^p\right]. \quad (4)$$

With the change of parameter p, the shape of potential is modified from the Gaussian potential $p = 2$ to the rectangular potential well $p = \infty$.

In this work, the systems of atomic impurities in spherical quantum dots characterized by finite oscillator (FO) and Gaussian potentials (Adamowski et al. (2000a;b); Kimani et al. (2008); Winkler (2004)) are investigated using the method of complex-coordinate rotation (Ho (1983); Reinhardt (1982)) in a finite-element discrete variable representation (FE DVR) (Balzer et al. (2010); Rescigno & McCurdy (2000)). The finite oscillator potential V_{FO} and Gaussian potential V_{G} are defined as

$$V_{\mathrm{FO}}(r) = -A\left(1 + \frac{B}{\sqrt{A}} r\right) \exp\left(-\frac{B}{\sqrt{A}} r\right) \quad (5)$$

and

$$V_{\mathrm{G}}(r) = -C \exp(-r^2/D^2), \quad (6)$$

where A and C are the confining strength of potentials, and the radii of dots are characterized inherently by $1/B$ and D for FO and Gaussian potentials, respectively. Figure 1 shows the examples of both potentials. The Gaussian potential being a special case of the power-exponential potential (Ciurla et al. (2002)) has a soft boundary of the potential. The one-electron energy spectrum for a Gaussian potential has been calculated by Adamowski et al. (Adamowski et al. (2000a;b)) using the variational method with Gaussian-type basis functions. The finite oscillator potentials as weakly confining potentials of quantum dots have been used to study the two-electron quantum dots by Winkler (Winkler (2004)), and later applied to few-electron quantum dots (Kimani et al. (2008)). Both potentials have r^2-dependence near the center of quantum dots, which is the typical character of harmonic oscillators. It should be noted that the impurity is not taken into account for the quantum dots in above-mentioned investigations.

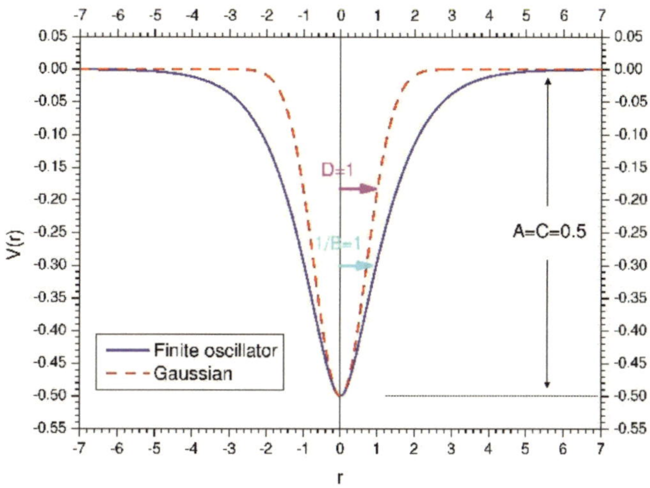

Fig. 1. Comparison of finite oscillator potential with Gaussian potential.

For the investigations of electronic structure and optical properties of atomic impurities in quantum dots, many efforts have been devoted to study hydrogenic impurity states in spherical quantum dots described by finite and infinite potential wells. The state energies of hydrogen impurity in spherical quantum dots with the infinite and finite well of rectangular potentials are explored by Chuu et al. (Chuu et al. (1992)), Yang et al. (Yang et al. (1998)), and Huang et al. (Huang et al. (1999)). Photoionization cross sections and oscillator strengths of hydrogenic impurities in spherical quantum dots are also obtained for the infinite and finite rectangular well models by Ham and Spector (Ham & Spector (n.d.)), Şahin (Şahin (2008)), and Stevanović (Stevanović (2010)). Recently, Lin and Ho (Lin & Ho (2011)) study the photoionization of hydrogen impurities in spherical quantum dots using the finite oscillator and Gaussian potentials. Chakraborty and Ho (Chakraborty & Ho (2011)) adopt the finite oscillator potential to describe the quantum dot for exploring the autoionization resonance states of helium impurities in quantum dots. In the present work, the alkali-metal atoms as impurities in the quantum dots are studied. On the basis of the finite oscillator and Gaussian models, the energy levels and photionization cross sections subject to the quantum confinement effect are illustrated.

The chapter is organized as follows. In Sec. 2, the FE DVR approach and complex-coordinate rotation method associated with the current work are described. The energy spectrum and photoionization cross sections varying with the different conditions of quantum dots for the lithium and sodium impurities are presented and discussed in Sec. 3. Section 4 summarizes this work and gives conclusions. Atomic units are used throughout unless otherwise noted.

2. Theoretical method

2.1 Finite-element discrete variable representation

The finite-element discrete variable representation (FE DVR) is a hybrid computation scheme taking advantage of the finite-element approach and the discrete variable representation to obtain the sparse kinetic-energy matrices and the diagonal representation of potential-energy matrices. Using this hybrid approach, the kinetic-energy matrix is block diagonal with matrix elements in compact expressions, and the potential-energy matrix elements are given by the potential values at grid points. This method has been implemented to investigate a variety of interesting physical problems, such as quantum-mechanical scattering problems (Rescigno & McCurdy (2000)), bright solitons in Bose-Einstein Condensates and ultracold plasmas (Collins et al. (2004)), non-equilibrium Greenąęs function calculations (Balzer et al. (2010)), and photoionization of impurities in quantum dots (Lin & Ho (2011)).

In the present work, the method of FE DVR which is detailed in references (Rescigno & McCurdy (2000)) and (Balzer et al. (2010)) is adopted to obtain the Hamiltonian matrix elements for atoms confined by quantum dots. Within the framework of FE DVR, the interval $[0,R_{max}]$ is divided into n_e finite elements, in which each element between $[x^i, x^{i+1}]$ is further subdivided by n_g Gauss quadrature points (see Fig. 2). Taking advantage of the standard Gauss-Lobatto points x_m and weights w_m (Michels (1963)), we define the generalized Gauss-Lobatto points,

$$x_m^i = \frac{1}{2}[(x^{i+1} - x^i)x_m + (x^{i+1} + x^i)], \tag{7}$$

and weights,

$$w_m^i = \frac{w_m}{2}(x^{i+1} - x^i). \tag{8}$$

It should be noted that $x_1^i = x^i$ and $x_{n_g}^i = x^{i+1}$ because $x_1 = -1$ and $x_{n_g} = 1$. In calculations, the integrals are approximated by Gauss-Lobatto quadrature,

$$\int_{x^i}^{x^{i+1}} \psi(r)dr \simeq \psi(x^i)w_1^i + \sum_{m=2}^{n_g-1} \psi(x_m^i)w_m^i + \psi(x^{i+1})w_{n_g}^i. \tag{9}$$

The wave functions are expanded in terms of local basis functions (see Fig. 2),

$$\chi_m^i(x) = \begin{cases} \dfrac{[f_{n_g}^i(x) + f_1^{i+1}(x)]}{\sqrt{(w_{n_g}^i + w_1^{i+1})}} & \text{for m=1 (bridge);} \\ \dfrac{f_m^i(x)}{\sqrt{w_m^i}} & \text{for else (element),} \end{cases} \tag{10}$$

where the Lagrange interpolating polynomials or so-called Lobatto shape functions $f_m^i(x)$ are given as

$$f_m^i(x) = \begin{cases} \prod\limits_{m' \neq m} \dfrac{(x - x_{m'}^i)}{(x_m^i - x_{m'}^i)} & \text{for } x^i \leq x \leq x^{i+1}; \\ 0 & \text{for } x < x^i \text{ or } x > x^{i+1}. \end{cases} \tag{11}$$

The bridge basis function $\chi_1^i(x)$ in Eq. (10) is in charge of connecting the adjacent elements to ensure the continuity of wave functions at end points of each finite element.

Based on the properties of the Lobatto shape functions and the approximation of Gauss-Lobatto quadrature for integrals, the matrix elements of kinetic-energy operator, $T = -\frac{1}{2}\frac{\mathrm{d}^2}{\mathrm{d}x^2}$, in FE DVR are evaluated by analytic formulas,

$$T_{m_1,m_2}^{i_1,i_2} = \langle \chi_{m_1}^{i_1} | T | \chi_{m_2}^{i_2} \rangle = \frac{1}{2}(\delta_{i_1,i_2} + \delta_{i_1,i_2\pm 1}) \int_0^\infty \mathrm{d}x \frac{\mathrm{d}}{\mathrm{d}x}\chi_{m_1}^{i_1}(x) \frac{\mathrm{d}}{\mathrm{d}x}\chi_{m_2}^{i_2}(x)$$

$$= \begin{cases} \frac{1}{2}\frac{\delta_{i_1,i_2}\tilde{T}_{m_1,m_2}^{i_1}}{\sqrt{w_{m_1}^{i_1} w_{m_2}^{i_2}}} & (m_1 > 1, m_2 > 1); \\ \frac{1}{2}\frac{(\delta_{i_1,i_2}\tilde{T}_{n_g,m_2}^{i_1} + \delta_{i_1,i_2-1}\tilde{T}_{1,m_2}^{i_2})}{\sqrt{w_{m_2}^{i_2}(w_{n_g}^{i_1} + w_1^{i_1+1})}} & (m_1 = 1, m_2 > 1); \\ \frac{1}{2}\frac{(\delta_{i_1,i_2}\tilde{T}_{m_1,n_g}^{i_1} + \delta_{i_1,i_2+1}\tilde{T}_{m_1,1}^{i_1})}{\sqrt{w_{m_1}^{i_1}(w_{n_g}^{i_2} + w_1^{i_2+1})}} & (m_1 > 1, m_2 = 1); \\ \frac{1}{2}\frac{(\delta_{i_1,i_2}(\tilde{T}_{n_g,n_g}^{i_1} + \tilde{T}_{1,1}^{i_1+1}) + \delta_{i_1,i_2-1}\tilde{T}_{1,n_g}^{i_2} + \delta_{i_1,i_2+1}\tilde{T}_{n_g,1}^{i_1})}{\sqrt{(w_{n_g}^{i_1} + w_1^{i_1+1})(w_{n_g}^{i_2} + w_1^{i_2+1})}} & (m_1 = m_2 = 1), \end{cases} \quad (12)$$

in which the term $\tilde{T}_{m_1,m_2}^i$ is defined as

$$\tilde{T}_{m_1,m_2}^i = \sum_m \frac{\mathrm{d}f_{m_1}^i(x_m^i)}{\mathrm{d}x} \frac{\mathrm{d}f_{m_2}^i(x_m^i)}{\mathrm{d}x} w_m^i. \quad (13)$$

According to Eq. (11), the first derivatives of the Lobatto shape functions at the quadrature points are given as

$$\frac{df_{m_1}^i(x_m^i)}{dx} = \begin{cases} \frac{1}{(x_{m_1}^i - x_m^i)} \prod_{m' \neq m_1, m} \frac{(x_m^i - x_{m'}^i)}{(x_{m_1}^i - x_{m'}^i)} & \text{for } m_1 \neq m; \\ \frac{1}{2w_{m_1}^i}(\delta_{m_1,n_g} - \delta_{m_1,1}) & \text{for } m_1 = m. \end{cases} \quad (14)$$

The matrix of the local potential-energy operator $V(x)$ in FE DVR has a diagonal representation with matrix element values equal to potential values at grid points, i.e.,

$$V_{m_1,m_2}^{i_1,i_2} = \int_0^\infty \mathrm{d}x \chi_{m_1}^{i_1}(x) V(x) \chi_{m_2}^{i_2}(x) = \delta_{i_1,i_2}\delta_{m_1,m_2}\tilde{V}_{m_1}^{i_1}, \quad (15)$$

with

$$\tilde{V}_m^i = \begin{cases} V(x_m^i) & \text{for } m > 1; \\ \frac{V(x_{n_g}^i)w_{n_g}^i + V(x_1^{i+1})w_1^{i+1}}{w_{n_g}^i + w_1^{i+1}} & \text{for } m = 1. \end{cases} \quad (16)$$

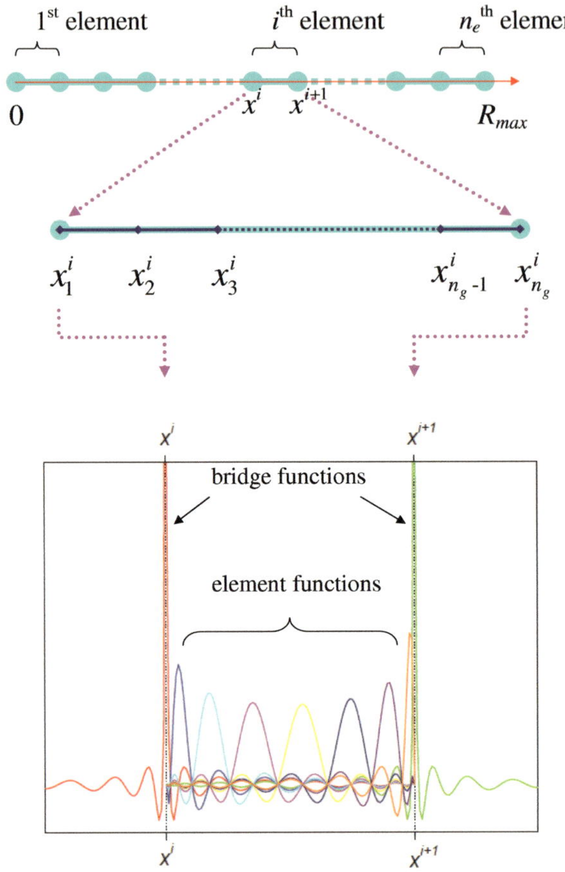

Fig. 2. Interval between 0 and R_{max} divided into n_e finite elements with n_g Gauss quadrature points and selected local basis functions (bridge and element functions) distributed in each finite element.

2.2 The method of complex coordinate rotation

The radial Schrödinger equation for the atomic impurity in spherical quantum dots is given as

$$\left[-\frac{1}{2}\frac{\mathrm{d}^2}{\mathrm{d}r^2} + V(r)\right]\phi(r) = E\phi(r), \tag{17}$$

where $V(r)$ is defined as

$$V(r) = \frac{l(l+1)}{2r^2} + U_a(r) + V_{\mathrm{QD}}(r), \tag{18}$$

where U_a is the atomic potential, and V_{QD} is given by V_{FO} for the confinement of finite oscillator potential (see equation (5)) or V_{G} for the Gaussian potential (see equation (6)). Within the framework of the complex scaling approach, the real coordinate r is transformed to complex coordinate z by the mapping

$$z = r\mathrm{e}^{\mathrm{i}\Theta}, \tag{19}$$

which turns the generalized Gauss-Lobatto points and weights in Eqs. (7) and (8) to be complex, i.e.,

$$x^i_m \to x^i_m e^{i\Theta}, \tag{20}$$

and

$$w^i_m \to w^i_m e^{i\Theta}. \tag{21}$$

The integrations of kinetic- and potential-energy matrix elements are performed along the complex path instead of the real axis. It turns out that the calculations of Eqs. (12) and (16) by the complex quadrature points and weights are equivalent to the usage of the real points and weights with the complex scaled operators. In other words, to obtain the complex scaled matrix elements, we can perform the transformation of the operators in advance,

$$-\frac{1}{2}\frac{d^2}{dr^2} \to -\frac{e^{-2i\Theta}}{2}\frac{d^2}{dr^2} \tag{22}$$

and

$$V(r) \to V(re^{i\Theta}), \tag{23}$$

followed by the implementation of integrals in real quadrature points and weights.

Through the standard diagonalization procedure, complex eigenvalues and eigenvectors are obtained. In the dipole approximation, the photoionization cross sections can be obtained by the optical theorem

$$\sigma(\omega) = \frac{4\pi\omega}{c}\mathrm{Im}(\alpha^-(\omega)), \tag{24}$$

where ω is the photon energy and c is the speed of light, i.e., the inverse fine-structure constant. The negative frequency component of the polarizability $\alpha^-(\omega)$ (Buchleitner et al. (1994); Rescigno & McKoy (1975)) calculated along the complex path C is given as

$$\alpha^-(\omega) = \int d\vartheta \int d\varphi \int_C dz\, z^2 \psi_0^\dagger(z,\vartheta,\varphi)\mu(z,\vartheta,\varphi)\psi^-(z,\vartheta,\varphi), \tag{25}$$

where μ is the component of the dipole operator along the direction of light polarization. The initial wave function ψ_0 with energy E_0 and the final scattered wave function ψ^- fulfill the equation

$$[H(z,\vartheta,\varphi) - E_0 - \omega]\,\psi^-(z,\vartheta,\varphi) = \mu(z,\vartheta,\varphi)\psi_0(z,\vartheta,\varphi). \tag{26}$$

3. Results and discussion

3.1 Lithium impurities in quantum dots

The method of complex coordinate rotation combined with the FE DVR approach are applied to investigate the state energies and photoionization cross sections of lithium impurities in the spherical quantum dots. The model potentials (Schweizer et al. (1999)), which are given as

$$U_a(r) = -\frac{1}{r}\left[\tilde{Z} + (Z - \tilde{Z}\exp(-a_1 r) + a_2 r\exp(-a_3 r))\right], \tag{27}$$

are adopted to simulate the alkali metals for the interaction of multi-electron core with the single valence electron. The parameters a_i ($i = 1,2,3$) of model potentials optimized by a least-square fit to experimental energies are listed in Table 1 for the lithium and sodium atoms. The energies of ground and first few excited states obtained by this model potential for the lithium atom are compared to the experimental data (Ralchenko et al. (2011)) in Table 2. Although the calculated ground-state energy of lithium atom is not as precise as the energy of excited states, the photoionization cross sections of free lithium atom as shown in Table 3 are in good agreement with other theoretical predictions.

Atom	$\bar{Z}$	Z	a_1	a_2	a_3
Li	1	3	3.395	3.212	3.207
Na	1	11	7.902	23.51	2.688

Table 1. Parameters of model potentials for lithium and sodium atoms.

	Theory		Experiment
	Present work	Sahoo & Ho	NIST
$1s^2 2s$	-0.197331	-0.198141	-0.198142
$1s^2 2p$	-0.130068		-0.130235
$1s^2 3s$	-0.074123		-0.074182
$1s^2 3p$	-0.057232		-0.057236

Table 2. Energies of ground and excited states for lithium model potential are compared with experimental values. Results of Sahoo & Ho refer to (Sahoo & Ho (2006)). Experimental data by NIST refer to (Ralchenko et al. (2011)).

ϵ	Present results	Sahoo & Ho	Peach et al.
0.01	1.568	1.470	1.565
0.03	1.638	1.551	1.640
0.05	1.653	1.575	1.659
0.10	1.557	1.500	1.571
0.50	0.568	0.562	0.580
1.00	0.218	0.217	0.218

Table 3. Photoionization cross sections (in units of Mb) of free lithium atom as functions of photoelectron energies ϵ (in atomic units). Results of Sahoo & Ho refer to (Sahoo & Ho (2006)). Data by Peach et al. refer to (Peach et al. (1988)).

In the present work, the energy levels of lithium impurities with the principal quantum number $n = 2$–3 and angular momentum quantum number $l = 0$–1 for the valence electron are calculated for the quantum dots modelled by the FO and Gaussian potentials. In Fig. 3, the $1s^2 2s$ and $1s^2 3s$ state energies of lithium impurities varying with the dot radii, $1/B$ and D for the FO and Gaussian potential, respectively, from 10^{-1} to 10^3 a.u. are displayed for the several confining strengths of potentials, A and C. Since the cases of $1/B = D = 0$ correspond to the free lithium atoms, the levels belonging to different confining strengths merge into one of the free lithium levels as the dot radii approach zero. With increasing the dot radii, the level energies are decreased until reaching a limit, which is equal to the energy of free lithium atom combined with the confining strength of potential, A or C. In other words, the total energy is then shifted down by an amount of A or C for the FO and Gaussian potentials, respectively. As long as the dot radii are small such that the confinement effect is negligible, the level energies are close to the energies of free atoms. The increased dot radii leading to the stronger confinement of quantum dots cause the wave function trapped into the inner region of a deeper potential well. As a result, the corresponding energies are decreased. For a specific confining strength of potential, $A = C = 0.5$ a.u., the energy variations of levels $1s^2 2s$, $1s^2 2p$, $1s^2 3s$, and $1s^2 3p$ with quantum dot radii ranging from 10^{-1} to 10^3 a.u. are shown in Fig. 4.

The rapidly downward shifts in $1s^2np$ levels caused by the confinement effect have analogy to the $1s^2ns$ levels.

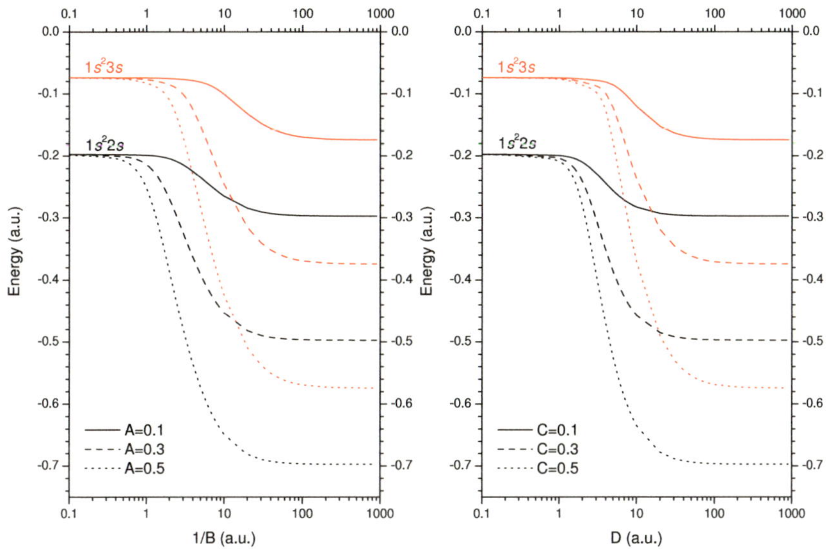

(a) Finite oscillator model (b) Gaussian model

Fig. 3. Energies of $1s^22s$ and $1s^23s$ states as functions of quantum dot radii ($1/B$ or D in atomic units) for several confining strengths of potentials (A or C in atomic units).

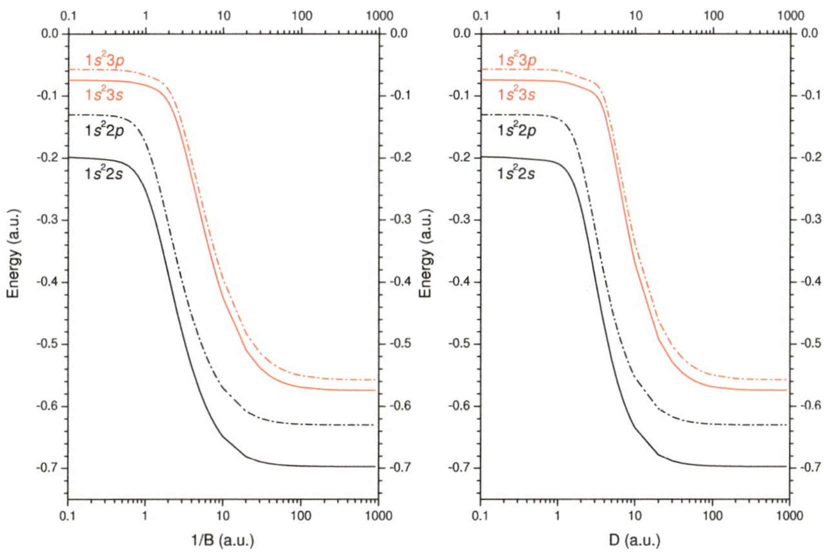

(a) Finite oscillator model (b) Gaussian model

Fig. 4. State energies of $1s^2nl$ ($n = 2$–3 and $l = 0$–1) as functions of quantum dot radii ($1/B$ or D) for the confining strengths of potentials $A = C = 0.5$ a.u.

The influence of quantum dot size on the photoionization cross sections of ground-state lithium impurities is demonstrated in Fig. 5 for the FO and Gaussian potentials with the confining strengths of potentials $A = C = 0.5$. The cross sections varying with the selected radii of quantum dots show the drastic change for photon energies near the ionization threshold. As observed in Fig. 5 for the both models of quantum dots, the cross sections of $1/B = D = 0.3$ are close to the data of the free lithium case ($1/B = D = 0$) for high photoelectron energies, but reduced for low photonelectron energies. With increasing the size of quantum dot, the cross sections are gradually enhanced, and reach maximum values at $1/B \sim 1.5$ and $D \sim 3.0$ for the FO and Gaussian model, respectively. As the dot radius is further increased, the cross sections are deceased from the maximum values gradually.

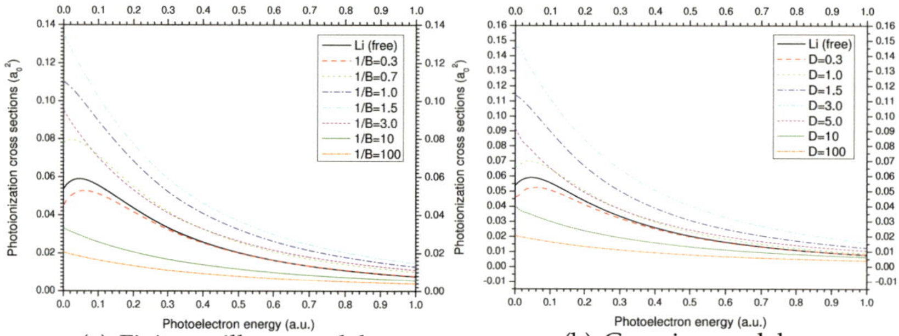

(a) Finite oscillator model (b) Gaussian model

Fig. 5. Photoionization cross sections as functions of photoelectron energies for several radii ($1/B$ and D in atomic units) of quantum dots with confining strengths of potentials $A = C = 0.5$.

One of striking properties due to the quantum confinement effect is the appearance of resonance-like profile in the photoionization cross sections as functions of quantum dot radii for a given photon energy. In Figs. 6, the cross sections as functions of confining strengths A and dot radii $1/B$ of the FO potentials are displayed for photon energies $\omega = 1$ and 3 a.u., respectively. For a given A, the occurrence of resonance-like structure demonstrates the constructive interference between the ground and continuum states due to the wave functions altered by the confinement effect of quantum dots. It is noticed that the peak of resonance-like profile rises with increasing the confining strength of potential. For the Gaussian potentials, the photoionization cross sections as functions of confining strengths C and dot radii D are shown in Fig. 7. The variation of cross sections with the confining strengths of potentials and dot radii resembles the results of the FO potentials, and the resonance-like profile of cross sections is also revealed. The numerical data of photoionization cross sections varying with the photoelectron energies are listed in Tables 4 and 5 for the FO and Gaussian potentials, respectively.

3.2 Sodium impurities in quantum dots

To investigate the state energies and photoionization cross sections of sodium impurities in the spherical quantum dots, we utilize the model potential in Eq. (27) with parameters given in Table 1 to describe the interaction of multi-electron core with the single valence electron for the sodium atom. The energies of ground and first few excited states calculated by this model potential for the sodium atom are compared to the experimental data (Ralchenko et al. (2011)) in Table 6.

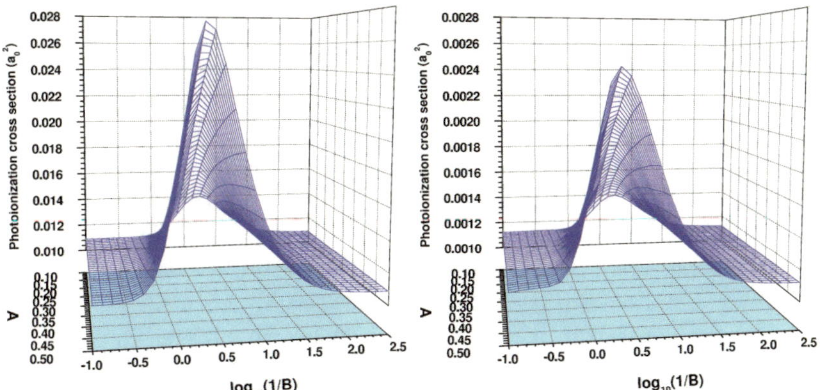

(a) Photon energy $\omega = 1$ a.u.

(b) Photon energy $\omega = 3$ a.u.

Fig. 6. Photoionization cross sections as functions of confining strengths (A in atomic units) and quantum dot radii ($1/B$ in atomic units) for finite oscillator potentials.

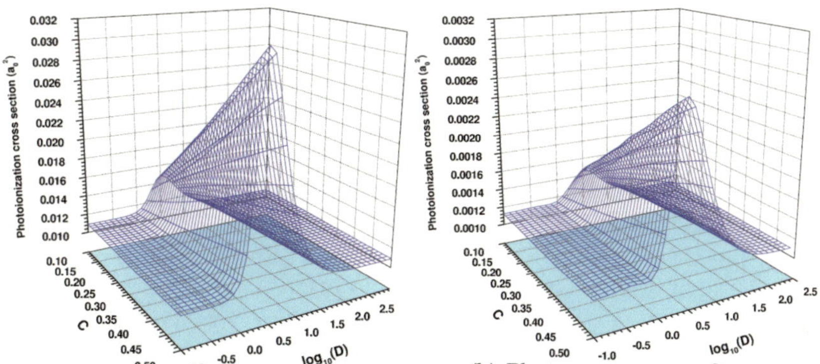

(a) Photon energy $\omega = 1$ a.u.

(b) Photon energy $\omega = 3$ a.u.

Fig. 7. Photoionization cross sections as functions of confining strengths (C in atomic units) and quantum dot radii (D in atomic units) for Gaussian potentials.

In Fig. 8, the energies of levels $2p^6 3s$ and $2p^6 4s$ of sodium impurities varying with the dot radii, $1/B$ and D for the FO and Gaussian potential, respectively, in between 10^{-1} and 10^3 a.u. are presented for the several confining strengths of potentials, A and C. The levels associated with different confining strengths of potentials merge into one of the free sodium levels for the dot radii approaching zero. On the contrary, the level energies split to different energy limits corresponding to the combined energy of free sodium atom with the confining strengths of potential as the radius of quantum dot is large. For a specific confining strength of potential, $A = C = 0.5$ a.u., the energy variations of levels $2p^6 3s$, $2p^6 3p$, $2p^6 4s$, and $2p^6 4p$ with quantum dot radii ranging from 10^{-1} to 10^3 a.u. are shown in Fig. 9.

In Fig. 10, the photoionization cross sections of ground-state sodium impurities in spherical quantum dots characterized by the FO and Gaussian models of confining strengths $A = C = 0.5$ are presented. The variation of cross sections with selected radii of quantum dots shows the influence of quantum confinement on the photoionization. As observed in Fig. 10 for the both model potentials with small parameters $1/B$ and D, the photoionization cross sections

ω_p	$1/B = 0.8$	$1/B = 3.0$	$1/B = 5.0$	$1/B = 10$
0.2	4.79447(-2)	6.26482(-2)	4.75992(-2)	3.13920(-2)
0.4	2.84476(-2)	3.70083(-2)	2.85878(-2)	1.96808(-2)
0.6	1.80247(-2)	2.36864(-2)	1.86717(-2)	1.32830(-2)
0.8	1.21964(-2)	1.61640(-2)	1.29792(-2)	9.46338(-3)
1.0	8.68848(-3)	1.15891(-2)	9.45430(-3)	7.02362(-3)
1.5	4.35280(-3)	5.87191(-3)	4.93428(-3)	3.78125(-3)
2.0	2.52575(-3)	3.43660(-3)	2.94528(-3)	2.29925(-3)
2.5	1.60966(-3)	2.20681(-3)	1.91726(-3)	1.51494(-3)
3.0	1.09482(-3)	1.51144(-3)	1.32609(-3)	1.05669(-3)
3.5	7.81176(-4)	1.08537(-3)	9.59280(-4)	7.69130(-4)
4.0	5.78273(-4)	8.08194(-4)	7.18347(-4)	5.78660(-4)

Table 4. Photoionization cross sections (in a_0^2) as functions of photoelectron energies ω_p (in atomic units) for finite oscillator potentials of $A = 0.3$ a.u. and $1/B = 0.8, 3.0, 5.0$ and 10 a.u. $a(b)$ denotes $a \times 10^b$.

ω_p	$D = 0.8$	$D = 3.0$	$D = 5.0$	$D = 10$
0.2	4.41468(-2)	7.40039(-2)	5.46382(-2)	3.12982(-2)
0.4	2.64195(-2)	4.42053(-2)	3.21495(-2)	1.96037(-2)
0.6	1.67271(-2)	2.81879(-2)	2.06879(-2)	1.32432(-2)
0.8	1.12819(-2)	1.90666(-2)	1.42416(-2)	9.44659(-3)
1.0	8.00689(-3)	1.35329(-2)	1.03097(-2)	7.01908(-3)
1.5	3.97977(-3)	6.70843(-3)	5.34077(-3)	3.78682(-3)
2.0	2.29879(-3)	3.87066(-3)	3.17933(-3)	2.30576(-3)
2.5	1.46226(-3)	2.46429(-3)	2.06729(-3)	1.52060(-3)
3.0	9.94378(-4)	1.67902(-3)	1.42909(-3)	1.06132(-3)
3.5	7.10053(-4)	1.20178(-3)	1.03348(-3)	7.72856(-4)
4.0	5.26265(-4)	8.92958(-4)	7.73777(-4)	5.81669(-4)

Table 5. Photoionization cross sections (in a_0^2) as functions of photoelectron energies ω_p (in atomic units) for Gaussian potentials of $C = 0.3$ a.u. and $D = 0.8, 3.0, 5.0$ and 10 a.u. $a(b)$ denotes $a \times 10^b$.

	Theory		Experiment
	Present work	Sahoo & Ho	NIST
$2p^6 3s$	-0.188860	-0.188857	-0.188858
$2p^6 3p$	-0.111520		-0.111600
$2p^6 4s$	-0.071672		-0.071578
$2p^6 4p$	-0.050985		-0.050934

Table 6. Energies of ground and excited states for sodium model potential are compared with experimental values. Results of Sahoo & Ho refer to (Sahoo & Ho (2006)). Experimental data by NIST refer to (Ralchenko et al. (2011)).

slightly deviate from the data of the free sodium case ($1/B = D = 0$). With increasing the size of quantum dots, the humps of cross sections are enlarged. For $1/B$ and D larger than

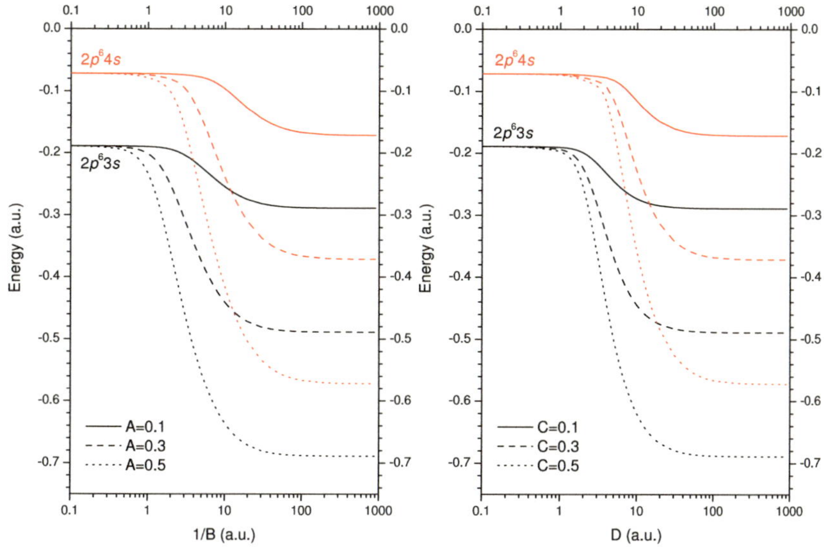

(a) Finite oscillator model (b) Gaussian model

Fig. 8. Energies of $2p^63s$ and $2p^64s$ states as functions of quantum dot radii ($1/B$ or D in atomic units) for several confining strengths of potentials (A or C in atomic units).

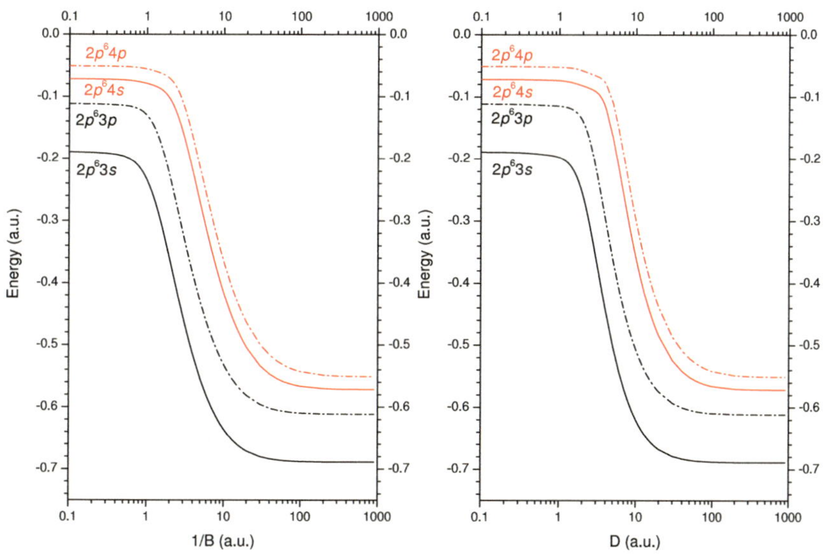

(a) Finite oscillator model (b) Gaussian model

Fig. 9. State energies of $2p^6nl$ ($n = 3$–4 and $l = 0$–1) as functions of quantum dot radii ($1/B$ or D) for the confining strengths of potentials $A = C = 0.5$ a.u.

5, the hump disappears and cross sections are reduced with the further increase of quantum dot radii. The existence of a Cooper minimum in the photoionization cross sections of free sodium atoms is well known (Cooper (1962); Marr & Creek (1968)). It is particular interesting to notice that the Cooper minimum is shifted back and forth and vanished eventually from the threshold energy with the change of quantum dot size.

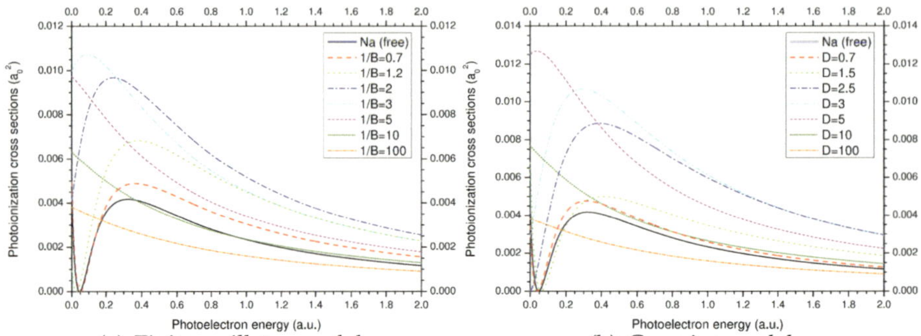

(a) Finite oscillator model (b) Gaussian model

Fig. 10. Photoionization cross sections as functions of photoelectron energies for several radii ($1/B$ and D in atomic units) of quantum dots with confining strengths $A = C = 0.5$.

Although the photoionization cross sections vary enormously and intricately with the size of quantum dots for photon energies near the threshold energy, the cross sections exhibit regular variation and resonance-like behavior for higher photon energies. In Figs. 11, the cross sections as functions of confining strengths A and dot radii $1/B$ of the FO potentials are displayed for photon energies $\omega = 1$ and 3 a.u., respectively. For a given confining strength of potential A, the resonance-like profile can be seen for photoionization cross sections varying with the dot radii $1/B$. The positions of resonance peak are shifted with the change of confining strength A. Because the photoionization cross sections are increased monotonically

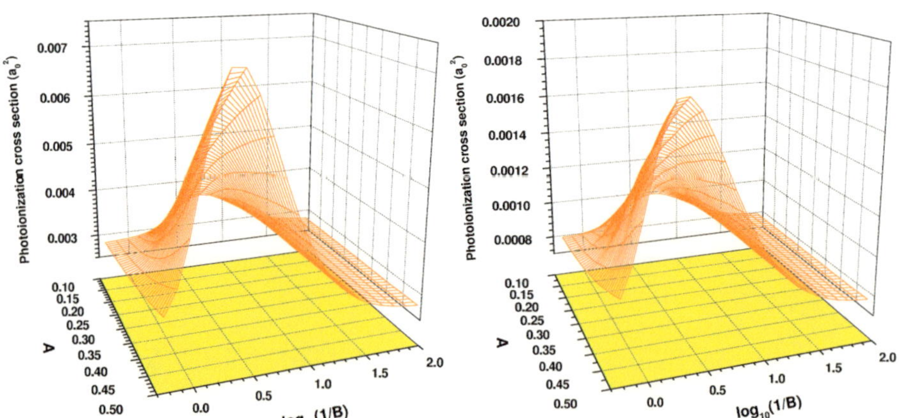

(a) Photon energy $\omega = 1$ a.u. (b) Photon energy $\omega = 3$ a.u.

Fig. 11. Photoionization cross sections as functions of confining strengths (A in atomic units) and quantum dot radii ($1/B$ in atomic units) for finite oscillator potentials.

with increasing the confining strength A for a given dot radius $1/B$, the peak of resonance rises with increasing the confining strengths of potentials.

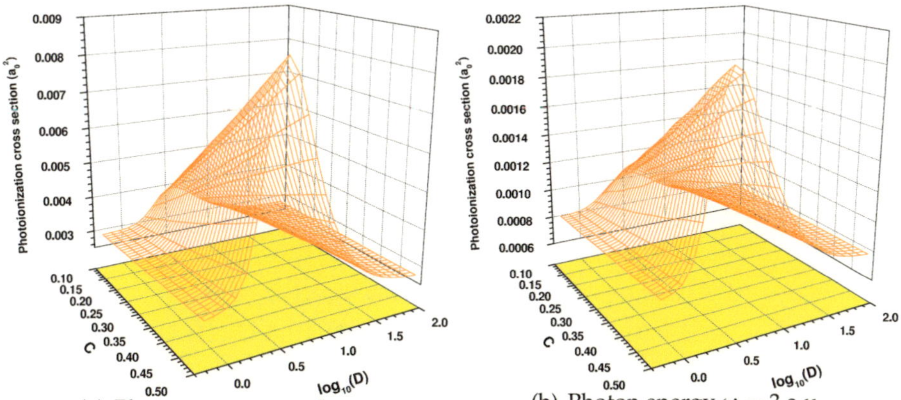

(a) Photon energy $\omega = 1$ a.u. (b) Photon energy $\omega = 3$ a.u.

Fig. 12. Photoionization cross sections as functions of confining strengths (C in atomic units) and quantum dot radii (D in atomic units) for Gaussian potentials.

For the Gaussian model, the photoionization cross sections as functions of confining strengths C and dot radii D are shown in Fig. 12. The variation of cross sections with the confining strengths of potentials and dot radii resembles the results of the FO potentials, and the resonance-like profile of cross sections is also revealed. To make the comparisons of FO model to the Gaussian model for the identical confining strengths $A = C$, the numerical data of photoionization cross sections varying with the photoelectron energies are listed in Tables 7 and 8 for the selected radii of quantum dots.

ω_p	$1/B = 0.8$	$1/B = 3.0$	$1/B = 5.0$	$1/B = 10$
0.2	3.74494(-3)	7.64326(-3)	7.91640(-3)	5.90620(-3)
0.4	4.44278(-3)	7.25065(-3)	6.51128(-3)	4.71764(-3)
0.6	3.81066(-3)	5.99769(-3)	5.19782(-3)	3.79446(-3)
0.8	3.15905(-3)	4.91032(-3)	4.21919(-3)	3.12203(-3)
1.0	2.64072(-3)	4.07371(-3)	3.50190(-3)	2.62573(-3)
1.5	1.80224(-3)	2.74326(-3)	2.38394(-3)	1.83387(-3)
2.0	1.32701(-3)	2.00153(-3)	1.75935(-3)	1.37624(-3)
2.5	1.02749(-3)	1.54056(-3)	1.36656(-3)	1.08106(-3)
3.0	8.23278(-4)	1.22982(-3)	1.09860(-3)	8.75958(-4)
3.5	6.76071(-4)	1.00775(-3)	9.05143(-4)	7.25869(-4)
4.0	5.65586(-4)	8.42125(-4)	7.59630(-4)	6.11833(-4)

Table 7. Photoionization cross sections (in a_0^2) as functions of photoelectron energies ω_p (in atomic units) for finite oscillator potentials of $A = 0.3$ a.u. and $1/B = 0.8, 3.0, 5.0$ and 10 a.u. $a(b)$ denotes $a \times 10^b$.

ω_p	$D = 0.8$	$D = 3.0$	$D = 5.0$	$D = 10$
0.2	3.93638(-3)	6.91812(-3)	9.30767(-3)	6.05882(-3)
0.4	4.41554(-3)	7.77475(-3)	7.57303(-3)	4.77588(-3)
0.6	3.70173(-3)	6.76452(-3)	5.94424(-3)	3.82507(-3)
0.8	3.02736(-3)	5.63360(-3)	4.75905(-3)	3.14297(-3)
1.0	2.50772(-3)	4.69180(-3)	3.91035(-3)	2.64251(-3)
1.5	1.68942(-3)	3.13333(-3)	2.62412(-3)	1.84642(-3)
2.0	1.23644(-3)	2.25964(-3)	1.92403(-3)	1.38660(-3)
2.5	9.54678(-4)	1.72367(-3)	1.48954(-3)	1.08981(-3)
3.0	7.64071(-4)	1.36737(-3)	1.19525(-3)	8.83429(-4)
3.5	6.27312(-4)	1.11558(-3)	9.83654(-4)	7.32305(-4)
4.0	5.24941(-4)	9.29355(-4)	8.24908(-4)	6.17421(-4)

Table 8. Photoionization cross sections (in a_0^2) as functions of photoelectron energies ω_p (in atomic units) for Gaussian potentials of $C = 0.3$ a.u. and $D = 0.8, 3.0, 5.0$ and 10 a.u. $a(b)$ denotes $a \times 10^b$.

4. Conclusions

The lithium and sodium impurities in spherical quantum dots are investigated using the method of complex-coordinate rotation in the finite-element discrete variable representation. Utilizing the FO and Gaussian potentials to mimic the environment of quantum dots, we study the energy spectra and photoionization of alkali metal impurities under the influence of quantum confinement effect. The level energies of impurities in the quantum dots are calculated for the both FO and Gaussian potentials in a variety of dot radii and confining strengths of potentials. The downward shift of impurity energy toward the combined energy of the free atom and the amplitude of the confining strength of potential is exhibited. The quantum confinement effect on the impurity energies due to the FO model is compared to the Gaussian model. The photoionization cross sections as functions of photoelectron energies are presented for the selected dot radii. The sensitivity of cross sections near the threshold energies to the dot radii demonstrates the significance of quantum confinement effect on the photoionization. The photoionization cross sections varying with different dot radii and confining strengths of potentials are given for specific photon energies. The enhancement of the constructive interference between the ground and continuum states due to the quantum confinement leads to the resonance-like profile for the cross sections varying with the dot radii at a given photon energy. The positions of resonance peak are associated with the confining strength. It is noted that the Cooper minimum existing in the photoionization cross sections of sodium impurities is shifted back and forth in energy positions and vanished eventually from the threshold because of the effect of quantum confinement.

This work is financially supported by the National Science Council of Taiwan.

5. References

Adamowski, J., Sobkowicz, M., Szafran, B. & Bednarek, S. (2000a). Electron pair in a gaussian confining potential, *Physical Review B* 62: 4234–4237.

Adamowski, J., Sobkowicz, M., Szafran, B. & Bednarek, S. (2000b). Erratum: Electron pair in a gaussian confining potential [phys. rev. b **62** , 4234 (2000)], *Physical Review B* 62: 13233–13233.

Alivisatos, A. P. (1996). Semiconductor clusters, nanocrystals, and quantum dots, *Science* 271: 933–937.

Balzer, K., Bauch, S. & Bonitz, M. (2010). Efficient grid-based method in nonequilibrium Green's function calculations: Application to model atoms and molecules, *Physical Review A* 81: 022510.

Buchleitner, A., Gremaud, B. & Delande, D. (1994). Wavefunctions of atomic resonances, *Journal of Physics B: Atomic, Molecular and Optical Physics* 27: 2663–2679.

Cao, Y. C. (2011). Impurities enhance semiconductor nanocrystal performance, *Science* 332: 48–49.

Chakraborty, S. & Ho, Y. K. (2011). Autoionization resonance states of two-electron atomic systems with finite spherical confinement, *Physical Review A* 84: 032515.

Chuu, D. S., Hsiao, C. M. & Mei, W. N. (1992). Hydrogenic impurity states in quantum dots and quantum wires, *Physical Review B* 46: 3898–3905.

Ciurla, M., Adamowski, J., Szafran, B. & Bednarek, S. (2002). Modelling of confinement potentials in quantum dots, *Physica E: Low-dimensional Systems and Nanostructures* 15: 261 – 268.

Collins, L. A., S, M., Kress, J. D., Schneider, B. I. & Feder, D. L. (2004). Time-dependent simulations of large-scale quantum dynamics, *Physica Scripta* T110: 408–412.

Connerade, J. P., Dolmatov, V. K., Lakshmi, P. A. & Manson, S. T. (1999). Electron structure of endohedrally confined atoms: atomic hydrogen in an attractive shell, *Journal of Physics B: Atomic, Molecular and Optical Physics* 32: L239–L245.

Cooper, J. W. (1962). Photoionization from outer atomic subshells. a model study, *Physical Review* 128: 681–693.

Costa, L. S., Prudente, F. V., Acioli, P. H., Neto, J. J. S. & Vianna, J. D. M. (1999). A study of confined quantum systems using the Woods-Saxon potential, *Journal of Physics B: Atomic, Molecular and Optical Physics* 32: 2461–2470.

Şahin, M. (2008). Photoionization cross section and intersublevel transitions in a one- and two-electron spherical quantum dot with a hydrogenic impurity, *Physical Review B* 77: 045317.

de Groot, S. R. & ten Seldam, C. A. (1946). On the energy levels of a model of the compressed hydrogen atom, *Physica* 12: 669–682.

Dolmatov, V. K., Baltenkov, A. S., Connerade, J. P. & Manson, S. T. (2004). Structure and photoionization of confined atoms, *Radiation Physics and Chemistry* 70: 417–433.

Ham, H. & Spector, H. N. (n.d.). Photoionization cross section of hydrogenic impurities in spherical quantum dots.

Ho, Y. K. (1983). The method of complex coordinate rotation and its applications to atomic collision processes, *Physics Reports* 99: 1 – 68.

Huang, Y.-S., Yang, C.-C. & Liaw, S.-S. (1999). Relativistic solution of hydrogen in a spherical cavity, *Physical Review A* 60: 85–90.

Kimani, P., Jones, P. & Winkler, P. (2008). Correlation studies in weakly confining quantum dot potentials, *International Journal of Quantum Chemistry* 108: 2763–2769.

Lin, C. Y. & Ho, Y. K. (2011). Photoionization cross sections of hydrogen impurities in spherical quantum dots using the finite-element discrete-variable representation, *Physical Review A* 84: 203407.

Marr, G. V. & Creek, D. M. (1968). The photoionization absorption continua for alkali metal vapours, *Proceedings of the Royal Society A* 304: 233–244.

Michalet, X., Pinaud, F. F., Bentolila, L. A., Tsay, J. M., Doose, S., Li, J. J., Sundaresan, G., Wu, A. M., Gambhir, S. S. & Weiss, S. (2005). Quantum dots for live cells, in vivo imaging, and diagnostics, *Science* 307: 538–544.

Michels, A., de Boer, J. & Bijl, A. (1937). Remarks concerning molecular interactions and their influence of the polarizability, *Physica* 4: 981–994.

Michels, H. H. (1963). Abscissas and weight coefficients for Lobatto quadrature, *Mathematics of Computation* 17: 237–244.

Peach, G., Saraph, H. E. & Seaton, M. J. (1988). Atomic data for opacity calculations. ix. the lithium isoelectronic sequence, *Journal of Physics B: Atomic, Molecular and Optical Physics* 21: 3669–3683.

Ralchenko, Y., Kramida, A. E., Reader, J. & Team, N. A. (2011). *NIST Atomic Spectra Database (ver. 4.1.0)* .
URL: *http://physics.nist.gov/asd3*

Reinhardt, W. P. (1982). Complex coordinates in the theory of atomic and molecular structure and dynamics, *Annual Review of Physical Chemistry* 33: 223–255.

Rescigno, T. N. & McCurdy, C. W. (2000). Numerical grid methods for quantum-mechanical scattering problems, *Physical Review A* 62: 032706.

Rescigno, T. N. & McKoy, V. (1975). Rigorous method for computing photoabsorption cross sections from a basis-set expansion, *Physical Review A* 12: 522–525.

Sahoo, S. & Ho, Y. K. (2006). Photoionization of Li and Na in Debye plasma environments, *Physics of Plasmas* 13: 063301.

Schedelbeck, G., Wegscheider, W., Bichler, M. & Abstreiter, G. (1997). Coupled quantum dots fabricated by cleaved edge overgrowth: From artificial atoms to molecules, *Science* 278: 1792–1795.

Schweizer, W., Faßbinder, P. & González-Férez, R. (1999). Model potentials for alkali metal atoms and Li-like ions, *Atomic Data and Nuclear Data Tables* 72: 33–55.

Stevanović, L. (2010). Oscillator strengths of the transitions in a spherically confined hydrogen atom, *Journal of Physics B: Atomic, Molecular and Optical Physics* 43: 165002.

Winkler, P. (2004). Electron interaction in weakly confining quantum dot potentials, *International Journal of Quantum Chemistry* 100: 1122–1130.

Xie, W. (2009). A study of two confined electrons using the Woods-Saxon potential, *Journal of Physics: Condensed Matter* 21: 115802.

Yang, C.-C., Liu, L.-C. & Chang, S.-H. (1998). Eigenstates and fine structure of a hydrogenic impurity in a spherical quantum dot, *Physical Review B* 58: 1954–1961.

8

Exciton States in Free-Standing and Embedded Semiconductor Nanocrystals

Yuriel Núñez Fernández[1,2], Mikhail I. Vasilevskiy[1],
Erick M. Larramendi[2] and Carlos Trallero-Giner[2,3]
[1]*Centro de Física, Universidade do Minho, Braga*
[2]*Facultad de Física and ICTM, Universidad de La Habana*
[3]*Departamento de Física, Universidade Federal de São Carlos*
[1]*Portugal*
[2]*Cuba*
[3]*Brazil*

1. Introduction

Semiconductor quantum dots (QDs), often referred to as "artificial atoms", have discrete energy levels that can be tuned by changing the QD size and shape. The existence of zero-dimensional states in QDs has been proved by high spectrally and spatially resolved photoluminescence (PL) studies Empedocles et al. (1996); Grundmann et al. (1995). Semiconductor QDs can be divided into two types, (1) epitaxially grown self-assembled dots (SAQDs) and (2) nanocrystals (NCs) surrounded by a non-semiconductor medium. Usually, SAQDs are obtained by using appropriate combinations of lattice mismatched semiconductors, taking advantage of the Stranski-Krastanov growth mode where highly strained 2D layers relax by forming 3D islands instead of generating misfit dislocations. SAQDs are robust and already integrated into a matrix appropriate for device applications Grundmann (2002). However, the size, shape and size distributions of the 3D islands are determined only by the strain related to the lattice mismatch of the specific heterojunction. Also, the density and the possibility of obtaining different nanocrystals over a given substrate have considerable limitations in this method.

Nanocrystal QDs have been produced by colloidal chemistry, melting, sputtering, ion implantation and some other techniques. An attractive feature of NCs is the possibility to control their electronic, optical and magnetic properties by varying their size, shape, surface characteristics and crystal structure, which is most efficiently achieved by using colloidal chemistry methods. These methods are known for the ability (i) to produce colloidal solutions of a broad variety of high quality semiconductor NCs of required size, (ii) to limit the size dispersion, and (iii) to control the NC surface Rogach (2008); Wang et al. (2005). Chemically grown NCs are more efficient light emitters than their bulk counterpart and even organic dyes. There are several reasons for this. First, quantum confinement of electronic states in QDs determines the transition energies and enhances the radiative transitions between conduction and valence bands. At the same time, it can be used to tune the luminescence wavelength and intensity, i.e., both the color and the brightness of the emission can be controlled. A second

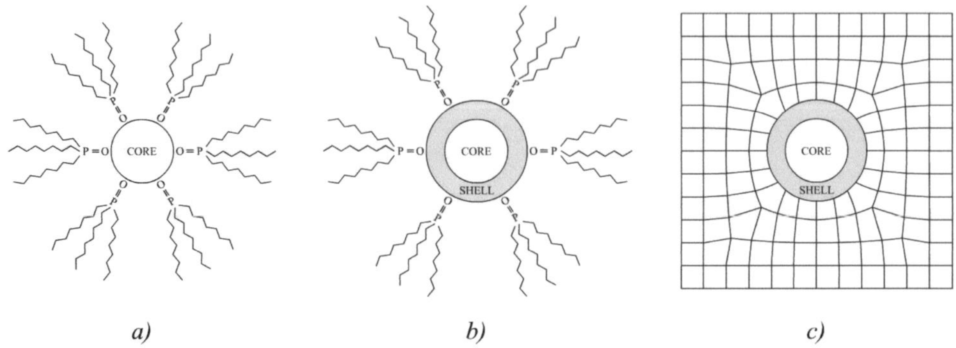

Fig. 1. Schematics of bare core (a) and core/shell (b) free standing and embedded (c) NC quantum dots.

effect, also characteristic of semiconductor nanoparticles, is not related to quantum physics but is purely geometrical. The spatial influence of defects acting as electronic traps is limited to the size of the host nanoparticle, whereas in bulk, nonradiative recombination sites can affect a much larger volume of material.

An obvious shortcoming of colloidal NCs, as compared to SAQDs, is that they are less stable and are not suitable for direct incorporation into electronic devices. One has to embed them into an appropriate matrix for device fabrication, especially for applications in the fields of optoelectronics and integrated optics. A possible approach consists in the integration of the colloidal chemistry methods with the epitaxial growth technology. The fabrication of high quality epitaxial films with embedded pre-fabricated NCs is a huge challenge. The successful integration of optically active colloidal NCs within an epitaxial structure has been demonstrated by combining the colloidal and molecular beam epitaxy (MBE) methods Woggon et al. (2005). It has been shown that core/shell nanoparticles (CSNPs) are more suitable for this purpose than bare core (e.g. CdSe) NCs. The luminescence properties and stability of CSNPs are generally better than those of single material nanocrystal QDs Rogach (2008). One of the earliest CSNP structures reported was CdSe/ZnS Dabbousi et al. (1997); Hines & Guyot-Sionnest (1996), which is at the same time the most intensively studied system to date. These particles show a very high photoluminescence (PL) quantum yield, which can be attributed to the better isolation of the electron-hole pair inside the dot from the surface recombination states. As well as NCs of a single semiconductor material, CSNPs are traditionally covered with trioctylphosphine oxide (TOPO) in order to prevent them from oxidation and to passivate dangling bonds at the semiconductor surface (See Fig. 1).

The nature of the medium surrounding a QD influences the quantum confinement effect and, consequently, the optical properties of these nanostructures. Already for colloidal NCs of a single II-VI material, the nature of the surface capping layer is important for the energy of the emitting states Jasieniak et al. (2011). For instance, exchange of TOPO with pyridine resulted in a a red shift of the order of 20-30 meV for CdSe NCs of 3.5-5.5 nm in diameter Luo et al. (2011). The transition energy and oscillator strength of the first excited state ($2S_{3/2}1S_e$) in these NCs can be strongly modified by their surface ligands and associated surface atomic arrangements Chen et al. (2011).

Introducing a shell also changes the energy spectrum of a colloidal QD. For CdSe/ZnS CSNPs, a red shift of the PL peak position and/or the absorption edge has been observed with increasing the thickness of the ZnS shell Baranov et al. (2003); Dabbousi et al. (1997); Talapin et al. (2001). The value of the red shift usually saturates for the shell thicknesses (d_s) above 3 monolayers (MLs) Baranov et al. (2003); Talapin et al. (2001), which may be an indication of the typical length scale of core wavefunction's penetration into the shell. Quite interesting results indirectly confirming this idea have been obtained for ternary core/shell/shell structures with the outer shell of CdSe. For a thin (1-2 ML) ZnS inner shell, there is a red shift of the PL peak observed for increasing thickness of the outer shell (d_{ss}), while for d_s =3 ML the peak position is practically independent of d_{ss} Gaponik et al. (2010). It means that the tail of the wavefunction inside the barrier (the first shell) is smaller than 3ML. Finally, when NCs are embedded into an epitaxial semiconductor matrix, a *blue* shift of the PL peak is observed, compared to the emission spectrum of the same NCs in organic solvent Rashad et al. (2010); Woggon et al. (2005). This effect is not straightforward to explain, because, in a simple view, the replacement of TOPO by a semiconductor matrix with a band structure similar to that of the QD materials should result in lower barriers at the interface and, consequently, in a weaker confinement and a red shift of the exciton transition. Therefore, some further effects have to be included into consideration, such as interface imperfection, surface charge or strain introduced by the matrix.

The purpose of this chapter is to provide a general yet simple theoretical description of the effects of surrounding media (shells and matrix) and interface characteristics on the exciton ground state in nanocrystal QDs that would be able to explain the above mentioned experimental results and could be applied to more sophisticated semiconductor structures based on NCs of an approximately spherical shape. Our approach is based on the effective mass approximation (EMA). Its advantages and shortcomings for calculations of the electronic properties of nanostructures are well known. On one hand, it reveals the underlying physics and clearly shows the effect of the material parameters on the observable properties, this is why it has been used by so many research groups since the beginning of the studies of nanocrystal QDs in the 80-s Brus (1984); Efros et al. (1996); Fomin et al. (1998); Miranda et al. (2006); Norris & Bawendi (1996); Pellegrini et al. (2005); Rolo et al. (2008); Vasilevskii et al. (1998). On the other hand, EMA is believed to overestimate the electron and hole confinement energies Jasieniak et al. (2011). Also, the scaling laws of these energies with the QD radius (R), obtained by fitting experimental Yu et al. (2003) and numerically calculated Delerue & Lannoo (2004) data, differ from the EMA predictions (R^{-2} in the strong confinement regime). Indeed, the EMA fails in the limit of very small clusters containing a hundred of atoms which should not even be called nanocrystals because their properties have more similarity with molecules than with crystals. Compared to the first version of the EMA theory for QDs Brus (1984), several improvements have been made, such as the consideration of finite barriers Norris & Bawendi (1996); Pellegrini et al. (2005) and the complex structure of the valence band of the underlying material Efros et al. (1996). As a result, it has been possible to assign several size-dependent transitions in measured optical spectra of CdSe Norris & Bawendi (1996) and CdTe Vasilevskii et al. (1998) QDs. Further improvement of the analytical description of QDs can be achieved by considering generalized boundary conditions allowing a discontinuity of the envelope functions at the interfaces, found to provide a better agreement with the results of *ab initio* numerical calculations Flory et al. (2008).

We apply the EMA approach to arbitrary centrosymmetric potentials, such as finite interface barriers due to band discontinuities or electric charges that can eventually accumulate at the NC/matrix interface. The EMA equations for electrons and holes (taking into account the complex valence band structure) will be formulated in terms of transfer matrices, allowing for the incorporation of generalized boundary conditions at all interfaces. The resulting matrix equations are solved numerically providing a rather simple and efficient tool for modeling different experimental situations and designing new complex QD-based nanostructures and optoelectronic devices with embedded nanoparticles as active optical components. The developed open-source software is available at http://sourceforge.net/projects/emaqdot. Finally, we present some calculated results concerning free-standing and embedded QDs and check them against experimental trends reported in the literature.

2. Exciton transition energy calculation

2.1 Basic equations

In the strong confinement regime ($R \ll a_{ex}$, a_{ex} is the exciton Bohr radius, and R is the core radius), the calculation of the lowest ($1S_{3/2}1S_e$) transition energy requires the electron ground state energy, E_e, the hole ground state energy, E_h, and the Coulomb interaction correction, E_c,

$$E_t = E_g + E_e + E_h + E_c$$

where E_g is the band gap energy of the core material. $E_e(E_h)$ is defined with respect to the bottom (top) of the conduction (valence) band.

For the spherically symmetric electron state $1S_e$, the envelope wavefunction can be written as $\Psi_e = \psi(r)/\sqrt{4\pi}r$ and the effective Schrödinger equation for $\psi(r)$ reads Efros et al. (1996); Norris & Bawendi (1996); Vasilevskii et al. (1998):

$$\psi''(r) + \frac{2m_e}{\hbar^2}\left[E_e - V_e(r)\right]\psi(r) = 0, \qquad (1)$$

where m_e is the electron effective mass and $V_e(r)$ is the potential acting on the electron. In order to obtain the electron ground state energy, we have to solve Eq. (1) together with the boundary conditions $\psi(0) = \psi(\infty) = 0$ and a matching condition at each interface of the heterostructure (see below).

Owing to the complex valence band structure of the involved semiconductor materials, the hole ground state is determined by the Luttinger Hamiltonian Luttinger (1956). In the centrosymmetric case, the radial part of the wavefunction is determined by two functions, $R_0(r)$ and $R_2(r)$, which satisfy the following system of differential equations Gelmont & Diakonov (1972):

$$(1+\beta)\left(\frac{d}{dr}+\frac{2}{r}\right)\frac{dR_0}{dr} + (1-\beta)\left(\frac{d}{dr}+\frac{2}{r}\right)\left(\frac{d}{dr}+\frac{3}{r}\right)R_2 + 4\frac{m_{lh}}{\hbar^2}\left(E_h - V_h(r)\right)R_0 = 0 \qquad (2)$$

$$(1-\beta)\left(\frac{d}{dr}-\frac{1}{r}\right)\frac{dR_0}{dr} + (1+\beta)\left(\frac{d}{dr}-\frac{1}{r}\right)\left(\frac{d}{dr}+\frac{3}{r}\right)R_2 + 4\frac{m_{lh}}{\hbar^2}\left(E_h - V_h(r)\right)R_2 = 0 \qquad (3)$$

where $V_h(r)$ is the potential acting on the hole,

$$\beta = \frac{\gamma_1 - 2\gamma}{\gamma_1 + 2\gamma} = \frac{m_{lh}}{m_{hh}}$$

and

$$m_{lh} = \frac{m_0}{\gamma_1 + 2\gamma}, \qquad m_{hh} = \frac{m_0}{\gamma_1 - 2\gamma}$$

are the light and heavy holes masses, respectively. Here, m_0 is the free electron mass and γ_1 and γ=$(2\gamma_2 + 3\gamma_3)$ /5 are the Luttinger parameters. They are constant within each material.

Finally, the Coulomb interaction energy is given by

$$E_c = -\frac{e^2 C}{\epsilon R} \tag{4}$$

where ϵ is the static dielectric constant of the QD material [1] and

$$C = \int_0^\infty t^2 \left[R_0^2(t) + R_2^2(t)\right] \left\{\frac{1}{t}\int_0^t \psi^2(s)ds + \int_t^\infty \frac{\psi^2(s)}{s}ds\right\} dt,$$

assuming that ψ, R_0 and R_2 are normalized according to:

$$\int_0^\infty \psi^2(t)dt = 1,$$

$$\int_0^\infty \left[R_0^2(t) + R_2^2(t)\right] t^2 dt = 1.$$

2.2 Boundary conditions

The differential equations presented above hold only inside each (e.g. core) material. At an interface between two materials, the following continuity conditions for the electron wavefunction should be applied:

$$\Psi_e \text{ and } \frac{1}{m_e}\frac{d\Psi_e}{dr} \text{ continuous.} \tag{5}$$

However, as it has been mentioned in the Introduction, the wavefunction $\psi(r)$ is just an *envelope* function and not necessarily must be continuous Flory et al. (2008); Laikhtman (1992). This issue has been widely discussed in the literature in relation to semiconductor heterostructures (see references in Laikhtman (1992)). Instead of (5), more general boundary conditions have been proposed, providing a better agreement with *ab initio* calculations for a number of III-V and II-VI compound heterostructures. A simplified version of such generalized boundary conditions that guarantees the continuity of the probability flux reads Rodina et al. (2002):

$$(m_e)^\alpha \Psi_e \quad \text{and} \quad \frac{1}{(m_e)^{\alpha+1}}\frac{d\Psi_e}{dr} \text{ continuous.} \tag{6}$$

where α is a phenomenological parameter. We shall also use these conditions (5) for interfaces between a semiconductor and TOPO. For $\alpha = 0$ the wavefunction is continuous while for $\alpha \neq 0$ it is not because of the difference in effective masses m_e at the interface.

At an ideal interface of two semiconductor materials of the same symmetry, the following continuity conditions for the hole envelope functions take place:

$$\begin{aligned} &R_0,\ R_2 \text{ continuous;} \\ &\frac{1}{m_{lh}}\frac{d}{dr}(R_0 + R_2) \text{ continuous;} \\ &\frac{1}{m_{hh}}\frac{d}{dr}(R_0 - R_2) \text{ continuous.} \end{aligned} \tag{7}$$

[1] For simplicity, we neglect the difference in the dielectric constant between different materials.

Although, in principle, these conditions should also be replaced by generalized ones, similar to Eqs. (6) Laikhtman (1992), we preferred to keep (7) in order to avoid additional free parameters.

2.3 Solution via transfer matrices

Since the equations are one-dimensional and the boundary conditions are linear, a transfer matrix formalism can be applied. This approach, borrowed from the optics of multilayer media Born & Wolf (1989), offers a convenient framework for linear problems and is straightforward to implement in a computer. If the potentials $V_e(r)$ and $V_h(r)$ are constant inside each material, Eqs. (1 - 3) can be solved explicitly. Considering one such material, two linearly independent solutions of (1) are $\cos(k_e r)$ and $\sin(k_e r)$ if

$$k_e = \sqrt{\frac{2m_e}{\hbar^2}\left(E_e - V_e\right)}$$

is real. If it is imaginary, $k_e = i\kappa_e$, then the solutions are $\exp(\pm\kappa_e r)$. For holes, if

$$k_h = \sqrt{\frac{2m_{hh}}{\hbar^2}\left(E_h - V_h\right)}$$

is real, the linearly independent solutions of Eqs. (2, 3) for the 2-vector $\begin{pmatrix} R_0 \\ R_2 \end{pmatrix}$ are

$$\begin{pmatrix} j_0(k_h r) \\ j_2(k_h r) \end{pmatrix}, \begin{pmatrix} j_0(k_l r) \\ -j_2(k_l r) \end{pmatrix}, \begin{pmatrix} y_0(k_h r) \\ y_2(k_h r) \end{pmatrix}, \begin{pmatrix} y_0(k_l r) \\ -y_2(k_l r) \end{pmatrix}, \tag{8}$$

where $k_l = \sqrt{\beta} k_h$, and j_ν, y_ν are the spherical Bessel functions of the first and second kind, respectively.

If $k_h = i\kappa_h$ with κ_h real, the solutions are:

$$\begin{pmatrix} i_0(\kappa_h r) \\ -i_2(\kappa_h r) \end{pmatrix}, \begin{pmatrix} i_0(\kappa_l r) \\ i_2(\kappa_l r) \end{pmatrix}, \begin{pmatrix} k_0(\kappa_h r) \\ -k_2(\kappa_h r) \end{pmatrix}, \begin{pmatrix} k_0(\kappa_l r) \\ k_2(\kappa_l r) \end{pmatrix}, \tag{9}$$

where $\kappa_l = \sqrt{\beta}\kappa_h$, and $i_\nu(z)$, $k_\nu(z)$ are the modified spherical Bessel functions of the first and third kind, respectively Abramowitz & Stegun (1970). Collecting the above solutions and combining them by using the boundary and matching conditions, one can obtain some transcendental equations for E_e and E_h. These transcendental equations are conveniently expressed in terms of transfer matrices, as shown below. For general centrosymmetric potentials, $V_e(r)$ and $V_h(r)$, the solutions of Eqs. (1-3) inside each material cannot be found analytically. The differential equations must be discretized using an appropriate numerical scheme. Then, our general approach still remains valid.

Let us suppose that $V_e(r)$, $V_h(r)$ = const in the following regions of the heterostructure: $0 < r < A_1$ (core) and $r > A_N$ (matrix far from the NC interface). Then the solutions in these regions are given by some particular combination of the above expressions (8) or (9). We consider a 2-vector, composed of the electron wavefunction and its derivative, $\mathbf{z}_e(r) = \begin{pmatrix} \psi(r) \\ \psi'(r) \end{pmatrix}$, which can be written explicitly for $r = A_1$ and $r = A_N$. These two vectors are connected by a transfer matrix, $\mathbf{T}_e$, that characterizes the region $r \in [A_1, A_N]$. It is constructed by multiplying the elementary transfer matrices describing the layers $A_i \leq r \leq A_{i+1}$, $(i = 2, 3, ..., N-1)$, and their interfaces. The details are given in Appendix I for a constant potential profile and in Appendix II for an arbitrary potential. Note that the case of infinite potential barrier at A_N requires a special analysis and is considered in Appendix III. Explicitly, we have for $r = A_1$:

$$\mathbf{z}_e(A_1) = c_1\mathbf{v}_1$$

where

$$\mathbf{v}_1 = \begin{pmatrix} \sin(k_eA_1) \\ k_e\cos(k_eA_1) \end{pmatrix} \qquad (E_e > V_e),$$

or

$$\mathbf{v}_1 = \begin{pmatrix} \sinh(\kappa_eA_1) \\ \kappa_e\cosh(\kappa_eA_1) \end{pmatrix} \qquad (E_e < V_e).$$

For $r = A_N$,

$$\mathbf{z}_e(A_N) = c_2\mathbf{v}_2 \qquad (E_e < V_e),$$

$$\mathbf{v}_2 = \begin{pmatrix} \exp(-k_eA_N) \\ -k_e\exp(-k_eA_N) \end{pmatrix}.$$

Connecting the points A_1 and A_N, we get

$$c_2\mathbf{v}_2 - c_1\mathbf{v}_3 = \mathbf{0}, \tag{10}$$

$$\mathbf{v}_3 = \mathbf{T}(A_1, A_N)\cdot\mathbf{v}_1.$$

The energy E_e is obtained from the equation,

$$\det\left(\left[-\mathbf{v}_3\ \mathbf{v}_2\right]\right) = 0. \tag{11}$$

Similarly, the corresponding 4-vectors for holes composed of $\begin{pmatrix} R_0 \\ R_2 \end{pmatrix}$ and their derivatives, are connected by a 4×4 transfer matrix $\mathbf{T}_h$ (see Appendix I). For $r = A_1$, we have:

$$\mathbf{z}_h(A_1) = c_1\mathbf{v}_1 + c_2\mathbf{v}_2,$$

where

$$\mathbf{v}_1 = \begin{pmatrix} j_0(k_hA_1) \\ j_2(k_hA_1) \\ k_hj_0'(k_hA_1) \\ k_hj_2'(k_hA_1) \end{pmatrix}, \mathbf{v}_2 = \begin{pmatrix} j_0(k_lA_1) \\ -j_2(k_lA_1) \\ k_lj_0'(k_lA_1) \\ -k_lj_2'(k_lA_1) \end{pmatrix} \qquad (E_h > V_h)$$

or

$$\mathbf{v}_1 = \begin{pmatrix} i_0(\kappa_hA_1) \\ -i_2(\kappa_hA_1) \\ \kappa_hi_0'(\kappa_hA_1) \\ -\kappa_hi_2'(\kappa_hA_1) \end{pmatrix}, \mathbf{v}_2 = \begin{pmatrix} i_0(\kappa_lA_1) \\ i_2(\kappa_lA_1) \\ \kappa_li_0'(\kappa_lA_1) \\ \kappa_li_2'(\kappa_lA_1) \end{pmatrix} \qquad (E_h < V_h).$$

For $r = A_N$,

$$\mathbf{z}_h(A_N) = c_3\mathbf{v}_3 + c_4\mathbf{v}_4$$

where

$$\mathbf{v}_3 = \begin{pmatrix} k_0(\kappa_hA_N) \\ -k_2(\kappa_hA_N) \\ \kappa_hk_0'(\kappa_hA_N) \\ -\kappa_hk_2'(\kappa_hA_N) \end{pmatrix}, \mathbf{v}_4 = \begin{pmatrix} k_0(\kappa_lA_N) \\ k_2(\kappa_lA_N) \\ \kappa_lk_0'(\kappa_lA_N) \\ \kappa_lk_2'(\kappa_lA_N) \end{pmatrix} \qquad (E_h < V_h).$$

Connecting the points $r = A_1$ and $r = A_N$, we have

$$c_3\mathbf{v}_3 + c_4\mathbf{v}_4 - c_1\mathbf{v}_5 - c_2\mathbf{v}_6 = \mathbf{0}, \tag{12}$$

$$\mathbf{v}_5 = \mathbf{T}_h(A_1, A_N) \cdot \mathbf{v}_1, \qquad \mathbf{v}_6 = \mathbf{T}_h(A_1, A_N) \cdot \mathbf{v}_2.$$

The energy E_h is determined by the equation

$$\det\left(\left[-\mathbf{v}_5 \ -\mathbf{v}_6 \ \mathbf{v}_3 \ \mathbf{v}_4\right]\right) = 0. \tag{13}$$

2.4 Too low barrier for holes

For embedded QDs, it is quite possible that the potential barriers provided by the semiconductor matrix are not sufficiently high (because of the small valence band-offset for II-VI semiconductors) to confine the carriers if the core radius is very small. Then one cannot consider the Coulomb interaction as a perturbation anymore and the confinement of the exciton as a whole should be considered. However, it can happen that the conduction band barrier is still quite high and the electrons still are in the strong confinement regime (i. e., $R \ll a_e$, a_e is the electron Bohr radius for the core material). Then the Coulomb interaction with the hole is still a small perturbation for the electron. Its ground state energy E_e and the wavefunction $\Psi_e(r)$ can be found as before. The above condition has profound physical consequences. The strongly localized electron shall keep the hole in the vicinity of the dot (otherwise it would be free to move into the matrix). This case can be called as "weak localization of the hole". We shall extend our formalism in order to include this case.

The Schrödinger equation for the electron-hole pair (exciton) is written as

$$\left[\hat{T}_e + V_e(\mathbf{r}_e) + \hat{T}_h + V_h(\mathbf{r}_h) + V_{eh}(\mathbf{r}_e, \mathbf{r}_h)\right] \psi_{eh} = E_{ex}\Psi_{ex}$$

where $\Psi_{ex} = \Psi_e(\mathbf{r}_e)\Psi_h(\mathbf{r}_h)$ (Ψ_h is a 2-vector), $\hat{T}_e$ ($\hat{T}_h$) represents the electron (hole) kinetic energy operator and V_{eh} is the electron-hole Coulomb interaction term. Multiplying by Ψ_e^* and integrating over the electron coordinates, $\mathbf{r}_e$, yields:

$$\left[\hat{T}_h + V_h(\mathbf{r}_h) + V_{\text{eff}}(\mathbf{r}_h)\right] \Psi_h = E_h\Psi_h \tag{14}$$

with

$$V_{\text{eff}} = \langle \Psi_e | V_{eh} | \Psi_e \rangle \,.$$

Using the well-known expansion,

$$\frac{1}{|\mathbf{r}_e - \mathbf{r}_h|} = \begin{cases} \frac{1}{r_e} \sum_{l=0}^{\infty} \left(\frac{r_h}{r_e}\right)^l P_l(\cos\theta) & \text{if } r_e \geq r_h \\ \frac{1}{r_h} \sum_{l=0}^{\infty} \left(\frac{r_e}{r_h}\right)^l P_l(\cos\theta) & \text{if } r_e < r_h \end{cases},$$

where P_l are Legendre polynomials and $\Psi_e(\mathbf{r}_e) = \psi(r_e)/\sqrt{4\pi}r_e$, we have:

$$V_{\text{eff}}(r_h) = -\frac{e^2}{\epsilon r_h} \int_0^{r_h} \psi^2(r_e) dr_e - \frac{e^2}{\epsilon} \int_{r_h}^{\infty} \frac{1}{r_e} \psi^2(r_e) dr_e. \tag{15}$$

For large r_h, the probability to find the electron there is small and $V_{\text{eff}}(r_h)$ behaves as $-e^2/\epsilon r_h$. On the other hand, for small r_h we have $V_{\text{eff}}(r_h) = V_{\text{eff}}(0) + O(r_h^2)$, where

$$V_{\text{eff}}(0) = -\frac{e^2 c_1^2}{2\epsilon} \left[\Gamma - Ci(2k_e R) + \ln(2k_e R)\right] - \frac{e^2}{\epsilon} \int_R^{\infty} \frac{\psi^2(r)}{r} dr,$$

Γ is the Euler-Mascheroni constant, $C_i(x)$ is the cos-integral function, and k_e, c_1 are the electron parameters inside the core (see above). Figure 2 shows the shape of the effective potential $V_{\text{eff}}(r_h)$.

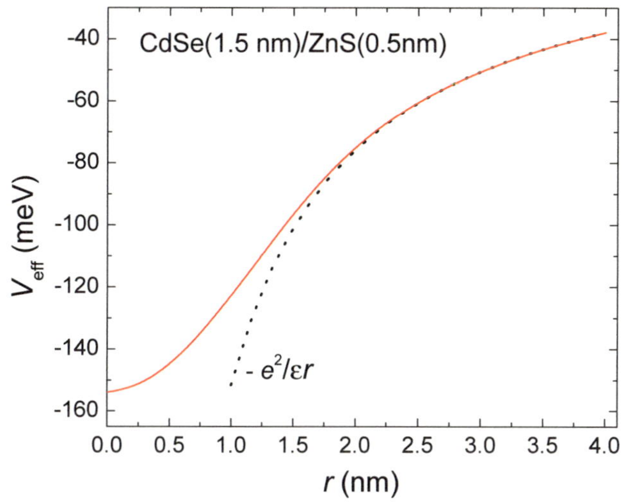

Fig. 2. Effective potential for holes (Eq. 15) produced by localized electron in $1S_e$ state.

This potential has to be added to V_h in Eqs. (2) and (3). Because of the non-explicit expression (15) for V_{eff} we have to solve (14) numerically. However, this does not rise any difficulty because this potential is also centrosymmetric (Appendix II). The asymptotic solutions for both $r_h \to 0$ and $r_h \to \infty$ are required. Near the origin, the potential $V_{\text{eff}}(r_h)$ is almost constant and the above mentioned solutions, $\mathbf{v}_1$ and $\mathbf{v}_2$, can be used (by putting $V_{\text{eff}}(r_h) \approx V_{\text{eff}}(0)$). In the same way, for large r_h the dominating terms of the asymptotic solutions behave as $\mathbf{v}_3$, $\mathbf{v}_4$ (replacing $V_{\text{eff}}(r_h)$ by 0).

Applying the continuity conditions similar to (7) and using the numerical transfer matrices described in Appendix II, the solutions of Eq. (14) near $r_h = 0$ are connected with the asymptotic ones at $r_h = A_N$, and the energy E_h is obtained.

3. Results and discussion

We applied the formalism described in the previous section to a set of "samples" that mimic free standing and embedded bare core and core/shell NCs. The parameters used in the calculations are listed in Table 1. The results are summarized below.

Bare core NCs

Concerning the size-dependent lowest transition energy in bare core NCs covered by TOPO Fig. 3 shows the results of our calculations in comparison with the experimental data of Murray et al. (1993); Yu et al. (2003) and those calculated assuming infinitely high barriers. A good agreement with the experiment is observed only for the calculations performed with finite barriers Pellegrini et al. (2005). We would like to point out that the complex valence band structure was neglected in the previous work considering finite barriers Pellegrini et al. (2005) and, consequently, an unrealistically small effective mass ($0.3m_0$) for heavy holes was

Parameter	$CdSe$	ZnS	$ZnSe$	$TOPO$
$m_e(m_0)$	0.119	0.22	0.16	1
$m_{hh}(m_0)$	0.82	0.61	0.495	1
$m_{lh}(m_0)$	0.262	0.23	0.177	1
E_g(eV)	1.75	3.78	2.7	5
VBO(eV)	0	0.99	0.22	1.63

Table 1. Parameters used in calculations for CdSe, ZnS, and ZnSe. Notation: VBO , E_g, m_e, m_{hh} and m_{lh} refer to the valence band offset, band gap, and effective masses of electrons, heavy holes and light holes, respectively. These values were taken from Li et al. (2009); Pellegrini et al. (2005); Schulz & Czycholl (2005); *Springer Materials The Landolt-Börnstein Database* (2011).

used. In the present work, we used the correct model for the valence band with realistic m_{lh} and m_{hh} and introduced an extra parameter $\alpha_{CdSe/TOPO} = -0.08$ to characterize the electron envelope function matching at the CdSe/TOPO interface. Barrier's heights used by Pellegrini et al. (2005) (1.63 eV) were obtained by subtracting semiconductor's E_g from the known energy (5 eV) of a certain electronic transition in TOPO and dividing the result equally between the CB and VB offsets, yielding the above rather low values. Most of the authors used much higher values for these barriers Dabbousi et al. (1997); Rashad et al. (2010) in order to obtain the correct trends for core-shell structures (see below) but then it is not possible to correctly reproduce the $E_t(R)$ dependence for bare core particles. Using the discontinuous envelope function can help to remedy this difficulty.

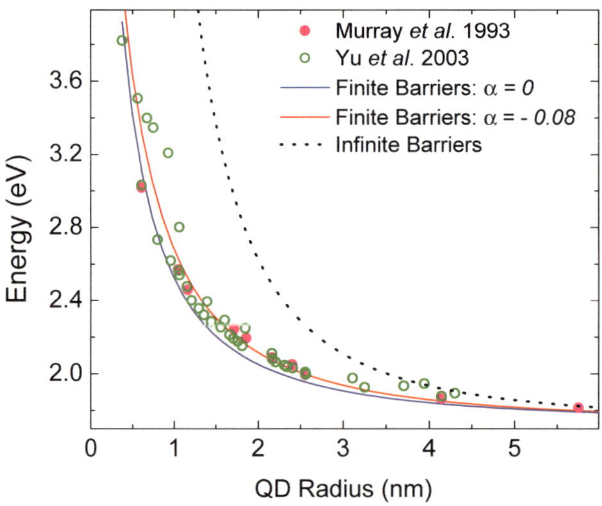

Fig. 3. Comparison between experimental and theoretical results for CdSe/TOPO NCs. Experimental values were taken from Refs. Murray et al. (1993); Yu et al. (2003). The lowest transition energy was calculated using finite barriers and either continuous ($\alpha = 0$) or discontinuous ($\alpha \neq 0$) boundary conditions for the electron wavefunction at the interface. For comparison, the results obtained assuming infinitely high barriers for both electron and hole are also presented.

Shell thickness effect in CdSe/ZnS CSNCs

As it has been said in the Introduction, the absorption and PL emission peaks in the spectra of CdSe/ZnS core/shell NCs are redshifted with respect to the bare core CdSe NCs of the same radius. The shift grows when the shell thickness is increased up to d_s=3 ML and then saturates. This effect has been observed by several groups Baranov et al. (2003); Dabbousi et al. (1997); Soni (2010). We focus on the absorption peak position because the emission peak is usually Stokes-shifted with respect to the former mostly because of the size distribution effects (larger dots emit stronger than the smaller ones in an ensemble of NCs Efros et al. (1996)). In our calculations of the lowest transition energy, the electron wavefunction at the CdSe/ZnS interface was considered continuous ($\alpha = 0$), while the ZnS/TOPO interface was characterized by an appropriate (non-zero) value of the electron wavefunction discontinuity parameter, $\alpha_{ZnS/TOPO}$. Fig. 4 shows a good agreement between the calculated and experimental results, obtained without using an unrealistically large heights of the ZnS/TOPO barriers (like 10 eV for holes in Ref. Dabbousi et al. (1997)). For comparison, the case $\alpha_{ZnS/TOPO} = 0$ is also presented, which lead to a blue shift in the transition energy when the shell thickness is increased, in direct contradiction with the experiments.

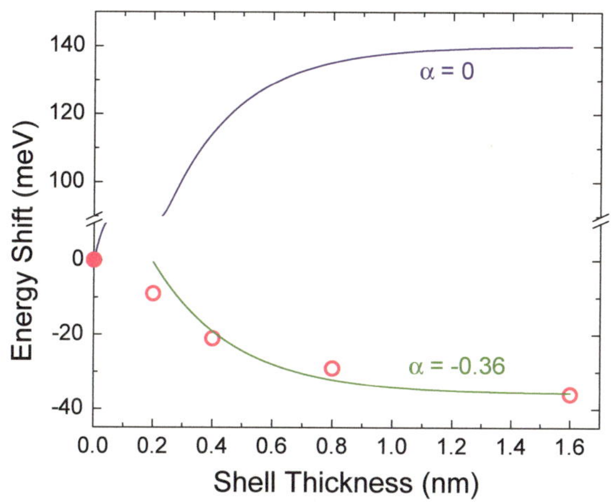

Fig. 4. Absorption peak shift in CdSe/ZnS CSNC, obtained experimentally by Dabboussi *et al.* (points) and calculated using present model with $\alpha_{ZnS/TOPO} = -0.36$ and $\alpha_{ZnS/TOPO} = 0$. The transition energy for the core/shell NCs is measured with respect to the CdSe bare core QD. The core size in both cases is $R = 2.0$ nm.

The effect of surface charge

As it has been pointed out in the Introduction, the utilization of colloidal NCs as active optical material in optoelectronic devices requires their incorporation into a high quality solid matrix. This process comprises the casting of the colloidal particles onto a substrate with the subsequent overgrowth of the matrix. For such embedded structures, it is relevant to consider the possibility of static charge accumulation at the nanoparticle/matrix interface Baccarani et al. (1978). Because of the incoherent incorporation of the nanoparticles into the crystalline matrix, the interface can create electronic trap states. Therefore the surface should be charged with a density σ_0 and a compensating space charge should be distributed in the matrix in the vicinity of the particle. The volume density of the latter can be assumed of the form,

$$\rho_m(r) = -\rho_0 \exp\left[-\frac{r-r_0}{l}\right] \qquad (r > r_0), \tag{16}$$

where $r_0 = R + d_s$ is the radius of the particle, l is a characteristic length (of the order of 1-2 nm), and ρ_0 is obtained from the charge neutrality condition:

$$\rho_0 = \frac{\sigma_0}{l\left[1 + 2l/r_0 + 2\left(l/r_0\right)^2\right]}.$$

Since epitaxial ZnSe normally is intrinsically n-type doped, we assume $\sigma_0 < 0$.

The electrostatic potential, $\varphi(r)$, is found by solving the Poisson equation with the charge density $\rho(r) = \sigma_0\delta(r - r_0) + \rho_m(r)$. The additional potential energy is:

$$U_c(r) = -e\varphi(r) = \begin{cases} AU_0 \frac{2l/r+1}{2l/r_0+1} \exp\left[-\frac{r-r_0}{l}\right] & (r \geq r_0) \\ AU_0 & (r < r_0) \end{cases}, \tag{17}$$

where $A = (l/r_0)\left(2l/r_0 + 1\right)\left[1 + 2l/r_0 + 2\left(l/r_0\right)^2\right]^{-1}$ is a dimensionless constant and the energy $U_0 = -4\pi r_0 \sigma_0 e/\epsilon$. Then the confinement potentials for electron and hole are obtained by considering $U_c(r)$ and the conduction and valence band offsets between the different materials forming each structure. The resulting band diagrams for two heterostructures (for the case of $\sigma_0 < 0$) are shown in Figure 5 a, b. Notice that, according to (17), if $l \ll r_0$ then the electron confinement decreases while the hole wavefunction becomes stronger localized owing to the surface charge effect. For large l the surface charge effect on the confinement can be neglected.

The results obtained for CdSe and CdSe/ZnS nanoparticles embedded in ZnSe are presented in Figs. 5-c, and 5-d. It can be seen that the surface charge effect on the hole energy is stronger for the bare core QD than for the CSNP. This is because in the latter, because of the presence of the ZnS shell, the hole penetrates less into the matrix (see Fig. 5 a, b) and hence a weaker interaccion with the charged interface is expected [2]. As a result, the surface charge effects on the electron and the hole nearly compensate each other in a bare core NC, while in a core/shell QD there is a significant net red shift of the exciton transition owing to the presence of the surface charge.

Free-standing *versus* embedded NCs

A blue shift in the PL emission when CdSe/ZnS CSNCs were embedded into a ZnSe crystalline matrix was observed in the work Larramendi (n.d.). Two samples studied in that work contained nanocrystals with different core size, $2R = 2.5$ and 3.0 nm, and $d_s = 0.5$ nm of ZnSe shell (2 MLs). A shift of the emission peak, of 68 and 33 meV, respectively, was observed (Fig. 6). We attempted to reproduce this effect in our calculations. Here we used $\alpha = 0$ for all (CdSe/ZnS, ZnS/TOPO, and ZnS/ZnSe) interfaces.

[2] For $r < R$ the hole does not interact with the charges outside by virtue of the Gauss law.

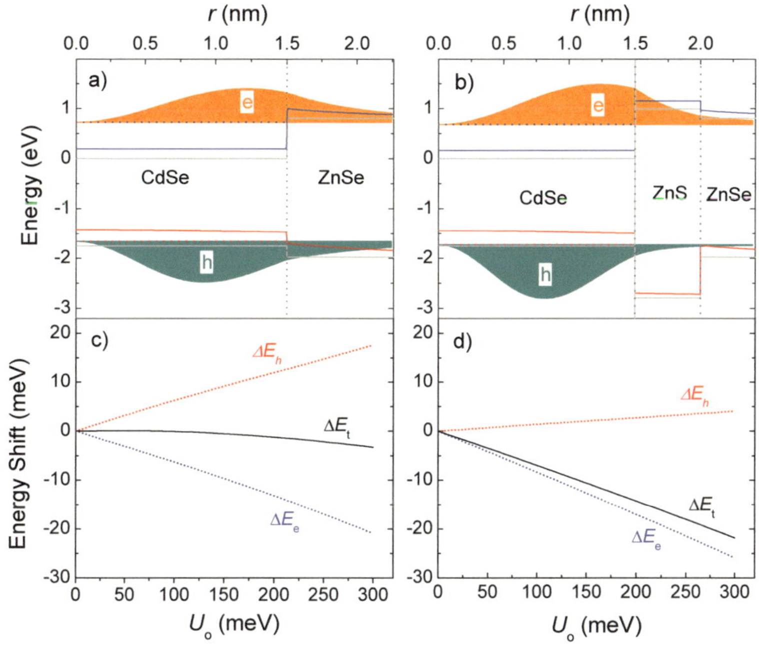

Fig. 5. Band energy diagrams for CdSe/ZnSe (a) and CdSe/ZnS/ZnSe (b) structures, with (full lines) and without (dashed lines) charge at the NC/matrix interface. Notice that the valence and conduction bands are shifted by a constant value (equal to AU_0) inside the particle. Also shown are the electron and hole probability densities for the case with surface charge. In panels c) and d) the electron (blue), the hole (red), and the transition (black) energy shifts (with respect to the case of $\sigma_0 = 0$) are presented *versus* the surface charge parameter U_0.

The results of our calculations are presented in Fig. 6. Essentially, the blue shift is obtained in a natural way, despite the higher barrier for TOPO-covered free-standing NCs. In order to qualitatively understand this result, we focus on Fig. 7, particularly on the electron wavefunctions shown in Fig. 7 d, which exemplifies the situation. For the free-standing NC, the higher barrier results in a smaller penetration length. This would favor larger energies. However, the larger difference in the effective masses on the left and on the right of the interface implies an abrupt change in the slope of the wavefunction. For the given set of parameters, the second effect (i.e. the change in the derivative of Ψ_e at the interface) dominates. Despite the longer tail of the electron wavefunction in the case of the embedded QD, the maximum of the probability density is located at a lower distance from the center and a blue shift is obtained (except for the smallest radii).

Let us point out that the situation depicted in the right panels of Fig. 7 corresponds to weak localization of the hole owing to its the Coulomb interaction with the strongly localized electron. Passing from strong to weak localization should affect the overlap integral between the electron and hole wavefunctions that determines the transition oscillator strength. In the ultimate limit of small core radius, the electron becomes delocalized and, consequently, the hole too. Then the overlap integral should decrease drastically. It could explain why the luminescence of the bare core NCs is weaker than that of the core shell NCs when they

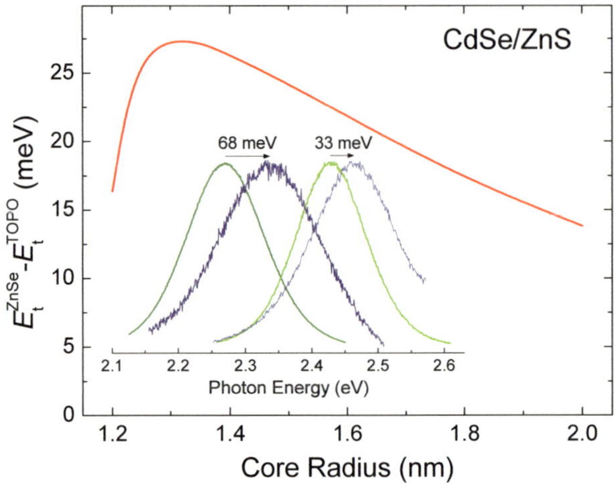

Fig. 6. Calculated transition energy shift *versus* core radius for CdSe/ZnS CSNPs embedded in ZnSe matrix ($d_s = 0.5$nm, $\sigma_0 = 0$). In the inset are displayed the PL bands measured for two samples of colloidal CdSe/ZnS NCs with different core size (2.5 and 3.0 nm), dispersed in toluene solution (green lines) and after embedding into epitaxial ZnSe layer (blue lines) [courtesy of Prof. U. Woggon and Dr. O. Schï£¡ps].

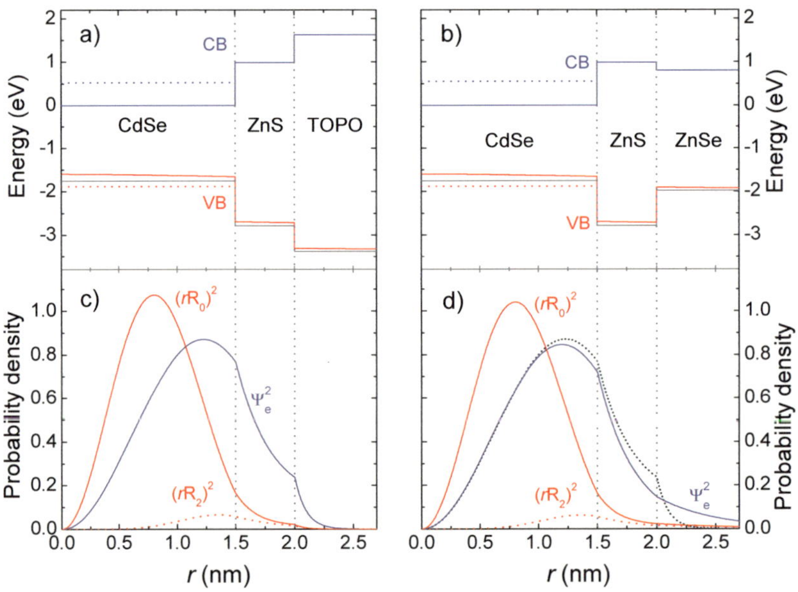

Fig. 7. Potential profiles and probability densities [electron (blue), $(rR_0)^2$ (red solid) and $(rR_2)^2$ (red dashed)] for CdSe/ZnS/TOPO nanoparticle (left panels) and CdSe/ZnS/ZnSe heterostructure (right panels). In the latter, the electron density of panel c) is also shown for comparison (black dotted line). In both cases $R = 1.5$ nm, $d_s = 0.5$ nm, $\sigma_0 = 0$, and $\alpha = 0$.

are embedded into the matrix Larramendi (n.d.). The threshold (in terms of the core size) between the different confinement regimes depends on several factors, such as the band offsets. These can be affected by the electric charge that can accumulate at the interface between the embedded QDs and the matrix. For instance, a positive surface charge would increase the electron barrier and decrease the one for the hole. The charge would also modify the potential profile seen by the confined electrons and holes, similar to the effect known for polycrystalline silicon Baccarani et al. (1978). It could bring about the hole confinement in the vicinity of the CdSe/ZnS interface, similar to what happens to 2D electrons in AlGaAs/GaAs heterostructures.

4. Conclusions

In summary, we have developed a transfer matrix approach to theoretically describe the effects of surrounding media and interface charge on the exciton ground state in nanocrystal QDs. It permits to explain a number of experimentally observed effects, including (i) the size-dependent lowest transition energy, (ii) the influence of shell thickness on the absorption and emission spectra of core/shell nanoparticles and (iii) the blue shift of the emission peak when NCs are embedded into an epitaxial matrix. For this, the utilization of generalized boundary conditions at nanoparticle/organic-ligand interfaces is required. We also showed that the Coulomb interaction between the electron and the hole can be important for the confinement of the latter. This is essential for the interpretation of the experimental results concerning colloidal CdSe NCs embedded in a semiconductor (e.g. ZnSe) host matrix because of the small valence band-offset between II-VI semiconductors. Note that, since the colloidal NCs are buried with random crystalline orientations and, consequently, are non-coherent to the host epitaxial matrix, the influence of the strain that could result from the lattice mismatch between the NCs and the matrix is not expected and therefore was not taken into account in our model. However, it could be incorporated using the standard theory of strained bulk semiconductors Bir & Pikus (1974). The same applies to the effects of the electron-hole exchange interaction and crystal field produced by hexagonal structure, which can be considered as perturbations and lead to the splitting of the $1S_{3/2}1S_e$ octet Efros et al. (1996), and the dielectric polarization effect owing the difference of the dielectric constant value between the NC material and TOPO, leading to a correction to the electron-hole Coulomb interaction Brus (1984). Thus, our approach provides a user-friendly tool to study different combinations of NC and surrounding materials and potential interesting physical effects, such as the crossover between strong and weak localization regimes for the QD hole.

5. Acknowledgements

The work was supported by the FCT (Portugal) under the grant PTDC/FIS/113199/2009. The authors are grateful to Prof. U. Woggon and Dr. O. Schï£¡ps for providing the photoluminescence measurement results presented in Fig. 6.

6. Appendix

Appendix I. Transfer matrix for a constant potential

Electrons

Let

$$\mathbf{z}(r) = \begin{pmatrix} \psi \\ \psi' \end{pmatrix} = \mathbf{M}(r) \cdot \mathbf{c}$$

where **c** is a 2-component constant vector, and

$$\mathbf{M}(r) = \begin{bmatrix} \cos(k_e r) & \sin(k_e r) \\ -k_e \sin(k_e r) & k_e \cos(k_e r) \end{bmatrix} \qquad (E_e > V_e)$$

or

$$\mathbf{M}(r) = \begin{bmatrix} \exp(\kappa_e r) & \exp(-\kappa_e r) \\ \kappa_e \exp(\kappa_e r) & -\kappa_e \exp(-\kappa_e r) \end{bmatrix} \qquad (E_e < V_e).$$

The key idea is to obtain $\mathbf{z}(r_2)$ given $\mathbf{z}(r_1)$:

$$\mathbf{z}(r_2) = \mathbf{T}(r_1, r_2) \cdot \mathbf{z}(r_1)$$

where $\mathbf{T}(r_1, r_2)$ is the transfer matrix. If r_1, r_2 are in the same material, the transfer matrix is $\mathbf{T}(r_1, r_2) = \mathbf{M}(r_2) \cdot \mathbf{M}^{-1}(r_1)$. Equivalently,

$$\mathbf{T}(r_1, r_2) = \begin{bmatrix} \cos(k_e \Delta r) & \frac{1}{k_e} \sin(k_e \Delta r) \\ -k_e \sin(k_e \Delta r) & \cos(k_e \Delta r) \end{bmatrix} \qquad (E_e > V_e)$$

or

$$\mathbf{T}(r_1, r_2) = \begin{bmatrix} \cosh(\kappa_e \Delta r) & \frac{1}{\kappa_e} \sinh(\kappa_e \Delta r) \\ \kappa_e \sinh(\kappa_e \Delta r) & \cosh(\kappa_e \Delta r) \end{bmatrix} \qquad (E_e < V_e).$$

At the interface Σ we have

$$\mathbf{z}(r^+) = \mathbf{T}_\Sigma \cdot \mathbf{z}(r^-),$$

$$\mathbf{T}_\Sigma = \begin{bmatrix} \beta_e^{-\alpha} & 0 \\ \frac{\beta_e^{-\alpha} - \beta_e^{\alpha+1}}{r} & \beta_e^{\alpha+1} \end{bmatrix},$$

and $\beta_e = m_e^+ / m_e^-$.

Holes

In this case

$$\mathbf{z}(r) = \begin{pmatrix} R_0 \\ R_2 \\ R_o' \\ R_2' \end{pmatrix} = \mathbf{M}(r) \cdot \mathbf{c},$$

where **c** is a 4-component constant vector and

$$\mathbf{M}(r) = \begin{bmatrix} j_0(k_h r) & j_0(k_l r) & y_0(k_h r) & y_0(k_l r) \\ j_2(k_h r) & -j_2(k_l r) & y_2(k_h r) & -y_2(k_l r) \\ k_h j_0'(k_h r) & k_l j_0'(k_l r) & k_h y_0'(k_h r) & k_l y_0'(k_l r) \\ k_h j_2'(k_h r) & -k_l j_2'(k_l r) & k_h y_2'(k_h r) & -k_l y_2'(k_l r) \end{bmatrix} \qquad (E_h > V_h)$$

or

$$\mathbf{M}(r) = \begin{bmatrix} i_0(\kappa_h r) & i_0(\kappa_l r) & k_0(\kappa_h r) & k_0(\kappa_l r) \\ -i_2(\kappa_h r) & i_2(\kappa_l r) & -k_2(\kappa_h r) & k_2(\kappa_l r) \\ \kappa_h i_0'(\kappa_h r) & \kappa_l i_0'(\kappa_l r) & \kappa_h k_0'(\kappa_h r) & \kappa_l k_0'(\kappa_l r) \\ -\kappa_h i_2'(\kappa_h r) & \kappa_l i_2'(\kappa_l r) & -\kappa_h k_2'(\kappa_h r) & \kappa_l k_2'(\kappa_l r) \end{bmatrix} \qquad (E_h < V_h).$$

The transfer matrix is $\mathbf{T}(r_1, r_2) = \mathbf{M}(r_2) \cdot \mathbf{M}^{-1}(r_1)$.

At the interface (Σ) we have:

$$\mathbf{T}_\Sigma = \begin{bmatrix} 1\,0 & 0 & 0 \\ 0\,1 & 0 & 0 \\ 0\,0 & \frac{1}{2}(\beta_l+\beta_h) & \frac{1}{2}(\beta_l-\beta_h) \\ 0\,0 & \frac{1}{2}(\beta_l-\beta_h) & \frac{1}{2}(\beta_l+\beta_h) \end{bmatrix},$$

and $\beta_l = \frac{m^+_{lh}}{m^-_{lh}}$, $\beta_h = \frac{m^+_{hh}}{m^-_{hh}}$.

Appendix II. Numerical transfer matrix

The solution of the first-order ordinary differential equations

$$\hat{L}[\mathbf{z}(t)] = 0$$

$$\mathbf{z}(t_1) = \mathbf{a}$$

where $\hat{L}$ is a linear operator, $\mathbf{a} = (a_1, a_2, ..., a_n)$, and $\mathbf{z} : \Re \to \Re^n$, can be expressed in terms of the solution of the auxiliary problems:

$$\hat{L}[\mathbf{u}_i(t)] = 0$$

$$\mathbf{u}_i(t_1) = \mathbf{E}_i$$

where $\mathbf{E}_i \in \Re^n$ is the i-th canonical vector, and $i = 1, 2, ..., n$. Since

$$\mathbf{a} = \sum_{i=1}^{n} \mathbf{E}_i a_i,$$

we have

$$\mathbf{z} = \sum_{i=1}^{n} \mathbf{u}_i a_i,$$

in particular,

$$\mathbf{z}(t_2) = \sum_{i=1}^{n} \mathbf{u}_i(t_2) a_i.$$

Taking into account that $z_i(t_1) = a_i$, and denoting $T_{ji} = (u_i)_j$,

$$\mathbf{z}(t_2) = \mathbf{T} \cdot \mathbf{z}(t_1).$$

Hence the transfer matrix $\mathbf{T}$ is given by the solutions $\mathbf{u}_i$ putted as column vectors.

Appendix III. Infinite potential barrier in A_N

For electron, the condition is now $\varphi(A_N) = 0$, hence the equation

$$\mathbf{v}_3[1] = 0 \tag{18}$$

instead of (10) determinates the electron energy E_e. The constant c_1 can be obtained from normalization.

For holes, we have now $R_0(A_N) = R_2(A_N) = 0$, hence the equation

$$\begin{bmatrix} v_5[1] & v_6[1] \\ v_5[2] & v_6[2] \end{bmatrix} \cdot \begin{pmatrix} c_1 \\ c_2 \end{pmatrix} = \begin{pmatrix} 0 \\ 0 \end{pmatrix} \tag{19}$$

replaces (12) and E_h is given by

$$\det\left(\begin{bmatrix} v_5[1] & v_6[1] \\ v_5[2] & v_6[2] \end{bmatrix}\right) = 0. \tag{20}$$

7. References

Abramowitz, M. & Stegun, I. A. (1970). *Handbook of Mathematical Functions*.

Baccarani, G., Riccó, B. & Spadini, G. (1978). Transport properties of polycrystalline silicon films, *Journal of Applied Physics* 49(11): 5565.
URL: *http://link.aip.org/link/JAPIAU/v49/i11/p5565/s1&Agg=doi*

Baranov, A., Rakovich, Y., Donegan, J., Perova, T., Moore, R., Talapin, D., Rogach, a., Masumoto, Y. & Nabiev, I. (2003). Effect of ZnS shell thickness on the phonon spectra in CdSe quantum dots, *Physical Review B* 68(16): 1–7.
URL: *http://link.aps.org/doi/10.1103/PhysRevB.68.165306*

Bir, G. L. & Pikus, G. E. (1974). *Symmetry and Strain-Induced Effects in Semiconductors*, Wiley, New York.

Born, M. & Wolf, E. (1989). *Principles of Optics*, Pergamon, Oxford.

Brus, L. E. (1984). Electron-electron and electron-hole interactions in small semiconductor crystallites: The size dependence of the lowest excited electronic state, *The Journal of Chemical Physics* 80(9): 4403.
URL: *http://link.aip.org/link/JCPSA6/v80/i9/p4403/s1&Agg=doi*

Chen, O., Yang, Y., Wang, T., Wu, H., Niu, C., Yang, J. & Cao, Y. C. (2011). Surface-Functionalization-Dependent Optical Properties of II-VI Semiconductor Nanocrystals., *Journal of the American Chemical Society* 133(43): 17504–12.
URL: *http://www.ncbi.nlm.nih.gov/pubmed/21954890*

Dabbousi, B. O., Rodriguez-Viejo, J., Mikulec, F. V., Heine, J. R., Mattoussi, H., Ober, R., Jensen, K. F. & Bawendi, M. G. (1997). (CdSe)ZnS Core-Shell Quantum Dots: Synthesis and Characterization of a Size Series of Highly Luminescent Nanocrystallites, *The Journal of Physical Chemistry B* 101(46): 9463–9475.
URL: *http://pubs.acs.org/doi/abs/10.1021/jp971091y*

Delerue, C. & Lannoo, M. (2004). *Nanostructures. Theory and Modeling*, Springer-Verlag: Berlin.

Efros, A. L., Rosen, M., Kuno, M., Nirmal, M., Norris, D. J. & Bawendi, M. (1996). Band-edge exciton in quantum dots of semiconductors with a degenerate valence band: Dark and bright exciton states, *Phys. Rev. B* 54: 4843–4856.
URL: *http://link.aps.org/doi/10.1103/PhysRevB.54.4843*

Empedocles, S. A., Norris, D. J. & Bawendi, M. G. (1996). Photoluminescence spectroscopy of single cdse nanocrystallite quantum dots, *Phys. Rev. Lett.* 77: 3873–3876.
URL: *http://link.aps.org/doi/10.1103/PhysRevLett.77.3873*

Flory, C. C., Musgrave, C. C. & Zhang, Z. (2008). Quantum dot properties in the multiband envelope-function approximation using boundary conditions based upon first-principles quantum calculations, *Physical Review B* 77(20): 1–13.
URL: *http://prb.aps.org/abstract/PRB/v77/i20/e205312*

Fomin, V. M., Gladilin, V. N., Devreese, J. T., Pokatilov, E. P., Balaban, S. N. & Klimin, S. N. (1998). Photoluminescence of spherical quantum dots, *Phys. Rev. B* 57: 2415–2425.
URL: *http://link.aps.org/doi/10.1103/PhysRevB.57.2415*

Gaponik, N., Hickey, S. G., Dorfs, D., Rogach, A. L. & Eychmüller, A. (2010). Progress in the light emission of colloidal semiconductor nanocrystals., *Small (Weinheim an der*

Bergstrasse, Germany) 6(13): 1364–78.
URL: *http://www.ncbi.nlm.nih.gov/pubmed/20564480*

Gelmont, B. L. & Diakonov, M. I. (1972). *Sov. Phys. Semicond.* 5: 1905.

Grundmann, M., Christen, J., Ledentsov, N. N., Böhrer, J., Bimberg, D., Ruvimov, ‡, S. S., Werner, P., Richter, U., Gösele, U., Heydenreich, J., Ustinov, V. M., Egorov, A. Y., Zhukov, A. E., Kop'ev, P. S. & Alferov, Z. I. (1995). Ultranarrow luminescence lines from single quantum dots, *Phys. Rev. Lett.* 74: 4043–4046.
URL: *http://link.aps.org/doi/10.1103/PhysRevLett.74.4043*

Grundmann, M. (ed.) (2002). *Nano-Optoelectronics. Concepts, Physics and Devices*, Springer-Verlag, Berlin.

Hines, M. A. & Guyot-Sionnest, P. (1996). Synthesis and characterization of strongly luminescing zns-capped cdse nanocrystals, *The Journal of Physical Chemistry* 100(2): 468–471.
URL: *http://pubs.acs.org/doi/abs/10.1021/jp9530562*

Jasieniak, J., Califano, M. & Watkins, S. E. (2011). Size-dependent valence and conduction band-edge energies of semiconductor nanocrystals., *ACS nano* 5(7): 5888–902.
URL: *http://www.ncbi.nlm.nih.gov/pubmed/21662980*

Laikhtman, B. (1992). Boundary conditions for envelope functions in heterostructures, *Phys. Rev. B* 46: 4769–4774.
URL: *http://link.aps.org/doi/10.1103/PhysRevB.46.4769*

Larramendi, E. M. (n.d.). To be published.

Li, Y., Walsh, A., Chen, S., Yin, W., Yang, J., Li, J., Da Silva, J., Gong, X. & Wei, S. (2009). Revised ab initio natural band offsets of all group IV, II-VI, and III-V semiconductors, *Applied Physics Letters* 94: 212109.
URL: *http://link.aip.org/link/?APPLAB/94/212109/1*

Luo, X., Liu, P., Truong, N., Farva, U. & Park, C. (2011). Photoluminescence Blue-Shift of CdSe Nano-Particles Caused By Exchange of Surface Capping Layer, *The Journal of Physical Chemistry C* pp. 20817–20823.
URL: *http://dx.doi.org/10.1021/jp200701x*

Luttinger, J. M. (1956). Quantum theory of cyclotron resonance in semiconductors: General theory, *Phys. Rev.* 102: 1030–1041.
URL: *http://link.aps.org/doi/10.1103/PhysRev.102.1030*

Miranda, R. P., Vasilevskiy, M. I. & Trallero-Giner, C. (2006). Nonperturbative approach to the calculation of multiphonon raman scattering in semiconductor quantum dots: Polaron effect, *Phys. Rev. B* 74: 115317.
URL: *http://link.aps.org/doi/10.1103/PhysRevB.74.115317*

Murray, C., Norris, D. & Bawendi, M. (1993). Synthesis and characterization of nearly monodisperse CdE (E= S, Se, Te) semiconductor nanocrystallites, *Journal of American Chemical Society* 115(4): 8706–8715.
URL: *http://cat.inist.fr/?aModele=afficheN&cpsidt=18785986*

Norris, D. & Bawendi, M. G. (1996). Measurement and assignment of the size-dependent optical spectrum in CdSe quantum dots, *Physical Review B* 53(24): 16338.
URL: *http://prb.aps.org/abstract/PRB/v53/i24/p16338_1*

Pellegrini, G., Mattei, G. & Mazzoldi, P. (2005). Finite depth square well model: Applicability and limitations, *Journal of Applied Physics* 97(7): 073706.
URL: *http://link.aip.org/link/JAPIAU/v97/i7/p073706/s1&Agg=doi*

Rashad, M., Paluga, M., Pawlis, a., Lischka, K., Schikora, D., Artemyev, M. V. & Woggon, U. (2010). MBE overgrowth of ex-situ prepared CdSe colloidal nanocrystals, *Physica Status Solidi (C)* 7(6): 1523–1525.
URL: *http://doi.wiley.com/10.1002/pssc.200983272*

Rodina, A., Alekseev, A., Efros, A., Rosen, M. & Meyer, B. (2002). General boundary conditions for the envelope function in the multiband k·p model, *Physical Review B* 65(12): 1–12.
URL: *http://link.aps.org/doi/10.1103/PhysRevB.65.125302*

Rogach, A. L. (ed.) (2008). *Semiconductor Nanocrystal Quantum Dots*, Springer-Verlag, Wien.

Rolo, A. G., Vasilevskiy, M. I., Hamma, M. & Trallero-Giner, C. (2008). Anomalous first-order raman scattering in iii-v quantum dots: Optical deformation potential interaction, *Phys. Rev. B* 78: 081304.
URL: *http://link.aps.org/doi/10.1103/PhysRevB.78.081304*

Schulz, S. & Czycholl, G. (2005). Tight-binding model for semiconductor nanostructures, *Phys. Rev. B* 72: 165317.
URL: *http://link.aps.org/doi/10.1103/PhysRevB.72.165317*

Soni, U. (2010). The Importance of Surface in Core-Shell Semiconductor Nanocrystals, *The Journal of Physical Chemistry* C 114: 22514.
URL: *http://pubs.acs.org/doi/abs/10.1021/jp1091637*

Springer Materials The Landolt-Börnstein Database (2011). Springer.

Talapin, D., Rogach, A., Kornowski, A., Haase, M. & Weller, H. (2001). Highly luminescent monodisperse CdSe and CdSe/ZnS nanocrystals synthesized in a hexadecylamine-trioctylphosphine oxide-trioctylphospine mixture, *Nano Letters* 1(4): 207–211.
URL: *http://pubs.acs.org/doi/abs/10.1021/nl0155126*

Vasilevskii, M., Akinkina, E., De Paula, A. & Anda, E. (1998). Effect of size dispersion on the optical absorption of an ensemble of semiconductor quantum dots, *Semiconductors* 32(11): 1229–1233.
URL: *http://www.springerlink.com/index/XW2V65248T002873.pdf*

Wang, X., Zhuang, J., Peng, Q. & Li, Y. (2005). A general strategy for nanocrystal synthesis, *Nature* 437(7055): 121–124.
URL: *http://www.nature.com/nature/journal/v437/n7055/suppinfo/nature03968_S1.html*

Woggon, U., Herz, E., Schöps, O., Artemyev, M. V., Arens, C., Rousseau, N., Schikora, D., Lischka, K., Litvinov, D. & Gerthsen, D. (2005). Hybrid epitaxial-colloidal semiconductor nanostructures., *Nano letters* 5(3): 483–90.
URL: *http://www.ncbi.nlm.nih.gov/pubmed/15755099*

Yu, W. W., Qu, L., Guo, W. & Peng, X. (2003). Experimental Determination of the Extinction Coefficient of CdTe, CdSe, and CdS Nanocrystals, *Chemistry of Materials* 15(14): 2854–2860.
URL: *http://pubs.acs.org/doi/abs/10.1021/cm034081k*

9

In-Gap State of Lead Chalcogenides Quantum Dots

Xiaomei Jiang*
Department of Physics,
University of South Florida, Tampa, FL
USA

1. Introduction

Lead selenide (PbSe) and lead sulfide (PbS) quantum dots (QDs) have many unique properties to make them promising materials for optoelectronic devices. Their bandgaps, ranging from 0.3~1.1eV, can be easily tuned via size control during synthesis, and their photo response in near infrared region promises their broad applications in bio-imaging [1], telecommunications [2], LEDs [3], lasers [4], photodetectors [5], and photovoltaic devices [6-8].

QDs are essentially nanocrystals consisting of tens to hundreds of atoms (Fig.1a). Due to the nanocrystal's small size (smaller than the exciton Bohr radius of the bulk semiconductor), strong quantum confinement results in discrete energy levels and bigger bandgaps compared with the respective bulk semiconductor (Fig. 1b)[9]. E_1, E_2 and E_3 stand for the first, second and third excitonic transitions, respectively. E_1 is the optical bandgap of QD, which is correlated with the size of QD, as shown in the absorption spectra of a series of PbSe QDs (Fig. 1c). QDs studied here are synthesized by colloidal chemistry [10, 11], where the QDs are kept in hexane or toluene as a suspension. The transmission electron micrograph (TEM) shows the highly monodisperse PbSe QDs with average size of 9.6 nm, and the single crystal structure was clearly shown [50]. The crystal structure for PbSe or PbS QDs is rock salt crystal, the same as its bulk semiconductor. The lattice constant for bulk PbS is about 5.9 Å, and 6.1 Å for PbSe [12]. In order to prevent coalescence of QDs, surface passivation by appropriate ligands/surfactants is necessary. Oleic acid is a common ligand for lead chalcogenides such as PbSe or PbS QDs [11]. The passivation layer (ligands) can also modify the optical and electronic properties of the QDs.

Unlike in their bulk semiconductors, enhanced Coulomb interaction in QDs results in much more tightly bonded excitons, and the fate of excitons in these quantum dots is of great relevance to their device applications. Although the properties of excitonic states have been thoroughly studied in the past decade, mostly employing transient spectroscopies [13-17], relatively less attention has been paid to the states within the

*Corresponding author

quantum dots bandgap. Conventionally, there are two types of in-gap states: one is the dark exciton state, which is due to the exchange splitting from confinement-enhanced exchange interaction [18, 19]; another type is trap state(s) associated with surface defects [20-22].

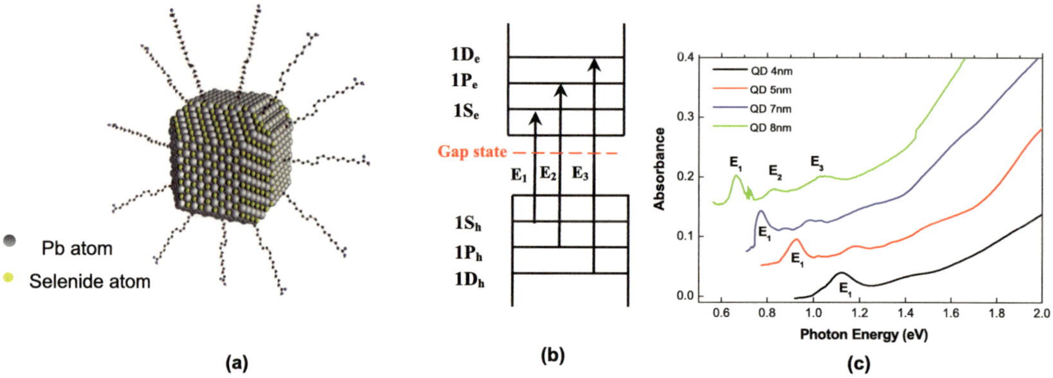

Fig. 1. (Color online) (a) Rock salt crystal structure of a PbSe quantum dot. (b) Quantized energy levels of a PbSe quantum dot (QD). The dashed line represents the gap state energy level found on IV-VI quantum dots. (c) Absorption spectra of variously sized PbSe QDs. E_1 is the optical gap of QD.

These in-gap states are of great importance since they affect the final destiny of excitons. We have previously reported a peculiar in-gap state that bears confinement dependence, with life time about 2 μs [23]. This single gap state (G.S) does not seem to fall into either one of the above conventional gap state categories. A detailed analysis of temperature dependence of photoluminescence (PL), absorption and photoinduced absorption (PA) reveals the unconventional G.S. is a new state of trapped exciton in QD film [24]. The importance of such gap state is illustrated below through the analysis of exciton loss mechanisms in QDs.

As can be seen in Figure 2, Auger recombination happens within sub-*ps*, followed by hot exciton cooling on the *ps* timescale [13]. Radiative recombination was surprisingly slow in these QDs, and was detected in the sub-*μs* range [25-27]. In contrast, relaxation to gap state(s) occurs much faster (<*ns*) [28]. Therefore, the final state of photogenerated carriers is likely to be the gap state. Due to its long lifetime, this state can be used to monitor charge transfer between QDs and polymers, similar to the case seen in π–conjugated polymer and fullerene systems [29, 30]. This feature also enables the investigation of charge transfer in hybrid composite of polymers and QDs using continuous wave photoinduced absorption (cw-PA) measurements of both constituents, and therefore provides a reliable and accurate study of charge transfer within the hybrid composite. This is especially useful when energy transfer is superimposed with charge transfer. As a matter of fact, the interplay of energy transfer and charge transfer in such composites has been one of the major obstacles that hinder the progress in QD / polymer solar cells. A thorough understanding of the gap state will help to identify photoinduced charge transfer between QDs and the polymer host, as will be illustrated in **2.3**.

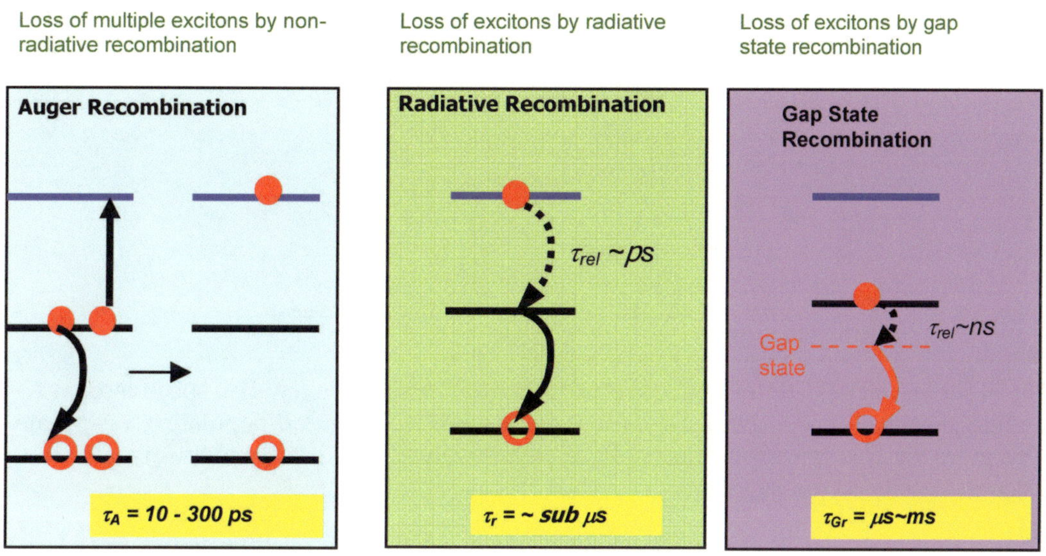

Fig. 2. (Color online) Schematic drawing of loss mechanism in lead chalcogenide QDs (Inspired by [51]).

2. Experimental observation of in-gap state in lead sulfide quantum dots

2.1 Experimental methodology

The main experimental method to study the in-gap state is Continuous Wave Photoinduced Absorption (cw-PA) spectroscopy. Continuous Wave Photoinduced Absorption (cw-PA) is also called pump & probe or photomodulation spectroscopy (Fig.3). A cw Ar^+ laser (pump), with its energy larger than the optical gap of the investigated material, excites the sample film and generates long-lived photoexcitations; a tungsten-halogen lamp is used to probe the modulated changes ΔT in transmission T among the interested energy range, usually the subgap regime. A lock-in amplifier is employed with an optical chopper for photomodulation. A series of solid-state photodetectors are coupled with light sources and optical components to span the detection range from UV to NIR. The advantages of cw-PA are that both neutral and charged excitations may be studied and there is no need to introduce dopants into the film [31].

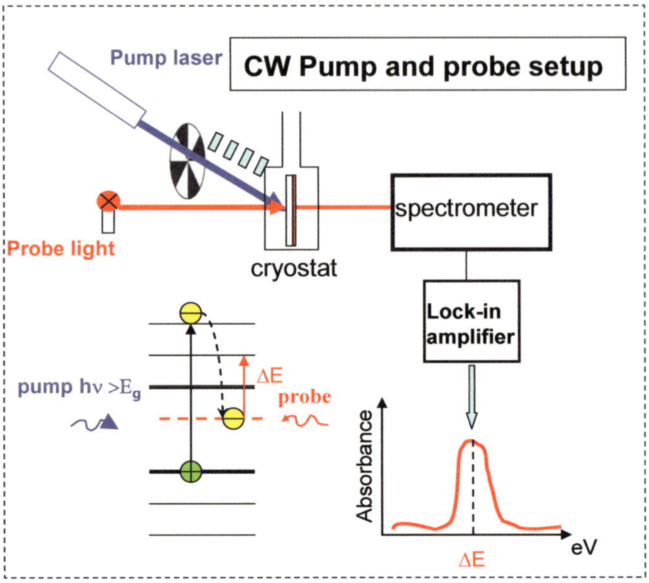

Fig. 3. (Color online) Continuous wave photoinduced absorption (cw-PA) spectroscopy. A regular absorption spectrum of a quantum dot is modified by pump-populating a gap state. This gives rise to a new absorption peak ΔE, whose intensity is directly proportional to the electron density on the gap state.

The pump (cw Ar^+ laser) excites the semiconductor sample with photons of an energy larger than the optical gap of the semiconductor (for example, E_g = 1.07eV for a 4nm PbS QD). The excited electrons thermalize into a long lived gap state(s) caused by defects in the semiconductors. This changes the absorption spectrum since now the transition ΔE becomes possible. A new peak arises in the spectrum at a wavelength commensurable with ΔE (for example ΔE = 0.33 eV for a 4nm PbS QD), as being schematically indicated in Figure 3. The important feature of this measurement is that the magnitude of this absorption peak is linearly proportional to the density of the electrons occupying this gap state:

$$PA = \frac{\Delta T}{T} = -\Delta \alpha d = \sigma n_e d \tag{1}$$

with n_e the density of photoexcitations, d the sample thickness, and σ the excited state optical cross section.

Previously several groups have been using cw-PA to investigate photophysics of CdSe and CdS quantum dots [32, 33]. Recently, we have applied this method to PbS and PbSe quantum dots system [23, 24, 34]. This methodology is very useful to study long-lived, in-gap states and their associated photoexcitations.

2.2 Spectral signature of charged species

We have measured cw photoinduced absorption (cw-PA) of PbS QD film in an energy range of interband electronic transitions at low temperature (10K) [34]. At photon energy near and above the QD bandgap, E_g, five photoinduced absorption (PA) peaks are clearly observed, and four photoinduced absorption bleaching (PB) valleys are also present **(Fig.4a)**. Figure 4b shows the five interband transitions involved. The close resemblance between PA spectrum (Fig. 4a, black circle) and the second derivative of the linear optical absorption (Fig. 4a, red line) strongly suggests that these steady state PAs may be caused by photoinduced local electric field, and therefore resemble the linear Stark effect [35]. Heretofore, there have been reports about QD charging associated with photoexcitation [36-38].

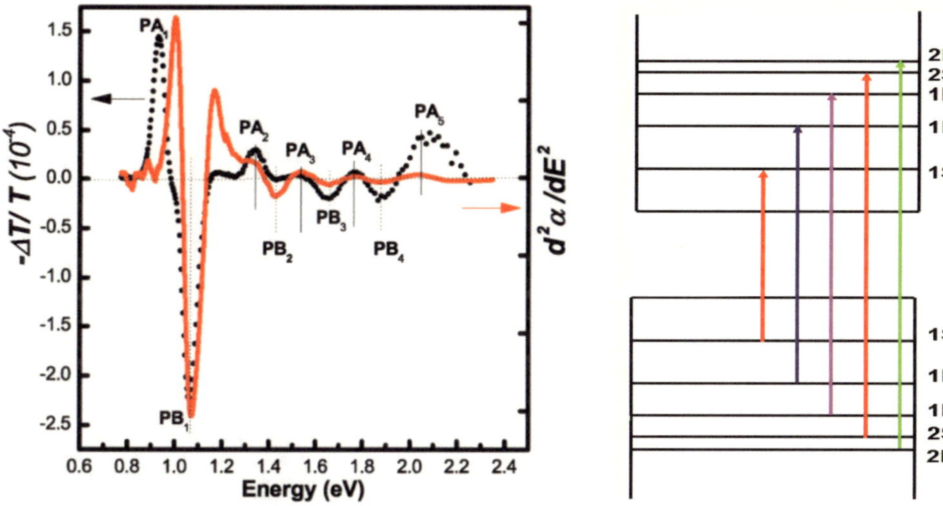

(a) (b)

Fig. 4. (Color online) (a) Photoinduced absorption (PA) spectrum (filled circle, black) of PbS nanocrystal (4.2 nm in diameter) film at T=10K and modulation frequency 400 Hz. Laser excitation is 488 nm with 150 mW/cm^2 intensity on film. The second derivative of the absorption spectrum (solid line, red) is also shown**.** PBs stand for photo bleaching bands ($\Delta\alpha<0$), whereas PAs are for photoinduced absorptions ($\Delta\alpha>0$). [Reprinted with permission from [34]. J. Zhang *et.al.*, *Appl. Phys. Lett.* 2008, 92, 14118. Copyright (2008), American Physical Society]. (b) A schematic drawing of the five interband transitions involved.

The photoinduced local electric field may be created by the trapped charges at the QD surface states. In cw-PA measurement, the laser is being modulated, which is equivalent to the modulation of the local electric field. As a result, the excitonic energy levels shifts and causes change of linear absorption, similar to that observed in electroabsorption spectrum [39]. Therefore, the electroabsorption (EA) feature above the QD bandgap is an indicator of charged species.

2.3 Spectral signature of neutral species

We have extended the cw-PA measurement to photon energy below the QD bandgap. A single PA band (called IR-PA) was observed in the near infrared range. The lifetime of this photoexcitation is about several microseconds, and the peak position of this band has correspondence with the QD size [23]. This photoexcitation was due to a transition from a gap state (G.S) to the second level of exciton excited state $1P_e$ (Fig. 1b). Two different scenarios could each partially explain our observations. Scenario 1 is that this G.S belongs to a certain trap state, and this could explain the large Stokes shift. Trap states in PbS or PbSe QDs have been previously observed [20, 40]. However, the characteristics of a conventional trap state do not completely go with the confinement-dependence, narrow emission band and short lifetime observed here. Furthermore, the temperature dependence of PL intensity (Fig. 5) shows no thermal activation behavior typical for trap state emission [40]. The lack of thermal activation also indicates there is negligible non-radiative recombination due to defects or aggregates in the films. Large Stokes shift from 100 to nearly 300 meV was observed [23]. The inset of Figure 5b plots the PL energy, E_{PL} vs. 1st excitonic peak (E_1). Dotted line shows the zero Stokes line (i.e., $E_{PL} = E_1$), which has a slope of 1.0. The linear fitting slope of $E_{PL} \sim E_1$ is 0.75, which is larger than 0.50, meaning the emission state or G.S is not fixed with respect to the bottom of the bulk conduction band, as previously reported for an in-gap hybrid state [41].

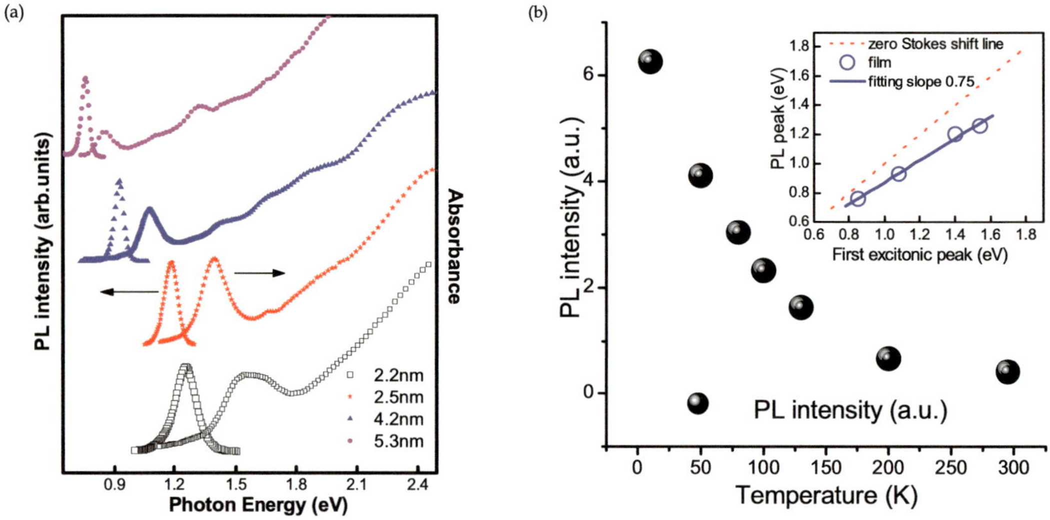

Fig. 5. (Color online) (a) PL (left) and absorption (right) spectra for four different sizes (2.2-5.3 nm in diameter) of PbS QD films on sapphire measured at $T = 10$ K. The baseline of the spectrum for each size was shifted vertically for clarity. (b) Temperature dependence of PL intensity for a 4.2 nm PbS QD film (black solid circle). The inset plots the PL energy, E_{PL} vs. first excitonic absorption, E_1 for the four sizes of PbS QDs. The red dotted line is zero Stokes shift line, and the blue solid line is a linear fit of experimental data (blue open circle). [Reprinted with permission from [24]. J E Lewis *et.al.*, *Nanotechnology* 2010, 21, 455402. doi:10.1088/0957-4484/21/45/455402. Copyright (2010), IOP Publishing Ltd].

We also rule out G.S to be a dark exciton state in PbS quantum dots, since the gap state is too 'deep' for dark exciton state from exchange splitting, which was calculated to be less than 10 meV below the lowest bright exciton for a 4.2 nm PbS QD [18], on the other hand, the activation energy of G.S was measured to be about 20 meV [24]. In terms of Stokes shift, even counting the total splitting due to exchange and intervalley interactions, the calculated value was less than 80 meV, whereas the Stokes shift we measured is 332 meV for this size QD [24]. These inconsistencies mean that G.S. is not the dark exciton state.

To further validate this claim, we measured the temperature dependences of PL energy, E_{PL} (solid circle, black) and the first excitonic absorption, E_1 (open triangle, red) of a 4.2 nm QD film (Fig. 6). Above T =50 K, a linear increase of dE_1/dT= 0.05 meV·K^{-1} was obtained from fitting of the absorption experimental data. On the other hand, a temperature coefficient of dE_{PL}/dT= 0.3 meV·K^{-1} was derived from fitting of E_{PL} data. $dE_1/dT << dE_{PL}/dT$ indicates that emission is not originated from a band edge splitting state such as a dark exciton state.

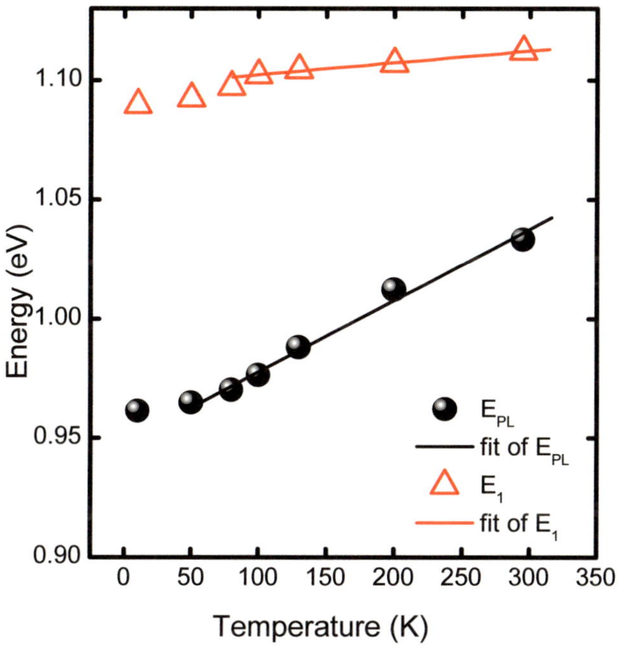

Fig. 6. (Color online) Temperature dependences of PL energy, E_{PL} (solid circle, black) and the first excitonic absorption E_1 (open triangle, red) of a 4.2 nm QD film. Black line is a linear fit for E_{PL} data and red line is a fit for E_1, at $T > 50$ K. [Reprinted with permission from [24]. J E Lewis *et.al.*, *Nanotechnology* 2010, 21, 455402. doi:10.1088/0957-4484/21/45/455402. Copyright (2010), IOP Publishing Ltd].

In summary, G.S is not a trap state based on lifetime and confinement dependence, nor it a dark exciton state based on its different characteristics as oppose to free excitons, i.e., the energy level within the bandgap, its temperature dependence and large Stokes shift. We therefore assign G.S a state for trapped exciton. Such a state, due to its long lifetime (~ several μs), is relevant to exciton dissociation and carrier extraction processes in QD/polymer composite, a material system potentially can be utilized for low-cost high efficiency solar cells [7,42].

3. Implication of in-gap state for quantum dots solar cells

Figure 7 is a schematic drawing of a QD/polymer composite. The absorption of photons by both moieties create excitons, with favorable type II ('staggered') energy level alignment, exciton dissociation could happen, with electron being transferred to the QD from the polymer, and holes to the polymer from the QD.

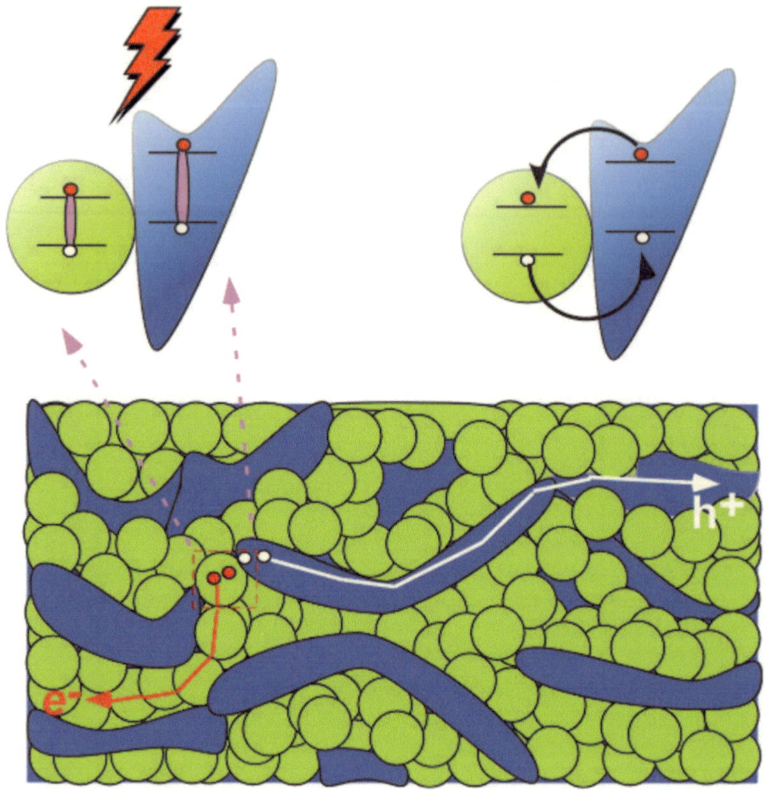

Fig. 7. (Color online) Schematic drawing of QD/polymer composite. QDs (green balls) were shown embedded in polymer matrix (blue). Upper left panel shows the creation of excitons upon light absorption; upper right panel shows the charge transfer process with type II ('staggered') energy level alignment between QDs and polymer.

Photoluminescence quenching has been primarily used to identify possible charge transfer between QD and polymer [32,43]. Figure 8a shows the absorption of a 2nm PbS QD film and PL of a poly(3-hexyl)thiophene (P3HT) film, both measured at T=10K. Because of the overlap between the absorption of QD (guest material) and the photoluminescence of polymer (host material), energy transfer (including Förster energy transfer and radiative energy transfer) can occur, which quenches the polymer PL. On the other hand, charge transfer between the polymer (host) and QD (guest) can also eliminate PL of the polymer (Fig. 8b). Unfortunately PL quenching itself cannot distinguish these two mechanisms.

Figure 8c shows the PL of QD/P3HT composite films with different weight ratio of QDs. It is clearly shown that energy transfer is quite efficient at high weight ratio of QDs, and the emission from QD came at a price of P3HT photoluminescence quench. However, comparing with the case of polymer/C_{60} composite, where <10% weight ratio of C_{60} completely quenches the PL of polymer [30], with 200% of QDs the magnitude of PL of P3HT was only reduced by a factor of 5 (inset of Fig. 8c). This indicates that the charge transfer (CT) is not efficient in PbS QD/P3HT composite, as previously reported in CdSe QD/polymer blend [32].

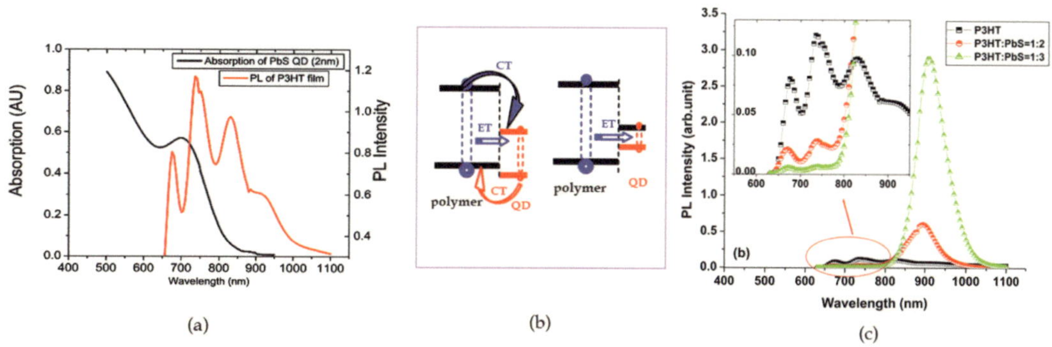

Fig. 8. (Color online) (a) PL spectrum of P3HT film (red line) and absorption spectrum of a 2nm PbS QD film (black line) measured at T=10K. (b) schematic drawing of energy transfer (ET) and charge transfer (CT) processes in QD/polymer composite when the energy level alignment is type II (staggered, left) and type I ('straddled', right). (c) PL spectra of neat P3HT film (black hall-filled square) and QD/P3HT composite with various weight ratios of QDs. The inset is a blow-up at short wavelength range.

Unlike in the case of CdSe and CdS QDs, with the spectral features we have discovered in PbS QDs, namely, IR-PA and EA (see sections 2.2&2.3 for details), we could qualitatively study charge transfer between PbS or PbSe QDs and polymers, without the complications from energy transfer which often occur in the composites. This is demonstrated in Figure 9. In Figure 9a, when the so-called type I alignment ("straddled") is present between energy levels of QDs and polymers, the energetics would be in favor of energy transfer, IR-PA signal should increase since now there are more excitons generated on polymer being transferred and eventually trapped on the gap state of QD. In this case, since negligible charges are added to the QDs, the EA feature at energy higher than the QD bandgap is expected to remain the same.

On the other hand, in Figure9b, when the energy alignment is type II ("staggered"), charge transfer could become more favorable, with hole being transferred to the polymer, the number of excitons originally trapped at the gap state would diminish, and therefore IR-PA signal is expected to decrease, and EA feature will increase due to enhanced QD charging.

Noticeably in both cases, PL quenching of polymer would occur, therefore the conventional way of measuring PL quenching alone is not sufficient to distinguish between whether or not charge transfer has occurred. Furthermore, in the QD/polymer composite, detection of the charge transfer process can be determined from PA measurements performed on the individual constituents (i.e. QDs and polymer), and the mixed composite. The relative change of ΔE in QDs and the change of polaron absorption in polymers would give direct insight into the charge transfer within the system. The possible overlap of the lower polaron PA band (P_1, usually at 0.5 eV) and IR-PA for QDs can be easily avoided by choosing different QD size and different polymers.

The fact that the IR-PA feature has strong temperature dependence does not prevent it being used to study relevant device physics at room temperature. A previous example in polymer/fullerene composite has shown polaron absorption, which almost vanishes at room temperature, could have implications for organic solar cells efficiency [44].

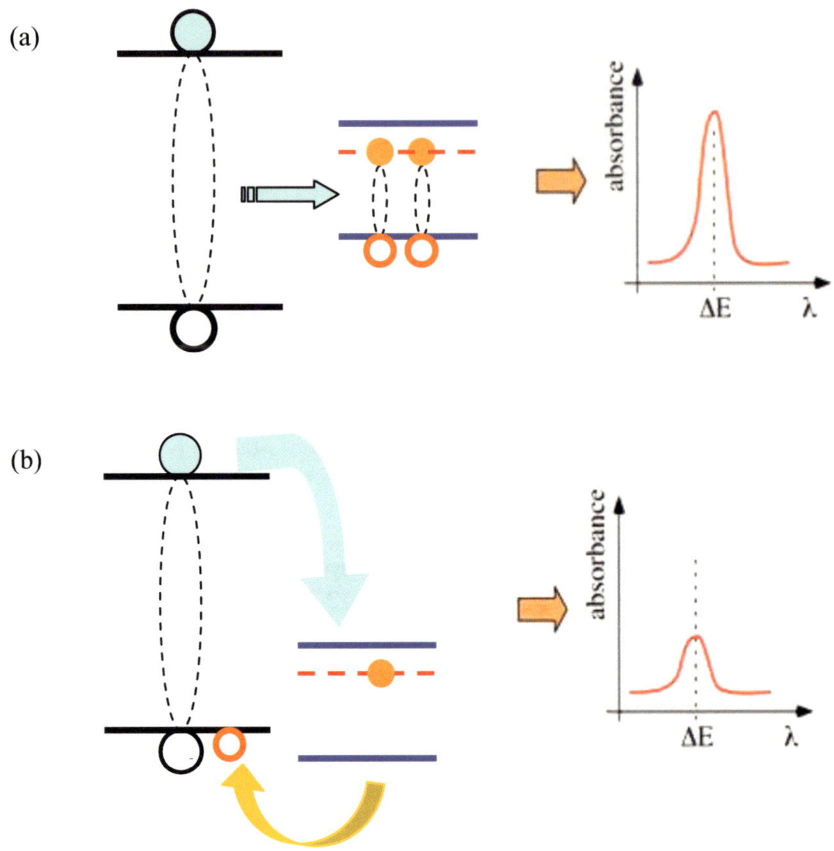

Fig. 9. (Color online) Schematic drawing of how energy transfer and charge transfer affect IR-PA and EA (not shown here). (a) When energy transfer dominates, the IR-PA signal increases, EA signal remains the same. (b) When charge transfer dominates, the IR-PA signal diminishes, accompanied by increased EA signal.

Figure 10a shows the PA spectra of neat QD and P3HT, as well as the QD/P3HT composite with 1:1 weight ratio. As can be seen, the polaron population (P_2) in P3HT shows slight increase, whereas the density of interchain excitons (IEX) was greatly enhanced due to morphology change of P3HT upon adding the QDs [45]. On the other hand, the EA signal of QDs remains the same in the composite film comparing with in neat QD film, with a slight increase of IR-PA signal, which means that charge transfer was inefficient between QD and polymer P3HT, combining with PL quenching (Fig. 11), we can draw the conclusion that energy transfer dominates the photoexcitation process in QD/polymer composite film. Similar result was recently reported [46]. Further improvement of charge transfer can be done by ligands manipulation of QD, i.e., ligand exchange with shorter surfactant groups [49], or ligand removal [43, 48, 49], to improve the interfacial properties between QD and polymer, and to facilitate charge transfer occurrence.

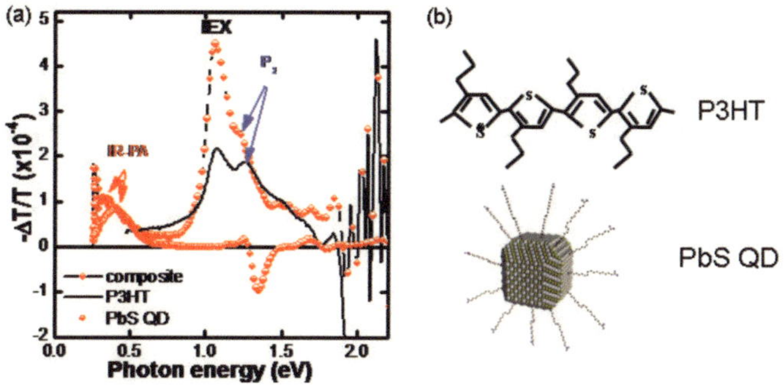

Fig. 10. (Color online) (a) Photoinduced absorption (PA) spectra of P3HT (black line), PbS QD (2.5nm) film (half-filled circle, red) and P3HT/PbS QD composite film (black line +red half-filled circle) measured at T =10K. The gap state is revealed as the near IR band (IR-PA), whereas the polaron absorption from P3HT is marked P_2, IEX stands for interchain exciton [45]; (b) Molecular structures of P3HT and PbS QD, (Courtesy of Dr. J. Lewis).

4. Conclusion

In conclusion, using photoinduced absorption (PA) spectroscopy, we have investigated the characteristics of a peculiar gap state (G.S) in films of PbS QDs with different sizes. Large Stokes shift was attributed to the difference from first excitonic absorption and emission from a gap state (G.S.) which bears quantum confinement dependence. A detailed analysis of temperature dependence of PL, absorption and photoinduced absorption reveals the unconventional G.S. is a new state of trapped exciton in QD film. This gap state is directly relevant to exciton dissociation and carrier extractions in this class of semiconductor quantum dots.

The spectral features of PA of PbS QD include an induced absorption band (IR-PA) at near infrared range, and electroabsorption peaks (EA) above the QD bandgap. Both features can be utilized to characterize charge transfer process between QD and conducting polymers such as poly (3-hexyl)thiophene (P3HT), a widely used electron donor in organic photovoltaics. The methodology developed in our work could separate the contributions of energy transfer from that of the charge transfer in QD/polymer composite, therefore solve the current difficulty of independently evaluating the role of charge transfer useful in hybrid photovoltaic devices built on QD/polymer mixture.

5. Acknowledgment

This work was sponsored by the ACS Petroleum Research Fund (PRF 47107-G10) and USF Grant No. NRG-R061717.

6. References

[1] Kim, S. *et.al., Nat. Biotechnol.* 2004, 22, 93–97.
[2] Steckel J S, Coe-Sullivan S, Bulovic V and Bawendi M G, *Adv. Mater.* 2003, 15, 1862.
[3] Rogach A L *et al., Angew. Chem. Int. Edn* 2008, 47, 6538.
[4] Eisler H J *et al., Appl. Phys. Lett.* 2002, 80, 4614.
[5] Konstantatos G *et al., Nature* 2006, 442, 180.
[6] McDonald S A *et al., Nat. Mater.* 2005, 4, 138.
[7] Jiang X *et al., J. Mater. Res.* 2007, 22, 2204.

[8] Luther J M *et al., Nano Lett.* 2008, 8, 3488.
[9] I. Kang and F. W. Wise, *J. Opt. Soc. Am. B* 1997, 14, 1632-1646.
[10] C. B. Murray, C. R. Kagan, and M. G. Bawendi, *Annu. Rev. Mater. Sci.* 2000, 30, 545–610.
[11] W. W. Yu, J. C. Falkner, B. S. Shih and V. L. Colvin, *Chem Mater* 2004, 16 (17), 3318-3322.
[12] D. Rached et al., *Physica B* 2003, 337, 394–403.
[13] Schaller R D, Agranovich V M and Klimov V I, *Nat.Phys.* 2005, 1, 189.
[14] Dementjev A and Gulbinas V, *Opt. Mater.* 2009, 31, 647.
[15] Dementjev A, Gulbinas V, Valkunas L and Raaben H, *Phys. Status Solidi* b 2004, 241, 945.
[16] Ellingson R J *et al Nano Lett.* 2005, 5, 865.
[17] Nair G, Geyer S M, Chang L Y and Bawendi M G, *Phys. Rev.* B 2008, 78,125325.
[18] An J M, Franceschetti A and Zunger A, *Nano Lett.* 2007, 7, 2129.
[19] Espiau de Lamaëstre R *et al., Appl. Phys. Lett.* 2006, 88, 181115.
[20] Kim D, Kuwabara T and Nakayama M, *J. Lumin.* 2006, 119/120, 214.
[21] Konstantatos G *et al., Nano Lett.* 2008, 8, 1446.
[22] Peterson J J and Krauss T D, *Phys. Chem. Chem. Phys.* 2006, 8, 3851.
[23] Zhang J and Jiang X, *J. Phys. Chem.* B 2008, 112, 9557.
[24] J E Lewis, SWu and X J Jiang, *Nanotechnology* 2010, 21, 455402.
[25] B. L. Wehrenberg, C. Wang, and P. Guyot-Sionnest, *J. Phys. Chem. B* 2002, 106, 10634-10640.
[26] Clark, S.W.H., J. M.; Wise, F. W., *J. Phys. Chem. C* 2007, 111, 7302.
[27] Du, H.C. et.al., *Nano letters* 2002, 2, 1321.
[28] M. Pope, C.E. Swenberg, *Electronic processes in Organic Crystals and Polymers.* 1999, Oxford Science Publications: Oxford.
[29] N.S. Sariciftci, L. Smilowitz, A.J. Heeger, F. Wudl, *Science* 1992, 258, 1474-1476.
[30] X. Wei, S.V. Frolov, Z.V. Vardeny, *Synth. Metal.* 78, 295-299 (1996).
[31] X. Wei and Vardeny, Handbook of Conducting Polymers II, Chapter 22, Marcel Dekker, Inc., New York, 1997.
[32] D. S. Ginger and N. C. Greenham, *Phys. Rev.* B 1999, 59, 10622.
[33] D. S. Ginger, N.C.G., *J. Appl. Phys* 2000, 87, 1361.
[34] J. Zhang, X. Jiang, *Appl. Phys. Lett.* 2008, 92, 14118.
[35] *Gerold U. Bublitz and Steven G. Boxer, Annu. Rev. Phys. Chem.* 1997, 48, 213–42.
[36] S. A. Empedocles and M. G. Bawendi, *Science* 1997, 278, 2114.
[37] K. Zhang, H. Chang, A. Fu, A. P. Alivisatos, and H. Yang, *Nano Lett.* 2006, 6, 843.
[38] T. D. Krauss, S. O'Brien, and L. E. Brus, *J. Phys. Chem. B* 2001, 105, 1725.
[39] E. J. D. Klem, L. Levina, and E. H. Sargent, *Appl. Phys. Lett.* 2005, 87, 053101.
[40] Turyanska L, Patané A and Henini M, *Appl. Phys. Lett.* 2007, 90, 101913.
[41] Fernée M J, Thomsen E, Jensen P and Rubinsztein-Dunlop H *Nanotechnology* 2006, 17 956.
[42] Kevin M. Noone et.al., *ACS Nano* 2009, 3, 1345–1352.
[43] X. Jiang, S. B. Lee, I. B. Altfeder, A. A. Zakhidov, R. D. Schaller, J. M. Pietryga, and V. I. Klimov, *Proc. Of SPIE* 2005, 5938, 59381F-1.
[44] T. Drori, C.-X. Sheng, A. Ndobe, S. Singh, J. Holt, and Z.V. Vardeny, *Phys. Rev. Lett.* 2008, 101, 037401.
[45] Österbacka R, An CP, Jiang XM, Vardeny ZV, *Science* 2000, 287, 839.
[46] Kevin M. Noone et.al., *Nano Lett.* 2010, 10, 2635–2639.
[47] S. Zhang, P. W. Cyr, S. A. McDonald, G. Konstantatos, and E. H. Sargent, *Appl. Phys. Lett.* 2005, 87, 233101-233103.
[48] D. V. Talapin and C. B. Murray, *Science* 2005, 310, 86-89.
[49] J. M. Luther, A. J. Nozik, et. al., *ACS Nano* 2008, 2, 271-280.
[50] G. Dedigamuwa et.al., *Appl Phys Lett* 2009, 95, 122107.
[51] R. D. Schaller and V. I. Klimov, *Phys. Rev. Lett.* 92, 186601, 2004.

10

Exciton Dynamics in High Density Quantum Dot Ensembles

Osamu Kojima
Kobe University
Japan

1. Introduction

Remarkable progress has been made in the fabrication of semiconductor quantum dots (QDs) using the self-assembling method in lattice-mismatched material systems; they are based on the Stranski-Krastanow growth mode (Goldstein et al., 1985). In this process, initial two-dimensional growth transforms into three-dimensional growth. Using the self-assembling technique, it is possible to fabricate semiconductor nanostructures in a continuous growth process in a vacuum. The self-assembled QDs grown on a semiconductor substrate can offer the possibility of realizing various interesting devices such as QD lasers, ultrafast optical switches, and solar cells (Arakawa & Sakaki, 1982; Huffaker et al., 1998; Prasanth et al., 2004; Bogaart et al., 2005; Martí et al., 2006; Oshima et al., 2008). To realize such QD devices, it is necessary to design the optical properties by controlling the exciton characteristics and to fabricate high-quality and high-density QDs. From this viewpoint, we have clarified that the photoluminescence (PL) characteristics of excitons in multiple stacked QDs fabricated by using the strain compensation technique (Akahane et al., 2002, 2008, 2011) can be controlled by changing the QD separations along the growth direction (Nakatani et al., 2008; Kojima et al., 2008). In this chapter, we introduce the control method of excitonic characteristics by using the overlap of electron envelope functions between QDs along the growth direction.

2. Sample structures

We used three samples in this study. For each sample, InAs self-assembled QDs with 30 periods was grown on an InP(311)B substrate, as shown in Fig. 1, by solid-source molecular beam epitaxy using a strain compensation technique(Nakatani et al., 2008; Kojima et al.,2008, 2010, 2011). After growing a 150-nm thick $In_{0.52}Al_{0.48}As$ buffer layer, 4-ML InAs QDs were deposited. The samples have $In_{0.5}Ga_{0.1}Al_{0.4}As$ spacer layers with thicknesses (*d*) of 20, 30, and 40 nm. Hereafter, we will call these samples *d*=*X* nm sample (*X*=20, 30, and 40). The spacer layer compensates the stress caused by the lattice mismatch to a QD layer. The QD density in each layer is $3.4 \times 10^{10}/cm^2$.

3. Control of optical characteristics of excitons in QD ensembles

Figure 2 shows the spacer-layer-thickness dependence of the PL spectra at 3.4 K (Kojima et al., 2008). The PL measurement was performed by using a mode-locked Ti:sapphire pulse laser. The emitted light was dispersed by a 32 cm single monochromator with a resolution of 1.0 nm. The PL intensity and peak energy clearly depend on the spacer layer thickness. With a decrease in the spacer layer thickness, the PL intensity decreases. The integrated intensity of PL band is plotted as a function of *d* in the inset of Fig. 2 for the clarification of the relation between the intensity of the PL band and *d*. The dotted line indicates the linear dependence of the intensity on *d*. The PL intensity is almost linearly proportional to *d*. It is well known that a decrease in the relative motion of electron and holes in systems of quantum wells or QDs causes an increase in the oscillator strength in the strong confinement regime because the oscillator strength is proportional to the probability of finding an electron and hole at the same position. Namely, since the overlapping of the envelope functions of the confined electron and hole is enlarged in each QD, the oscillator strength of QD excitons is approximately inversely proportional to the confinement volume (Takagahara, 1987; Kayanuma, 1988).

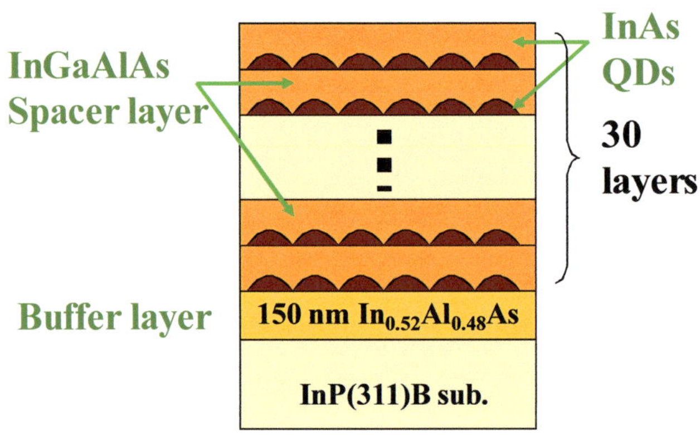

Fig. 1. Schematic of the sample structure.

Here, we discuss the origin of the confinement volume change. There are two possible factors for causing the expansion: (i) the creation of larger QDs owing to a decrease in *d* and (ii) the growth-direction elongation of the envelope function of confined carriers. It has been reported that larger-sized QDs are created in the case of the thinner capping layer (Xie et al., 1995; Saito et al., 1998; Persson et al., 2005). Indium atoms have a strong tendency to segregate at the surface when GaAs is deposited over InAs at temperature ~ 500 °C (Brandt et al., 1992), so that the capped InAs QD size tends to be small by a thick cap layer (Xie et al., 1995; Inoue et al., 2008). In addition, the thinner cap layer results in less compressive stress in the QDs (Saito et al., 1998; Persson et al., 2005), leading the strain-relaxed larger QDs. If a change in *d* causes this QD size variation due to the indium segregation or strain reduction, the magnitude of overlap of the envelope functions of confined electron and holes varies according to the change in the QD size. Since the oscillator strength depends on the QD size as described above, the PL intensity is expected to correspond to the variation of *d*. However, such possibility of the change in the QD size is denied as follows. The indium segregation is suppressed when the InAs QDs are capped by the indium alloys (Kim et al., 2003) so that in our samples, InGaAlAs spacer layer suppresses the indium segregation. Moreover, we employed the strain compensation technique in the QD growth, which means that the compressive strain to QDs is independent of *d* is practically independent. Therefore, we can focus on the growth-direction elongation of the envelope functions of confined carriers.

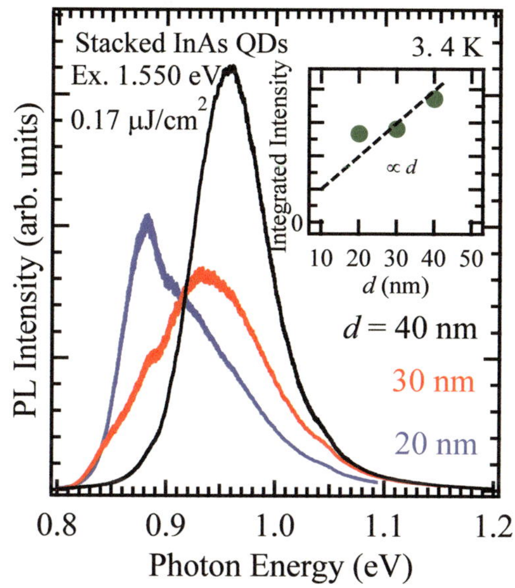

Fig. 2. The spacer-layer-thickness dependence of the PL spectrum in the stacked InAs QDs with *d*=20, 30, and 40 nm at 3.4 K. The inset shows the d dependence of the integrated intensity of the PL bands. The dotted line denotes the linear dependence of the intensity on *d*.

The heavy holes forming excitons are strongly localized in each of QDs (Saito et al., 2005) because of the heavier mass. On the other hand, since the effective mass of an electron is less than that of a hole, the electron envelope functions are sensitive to the quantum confinement effect. When the spacer layer thickness decreases, the electrons approach each other along the growth direction, and the electron envelope functions overlap within the spacer layer. Consequently, it is considered that the electron envelope functions in the samples with 30 QD layers separated by thinner spacer layers interconnect weakly along the growth direction owing to the overlap. This interconnection results in the lowering of the oscillator strength because of the reduction in the magnitude of the overlap integral between the electron and hole envelope functions.

The interconnection of electron envelope functions induces the lower-energy shift and broadening of PL band, as shown in Fig. 2. In Fig. 2, since the electrons in the *d*=40 nm sample are practically isolated within each QD, we can conclude that the PL band at around 0.95 eV in the *d*=40 nm sample is typical for our QDs. The peak energy shift in the thinner *d* samples comes from the reduction in the confinement effects by the expansion of the confinement volume owing to the interconnection of electron envelope functions. In addition, the PL band broadens and the shape becomes asymmetric owing to the appearance of a new PL band at low energy side below 0.9 eV in Fig. 2. In the *d*=20 nm sample, the PL band clearly shows two components. As described above, the PL band around 0.95 eV results from the uncoupled QDs. Thus, the PL band below the energy of 0.9 eV arises from the interconnection of the electron envelope function.

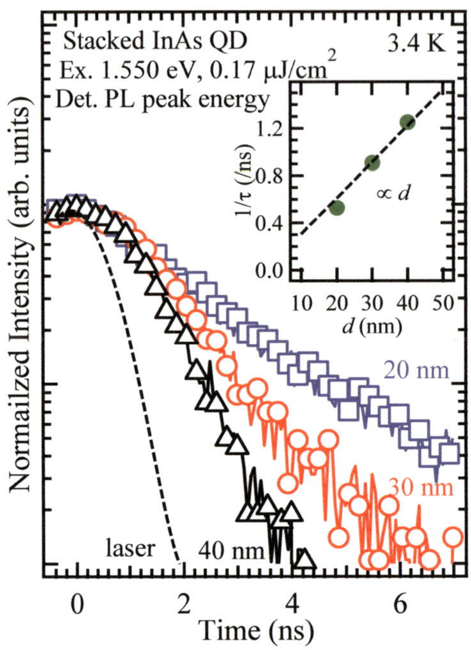

Fig. 3. The PL decay profiles observed at the peak energy in each sample. The dotted curve indicates the laser profile. The inset indicates the relation between $1/\tau$ and *d*. The dotted line denotes the linear dependence of $1/\tau$ on *d*.

When the above consideration is correct, the PL decay time should be inversely proportional to the spacer layer thickness. The spacer-layer-thickness dependence of the PL decay profile observed at a peak energy in each sample is shown in Fig. 3 (Kojima et al., 2008). The PL decay profiles were measured by a time-correlated single-photon counting method with a time resolution of 0.8 ns. The excitation source was a mode-locked Ti:sapphire pulse laser delivering 110 fs pulses with a repetition rate of 4 MHz. We used a pulse picker to reduce the laser repetition rate from 80 to 4 MHz. The excitation photon energy was 1.550 eV and the excitation density was 0.17 $\mu J/cm^2$. PL was dispersed by a 27-cm single monochromator with a resolution of 1.0 nm and detected by a time-to-amplitude converter system with the use of a liquid-nitrogen-cooled InP/InGaAsP photomultiplier. All profiles are normalized by the PL intensity observed at 0 ns. The dotted curve shows a laser profile. The PL decay profile has a single component and it clearly depends on *d*. The evaluated PL decay times τ_d obtained by fitting with a single exponential function are 1.9, 1.1, and 0.8 ns in d=20, 30, and 40 nm, respectively. This result is very different from that observed in the case of QD molecules. In the case of excitons in the QD molecules, the PL decay time decreases with a decrease in the interdot distance because of the superradiance effect (Bardot et al., 2005). In the inset of Fig. 3, $1/\tau_d$ was plotted as a function of *d*. The dotted line indicates the linear dependence of $1/\tau_d$ on *d*. τ_d of the QD excitons is inversely proportional to the oscillator strength *f* as described by the following equation (Andreani et al., 1999; Hours et al., 2005):

$$f = \frac{1}{\tau}\frac{3\pi\varepsilon_0}{n}\frac{2mc^3}{e^2\omega_0^2} \tag{1}$$

where ω_0 is the optical transition frequency, n is the refractive index, and m is the free-electron mass. Hence, the relation between τ and d strongly supports our consideration that the elongation of the electron envelope function results in the lowering of the oscillator strength.

4. Temperature dependence of excitons in QD ensembles

The reduced oscillator strength of excitons in QD ensembles with interconnection effects may not be suitable for light-emission devices such as LEDs, laser devices, and so on without a further increase in the number of the stacking layer. However, some devices such as quantum information devices and solar cells require a long exciton lifetime. Thus, the controllable long exciton lifetime described in previous section is considered to be a noteworthy property of interconnected QDs for realizing novel devices. Here, we focus on a carrier dynamics depending on the temperature in order to clarify the effects of the interconnection in vertically aligned QD ensembles.

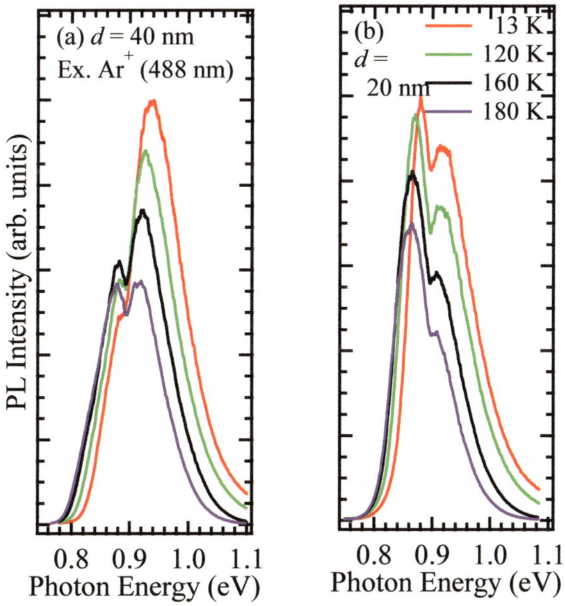

Fig. 4. Temperature dependence of the PL spectra in (a) the d= 40 nm and (b) the d=20 nm samples. The dip at around 0.9 eV originates from the hydroxy group in the optical fiber.

Figure 4 (a) and 4 (b) show the temperature dependence of the PL spectra in the d=40 nm and 20 nm samples, respectively (Kojima et al., 2010). The PL measurement was performed by using the 488 nm line of a CW Ar^+ laser. The excitation density was kept at 10 W/cm^2. The emitted light was dispersed by a 32-cm single monochromator with a resolution of 1.0 nm and was detected by using a liquid-nitrogen-cooled InGaAs-photodiode array. All spectra were normalized by the maximum intensity of the spectra at 13 K in each sample. The dip at around

0.9 eV is due to absorption by the hydroxy group in the optical fiber. The decrease in the PL intensity with an increase in the temperature in the d=20 nm sample is same as that in the d=40 nm sample. This result indicates that the nonradiative recombination process induced by an increase in the temperature is similar in both the samples.

In order to clarify the difference between the temperature dependences in both samples, we measured the PL decay profiles at various temperatures, as shown in Fig. 5. All the profiles were recorded at the PL peak energy. The PL decay times of both samples increase with the temperature. In Fig. 5 (c), the PL decay times of both samples are plotted as a function of temperature. The increase in PL decay time indicates that the growth by the strain compensation technique can suppress the generation of nonradiative centers with QD stacking. However, the temperature dependence of the PL decay time in the d=20 nm sample is different from that in the d=40 nm sample.

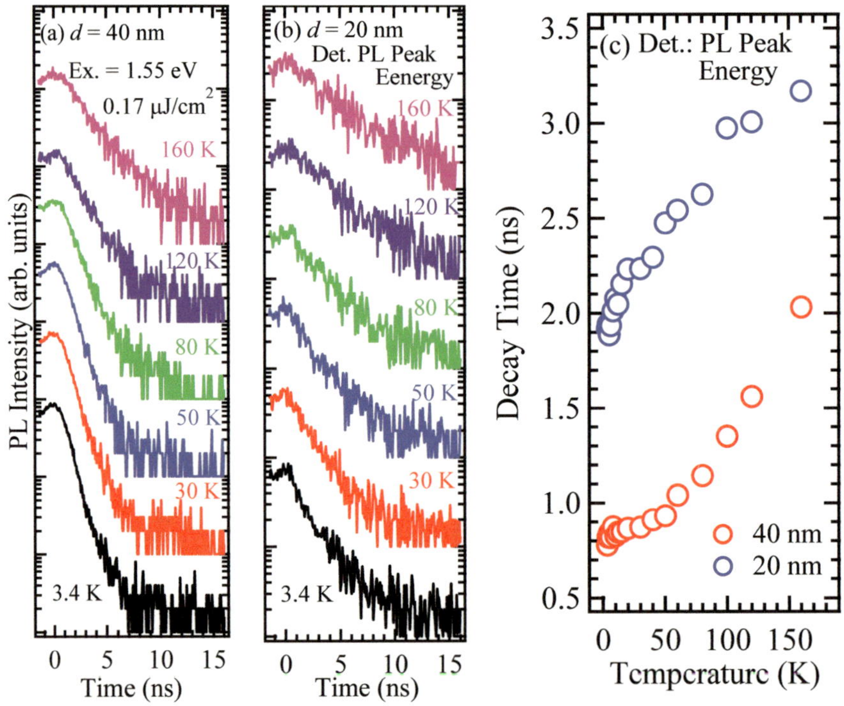

Fig. 5. Temperature dependence of the PL decay profiles in (a) d=40 nm and (b) d=20 nm samples. Each profile was recorded at the PL peak energy. (c) The PL decay times in both the samples are plotted as a function of temperature.

The increase in the PL decay time with the temperature is generally explained by the thermal dissociation of excitons into the electron-hole pairs, in which the excitons escape from the QDs into the spacer layer and/or the upper subband levels of electron and holes via thermionic emission (Wang et al., 1994; Yu et al., 1996; Fiore et al., 2000; Hostein et al., 2008), which is described as lateral coupling model. We considered that the temperature dependence of the PL decay time in the d=40 nm sample can be explained by this lateral coupling model. However, the temperature dependence of the PL decay time in the d=20 nm sample should include another factor, namely, the interconnection effect of the electron envelope functions.

In Fig. 6, the PL decay times in the low temperature region were plotted as a function of temperature. The PL decay time in the *d*=40 nm sample shows the almost constant value. This is the typical result for localized excitons. On the other hand, the PL decay time in the *d*=20 sample indicates the $T^{0.5}$ dependence as shown by the solid curve, which is similar property of the excitons in the quantum wires (Akiyama et al., 1994). This is an evidence of the QD interconnection. Therefore, the difference of the temperature dependence of the PL decay time in Fig. 5(c) arises from the dimensionality of excitons.

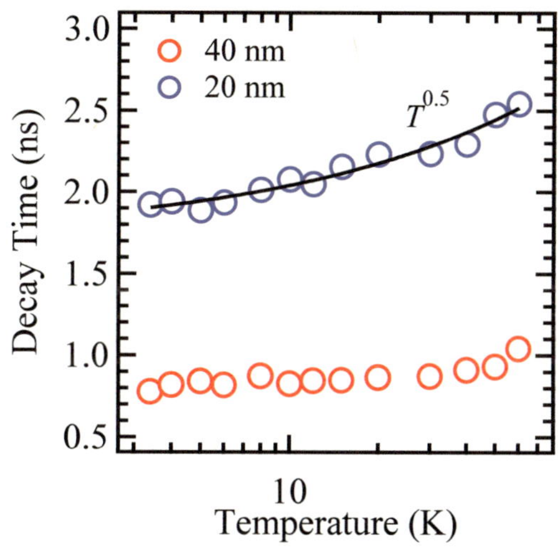

Fig. 6. Temperature dependence of the PL decay time around low temperature region.

Figure 7 shows the detection-energy dependence of the PL decay time in the *d*=40 nm sample measured at various temperatures (Kojima et al., 2011). For reference, the PL spectrum at 3.4 K is depicted. Even at 50 K, corresponding to 4.3 meV, the PL decay time depends on the detection energy; the lateral coupling occurs. Moreover, the increase factor of the decay time is enhanced at 100 K, which indicates the enhancement of the lateral coupling due to thermal activation of carriers.

If the lateral coupling originates only from the exciton/carrier transfer due to the thermal dissociation, the 50 K temperature is considered to be insufficiently to cause the lateral coupling. Therefore, it was deduced that the hole injection from the spacer layers to QDs results in the lateral coupling-like behaviour in the detection-energy dependence at lower temperature region. Assuming the strain distribution calculated by Grundmann in and around a pyramidal InAs QD (Grundmann et al., 1995), there exist lateral potentials for electrons and holes in the vicinity of a QD. The potential for holes increases close to the QD and gives rise to a barrier for the capture of holes from the wetting layer (WL) in QD. On the other hand, the potential for electrons drops monotonically due to the weak influence. Therefore, the excited carriers in the WLs and the spacer layers lead to the transfer of an impaired hole into the QDs (Adler et al., 1996). In the lower temperature region, lateral coupling-like behaviour arises from the hole injection from the spacer layers. In our QD systems, there are two possible reasons causing the lateral coupling at lower temperature region: the high QD density and the decrease in the potential height due to strain compensation. The high QD density enables to transfer in-plane direction in comparison with the vertical direction, and the strain compensation reduces the potential height induced by the lattice-mismatch strain. These two factors will lead the lateral coupling in the lower temperature comparably.

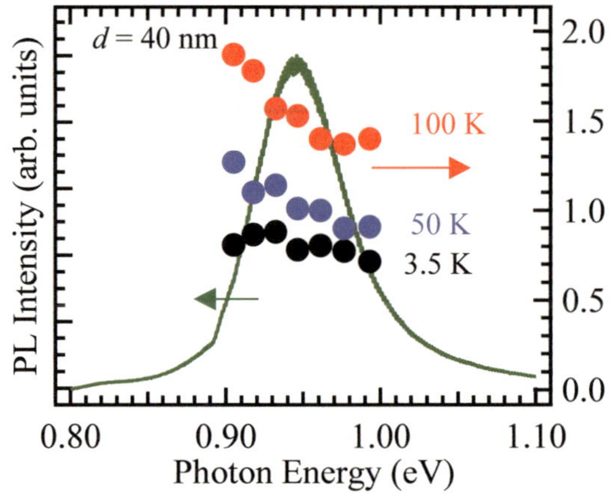

Fig. 7. Temperature dependence of the PL decay time and PL spectrum at 3.5 K in the *d*=40 nm sample.

To reveal the effects of the lateral coupling on the intraband transition process, we measured the excitation-energy dependence of the PL spectrum systematically. In measurements of the excitation-energy dependence, the excitation light was produced by combination of a 100-W Xe lamp and a 32-cm single monochromator with a resolution of 5 nm. The emitted light was dispersed by a 32-cm single monochromator with a resolution of 1.0 nm. In Fig. 8, the excitation-energy dependence of the PL intensity monitored at the PL peak energy

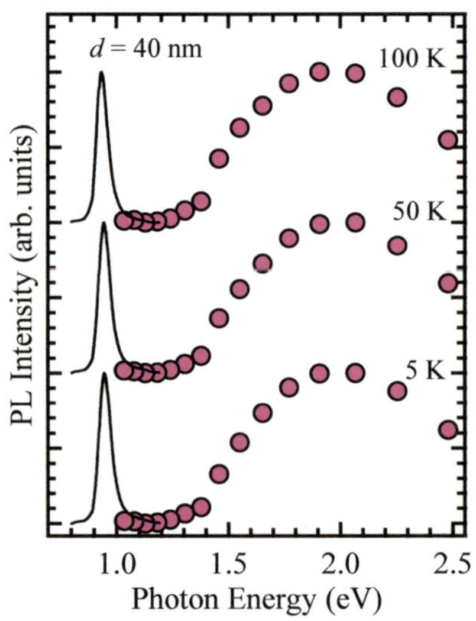

Fig. 8. Excitation-energy dependence of the PL peak intensity measured at various temperatures (circles) in the *d* = 40 nm sample. The PL spectrum at each temperature is also shown.

measured at various temperatures is depicted. All the profiles were normalized by the maximum intensity. All profiles show the maximum intensity around 1.9 eV. The bandgap energy of the InP substrate is 1.424 eV at 1.6 K and the $In_{0.5}Ga_{0.1}Al_{0.4}As$ spacer layer is around 1.357 eV at room temperature (Madelung, 2004). Therefore, the maximum intensity around 1.9 eV is attributed not to the resonant carrier injection from the spacer layers and substrates but to the higher-order excitons in InAs QDs. This demonstrates the existence of the above barrier exciton states around these energy regions. The profile of the excitation-energy dependence of the PL peak intensity hardly changes with the temperature. This result indicates that the lateral coupling is negligible for the intraband relaxation process in this sample.

Next, we discuss the relation between the interconnection effects along the growth direction and the lateral coupling. We performed the same experiments in the *d*=20 nm sample which has a longer exciton lifetime due to the interconnection effect as shown in Fig. 9. The increase factor of the PL decay time depending on the temperature is much larger than that in the *d*=40 nm sample. This result indicates that the lateral coupling effect affects the carrier relaxation process in the *d*=20 nm sample. Therefore, intraband relaxation process will change with temperature.

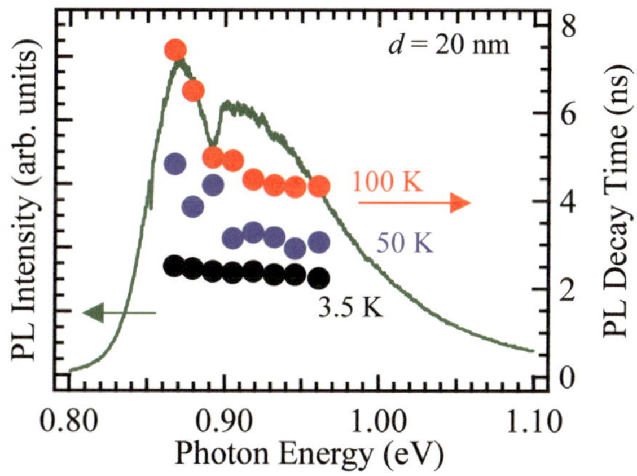

Fig. 9. Temperature dependence of the PL decay time and PL spectrum at 3.5 K in the *d*=20 nm sample.

Figure 10 shows the excitation-energy dependence of the PL peak intensity measured at various temperatures in the *d*=20 nm sample (Kojima et al., 2011). The PL spectrum at each temperature is also shown. The profiles show the peak around 1.5 eV. The difference of the peak energy between the *d*=40 nm and *d*=20 nm samples comes from that of the lowest exciton energy. While the dependence of the PL intensity in the *d*=40 nm sample hardly changes, that in the *d*=20 nm sample clearly changes around the higher energy over 1.6 eV. This change is related to the exciton lifetime. As mentioned above, the exciton lifetime in the *d*=40 nm (*d*=20 nm) sample is 0.8 (1.9) ns. When the exciton lifetime is shorter than the transfer time under the lateral coupling conditions, the transfer process does not have a sufficient effect on the intraband relaxation process. On the other hand, in the case that the exciton lifetime is longer than the transfer time, the transfer process changes the intraband relaxation process, because it will be difficult for the transferred carriers to relax into the

QDs because of the occupied states. Therefore, the carriers generated in the smaller QDs with the larger transition energies strongly affects the larger QDs with the smaller transition energies. However, for application of the QDs with longer exciton lifetime to optical devices, especially photo receiving devices such as the photodetectors or solar cells, the exciton transfer time is considered to be not as fast as the deteriorating carrier extraction.

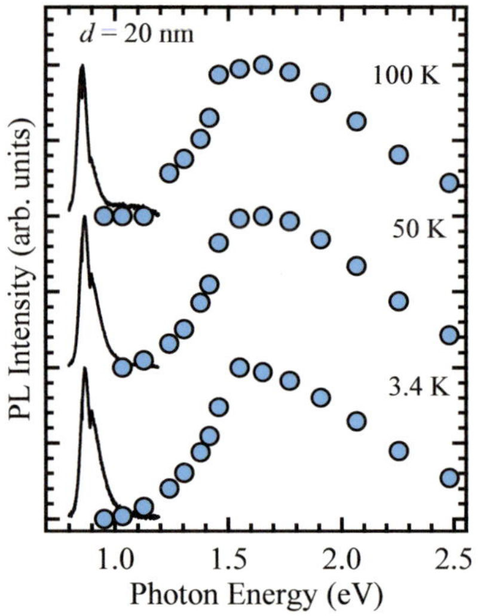

Fig. 10. Excitation-energy dependence of the PL peak intensity measured at various temperatures (circles) in the d = 20 nm sample. The PL spectrum at each temperature is also shown.

5. Conclusion

We have investigated the PL characteristics of excitons in multilayer stacked QDs with different spacer layer thicknesses. We found that the intensity of the PL band decreases with a decrease in spacer layer thickness. The PL spectra in the thinner spacer layer sample indicate the elongation of electron envelope functions along the growth direction. Moreover, from the PL decay time, it is revealed that the elongation of the electron envelope functions induces the lowering of an oscillator strength, leading to the lengthening of the PL decay time. This result suggests the interconnection of QDs along the growth direction via the overlap of the electron envelope functions. It is concluded that the PL characteristics of stacked QDs can be controlled by altering the spacer layer thickness through the variation of the exciton oscillator strength.

In addition, we have investigated the effects of temperature on the PL characteristics of excitons in the d=40 nm and d=20 nm samples. We found that the decrease in the PL intensity in the d=20 nm sample with interconnection effect is similar to that in the d=40 nm sample. To clarify the effect of the interconnection in the d=20 nm sample, we examined the temperature dependence of the PL decay time. The PL decay profiles, which show the increase in the PL decay time with temperature, indicated the suppression of nonradiative recombination paths caused during the QD and spacer layer growth processes. The increase in the PL decay time arises from the thermal delocalization.

However, the temperature dependence of the PL decay time in the two samples is different. In order to reveal the discrepancy in the temperature dependence of the PL characteristics, we examined the detection energy dependence of the PL decay time. The PL decay times of both samples clearly depend on the detection energy; this indicates the lateral coupling between the QDs. As the temperature increases, the excitons transfer from smaller QDs to larger ones. This affects the exciton relaxation process. However, in the d=20 nm sample, the vertical interaction in addition to the lateral interaction strongly affects the excitonic process, and therefore, the temperature dependence of the PL decay time differs from that in the d=40 nm sample.

Finally, we investigated the effect of the lateral coupling, namely the exciton/carrier transfer process in the in-plane direction, on the intraband relaxation process of photoexcited carriers in d=20 nm and d=40 nm samples. The detection-energy dependence of the PL decay time indicates the in-plane interaction between QDs even at 50 K in both the samples. In the excitation-energy dependence of the PL intensity, while the transfer process hardly changes the intraband relaxation process in the d=40 nm sample, that in the d=20 nm sample changes the intraband relaxation process. Because the exciton lifetime in the d=20 nm sample is longer than that in the d=40 nm sample, this change depends on the exciton lifetime.

These findings suggest that the interconnection of QDs along the growth direction via the overlapping of electron envelope functions occurs at high temperatures. These may aid the development of some functional devices by using QDs. In particular, they will be advantageous for the devices based on the so-called QD superlattice.

6. Acknowledgment

We would like to thank Dr. K. Akahane from National Institute of Information and Communications Technology, Japan, Prof. O. Wada, Prof. T. Kita, Mr. H. Nakatani, and Mr. M. Mamizuka from Kobe University for their fruitful discussions. These works were partially supported by the Grant-in-Aid for the Scientific Research and from the Ministry of Education, Culture, Sports, Science, and Technology of Japan (No.23656050).

7. References

Adler, F.; Geiger, M.; Bauknecht, A.; Scholz, F.; Schweizer, H.; Pilkuhn, M. H.; Ohnesorge, B. & Forchel, A. (1996) Optical transitions and carrier relaxation in self assembled InAs/GaAs quantum dots. *Journal of Applied Physics* Vol. 80, No. 7, pp. 4019-4026, ISSN 0021-8979

Akahane, K.; Ohtani, N.; Okada, Y. & Kawabe, M. (2002) Fabrication of ultra-high density InAs-stacked quantum dots by strain-controlled growth on InP(311)B substrate. *Journal of Crystal Growth* Vol. 245, No. 1-2, pp. 31-36, ISSN 0022-0248

Akahane, K.; Yamamoto, N. & Tsuchiya, M. (2008) Highly stacked quantum-dot laser fabricated using a strain compensation technique. *Applied Physics Letters* Vol. 93, No. 4, pp. 041121-1-3, ISSN 0003-6951

Akahane, K.; Yamamoto, N. & Kawanishi, T. (2011) Fabrication of ultra-high-density InAs quantum dots using the strain-compensation technique. *Physica Status Solidi A* Vol. 208, No. 2, pp. 425-428, ISSN 1862-6300

Akiyama, H.; Koshiba, S.; Someya, T.; Wada, K.; Noge, H.; Nakamura, Y.; Inoshita, T.; Shimizu, A. & Sakaki, H. (1994) Thermalization effect on radiative decay of excitons in quantum wires. *Physical Review Letters* Vol. 72, No. 6, pp. 924-927, ISSN 1079-7114

Andreani, L. C.; Panzarini, G. & Gerard, J.-M. (1999) Strong-coupling regime for quantum boxes in pillar microcavities: Theory. *Physical Review B* Vol. 60, No. 19, pp. 13276-13279, ISSN 1098-0121

Arakawa, Y. & Sakaki, H. (1982) Multidimensional quantum well laser and temperature dependence of its threshold current. *Applied Physics Letters* Vol. 40, No. 11, pp. 939-941, ISSN 0003-6951

Bardot, C.; Schwab, M.; Bayer, M.; Farad, S.; Wasilewski, Z. & Hawrylak, P. (2005) Exciton lifetime in InAs/GaAs quantum dot molecules. *Physical Review B* Vol. 72, No. 3, pp. 035314-1-7, ISSN 1098-0121

Bogaart, E. W.; Nötzel, R.; Gong, Q.; Haverkort, J. E. M. & Wolter, J. H. (2005) Ultrafast carrier capture at room temperature in InAs/InP quantum dots emitting in the 1.55 μm wavelength region. *Applied Physics Letters* Vol. 86, No. 17, pp.173109-1-3, ISSN 0003-6951

Brandt, O.; Tapfer, L.; Ploog, K.; Bierwolf, R. & Hohenstein, M. (1992) Effect of In segregation on the structural and optical properties of ultrathin InAs films in GaAs. *Applied Physics Letters* Vol. 61, No. 23, pp.2814-2816, ISSN 0003-6951

Fiore, A.; Borri, P.; Langbein, W.; Hvam, J. M.; Oesterie, U.; Houdré, R.; Stanley, R. P. & Ilegems, M. (2000) Time-resolved optical characterization of InAs/InGaAs quantum dots emitting at 1.3 μm. *Applied Physics Letters* Vol. 76, No. 23, pp.3430-3432, ISSN 0003-6951

Goldstein, L.; Glas, F.; Marzin, J. Y.; Charasse, M. N. & Le Roux, G. (1985) Growth by molecular beam epitaxy and characterization of InAs/GaAs strained-layer superlattices. *Applied Physics Letters* Vol. 47, No. 10, pp.1099-1101, ISSN 0003-6951

Grundmann, M.; Stier, O. & Bimberg, D. (1995) InAs/GaAs pyramidal quantum dots: Strain distribution, optical phonons, and electronic structure. *Physical Review B* Vol. 52, No. 16, pp. 11969-11981, ISSN 1098-0121

Hostein, R.; Michon, A.; Beaudoin, G.; Gogneau, N.; Patriache, G.; Marzin, J.-Y.; Robert-Phillip, I.; Sagnes, I. and Beveratos, A. (2008) *Applied Physics Letters* Vol. 93, No. 7, pp.073106-1-3, ISSN 0003-6951

Hours, J.; Senellart, P.; Peter, E.; Cavanna, A. & Bloch, J. (2005) Exciton radiative lifetime controlled by the lateral confinement energy in a single quantum dot. *Physical Review B* Vol. 71, No. 16, pp. 161306-1-4, ISSN 1098-0121

Huffaker, D. L.; Park, G.; Zou, Z.; Shchekin, O. B. & Deppe D. G. (1998) 1.3 μm room-temperature GaAs-based quantum-dot laser. *Applied Physics Letters* Vol. 73, No. 18, pp. 2564-2566, ISSN 0003-6951

Inoue, T.; Kita, T.; Wada, O.; Konno, M.; Yaguchi, T. & Kamino, T. (2008) Electron tomography of embedded semiconductor quantum dot. *Applied Physics Letters* Vol. 92, No. 3, pp. 031902-1-3, ISSN 0003-6951

Kayanuma, Y. (1988) Quantum-size effects of interacting electrons and holes in semiconductor microcrystals with spherical shape. *Physical Review B* Vol. 38, No. 14, pp. 9797-9805, ISSN 1098-0121

Kim, J. S.; Lee, J. H.; Hong, S. U.; Han, W. S.; Kwack, H.-S. & Oh, D. K. (2003) Influence of InGaAs overgrowth layer on structural and optical properties of InAs quantum dots. *Journal of Crystal Growth* Vol. 255, No. 1-2, pp. 57-62, ISSN 0022-0248

Kojima, O.; Nakatani, H.; Kita, T.; Wada, O.; Akahane, K. & Tsuchiya, M. (2008) Photoluminescence characteristics of quantum dots with electronic states interconnected along growth direction. *Journal of Applied Physics* Vol. 103, No. 11, pp. 113504-1-5, ISSN 0021-8979

Kojima, O.; Nakatani, H.; Kita, T.; Wada, O. & Akahane, K. (2010) Temperature dependence of photoluminescence characteristics of excitons in stacked quantum dots and quantum dot chains. *Journal of Applied Physics* Vol. 107, No. 11, pp. 073506-1-4, ISSN 0021-8979

Kojima, O.; Mamizuka, M.; Kita, T.; Wada, O. & Akahane, K. (2011) Intraband relaxation process in highly stacked quantum dots. *Physica Status Solidi C* Vol. 8, No. 1, pp. 46-49, ISSN 1610-1642

Madelung, O. (2004) *Semiconductors: Data Handbook*, Springer-Verlag, ISBN 978-3540404880

Martí, A.; Antolín, E.; Stanley, C. R.; Farmer, C. D.; López, N.; Díaz, P.; Cánovas, E.; Linares, P. G. & Luque, A. (2006) Production of photocurrent due to intermediate-to-conduction-band transitions: a demonstration of a key operating principle of the intermediate-band solar cell. *Physical Review Letters* Vol. 97, No. 24, pp. 247701-1-4, ISSN 1079-7114

Nakatani, H.; Kita, T.; Kojima, O.; Wada, O.; Akahane, K. & Tsuchiya, M. (2008) Photoluminescence dynamics of coupled quantum dots. *Journal of Luminescence* Vol. 128, No. 5-6, pp. 975-977, ISSN 0022-2313

Oshima, R.; Takata, A. & Okada, Y. (2008) Strain-compensated InAs/GaNAs quantum dots for use in high-efficiency solar cells. *Applied Physics Letters* Vol. 93, No. 8, pp. 083111-1-3, ISSN 0003-6951

Persson, J.; Håkanson, U.; Johansson, M. K.-J.; Samuelson, L. & Pistol, M.-E. (2005) Strain effects on individual quantum dots: Dependence of cap layer thickness. *Physical Review B* Vol. 72, No. 8, pp. 085302-1-2, ISSN 1098-0121

Prasanth, R.; Haverkort, J. E. M.; Deepthy, A.; Bogaart, E. W.; van der Tol, J. J. G. M.; Patent, E. A.; Zhao, G.; Gong, Q.; van Veldhoven, P. J.; Nötzel, R. & Wolter, J. H. (2004) All-optical switching due to state filling in quantum dots. *Applied Physics Letters* Vol. 84, No. 20, pp. 4059-4061, ISSN 0003-6951

Saito, H.; Nishi, K. & Sugou, S. (1998) Influence of GaAs capping on the optical properties of InGaAs/GaAs surface quantum dots with 1.5 μm emission. *Applied Physics Letters* Vol. 73, No. 19, pp. 2742-2744, ISSN 0003-6951

Saito, T.; Nakaoka, T.; Kakitsuka, T.; Yoshikuni, Y. & Arakawa, Y. (2005) Strain distribution and electronic states in stacked InAs/GaAs quantum dots with dot spacing 0–6 nm. *Physica E: Low-dimensional Systems and Nanostructures* Vol. 26, No. 1-4, pp. 217-221, ISSN 1386-9477

Takagahara, T. (1987) Excitonic optical nonlinearity and exciton dynamics in semiconductor quantum dots. *Physical Review B* Vol. 36, No. 17, pp. 9293-9296, ISSN 1098-0121

Wang, G.; Fafard, S.; Leonard, D.; Bowers, J. E.; Merz, J. L. & Petroff, P. M. (1994) Time-resolved optical characterization of InGaAs/GaAs quantum dots. *Applied Physics Letters* Vol. 64, No. 21, pp. 2815-2817, ISSN 0003-6951

Xie, Q.; Chen, P.; Kalburge, A.; Ramachandran, T. R.; Nayfonov, A.; Konkar, A. & Madhukar, A. (1995) Realization of optically active strained InAs island quantum boxes on GaAs(100) via molecular beam epitaxy and the role of island induced strain fields. *Journal of Crystal Growth* Vol. 150, No. 1, pp. 357-363, ISSN 0022-0248

Yu, H.; Lycett, S.; Roberts, C. & Murray, R. (1996) Time resolved study of self-assembled InAs quantum dots. *Applied Physics Letters* Vol. 69, No. 26, pp. 4087-4089, ISSN 0003-6951

Section 2

Transport and Eletronics Properties of Quantum Dot Systems

11

Electron Transport Properties of Gate-Defined $GaAs/Al_xGa_{1-x}As$ Quantum Dot

Dong Ho Wu and Bernard R. Matis
Naval Research Laboratory, Washington, DC, USA

1. Introduction

In this chapter we explore transport properties of lateral, gate defined quantum dots in $GaAs/Al_xGa_{1-x}As$ heterostructures. The term "quantum dot" as defined here refers to small regions of charge carriers within a 2-dimensional electron gas (2DEG), established via electrically biased surface gates used to isolate the charge carriers from the rest of the 2DEG, which are confined to length scales on the order of nanometers. While there are several other forms of quantum dots, including colloidal and self-assembled dots, in this chapter, however, we consider only gate defined quantum dots.

Recent advancements in the research areas of quantum dot (QD) and single electron transistors (SET) have opened up an exciting opportunity for the development of nanostructure devices. Of the various devices, our attention is drawn in particular to detectors, which can respond to a single photon over a broad frequency spectrum, namely, microwave to infrared (IR) frequencies. Here, we report transport measurements of weakly coupled double quantum dots, fabricated on a GaAs/AlGaAs 2-dimensional electron gas material, under the influence of external fields at 110GHz. In this experiment, transport measurements are carried out for coupled quantum dots in the strong-tunneling Coulomb blockade (CB) regime. We present experimental results and discuss the dependence on quantum dot size, 2DEG depth, fabrication techniques, as well as the limitations in developing a QD photon detector for microwave and IR frequencies, whose noise equivalent power (NEP) can be as sensitive as 10^{-22} W/Hz$^{1/2}$.

The charging energy E_C of a quantum dot is the dominant term in the Hamiltonian and is inversely related to the self capacitance of the dot C_{dot} according to $E_C = e^2/C_{dot}$. The temperature of the charge carriers within the 2DEG must be kept below a certain value, namely K_BT, so that the thermal energy of the electrons does not exceed the charging energy E_C of the dot. Keeping the temperature below the K_BT limit prevents electrons from entering or leaving the dot at random, thereby allowing one to control the number of electrons in the dot. In order to raise the operating temperature T of the single photon detector we must also raise the charging energy E_C, which is accomplished by decreasing C_{dot}. Since C_{dot} is directly related to the dimensions of the quantum dot our focus was directed at decreasing the overall size of the quantum dots. For smaller gate-defined quantum-dots the inclusion of shallower 2DEG is necessary.

However the experiments that we carried out to determine the effect of 2DEG depth on lateral gate indicated that leakage currents within a GaAs/AlGaAs heterostructure increased dramatically as the 2DEG depth became shallower. At this moment the leakage current in shallower 2DEG materials is one of the most significant technical challenges in achieving higher operating temperature of the single photon detector.

2. Gate-defined quantum-dots

2.1 2-dimensional electron gas (2DEG)

In contrast to colloidal and self-assembled quantum dots, which are physically well defined small dots separated from other media, the gate-defined quantum dot means charge carriers (either electrons or holes) confined in a small region, which is formed by electrically biased gates surrounding the region. First the charge carriers are confined within the so-called 2-dimentional electron gas (2DEG) material, which is typically made of GaAs/AlxGa1-xAs heterostructure. Figure 1 shows an example of the vertical profile of a 2DEG heterostructure and the corresponding energy-band diagram.

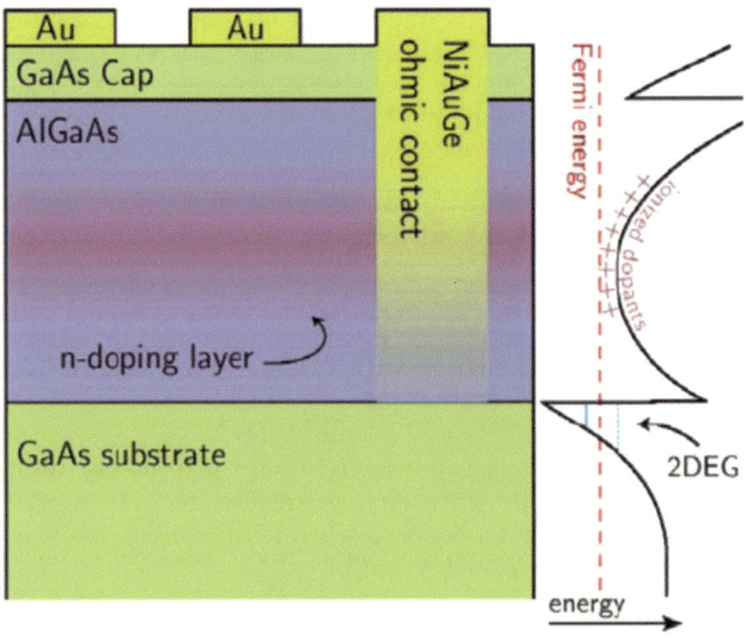

Fig. 1. An example of a the vertical profile of the GaAs/AlGaAs 2DEG heterostructure (Left) and the corresponding energy-band diagram (Right).[Ref.1] Included in the vertical profile are the patterned metallic surface gates (Au) that define the nanostructure devices and the ohmic contacts (NiAuGe), which when annealed penetrate through the top layers of the structure and make electrical contact to the 2DEG.

One layer of AlGaAs in particular contains a region of n-type dopants, either as a single layer (so called δ-doped layer) or homogeneously doped (modulation doping). In our case the dopants are Si atoms, which are deposited within an AlGaAs layer and are separated away from the 2DEG by an undoped AlGaAs spacer layer. The spacer layer which is typically 10 - 100 nm thick is to minimize the effect of scattering from the dopants; the various layer thicknesses can be modulated to vary the properties of the 2DEG.

At low temperature each Si atom produces a free electron as the electrons become thermally ionized [2, 3]. The offset in the conduction bands between GaAs and AlGaAs results in each free electron migrating toward the energetically favorable GaAs substrate layer. The charge carriers still feel the electrostatic attractive forces from the ionized donor atoms, however, and ultimately become trapped at the interface between the GaAs layer and an undoped AlGaAs layer. These trapped electrons are called 2-dimentional electron gas (2DEG). As the temperature decreases to very cold temperatures (< 1 K) the thermal smearing of the vertical "z" profile of the 2DEG becomes less pronounced as the electrons occupy only the lowest energy levels up to the Fermi Energy, resulting in a very clean glass of electrons confined within a 2-dimensional plane.

Because the lattice constants of GaAs and AlGaAs are only slightly different (~7% mismatch) the interface is essentially defect free. Because of this defect free interface and the separation of the 2DEG from the Si dopants 2DEG can have high electron mobility, $\mu_e \sim 10^5$ - 10^7 cm^{2}V/s, and long mean free paths, ℓ ~ 1 - 1000 nm. These properties are often exploited for quantum dot devices which require coherent and ballistic electron transport behavior.

2.2 Gate-defined quantum-dot

The local electron density within the 2DEG can be manipulated by placing electrodes on GaAs cap surface, as shown in Figure 1. When a negative bias voltage is applied to the electrodes the negatively charged gates repel electrons in the 2DEG. If the negative field strength is strong enough all electrons beneath the electrodes will be fully depleted. The electrodes can be lithographically arranged over an area with a certain geometric shape, such as a circular disk. An example is shown in Figure 2.

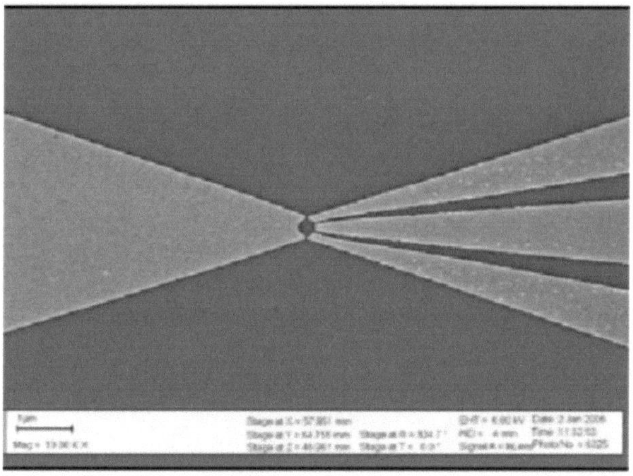

Fig. 2. An example of gate-defined quantum dot. The four gates are fabricated by the e-beam lithography, and surround a circular disk in the center, which becomes the quantum dot when the four gates are negatively biased. The lithographically defined circular-disk size is about 250 nm in diameter. However the actual size of the quantum dot depends on the strength of the bias voltage. Also the shape of the quantum dot depends on how the bias voltage is applied to each gate, and it can be deviated from the disk shape.

With strong enough negative bias voltages applied to the electrodes, electrons confined inside the area (e.g. circular disk) will be isolated from the rest of the electrons in the 2DEG. These isolated electrons in the area (e.g. disk) are called the gate-defined quantum-dot and the rest of the electrons in the 2DEG are called the reservoir. While the configuration of the gates influence the overall shape and determine the maximum size of quantum dot, actual shape and size of the isolated electron puddle (i.e. quantum dot) are dependent upon the strength of the negative bias voltage applied to each gate. The gap between gates is often called the quantum point contact (QPC) and is typically a few tens of nm. It pinches off electrons when the negative bias voltage is applied to the gates. The QPCs can individually tune the potential barriers between the dot and the reservoirs, and hence control the tunneling rate from the leads and the dot. The transport through a quantum dot can be divided into two categories, "open" and "closed," depending upon the conductance of the QPCs. For strong coupling, the conductance $G > e^2/h$, where each QPC passes one or more modes, the dot is considered "open." In an "open" dot electrons are classically allowed to travel through the dot from one reservoir to the other. For weak coupling, $G < e^2/h$, where each QPC is set to pass less than one fully transmitting mode, the dot is considered "closed." If the bias voltage is large enough the electrons near the quantum point contacts are completely pinched off, making the quantum dot to be "closed" or isolated from the reservoir. However electrons can tunnel through the "closed" quantum dot, allowing very small currents. Therefore the conductance is orders of magnitude lower than that of 2DEG.

Quantum Dots are often referred to as zero-dimensional systems, as the electronic motion is entirely restricted in all directions. The size of quantum dot is typically smaller than a few hundred nano-meters in diameter. As electrons are confined within such a length scale the spacing between each quantum energy level of the electron becomes very pronounced when the temperature of the quantum dot drops below 4.2 K and thermal smearing is very much reduced. Because of these well defined quantum energy levels of electrons within the quantum dot the tunneling currents through the quantum dot exhibit the characteristic Coulomb blockade effect.

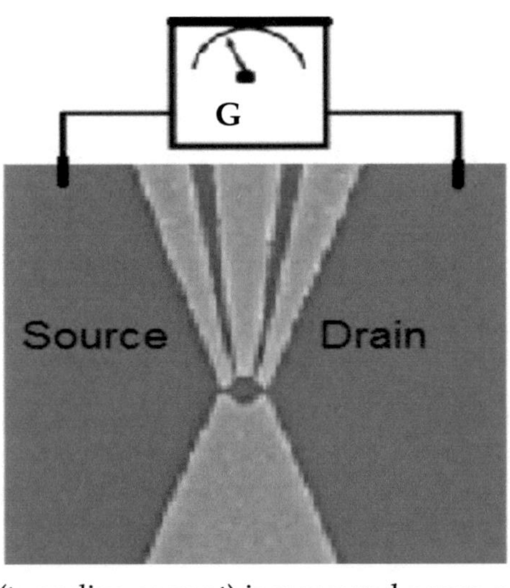

Fig. 3. The conductance G (tunneling current) is measured across a quantum dot device, from the source to drain reservoirs.

The Coulomb blockade occurs due to the fact that conduction through the dot is prevented for most settings of the electrostatic gates simply because the available energy levels within the dot are not in alignment with the Fermi levels in the source and drain (i.e. reservoir). An electron is unable to tunnel into the dot if the energy needed to add an additional electron (from N to N + 1 electrons) is above the Fermi Energy in the source. Similarly an electron is unable to tunnel out of the dot if the energy carried by that electron is less than the Fermi Energy in the drain. If electrons have enough energy to tunnel into the dot and then tunnel out of the dot, the measured conductance displays a large conductance spike, which indicates tunneling currents. This is known as a Coulomb blockade peak.

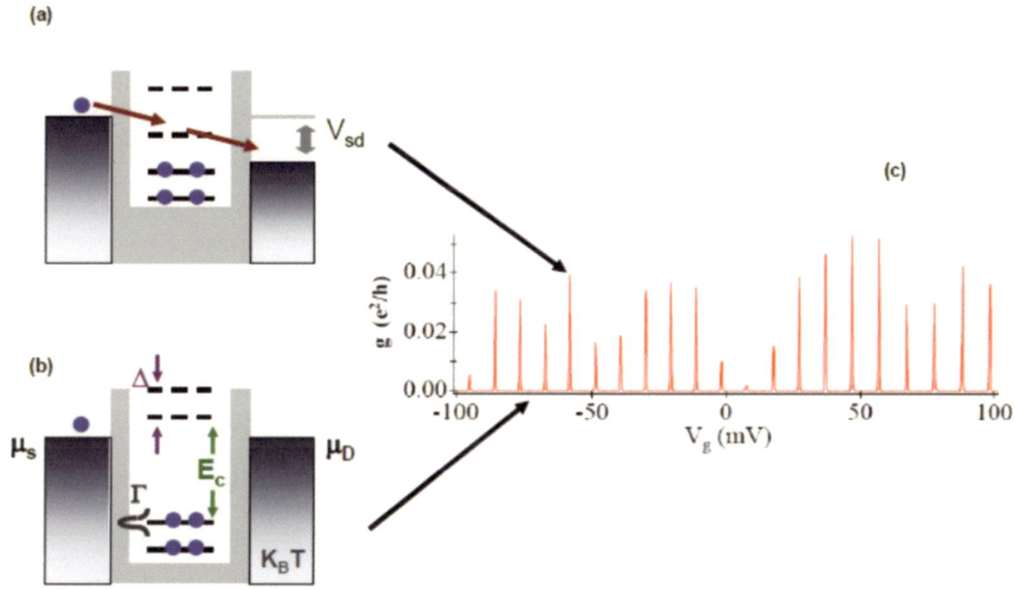

Fig. 4. (a) and (b): Coulomb blockade energy diagrams. (a) The energy level in the quantum dot allows the electron to tunnel through the dot. The tunneling currents produce a Coulomb blockade peak. (b) The energy level in the quantum dot is higher than the electron's energy so that the electron cannot tunnel through the quantum dot. Therefore the conductance is zero, and there is no Coulomb blockade peak in (c). As a gate voltage is swept the energy levels of the dot are raised and lowered, making the quantum energy levels to move in and out of alignment with the chemical potentials (μ_S and μ_D) of the source and drain, resulting in large spikes in the conductance. The interval and the sharpness of Coulomb blockage peaks are determined by the quantum energy level spacing Δ and the finite thermal broadening Γ for a given temperature T.

For the tunneling currents and the Coulomb blockade five separate energy parameters need to be considered, including the source-drain voltage V_{sd}, the chemical potentials of source μ_S and drain μ_D, the charging energy E_C and the thermal energy of charge carriers K_BT. For the conductance measurement a small source-drain voltage Vsd, which is typically limited to be less than a few μV so as not to impart energy to the electrons greater than the thermal energy, is held across the dot. The source-drain voltage results in the chemical potential difference between the chemical potentials of source and drain so that $eV_{sd} = \mu_S - \mu_D$. The charging energy E_C is an additional Coulomb energy that is needed to add an additional electron to the quantum dot, and can be expressed as

$$E_C = e^2/2C_{dot}. \quad (1)$$

Here C_{dot} is the self-capacitance of the quantum dot. At temperature T an electron has the thermal energy K_BT. If the thermal energy becomes comparable or larger than the charging energy it causes the electron randomly to tunnel through the quantum dot, and also results in a thermal broadening larger than the energy level spacing Δ. Then the quantum dot will not be functional, as the electron is no longer controllable by the gate bias voltage. Hence it is very important to keep the quantum dot at very low temperatures so that its thermal energy is well below the charging energy (i.e. $E_C > K_BT$).

The quantum dot can manipulate the flow of an individual electron by controlling the gate bias voltage. As shown in Figure 5, a quantum dot with a capacitively coupled gate can be used as a single electron transistor. The bias voltage applied to the gate raises or lowers the energy level of the dot so that each single electron can tunnel through the quantum dot. Such a device is called a single electron transistor. We utilized the single electron transistor for our quantum dot single photon detector.

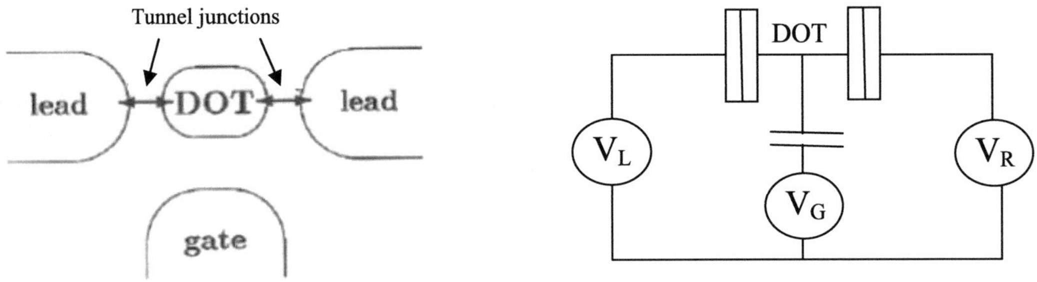

(a) (b)

Fig. 5. (a) A schematic diagram of a single electron transistor made of a quantum dot. (b) An equivalent circuit for a single electron transistor.

3. A single photon detector based on coupled double quantum dots

The quantum energy levels as well as the level spacing Δ can be adjusted by controlling the physical parameters of the quantum dot. A photon can change the energy level of a quantum dot, which leads to electron tunneling through the quantum dot. This is known as photon assisted tunneling in a quantum dot. In 2000 Komiyama and his coworkers exploited this property and developed a detector, which can detect a single photon at far-infrared frequencies. The quantum dot size that they used in the experiments was about 500 nm in diameter fabricated on a 100 nm thick 2DEG substrate. Their large quantum dots resulted in a large self capacitance C_{dot} and a small charging energy $E_C = e^2/2C_{dot}$. Hence their detectors had to be operated at 100 mK or below, which made the detector less practical. We attempted to adopt their quantum dot detector technology and raise the charging energy and the operating temperature by reducing the quantum dot size.

3.1 A shallow 2-dimentional electron gas for quantum dot single photon detector

For the design of our quantum dot detector we have performed numerical calculations. The calculations indicate that our detector should be fabricated on a shallow 2 dimensional electron gas (2DEG) substrate in order to achieve an operating temperature above 4 K. [4] As shown in Figure 6 the depth of the 2DEG that we have used for the detector is about 40 nm. This is much shallower in comparison with the Komiyama group's 2DEG, which was buried approximately 100 nm beneath the un-doped GaAs cap layer.

The high-mobility GaAs/$Al_{0.24}Ga_{0.76}As$ heterostructure crystal was grown by molecular beam epitaxy in the [001] direction. The heterostructure layers were deposited on an n-type GaAs substrate, carried a 5000Å thick GaAs buffer layer, a non-inverted heterostructure (500 Å thick GaAs/ 140Å thick $Al_{0.24}Ga_{0.76}As$), a δ-doped barrier layer (250 Å thick $Al_{0.24}Ga_{0.76}As$), and a δ-doped GaAs cap layer (10 Å thick). The silicon n-type dopants (level $6\times10^{18}/cm^3$) provide the excess charge carriers (target value was $6x10^{11}/cm^2$ at room temperature), which constitute a 2 dimensional electron gas (2DEG) at the hetero-interface 400 Å below the wafer surface and 140 Å from the dopant atoms.

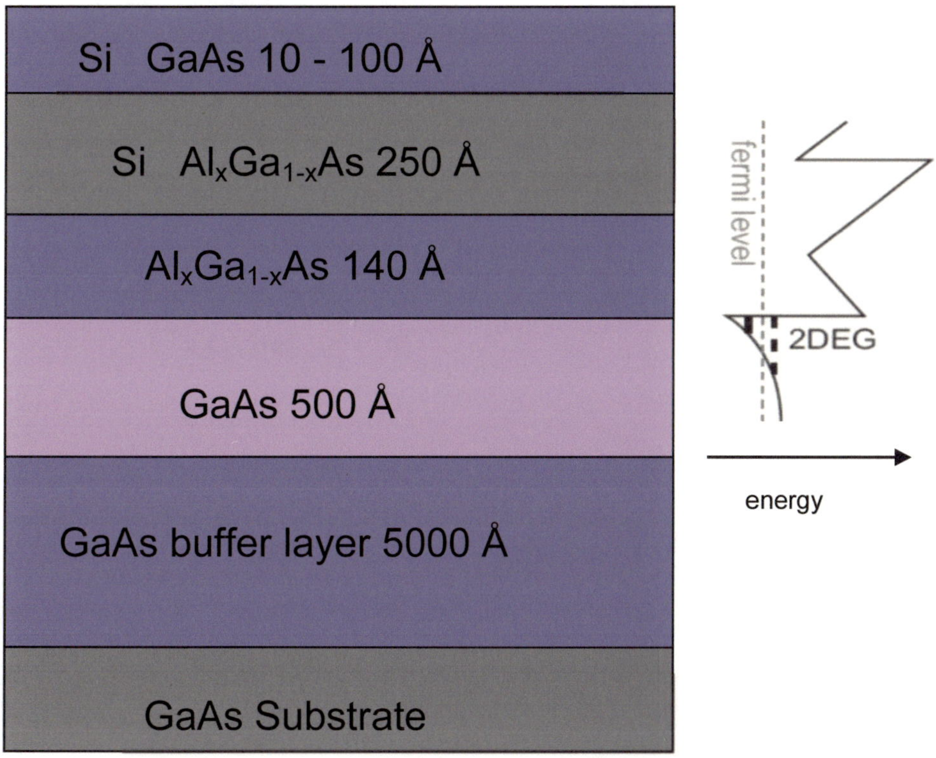

Fig. 6. The 2DEG structure we have used for our double-dot detector. For this work the 2DEG depth was a shallow 40 nm beneath the surface of the wafer. Here, x = 0.24.

For the characterization of 2DEG as well as for the quantum dot device good ohmic contacts should be made on the GaAs cap layer, as illustrated in Figure 1. A good ohmic contact has a non-zero internal resistance R_c that obeys Ohm's law for all current densities of interest. The contact should work at the lowest temperatures reached in quantum dot experiments where thermionic currents are negligible, but tunnel currents are allowed [5-7]. Fabrication of good ohmic contacts is not always trivial. The standard process includes depositing metals onto the surface and then annealing them into the wafer in order to make electrical contact to the 2DEG. We have used GaAs/AlGaAs heterostructures with several different 2DEG depths ranging from a shallow 40 nm to a deeper 160 nm. In each case a separate ohmic recipe had to be developed.

After several steps of cleaning procedures we carried out acid wet etch in order to remove any GaAs oxide layer that has formed on the surface. [8] Then we performed metal deposition on the 2DEG substrate. With the following recipes we have achieved low resistance, good ohmic contacts at cryogenic temperatures.

For shallow 2DEG	For 160nm deep 2DEG
1. 5 nm Ni	1. 5 nm Ni
2. 40 nm Ge	2. 125 nm Ge
3. 80 nm Au	3. Wait 30 min.
4. Wait 30 min.	4. 250 nm Au
5. 35 nm Ni	5. 50 nm Ni
6. 30 nm Au	6. Wait 30 min.
	7. 50 nm Ge
	8. 100 nm Au
	9. 35 nm Ni

Table 1. Ohmic contact recipes for two different 2DEG depths that have given low resistance ($<100\ \Omega$) at low temperature when annealed above the eutectic point.

The first Ni layer acts as a wetting layer and enhances the uniformity of the contacts; 5 nm is enough as this layer should not be thick. Otherwise it may prevent the other elements from penetrating into the wafer. The 2:1 ratio of Au:Ge forms a eutectic mixture, which is the ratio of two substances with the lowest melting point (a 2:1 ratio is essentially 88% Au and 12% Ge by weight with the melting point of this eutectic at ~ 380 °C). Each metal was evaporated one at a time. The second Ni layer acts as a barrier for the top layers of metals. The metalized 2DEG substrate is then submersed in Acetone for liftoff, and then rinsed with IPA and DI-H_2O. Finally it was dried by blowing dry N_2 gas.

In order to make electrical contact to the 2DEG the metals must be annealed into the substrate after the liftoff process. For the annealing we used AS-One 150 Rapid Thermal

Annealer (RTA) from ANNEALSYS with the following annealing procedure: Start with a ramp to 260 °C at 10% power to drive off any moisture from the chip. The pyrometer target is set at 510 °C for 100 seconds for the 160 nm deep 2DEG's, and 450 °C for 100 seconds for the 40, 43, and 90 nm deep 2DEG's.

The resulting R_c resistances for each contact are on the order of tens of kΩ at room temperature and decrease to a value on the order of kΩ at 4.2 K for the 160 nm deep 2DEG. For the shallower 2DEGs the contact resistances are even lower; they are on the order of kΩ at room temperature.

Low temperature measurements for the ohmic contacts and the 2DEG mobility characterizations are carried out using a Physical Property Measurement System (Quantum Design). The system is also equipped with a 9 T superconducting magnet and is capable for carrying out our Hall Effect measurements. Figure 7 shows the temperature dependence of both good and poor Ohmic contacts that were measured at zero field.

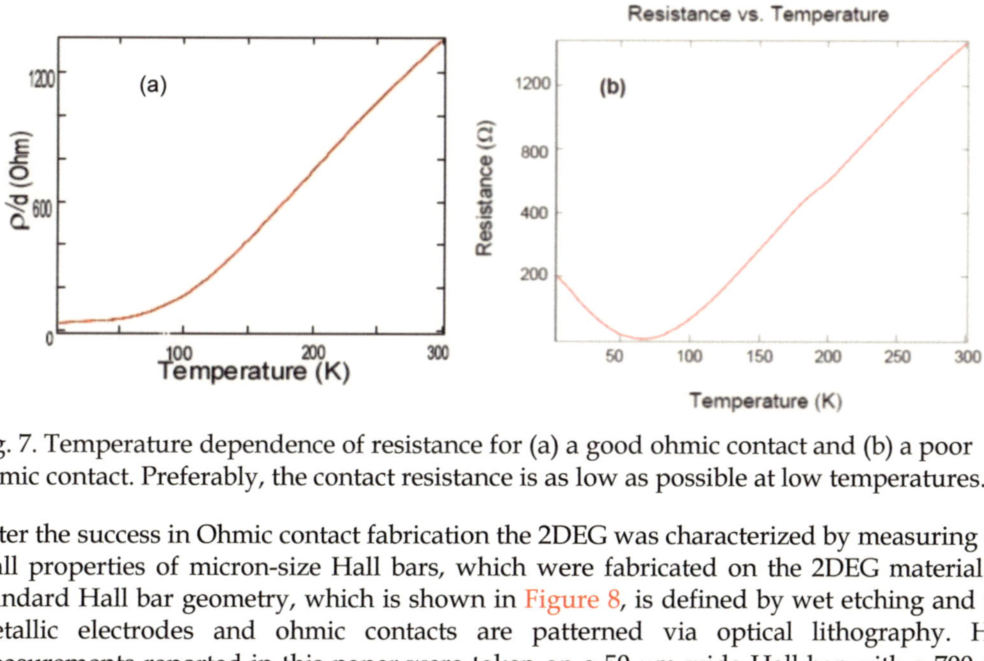

Fig. 7. Temperature dependence of resistance for (a) a good ohmic contact and (b) a poor ohmic contact. Preferably, the contact resistance is as low as possible at low temperatures.

After the success in Ohmic contact fabrication the 2DEG was characterized by measuring the Hall properties of micron-size Hall bars, which were fabricated on the 2DEG material. A standard Hall bar geometry, which is shown in Figure 8, is defined by wet etching and the metallic electrodes and ohmic contacts are patterned via optical lithography. Hall measurements reported in this paper were taken on a 50 μm wide Hall bar with a 700 μm distance between longitudinal taps. Electrical contact is made with the 2DEG by lithographically patterned Ni-Au-Ge Ohmic contacts, which when annealed at temperatures above 400 degrees Celsius provide for low resistive transport into and out of the 2DEG at cryogenic temperatures.

Two different Hall bars were fabricated, with and without an overlaying Si_3N_4 (silicon nitride) dielectric layer, which was tested to shield the 2DEG along the mesa edge from unwanted field effects caused by voltage biased leads. For the characterization of ohmic contacts we used a standard Van der Pauw experimental configuration. As shown in Figure 6, the resistivity decreased with temperature monotonically indicating the correct Ohmic contact behavior.

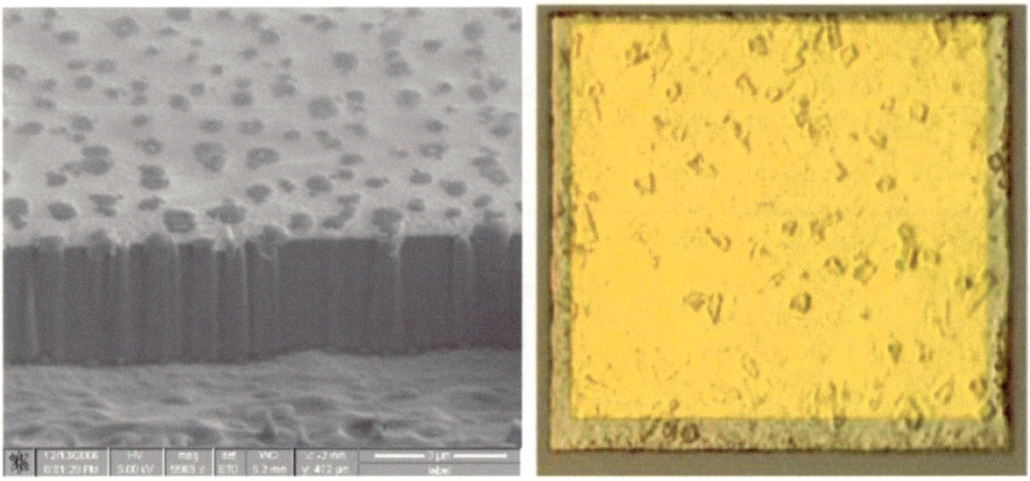

(a) (b)

Fig. 8. SEM and optical images taken of an annealed, low resistance ohmic contact. (a) The cross-sectional SEM image, taken after a Focused Ion Beam (FIB) cut into the contact, shows the puncturing of the deposited metal into the host GaAs/AlGaAs wafer. (b) An optical micrograph of a contact measured to have less than 1 kΩ of resistance at 4.2 K. The dimensions of the contact are 200 μm x 200 μm. The smaller gold square is additional metal deposited during the last optical lithography step (large gate pads) to help in wire-bonding.

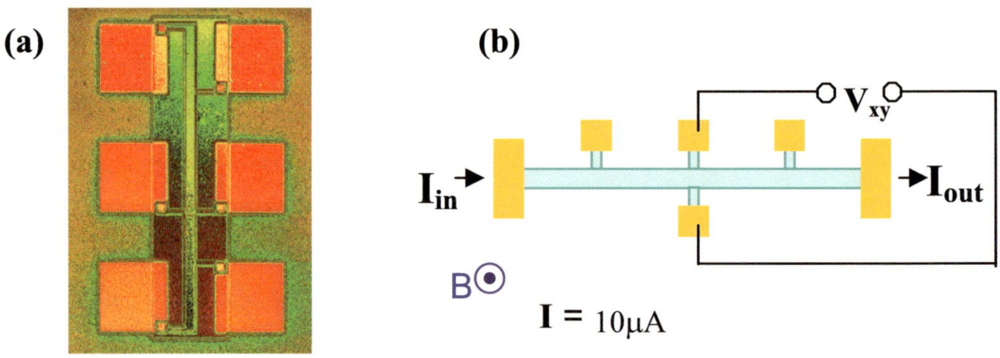

Fig. 9. (a) Optical micrograph of a micron-size Hall bar fabricated on the 2DEG material, and (b) the schematic diagram of the Hall bar and its characterization. Note that a magnetic field *B* perpendicular to the Hall bar is applied during the characterization, and a 10 μA current was used.

When a magnetic field is applied to 2DEG, electrons moving within the 2-dimensional system experience a Lorentz force that pushes them into circular orbits. Since in the 2-dimensional system only certain orbits (or energy states) are quantum mechanically allowed, the energy levels of the circular orbits are quantized, just as in the discrete set of allowed energy levels in an atom. These quantized energy states, or *Landau levels*, can be expressed as

$$E_j = (j - ½)\, h\, e\, B\, /\, (2\, \pi\, m) \tag{2}$$

Here, j is an integer, h the Planck's constant, e the fundamental electron charge, and m the electron mass. Assuming a fixed electron density n for a 2-dimensional system, at low temperatures all electrons occupy the lowest allowable energy state, or Landau level, filling it only partially. As the field B is swept toward zero the capacity for each Landau level to hold each electron decreases according to

$$N = eB / h \tag{3}$$

where N is the number of orbits that can be packed per Landau level into each cm^2 of the system. At various points along the magnetic field all electrons fill up an exact number of Landau levels with all higher energy states remain empty. When this occurs the B-field is quantized and can be expressed as

$$B = (n\,h / e) / j, \tag{4}$$

where n is the electron density for a given state. Then the magneto-resistance -- resistance measured along the initially supplied current path -- drops and the Hall resistance R_H becomes quantized as

$$R_H = B / (n\,e) = h / (j\,e^2). \tag{5}$$

The first expression is just the classical Hall resistance while the second expression comes from substituting the values for B into the first expression. From this equation it is possible to extract the charge carrier density of the material by examining the periodicity of the plateaus in the quantum Hall effect measurement.

Our Hall resistance measurements were carried out on a patterned Hall bar shown in Figure 9. A drive current of 10 μA, which was the minimum current setting available on our Physical Properties Measurement System at a frequency of 30 Hz was supplied across the length of the Hall-bar, and a magnetic field B was applied along the direction perpendicular to both the current path and the measured V_H direction. A 9 Tesla superconducting magnet was used to generate the field, though for safety purposes the magnet was only ramped to 7 T in each direction. The measurements performed at 1.7 K

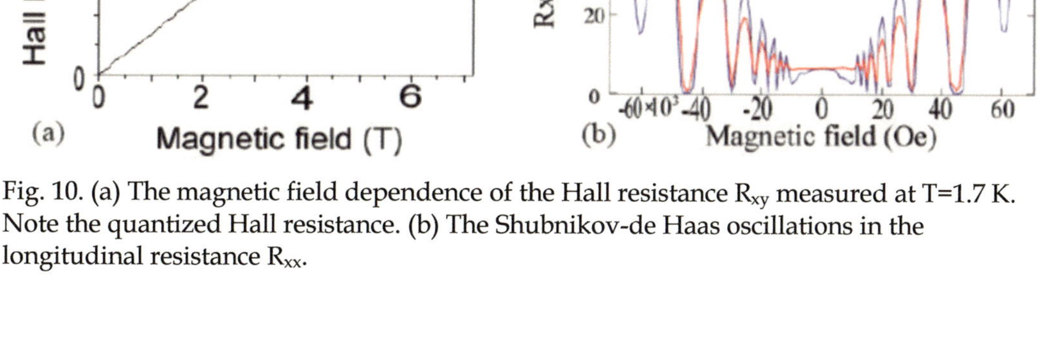

Fig. 10. (a) The magnetic field dependence of the Hall resistance R_{xy} measured at T=1.7 K. Note the quantized Hall resistance. (b) The Shubnikov-de Haas oscillations in the longitudinal resistance R_{xx}.

show that the magnetic field dependence of the Hall resistance is quantized to $R_{xy} = h/je^2$ in our 2DEG material. Also, the longitudinal resistance R_{xx} measured as a function of magnetic field exhibits the characteristic Shubnikov-de Haas oscillations, as shown in Figure 10.

2DEG Property	Symbol	Value	Units
Charge carrier density	n	5.0×10^{11}	cm^{-2}
Charge carrier mobility	μ	3.0×10^{5}	cm^2/Vs
Effective mass	m^*	0.067	$m_e = 9.1\times10^{-28}g$
Spin degeneracy	g_S	2	
Valley degeneracy	g_V	1	
Density of states	$\rho(E) = g_S g_V m^*/(h^2/2\pi)$	2.8×10^{10}	$cm^{-2}meV^{-1}$
Landau level spacing	$1/\rho(E)$	3.57	$\mu eV \mu m^2$
Fermi wave vector	$k_F = (4\pi n/g_S g_V)^{1/2}$	1.8×10^{6}	cm^{-1}
Fermi Energy	$E_F = (hk_F/2\pi)^2/2m^*$	17.88	meV
Fermi wavelength	$\lambda_F = 2\pi/k_F$	35	nm
Fermi velocity	$v_F = (hk_F/2\pi)/m^*$	3.07×10^{7}	cm/s
Scattering time	$\tau = m^* \mu/e$	11	ps
Mean free path	$\ell = v_F \tau$	3.5	μm
Cyclotron radius	$r_C = (hk_F/2\pi)/eB$	26	$nm/B^{1/2}$

Table 2. Typical parameters of our 40 nm deep 2DEG formed in a GaAs/AlGaAs heterostructure. The unit of B is in Tesla.

From the periodicity of the plateaus in Figure 10 (a) and (b), the 2DEG charge carrier density n was estimated to be about 5.0×10^{11} charges/cm^2 while the charge carrier mobility μ was estimated to be about 3.0×10^5 cm^{2}V/s. These two parameters were then used to obtain for example the Fermi Energy E_F, mean free path ℓ, Fermi wavelength λ_F, and effective mass m*. Table 2 lists the various properties that were calculated for one of our shallower (40 nm thick) 2DEGs.

3.2 Fabrication of gates on a shallow 2DEG and gate-defined double quantum dots

As our heterostructure material showed the typical 2DEG behavior, we fabricated quantum point-contact (QPC) devices to see further 2DEG behavior in another nano-device form. Moreover, we did this to test our device fabrication technique. The gap on the QPC was set at 250 nm, shown in Figure 11. We tested the device at 4.2 K using AC lock-in techniques and found that the device did indeed exhibit quantized resistance behavior on account of the quantized transverse electron momentum through the QPC.

A Quantum Point Contact is defined as a short one-dimensional channel that is connected adiabatically to large source and drain reservoirs and that supports one or more wave modes. Our QPCs were made by electron beam lithography where two small metallic electrodes are patterned to form a small gap between them (100 nm – 1 μm in a typical QPC experiment). When the device is very cold and the negative bias voltage applied to it is strong enough to fully deplete electrons in the local 2DEG underneath, the electrons within the 2DEG are forced through a narrow constriction having now been permitted to move in only one direction. The width of the channel can be controlled by adjusting the gate voltages and can be made small enough to be comparable to the Fermi wavelength of the electrons (~40 nm). When the wavelength of the electrons is on the order of or greater than the characteristic size of the system quantum effects become pronounced. Here, since the Fermi wavelength is comparable to the width of the QPC's narrow constriction quantum effects are observable. Figure 11 shows examples of QPC's while Figure 12 shows a quantized resistance obtained from a QPC shown in Figure 11 (a), which indicates the quantization of the conductance in the QPC.

(a) (b)

Fig. 11. Two Scanning Electron Micrographs (SEM) of QPC's used to restrict the electrons in the 2DEG to motion in one direction. (a) QPC with a ~250 nm wide gap. (b) QPC with a ~300 nm wide gap. Note the slight bend at the edges of the electrodes in (a), which is due to the *proximity effect*, a result of secondary backscattering electrons in the electron beam lithography writer.

If the negative voltages on the QPC electrodes are made sufficiently strong so that the first subband is above the Fermi level, then the electrons can only tunnel across from one reservoir to the other, and the QPC then acts as a tunnel barrier. A good review of the theory regarding quantum point contacts can be found in [9]. The conductance is calculated starting with the simple Hamiltonian

$$H = p_x^2/2m^* + eV(x) + p_y^2/2m^* \tag{6}$$

Here, V(x) is the confining potential from the gate electrodes in the lateral direction. In this Hamiltonian V(y), the potential in the longitudinal direction that describes the transition from the 2DEG reservoirs to the constriction, is not included, assuming the one-dimensional electron's motion in the x-direction. V(x) takes on a parabolic form in the lateral direction

$$V(x) = \tfrac{1}{2}\, m^*\omega_o^2 x^2. \tag{7}$$

The solutions to the Schrödinger equation with this V(x) can be written in the Energy eigenvalue form

$$E_n = (n-1/2)\hbar\omega_o + \hbar^2 k_y^2/2m^*. \tag{8}$$

Here n is an integer (n = 1, 2,...). The conductance of the QPC can be calculated using Landauer-Buttiker formalism if the transmission probability are known, and is given by (see ref.[9] for details)

$$G = \frac{2e^2}{h} \sum_{n=1}^{N} T_n(E) \tag{9}$$

Essentially, the summation is over all modes of the QPC and $T_n(E)$ represents the transmission probability of each individual mode. For small V_{sd} values this can be simplified by making the approximation $T_n(E) = T_n(E_F)$. If there is no backscattering from the QPC (although this assumption is not realistic), $\Sigma T_n = N$, where N is an even number integer (N = 0, 2, 4,...) for the case of no applied magnetic field, representing each fully occupied subband. Then the conductance of a QPC can be written as

$$G = \frac{2e^2}{h} N \tag{10}$$

The conductance G of a QPC is quantized in units of $2e^2/h$ depending on the number of modes accessible in the device.

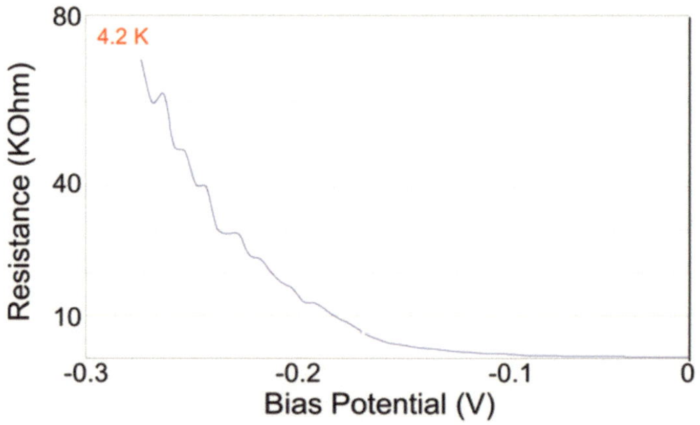

Fig. 12. Quantized resistance obtained from the quantum point contact shown in Figure 11 (a), which indicates the quantized conductance of the QPC at 4.2K.

The quantized conductance also can be seen from a quantum point contact formed by a gap between the gates, which are fabricated to define the quantum dot. An example of a single quantum dot is shown in Figure 2 and also in the inset of Figure 13. The conductance measurements at 80 mK through a quantum point contact formed by the gates 4 and 8 exhibits a well defined quantized conductance.

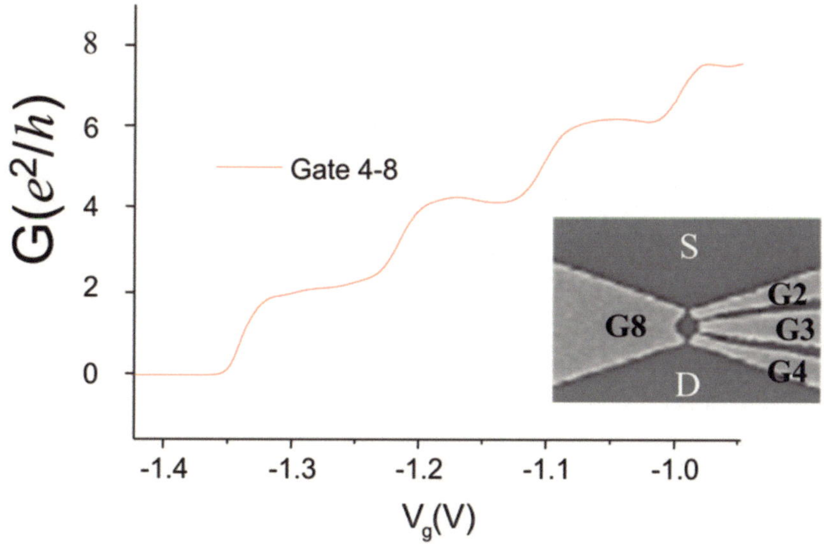

Fig. 13. The conductance G through a quantum point contact between the gates G4 and G8 is quantized in units of e^2/h, with the multiple of 2 arising from spin degeneracy. Inset: a single quantum dot.

3.3 Weakly coupled double quantum dot for a single photon detection application

For a small number of electrons in the quantum dot it is possible to calculate many-electron wave-functions and energy states. The many-body spectrum at zero magnetic field is then governed by the quantum confinement energy E_q and the charging energy E_c. For the simplest case of two parabolically confined electrons these parameters may be expressed in terms of l_o, which is related to a characteristic frequency Ω_0 determined by the electrostatic environment, as

$$l_o = (\hbar / m\,\Omega_0)^{½}. \quad (11)$$

The confinement length of the harmonic oscillator can be expressed as

$$E_q = \hbar^2 / (m * l_o^2)$$

and the charging energy as

$$E_c = e^2 / 2C \propto e^2 / (4\pi\mu\varepsilon\, l_o).$$

If one uses the quantum dot as a photon detector, the characteristic frequency Ω_0 is related to the frequency of the photon absorbed by the quantum dot. This means that the photon frequency of the quantum dot detector can be tuned by adjusting the electrostatically defined quantum dot size. When a photon is absorbed by the dot its energy level is shifted resulting in a pair of excited electrons and holes. The excited charge can tunnel to the electron reservoir (i.e. outside of the quantum dots), resulting in the conductance-resonance peak shift. The variation of conductance can be detectable when the quantum dot absorbs even a single photon. As demonstrated by Komiyama and his coworkers [10, 11], such photon detection can be achieved using a single quantum dot or weakly-coupled double quantum-dots. Since the photon detection using double quantum-dots seemed to be more practical than that of a single quantum dot, which can be achieved by applying a considerable magnetic field (3.4 - 4.15 Tesla) to the quantum dot, we adopted the double quantum-dot technique.

3.4 Double quantum dot photon-detector

Similar to Komiyama's double quantum dot detector, our single photon detector consists of double quantum dots in a parallel geometry that is defined by metallic electrodes deposited on the 2DEG substrate surface. In this experiment we have used several different 2DEG substrates with the 2DEG depth ranging from 40 nm to 160 nm, and fabricated more than several hundred devices. Figure 14(a) shows the gate electrodes and the Ohmic contacts with the quantum dots located at the center of the white frame. Figure 14 (b) shows another SEM picture of the double quantum dots, which is a magnified view of the center part of Figure 14(a). The lower quantum dot (QD1) acts as a photon absorber and the upper quantum dot (QD2) functions as a single electron transistor.

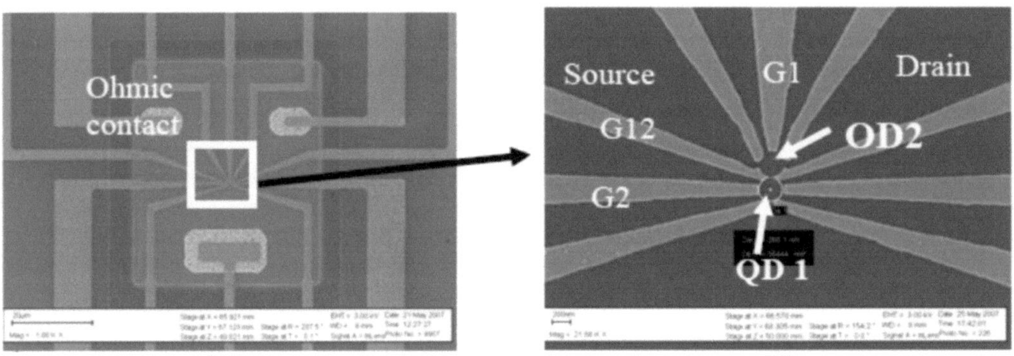

(a) (b)

Fig. 14. (a) Overall SEM view of the NRL double quantum dot detector. (b) SEM picture of the electrode defined double quantum dots.

The gate electrodes were defined via e-beam lithography after which we deposited a 50 Å thick Cr layer (acts as a wetting layer) on the surface of a GaAs/AlGaAs heterostructure and then a 150 Å thick Au layer on top of the Cr layer. The diameters of QD1 and QD2, as defined by the surrounding electrodes, are roughly 250 nm with the diameter of the SET dot (QD2) slightly smaller than that of the absorber dot (QD1). As mentioned earlier, the actual size of the quantum dot is dependent upon the strength of the negative bias voltage applied to the gates: the stronger the bias voltage the smaller the quantum dot. As the capacitance and the electrochemical potential of the quantum dot are closely related to the number of isolated electrons, one can control the capacitance and the electrochemical potential by adjusting the gate voltage. The plunger gate G_1 shown in Figure 14 provides experimental control of the SET dot's self capacitance (C_1) and electrochemical potential (μ_1), and the pair of gates labeled G_2 control the absorber dot's self capacitance (C_2) and electrochemical potential (μ_2). The electrodes labeled G_{12} control the potential barrier that couples the SET dot and the absorber dot.

As the absorber (QD1) and the SET (QD2) are weakly coupled by the voltage on gates G12, the excited energy level of the absorber alters the energy levels of the SET. If an energy level of the SET aligns within the energy levels of the source and the drain, electrons begin to flow through the SET. These excited energy levels of the absorber and the SET are in what are referred to as meta-stable states, which survive typically on the order of or less than a few milli-seconds. This short meta-stable state is due to the fact that a finite probability exists that an electron from one of the large 2DEG reservoirs "hops" onto the absorber dot. This results in a change in the energy of the absorber dot, which can affect the energy level matching between the SET dot and source and drain, since the two dots are electrostatically coupled. Since the electron mobility is very high (3.0×10^5 cm^2/Vs) and the electron density is very large ($5.0\text{x}10^{11}$ cm^{-2}) in the 2DEG, a significantly large number (~10^6 - 10^7) of electrons can flow through the SET within the short time of a meta-stable state, resulting in electric currents on the order of a pico-ampere. By employing a lock-in technique one can readily measure such currents. This operating principle is somewhat analogous to the photomultiplier tube as a single photon triggers a measureable electron flow in the detector.

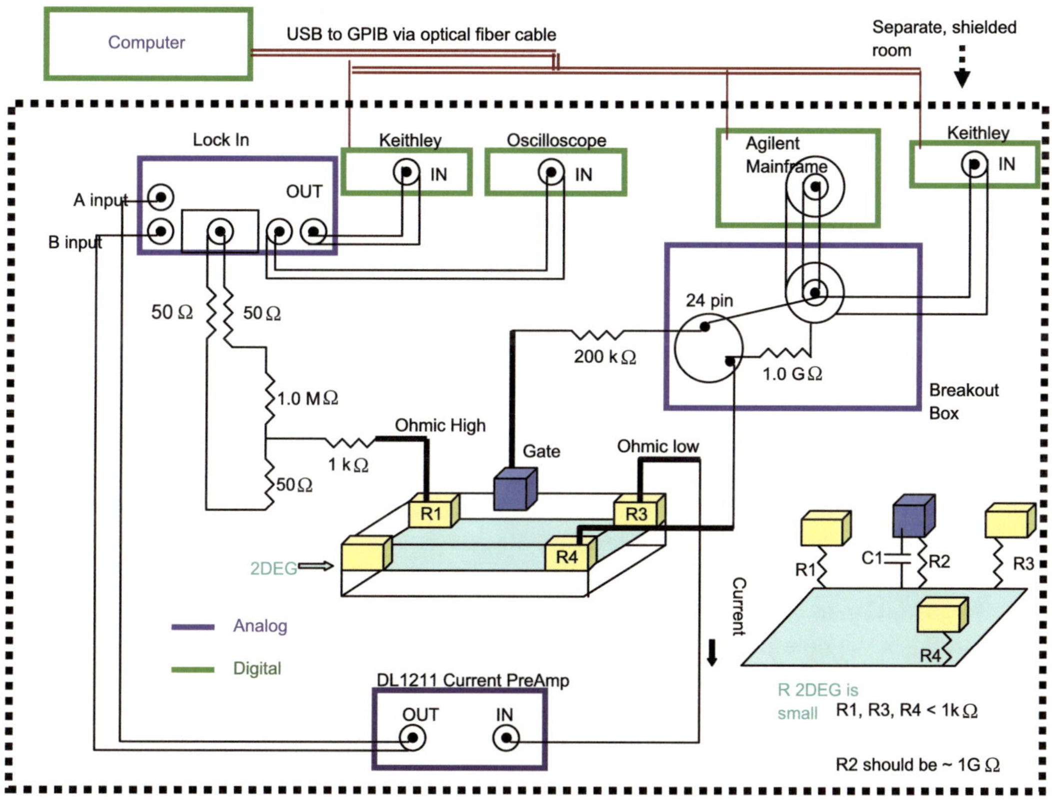

Fig. 15. A schematic diagram of our experimental setup to measure low current signals from a device. This diagram shows the wiring between the electronic equipment and the gate and Ohmic pads on a device. A lock-in sources an AC signal (~ 0.1 V) and a simple voltage divider circuit 10^5:1 is connected to an Ohmic contact on the high side and is used to supply the μV drop across the device. An Agilent E5270B Mainframe supplies the gate voltages through a homemade breakout box. Final data values can be read off of the Keithley digital multi-meters (DMM) and oscilloscope. The computer that collects data is housed in a separate room and is connected to the equipment via an optical fiber.

The electrons (~10^6 -10^7) flowing through the SET result in an electric current, which is on the order of a pico-Ampere (10^{-12}) or less. [1, 3, 12-14] In order to measure this weak current one should carefully design the experimental set up. Since electrical noises can induce currents much larger than pico-Ampere, it is necessary to minimize ambient electrical noises, which usually can be achieved by carrying out the measurements within a shielded room, and also by employing a lock-in technique. [15-17] An example is shown in Figure 15. A measurement includes the application of a source-drain voltage (or current source) over the device, or part of the device, and measuring the resulting current or voltage signal as a function of various parameters, such as the negative voltages applied to the depletion gates, temperature, electromagnetic fields, etc.

During the measurements it is important to keep the current and voltage across a device small enough in order to maintain the device temperature sufficiently cold. The energy associated with the voltage drop across the source and the drain, V_{sd}, should not exceed the electron thermal energy, K_BT_e, within the 2DEG. If $eV_{sd} > K_BT_e$, then the electrons within the Fermi reservoirs may enter or leave the quantum dots at random and/or the higher energy charge states may become allowed within the dots. Then the electron flow through the quantum dot cannot be controlled. Therefore, it is necessary to limit the voltage drop across a device such that $eV_{sd} < K_BT$, where K_B is the Boltzmann's constant. For example, V_{sd} should be less than 345 μV for T_e = 4 K, or 8.62 μV for T_e = 100 mK.

3.5 Photon detection

A schematic diagram of our photon detection setup is shown in Figure 16. For this demonstration we used a double quantum-dot detector fabricated on 100 nm 2DEG, and employed an HP 85105 millimeter-wave controller and an HP W85104A test-set module, which were attached to an HP8510C Vector Network Analyzer. The millimeter-wave signal was sent through micro-coaxial cable (Lakeshore Type C cable). To modulate the signal we split the micro-coaxial cable and made two sets of dipole antennae, which face each other across an optical beam chopper (Stanford Research SR540) set to produce 1 - 2 Hz modulation. However, at the low modulation frequency, the chopper's blade did not rotate smoothly resulting in irregular modulation so much that the modulation interval was highly irregular. Also, we note that the millimeter-wave signal was highly attenuated through the micro-coaxial cables, as well as through the dipole-antenna to dipole-antenna coupling. We estimate the attenuation rate was much more than 5dB per foot for the micro-coaxial cable and the coupling efficiency through the dipole-antenna coupling to be less than 10%. Since the initial mm-wave input from the W85104A was approximately 50 μW, and the transmission efficiency of the millimeter-wave photon through the coaxial cable and the dipole-antenna coupling was extremely poor, we think that the millimeter-wave signal radiated onto the double dot detector was sub-microwatts.

We measured the temperature dependent conductance as well as the bias voltage dependent conductance of our double dot device. The experiments indicate that although the millimeter-wave signal power was very weak our double dot device could detect the signal (Figure 17). However, the results may not indicate single photon detection. We think that our double dot detector could detect a few millimeter-wave photons at 100 mK.[18]

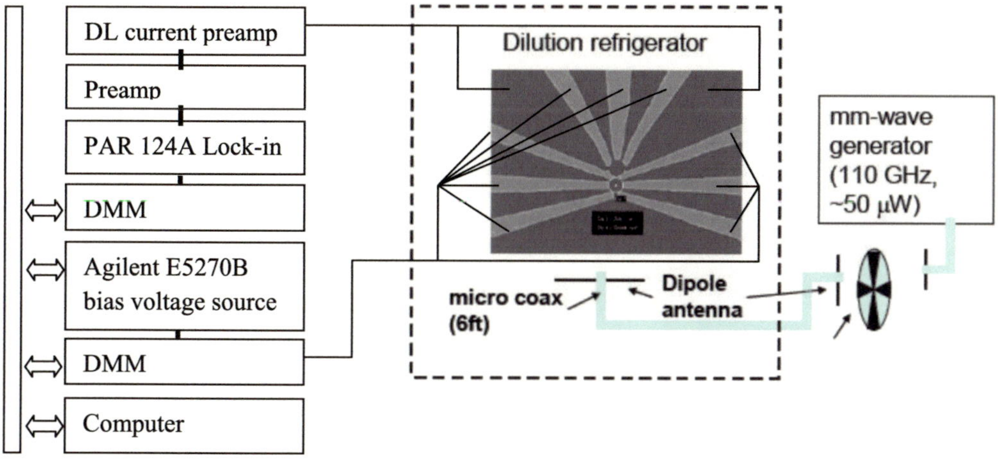

Optical BUS

Fig. 16. Experimental set up for the millimeter-wave photon detection with our double dot detector.

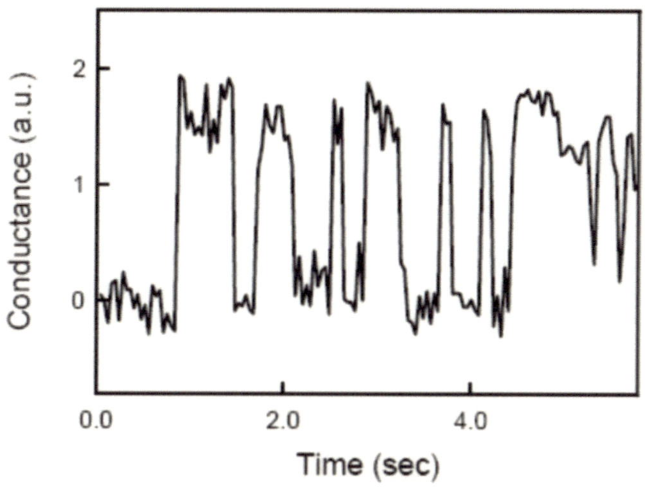

Fig. 17. The conductance variation as the double dot device detects millimeter-wave photons

While experiments indicate that it was possible to detect some photons at 110 GHz with the double quantum-dot structure shown in Figure 14, the detection efficient was very poor. We suspect that the inefficiency was largely due to the gate G2, which was supposed to function as an antenna. With its improper shape as an antenna, it did not efficiently pick up photons. Later we modified the double quantum dot detector, implementing bow-tie antenna geometry for the gate G2, as shown in Figure 18.

Also we attempted to reduce the quantum dot size in order to detect photons at an elevated temperature. As discussed earlier, a quantum dot detector should be operated at a temperature T that has a thermal energy K_BT below the charging energy of the quantum dot. The charging energy E_C is given as $E_C = e^2/C_{dot}$ where C_{dot} is the self capacitance of the quantum dot, which is proportional to the size of quantum dot. Hence, by reducing the quantum dot size, one can raise the charging energy E_C, as well as the operating temperature T of the quantum dot detector.

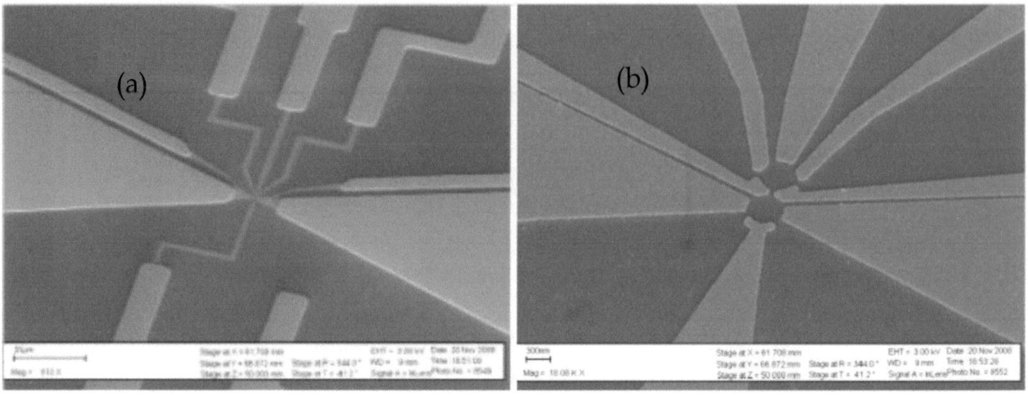

Fig. 18. SEM images of a modified double quantum-dot detector. (a) Overall view. (b) Magnified view of the center of picture (a), showing double quantum dots and a bow-tie antenna. The lower quantum dot is coupled to a dipole (bow-tie) antenna fabricated to absorb THz frequencies, while the upper dot acts as a single electron transistor.

As a rule of thumb the depletion length that is the lateral depletion of the electrons around the gates is roughly equal to the depth of the 2DEG. This means that, for a smaller quantum dot, we need to use a shallower 2DEG substrate. For our experimental demonstration of a photon detector at temperatures considerably higher than 100 mK, we have fabricated about a hundred double quantum-dot detectors on 2DEG substrates with thickness ranging from 40 nm to 160 nm.

4. Leakage currents in GaAs/AlGaAs heterostructures

Our experiments revealed that the quantum dot detectors fabricated on a shallow 2DEG suffered from problems associated with overwhelming leakage currents. Within a GaAs/AlGaAs heterostructure the leakage currents increased dramatically as the 2DEG depth became shallower. Since the leakage currents dominate, it was not possible to obtain any discernible signal from the quantum dot detector. Also the leakage currents caused severe damage to the quantum dot gates, often resulting in a short circuit on the gates.

Measurements were performed to determine the currents that flow between the 2DEG and a laterally defined depletion gate on the wafer's surface. Current is measured as a function of the voltage applied to the gate when the gate is biased with respect to the 2DEG underneath. While we expect that the current should be zero ideally or much less than pico-Amperes, the actual leakage current measured is orders of magnitude larger than anticipated. Even our numerical calculation, which was performed along with our experimental efforts, indicates that the leakage current is substantially larger than previously expected for shallow 2DEG wafers. Unless we find a way to prevent this large leakage it may lead to a limit for the maximum operating temperature obtainable for our quantum-dot photon detector.

Figure 19 shows a simple setup used to test the leakage currents within our GaAs/AlGaAs 2DEG substrate. The leakage current measurements were performed for several different 2DEGs with depths of 40 nm, 43 nm, 90 nm, and 160 nm. As shown in Figure 19, the voltage potential is applied directly to a gate and to an Ohmic contact in the reservoir. Any leakage current between the gate and the 2DEG is measured by a current amplifier (DL1211). The DL1211 converts the measured current signal into a voltage signal, which is read by the Keithley multimeter.

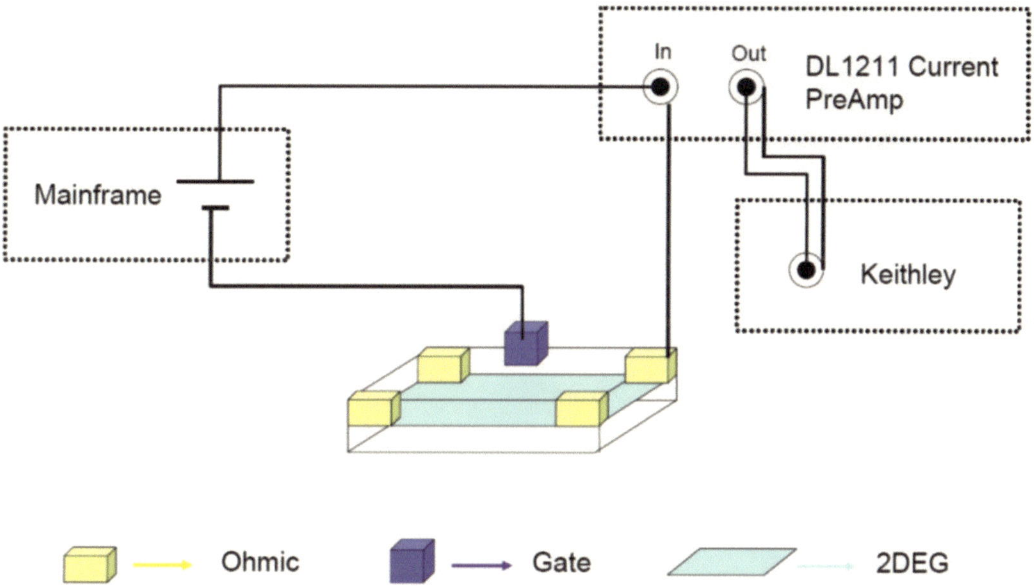

Fig. 19. A schematic experimental setup for the measurement of leakage currents in our 2DEG wafers. The voltage potential is supplied by an Agilent E5270B Precision Measurement Mainframe.

The leakage currents increased 6 orders of magnitude when the 2DEG depth was varied from 160 nm to 40 nm. Some of the shallowest (40 nm) 2DEG substrates generated leakage currents as high as tens of micro-Amperes, while the leakage current from the thickest one was much smaller, less than a pico-Ampere. When leakage currents are as high as several micro-Amperes the electron flow in and around the quantum dots cannot be controlled, and it is impossible to obtain any meaningful signal from the quantum dot detector. Our experiments further revealed that the strength of leakage currents vary depending on the individual 2DEG substrate. In other words, when we measure the leakage currents from two different 40 nm thick 2DEG substrates, we obtain very inconsistent results. This suggests us that the leakage current problem may not be entirely due to the intrinsic property of a shallow 2DEG, rather it may suggest that the problem is related to the defects in the 2DEG substrate.

Figure 20 shows an example of our leakage current data obtained from a 43 nm deep 2DEG at 4.2 K. The leakage currents increased linearly with the gate potential, and could reach as high as several hundred nano-Amperes. In order to investigate the origin of this leakage current we performed numerical simulations, in which we assumed the leakage currents were through a 1-dimensional barrier. In this simplified model only the tunneling across a Schottky barrier from a metal to a GaAs layer under a reverse gate bias was considered. The model is so simplified that we did not include the charge carrier interaction with the heterostructure, such as how the charge carriers, after having passed the Schottky barrier, travel through several regions of GaAs and AlGaAs, including a heavily doped AlGaAs layer, before reaching the 2DEG. The inclusion of these would make the simulations more realistic. However it will require implementing a lot more difficult calculations in the simulations. So we only considered the problem only the top layer of GaAs. We know that this simulation is not realistic, but we think that it will give us some insights about the leakage currents.

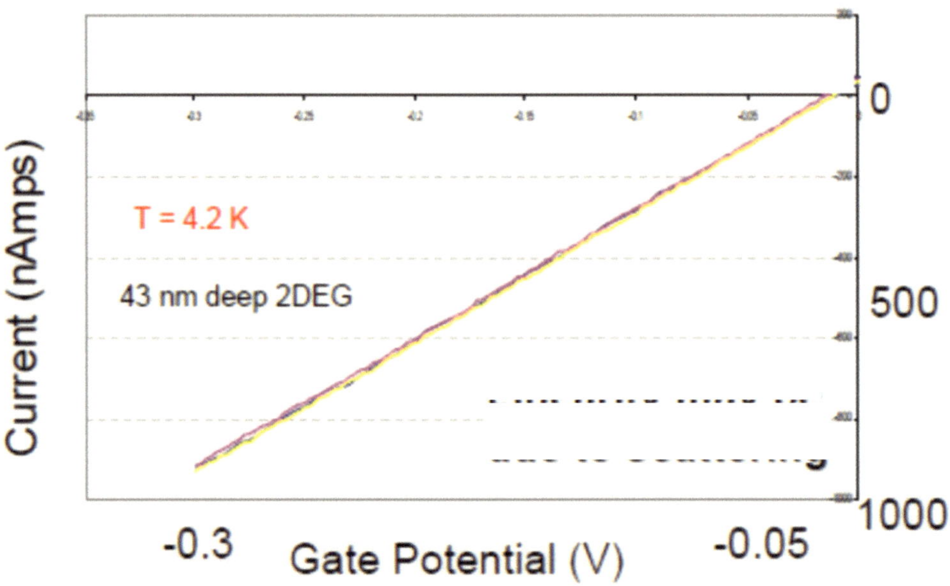

Fig. 20. Leakage current data taken from a 43 nm deep 2DEG at 4.2 K. Leakage currents for this shallow 2DEG structure can be as high as hundreds of nano-Amperes.

The simulations indicate that a shallower 2DEG leads to a larger leakage current, and the leakage current can exceed a thousand nano-Amperes. These results are at least qualitatively consistent with our experimental results. A simulated result that shows the leakage current as a function of gate voltage is presented in Figure 21. Apparently the exponential relationship between the leakage current and the bias voltage is not consistent with our experiments. (See Figure 20) The discrepancy may be due to the fact that our model is too simple and does not reflect realistic conditions, for instance the scattering that the charge carriers experience, due to the Si dopants, as they pass from the lateral surface gates to the 2DEG, and the effect of lattice mismatching between GaAs and AlGaAs.

In order to construct a device using quantum dots, one should minimize the leakage current since it not only prevents the proper control of the quantum dot but also sometimes leads to physical damage to the quantum dot. The gates surrounding the quantum dot are very small, typically less than 20 nm thick and a few tens of nm wide. Around the quantum point contacts formed by the gates the cross-sections of these metal-structures can be as small as a few tens of nm^2. When the leakage current exceeds several hundred nano-Amperes, the current density near the quantum point contacts can exceed 10^{10} - 10^{11} A/m^2. The current density may become high enough to fuse metallic structures near the quantum point contacts. This is what presumably happened to some of our quantum dot devices. An example is shown in Figure 22.

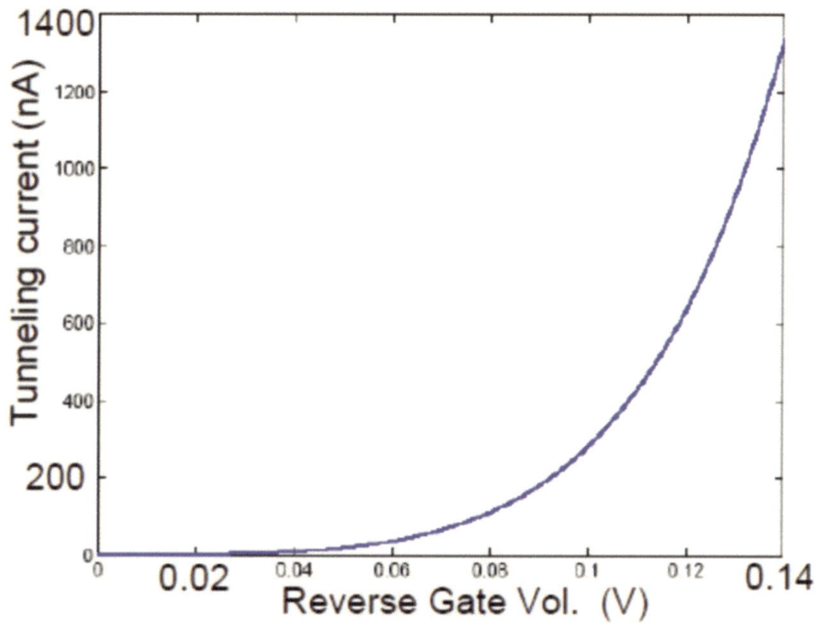

Fig. 21. An example of a numerical calculation showing the leakage current across a Schottky barrier from a metal to a GaAs layer under a reverse bias.

Our attempt to raise the operating temperature of a quantum dot photon detector did not succeed. The leakage current in shallower 2DEG materials remains one of the most significant technical challenges in achieving higher operating temperatures for single photon detectors. The origin of leakage currents in 2DEG substrates and a method to avoid them are topics for future research.

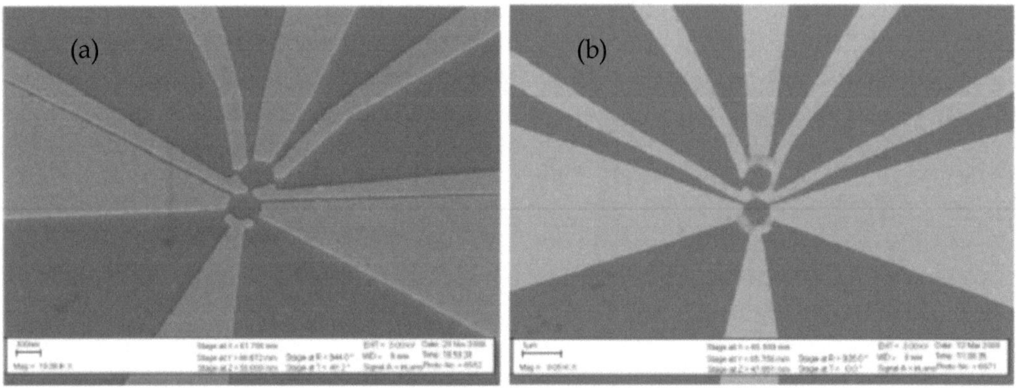

Fig. 22. SEM images of (a) BEFORE and (b) AFTER testing the quantum dots, which involves a large leakage current flowing through the device. The damage to the thin nanostructure gates near the center of the picture in (b) is due to the large leakage current that passes through the very small nanostructure (e.g. quantum point contacts between the gates).

5. References

[1] W. G. van der Wiel, S. DeFranceschi, J. M. Elzerman, T. Fujisawa, S. Tarucha, L. P. Kouwenhoven, Electron transport through double quantum dots, *Reviews of Modern Physics* 75 (2003).

[2] T. J. Thornton, M. Pepper, H. Ahmed, D. Andrews, G. J. Davies, One-Dimensional conduction in the 2D electron gas of a GaAs/AlGaAs Heterojunction, *Phys. Rev. Lett.* 56, 11 (1986).

[3] NATO ASI Series, Quantum Transport in Semiconductor Submicron Structures, *Series E: Appl. Sci.* 326 (1996)

[4] Vurgaftman, J. R. Meyer, D. H. Wu, K. Bussmann, and B. T. Jonker, Spectral simulation of GaAs and InAs quantum-dot terahertz detectors designed for higher-temperature operation, *Journal of Applied Physics* 100, 064509 (2006).

[5] P. O'connor, A. Dori, M. Feuer, R. Vounckx, Gold-germanium-based ohmic contacts to the two-dimensional electron gas at selectively doped semiconductor heterointerfaces, *IEEE Transactions on Electron Devices* ED-34, 4 (1987).

[6] M. Furno, F. Bonani, G. Ghione, Transfer matrix method modeling of inhomogeneous Schottky barrier diodes on silicon carbide, *Solid-State Electronics* 51, 466-474 (2007).

[7] J. Crofton, P. A. Barnes, A comparison of one, two, and three band calculations of contact resistance for a GaAs ohmic contact using the WKB approximation and a numerical solution to the Schrodinger equation, *J. Appl. Phys.* 69, 11 (1991).

[8] D. A. Muller, T. Sorsch, S. Moccio, F. H. Baumann, K. Evans-Lutterdot, G. Timp, The electronic structure at the atomic scale of ultrathin gate oxides, *Nature* 399, 758 (1999)

[9] H. van Houten, C. W. J. Beenakker, B. J. van Wees, Quantum point contacts *Nanostructured Systems*, M. A. Reed, Ed. (Academic Press, San Diego, 1992) 35, pp. 9-112 (1992)

[10] O. Astafiev, S. Komiyama, T. Kutsuwa, V. Antonov, Y. Kawaguchi, K. Hirakawa, Single-photon detector in the microwave range, *Applied Physics Letters* 80, 22 (2002).
[11] S. Komiyama, O. Astafiev, V. Antonov, T. Kutsuwa, H. Hirai, A single-photon detector in the far-infrared range, *Nature* 403, 405 (2000).
[12] L. P. Kouwenhoven, C. M. Marcus, P. L. McEuen, S. Tarucha, R. M. Westervelt, N. S. Wingreen, Electron transport in quantum dots, *Proceedings of the Advanced Study Institute on Mesoscopic Electron Transport* (1997).
[13] C. Livermore, C. H. Crouch, R. M. Westervelt, K. L. Campman, A. C. Gossard, The Coulomb Blockade in Coupled Quantum Dots, *Science* 274, 1332 (1996)
[14] L. P. Kouwenhoven, C. M. Marcus, P. L. McEuen, S. Tarucha, R. M. Westervelt, N. S. Wingreen, Electron transport in quantum dots, *Mesoscopic Electron Transport*, Kluwer Academic, Dordrecht (1997)
[15] F. Pobell, Matter and Methods at Low Temperatures, 2nd Edition (2009)
[16] H. W. Ott, Noise Reduction Techniques in Electronic Systems, 2nd Edition (1988)
[17] R. C. Richardson, E. N. Smith, Experimental Techniques in Condensed Matter Physics at Low Temperatures, Advanced Book Classics (1998)
[18] D. H. Wu, B. R. Matis, K. Bussmann, Scalable single photon detector for terahertz and infrared applications, *International Journal of High Speed Electronics and Systems* 18, 4 (2008)
[19] B. J. van Wees, H. van Houten, C. W. J. Beenakker, J. G. Williamson, L. P. Kouwenhoven, D. van der Marel, C. T. Foxon, Quantized conductance of point contacts in a two-dimensional electron gas, *Phys. Rev. Lett.* 60, 9 (1988).
[20] T. H. Oosterkamp, T. Fujisawa, W. G. van der Wiel, K. Ishibashi, R. V. Hijman, S. Tarucha, L. P. Kouwenhoven, Microwave spectroscopy of a quantum-dot molecule, *Nature* 395, 873 (1998).
[21] L. P. Kouwenhoven, S. Jauhar, J. Orenstein, P. L. McQueen, Y. Nagamune, J. Motohisa, H. Sakaki, Observation of photon-assisted tunneling through a quantum dot, *Phys. Rev. Lett.* 73, 25 (1994).
[22] K. Ikushima, Y. Yoshimura, T. Hasegawa, S. Komiyama, T. Ueda, K. Hirakawa, Photon-counting microscopy of terahertz radiation, *App. Phys. Lett.* 88, 152110 (2006).
[23] S. M. Cronenwett, S. M. Maurer, S. R. Patel, C. M. Marcus, C. I. Duruoz, J. S. Harris Jr., Mesoscopic Coulomb blockade in one-channel quantum dots, *Phys. Rev. Lett.* 81, 26 (1998).
[24] S. Gustavsson, M. Studer, R. Leturcq, T. Ihn, K. Ensslin, D. C. Driscoll, A. C. Gossard, Frequency-selective single-photon detection using a double quantum dot, *Phys. Rev. Lett.* 99, 206804 (2007).
[25] S. R. Patel, S. M. Cronenwett, D. R. Stewart, A. G. Huibers, C. M. Marcus, C. I. Duruoz, J. S. Harris Jr., K. Campman, A. C. Gossard, Statistics of Coulomb Blockade Peak Spacings, *Phys. Rev. Lett.* 80, 20 (1998)
[26] M. H. Devoret, R. J. Schoelkopf, Amplifying quantum signals with the single-electron transistor, *Nature* 406, 1039 (2000)
[27] H. L. Stormer, *Rev. Mod. Phys.* 71, 875 (1999)
[28] M. Pioro-Ladriere, J. H. Davies, A. R. Long, A. S. Sachrajda, L. Gaudreau, P. Zawadzki, J. Lapointe, J. Gupta, Z. Wasilewski, S. Studenikin, Origin of switching noise in GaAs/Al$_x$Ga$_{1-x}$As lateral gated devices, *Phys. Rev. B* 72, 115331 (2005)

[29] H. Hashiba, V. Antonov, L. Kulik, A. Tzalenchuk, P. Kleinschmid, S. Giblin, S. Komiyama, Isolated quantum dot in application to terahertz photon counting, *Phys. Rev. B* 73, 081310(R) (2006)
[30] S. Kim, M. S. Sherwin, J. D. Zimmerman, A. C. Gossard, P. Focardi, D. H. Wu, Room temperature terahertz detection based on electron plasma resonance in an antenna-coupled GaAs MESFET, *Appl. Phys. Lett.* 92, 253508 (2008)

12

Tunneling Atomic Force Microscopy of Self-Assembled In(Ga)As/GaAs Quantum Dots and Rings and of GeSi/Si(001) Nanoislands

Dmitry Filatov[1], Vladimir Shengurov[1], Niyaz Nurgazizov[2], Pavel Borodin[2] and Anastas Bukharaev[2]
[1]*Technical Physics Research Institute, N.I. Lobachevskii University of Nizhny Novgorod,*
[2]*E.K. Zavoisky Kazan' Physical-Technical Institute, Kazan' Scientific Centre, Russian Academy of Sciences*
Russia

1. Introduction

Scanning Tunnelling Microscopy (STM) has been being used for the investigation of the morphology and of the atomic structure of the semiconductor nanostructures extensively since early 1990-s (Medeiros-Ribeiro et al., 1998). More recently, STM in Ultra High Vacuum (UHV) has been applied also to the investigation of the spatial and energy distributions of the local density of states (LDOS) in the quantum semiconductor heterostructures. For example, Cross-Sectional STM (X-STM) has been applied to the visualization of the envelope wavefunctions of the quantum confined states in the GaSb/InAs(001) quantum wells (QWs) (Suzuki et al., 2007) and in the self-assembled InAs/GaAs(001) quantum dots (ODs) (Grandidier et al., 2000). Also, the surface InAs/GaAs(001) QDs grown by Molecular Beam Epitaxy (MBE) have been investigated by UHV STM *in situ* (Maltezopoulos et al., 2003). The peaks related to the quantum confined states in the QDs have been observed in the differential conductivity $\sigma_d = dI_t/dV_g$ spectra of the STM tip contact to the QDs (here I_t is the tip current and V_g is the gap voltage). The $\sigma_d(x, y)$ images of the ODs (x and y are the tip coordinates on the sample surface) recorded at the values of V_g corresponding to the peaks in the $\sigma_d(V_g)$ spectra correlated with the probability density patterns $|\chi(x, y)|^2$ where $\chi(x, y)$ are the lateral components of the electron quantum confined states envelope wavefunctions calculated for the pyramidal InAs/GaAs(001) ODs defined by the (101) facets (Stier et al., 1999).

The present chapter is devoted to the investigation of the electronic states in the self-assembled semiconductor nanostructures [namely, the InAs/GaAs(001) QDs, the InGaAs/GaAs(001) quantum rings (QRs), and the GeSi/Si(001) nanoislands] by Tunnelling Atomic Force Microscopy (AFM). The samples with the surface self-assembled semiconductor nanostructures were scanned across by a conductive Si AFM probe covered by a conductive coating (Pt, W_2C, or a diamond-like film) in the contact mode. The bias voltage V_g was applied between the AFM probe and the sample. Simultaneously with the acquisition of the topography $z(x, y)$, the $I-V$ curves of the probe-to-sample contact $I_t(V_g)$ were acquired in each point in the scans.

Earlier, Tunnelling AFM has been applied mainly to the characterization of the local electrical properties of the thin dielectric films on the conductive substrates (Yanev et al., 2008). Also, Tunnelling AFM in UHV has been applied to the tunnel spectroscopy of individual Au nanoclusters in the ultrathin SiO_2/Si films (Zenkevich et al., 2011). The present chapter summarizes a series of the original studies where the authors have applied Tunnelling AFM to the mapping of the LDOS in the self assembled semiconductor nanostructures for the first time (Filatov et al., 2010, 2011, Borodin et al., 2011). The main advantage of Tunnelling AFM compared to UHV STM is that the former allows the *ex situ* investigation of the surface semiconductor nanostructures covered by a native oxide layer during the sample transfer from the growth setup to the AFM one through the ambient air. This makes the STM studies of these samples hardly possible.

Another distinctive feature of the studies present in this chapter is that the samples with InGaAs/GaAs(001) QDs and QRs have been grown by Atmospheric Pressure Metal Organic Vapour Phase Epitaxy (AP-MOVPE). In most studies reported in the literature, the QDs grown by MBE or by Low Pressure (LP) MOVPE have been investigated. Nevertheless, the investigations of the electronic properties of the InAs/GaAs(001) QDs grown by AP MOVPE are of a considerable interest because this growth method is more promising for the commercial device production due to its lower cost and higher productivity as compared to MBE and LP-MOVPE.

Also, the structures with the self-assembled GeSi/Si(001) nanoislands studied in the present chapter were grown by a novel technique of Sublimation MBE (SMBE) in GeH_4 ambient. In this method, the Si layers are grown from an ordinary sublimation source in UHV. To deposit Ge, GeH_4 is introduced in the growth chamber at the pressure of $\sim 10^{-2} \div 10^{-4}$ Torr and undergo pyrolysis on the heated substrate. So far, this method is some hybrid between the conventional MBE from the sublimation source and LP VPE. Again, in the majority of works, the GeSi/Si(001) nanoislands grown by MBE have been studied (Berbezier & Ronda, 2009). The main advantage of the hybrid technique of SMBE in GeH_4 ambient as compared to the ordinary VPE of Si and Ge from silanes and germanes, respectively is that SMBE allows growing the Si layers of high crystalline quality and purity at relatively low temperatures (450 ÷ 500°C) keeping the high enough growth rates. In addition, SMBE offers a broader choice of the doping impurities as well as a wider range of their concentrations achievable than VPE. On the other hand, the deposition of Ge from a gaseous precursor provides higher uniformity of the nominal thickness of the deposited Ge layer d_{Ge} over the substrate surface.

Having applied Tunnel AFM to the investigation of the LDOS in the self assembled semiconductor nanostructures described above, we have observed the patterning of the probe current images $I_t(x, y)$ of the InAs/GaAs(001) QDs and of the InGaAs/GaAs(001) QRs as well as the peaks in the $\sigma_d(x, y)$ spectra of the contact of the AFM probe to the sample surface attributed respectively to the spatial and energy distributions of the LDOS in the quantum heterostructures. The results of the LDOS mapping by Tunnelling AFM have been compared to the results of the calculations of the probability density patterns $|\chi(x, y)|^2$ reported in the literature. The Tunnel AFM data allowed the identification of the quantum confined states in the InAs/GaAs(001) QDs grown by AP MOVPE the interband optical transitions between which are manifested in the photosensitivity (PS) spectra of the QD structures. Finally, the direct measurements of the LDOS spectrum in the self assembled $Ge_xSi_{1-x}/Si(001)$ nanoislands demonstrated the type I conduction band alignment in them at $x < 0.45$.

2. Self assembled InAs/GaAs(001) quantum dots

2.1 Growth and characterization

In this subsection, the details of the growing the samples for the Tunnelling AFM investigation by AP MOVPE and of their characterization are presented.

The InAs/GaAs(001) QD structures for the tunnel spectroscopy of the LDOS in the conduction band were grown on the n^+-GaAs(001) substrates by Dr. B. N. Zvonkov in Research Institute for Physics and Technology, N. I. Lobachevskii University of Nizhny Novgorod, Russia using a homemade setup for AP MOVPE from trimethylgallium, trimethylindium, and AsH_3. The schematic of the QD structures for the Tunnelling AFM investigations is shown in Fig. 1, *a*. The substrates were misoriented from (001) by 3 ÷ 5° towards <110>. The donor concentration in the substrate material was ~10^{18} cm^{-3}. The GaAs buffer layers with the thickness $d_b \approx 200$ nm were doped by Si heavily up to the donor concentration ~ 10^{18} cm^{-3} using pulsed laser sputtering of a bulk Si target placed in the MOVPE reactor. More detains on this growth technique can be found elsewhere (Karpovich et al., 2004a). The 3 nm thick intentionally undoped GaAs spacer layers were grown between the n^+-GaAs buffers and the InAs ODs. The latter were grown at the substrate temperature T_g = 530°C, the nominal thickness of the InAs layers d_{InAs} was ≈ 5 monolayers (ML).

The morphology of the grown samples was first examined by ambient air AFM using NT MDT® Solver Pro™ instrument in Contact Mode. The NT MDT® CSG-01 silicon AFM probes were used. The curvature radii of the probe tips R_p were < 10 nm (according to the vendor's specifications).

The morphology of the surface QD arrays was characterized quantitatively by the following parameters:

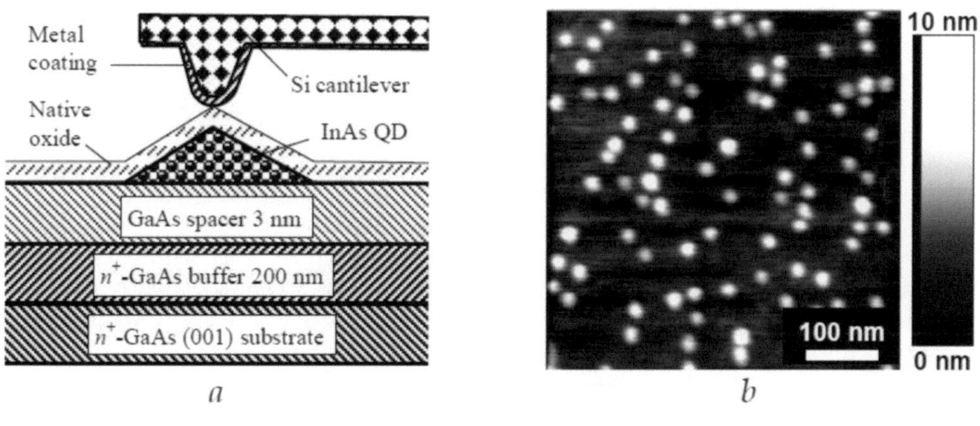

Fig. 1. The schematic (*a*) and an ambient air AFM image (*b*) of a structure with the InAs/*n*-GaAs/ n^+-GaAs(001) surface QDs.

- The average height of the QDs <*h*>
- The averaged diameter of the QDs <*D*>. The value of D for given QD was defined as $D = P/2\pi$ where P was the perimeter of the AFM image of the QD measured at the level of 0.1 <*h*> above the surface of the wetting layer
- The surface density of the QDs N_s

The morphological parameters of the QD arrays listed above were determined by the digital processing of the AFM data. In order to identify the QDs on the surface of the wetting layer, a threshold particle recognition algorithm has been applied.

The key issue in the investigations of the morphology of the surface InAs/GaAs(001) QDs by AFM was the effect of convolution (Bukharaev et al., 1999) originating from a relatively large AFM probe tip radius R_p (~ 10 nm) as compared to the typical sizes of the InAs/GaAs(001) QDs defined by the (101) facets ($D = 12 \div 18$ nm, $h = 5 \div 6$ nm). In order to extract the actual size and shape of the InAs/GaAs(001) QDs grown by AP MOVPE from the AFM data, we have applied the digital processing of the AFM images using an original software for the correction of the convolution artifacts (so called "deconvolution"). This software utilizes the "virtual AFM" algorithm (Bukharaev et al., 1998).

In order to apply this algorithm to the correction of the convolution artifacts, one needs to know the exact geometry of the actual probe tip used in the experiment. The specifications provided by the AFM cantilevers' vendors appear to be insufficient often. The actual probe tip shape may be determined by the measuring of a standard sample with the geometry known *a priori*. However, the theory of the convolution artifacts sets up an essential requirement on the geometrical parameters of the standard sample to be close to these ones of the samples under study [i. e. the InAs/GaAs(001) QDs in our case]. In order to evaluate the AFM tip shape, we have used a structure with the surface self assembled GeSi/Si(001) pyramidal-shaped nanoislands (Medeiros-Ribeiro et al., 1998) as a standard sample. The structure was grown by Dr. A. V. Novikov, Institute for Physics of Microstructures, Russian Academy of Sciences (Nizhny Novgorod, Russia) using standard MBE. The self assembled GeSi/Si(001) pyramidal-shaped nanoislands are defined by the (105) facets, their heigt can be extracted just from the AFM data directly. Note that the convolution artifacts do not affect the acuracy of the measurements of the heights h of the GeSi/Si(001) nanoislands as well as of the InAs/GaAs(001) QDs. So far, all the parameters needed to determine the actual probe shape can be determined from a single AFM scan of a single GeSi/Si(001) pyramidal island.

Another ulitmate requrement is that the hieght of the topographic elements on the standard sample should exceed the heihgts of the invesitgated objects [namely, the InAs/GaAs(001) QDs]. The self assembled GeSi/Si(001) pyramid islands satisfy this requirement as thier height h may reach ~ 10 nm when grown in the appropriate conditions, that is well enough for the InAs/GaAs(001) QDs (typically, $h = 5 \div 6$ nm). So far, the GeSi/Si(001) pyramid nanaoislands appear to be a good natural standard for the measurement of the probe shape in the particular case of the deconvolution of the AFM images of the self-assembled InAs/GaAs(001) QDs.

The raw images of the QDs seemd round, their lateral sizes D were 30 to 50 nm (Fig. 2, *a*). In the AFM images after the deconvolution (Fig. 2, *b*) the (101) faceting of the InAs/GaAs(001) QDs grown by AP MOVPE is seen clearly.

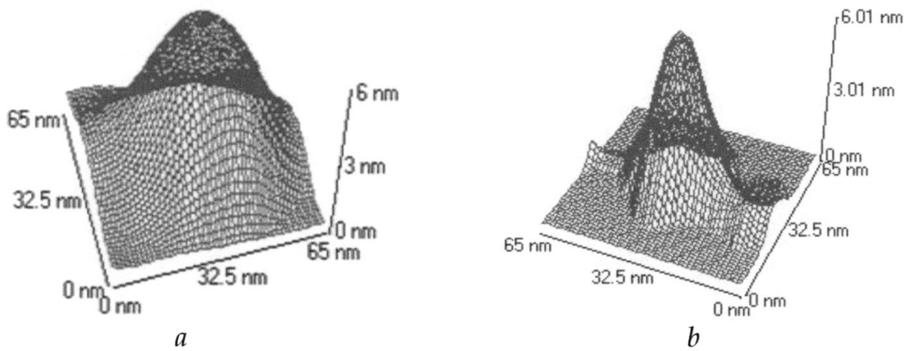

Fig. 2. The AFM images of a surface InAs/GaAs(001) QD grown by AP MOVPE: as measured (*a*) and after the deconvolution (*b*).

The base sides were directed along <110>, their length $b = 14 \div 18$ nm like the ones grown by MBE (Ledentsov et al., 1999, Maltezopoulos et al., 2003). The top of the pyramid was truncated slightly (Fig. 2, *b*) that could be explained noting that the top of the pyramid is a concentrator of the tensile strain. So far, the truncation of the pyramid reduces the overall elastic energy of the QD. Similar shape of the QD tops has been observed by UHV STM (Maltezopoulos et al., 2003). Therefore, the aspect ratio A_R of the InAs/GaAs(001) QDs is less than 1:2 slightly.

The size quantization energy spectrum of the InAs/GaAs(001) QDs grown by AP MOVPE has been examined by photoluminescence (PL) spectroscopy at 77K and by the photoelectric spectroscopy at 300K. The structures for the PL spectroscopy have been grown on the semi-insulating GaAs(001) substrates, the buffer layers have been undoped intentionally. The InAs QDs in these structures have been capped by a 30 nm thick GaAs cladding layer. The structures for the photoelectric measurements had the same design except the substrates were form n^+-GaAs. The InAs in all three types of the structures (for Tunnelling AFM, for PL, and for the photoelectric spectroscopy) have been grown in the same conditions (T_g = 530°C, d_{InAs} = 5 ML). The details of the measurement techniques as well as the analysis of the experimental results can be found elsewhere (Karpovich et al., 2004b).

2.2 The conduction band states

In this subsection, the results of the Tunnelling AFM investigations of the quantum confined electron states in the surface InAs QDs are presented and discussed. Also, the results of the studies of the laterally coupled surface InAs QDs are presented.

The Tunnelling AFM studies were carried out at 300K using Omicron® MultiProbe P™ UHV system equipped by Omicron® UHV VT AFM/STM. A typical topographic image $z(x, y)$ and the probe current one $I_t(x, y)$ of an InAs/GaAs(001) QD sample are presented in Fig. 3, *a* & *b*, respectively. An increasing of I_t every time the AFM tip encounters the QD surface had been observed. This observation has been attributed to the electron tunnelling between the AFM tip and the conductive buffer layer through the quantum confined states in the QDs (a qualitative band diagram of a contact of a Pt coated AFM tip to an InAs/GaAs/n^+-GaAs biased negatively is presented in Fig. 4, *b*).

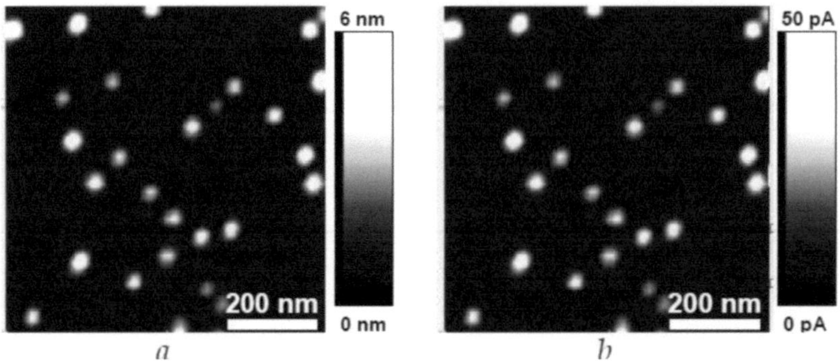

Fig. 3. The topographic (*a*) and the probe current (*b*) images of the surface InAs/GaAs/n^+-GaAs(001) QDs. V_g = -3.7 V. Reproduced from (Filatov et al., 2010) under license by IoP Publishing Ltd.

The differential conductivity spectra $\sigma_d(V_g) = dI_t/dV_g$ of the tunnel contact of a Pt coated AFM tip to different points of the surface an InAs/GaAs/n^+-GaAs(001) QD are presented in Fig. 4. , *a*. The $\sigma_d(V_g)$ spectra have been calculated from the measured $I_t(V_g)$ curves of the probe-to-sample contact by the numerical differentiation with the nonlinear smoothing. The points on the QD surface where the respective $I_t(V_g)$ curves had been measured are marked in Fig. 5, *a*. The peaks observed in the $\sigma_d(V_g)$ spectra were attributed to the tunnelling of the electrons between the metallic tip coating and the conductive substrate through the quantum confined states in the QDs (Fig. 4. , *b*). The native oxide on the sample surface formed a potential barrier, the second triangle potential barrier was formed by the depletion layer of the contract of the metal tip coating to the GaAs/n^+-GaAs. In the UHV STM studies (Maltezopoulos et al., 2003) the first potential barrier was formed by the vacuum gap between the metal STM probe and the sample surface.

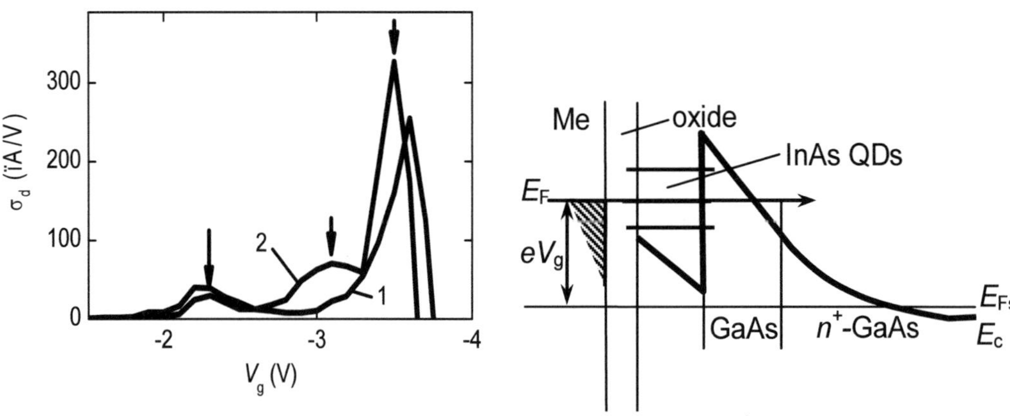

Fig. 4. The differential conductivity spectra $\sigma_d(V_g)$ (*a*) and the qualitative band diagram (*b*) of a negatively biased contact of a metal coated AFM tip to an InAs QD on the *n*-GaAs/n^+-GaAs (001) substrate. The curve numbers denote the points of the $I-V$ curves' measurements marked in Fig. 5, *a*. Reproduced from (Filatov et al., 2010) with permission from ©Pleiades Publishing, Ltd.

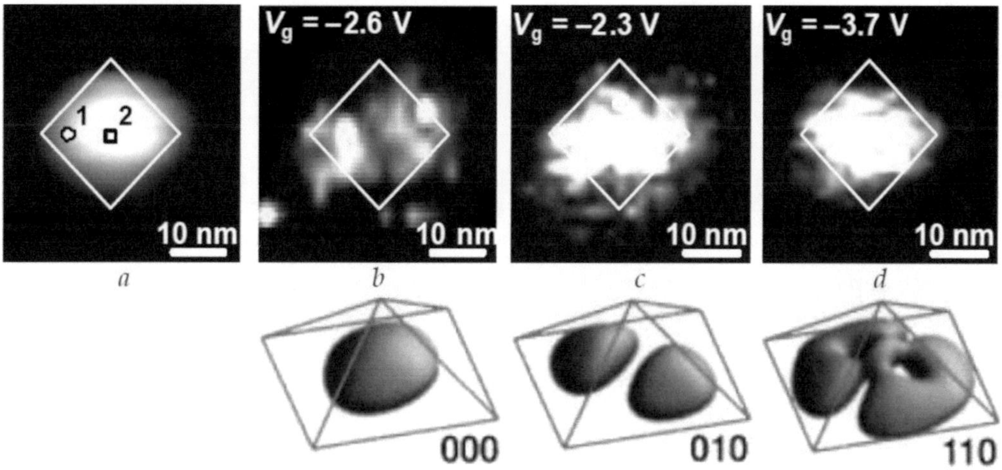

Fig. 5. The UHV AFM (*a*) and the probe current (*b* – *d*) images of an InAs QD on the *n*-GaAs/n^+-GaAs(001) substrate. Reproduced from (Filatov et al., 2010) with permission from ©Pleiades Publishing, Ltd. Below, the surfaces of the equal probability density $|\chi(x, y, z)|^2$ = 0.65 calculated for several quantum confined electron states $| n_1n_2n_3>$ in a pyramidal InAs/GaAs(001) QD with the base side length b = 16 nm are presented. Reproduced partly from (Stier et al., 1999) with permission from ©American Physical Society.

The AFM and the probe current images of an individual InAs/n^+-GaAs QD are presented in Fig. 5. Note that because of the convolution effect, the AFM image of the QD is rounded and enlarged as compared to the actual QD size revealed using the deconvolution (b = 14 ÷ 16 nm, see Sec. 2.1 above).

Following (Maltezopoulos et al., 2003), the patterns of the $I_t(x, y)$ images at certain values of V_g corresponding to the maxima of the $\sigma_d(V_g)$ spectra in Fig. 4, *a* have been related to the spatial distribution of the LDOS in the (x, y) plane:

$$\rho_E(x,y) \propto \sum_{n_1n_2n_3=0}^{N_1N_2N_3} \left|\chi_{n_1n_2n_3}(x,y)\right|^2 . \tag{1}$$

Here the envelope wavefunctions $\chi(x, y)$ were considered to be spin-independent and twofold spin-degenerated. The summation in (1) was taken over the states below the Fermi level in the probe coating material E_F. In other words, in the case of the QDs grown on the *n*-GaAs/n^+-GaAs(001) substrate the energies of the respective quantum confined states must satisfy the following condition:

$$E_{n_1n_2n_3} < E_{Fs} + eV_g , \tag{2}$$

where E_{Fs} is the Fermi level in the n^+-GaAs buffer (see Fig. 4. , *b*). Condition (2) defines the upper limits of the summation N_1, N_2, and N_3 in (1).

Again, following (Maltezopoulos et al., 2003), the tunnel current images of the QDs $I_t(x, y)$ were compared to the probability density patterns $|\chi(x, y, z)|^2$ = *const* calculated for the quantum confined electron states in the pyramidal InAs/GaAs(001) QDs (Stier et al., 1999).

The probe current images of the QDs obtained by Tunnelling AFM were more noisy than the ones obtained by STM in UHV (Maltezopoulos et al., 2003) that was attributed to the nonuniformity of the thickness of the native oxide covering the QDs. Nevertheless, several electron quantum confined states in the QDs were identified, the respective images of the $|\chi(x, y, z)|2 = 0.65$ surfaces (Stier et al., 1999) are shown in Fig. 5.

In order to associate the spectral positions of the peaks in the $\sigma_d(V_g)$ spectra with the quantum confined level energies $E_{n_1n_2n_3}$, one must take into account the partial voltage drop on the depletion layer between the QDs and the n^+-GaAs buffer layer as well as the one on the surface states at the semiconductor/native oxide interface. Following (Suzuki et al., 2007), we have applied a simple one-dimensional model based on the solution of one-dimensional Poisson's equation (Feenstra & Stroscio, 1987) to account for the voltage drop on the depletion layer of the probe-to-sample contact. In order to account for the surface charge density on the surface states on the interface between the sample surface and the native oxide, we have applied Hasegawa's model (Hasegawa & Sawada, 1983). The calculations have shown that approximately ½ of V_g drops on the surface states.

The InAs/GaAs (001) surface QD structures grown by AP MOVPE are featured by the presence of a considerable number of the QDs arranged in the pairs along the growth steps in a close proximity to each other (see Fig. 1, *b*) that may result in a considerable overlap of the envelope wavefunctions of the quantum confined states in the adjacent QDs. An example of the UHV AFM image as well as a series of the probe current images of a pair of the laterally coupled surface InAs/GaAs(001) QDs are presented in Fig. 6.

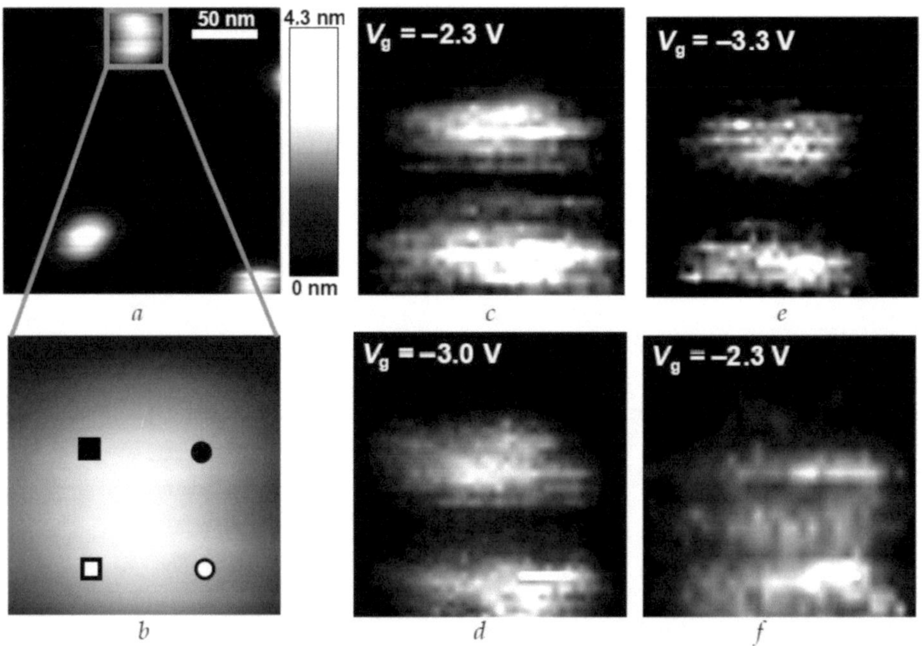

Fig. 6. The UHV AFM (*a*, *b*) and the probe current (*c* – *f*) images of the laterally coupled surface InAs/GaAs(001) QDs. The symbols in fig. (*b*) mark the points of the measurement of the respective tunnel spectra presented in Fig. 7.

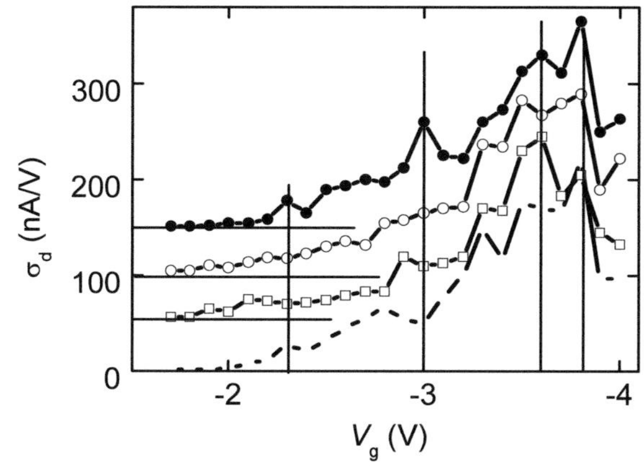

Fig. 7. The differential conductivity spectra $\sigma_d(V_g)$ of the tunnel contact of a Pt coated AFM tip to the laterally coupled InAs/n-GaAs/n^+-GaAs(001) QDs. The points of measurement of the initial $I-V$ curves are denoted in Fig. 6, *b* by the respective symbols.

The differential conductivity spectra $\sigma_d(V_g)$ of the tunnel contact of a Pt coated AFM tip to the laterally coupled InAs/n-GaAs/n^+-GaAs(001) QDs are presented in Fig. 7. The asymmetry of the current images in Fig. 6, $c-f$ (as compared to the ones of the single QDs, see Fig. 5, $b-d$), which had varied with increasing V_g along with the splitting of the peaks related to the quantum confined states in the coupled QDs (Fig. 7) were attributed to the hybridization of the quantum confined states in the laterally coupled QDs. Similar patterns of the tunnel current images and tunnel spectra as well as of their dynamics with varying V_g have been observed while studying the hybridization of the quantum confined states in the GaSb/InAs(001) double symmetric QWs by X-STM in UHV (Suzuki et al., 2007).

2.3 The valence band states

In this subsection, the results of the Tunnelling AFM studies of the hole quantum confined states in the InAs/GaAs(001) QDs are presented. It is worth noting that the authors of the present chapter had applied Scanning Probe Microscopy technique for the studying of the valence band states in the InAs/GaAs(001) self assembled QDs for the first time.

The InAs/GaAs(001) QD samples for the investigation of the valence band states in the InAs QDs by Tunneling AFM have been grown on the p^+-GaAs(001) substrates. The GaAs buffer layers were also doped heavily by Zn from diethylzinc up to the acceptor concentration $N_A \sim 10^{18}$ cm^{-3}. The intentionally undoped 3 nm thick GaAs spacer layers were grown prior to the deposition of InAs, as in the InAs/GaAs/n^+-GaAs(001) QD structures for the investigations of the electron states described in the previous subsection. The technological parameters of the process of growing the InAs QDs were the same, as in the case of the structures grown on the n^+-GaAs substrates: T_g = 530°C, d_{InAs} = 5 ML.

The differential conductivity spectra $\sigma_d(V_g)$ of the tunnel contact of a Pt coated AFM tip to an InAs QD on the GaAs/p^+-GaAs (001) substrate are presented in Fig. 8, *a*. The oscillations of the $\sigma_d(V_g)$ spectra have been attributed to the tunneling from the valence band states in the p^+-GaAs buffer to the free states above the Fermi level in the metal tip coating through the quantum confined hole states in the InAs QD (Fig 8, *b*).

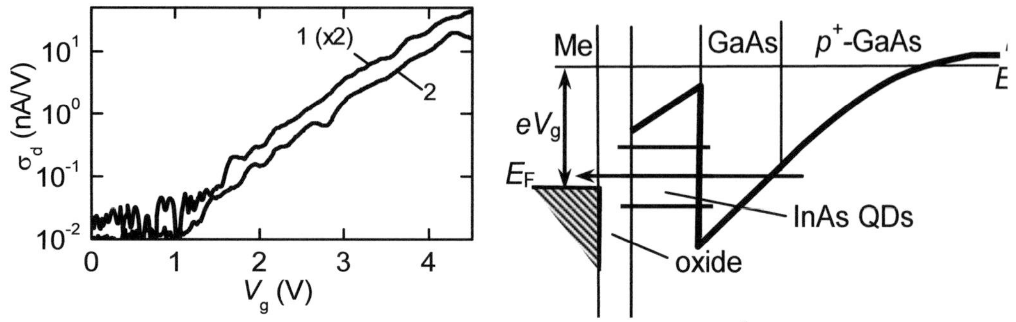

Fig. 8. The differential conductivity spectra $\sigma_d(V_g)$ (*a*) and the qualitative band diagram (*b*) of the positively biased contact of a Pt coated AFM tip to an InAs QD on the GaAs/p^+-GaAs (001) substrate. The curve numbers denote the points of the initial $I - V$ curves' measurements shown in Fig 9, *a*. Reproduced partly from (Filatov et al., 2010) under license from ©IoP Publishing, Ltd.

Comparing the $\sigma_d(V_g)$ spectra of the QDs on the p^+-GaAs (Fig. 8, *a*) with the ones of the QDs grown on n^+-GaAs (Fig 4, *a*), one can note that the peaks related to the tunnelling via the quantum confined electron states in the QDs were well resolved in the case of the InAs/n+-GaAs QDs while a nearly exponential curves with the weak oscillations only have been observed in the InAs/p^+-GaAs QDs(Fig. 8, *a*). This observation could be explained noting that according to (Stier et al., 1999), the energy spacing between the electron levels in the InAs/GaAs(001) QDs with the base size length $b = 15 \div 20$ nm is ~100 meV while the one for the hole levels is $10 \div 20$ meV only. As a result, the peaks in the $\sigma_d(V_g)$ spectra related to the quatum confined hole states in the InAs QDs were resolved poorly because of the thermal and structural broadening. Note also that the spectral spacing between the peaks in the $\sigma_d(V_g)$ spectra ascribed to the quantum confined hole states in Fig. 8, *a* was much less than the one for the conduction band states (Fig. 4, *a*), that agrees with proposed interpretation of the QDs' tunnel spectra as well.

The AFM and the probe current images of an individual InAs QD on the GaAs/p^+-GaAs(001) substrate are presented in Fig. 9. The probe current images $I_t(x, y)$ of the QD measured at different values of V_g were related to the lateral spatial distribution of the LDOS of the quantum confined hole states in the QD $\rho_E(x, y)$ at the respective vaules of E. In the case of the InAs QDs grown on the GaAs/p^+-GaAs(001) substrate, the hole quantum confined states, the energy of which satisfies the following condition:

$$E_{Fs} - eV_g < E_{n_1 n_2 n_3} < E_{Fs} , \tag{3}$$

where E_{Fs} is the Fermi level energy in the p^+-GaAs buffer (see the band sketch in Fig. 8, *a*) can manifest themselves in the probe current images $I_t(x, y)$.

Again, the maps of the probe current $I_t(x, y)$ have been compared to the equal probability density patterns $|\chi(x, y, z)|^2 = const$ calculated for several lower quantum confined hole states (Stier et al., 1999). Although the quality of the probe current images was not so good, several quantum confined states have been identified (Fig. 9).

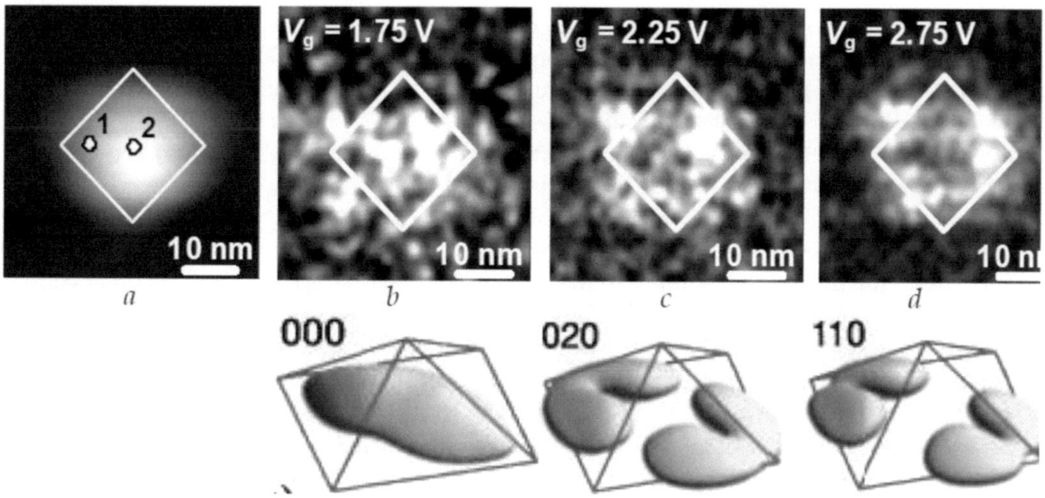

Fig. 9. The AFM (*a*) and the probe current (*b* – *d*) images of an InAs QD on the GaAs/p^+-GaAs(001) substrate. Reproduced partly from (Filatov et al., 2010) under license from ©IoP Publishing, Ltd. Below the surfaces of equal probability density $|\chi(x, y, z)|^2 = 0.65$ calculated for several quantum confined hole states $|n_1n_2n_3>$ in a pyramidal InAs/GaAs(001) QD with the base side length b = 16 nm are presented. Reproduced partly from (Stier et al., 1999) with permission from ©American Physical Society.

Having identified the quantum confined electron and hole states in the InAs/GaAs(001) QDs grown by AP MOVPE, we succeeded to identify the states the interband optical transition between which are manifested in the PS spectra of the QD structures grown by AP MOVPE.

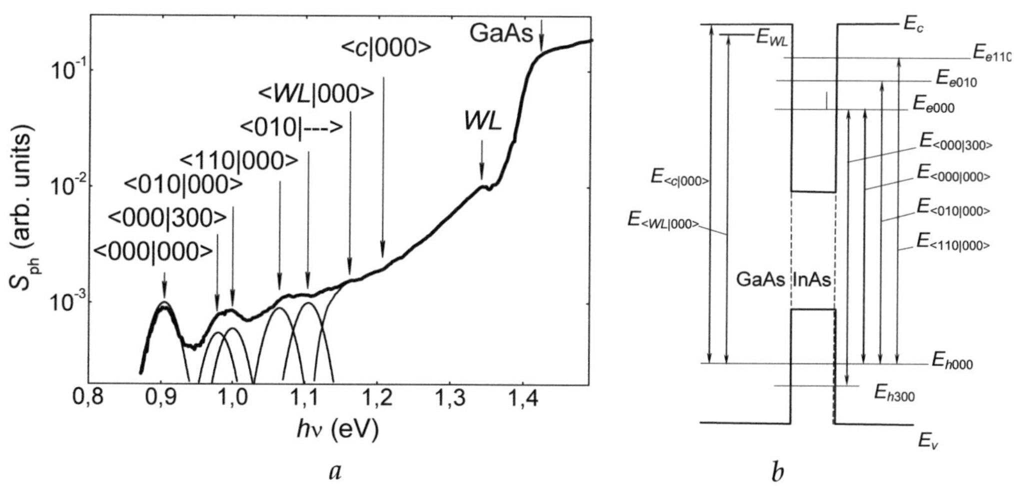

Fig. 10. The photosensitivity spectrum (*a*) and the band diagram (*b*) (300K) of an InAs/GaAs(001) QD structure grown by AP MOVPE. Reproduced from (Filatov et al., 2010) with permission from ©Pleiades Publishing Ltd.

An example of the PS spectrum $S_{ph}(h\nu)$ of an InAs/GaAs(001) QD structure measured by the photovoltage spectroscopy in a liquid electrolyte (Karpovich et al., 2004b) is presented in Fig. 10, *a*. The QDs in this particular structure were grown in the same conditions as in the samples for the Tunnelling AFM investigstions described above. Several peaks related to the interband optical transitions between the quantum confined electron and hole states in the QDs are present in the PS spectrum. Also, the PS bands related to the transitions from the ground hole state in the QD |000> to the ground electron states in the InAs wetting layer |WL> and to the conduction band states in GaAs have been observed. The band diagram of an InAs/GaAs QD structure with the transitions manifested in the PS spectrum (Fig. 10, *a*) shown schematically is presented in Fig. 10, *b*. This diagram was based on the best fit between the interband transition energies extracted from the PS spectrum in Fig. 10, *a* and the ones calculated from the data on the quantum confined level energies in the pyramidal InAs/GaAs(001) QDs as the function of the QD base size *b* (Stier et al., 1999). The best fit was found at $b \approx 16$ nm that is consistent with the ambient air AFM data obtained using the deconvolution (see Sec. 2.1). Note that many possible transitions are not manifested in the PS spectrum, because the overlap integrals between the envelopes of the respective electron and hole states are close to zero (Stier et al., 1999). Only the transitions between the states the overlap integrals for which are close to unity are manifested in the PS spectrum.

We have failed to identify one of the higher energy hole states marked as |---> in Fig. 10 because the data for the respective hole energy band are not presented in the publication by (Stier et al,. 1999). In the other aspects, the data on the morphology, on the electronic structure, and on the energy spectrum of the InAs/GaAs(001)QDs grown by AP MOVPE provided by ambient air and Tunnelling AFM, by the PL spectroscopy, and by the PS one appeared to be consistent with each other as well as with the results of the theoretical calculations reported in the literature (Stier et al., 1999).

3. InGaAs/GaAs(001) quantum rings

3.1 Growth and characterization

In this subsection, the details on the growth of the InAsGa/GaAs(001) QRs by AP MOVPE as well as of their characterization are presented.

The procedure of growing the self-assembled InAsGa/GaAs(001) QRs by AP MOVPE was different from that used in the standard MBE. Usually, in order to grow the InGaAs/GaAs(001) QRs by standard MBE, the InAs/GaAs(001) QDs are capped by a thin GaAs cladding layer with $d_c \approx \langle h \rangle$, and then the structures are annealed at the temperature $T_A \approx 600 \div 630$°C for several tens of minutes (Lorke et al., 2002). However, the QRs grown by AP MOVPE have been proven to form just during capping of the InAs QDs by the GaAs cladding layer (Baidus' et al., 2000). This phenomenon had been attributed to the enhanced surface diffusion of In adatoms in the AP MOVPE process as compared to MBE because in the former case the surface dangling bonds are passivated by hydrogen residual from the cracking of the metal organic compounds and/or arsine.

The InGaAs/GaAs(001) QR structures for the Tunnelling AFM investigations described in the present chapter have been grown on the n^+-GaAs(001) substrates misoriented by 5° towards <110>.

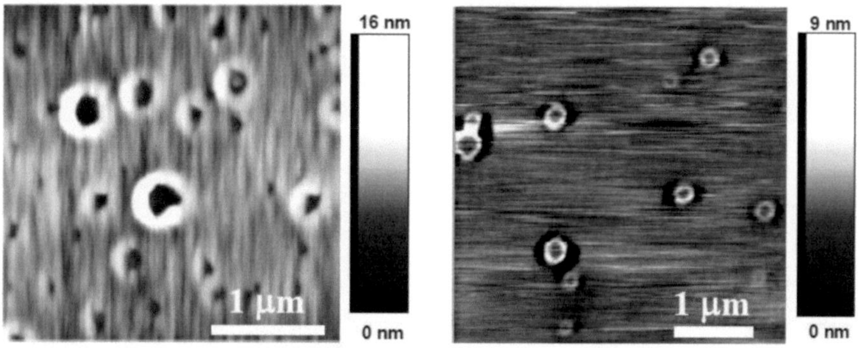

a *b*

Fig. 11. The ambient air AFM image of an InGaAs/GaAs(001) QR structure (*a*); a liquid AFM image of an InGaAs/GaAs(001) QR structure with a 30 nm thick GaAs cladding layer removed by selective wet etching. Reproduced from (Filatov et al., 2011) with permission from ©Pleiades Publishing, Ltd.

The 3 nm thick intentionally undoped GaAs spacer layers have been grown on the n^+-GaAs buffer layers. Then, the InAs/GaAs QDs were grown at T_g = 530°C and capped by a GaAs cladding layer with the thickness d_c = 10 nm at T_g = 600°C. The nominal thickness of the InAs layerd_{InAs} was ≈ 5 ML. The ambient air AFM measurements have demonstrated the InGaAs QRs to form during capping the InAs QDs at elevated temperatures (Fig. 11, *a*). The QRs of various diameters D = 150 ÷ 400 nm have been observed on surface (Fig. 11, *a*). As it had been found earlier (Karpovich et al., 2004b), the structures with the surface InAs/GaAs(001) QDs grown by AP MOVPE are features by a number of the relaxed InGaAs clusters present on the surface (Fig. 12, *b*). These clusters are formed via the coalescence of the smaller InAs coherent QDs during growth, which, in turn, has been attributed to the enhanced surface diffusion in the AP MOVPE process. Within this paradigm, the formation of the smaller QRs has been attributed to the transformation of the smaller coherent InAs/GaAs QDs during capping while the formation of the larger QRs has been attributed to the transformation of the larger relaxed InGaAs clusters.

Also, the structure with d_c ≈ 30 nm was grown on a semi insulating GaAs(001) substrate in order to examine the optical properties of the QRs by PL spectroscopy. In order to reveal the morphology of the overgrown QRs, we have employed the removal of the GaAs cladding layer by wet selective etching followed by AFM investigation in liquid (Karpovich et al., 2004a). The application of liquid AFM was to avoid the probe fadeouts due to the residual etching solution contamination (Fig. 11, *b*).

The PL (77 K) and PS (300 K) spectra of the structures with the thickness of the GaAs cladding layer d_c ≈ 30 and 10 nm, respectively, are presented in Fig. 12, *a*. A peak of the edge PL of GaAs with the maximum energy $h\nu_m$ ≈ 1.51 eV (GaAs) and a peak with $h\nu_m$ ≈ 1.41 eV (WL) originating from the interband radiative transitions between the ground quantum confined states in the InAs/GaAs(001) wetting layer are present in the PL spectrum. Also, a broad PL band with $h\nu_m$ ≈ 1.2 eV attributed to the ground state transitions in the InGaAs/GaAs(001) QRs have been observed. This band has a longer wavelength shoulder and can be decomposed into two Gaussian components with $h\nu_m$ ≈ 1.2 and $h\nu_m$ ≈ 1.1 eV related to the smaller and larger QRs, respectively.

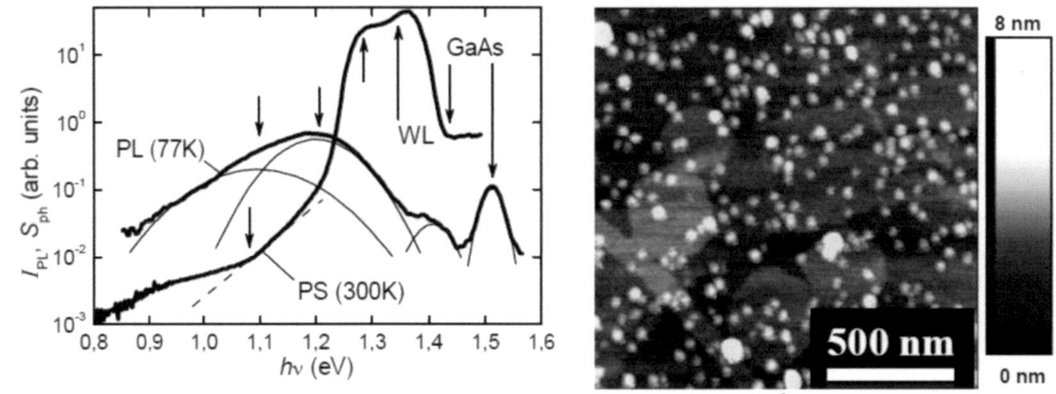

a *b*

Fig. 12. The PL (77 K) and photosensitivity (300K) spectra of the InGaAs/GaAs(001) QR structures with the 30 nm and 10 nm cladding layers, respectively (*a*). Reproduced from (Filatov et al., 2011) with permission from ©Pleiades Publishing Ltd.; an AFM image of a surface InAs/GaAs(001) QDs (*b*).

Also, the PS bands with the edges at $h\nu \approx 1.36$ eV (WL), 1.29 eV, and 1.09 eV corresponding to the above PL peaks have been observed in the PS spectrum (Fig. 12, *a*). The difference in the values of the PL peak energies and the band edges ones in the PS spectrum could be related to the different in d_c, and, in turn, to the difference in the morphology, in the composition, and in the elastic strain of the QR material in these structures. Note that since the structure was photoexcited through the substrate, playing a role of a low-pass optical filter, the PS spectrum was truncated at $h\nu \approx 1.43$ eV that is the optical gap energy of n^{+}-GaAs. The PS band with the edge at $h\nu \approx 0.8$ eV was related, most likely, to the impurity PS from the defect complexes in GaAs (Karpovich et al., 2004b).

The In molar fraction x in the QR material ($In_xGa_{1-x}As$) was estimated from the PL and PS spectra to be $0.35 \div 0.4$. The estimates were made by the best fit between the energies of the PL peak and of the PS band edges related to the ground state transitions in the QRs and the calculated values of the ground state transitions in the QRs E_0, x being the fitting parameter. The values of E_0 as a function of the QRs' sizes determined form the AFM images and of x were calculated by solving the Schrödinger's equation in the effective mass approximation for a model potential of a flat gasket-shaped ring (rectangular in the cross-section) with potential wells of finite height (Barticevic et al., 2000). The ring material was assumed to be pseudomorphic to the GaAs matrix. The materials parameters (the effective masses of electron and holes masses, the band offsets, etc.) were taken from (Stier et al., 1999).

3.2 Tunneling AFM investigations

In this subsection, the results of the investigations of the electronic structure of the self assembled InGaAs/GaAs(001) QRs by Tunnelling AFM are presented. Again, it is worth noting that the authors of the present chapter had applied Scanning Probe Microscopy technique for the studying of the LDOS in the self assembled semiconductor QRs for the first time.

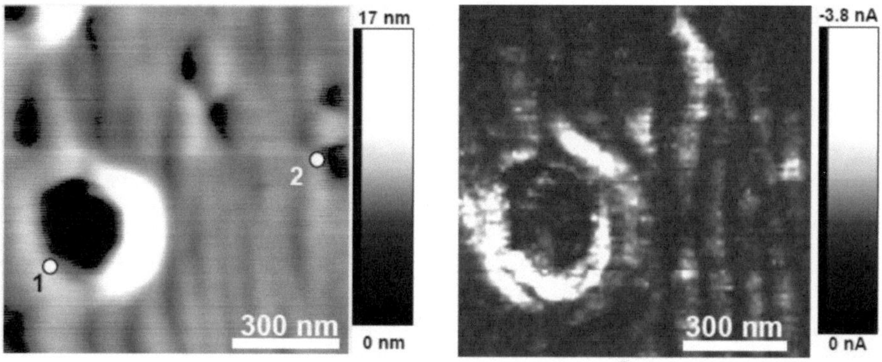

a *b*

Fig. 13. The UHV AFM (*a*) and the probe current (*b*) images of an InGaAs/GaAs(001) QR structure. V_g = -1.37 V. The figures (1) and (2) mark the points of measurement of the $I-V$ curves (the respective $\sigma_d(V_g)$ spectra are presented in Fig. 14). Reprinted from (Filatov et al., 2010) under license by IoP Publishing Ltd.

The UHV AFM and the probe current images of an InGaAs/GaAs(001) QR structure are presented in Fig. 13. The QRs are manifested in the probe current image $I_t(x, y)$ by the increased probe current I_t that was related to the tunnelling of the electrons from the probe coating material into the n^+-GaAs buffer through the quantum confined states in the InGaAs/GaAs(001) QRs (Fig 4, *b*). The differential conductivity spectra $\sigma_d(V_g)$ of the contact of a Pt coated AFM tip to the surface of the InGaAs QRs on the n-GaAs/n^+-GaAs (001) substrate are presented in Fig. 14. The peaks related to the quantum confined states in the QRs are well expressed in the $\sigma_d(V_g)$ spectrum of the smaller QR (Fig. 14, *b*). However, the features, which could be related to the quantum size effect are present in the $\sigma_d(V_g)$ spectrum of the larger QR as well (Fig. 14, *a*). So far, the larger QRs could be classified as the quantum size structures as well in spite of their relatively large sizes. The calculations have shown that the size quantization in z direction (normal to the substrate) affects the energy spectrum of the electrons and holes in QRs most.

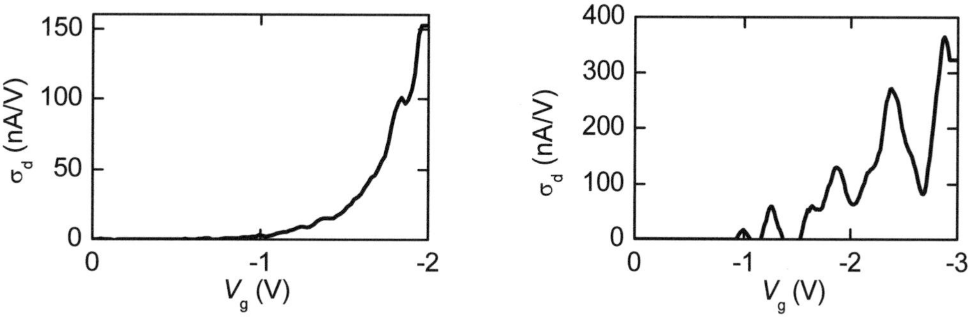

a *b*

Fig. 14. The differential conductivity spectra $\sigma_d(V_g)$ of the contact of a Pt coated AFM tip to the surface of the InGaAs QRs of larger (*a*) and smaller (*b*) size on the n-GaAs/n^+-GaAs (001) substrate. The points of measurement of the initial $I-V$ curves are marked in Fig. 13, *a*. Reproduced from (Filatov et al., 2011) with permission from ©Pleiades Publishing, Ltd.

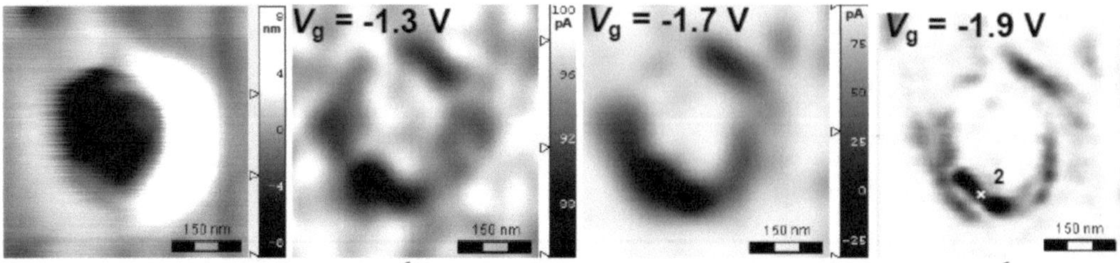

Fig. 15. The AFM (*a*) and the inverted probe current (*b* – *d*) images of an InGaAs QR on the *n*-GaAs/*n*+-GaAs(001) substrate. Reproduced partly from (Filatov et al., 2011) with permission from ©Pleiades Publishing, Ltd.

The AFM images and the inverted probe current images of an individual InGaAs/GaAs(001) QR measured at different values of V_g are presented in Fig. 15. The probe current images of the QR change with increasing V_g. The most important result of the studies presented in this section was the observation of the angular patterning in the probe current images of the QRs. Ideally, in the circularly symmetric potential (Curie group of symmetry C_∞), the angular dependence of the envelope eigenfuntions of the quantum confined states is expressed in the polar coordinates (r, φ) by the term $\exp(-il\varphi)$ where i is the imaginary unity and $l = 0, \pm 1, \ldots$ is the angular quantum number. As $|\exp(-il\varphi)|^2 = 1$ for any l and φ, no angular dependence of I_t should be observed.

The observation of the angular patterning in the current images of the QRs has been attributed to the asymmetry of the QRs' shape. In addition, one should take into account the effect of the piezoelectric field in the strained InGaAs, which also reduces the potential symmetry from C_∞ downto C_{2v} even in a perfectly round QR (Stier et al., 1999).

In order to account for the effect of the effect of the strain-induced piezoelectric field on the angular pattern of the LDOS in the InGaAs/GaAs(001) QRs, we have applied a simple model in the framework of the first-order perturbation theory. The lateral part of the spin-independent eigenfunction for the model flat gasket finite well height potential (Barticevic et al., 2000) can be written in the form:

$$\chi_{\mathrm{ml}}(\mathrm{r},\varphi) = \chi_{\mathrm{ml}}(\mathrm{r})\exp(-\mathrm{i}l\varphi), \tag{4}$$

where $\chi_{ml}(r)$ is the radial part of $\chi_{ml}(r, \varphi)$ expressed via the Bessel functions and m is the radial quantum number. As the perturbation piezoelectric potential has the C_{2v} symmetry, an additional integral of motion, parity of states, appears. So, one can write the angular components of the correct envelopes of the zero-order approximation as

$$\begin{aligned} \chi_0^{(0)}(\varphi) &= \frac{1}{\sqrt{2\pi}} \\ \chi_{\mu+}^{(0)}(\varphi) &= \frac{1}{\sqrt{\pi}}\cos(\mu\varphi), \\ \chi_{\mu-}^{(0)}(\varphi) &= \frac{1}{\sqrt{\pi}}\sin(\mu\varphi) \end{aligned} \tag{5}$$

where $\mu = |l| = 1, 2, \ldots, |m-1|$.

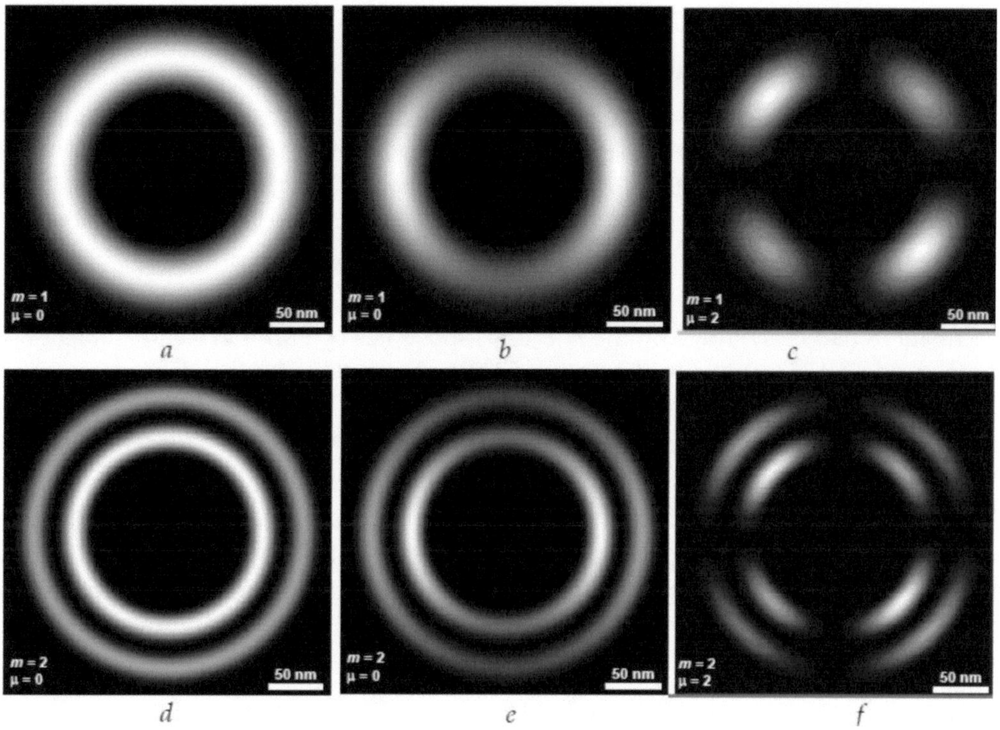

Fig. 16. The model maps of the spatial distribution of the lateral component of the probability density of the envelope wavefunction of the quantum confined states $|\chi_{nl}(r, \varphi)|^2$ in the plane of an $In_{0.35}Ga_{0.65}As$/GaAs(001) for different values of m and μ. (*a, d*): no perturbation potential; (*b, c, e, f*): the perturbation potential $V_0\cos(2\varphi)$ is imposed. Reproduced from (Filatov et al., 2011) with permission from ©Pleiades Publishing Ltd.

The model perturbation potential to account for the effect of the piezoelectric field has been selected in the form $V(\varphi)= V_0\cos(2\varphi)$. Such a model potential satisfies the symmetry C_{2v} and is convenient for the calculations of the matrix elements which appeared to be equal to zero except the $\langle ml \mid ml\pm2\rangle$ ones.

The model maps of $|\chi_{nl}(r, \varphi)|^2$ calculated for the $In_{0.35}Ga_{0.65}As$/GaAs(001) QRs with the outer and the inner radii of 70 and 140 nm, respectively, are presented in Fig. 16. When no perturbation potential is imposed, no angular patterning of the probability density maps takes place (Fig. 16, *a*&*d*). However, the model $|\chi_{nl}(r, \varphi)|^2$ maps calculated in the first order of the perturbation theory for V_0 = 5 meV demonstrate well expressed angular patterning. Note that the probability density images keep the C_{2v} symmetry regardless to the number of knots in the circle, in accordance with Curie's theorem.

Comparing the calculated $|\chi_{nl}(r, \varphi)|^2$ patterns with the the experimental probe current images of the QRs (cf, for example, Fig. 16, *c* & Fig. 15, *b*; Fig. 16, *f* & Fig. 15, *d*) demonstrate that in general the model presented above describes the qualitative features of the experimental current images more or less accurately. In turn, this could be considered as an evidence for the asymmetry of the quantum confining potential in the QRs as the origin of the angular patterning of these ones.

4. GeSi/Si(001) nanoislands

4.1 Growth and characterization

In this section, the results of the investigation of the LDOS in the self-assembled GeSi/Si(001) self assembled nanoislands by Tunnelling AFM are presented. Again, it should be stressed here that the authors of the present chapter had applied Scamming Probe Microscopy to the investigation of the LDOS in the self-assembled GeSi/Si(001) nanoislands for the first time. (Borodin et al., 2011).

The heterostructures with the surface GeSi/Si(001) nanoislands for the Tunnelling AFM investigations had been grown on the p^+-Si(001) substrates by SMBE in GeH_4 ambient using a home-made UHV setup (Svetlov et al., 2001). The qualitative band diagrams of the structures for the Tunnelling AFM investigations are presented in Fig. 19, *b* & *c*. The buffer layers of ≈ 200 nm in thickness were doped heavily by boron up to the hole concentration ~ 10^{18} cm^{-3}. More details on the growth technique can be found elsewhere (Filatov et al., 2008a, 2008b).

The topography of the samples destined to the Tunnelling AFM investigations was examined first by ambient air AFM. The AFM images of the structures with the surface GeSi/Si(001) nanoislands grown at various growth temperatures T_g are presented in Fig. 17. A dense array of the pyramid-shaped nanoislands ($N_s \sim 10^{11}$ cm^{-2}) has been observed on the surface of the sample grown at T_g = 500°C (Fig. 17, *a*). The (105) faceting of the pyramid nanoislands was resolved poorly, again, due to the convolution effect. A bimodal size distribution of the nanoislands took place on the surface of the sample grown at T_g = 600°C (Fig. 17, *b*). Along with the dome-shaped GeSi islands, a small number of larger so called super dome islands (Kamins et al., 1999) has been observed on the surface. The super dome islands originate from the coalescence of the smaller dome ones. It should be noted that in the structures grown by MBE the formation of the super dome islands has been observed in the process of the post-growth annealing (Kamins et al, 1999). In contrary, in SMBE in GeH_4 ambient the formation of the super dome islands has been observed just during growth (Filatov et al., 2008a, 2008b), that was attributed to Ostwald ripening.

The surface density of the islands N_s decreased and their sizes increased with increasing T_g while the bimodal size distribution remained in the structures grown at T_g = 700°C and 800°C (Fig. 17, *c* & *d*), in accordance with Lifshits-Sloyzov-Wagner theory.

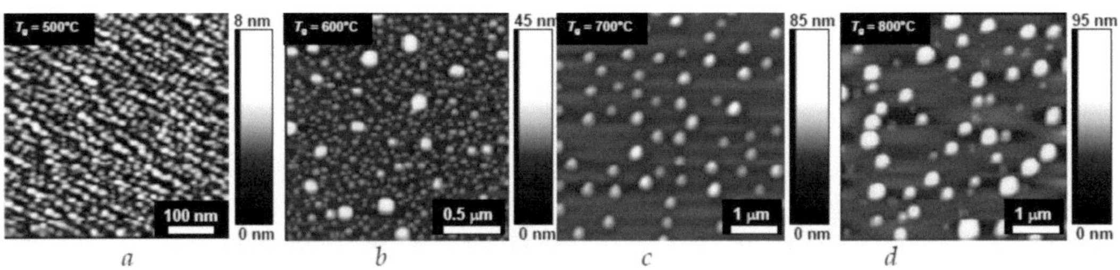

Fig. 17. The ambient air AFM images of the surface GeSi/Si(001) nanoislands grown by SMBE in GeH_4 ambient at various growth temperatures T_g.

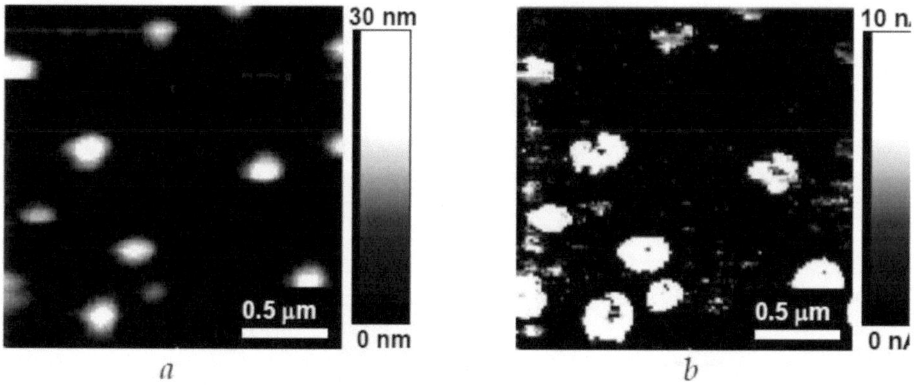

Fig. 18. The UHV AFM (*a*) and the probe current (*b*) images of the GeSi/Si(001) nanoislands. V_g = 2.0 V. Reproduced from (Borodin et al., 2011) with permission from ©Pleiades Publishing, Ltd.

The composition of the GeSi/Si(001) nanoislands' material has been examined by PL spectroscopy (Filatov et al., 2008a, 2008b) and by Confocal Raman Microscopy (CRM, Mashin et al., 2010). The GeSi/Si(001) structures grown on the *p*-Si substrates with the nanoislands grown in the same conditions as the surface ones destined to the Tunnelling AFM studies but capped with the 40 nm thick cladding Si layers had been used for the optical investigations. It has been found that the nanoislands grown within T_g = 600 ÷ 800°C consisted of Ge_xSi_{1-x} alloy with the Ge molar fraction x decreasing from ≈0.55 downto ≈ 0.25 with T_g increasing from 600°C up to 800°C (Filatov et al., 2008b, Mashin et al., 2010).

4.2 Tunneling AFM investigations

The UHV AFM and the probe current images of the surface GeSi/Si(001) nanoislands grown at T_g = 700°C are presented in Fig. 18, *a* and *b*, respectively. The GeSi islands of various sizes and shapes have been observed on the sample surface (Fig. 18, *a*). The smaller islands were dome shaped while the larger ones (super dome islands) were shaped as the truncated pyramids defined by the (101) facets.

The spots of increased probe current I_t in the probe current image (Fig. 18, *b*) were located at the places corresponding to GeSi islands. The increased probe current has been related to the tunnelling of the electrons from the electronic states in the valence band of the GeSi islands into the free states above the Fermi level E_F in the Pt AFM tip coating through the native oxide layer covering the island surface. The band diagram of a positively biased contact between the Pt coated AFM tip and a surface GeSi/*p*-Si/*p*+-Si island is shown in Fig. 19, *c*. It should be noted that the sizes of the spots of increased I_t in the probe current image (Fig. 18, *b*) are larger than the sizes of the topographic images of the respective islands in Fig. 18, *a*. This has been attributed to the convolution effect due to the relatively large radius of the curvature of the AFM tip R_p. Typical values of R_p for NT MDT® NSG-01 Pt coated AFM probes are ≈ 35 nm, according to the vendor's specifications.

The differential conductivity spectra $\sigma_d(V_g)$ of the contact of a Pt coated AFM tip to the structure with the surface pyramid Ge/Si(001) nanoislands grown at T_g = 500°C are presented in Fig. 19, *a*.

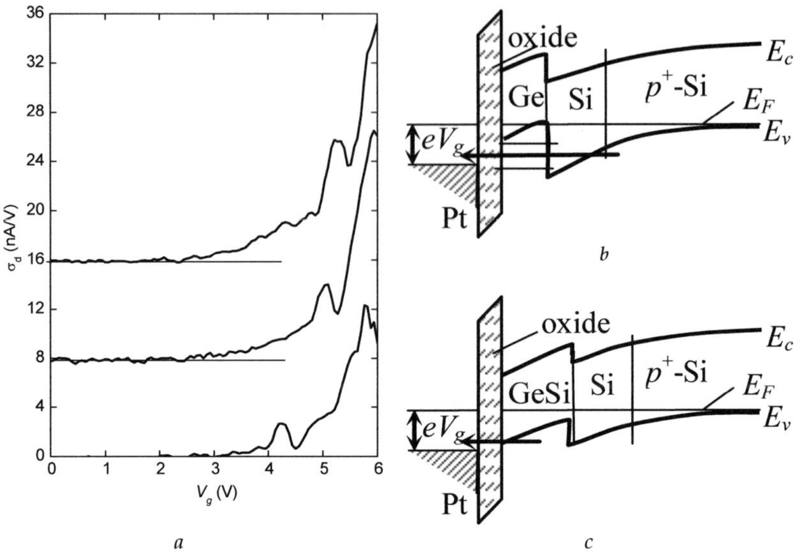

Fig. 19. The differential conductivity spectra $\sigma_d(V_g)$ of the contact of a Pt coated AFM tip to the structure with the surface pyramid Ge/Si(001) nanoislands grown at T_g = 500°C (*a*), the qualitative band diagrams of the positively biased contact s of the Pt coated AFM tip to the pyramid islands (*b*) and to the dome-shaped or the super-dome islands (*c*).

Again, as in the cases of the InAs/GaAs(001) QDs and of the InGaAs/GaAs(001) QRs (see Sec. 2 & 3 above), the differential conductivity spectra $\sigma_d(V_g)$ have been calculated from the measured $I_t(V_g)$ curves by numerical differentiation with non-linear smoothing. The spectra presented in Fig. 19, *a* were the results of the averaging of 25 $I_t(V_g)$ curves measured in the different areas on the sample surface 5 × 5 pixels by size. The peaks in the $\sigma_d(V_g)$ spectra were ascribed to the tunneling of the electrons from the valence band states in the p^+-Si buffer through the quantum confined hole states in the Ge/Si(001) pyramid islands into the free states above the Fermi level in the Pt AFM tip coating (Fig. 19, *b*).

We have failed to obtain the probe current images, which could be associated to the probability density patterns $|\chi(x, y)^2|$ of the quantum confined hole states in the pyramid Ge/Si(001) nanoislands. The most probable cause for this, in our opinion, were too small sizes (b = 50 ÷ 80 nm and h = 5 ÷ 8 nm) and too large surface density ($N_s \sim 10^{11}$ cm^{-2}) of the Ge nanoislands on the surface of this particular sample as compared to the AFM probe tip dimensions ($R_p \approx 35$ nm). As a result, the islands have been resolved poorly in the UHV AFM images due to the convolution artifact unlike the ambient air AFM (Fig. 17, *a*) measured with NT MDT® CSG-01 probes without metal coating ($R_p < 10$ nm, according to the vendor's specifications).

The differential conductivity spectra $\sigma_d(V_g) = dI_t/dV_g$ and the normalized differential conductivity spectra $(dI_t/dV_g)/(I_t/V_g)$ of the contact of the Pt coated AFM probe tip to the GeSi/Si(001) nanoislands grown at various temperatures and to the sample surface between the islands (WL) are presented in Fig. 20. The points on the sample surface where the initial $I_t(V_g)$ curves have been measured are marked in Fig. 21. The spectra presented in Fig. 20 were the results of the averaging of 25 $I_t(V_g)$ curves measured in the spots on the sample surface 5 × 5 pixels by size.

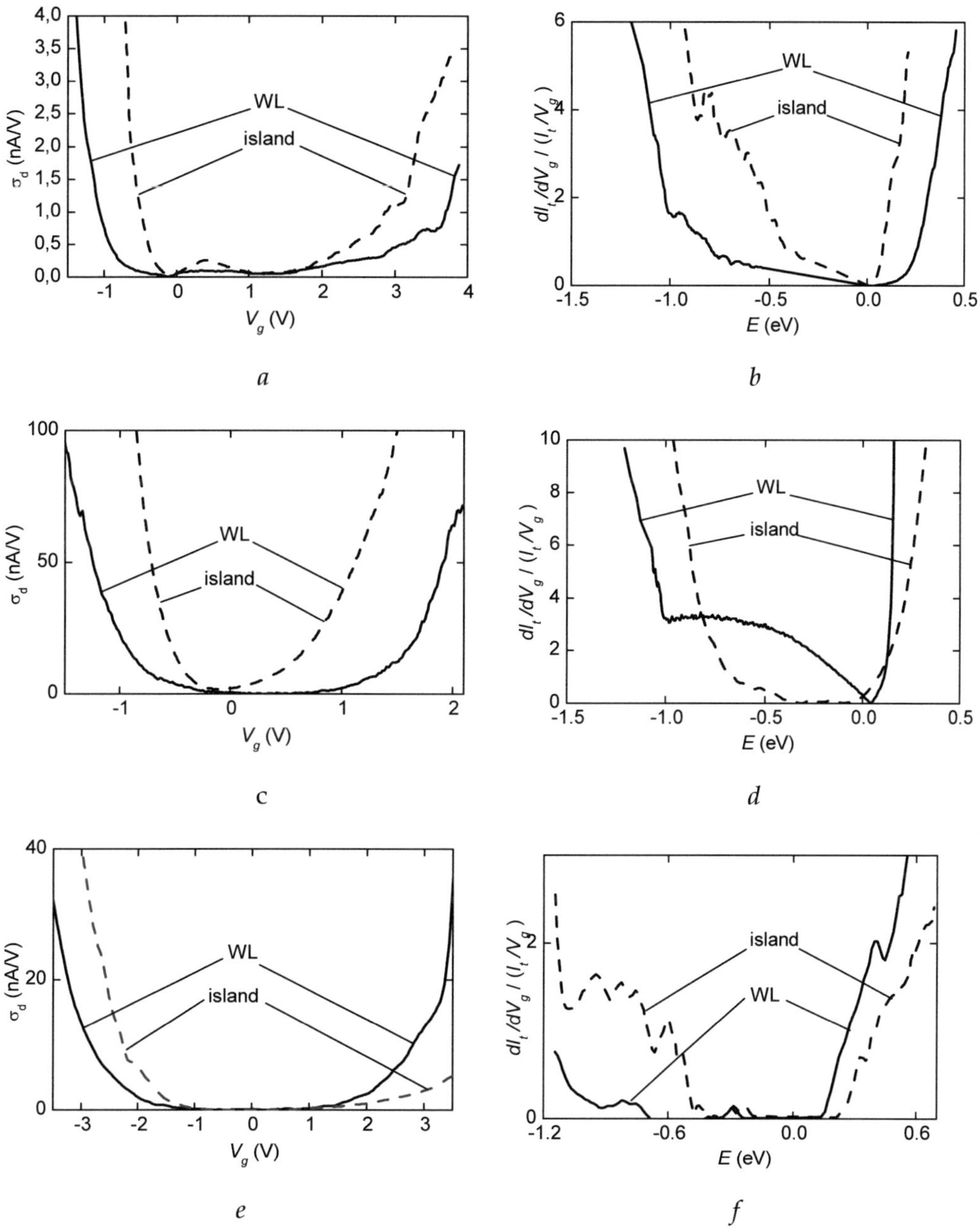

Fig. 20. The differential conductivity spectra $\sigma_d(V_g)$ (*a, c, e*) and the normalized differential conductivity spectra $(dI_t/dV_g)/(I_t/V_g)$ (*b, d, f*) of the contact of the Pt coated AFM probe tip to the GeSi/Si(001) nanoislands and to the sample surface between the islands (WL). The points on the sample surface where the initial $I_t(V_g)$ curves have been measured are denoted in Fig. 21. Reproduced partly from (Borodin et al., 2011) with permission from ©Pleiades Publishing, Ltd.

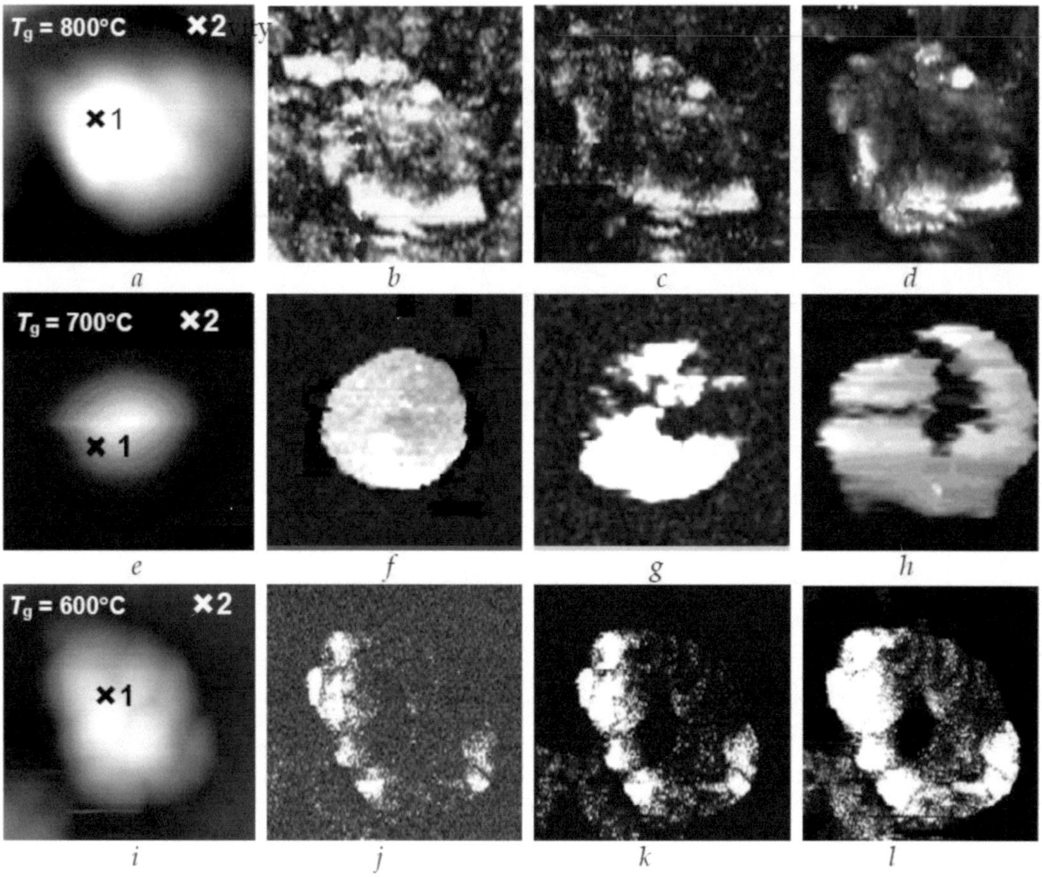

Fig. 21. The AFM (*a, e, i*) and the runnel current (*b – d, f – h, j – l*) images of the individual GeSi/Si(001) nanoislands grown by SMBE in GeH_4 ambient at various growth temperatures T_g. The points of measurements of the $I_t(V_g)$ curves the derivatives of which are presented in Fig. 20 are denoted in the AFM images (*a, e, i*). Reproduced partly from (Borodin et al., 2011) with permission from ©Pleiades Publishing, Ltd.

According to the theory of the tunnel spectroscopy, the normalized differential conductivity spectra $(dI_t/dV_g)/(I_t/V_g)$ of the contact of a metal STM tip to a sample is proportional to the LDOS at the sample surface. The relationship of the energies in the LDOS spectra to V_g, again, has been established taking into account the partial drop of V_g on the depletion layer of the tip-to-sample contact on the base of one-dimensional Poisson's equation (Feenstra & Stroscio, 1987). Unlike the case of the InAs/GaAs(001) QDs considered above in Sec. 2, the charge on the surface states on the boundary of the semiconductor with the native oxide has been neglected because the SiO_2/Si and GeO_2/Ge interfaces are known to be featured by low density of the surface states.

As it is evident from Fig. 20, tunnel spectra of the GeSi/Si(001) nanoislands grown at T_g = 700 and 800°C demonstrate the I-type conduction band alignment. Traditionally, the GeSi/Si(001) heterostructures had been being considered to be the II-type ones, i. e. the GeSi layer had been being treated as a potential barrier for the electrons with respect to the ones in Si (Berbezier & Ronda, 2009). However, there has been an increasing number of publications, both theoretical and experimental, where the Ge_xSi_{1-x}/Si(001) heterostructures have been reported to be of the I type ones at low enough values of x, i. e. the GeSi layers were proven to be the potential wells for the electrons. More recent 30-band $k{\cdot}p$ calculations (El Kurdi et al., 2006) have demonstrated the pseudomorphic Ge_xSi_{1-x}/Si(001) heterolayers to be the ones of the I type when $0.05 < x < 0.45$. In contrary, the GeSi/Si(001) nanoislands were reported to be always the II type heterostructures within the whole range of $0 < x < 1$ due to the nonuniform tensile strain of Si near the tops and the bottoms of the islands (El Kurdi et al., 2006). Anyway, the magnitude of the conduction band offset ΔE_c do not exceed several meV.

The $I_t(x, y)$ images of the GeSi/Si(001) islands were found to depend on V_g. At lower V_g corresponding to the extraction of the electrons From the electron states near the top of the valence band in the GeSi islands (see Fig. 19, *b*), the probe current images had more or less round shape (Fig. 21, *b, f, j*). At higher V_g the current image patterns transformed into the ones having a twofold-like symmetry Fig. 21, *c, g, k*). With further increasing of V_g, corresponding to the extraction of the electrons from the electron states near the Si valence band edge, the current images took a 4-fold-like symmetry Fig. 21, *d, h, l*) that was attributed to the elastic strain relaxation at the pyramidal island edges.

The values of ΔE_c at the Ge_xSi_{1-x}/Si(001) nanoislands' heterointerface determined from the tunnel spectra of the individual islands presented in Fig. 20 are plotted *vs* the averaged Ge molar fraction in the nanoisland material <*x*> determined by PL spectroscopy and CRM (Filatov et al., 2008b, Mashin et al., 2010) in Fig. 22, *a*. Also, the theoretical dependence of ΔE_c on <*x*> calculated for a strained Ge_xSi_{1-x}/Si(001) heterosturcture at 300K according to (Aleshkin & Bekin, 1997) is presented in Fig. 22, *a*. The theory predicts the strained Ge_xSi_{1-x}/Si(001) heterostructures to be the ones of the I type when $0 < x < 0.44$, the conduction band minima in the strained GeSi are related to the Δ_4 valleys. The experimental data of the tunnel spectroscopy agree qualitatively with the theory by (Aleshkin & Bekin, 1997) as well as the one by (El Kurdi et al., 2006), although the magnitudes of ΔE_c at the respective values of <*x*> are larger than the ones predicted by the theory.

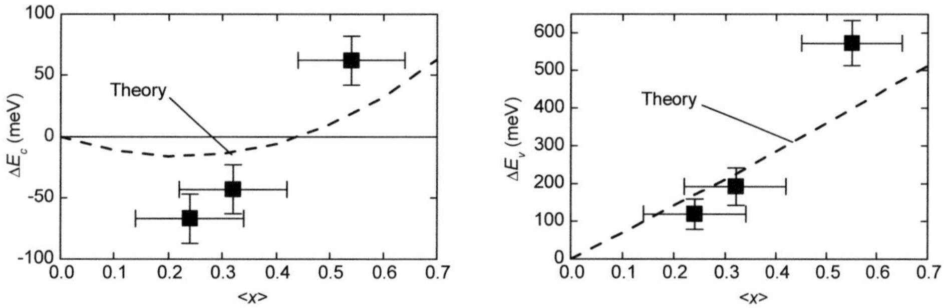

a *b*

Fig. 22. The dependencies of the conduction band offset ΔE_c (*a*) and of the valence band one ΔE_v (*b*) in the surface Ge_xSi_{1-x}/Si(001) nanoislands on the averaged Ge molar fraction in the island material <>. The theoretical curves have been calculated for the strained Ge_xSi_{1-x}/Si(001) layers according to (Aleshkin & Bekin, 1997).

This disagreement could be attributed probably to the partial strain relaxation in the surface GeSi/Si(001) nanoislands as compared to the pseudomorphic GeSi/Si(001) heterostructures or to the coherent GeSi/Si(001) nanoislands embedded into the single crystal Si.

Also, the values of the valance band offsets at the Ge_xSi_{1-x}/Si(001) nanoislands' heterointerface ΔE_v determined from the tunnel spectra presented in Fig. 20 are plotted *vs* $<x>$ along with the ones calculated according to (Aleshkin & Bekin, 1997) in Fig. 22, *b*. A rather good agreement between the calculated and measured values of ΔE_v supports the conclusions about the type of the conduction band alignment in the self assembled GeSi/Si(001) nanoislands derived out from the tunnel spectroscopy data. Note that unlike the tunnel spectra of the Ge/Si(001) pyramid islands (Fig. 19, *a*), the tunnel spectra of the larger GeSi/Si(001) islands with lateral sizes $D > 100$ nm and height $h > 20$ nm (Fig. 20) do not exhibit any features, which could be attributed to the size quantization. This is not surprising taking into account the relatively large sizes of the islands grown at $T_g \geq 600$°C as compared to the de Broglie wavelength for the holes in GeSi alloy at 300K.

5. Conclusion

The results presented in this chapter demonstrate applicability of Tunneling AFM to the *ex situ* investigations of the LDOS of the quantum confined states in the InGaAs/GaAs and GeSi/Si nanostructures covered by the native oxide. The observed patterns of the probe current images agree with the results of the quantum confined states eigenfunction calculations reported in the literature. Besides, the application of Tunnelling AFM technique have brought some fundamental results on the basic properties of various nanostructures. Particularly, the angular patterning of the current images of the InGaAs/GaAs(001) quantum rings was related to the asymmetry of the confining potential. Also, the self assembled Ge_xSi_{1-x}/Si(001) nanoislands have been proven to be the I type heterostructures at $x < 0.45$.

6. Acknowledgement

The authors gratefully acknowledge the financial support from Russian Foundation of Basic Research (RFBR 10-02-90738-mob_st) and from Ministry of Science and Education, Russian Federation (NK-346P-25 and RNP 2.1.1.3615).

7. References

Baidus', N.V., Zvonkov, B.N., Filatov, D.O., Guschina, Yu.Yu., Karpovich, I.A. & Zdoroveischev, A.V. (2000). Investigaton of the overgrowth of the InAs nanoislands in the GaAs/lnAs quantum dot heterostructures grown by Vapor Phase Epitaxy. *J. Surf. Investigations: Physics, Chemistry & Mechanics*, No. 7 (July 2000), pp. 71-75, ISSN 1027-4510.

Barticevic, Z., Pacheco, M. & Latge, A. (2000). Quantum rings under magnetic fields: Electronic and optical properties. *Phys. Rev. B*, Vol. 62, No. 11 (September, 2000), pp. 6963-6966, ISSN 1098-0121.

Berbezier, I. & Ronda, A. (2009). SiGe nanostructures. *Surf. Sci. Rep.*, Vol.64, No. 2 (February 2009), pp. 47-98, ISSN 0167-5729.

Borodin, P.A., Bukharaev, A.A., Filatov, D.O., Isakov, M.A., Shengurov, V.G., Chalkov, V.Yu. & Denisov, Yu.A. (2011). Investigation of the local density of states in self-assembled GeSi/Si(001) nanoislands by Combined Scanning Tunneling and Atomic Force Microscopy. *Semicond.*, Vol.45, No.3 (March 2011), pp.403–407, ISSN 1063-7826.

Bukharaev, A.A., Berdunov, N.V., Ovchinnikov, D.V. & Salikhov, K.M. (1998). Three-dimensional probe and surface reconstruction for Atomic Force Microscopy using a deconvolution algorithm. *Scanning Micros.*, Vol. 12, No. 1 (January 1998), pp. 225-234, ISSN 0891-7035.

El Kurdi, M., Sauvage, S., Fishman, G. & Boucaud, P. (2006). Band-edge alignment of SiGe/Si quantum wells and SiGe/Si self-assembled islands. *Phys. Rev. B*, Vol. 73, No. 11 (May 2006), p. 195327 (9 p.), ISSN 1098-0121.

Feenstra, R.M. & Stroscio, J.A. (1987). Tunneling spectroscopy of the GaAs (110) surface. *J. Vac. Sci. Technol. B* Vol. 5, No. 4 (April 1987), pp. 923-928, ISSN 0734-211X.

Filatov, D.O., Kruglova, M.V., Isakov, M.A., Siprova, S.V., Marychev, M.O., Shengurov, V.G., Chalkov, V.Yu. & Denisov, S.A. (2008a). Morphology and photoluminescence of self-assembled GeSi/Si nanoclusters grown by Sublimation Molecular-Beam Epitaxy in a GeH_4 ambient. *Bull. RAS: Physics*, Vol. 72, No. 2 (February 2008), pp. 249-252, ISSN 1062-8738.

Filatov, D.O., Kruglova, M.V., Isakov, M.A., Siprova, S.V., Marychev, M.O., Shengurov, V.G., Chalkov, V.Yu. & Denisov, S.A. (2008b). Photoluminescence of GeSi/Si nanoclusters formed by Sublimation Molecular-Beam Epitaxy in GeH_4 medium. *Semicond.*, Vol. 42, No. 9 (September 2008), pp. 1098-1103, ISSN 1063-7826.

Filatov, D.O., Lapshina, M.A., Isakov, M.A., Borodin, P.A. & Bukharaev, A.A. (2010). Tunnelling AFM study of the local density of states in the self assembled In(Ga)As/GaAs(001) quantum dots and rings. *J. Phys. Conf. Ser.* Vol. 245 (October, 2010), p. 012017 (4 p.), ISSN 1742-6596.

Filatov, D.O., Borodin, P.A. & Bukharaev, A.A. (2011). Study of local density of electron states in InGaAs/GaAs quantum rings by Combined STM/AFM. (2011). *J. Surf. Investigation. X-ray, Synchrotron, & Neutron Techniques*, Vol. 5, No. 3 (March 2011), pp. 547–553, ISSN 1027-4510.

Grandidier, B., Niquet, Y. M., Legrand, B., Nys, J. P., Prieste, C., Stievenard, D., Gerard, J.M. & Thierry-Mieg, V. (2000). Imaging the wave-function amplitudes in cleaved semiconductor quantum boxes. *Phys. Rev. Lett.*, Vol. 85, No. 5 (July 2000), pp. 1068-1073, ISSN 0031-9007.

Hasegawa, H. & Sawada, T. (1983). On the electrical properties of compound semiconductor interfaces in metaliinsulatorisemiconductor structures and the possible origin of interface states. *Thin Solid Films*, Vol. 103, No. 1-3 (January 1983), pp. 119-140, ISSN 0040-6090.

Kamins, T.I., Medeiros-Ribeiro, G., Ohlberg, D.A.A. & Stanley Williams, R. (1999). Evolution of Ge islands on Si(001) during annealing. *J. Appl. Phys.*, Vol. 85, No. 2 (February 1999), pp. 1159-1162, ISSN 1089-7550.

Karpovich, I.A., Zvonkov, B.N., Baidus', N.V., Tikhov, S.V. & Filatov, D.O. (2004a). Tuning the energy spectrum of the InAs/GaAs quantum dot structures by varying the thickness and composition of a thin double GaAs/InGaAs cladding layer. In:*Trends in Nanotechnology Research*, E.Dirote, pp.173-208, Nova Science, ISBN1-59454-091-8, New York.

Karpovich, I.A., Baidus', N.V., Zvonkov, B.N., Filatov, D.O., Levichev, S.B., Zdoroveishev, A.V. & Perevoshikov, V.A. (2004b). Investigation of the buried InAs/GaAs quantum dots by SPM combined with selective chemical etching *Phys.Low-Dim.Struct.*, No. 3/4 (March 2004), pp. 341-345, ISSN 0204-3467.

Lorke, A., Blossey, R., Garcia, J.M., Bichler, M. & Abstreiter, G. (2002). Morphological transformation of $In_yGa_{1-y}As$ islands, fabricated by Stranski–Krastanov growth. *Mater. Sci. Eng. B*, Vol. 88, No. 2-3 (January 2002), pp. 225-229, ISSN 0921-5107.

Maltezopoulos, T., Bolz, A., Meyer, C., Heyn, C., Hansen, W., Morgenstern, M. & Wiesendanger, R. (2003). Wave-function mapping of InAs quantum dots by Scanning Tunneling Spectroscopy. *Phys. Rev. Lett.*, Vol. 91, No. 19 (November 2003), p. 196804 (4 pp.), ISSN 0031-9007.

Mashin, A.I., Nezhdanov, A.V., Filatov, D.O., Isakov, M.A., Shengurov, V.G., Chalkov, V.Yu. & Denisov, S.A. (2010). Confocal Raman Microscopy of Self-assembled GeSi/Si(001) islands. *Semicond.*, Vol. 44, No. 11 (November 2010), pp. 1504–1510, ISSN 1063-7826.

Medeiros-Ribeiro, G., Bratkovski, A.M., Kamens, T.I., Ohlberg, D.A.A. & Stanley Williams, R. (1998). Shape transition of germanium nanocrystals on a silicon (001) surface from pyramids to domes. *Science*, Vol. 279, No. 5349 (January 1998), pp. 353-355, ISSN 1095-9203.

Stier, O., Grundmann, M. & Bimberg, D. (1999). Electronic and optical properties of strained quantum dots modeled by 8-band $k{\cdot}p$ theory. *Phys. Rev. B*, Vol. 59, No. 8 (February 1999), pp. 5688-5701, ISSN 1098-0121.

Suzuki, K., Kanisawa, K., Janer, C., Perraud, S., Takashina, K., Fujisawa, T. & Hirayama, Y. (2007). Spatial imaging of two-dimensional electronic states in semiconductor quantum wells. *Phys.Rev.Lett.*, Vol.98, No.13 (March 2007), p.136802 (4 p.), ISSN 0031-9007.

Svetlov, S.P., Shengurov, V.G., Tolomasov, V.A., Gorshenin, G.N. & Chalkov, V.Yu. (2001). A sublimation silicon Molecular Beam Epitaxy system. *Instrum. Exp. Tech.*, Vol. 44, No. 5 (May 2001), pp. 700-703, ISSN 1608-3180.

Yanev, V.; Rommel, M.; Lemberger, M.; Petersen, S.; Amon, B.; Erlbacher, T.; Bauer, A.J.; Ryssel, H.; Paskaleva, A.; Weinreich, W.; Fachmann, C.; Heitmann, J. & Schroeder, U. (2008). Tunneling atomic-force microscopy as a highly sensitive mapping tool for the characterization of film morphology in thin high-*k* dielectrics. *Appl. Phys. Lett.*, Vol. 92, No. 25 (June 2008), pp. 2910-2912, ISSN 0003-6951.

Zenkevich, A., Lebedinskii, Yu., Antonov, D., Gorshkov, O. & Filatov, D. (2011). Structure and electron transport in the ultrathin nanocomposite SiO_2/Si films with the metal nanoclusters. In: *Advances in Diverse Industrial Applications of Nanocompositess, B.* Reddy, pp. 317-340, InTech, ISBN 978-953-307-202-9, Vienna.

13

Quantum Injection Dots

Eliade Stefanescu
Center of Advanced Studies in Physics of the Romanian Academy
Romania

1. Introduction

In optoelectronics, quantum dots are essential elements for coupling a device to an electromagnetic field in the infrared domain of the frequency spectrum. Such a dot is a small semiconductor region, with a forbidden band specific to a given application, embedded in the active i-region of the p-i-n junction of a laser structure (1), or in the sensitive i-region of the p-i-n junction of a photovoltaic structure (2). These quantum dots have the advantage of

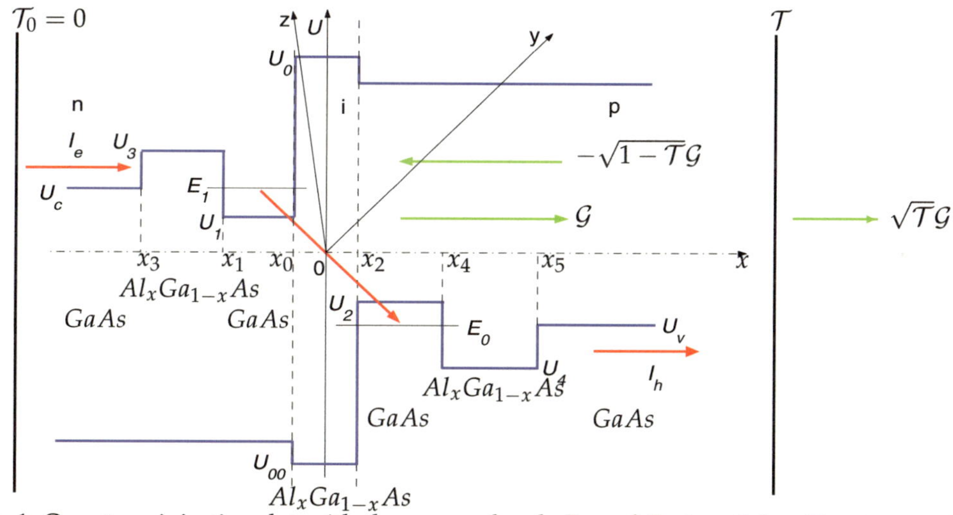

Fig. 1. Quantum injection dot with the energy levels E_1 and E_0, in a Fabry-Perot resonator with transmission coefficients of the mirrors $\mathcal{T}_0 = 0$ and $\mathcal{T} > 0$. By quantum transitions between these levels, a super radiant electromagnetic field with two counter-propagating waves of amplitudes $\mathcal{G}$ and $\sqrt{1-\mathcal{T}}\mathcal{G}$ is generated.

being feasible for a large class of applications (transition frequencies). However, they have the disadvantage that, being embedded in a bulk region, any injection or photovoltaic process includes transport processes through the active (sensitive) region, which are very dissipative. In such a system of quantum dots, an injected electron in the bulk active region of a laser structure has a large probability to recombine with an electron hole before reaching a quantum dot, where this electron becomes an active one, contributing to the laser field generation. Similarly, an optically excited electron in a quantum dot of a photo-voltaic bulk active region, has a large probability to meet another quantum dot, and to decay in the ground state of this quantum dot before reaching the quasi-neutral zone of the n-region, where this electron brings its contribution to the generated current.

In this chapter, we deal with a different kind of quantum dots (figure 1), as basic elements of a new class of power optoelectronic devices, devoted to the conversion of the environmental heat into usable energy (3–6). Such a quantum dot is substantially different from a bulk quantum dot (1; 2). These dots are deployed in double arrays of donor-acceptor pairs of impurity atoms placed in quantum wells at the two sides of the i-region of an n-i-p junction. The quantum dot density has an appropriate value to include the whole internal field of the semiconductor junction in the quantum dot region, i.e. between the two impurity arrays forming this region. Otherwise, spatial or mobile charge layers arise at the external sides of the two quantum wells, altering the distance $E_1 - E_0$ between the two ground energy levels of these wells. Thus, in a process of current injection, or of an optical current generation, the electrons traverse the internal field region by quantum transitions between the two levels of interest (see figure 1).

In section 2, we derive explicit expressions for electric potentials, energy levels in these potentials, wave-functions, the quantum dot density, and transition dipole moments, which determine the strength of the coupling of a quantum dot to a quasi-resonant electromagnetic field, and to the dissipative environment. We obtain operation conditions for the characteristics of the separation barriers.

In section 3, we consider the dissipative couplings of a quantum injection dot. We describe the dissipative dynamics of such a quantum dot by a master equation for a system of Fermions, coupled to a complex environment of other Fermions, Bosons, and the free electromagnetic field (7). This equation depends on analytic dissipative coefficients, describing correlated transitions of the system particles with the environmental particles, transitions stimulated by thermal fluctuations of the self-consistent field of the environmental Fermions, and the non-Markovian dynamics induced by these fluctuations. We derive explicit expressions of the dissipation coefficients as functions of universal constants and physical properties of the semiconductor structure: effective masses of electrons and electron holes, concentrations of donors and acceptors in the conduction regions, transition frequency, dipole moments, crystal density, elasticity coefficient, geometrical characteristics of the semiconductor structure, and temperature.

In section 4, we consider the dissipative dynamics of an electromagnetic field interacting with the quasi-free electrons of a semiconductor structure. By a method previously used in (7), we obtain a master equation with coefficients depending on frequency and the effective masses, transition dipole moments, and densities of states of these electrons. We derive field equations coupled to the polarizations of the system of active Fermions, with explicit expressions of the coupling and absorption coefficients, as functions of the physical properties of the active quantum dot system, and characteristics of the dissipative environment.

In section 5, we derive equations for the density matrix elements of a quantum injection dot interacting with a quasi-resonant electromagnetic field. We obtain equations for the amplitude mean-values of the forward and backward electromagnetic waves, propagating in a Fabry-Perot cavity which includes a system of such quantum dots. We derive optical equations for a system of quantum injection dots in a resonant Fabry-Perot cavity, with an additional term in the population equation, for describing a current injection in the semiconductor structure (3–5).

In section 6, we present the concept of quantum heat converter, as a device based on systems of quantum injection dots. This device is conceived in two versions: (1) longitudinal quantum heat converter, where the electromagnetic field propagates in the direction of the injected current, i.e. perpendicularly to the semiconductor chip (3; 4), and (2) transversal quantum heat converter, where the electromagnetic field propagates perpendicularly to the injected current, i.e. in the plane of the semiconductor chip (3; 5). Any of these versions could be preferred in specific applications. We derive analytical expressions of the super radiant power, as a function of the injected current, dissipative coefficients, coupling coefficient, number of super radiant transistors, transmission coefficient of the output mirror, and the geometrical characteristics of the device. We get operation conditions for these parameters. We describe the super radiant dissipative dynamics of a quantum injection dot, when a step current is injected in the device.

In this chapter, we present an analytical model depending only on material and geometrical characteristics, temperature, and universal constants.

2. Physical model

The essential problem of any optoelectronic device is the coupling of an electric current to an electromagnetic field. As a function of the roles played by these physical quantities into the input-output characteristic, one can conceive two kinds of devices. In the conventional optoelectronics, a device with the electric current as an input and the electromagnetic field as an output, if this field is not coherent, is called LED. If the output field is coherent, it is called laser. Conversely, if the electric current is the output, depending on an input field, the device is called photodiode, or photovoltaic cell if it is a power device, devoted to the electric energy production.

The new field of the heat conversion optoelectronics, includes two similar kinds of device. In this case, the radiant device is called quantum heat converter (3–5), while the photovoltaic device is called quantum injection system (6). Both kinds of such devices are based on an electron-field interaction with a potential V in the total Hamiltonian

$$H = H_0^S + H^F + V, \tag{1}$$

including a term for an electron with the electric charge $-e$ in an electromagnetic field with the vector potential $\vec{A}$,

$$H_0^S + V = \frac{(\vec{p} + e\vec{A})^2}{2M} + U(\vec{r}), \tag{2}$$

and the Hamiltonian of this field which, for the system represented in figure 1, is of the form

$$H^F = \hbar\omega(a_+^+ a_+ + a_-^+ a_- + 1), \tag{3}$$

where $a_+ - a_+^+$, $a_- - a_-^+$ are the creation-annihilation operators of the two counter-propagating waves. In equation (2), we distinguish the Hamiltonian of the electron in the potential $U(\vec{r})$ of an active quantum dot

$$H_0^S = \frac{\vec{p}^2}{2M} + U(\vec{r}) = \sum_i \varepsilon_i c_i^+ c_i, \tag{4}$$

and the interaction potential

$$V = \frac{e}{M}\vec{p}\vec{A}, \tag{5}$$

while the term in $\vec{A}^2$ is negligible in the non-relativistic approximation. This potential depends on the electron momentum

$$\vec{p} = \mathrm{i}M\sum_{ij}\omega_{ij}\vec{r}_{ij}c_i^+c_j, \tag{6}$$

and the vector potential

$$\vec{A} = \frac{\hbar}{e}\vec{K}\left(a_+e^{ikx} + a_+^+e^{-ikx} + a_-e^{-ikx} + a_-^+e^{ikx}\right), \tag{7}$$

of the electric field

$$\vec{E} = \mathrm{i}\frac{\hbar\omega}{e}\vec{K}\left(a_+e^{ikx} - a_+^+e^{-ikx} + a_-e^{-ikx} - a_-^+e^{ikx}\right). \tag{8}$$

In these expressions, M is the electron mass, $c_i^+ - c_j$ are Fermion operators, ω_{ij} are transition frequencies of the active electron, $\vec{r}_{ij}$ are dipole moments, ω is the frequency of the field, and

$$\vec{K} = \vec{1}_y\sqrt{\alpha\frac{\lambda}{\mathcal{V}}} \tag{9}$$

is a vector in the y-direction of this field, depending on the wavelength λ, the fine structure constant $\alpha = \frac{e^2}{4\pi\varepsilon_0\hbar c} \approx \frac{1}{137}$, and the quantization volume $\mathcal{V}$.

From (5) and (6), we notice that the electron-field interaction of the system depends on the transition dipole moment

$$\vec{r}_{01} = \vec{r}_{10} = \int_{\mathcal{V}_s}\Psi_0\vec{r}\Psi_1 d^3\vec{r}, \tag{10}$$

where

$$\Psi_1(x,y,z) = \psi_1(x)\phi_1(y)\chi_1(z) \tag{11a}$$
$$\Psi_0(x,y,z) = \psi_0(x)\phi_0(y)\chi_0(z) \tag{11b}$$

are eigenfunctions of the Hamiltonian (4), for the ground states of the two quantum wells. For the potential $U(\vec{r})$, we distinguish seven regions, of four $GaAs$ quantum wells, and three $Al_xGa_{1-x}As$ potential barriers, determined by the impurity concentrations of these regions (see figure 1). For the two thick conduction regions with the potentials U_c and U_v, we use a three-dimensional model, with a quantization volume $\mathcal{V}_n = 1/N_D$ for the n-region of donor concentration N_D, and $\mathcal{V}_p = 1/N_A$ for the p-region of acceptor concentration N_A. In these quantization volumes, for an electron with the effective mass M_n, and an electron hole with the effective mass M_p, we consider the densities of states (8)

$$g^{(n)}(E_\alpha) = \mathcal{V}_n\frac{\sqrt{2}M_n^{3/2}}{\pi^2\hbar^3}\sqrt{E_\alpha} \tag{12a}$$

$$g^{(p)}(E_\alpha) = \mathcal{V}_p\frac{\sqrt{2}M_p^{3/2}}{\pi^2\hbar^3}\sqrt{E_\alpha}, \tag{12b}$$

and the occupation probabilities of these states with the kinetic energies E_α:

$$f^{(n)}(E_\alpha) = \frac{1}{e^{(U_c+E_\alpha)/T}+1} \approx e^{-(U_c+E_\alpha)/T} \tag{13a}$$

$$f^{(p)}(E_\alpha) = \frac{1}{e^{(-U_v+E_\alpha)/T}+1} \approx e^{-(-U_v+E_\alpha)/T}, \tag{13b}$$

where approximate expressions are taken into account for the usual case of a non-degenerate semiconductor. Considering the integral of the number of particles occupying the states of a quantization volume, one gets the two potentials

$$U_c(T) = T \ln \frac{N_c(T)}{N_D}, \qquad N_c(T) = 2\left(\frac{\sqrt{M_n T/2\pi}}{\hbar}\right)^3 \tag{14a}$$

$$U_v(T) = -T \ln \frac{N_v(T)}{N_A}, \qquad N_v(T) = 2\left(\frac{\sqrt{M_p T/2\pi}}{\hbar}\right)^3. \tag{14b}$$

For the very thin layers of the quantum wells and potential barriers, we use a two-dimensional model. For a quantization area A_e, in n and p regions, one gets the densities of states

$$g^{(1)} = A_e \frac{M_n}{\pi\hbar^2} \tag{15a}$$

$$g^{(2)} = A_e \frac{M_p}{\pi\hbar^2}. \tag{15b}$$

By using the Fermi-Dirac distribution in the particle number integral, we obtain expressions similar to (14) for the potentials of the two $GaAs$ quantum wells, as a function of the surface quantum dot density N_e

$$U_1(T) = -T \ln\left(e^{\frac{\pi\hbar^2 N_e}{M_n T}} - 1\right) \tag{16a}$$

$$U_2(T) = T \ln\left(e^{\frac{\pi\hbar^2 N_e}{M_p T}} - 1\right). \tag{16b}$$

Similar expressions are obtained for the separation barriers, as functions of the donor and acceptor arrays with concentrations N_3, N_4 embedded in the very thin $Al_xGa_{1-x}As$-layers of these barriers,

$$U_3(T) = -T \ln\left(e^{\frac{\pi\hbar^2 N_3}{M_n T}} - 1\right) \tag{17a}$$

$$U_4(T) = T \ln\left(e^{\frac{\pi\hbar^2 N_4}{M_p T}} - 1\right), \tag{17b}$$

and for the potential barrier U_0 between the two quantum wells,

$$U_0(T) = -T \ln\left(e^{\frac{\pi\hbar^2 N_0}{M_n T}} - 1\right), \tag{18}$$

with a slight donor concentration N_0, controlling this potential. For these wells, we consider harmonic wave-functions with exponential tails in the neighboring barriers

$$\psi_1(x) = A_1 \cos\left[k_1(x_0 - x) - \arctan\frac{\alpha_1}{k_1}\right], \quad x_1 \le x \le x_0 \tag{19a}$$

$$\psi_1(x) = A_1\sqrt{\frac{E_1 - U_1}{U_0 - U_1}}e^{-\alpha_1(x - x_0)}, \quad x_0 \le x \le x_2 \tag{19b}$$

$$\psi_1(x) = A_1\sqrt{\frac{E_1 - U_1}{U_3 - U_1}}e^{-\alpha_3(x_1 - x)}, \quad x_3 \le x \le x_1 \tag{19c}$$

and

$$\psi_0(x) = A_0 \cos\left[k_0(x - x_2) - \arctan\frac{\alpha_0}{k_0}\right], \quad x_2 \le x \le x_4 \tag{20a}$$

$$\psi_0(x) = A_0\sqrt{\frac{U_2 - E_0}{U_2 - U_{00}}}e^{-\alpha_0(x_2 - x)}, \quad x_0 \le x \le x_2 \tag{20b}$$

$$\psi_0(x) = A_0\sqrt{\frac{U_2 - E_0}{U_2 - U_4}}e^{-\alpha_4(x - x_4)}, \quad x_4 \le x \le x_5, \tag{20c}$$

while the tails beyond these barriers are neglected. These wave-functions depend on the wave-numbers

$$k_1 = \frac{1}{\hbar}\sqrt{2M_n(E_1 - U_1)} \tag{21a}$$

$$k_0 = \frac{1}{\hbar}\sqrt{2M_p(U_2 - E_0)}, \tag{21b}$$

attenuation coefficients

$$\alpha_1 = \frac{1}{\hbar}\sqrt{2M_n(U_0 - E_1)} \tag{22a}$$

$$\alpha_0 = \frac{1}{\hbar}\sqrt{2M_p(E_0 - U_{00})} \tag{22b}$$

$$\alpha_3 = \frac{1}{\hbar}\sqrt{2M_n(U_3 - E_1)} \tag{22c}$$

$$\alpha_4 = \frac{1}{\hbar}\sqrt{2M_p(E_0 - U_4)}, \tag{22d}$$

and normalization coefficients

$$A_1 = \sqrt{2}\left[x_0 - x_1 + \frac{\hbar}{\sqrt{2M_n}}\left(\frac{1}{\sqrt{U_0 - E_1}} + \frac{1}{\sqrt{U_3 - E_1}}\right)\right]^{-1/2} \tag{23a}$$

$$A_0 = \sqrt{2}\left[x_4 - x_2 + \frac{\hbar}{\sqrt{2M_p}}\left(\frac{1}{\sqrt{E_0 - U_{00}}} + \frac{1}{\sqrt{E_0 - U_4}}\right)\right]^{-1/2}, \tag{23b}$$

while the energy eigenvalues are given by the equations:

$$E_1 - U_1 = \frac{\hbar^2}{2M_n(x_0 - x_1)^2} \left(\arctan \sqrt{\frac{U_0 - E_1}{E_1 - U_1}} + \arctan \sqrt{\frac{U_3 - E_1}{E_1 - U_1}} \right)^2 \tag{24a}$$

$$U_2 - E_0 = \frac{\hbar^2}{2M_p(x_4 - x_2)^2} \left(\arctan \sqrt{\frac{E_0 - U_{00}}{U_2 - E_0}} + \arctan \sqrt{\frac{E_0 - U_4}{U_2 - E_0}} \right)^2 . \tag{24b}$$

We take the two energy eigenvalues as $E_1 = U_c$, $E_0 = U_v$. In this case, the whole internal potential of the n-i-p junction is included on the distance d between the two charge layers of the quantum dot region, which means a quantum dot surface density

$$N_e = \varepsilon \frac{E_1 - E_0}{e^2 d}, \tag{25}$$

while this distance can be approximated as

$$d = \frac{1}{2}(x_2 - x_0 + x_4 - x_1). \tag{26}$$

The energy levels $E_1 = U_c$ and $E_0 = U_v$ can be obtained from (14), as functions of the donor and acceptor concentrations N_D and N_A of the conduction regions. By choosing appropriate values for the separation and quantum dot barriers, from (17) and (18) one gets the surface concentrations N_3, N_4 and N_0 of these barriers. With the widths $x_1 - x_3$ and $x_5 - x_4$, the separation barrier must have a higher penetrability P than the necessary value to provide the injected current I through the device area A_D, which means that the density of this current must be smaller than the thermal current $\frac{1}{6}eN_D v_T P$ emergent from a unit volume with the thermal velocity $v_T = \sqrt{T/M_n}$, and crossing the barrier. We get the conditions

$$\alpha_3(x_1 - x_3) < \frac{1}{2} \ln \left(\frac{eN_D A_D}{6I} \sqrt{\frac{T}{M_n}} \right) \tag{27a}$$

$$\alpha_4(x_5 - x_4) < \frac{1}{2} \ln \left(\frac{eN_A A_D}{6I} \sqrt{\frac{T}{M_p}} \right) . \tag{27b}$$

Thus, a quantum injection dot is a two-level system, with the energy levels E_0 and E_1, in a quantization volume shaped as a parallelepiped with the basis area $\mathcal{A} = \frac{1}{N_e}$ and the height $x_5 - x_3$. In this volume, we consider the two wave-functions (19)-(20) for the coordinate x, while, for the coordinates y and z in the plane of the quantum dot array, we consider the wave-functions $\phi_1(y)$, $\phi_0(y)$ and $\chi_1(y)$, $\chi_0(y)$, describing a thermal motion with an energy mean-value T. For a longitudinal device, when the electromagnetic field propagates in the direction x of the injected current, the electric component E_y of this field is coupled with the component y_{01} of the transition dipole moment of the system between thermal states. For a transversal device, when the electromagnetic field propagates in a direction y perpendicular to the direction x of the injected current, i.e. in the plane of the quantum dot array, the electric component E_x of this field is coupled with the transition dipole moment x_{01} between the two states (19)-(20) of the system. These dipole moments also determine the dissipative couplings of the quantum dot. They essentially depend on the width $x_2 - x_0$ of the quantum dot barrier of height U_0, which determines the overlap of the two wave-functions $\psi_1(x)$ and $\psi_0(x)$. When this width is chosen for reasonable values of the electron-field and dissipative couplings, and N_3, N_4 for reasonable values of the separation barriers satisfying the conditions (27), from equations (16) and (24)-(26) the geometrical characteristics $x_0 x_1, x_2, x_3, x_4$ are obtained.

As an example, for a concentration $N_D = N_A = 3.16 \times 10^{16}\ cm^{-3}$ of a super radiant junction, working at temperature $T = 10\ ^0C$, we get a transition frequency $E_1 - E_0 = 0.1866\ eV$. A quantum dot $Al_{0.37}Ga_{0.63}As$-barrier of $U_0 = 0.5\ eV$ is obtained for a surface concentration of $N_0 = 6.4243 \times 10^6\ m^{-2}$ donors, and separation barriers of $U_3 - U_c = U_v - U_4 = 0.05\ eV$ are obtained for the surface concentrations of these barriers $N_3 = 8.01 \times 10^{13}\ m^{-2}$ and $N_4 = 2.552 \times 10^{13}\ m^{-2}$. For a width $x_2 - x_0 = 5.5\ nm$ of the quantum dot barrier, we get a quantum dot surface concentration $N_e = 1.476 \times 10^{16}\ m^{-2}$ and the widths of the two quantum dot wells $x_0 - x_1 = 4.189\ nm$ and $x_4 - x_2 = 1.576\ nm$, while separation barriers with widths $x_1 - x_3 = 10\ nm$ and $x_5 - x_4 = 3\ nm$ satisfy the conditions (27) for an injected current $I = 45\ A$ in a device with an area $A_D = 4\ cm^2$.

3. Dissipative dynamics of quantum injection dots

We consider a quantum injection dot with two energy levels E_1, E_0, coupled to a super radiant field and a complex dissipative environment of a semiconductor structure as it is represented in figure 1: (1) the quasi-free electrons of the n-conduction region $x < x_3$, (2) the quasi-free holes of the p-conduction region $x > x_5$, (3) the phonons of the crystal at temperature T, and (4) the photons of the free electromagnetic field existing at temperature T (see figure 2). In (3)

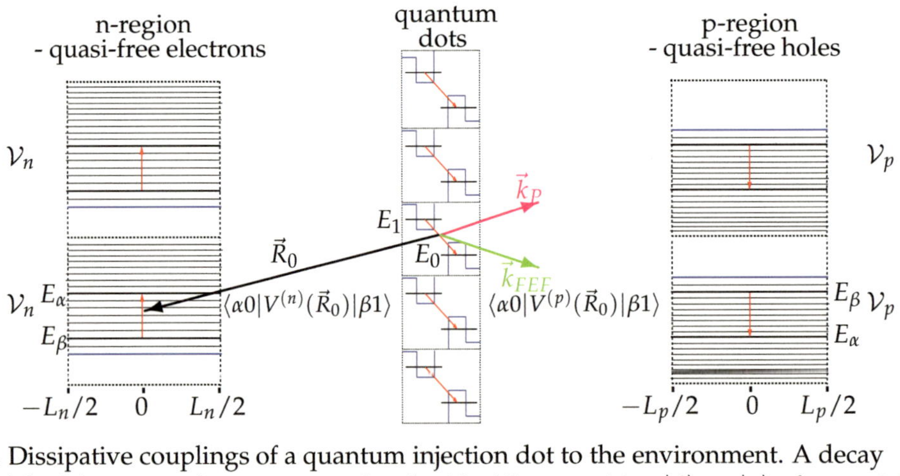

Fig. 2. Dissipative couplings of a quantum injection dot to the environment. A decay $|1\rangle \to |0\rangle$ of the active electron is correlated with: (1) a transition $|\beta\rangle \to |\alpha\rangle$ of a quasi-free electron in a quantization volume $\mathcal{V}_n$, (2) a transition $|\beta\rangle \to |\alpha\rangle$ of a quasi-free hole in a quantization volume $\mathcal{V}_p$, (3) a phonon creation with a wave vector $\vec{k}_P$, and (4) a photon creation with a wave vector $\vec{k}_{FEF}$.

we showed that, for a semiconductor structure with the characteristics mentioned at the end of the preceding section, the decay rate corresponding to the phonon environment is dominant, the decay rate due to the conduction electrons and holes is smaller, while the decay rate given by the free electromagnetic field is negligible.

We consider a system of interest, including a system of Fermions S with the Hamiltonian H_0^S and an electromagnetic field F with the Hamiltonian H^F, in a dissipative environment with the Hamiltonian H^E. Taking into account a potential of interaction V between the Fermion system S and the field F, and a system-environment potential V^E, the dynamics of the total system is described by an equation of motion of the form:

$$\frac{d}{dt}\tilde{\chi}(t) = -\frac{i}{\hbar}\left[\varepsilon\tilde{V}(t) + \varepsilon\tilde{V}^E(t), \tilde{\chi}(t)\right]. \tag{28}$$

In this equation, tilde denotes operators in the interaction picture, e.g.

$$\tilde{\chi}(t) = e^{\frac{i}{\hbar}(H^E + H_0^S + H^F)t}\chi(t)e^{-\frac{i}{\hbar}(H^F + H_0^S + H^E)t}. \tag{29}$$

According to a general procedure disclosed in (9), we take a total density of the form

$$\tilde{\chi}(t) = R \otimes \tilde{\rho}(t) + \varepsilon\tilde{\chi}^{(1)}(t) + \varepsilon^2\tilde{\chi}^{(2)}(t) + ... \;, \tag{30}$$

where $\rho(t)$ is the reduced density matrix of the system of interest, while R is the density matrix of the dissipative environment at the initial moment of time, $t = 0$, the time-evolution of the environment being described by the higher-order terms $\tilde{\chi}^{(1)}(t), \tilde{\chi}^{(2)}(t), ...$ The parameter ε is introduced to handle the orders of the terms in this series expansion, and is set to 1 in the final results. The reduced density of the system is

$$\tilde{\rho}(t) = Tr_E\{\tilde{\chi}(t)\}, \tag{31}$$

while the higher-order terms of the total density have the property:

$$Tr_E\{\tilde{\chi}^{(1)}\} = Tr_E\{\tilde{\chi}^{(2)}\} = ... \; = 0. \tag{32}$$

If initially the environment is in the equilibrium state R, the density matrix of the total system is of the form $\chi(0) = R\rho(0)$. We suppose that at time $t = 0$, due to the interaction V of the system of Fermions with the electromagnetic field, or due to a non-equilibrium initial state $\rho(0) \neq \rho_T$, a time-evolution begins, while the reduced density satisfies a quantum dynamical equation of the form

$$\frac{d}{dt}\tilde{\rho}(t) = \varepsilon\tilde{B}^{(1)}(\tilde{\rho}(t), t) + \varepsilon^2\tilde{B}^{(2)}(\tilde{\rho}(t), t) + ... \;. \tag{33}$$

From the dynamic equation (28), with expressions (30)-(33), we obtain the quantum master equation

$$\begin{aligned}
\frac{d}{dt}\rho(t) = & -\frac{\mathrm{i}}{\hbar}[H, \rho(t)] - \mathrm{i}\sum_{ij}\zeta_{ij}[c_i^+ c_j, \rho(t)] \\
& + \sum_{ij}\lambda_{ij}([c_i^+ c_j \rho(t), c_j^+ c_i] + [c_i^+ c_j, \rho(t) c_j^+ c_i]) \\
& + \sum_{ijkl}\zeta_{ij}\zeta_{kl}\int_{t-\tau}^{t}[c_i^+ c_j, e^{-\mathrm{i}[\phi(t') + \frac{1}{\hbar}H_0^S(t-t')]}[c_k^+ c_l, \rho(t')]e^{\mathrm{i}[\phi(t') + \frac{1}{\hbar}H_0^S(t-t')]}]dt',
\end{aligned} \tag{34}$$

where the coefficients

$$\zeta_{ij} = \frac{1}{\hbar}\sqrt{\frac{1}{Y^F}\int_{(\alpha)}\langle\alpha i|(V^F)^2|\alpha j\rangle f_\alpha^F(\varepsilon_\alpha) g_\alpha^F(\varepsilon_\alpha) d\varepsilon_\alpha} \tag{35}$$

describe transitions stimulated by the thermal fluctuations of the self-consistent field of the Y^F environmental Fermions in a certain quantization volume - hopping potential (10), $\phi(t')$ is a phase fluctuation operator, while the coefficients

$$\lambda_{ij} = \lambda_{ij}^F + \lambda_{ij}^B + \gamma_{ij} \tag{36}$$

describe correlated transitions of the system Fermions with environment particles, including explicit terms for Fermions, Bosons, and photons of the free electromagnetic field. These terms depend on the dissipative two-body potentials V^F, V^B, the densities of the environment states $g^F(\varepsilon_\alpha), g^B(\varepsilon_\alpha)$, the occupation probabilities of these states $f^F(\varepsilon_\alpha), f^B(\varepsilon_\alpha)$, and temperature T. For a rather low temperature, $T \ll \varepsilon_{ji},\ j > i$, these terms become

$$\lambda_{ij}^F = \frac{\pi}{\hbar}|\langle\alpha i|V^F|\beta j\rangle|^2[1 - f^F(\varepsilon_{ji})]g^F(\varepsilon_{ji}) \tag{37a}$$

$$\lambda_{ji}^F = \frac{\pi}{\hbar}|\langle\alpha i|V^F|\beta j\rangle|^2 f^F(\varepsilon_{ji})g^F(\varepsilon_{ji}), \tag{37b}$$

for the Fermion environment,

$$\lambda_{ij}^B = \frac{\pi}{\hbar}|\langle\alpha i|V^B|\beta j\rangle|^2[1 + f^B(\varepsilon_{ji})]g_\alpha^B(\varepsilon_{ji}) \tag{38a}$$

$$\lambda_{ji}^B = \frac{\pi}{\hbar}|\langle\alpha i|V^B|\beta j\rangle|^2 f^B(\varepsilon_{ji})g_\alpha^B(\varepsilon_{ji}) \tag{38b}$$

for the Boson environment, and

$$\gamma_{ij} = \frac{2\alpha}{c^2\hbar^3}\left|\vec{r}_{ij}\right|^2 \varepsilon_{ji}^3(1 + \frac{1}{e^{\varepsilon_{ji}/T} - 1}) \tag{39}$$

for the Boson environment of the free electromagnetic field, where $\vec{r}_{ij}$ are the transition dipole moments. The dissipative terms of the master equation (34) with coefficients (36)-(39) describe correlated transitions of the system and the environment particles, with energy conservation, $\varepsilon_{ji} = \varepsilon_{\alpha\beta}$, in agreement with the quantum-mechanical principles and the detailed balance principle (11). The non-Markovian part of this equation takes into account the fluctuations of the self-consistent field of the environment Fermions, with the coefficients (35).

A significant component of the dissipative dynamics comes from the Coulomb interaction of the active electrons, mainly located in the interval (x_3, x_5), with the conduction electrons and holes in the conduction regions $(-\infty, x_3)$ and $(x_5, +\infty)$, respectively (figure 1). We use the notations $\vec{r}$ for the position vector of an active electron, and $\vec{R}_0 + \vec{R}$ for the position vector of a dissipative electron (hole), where $\vec{R}_0$ is the position vector of an arbitrary n (p) cluster, and $\vec{R} = \vec{1}_x X + \vec{1}_y Y + \vec{1}_z Z$ is the position of an electron (hole) in this cluster (figure 3). In this case, the Coulomb potential in a first-order approximation of the two-body term $\vec{R}\vec{r} = Xx + Yy + Zz$ is

$$V^C(\vec{R}, \vec{r}) = \frac{\alpha\hbar c}{|\vec{R}_0 + \vec{R} - \vec{r}|} \approx \frac{\alpha\hbar c}{|\vec{R}_0|}\left(1 + \frac{Xx + Yy + Zz}{\vec{R}_0^2}\right). \tag{40}$$

From this expression, only the second term, bilinear in the coordinates of an active electron and of an electron (hole) of the environment, yields contributions in the two-body transition matrix elements of the decay (excitation) rates (37):

$$V^F(\vec{R}_0, \vec{R}, \vec{r}) \doteq V^{(n)}(\vec{R}_0, \vec{R}, \vec{r}) = -V^{(p)}(\vec{R}_0, \vec{R}, \vec{r}) = \frac{\alpha\hbar c}{|\vec{R}_0|^3}(Xx + Yy + Zz). \tag{41}$$

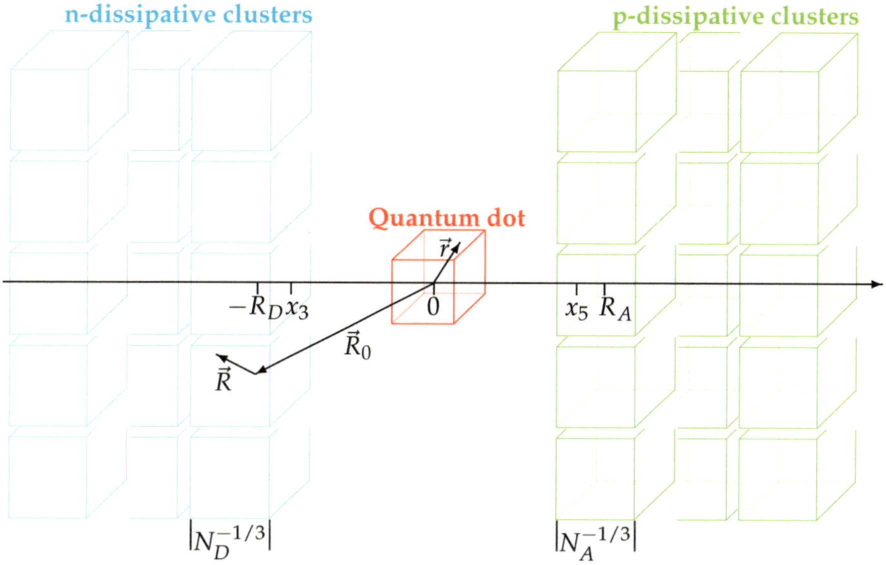

Fig. 3. The electron of a quantum injection dot is coupled to the quasi-free electrons of the n-dissipative clusters (n-region) and quasi-free holes of the p-dissipative clusters (p-region) by an electric dipole-dipole interaction: $V^F(\vec{R}_0, \vec{R}, \vec{r}) = \frac{\alpha\hbar c}{|\vec{R}_0|^3}\vec{R}\vec{r}$.

From the wave-functions derived in the preceding section, we obtain the dipole moment of a quantum dot:

$$x_{01}^{(\Psi)} = c_{01}^{(x)} \left(\frac{x_2 - x_0}{2} - \frac{1}{\alpha_0 - \alpha_1} \right) \tag{42a}$$

$$y_{01}^{(\Psi)} = z_{01}^{(\Psi)} = c_{01}^{(x)} \frac{\hbar}{2\sqrt{M_n T}} \tag{42b}$$

$$y_{10}^{(\Psi)} = z_{10}^{(\Psi)} = c_{01}^{(x)} \frac{\hbar}{2\sqrt{M_p T}}, \tag{42c}$$

as a product of the overlap function

$$c_{01}^{(x)} = \frac{A_1 A_0}{\alpha_0 - \alpha_1} \sqrt{\frac{(E_1 - U_1)(U_2 - E_0)}{(U_0 - U_1)(U_2 - U_{00})}} \left(e^{-\alpha_1(x_2 - x_0)} - e^{-\alpha_0(x_2 - x_0)} \right). \tag{43}$$

and a quantity that we call the state separation distance. At the same time, with the initial and the final energies $E_\beta = T/2$, $E_\alpha = E_\beta + \varepsilon_{10}$, we obtain the dipole moments for the n-zone

$$X_{\alpha\beta}^{(n)} = Y_{\alpha\beta}^{(n)} = Z_{\alpha\beta}^{(n)} = \frac{\hbar}{\varepsilon_{10}} \sqrt{\frac{2\varepsilon_{10} + T}{M_n}} \approx \sqrt{\frac{2\hbar}{M_n \omega_0}}, \tag{44}$$

and for the p-zone,

$$X_{\alpha\beta}^{(p)} = Y_{\alpha\beta}^{(p)} = Z_{\alpha\beta}^{(p)} = \frac{\hbar}{\varepsilon_{10}} \sqrt{\frac{2\varepsilon_{10} + T}{M_p}} \approx \sqrt{\frac{2\hbar}{M_p \omega_0}}, \tag{45}$$

where we used the notations $\varepsilon_{10} \doteq \hbar\omega_0 \doteq E_1 - E_0$. With the quantum dot dipole moments (42), and the environment dipole moments (44)-(45), we calculate the matrix elements of the two-body potential (41). With these matrix elements, from (37) with the densities of states (12) and the occupation probabilities of these states (13), one obtains the dissipative coefficients for the coupling of a quantum dot with a dissipative cluster. By integration over all clusters of both hemispheres of the n and p conduction regions, with the quantization volumes $\mathcal{V}_n = \frac{1}{N_D}$ and $\mathcal{V}_p = \frac{1}{N_A}$ as differential volumes in these integrals, we obtain the dissipation coefficients (3):

$$\lambda_{01}^{(n)} = \frac{4\alpha^2 c^2 \sqrt{2M_n}(\varepsilon_{10} + \frac{T}{2})|c_{01}^{(x)}|^2 \mu_{01}^2}{3\left(\frac{N_D^{-1/3}}{2} - x_3\right)^3 \varepsilon_{10}^{3/2}(e^{-(U_c+\varepsilon_{10})/T} + 1)} \tag{46a}$$

$$\lambda_{10}^{(n)} = \frac{4\alpha^2 c^2 \sqrt{2M_n}(\varepsilon_{10} + \frac{T}{2})|c_{01}^{(x)}|^2 \mu_{01}^2}{3\left(\frac{N_D^{-1/3}}{2} - x_3\right)^3 \varepsilon_{10}^{3/2}(e^{(U_c+\varepsilon_{10})/T} + 1)} \tag{46b}$$

$$\lambda_{01}^{(p)} = \frac{4\alpha^2 c^2 \sqrt{2M_p}(\varepsilon_{10} + \frac{T}{2})|c_{01}^{(x)}|^2 \mu_{01}^2}{3\left(\frac{N_A^{-1/3}}{2} + x_5\right)^3 \varepsilon_{10}^{3/2}(e^{-(-U_v+\varepsilon_{10})/T} + 1)} \tag{46c}$$

$$\lambda_{10}^{(p)} = \frac{4\alpha^2 c^2 \sqrt{2M_p}(\varepsilon_{10} + \frac{T}{2})|c_{01}^{(x)}|^2 \mu_{01}^2}{3\left(\frac{N_A^{-1/3}}{2} + x_5\right)^3 \varepsilon_{10}^{3/2}(e^{(-U_v+\varepsilon_{10})/T} + 1)}, \tag{46d}$$

where

$$\mu_{01}^2 = \left(\frac{x_2 - x_0}{2} - \frac{1}{\alpha_0 - \alpha_1} + \frac{\hbar}{\sqrt{M_n T}}\right)\left(\frac{x_2 - x_0}{2} - \frac{1}{\alpha_0 - \alpha_1} + \frac{\hbar}{\sqrt{M_p T}}\right). \tag{47}$$

is the square of the separation distance of the two states $\Psi_0(\vec{r})$ and $\Psi_1(\vec{r})$. From (35), by similar calculations we obtain the fluctuation coefficients of a quantum dot in the self-consistent field of dissipative clusters:

$$\left[\zeta_{11}^{(n)}\right]^2 = \frac{\alpha^2 c^2 M_n^{3/2} T^{3/2}}{360\pi\sqrt{2\pi}\hbar^3} \cdot \frac{N_D^{1/3}[A_1^2(x_0^3 - x_1^3) + \frac{1}{Ne}]}{N_e N_c \left(\frac{N_D^{-1/3}}{2} - x_3 + \frac{x_0 + x_1}{2}\right)^5} \tag{48a}$$

$$\left[\zeta_{11}^{(p)}\right]^2 = \frac{\alpha^2 c^2 M_p^{3/2} T^{3/2}}{360\pi\sqrt{2\pi}\hbar^3} \cdot \frac{N_A^{1/3}[A_1^2(x_0^3 - x_1^3) + \frac{1}{Ne}]}{N_e N_v \left(\frac{N_A^{-1/3}}{2} + x_5 - \frac{x_0 + x_1}{2}\right)^5} \tag{48b}$$

$$\left[\zeta_{00}^{(n)}\right]^2 = \frac{\alpha^2 c^2 M_n^{3/2} T^{3/2}}{360\pi\sqrt{2\pi}\hbar^3} \cdot \frac{N_D^{1/3}[A_0^2(x_4^3 - x_2^3) + \frac{1}{Ne}]}{N_e N_c \left(\frac{N_D^{-1/3}}{2} - x_3 + \frac{x_4 + x_2}{2}\right)^5} \tag{48c}$$

$$\left[\zeta_{00}^{(p)}\right]^2 = \frac{\alpha^2 c^2 M_p^{3/2} T^{3/2}}{360\pi\sqrt{2\pi}\hbar^3} \cdot \frac{N_A^{1/3}[A_0^2(x_4^3 - x_2^3) + \frac{1}{Ne}]}{N_e N_v \left(\frac{N_A^{-1/3}}{2} + x_5 - \frac{x_4 + x_2}{2}\right)^5}. \tag{48d}$$

The Markovian dissipative coefficients (46) describe a very strong, exponential decrease of the decay rate with the width $x_2 - x_0$ of the quantum dot barrier, given by the square of the overlap function (43), and a strong decrease with the distances x_3 and x_5 of the quantum dot separation from the two conduction regions, which enter at the denominators with power 3. The non-Markovian coefficients (48) describe a very strong decrease of the fluctuation rate with the separation distances x_3 and x_5, which enter at the denominators with power 5.

The coupling of a quantum dot electron with a vibrational mode α is described by the potential matrix element (3):

$$V_{01\alpha}^{EP} = V_{10\alpha}^{EP} = -\hbar\omega_0^{3/2} \frac{M\vec{r}_{01}\vec{1}_\alpha}{\sqrt{\mathcal{M}\hbar/2}}, \tag{49}$$

where $\mathcal{M}$ is the mass of the phonon quantization volume $\mathcal{V}_\mathcal{P}$. For the density of phonon states of energy $\hbar\omega_0$, we obtain an expression similar to (12):

$$g_P(\hbar\omega_0) = \mathcal{V}_\mathcal{P} \frac{\sqrt{2}\mathcal{M}^{3/2}}{\pi^2\hbar^3} \sqrt{\hbar\omega_0}. \tag{50}$$

We consider the sound velocity v from the phonon wavelength expressions

$$\lambda_P = \frac{v}{\nu} = \frac{2\pi v}{\omega} = \frac{2\pi\hbar v}{\varepsilon_{10}}, \qquad \lambda_P \equiv \frac{2\pi}{k_P} = \frac{2\pi\hbar}{\sqrt{2\mathcal{M}\varepsilon_{10}}}, \tag{51}$$

and the crystal density

$$D \equiv \frac{\mathcal{M}}{\mathcal{V}_\mathcal{P}} = \frac{2\pi^2\hbar^2}{\mathcal{V}_\mathcal{P}\lambda_P^2\varepsilon_{10}}. \tag{52}$$

With (49)-(52), from (38) we obtain the decay (excitation) rates

$$\lambda_{01}^P = \frac{E_e^2\varepsilon_{10}^5}{\pi\hbar^6 c^4 v^3 D} \cdot \frac{|c_{01}^{(x)}|^2\mu_{01}^2}{1 - e^{-\varepsilon_{10}/T}} \tag{53a}$$

$$\lambda_{10}^P = \frac{E_e^2\varepsilon_{10}^5}{\pi\hbar^6 c^4 v^3 D} \cdot \frac{|c_{01}^{(x)}|^2\mu_{01}^2}{e^{\varepsilon_{10}/T} - 1}, \tag{53b}$$

where $E_e = Mc^2$ is the rest energy of the electron, and v is the sound velocity, which can be calculated from the Young elasticity coefficient E and the crystal density D:

$$v \approx \sqrt{\frac{E}{D}}. \tag{54}$$

We notice that both systems of dissipation coefficients (46) and (53) are proportional to the squares of the state separation distance and overlap function. Expressions (53) describe a very strong dependence of the decay rates on the transition energy ε_{10}, being proportional to this energy with power 5. However, they are valid for phonon wavelengths much larger than the distance between the atoms of the crystal lattice. Otherwise, the number of the density modes can no more be treated as a quasi-continuous function of frequency, and the probability of any non-resonant interaction vanishes (Mösbauer effect). We also found that for the rather low transition energies specific to the quantum injection dots, the decay rate due to the phonon coupling is rather low, e.g. for the structure presented at the end of the preceding section with $\varepsilon_{10} = 0.1866\ eV$, we got $\lambda_{01}^P = 2 \times 10^7\ s^{-1}$. As we found by direct calculations (3), for the quantum injection dots, which are separated by potential barriers from the conduction electrons, the decay rate due to these electrons is much lower than the decay rate due to the coupling to the phonons of the crystal lattice vibrations.

4. Dissipative dynamics of electromagnetic field

The operation of a semiconductor device with quantum dots interacting with a quasi-resonant electromagnetic field is based on the transparency of the host semiconductor structure, with a band to band transition frequency much higher than the quantum dot transition frequency. That means that this electromagnetic field is absorbed by the host semiconductor structure mainly by intra-band transitions, essentially depending on overlap functions of thermal states with excited states populated (depopulated) by these transitions.

The dynamics of this electromagnetic field, with a potential $\tilde{V}(t)$ of interaction with a quantum dot and a potential $\tilde{V}^E(t)$ of interaction with a conduction electron (hole), is described by a system of equations of the form (28), (30), (31) (33), which, in the second-order approximation, provides

$$\frac{d}{dt}\tilde{\rho}(t) = \tilde{B}^{(1)}[\tilde{\rho}(t), t] + \tilde{B}^{(2)}[\tilde{\rho}(t), t] \tag{55a}$$

$$\tilde{B}^{(1)}[\tilde{\rho}(t), t] = -\frac{\mathrm{i}}{\hbar} Tr_E[\tilde{V}(t) + \tilde{V}^E(t), R \otimes \tilde{\rho}(t)] \tag{55b}$$

$$\tilde{\chi}^{(1)}(t) = \int_0^t \left\{ -\frac{\mathrm{i}}{\hbar}[\tilde{V}(t') + \tilde{V}^E(t'), R \otimes \tilde{\rho}(t')] - R \otimes \tilde{B}^{(1)}[\tilde{\rho}(t'), t'] \right\} dt' \tag{55c}$$

$$\tilde{B}^{(2)}[\tilde{\rho}(t), t] = -\frac{\mathrm{i}}{\hbar} Tr_E[\tilde{V}(t) + \tilde{V}^E(t), \tilde{\chi}^{(1)}(t)]. \tag{55d}$$

The interaction potential $\tilde{V}(t)$ is obtained from (5)-(6), while the dissipative potential $\tilde{V}^E(t)$ is given by the similar expressions

$$V^E = \frac{e}{M}\vec{P}\vec{A} \tag{56}$$

and

$$\vec{P} = \mathrm{i}M \sum_{\alpha\beta} \omega_{\alpha\beta} \vec{R}_{\alpha\beta} c_\alpha^+ c_\beta. \tag{57}$$

From (5)-(7) and (56)-(57), we derive expressions of these potentials depending only on the positive transition frequencies $\omega_{ji}(j > i)$ and $\omega_{\alpha\beta}(\alpha > \beta)$, and take into account the so called "rotating-wave approximation", which includes only conservative processes, when an electron excitation is correlated only with a photon annihilation, while an electron decay is correlated only with a photon creation:

$$V = \mathrm{i}\sum_{j>i} \hbar\omega_{ji}\vec{K}\vec{r}_{ij} \left[c_j^+ c_i \left(a_+ e^{\mathrm{i}kx} + a_- e^{-\mathrm{i}kx} \right) - c_i^+ c_j \left(a_+^+ e^{-\mathrm{i}kx} + a_-^+ e^{\mathrm{i}kx} \right) \right] \tag{58}$$

$$V^E = \mathrm{i}\sum_{\alpha>\beta} \hbar\omega_{\alpha\beta}\vec{K}\vec{R}_{\alpha\beta} \left[c_\alpha^+ c_\beta \left(a_+ e^{\mathrm{i}kx} + a_- e^{-\mathrm{i}kx} \right) - c_\beta^+ c_\alpha \left(a_+^+ e^{-\mathrm{i}kx} + a_-^+ e^{\mathrm{i}kx} \right) \right]. \tag{59}$$

We consider the time-dependent expressions of these operators in the interaction picture,

$$\tilde{a}(t) = ae^{-\mathrm{i}\omega t}, \qquad \tilde{a}^+(t) = a^+ e^{\mathrm{i}\omega t} \tag{60}$$

$$\tilde{c}_i^+(t)\tilde{c}_j(t) = c_i^+ c_j e^{-\mathrm{i}\omega_{ji} t}, \qquad \tilde{c}_j^+(t)\tilde{c}_i(t) = c_j^+ c_i e^{\mathrm{i}\omega_{ji} t} \tag{61}$$

$$\tilde{c}_\beta^+(t)\tilde{c}_\alpha(t) = c_\beta^+ c_\alpha e^{-\mathrm{i}\omega_{\alpha\beta} t}, \qquad \tilde{c}_\alpha^+(t)\tilde{c}_\beta(t) = c_\alpha^+ c_\beta e^{\mathrm{i}\omega_{\alpha\beta} t}, \tag{62}$$

and take equations (55) for the mean-values of the electron operators. We retain only the slowly time varying terms, obtained from the resonance condition $\omega_{\alpha\beta} = \omega$, while the rapidly varying terms are neglected. We consider the summations over the environmental states as integrals over a quasi-continuum of states, with the densities $g(\varepsilon_\alpha)$ and $g(\varepsilon_\beta)$ and occupation probabilities $f(\varepsilon_\alpha)$ and $f(\varepsilon_\beta)$, and neglect the thermal energies $\varepsilon_\beta \sim T$ in comparison with the transition energy $\hbar\omega$. We obtain the quantum master equation

$$\begin{aligned}\frac{d}{dt}\rho(t) =& -i\omega[a_+^+a_+ + a_-^+a_-, \rho(t)] \\ &+ \sum_{j>i} \omega_{ji}\vec{K}\vec{r}_{ij} \left[\langle c_j^+ c_i \rangle (a_+ e^{ikx} + a_- e^{-ikx}) - \langle c_i^+ c_j \rangle (a_+^+ e^{-ikx} + a_-^+ e^{ikx}), \rho(t) \right] \\ &+ \Lambda \left\{ [a_+\rho(t), a_+^+] + [a_+, \rho(t)a_+^+] + [a_-\rho(t), a_-^+] + [a_-, \rho(t)a_-^+] \right\},\end{aligned} \tag{63}$$

with the dissipation coefficient

$$\Lambda = \pi\hbar\omega^2 g(\hbar\omega)(\vec{K}\vec{R}_{\alpha\beta})^2, \tag{64}$$

depending on quantities which, according to (9), (44), and (12), are of the form

$$K = \sqrt{\alpha\frac{\lambda}{\mathcal{V}}} \tag{65}$$

$$R_{\alpha\beta} = \sqrt{\frac{2\hbar}{M\omega}} \tag{66}$$

$$g(\hbar\omega) = \mathcal{V}^S \frac{\sqrt{2}M^{3/2}}{\pi^2\hbar^3}\sqrt{\hbar\omega}. \tag{67}$$

Unlike the master equation for an electromagnetic field mode derived in (12), we considered explicit expressions of the electron-field potential of interaction, and neglected the thermal energy in comparison with the transition energy. With (65)-(67), the dissipation coefficient (64) takes a form

$$\Lambda = \Omega\frac{\mathcal{V}^S}{\mathcal{V}} \tag{68}$$

depending on the quantity

$$\Omega = 4\alpha\sqrt{\frac{2Mc^2\omega}{\hbar}}, \tag{69}$$

and the two quantization volumes, $\mathcal{V}$ of the electromagnetic field and $\mathcal{V}^S$ of the dissipative electron system. We consider a quantization volume $\mathcal{V}^S = V_n = \frac{1}{N_D}$ for an n-type region, or $\mathcal{V}^S = V_p = \frac{1}{N_A}$ for a p-type region. From physical point of view, a quantization volume of the electromagnetic field $\mathcal{V}$ means a measuring process corresponding to a confinement in this volume. The electromagnetic field can not be quantized in a volume $\mathcal{V}^S$, but in a much larger one, with much larger dimensions than the field wavelength λ. For an electromagnetic field, we consider a unit quantization volume $\mathcal{V} = 1_V = 1_L^3 = 1m^3$, because, in this case, the radiation density is equal to the electromagnetic field density times the light velocity, $S = w_E c\ [W/m^2]$, where $w_E = \frac{\varepsilon_0 E^2}{2}\ [J/m^3]$ is calculated with this quantization volume in the expression (8)-(9) of the field.

We notice that the master equation (63) describes an electromagnetic field quantized in a unit volume $\mathcal{V} = 1m^3$ in interaction with an electron system occupying a much smaller volume $\mathcal{V}^S = V^{(n)}, V^{(p)}$. A system of N dissipative electrons can be taken into account by multiplying the dissipative coefficient Λ with N. However, in such a description, the electromagnetic field is considered of a constant amplitude inside the quantization volume $\mathcal{V} = 1_V$, while, in fact, this amplitude undertakes a spatial variation due to the interaction with the system quantized in a volume $\mathcal{V}^S \ll \mathcal{V}$, i.e. propagation characteristics as the absorption coefficient and the refractive index inside the field quantization volume are not taken into account. We take into account the spatial variation of the electromagnetic field by considering this field as being given by an x dependent density matrix, as product of density matrices for the two counter-propagating waves, $\rho(x,t) = \rho_+(x,t)\rho_-(x,t)$, and taking the dissipative terms as integrals over the paths traveled by these waves. Considering a distribution of N_D dissipative clusters over the the thickness L_D of the device, from the master equation (63), we get

$$\frac{d}{dt}\rho_+(x,t) = -i\omega[a_+^+ a_+, \rho_+(x,t)] + \sum_{j>i} \omega_{ji}\vec{K}\vec{r}_{ij}\left[\langle c_j^+ c_i\rangle a_+ e^{ikx} - \langle c_i^+ c_j\rangle a_+^+ e^{-ikx}, \rho_+(x,t)\right] \quad (70a)$$
$$+\frac{\Omega_D}{1_L}\int_0^x \left\{[a_+\rho_+(x',t'), a_+^+] + [a_+, \rho_+(x',t')a_+^+]\right\} e^{-ik(x-x')}dx'$$
$$\frac{d}{dt}\rho_-(x,t) = -i\omega[a_-^+ a_-, \rho_-(x,t)] + \sum_{j>i} \omega_{ji}\vec{K}\vec{r}_{ij}\left[\langle c_j^+ c_i\rangle a_- e^{-ikx} - \langle c_i^+ c_j\rangle (a_-^+ e^{ikx}, \rho_-(x,t)\right] \quad (70b)$$
$$+\frac{\Omega_D}{1_L}\int_x^{L_D} \left\{[a_-\rho_-(x',t'), a_-^+] + [a_-, \rho_-(x',t')a_-^+]\right\} e^{-ik(x'-x)}dx',$$

depending on the dissipative coefficient

$$\Omega_D = \Omega\frac{1_L^2 L_D}{1_L^3} = \Omega\frac{L_D}{1_L}, \quad (71)$$

obtained by summation over the dissipative clusters with the volume $1_L^2 L_D$ in the quantization volume 1_L^3. The exponential factors in the integrals describe the delay of the field propagating from the coordinate x' of a dissipative element to the coordinate x of the density matrix of this field at this coordinate. These equations describe the dissipative dynamics of a forward electromagnetic wave, propagating from $x' = 0$ to $x' = x$, and of a backward electromagnetic wave, propagating from $x' = L_D$ to $x' = x < L_D$. With these equations, we calculate the mean-values of the field operators

$$a_+(x,t) = \langle a_+\rangle = Tr\{a_+\rho_+(x,t)\} = \mathcal{A}_+(x,t)e^{-i\omega t} \quad (72a)$$
$$a_-(x,t) = \langle a_-\rangle = Tr\{a_-\rho_-(x,t)\} = \mathcal{A}_-(x,t)e^{-i\omega t}. \quad (72b)$$

For an array of two-level systems with the coordinate x, interacting with the electromagnetic field with the frequency $\omega \approx \omega_0$, we define the time slowly-varying amplitude of the polarization $\mathcal{S}(x,t)$ by the relations

$$\langle c_i^+ c_j\rangle = \rho_{ji}(x,t) = \frac{1}{2}\mathcal{S}(x,t)e^{-i\omega t}, \quad (73)$$
$$\mathcal{S}(x,t) = \mathcal{S}_+(x,t)e^{ikx} + \mathcal{S}_-(x,t)e^{-ikx}, \quad (74)$$

where $\mathcal{S}_+(x,t), \mathcal{S}_-(x,t)$ are slowly-varying in space and time amplitudes of the polarization induced by the two counter-propagating waves of the field. Having in view Heisenberg's uncertainty principle

$$\Delta k\Delta x \geq \frac{1}{2}, \qquad \Delta\omega\Delta t \geq \frac{1}{2}, \quad (75)$$

we notice that, in equations (70), this relation selects only the close terms, with $x' \approx x$, while the farer terms in $x - x'$ are washed up by the uncertainty Δk in the oscillating functions under the x'-dependent integrals. By definition, these x-dependent integrals describe the attenuation of an electromagnetic wave squeezed in the x-domain, $\Delta x = 0$. We take into account a finite uncertainty Δx, in the vicinity of x, by extending these x'-integrals with the half-width $\Delta x/2$. We obtain the field equations

$$\frac{d}{dt}\mathcal{A}_+(x,t) = -\frac{1}{2}\omega_0\vec{K}\vec{r}_{01}\mathcal{S}_+(x,t) - \frac{\Omega_D}{1_L}\int_0^{x-\Delta x/2}\mathcal{A}_+(x',t')e^{-\mathrm{i}[k(x-x')-\omega(t-t')]}dx' - \frac{\Omega_D}{1_L}\int_{x-\Delta x/2}^{x+\Delta x/2}\mathcal{A}_+(x',t')e^{-\mathrm{i}[k(x-x')-\omega(t-t')]}dx' \quad (76a)$$

$$\frac{d}{dt}\mathcal{A}_-(x,t) = -\frac{1}{2}\omega_0\vec{K}\vec{r}_{01}\mathcal{S}_-(x,t) + \frac{\Omega_D}{1_L}\int_{x+\Delta x/2}^{L_D}\mathcal{A}_-(x',t')e^{-\mathrm{i}[k(x'-x)-\omega(t-t')]}dx' + \frac{\Omega_D}{1_L}\int_{x-\Delta x/2}^{x+\Delta x/2}\mathcal{A}_-(x',t')e^{-\mathrm{i}[k(x'-x)-\omega(t-t')]}dx'. \quad (76b)$$

We notice that these are non-local in space equations including retarded contributions of the dissipation processes along the distance $|x - x'|$, and absorption processes in the vicinity Δx of the coordinate x. We consider the quantities under these integrals as spectral lines integrated over half-widths, $k\Delta x - \omega\Delta t = (k + \Delta k)\Delta x - (\omega + \Delta\omega)\Delta t = \Delta k\Delta x - \Delta\omega\Delta t = \frac{\pi}{6} + \frac{\pi}{6} > \frac{1}{2} + \frac{1}{2}$. By integrating the first integral of the first equation two times by parts, for a large distance $x - x'$ we obtain

$$\begin{aligned}\int_0^{x-\Delta x/2}\mathcal{A}_+(x',t')e^{-\mathrm{i}[k(x-x')-\omega(t-t')]}dx' &= \frac{1}{\mathrm{i}k}\mathcal{A}_+(x',t')e^{-\mathrm{i}[k(x-x')-\omega(t-t')]}\Big|_0^{x-\Delta x/2} \\ &\quad - \frac{1}{\mathrm{i}k}\int_0^{x-\Delta x/2}\frac{d}{dx'}\mathcal{A}_+(x',t')e^{-\mathrm{i}[k(x-x')-\omega(t-t')]}dx' \\ &= \left[-\frac{\mathrm{i}}{k}\mathcal{A}_+(x,t) + \frac{1}{k^2}\frac{d}{dx}\mathcal{A}_+(x,t)\right]\left[\cos\left(\frac{k\Delta x - \omega\Delta t}{2}\right) - \mathrm{i}\sin\left(\frac{k\Delta x - \omega\Delta t}{2}\right)\right] \\ &= -\frac{1 + \mathrm{i}\sqrt{3}}{2k}\mathcal{A}_+(x,t) + \frac{\sqrt{3} - \mathrm{i}}{2k^2}\frac{d}{dx}\mathcal{A}_+(x,t),\end{aligned} \quad (77)$$

while, taking into account that on a vary short distance Δx the field amplitude is practically constant, the second integral of this equation takes a simple form

$$\begin{aligned}\int_{x-\Delta x/2}^{x+\Delta x/2}\mathcal{A}_+(x',t')e^{-\mathrm{i}[k(x-x')-\omega(t-t')]}dx' &= \frac{1}{\mathrm{i}k}\mathcal{A}_+(x',t')e^{-\mathrm{i}[k(x-x')-\omega(t-t')]}\Big|_{x-\Delta x/2}^{x+\Delta x/2} \\ &= \frac{2}{k}\mathcal{A}_+(x,t)\sin\left(\frac{k\Delta x - \omega\Delta t}{2}\right) = \frac{1}{k}\mathcal{A}_+(x,t).\end{aligned} \quad (78)$$

These terms describe a slight variation of the wave-vector k, $k' = k + \kappa$, which means that the amplitude of the mean-value of the field operator takes a form

$$\mathcal{A}_+(x,t) = \tilde{\mathcal{A}}_+(x,t)e^{\mathrm{i}\kappa x}. \quad (79)$$

Taking into account that $\frac{\Omega_D}{ck^2 1_L} \ll 1$, while $\kappa = \frac{k}{1+\frac{2ck^2 1_L}{\sqrt{3}\Omega_D}} \ll k$, we get a field equation

$$\frac{\partial}{\partial t}\mathcal{A}_+(x,t) + c\left[\frac{\partial}{\partial x}\mathcal{A}_+(x,t) + \alpha'\mathcal{A}_+(x,t)\right] = -\frac{1}{2}\omega_0\vec{K}\vec{r}_{01}\mathcal{S}_+(x,t), \quad (80)$$

with an absorption coefficient

$$\alpha' = \frac{2\alpha c}{1_L}\sqrt{\frac{2M}{\hbar\omega}}\frac{L_D}{1_L}. \tag{81}$$

In the following calculations, we are interested in a form of this equation in a cavity with the length L_D,

$$\frac{d}{dt}\mathcal{A}_+(x,t) = -\gamma_F \mathcal{A}_+(x,t) - \frac{1}{2}\omega_0 \vec{K}\vec{r}_{01}\mathcal{S}_+(x,t), \tag{82}$$

with a field decay rate

$$\gamma_F = c\alpha' = \frac{2\alpha c^2}{1_L^2}\sqrt{\frac{2M}{\hbar\omega}}L_D, \text{ or } 1_L\gamma_F = \frac{2\alpha c^2}{1_L}\sqrt{\frac{2M}{\hbar\omega}}L_D, \tag{83}$$

which we call field decay velocity. In section 6, it will be shown that the field decay velocity $1_L\gamma_F$ describes the field loss by dissipation, as the quantity $\mathcal{T}c$ describes the electromagnetic energy loss by radiation through the output mirror with the transparency $\mathcal{T}$. We notice that the decay rate of an electromagnetic field in a cavity is proportional to the length of this cavity. For a semiconductor chip with the thickness $L_D = 2\ mm$, we considered in our calculations in (3), from (83) we get a field decay rate $\gamma_F = 2.05 \times 10^7\ s^{-1}$, which is in agreement with the empirical values $\gamma_F = 10^7,\ 10^8\ s^{-1}$, we considered in these calculations. It is interesting that, in this model, the decay rate does not depend on the concentration of the dissipative clusters, since an increase of this concentration means a decrease of the density of states in every cluster, which, in this way, becomes smaller. These two variations cancel exactly one another in the final result. By taking into account the spreading of a dissipative electron wave-function beyond the the boundaries of its cluster due to the thermal motion, one obtains a lower value of the decay rate, but with an increase with the concentration of these clusters.

5. Optical equations for a system of quantum injection dots

From the quantum master equation (34), we derive optical equations for a two-level system. In the approximation of the slowly varying amplitudes, we consider the non-diagonal matrix elements

$$\rho_{10}(t) = \rho_{01}^*(t) = \frac{1}{2}\left[\mathcal{S}_+(t)e^{ikx} + \mathcal{S}_-(t)e^{-ikx}\right]e^{-i\omega t}, \tag{84}$$

and the population difference

$$w(t) = \rho_{11}(t) - \rho_{00}(t), \text{ with the normalization condition} \tag{85a}$$

$$1 = \rho_{11}(t) + \rho_{00}(t). \tag{85b}$$

Calculating the matrix elements of the two-level system, and averaging over the field states, from the master equation (34) we get:

$$\begin{aligned}\frac{d}{dt}\rho_{10}(t) = & -[\lambda_{01} + \lambda_{10} + i(\omega_0 + \zeta_{11} - \zeta_{00})]\rho_{10}(t) \\ & +\vec{K}\left[(\langle a_+\rangle + \langle a_-^+\rangle)e^{ikx} + (\langle a_+^+\rangle + \langle a_-\rangle)e^{-ikx}\right]\omega_0\vec{r}_{10}[\rho_{00}(t) - \rho_{11}(t)] \\ & +(\zeta_{11} - \zeta_{00})^2\int_{t-\tau}^{t}\rho_{10}(t')e^{-i[\phi_{10}(t')+\omega(t-t')]}dt'\end{aligned} \tag{86a}$$

$$\begin{aligned}\frac{d}{dt}\rho_{11}(t) = & -\frac{d}{dt}\rho_{00}(t) = 2[\lambda_{10}\rho_{00} - \lambda_{01}\rho_{11}] \\ & +\vec{K}\left[(\langle a_+\rangle + \langle a_-^+\rangle)e^{ikx} + (\langle a_+^+\rangle + \langle a_-\rangle)e^{-ikx}\right]\omega_0\vec{r}_{10}[\rho_{10}(t) + \rho_{01}(t)].\end{aligned} \tag{86b}$$

From the expression (8) of the quantized electric field $\vec{E}$ in the plane-wave approximation, and mean-values of the annihilation operators of the form

$$\langle a_+ \rangle = \bar{a}_+(t)e^{-\mathrm{i}\omega t} \tag{87a}$$

$$\langle a_- \rangle = \bar{a}_-(t)e^{-\mathrm{i}\omega t}, \tag{87b}$$

we get the mean-value of this field,

$$\langle \vec{E} \rangle = \frac{1}{2}\left[\vec{\mathcal{E}}(t)e^{-\mathrm{i}\omega t} + \vec{\mathcal{E}}^*(t)e^{\mathrm{i}\omega t}\right], \tag{88}$$

with the time slowly-varying amplitude

$$\vec{\mathcal{E}}(t) = \vec{\mathcal{E}}_+(t)e^{\mathrm{i}kx} + \vec{\mathcal{E}}_-(t)e^{-\mathrm{i}kx}, \tag{89}$$

while the amplitudes of the two counter-propagating waves are

$$\vec{\mathcal{E}}_+(t) = 2\mathrm{i}\frac{\hbar\omega}{e}\vec{K}\bar{a}_+(t) \tag{90a}$$

$$\vec{\mathcal{E}}_-(t) = 2\mathrm{i}\frac{\hbar\omega}{e}\vec{K}\bar{a}_-(t). \tag{90b}$$

In this description we neglect the variation of the amplitudes inside the cavity, by taking into account these two amplitudes only as mean-values over the space coordinate, related by the boundary condition for the output mirror of transmission coefficient $\mathcal{T}$:

$$\vec{\mathcal{E}}_-(t) = -\sqrt{1-\mathcal{T}}\vec{\mathcal{E}}_+(t). \tag{91}$$

With the notations

$$\vec{g} = \frac{e}{\hbar}\vec{r}_{10} \tag{92}$$

for the coupling coefficient,

$$\gamma_\perp = \lambda_{01} + \lambda_{10} \tag{93}$$

for the dephasing rate,

$$\gamma_\parallel = 2(\lambda_{01} + \lambda_{10}) \tag{94}$$

for the decay rate,

$$\gamma_n = |\zeta_{11} - \zeta_{00}| \tag{95}$$

for the fluctuation rate of the self-consistent field, and

$$w_T = -\frac{\lambda_{01} - \lambda_{10}}{\lambda_{01} + \lambda_{10}}, \tag{96}$$

from (84)-(91) we obtain equations for the slowly-varying amplitudes

$$\begin{aligned}\frac{d}{dt}\mathcal{S}_+(t) = {} & -[\gamma_\perp + \mathrm{i}(\omega_0 + \gamma_n - \omega)]\mathcal{S}_+(t) + \mathrm{i}\vec{g}\vec{\mathcal{E}}_+(t)w(t) \\ & + \gamma_n^2 \int_{t-\tau}^{t} \mathcal{S}_+(t')e^{-\mathrm{i}[\phi_{10}(t') + \omega(t-t')]}dt' \end{aligned} \tag{97a}$$

$$\frac{d}{dt}w(t) = -\gamma_\parallel[w(t) - w_T] + (2-\mathcal{T})\mathrm{i}\vec{g}\frac{1}{2}\left[\vec{\mathcal{E}}_+^*(t)\mathcal{S}_+(t) - \vec{\mathcal{E}}_+(t)\mathcal{S}_+^*(t)\right]. \tag{97b}$$

In equation (97b) we have taken into account that the term

$$\Phi_+(t) = \mathrm{i}g\frac{1}{2}\left[\vec{\mathcal{E}}_+^*(t)\mathcal{S}_+(t) - \vec{\mathcal{E}}_+(t)\mathcal{S}_+^*(t)\right] \tag{98}$$

is a particle flow due to the forward electromagnetic wave propagating in the cavity, while

$$\Phi_-(t) = \mathrm{i}\vec{g}\frac{1}{2}\left[\vec{\mathcal{E}}_-^*(t)\mathcal{S}_-(t) - \vec{\mathcal{E}}_-(t)\mathcal{S}_-^*(t)\right] \tag{99}$$

is a particle flow due to the backward electromagnetic wave, which means that the two flows satisfy the boundary condition for the energy flow of the electromagnetic field

$$\Phi_-(t) = (1-\mathcal{T})\Phi_+(t). \tag{100}$$

At the same time, calculating the mean-value of the field operator a, averaging over the states of the two-level system, and taking into account the relation

$$\langle c_i^+ c_j\rangle = \rho_{ji}(t), \tag{101}$$

from equation (34) we get the field equation

$$\frac{d}{dt}\langle a_+\rangle = -\mathrm{i}\omega\langle a_+\rangle + \vec{K}\omega_0\vec{r}_{10}[\rho_{10}(t) - \rho_{01}(t)]e^{-\mathrm{i}kx}. \tag{102}$$

Thus, with (84), (87) and (90), we get a field equation for slowly-varying amplitudes

$$\frac{d}{dt}\vec{\mathcal{E}}_+(t) = -\mathrm{i}\omega_0\frac{\hbar\omega}{e}\vec{K}(\vec{K}\vec{r}_{10})\mathcal{S}_+(t). \tag{103}$$

We consider this equation for the components $u(t)$ and $v(t)$ of the polarization amplitude

$$\mathcal{S}_+(t) = u(t) - \mathrm{i}v(t), \tag{104}$$

and $\mathcal{F}(t)$ and $\mathcal{G}(t)$ of the electromagnetic field

$$\mathcal{E}_+(t) = \mathcal{F}(t) + \mathrm{i}\mathcal{G}(t), \tag{105}$$

and take into account the field dissipation described by the dissipation rate γ_F. We get

$$\frac{d}{dt}\mathcal{F}(t) = -\gamma_F\mathcal{F}(t) - g\frac{\hbar\omega_0}{2\varepsilon\mathcal{V}}v(t) \tag{106a}$$

$$\frac{d}{dt}\mathcal{G}(t) = -\gamma_F\mathcal{G}(t) - g\frac{\hbar\omega_0}{2\varepsilon\mathcal{V}}u(t). \tag{106b}$$

We consider these equations for the electromagnetic energy in the quantization volume $\mathcal{V}$, and introduce the energy flow through the surface $\mathcal{A}$ of this volume:

$$\frac{d}{dt}\left[\mathcal{V}\frac{1}{2}\varepsilon\mathcal{F}^2(t)\right] = -\mathcal{T}c\frac{1}{2}\varepsilon\mathcal{F}^2(t)\mathcal{A} - \gamma_F\mathcal{V}\varepsilon\mathcal{F}^2(t) - g\frac{\hbar\omega_0}{2}\mathcal{F}v(t) \tag{107a}$$

$$\frac{d}{dt}\left[\mathcal{V}\frac{1}{2}\varepsilon\mathcal{G}^2(t)\right] = -\mathcal{T}c\frac{1}{2}\varepsilon\mathcal{G}^2(t)\mathcal{A} - \gamma_F\mathcal{V}\varepsilon\mathcal{G}^2(t) - g\frac{\hbar\omega_0}{2}\mathcal{G}u(t). \tag{107b}$$

At the same time, from (97b) with (104) and (105), we derive the equation for the population difference (85a), and introduce the particle flow $\mathcal{I}$ in a two-level system, due to the electric current $I = eA_D N_e \mathcal{I}$ injected in the device:

$$\frac{d}{dt}w(t) = -\gamma_F[w(t) - w_T] + 2\mathcal{I} + (2-\mathcal{T})g[\mathcal{F}(t)v(t) + \mathcal{G}(t)u(t)] \tag{108}$$

From (107) and (108) with (85), we get an equation of energy conservation:

$$\begin{aligned}\hbar\omega_0\mathcal{I} = \frac{d}{dt}\left\{\hbar\omega_0\rho_{11}(t) + (2-\mathcal{T})\mathcal{V}\frac{1}{2}\varepsilon[\mathcal{F}^2(t) + \mathcal{G}^2(t)]\right\} + \gamma_\parallel\left[\rho_{11}(t) - \frac{1+w_T}{2}\right]\hbar\omega_0 \\ +(2-\mathcal{T})(\mathcal{T}c\frac{\mathcal{A}}{\mathcal{V}} + 2\gamma_F)\mathcal{V}\frac{1}{2}\varepsilon[\mathcal{F}^2(t) + \mathcal{G}^2(t)].\end{aligned} \tag{109}$$

This equation describes the transition power $\hbar\omega_0\mathcal{I}$ of the active system providing the energy transfer processes involved in the dissipative super radiant decay: (1) the energy variation of the electron-field system, (2) the dissipative decay of the electron energy, proportional to $\gamma_\parallel$, (3) the radiation of the field energy, proportional to the light velocity c and the transmission coefficient $\mathcal{T}$ of the output mirror, and (4) the dissipation of the field energy, proportional to γ_F. In this equation, both waves leaving the quantum system and propagating in the cavity, the forward wave with an amplitude coefficient 1 and the backward wave with an amplitude coefficient $\mathcal{R} = 1 - \mathcal{T}$, are taken into account with the coefficient $1 + \mathcal{R} = 2 - \mathcal{T}$.

From the polarization equation (97a) with (104) and (105), the population equation (108), and the field equations (107), we obtain the equations of the slowly varying amplitudes of the system:

$$\begin{aligned}\frac{d}{dt}u(t) = &\quad -\gamma_\perp[u(t) - \delta\omega v(t)] - g\mathcal{G}(t)w(t) \\ &+\gamma_n^2\int_{t-\tau}^{t}\{u(t')\cos[\phi_n(t') + (\omega-\omega_0)(t-t')] + v(t')\sin[\phi_n(t') + (\omega-\omega_0)(t-t')]\}dt'\end{aligned} \tag{110a}$$

$$\begin{aligned}\frac{d}{dt}v(t) = &\quad -\gamma_\perp[v(t) + \delta\omega u(t)] - g\mathcal{F}(t)w(t) \\ &+\gamma_n^2\int_{t-\tau}^{t}\{v(t')\cos[\phi_n(t') + (\omega-\omega_0)(t-t')] - u(t')\sin[\phi_n(t') + (\omega-\omega_0)(t-t')]\}dt'\end{aligned} \tag{110b}$$

$$\frac{d}{dt}w(t) = -\gamma_\parallel[w(t) - w_T] + 2\mathcal{I} + (2-\mathcal{T})g[\mathcal{G}(t)u(t) + \mathcal{F}(t)v(t)] \tag{110c}$$

$$\frac{d}{dt}\mathcal{F}(t) = -\frac{1}{2}\mathcal{T}c\frac{\mathcal{A}}{\mathcal{V}}\mathcal{F}(t) - \gamma_F\mathcal{F}(t) - g\frac{\hbar\omega_0}{2\varepsilon\mathcal{V}}v(t) \tag{110d}$$

$$\frac{d}{dt}\mathcal{G}(t) = -\frac{1}{2}\mathcal{T}c\frac{\mathcal{A}}{\mathcal{V}}\mathcal{G}(t) - \gamma_F\mathcal{G}(t) - g\frac{\hbar\omega_0}{2\varepsilon\mathcal{V}}u(t), \tag{110e}$$

where $\phi_n(t') \equiv \phi_{01}(t') \equiv -\phi_{10}(t')$ is the phase fluctuation with a fluctuation time $\tau_n = 1/\gamma_n$, and

$$\delta\omega = \frac{\omega - \omega_0 - \gamma_n}{\gamma_\perp} \tag{111}$$

is the relative atomic detuning. In these equations, the coupling of the electron system to the electromagnetic field is described by a coupling coefficient for the dipole interaction $g = \vec{g}\vec{1}_E$. These equations also describe a dissipative decay of the electron system by the coefficients $\gamma_\parallel$

and $\gamma_\perp$, non-Markovian effects by time-integrals proportional to the fluctuation coefficient γ_n^2 in the polarization equations (110a) and (110b), a decrease of the electron-field coupling due to the field radiation by the term proportional to the coefficient $(2 - \mathcal{T})$ in (110c), and a decrease of field by the radiation terms proportional to the product $c\mathcal{T}$ in (110d) and (110e), and by the terms proportional to the decay rate γ_F.

6. Superradiant quantum injection dots and heat conversion

The dynamic equations (110) take a simpler form in a stationary regime when the derivatives with time become zero and the polarization variables can be taken out from the integrals. Considering an integration over a fluctuation time $\tau_n = 1/\gamma_n$, we get a time oscillating term, generated by the fluctuations of the environment particles. In the Markovian approximation, when these oscillations are neglected, we get the steady-state equations:

$$-\gamma_\perp[u - \delta\omega v] - g\mathcal{G}w = 0 \tag{112a}$$
$$-\gamma_\perp[v + \delta\omega u] - g\mathcal{F}w = 0 \tag{112b}$$
$$-\gamma_\parallel(w - w_T) + 2\mathcal{I} + (2 - \mathcal{T})g(\mathcal{G}u + \mathcal{F}v) = 0 \tag{112c}$$
$$-\Gamma_F\mathcal{F} - Gv = 0 \tag{112d}$$
$$-\Gamma_F\mathcal{G} - Gu = 0, \tag{112e}$$

where

$$G = g\frac{\hbar\omega_0}{2\varepsilon\mathcal{V}} \tag{113a}$$
$$\Gamma_F = \frac{1}{2}\mathcal{T}c\frac{\mathcal{A}}{\mathcal{V}} + \gamma_F. \tag{113b}$$

From the system of equations (112), for the resonance case ($\delta\omega = 0$), we calculate the flow density of the electromagnetic energy radiated by the device:

$$S = \mathcal{T}c\frac{1}{2}\varepsilon(\mathcal{F}^2 + \mathcal{G}^2). \tag{114}$$

We get

$$S = \frac{\frac{\hbar\omega_0}{(2-\mathcal{T})\mathcal{A}}}{1 + \frac{2\gamma_F\mathcal{V}}{\mathcal{T}c\mathcal{A}}}\left[\mathcal{I} - \left(-w_T\frac{\gamma_\parallel}{2} + \frac{\frac{1}{2}\mathcal{T}c\frac{\mathcal{A}}{\mathcal{V}} + \gamma_F}{\frac{g^2\hbar\omega_0}{\gamma_\perp\gamma_\parallel\varepsilon\mathcal{V}}}\right)\right]. \tag{115}$$

This expression of the flow density S has a nice physical interpretation, being proportional to the product of the transition energy $\hbar\omega_0$, divided to the radiation area of a quantum dot $\mathcal{A}$, with the difference between the particle flow $\mathcal{I}$ and a threshold value depending on coupling, radiation, and dissipation coefficients. This expression is valid when the quantization volume $\mathcal{V}$ of the field corresponds to the electromagnetic energy delivered by the whole system of $N_e N_t$ quantum dots to a volume unit, which means

$$\mathcal{V}[m^3] = \frac{1}{N_e[m^{-2}]N_t[m^{-1}]}, \tag{116}$$

where $N_e[m^{-2}]$ is the number of quantum dots per area unit, and $N_t[m^{-1}]$ is the number of super radiant junctions per length unit.

In a first approximation, we neglect the temperature variation due to the heat transfer through the semiconductor structure. To take into account this temperature variation, one has to make corrections of the parameters, to obtain the same transition frequency on the whole chain of super radiant junctions.

Such a device can be realized in two versions schematically represented in figure 4: (a) a longitudinal device, with the two mirrors M_1 and M_2 made on the two surfaces in the plane of the chip, of transmission coefficients $\mathcal{T}_0 = 0$ and $\mathcal{T} > 0$, which form a Fabry-Perot cavity coupling a super radiant mode that propagates in the x-direction of the injection current; (b) a transversal device, with the two mirrors M_1 and M_2 made on two lateral surfaces of the chip, of transmission coefficients $\mathcal{T}_0 = 0$ and $\mathcal{T} > 0$, which form a Fabry-Perot cavity coupling a super radiant mode that propagates in the y-direction, perpendicular to the injection current. While in version (a) the roles of the mirrors M_1 and M_2, and of the injection electrodes $\mathcal{E}_1$ and $\mathcal{E}_2$, are played by the same metalizations, made on the two surfaces in the plane of the chip, in version (b) the mirror metalizations M_1 and M_2, which are made on two lateral surfaces, are different from the electrode metalizations $\mathcal{E}_1$ and $\mathcal{E}_2$.

The two devices have the same structure, including layers of $GaAs$, with a narrower forbidden band and a heavier doping, for the quantum wells, and layers of $Al_xGa_{1-x}As$, with a larger forbidden band and a lighter doping, for the potential barriers. The margins of these bands are determined by the concentrations of the donors (acceptors) embedded in the semiconductor layers. For a longitudinal device (figure 4a), the $\bar{N}_t$ (dimensionless) quantum dots in the x-direction, radiate through an area $\dfrac{1}{N_e[m^{-2}]}$, which means

$$\mathcal{A}_L[m^2] = \frac{1}{N_e[m^{-2}]\bar{N}_t}, \tag{117}$$

while for a transversal device (see figure 4b), the $\sqrt{N_e[m^{-2}]A_D[m^2]}$ quantum dots in the y-direction, radiate through an area $\dfrac{L_D[m]}{\bar{N}_t}\dfrac{1}{\sqrt{N_e[m^{-2}]}}$, which means

$$\mathcal{A}_T[m^2] = \frac{L_D[m]}{N_e[m^{-2}]\bar{N}_t\sqrt{A_D[m^2]}}. \tag{118}$$

With the radiation area $\mathcal{A}_L$ ($\mathcal{A}_T$) of a quantum dot, from (115) we derive the flow density S_L (S_D), and the total flow of the electromagnetic field radiated by the device in the two versions:

$$\Phi_L = A_D S_L \tag{119a}$$

$$\Phi_T = L_D\sqrt{A_D}S_T. \tag{119b}$$

We obtain

$$\Phi_L = \frac{\bar{N}_t}{(2-\mathcal{T})(1+2\frac{1_L\gamma_F}{\mathcal{T}c})} \cdot \frac{\hbar\omega_0}{e}(I - I_{0L}) \tag{120a}$$

$$\Phi_T = \frac{\bar{N}_t}{(2-\mathcal{T})(1+2\frac{1_L\gamma_F}{\mathcal{T}c}\frac{A_D^{1/2}}{L_D})} \cdot \frac{\hbar\omega_0}{e}(I - I_{0T}), \tag{120b}$$

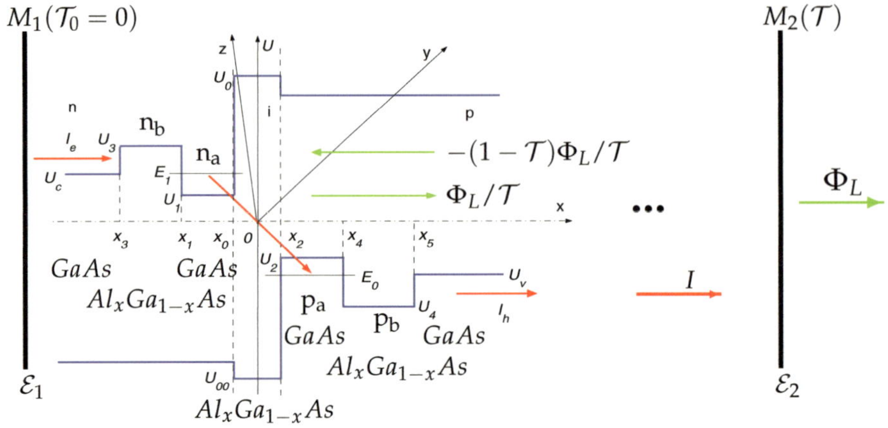

(a) Longitudinal super radiant device with the Fabry-Perot cavity oriented in the x-direction of the injected current $I = I_e = I_h$, i.e. perpendicular to the semiconductor layers.

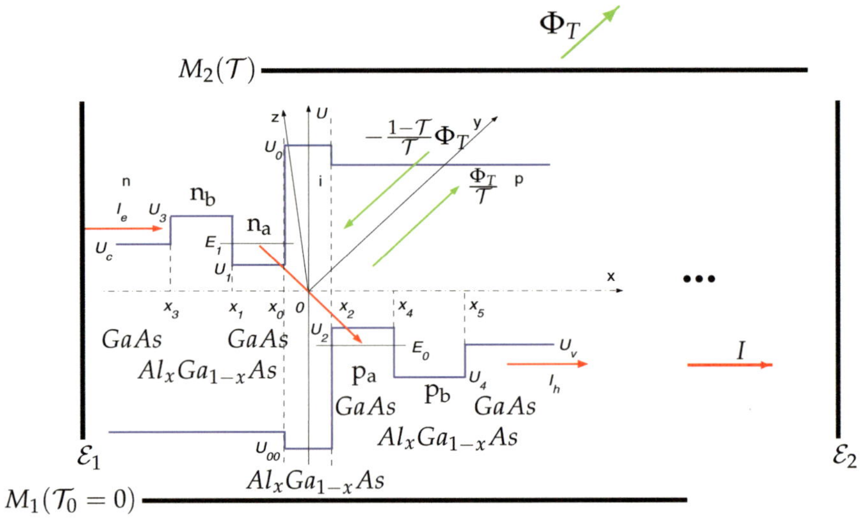

(b) Transversal super radiant device with the Fabry-Perot cavity oriented in the y-direction, perpendicular to the injected current $I = I_e = I_h$, i.e. in the plane of the semiconductor layers.

Fig. 4. Dissipative super radiant n-i-p device with two injection electrodes $\mathcal{E}_1$ and $\mathcal{E}_2$ and a Fabry-Perot cavity with the mirrors M_1 and M_2 of transmission coefficients $\mathcal{T}_0 = 0$ and $\mathcal{T}$, respectively, in two possible versions (a) and (b).

as a function of the injected current I and the threshold currents

$$I_{0L} = \frac{1}{2} e N_e A_D \gamma_{\parallel} \left[-w_T + \frac{\varepsilon \gamma_{\perp}}{g_L^2 \hbar \omega_0 N_e \bar{N}_t} (\mathcal{T} c + 2 \cdot 1_L \gamma_F) \right] \tag{121a}$$

$$I_{0T} = \frac{1}{2} e N_e A_D \gamma_{\parallel} \left[-w_T + \frac{\varepsilon \gamma_{\perp}}{g_T^2 \hbar \omega_0 N_e \bar{N}_t} (\mathcal{T} c \frac{L_D}{A_D^{1/2}} + 2 \cdot 1_L \gamma_F) \right], \tag{121b}$$

which depends on the field decay velocity (83). The threshold current is proportional to the threshold population, which includes three terms for the three dissipative processes that must be balanced by current injection for creating a coherent electromagnetic field: (1) a term $-w_T$, for a population inversion, (2) a term proportional to the light velocity c and the transmission coefficient $\mathcal{T}$, for the field radiation, and (3) a term proportional to decay rate γ_F, for the field dissipation.

From (112c) and (120), we notice that when the injection current $I = eN_eA_D\mathcal{I}$ is under the threshold value I_{0L} (I_{0T}), the radiation field is $\mathcal{F} + i\mathcal{G} = 0$, while the population difference w increases with this current. When the injection current I reaches the threshold current I_{0L} (I_{0T}), the population difference w reaches the radiation value

$$w_R = \frac{\mathcal{T}c\frac{A}{\mathcal{V}} + 2\gamma_F}{\frac{g^2\hbar\omega_0}{\gamma_\perp \varepsilon \mathcal{V}}}. \tag{122}$$

Increasing the injection current I beyond the threshold value I_{0L} (I_{0T}), the population difference keeps this value, while the super radiant field and the polarization ($u = -\frac{g}{\gamma_\perp}w_R\mathcal{G}, v = -\frac{g}{\gamma_\perp}w_R\mathcal{F}$) increases. However, the polarization (u, v) can not increase indefinitely, being constrained by the condition of the Bloch vector length $(2-\mathcal{T})(u^2 + v^2) + w^2 \leq w_T^2$. For the maximum value (u_M, v_M) of the polarization, while $u_M^2 + v_M^2 = (w_T^2 - w_R^2)/(2-\mathcal{T})$, the super radiant field reaches its maximum flow density

$$S_M = \frac{\mathcal{T}c\varepsilon}{2(2-\mathcal{T})}\left[w_T^2 \frac{g^2\frac{\hbar^2\omega_0^2}{\varepsilon^2\mathcal{V}^2}}{\left(\mathcal{T}c\frac{A}{\mathcal{V}} + 2\gamma_F\right)^2} - \frac{\gamma_\perp^2}{g^2}\right]. \tag{123}$$

From this equation with equation (115) for $S = S_M$, we get the value $I_M = eN_eA_D\mathcal{I}_M$ of the injection current producing the maximum flow of the electromagnetic energy. Increasing the injection current beyond this value, the polarization (u, v) will not increase any more, but the population will increase, leading to a rapid decrease of the polarization. Neglecting the current increase from I_M to the value I'_M when the polarization vanishes, from equation (112c) with $w = -w_T$ and $u = v = 0$, we get a simple, approximate expression $I_M \approx I'_M = \frac{1}{2}eN_eA_D\gamma_\parallel(-w_T - w_T)$, which can be compared with (121). From the operation condition $I_{0L}, I_{0T} < I_M$, we get conditions for the coupling, dissipation, and radiation coefficients:

$$w_{IL} = \frac{\varepsilon_0\gamma_\perp}{g_L^2\hbar\omega_0 N_e\bar{N}_t}(\mathcal{T}c + 21_L\gamma_F) < -w_T \approx 1 \tag{124a}$$

$$w_{IT} = \frac{\varepsilon_0\gamma_\perp}{g_T^2\hbar\omega_0 N_e\bar{N}_t}(\mathcal{T}c\frac{L_D}{A_D^{1/2}} + 21_L\gamma_F) < -w_T \approx 1. \tag{124b}$$

From equations (46) and (53), with (47) and (43), we notice that the dephasing and decay rates (93) and (94) strongly depend on the i-layer thickness $x_2 - x_0$. In figure 5a, we represent the decay rates $\gamma_\parallel^P, \gamma_\parallel^E, \gamma_\parallel^{EM}$ for the three dissipative couplings, with the phonons of the crystal vibrations, the conduction electrons and holes, and the free electromagnetic field, $\gamma_\parallel = \gamma_\parallel^P + \gamma_\parallel^E + \gamma_\parallel^{EM}$. We also represent the fluctuation rate (95) with the components (48) for the two neighboring conduction regions n and p, $\gamma_n^2 = \left[\gamma_n^{(n)}\right]^2 + \left[\gamma_n^{(p)}\right]^2$. In figure

5b, we represent the coupling coefficients for a longitudinal and a transversal structure. For all these coefficients, we get quasi-exponential variations with the i-layer thickness $x_2 - x_0$. We notice that, in these structures, the phonon decay rate $\gamma_{\|}^{P}$, which is unavoidable, dominates the electric decay rate $\gamma_{\|}^{E}$, which depends on the separation barriers, while the electromagnetic decay rate $\gamma_{\|}^{EM}$ is negligible. It is remarkable that the decay rate of a quantum

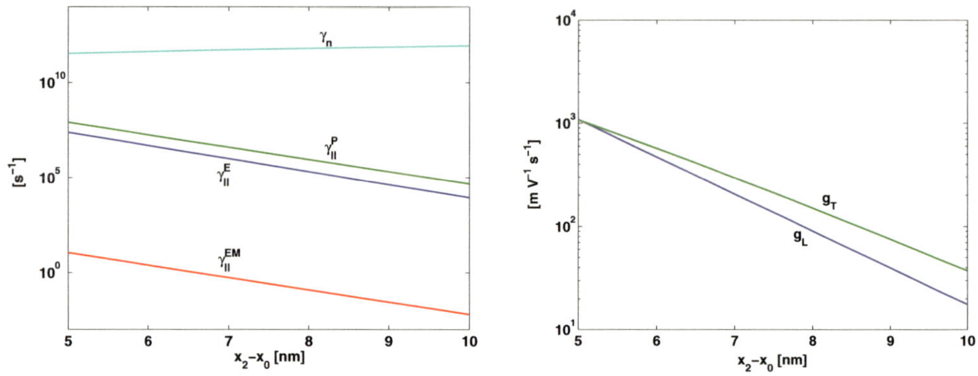

(a) (b)

Fig. 5. (a) The dependence of the dissipative coefficients on the i-zone thickness; (b) The dependence of the coupling coefficients on the i-zone thickness, for a longitudinal and a transversal structure.

injection dot, with a value around $10^7 s^{-1}$, is significantly lower than the decay rates of other $GaAs$ structures, which are at least somewhere around $10^{12} s^{-1}$ (13). From figure 5b, we notice that, although the two coupling coefficients are calculated with completely different dipole moments, g_L with (42b)-(42c), and g_T with (42a), the values of these coefficients are approximately equal for small values of $x_2 - x_0$, and keep near values for thicker i-zones.

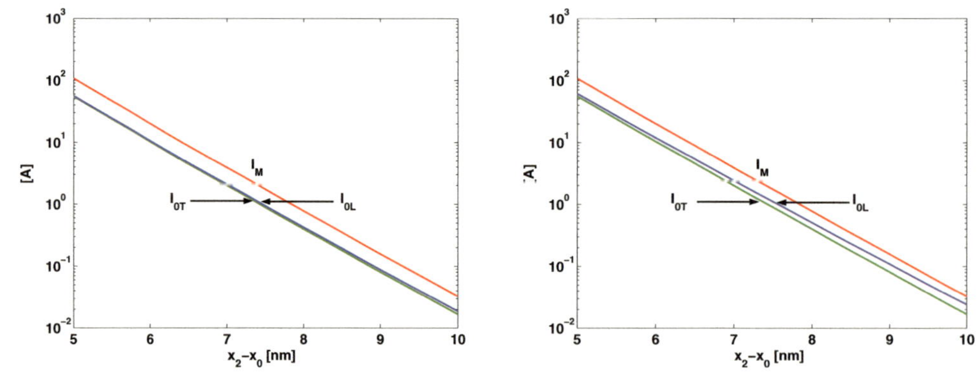

(a) (b)

Fig. 6. The dependence of the threshold currents on the thicknesses of the i-zone for two values of the transmission coefficient of the output mirror: (a) $\mathcal{T} = 0.1$; (b) $\mathcal{T} = 0.5$.

From equations (121), we notice that the dissipative rates $\gamma_{\|}, \gamma_{\perp}$ and the coupling coefficient g_L (g_T), determine the threshold current I_{0L} (I_{0T}). In figure 6a we represent the dependence of these currents in comparison with the maximum current I_M, for two values of the output transmission coefficient $\mathcal{T}$. We notice that the operation condition $I_{0L}, I_{0T} > I_M$ is satisfied for both values of these coefficients. This property can be understood from the analytical expressions (121) or (124), having in view that the dephasing rate $\gamma_\perp$ and the square of the coupling coefficient g_L (g_T) are proportional to the square of the dipole moment, which means that the operation condition does not depend on this moment. Since the quantum

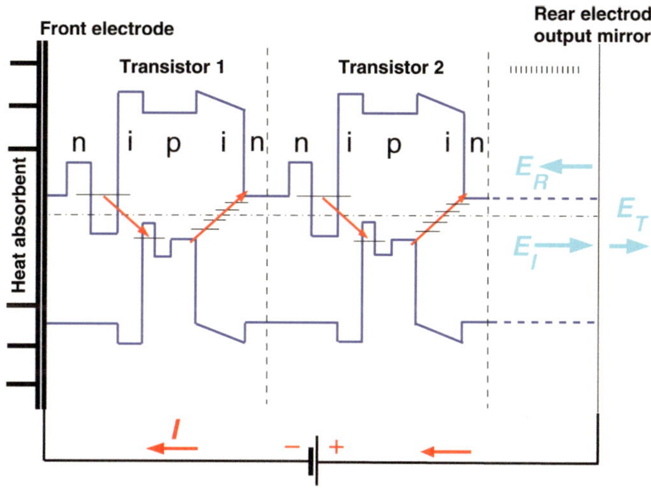

Fig. 7. Quantum heat converter, as a packet of super radiant transistors, thermally coupled to a heat absorbent. While a current I is injected in the device, an electromagnetic flow is obtained, mainly on the account of the heat absorption.

dot density N_e is determined by physical conditions, according to (25), the threshold current (121) can be controlled only by the number of super radiant transistors $\bar{N}_t$ in the structure. In our calculations we considered a number of superadiant transistors $\bar{N}_t = 1045$. While the heat propagates from the heat absorbent (see figure 7) throughout the semiconductor structure, a portion of this heat is absorbed by every super radiant transistor, producing a temperature decrease from the front electrode to the rear one. In figure 8a we represent the electric power and the radiation power as functions of the injected current, for a longitudinal and a transversal configuration of the device. A radiation power arises only when the injection current exceeds a threshold value. From (121a) and (121b), we notice that, due to the factor $\frac{L_D}{A_D^{1/2}}$ in the radiation term of the population inversion, the threshold current of a transversal device is lower than that of a longitudinal one. However, due the same factor at the denominator of (120b), the increase of the radiation power with the injection current is lower for a transversal device than for a longitudinal one. In figure 8b the total temperature variation in the semiconductor structure is represented. We notice that a rather high power of 200 W, that means 0.500 MW from an active area of 1 m^2, can be obtained at a rather low temperature difference of about 7 ^0C.

The radiation power of a transversal device becomes much higher by increasing the transmission coefficient from $\mathcal{T} = 0.1$ to $\mathcal{T} = 0.5$ and the transition dipole moment by diminishing the thickness of the i-zone from $x_2 - x_0 = 6.5\ nm$ to $x_2 - x_0 = 6\ nm$ as is represented in figure 9. In this case, the threshold current of the transversal device becomes significantly lower than that of the longitudinal one. The threshold current of the longitudinal device is sgnificantly lowered by decreasing the transmission coefficient from $\mathcal{T} = 0.5$ to $\mathcal{T} = 0.2$ as is represented in figure 10.

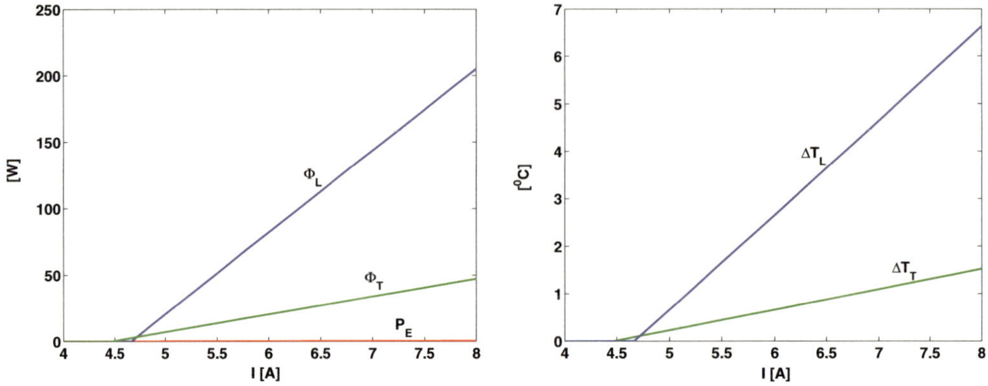

(a) (b)

Fig. 8. (a) The radiation powers Φ_L and Φ_T and the electric power P_E as functions of the injection current I, for $x_2 - x_0 = 6.5\ nm$, $\mathcal{T} = 0.1$, and $\gamma_F = 10^7\ s^{-1}$; (b) The temperature variations ΔT_L, ΔT_T as functions of the injection current I.

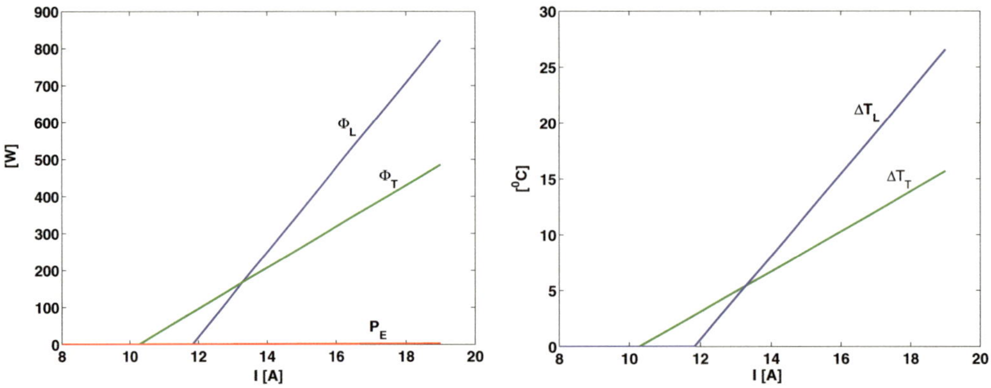

(a) (b)

Fig. 9. The radiation powers Φ_L and Φ_T, the electric power P_E, and the temperature variations ΔT_L, ΔT_T as functions of the injection current, for $x_2 - x_0 = 6\ nm$, $\mathcal{T} = 0.5$, and $\gamma_F = 10^7\ s^{-1}$.

It is remarkable that in the three cases represented in figures 8-10 the electric power dissipated in the device by the injection current I is much lower than the super radiant power. This is because, as one can notice also from (120), the super radiant power produced by the injected current corresponds to the high transition energy $\hbar\omega_0$ between the two zones n and p, while the power electrically dissipated by this current corresponds to a very low potential difference $U_c - U_{c1}$, necessary for carrying this current through the two rather thin highly conducting zones n and p (figure 7b). The difference between these two powers is obtained by heat absorption, when the electrons are excited from the lower potential of the p-zone to the higher

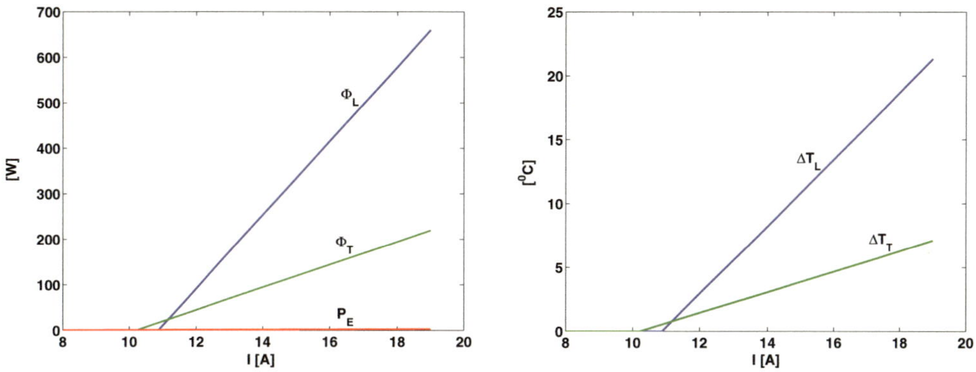

(a) (b)

Fig. 10. The radiation powers Φ_L and Φ_T, the electric power P_E, and the temperature variations ΔT_L, ΔT_T as functions of the injection current, for $x_2 - x_0 = 6\ nm$, $\mathcal{T} = 0.2$, and $\gamma_F = 10^7\ s^{-1}$.

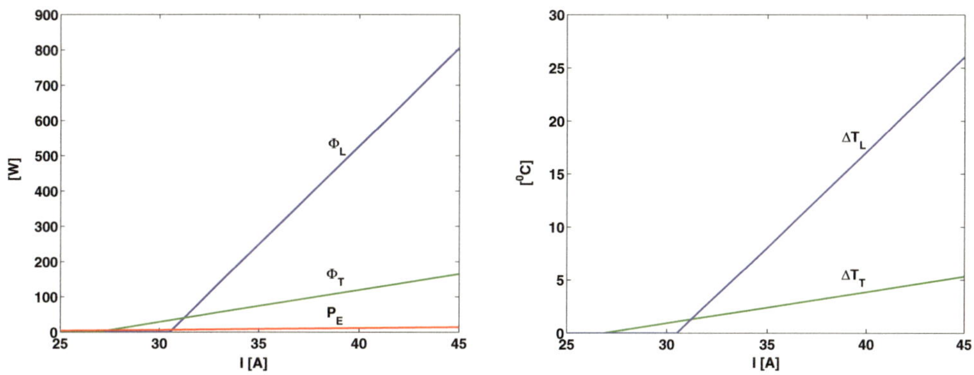

(a) (b)

Fig. 11. The radiation powers Φ_L and Φ_T, the electric power P_E, and the temperature variations ΔT_L, ΔT_T as functions of the injection current, for $x_2 - x_0 = 5.5\ nm$, $\mathcal{T} = 0.5$, and $\gamma_F = 10^8\ s^{-1}$.

potential of the n-zone of the base-collector junction. In figure 11 we consider a much larger decay rate of the electromagnetic field, $\gamma_F = 10^8\ s^{-1}$ instead of $\gamma_F = 10^7\ s^{-1}$, when the operation conditions (124) are also satisfied. In this case, we also obtain a high radiation power, but with a higher injection current, which, however, does not produce an important electrical power P_E, dissipated in the device.

We study the time evolution of a quantum heat converter, by solving the time dependent system of equations (110), for a step current injected at the initial $t = 0$, and a fluctuation that arises at a certain time $t > 0$. Non-Markovian fluctuations are time-evolutions of polarization, population and field due to the self-consistent field of the environment particles that, in our case, are the quasi-free electrons and holes in the conduction regions of the device. In figure 12, we represent the dynamics of a longitudinal device with a thickness of the i-zone $x_2 - x_0 = 5.5\ nm$ and a transmission coefficient of the output mirror $\mathcal{T} = 0.1$, while the threshold current is $I_{0L} = 24.1149\ A$ and the maximum current is $I_M = 46.0995\ A$. We consider a step current $I = 45\ A$ injected at time $t = 0$. In the Markovian approximation, the super radiant power $\Phi_L(t)$ of a longitudinal quantum heat converter is generated as in figure 12a, while the population $w(t)$ and polarization variables $u(t)$, $v(t)$ have the time-evolutions represented in figure 12b. The sudden jumps of the polarization variables in figure 12b, are detailed in

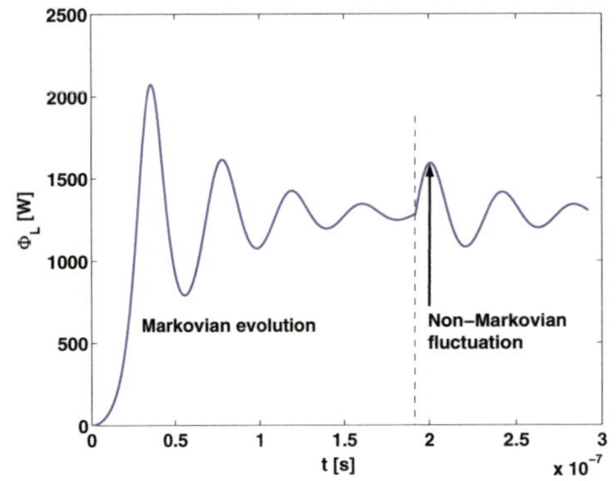

(a)

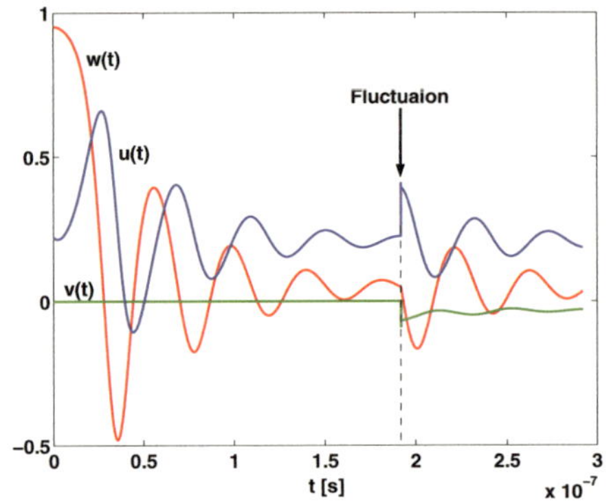

(b)

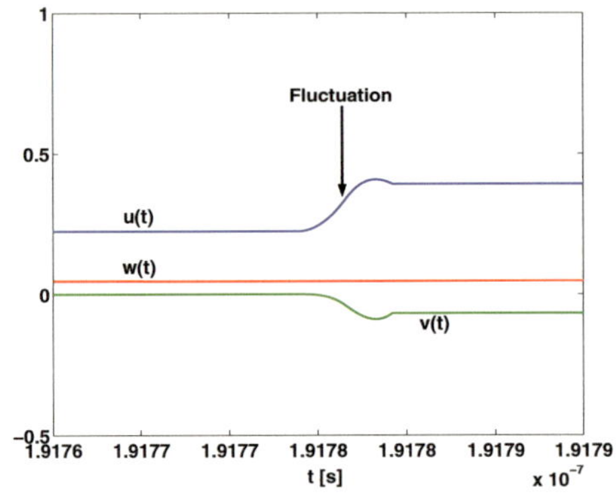

(c)

Fig. 12. Dynamics of a longitudinal super radiant device with $x_2 - x_0 = 5.5\ nm$ and $\mathcal{T} = 0.1$ when a step current of $I = 45\ A$ is injected in the device: (a) super radiant power; (b) polarization and population; (c) polarization fluctuation in a short timescale.

figure 12c, in a short timescale. At $t = 0$, the population increases from the equilibrium value w_T for the temperature T, to $w(0) = w_T + 2I/(eN_eA_D\gamma_\parallel)$ and, after that, while the radiation field increases, the population decreases tending to an asymptotic value. With an appropriate choice of the phase of the initial polarization, $v(0) = 0$, $u(0)$ takes a value corresponding to the maximum value $-w_T$ of the Bloch vector, which is $u(0) = \sqrt{[w_T^2 - w^2(0)]/(2-\mathcal{T})}$. In the Markovian approximation, the electromagnetic power is growing to a certain value, and after a short oscillation tends to the asymptotic value that according to (120a) is $\Phi_L = 1.2843 \times 10^3\ W$. However, in the non-Markovian approximation, random fluctuations of the polarization, population, and field arise. These fluctuations are described by the time integrals in the polarization equations (110a) and (110b) depending on the time-dependent phase term $\phi_n(t')$, with a mean-value of the fluctuation time $\tau_n = 1/\gamma_n$. From figure 5a, we notice that the fluctuation rate γ_n is four orders higher than the decay rate $\gamma_\parallel$, corresponding to the timescale of the Markovian processes. In equations (110a) and (110b), we take a positive fluctuation with a duration tn $\tau_n = 2.6305 \times 10^{-12} s$, followed by a negative one with the same duration. In figure 12c such a fluctuation is represented in a short timescale, specific to the non-Markovian fluctuations, while in figures 12a and 12b it is represented in a long timescale specific to the Markovian processes. We notice that, while the polarization variables $u(t)$ and $v(t)$, which depend on the transition elements of the density matrix, undertake considerable variations in a fluctuation time, for population and super radiant field these variations only initialize long time oscillations.

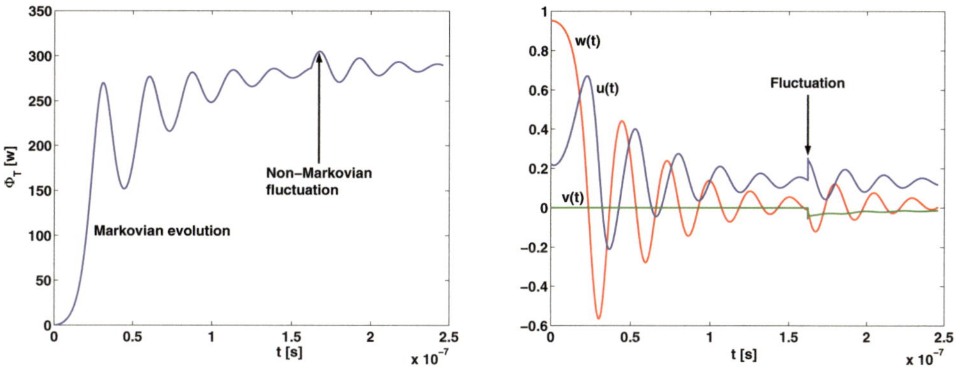

(a) (b)

Fig. 13. Dynamics of a transversal super radiant device with $x_2 - x_0 = 5.5\ nm$ and $\mathcal{T} = 0.1$ when a step current of $I = 45\ A$ is injected: (a) Superradiant power; (b) Population and polarzation.

In figure 13, we represent the dynamics of the transversal device with the same semiconductor structure and injected current, while the threshold current takes a lower value $I_{0T} = 23.4528\ A$. This decrease of the threshold current for a transversal device, in comparison with a longitudinal one with the same semiconductor structure, is obtained due to the field amplification on the longer path of the field propagation in the plane of the quantum dot layers, which is described by the term $c\mathcal{T}\frac{L_D}{A_D^{1/2}}$ in equations (121). However, this small difference is not very significant, since, according to equation (83), a longer propagation path leads also to a higher decay rate of the field, i.e. to an increase of the dissipative term $1_L\gamma_F$. We notice that, while the radiation power is lower, this device is much less sensitive to the thermal fluctuations described by the non-Markovian term. This decrease of the radiation power for a transversal device, in comparison with a longitudinal one with the same semiconductor structure, is obtained due to the factor $\frac{A_D^{1/2}}{L_D}$ at the denominator of equation (120b). An essential advantage of a transversal quantum heat converter, in comparison with a longitudinal one, consists in injection electrodes as zero-transmission mirrors, i.e. these electrodes are thick metalizations, providing an uniform current injection in the device. For an uniform current injection, a longitudinal quantum heat converter needs a special output structure, as a high transmission output Fabry-Perot cavity (4). Although for a transversal device we obtained a lower radiation power than a longitudinal one, it could be advantageous for some applications: for instance to obtain a powerful radiation device, as a stack of many transversal quantum heat converters. Another application could be an electric generator with the three semiconductor devices of the system, transversal quantum heat converter, quantum injection system, and total quantum injection system, in the same plane, eventually stuck on the same pad.

In figures 12 and 13, we considered a positive fluctuation followed by a negative one, which means an integration over a first interval of time $\tau_n = 1/\gamma_n$ with a phase $\phi_n = 0$ followed by an integration over a second interval of time τ_n with a phase $\phi_n = \pi$ in the polarization equations (110a) and (110b).

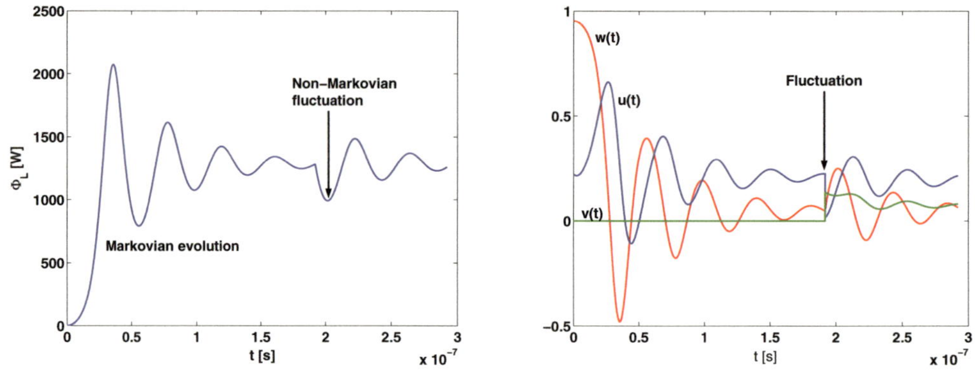

(a) (b)

Fig. 14. Dynamics of a longitudinal super radiant device with a negative fluctuation ($\phi_n = \pi$), followed by a positive one ($\phi_n = 0$).

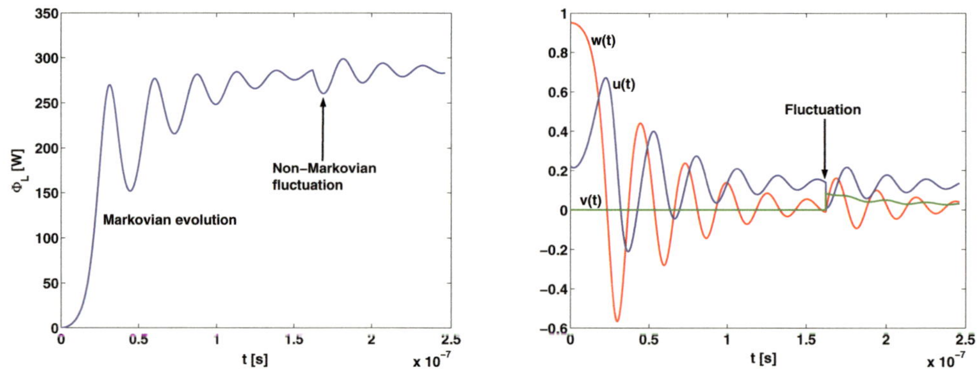

(a) (b)

Fig. 15. Dynamics of a transversal super radiant device with a negative fluctuation ($\phi_n = \pi$), followed by a positive one ($\phi_n = 0$).

Changing the phases of the fluctuations, i.e. taking a negative fluctuation followed by a positive one (figures 14 and 15), we get similar evolutions but with opposite signs. Obviously, the realistic evolution of a device is the result of the random phases ϕ_n, arising during the whole evolution of the system. Thus, the system dynamics takes a noisy form, with the polarization undertaking rapid variations during a fluctuation time, while the population and the super radiant field are only initialized into slow oscillations.

7. Conclusions

We presented a new kind of quantum dots, with a quantum well for electrons in the n-region of an n-i-p heterostructure, and a quantum well for holes in the p-region of this structure. These quantum wells are separated from the two n and p conduction regions by transparent potential barriers, and separated from one another by the potential barrier of the i-region. Such a quantum dot, we call "quantum injection dot", can be compared with a conventional quantum dot, as a small semiconductor region with a narrower forbidden band in a much larger i-region of an n-i-p semiconductor structure. Quantum injection dots have mainly been conceived for conversion of environmental heat into usable energy, while the conventional quantum dots are mainly used in information technology. A quantum injection dot differs in many respects from a conventional quantum dot, where the quantum well for holes is placed under the quantum well for electrons, as conduction and valence bands of the same semiconductor region: (1) while a quantum injection dot is supplied with electrons and holes from the two conduction regions by quantum tunneling through the n and p separation barriers, without any energy increase, a conventional quantum dot is supplied with electrons and holes only by providing a substantial energy, necessary to raise these electrons and holes from the n and p conduction regions to the conduction and valence bands of the i-region, from where they fall in the two potential wells of the quantum dot; (2) a quantum injection dot provides the electron transfer from the n-region to the p-region only by quantum transitions, while the electron transfer provided by a conventional quantum dot includes additional transport processes from the two n and p regions to the i-region where this quantum dot is located; (3) a quantum injection dot is a one-electron normalized two-level quantum system, while a conventional quantum dot is a confinement semiconductor region where many electrons and holes are simultaneously present to provide a larger probability for the super radiant transitions. In comparison with a conventional quantum dot, a quantum injection dot is much less dissipative, and, due to its simpler structure, enables a much higher packing degree in a semiconductor structure.

We studied a system of quantum injection dots by using the available means of quantum mechanics: (1) we calculated wave-functions, dipole moments, and eigenvalue equation for energy; (2) we derived equations for the dissipative super radiant dynamics of the system; (3) we obtained analytical coefficients depending only on physical characteristics and universal constants, without any phenomenological parameter. In the dynamics of a quantum dot system, we distinguish five dissipative processes: (1) correlated transitions with phonons of the crystal lattice vibrations, which is the dominant dissipation process (2) correlated transitions with quasi-free electrons and holes in the conduction regions, (3) correlated transitions with the quasi-free electromagnetic field, which are negligible, (4) transitions stimulated by the thermal fluctuations of the self-consistent field of the electrons and holes in the conduction regions, (5) non-Markovian processes induced by these fluctuations. However, we found that the fluctuation time is much shorter than the decay time, which means that the system is in fact quasi-Markovian, while the non-Markovian fluctuations manifest themselves only as a noise. For the propagation of the electromagnetic field throughout the semiconductor structure, by taking into account the dissipative interaction with the quasi-free electrons an holes in the conduction regions, we obtained an analytical expression of the field decay rate as a function of effective masses, frequency, and propagation path.

We studied a device converting environmental heat into coherent electromagnetic energy in two versions: (1) longitudinal quantum heat converter, with the electromagnetic field propagating in the direction of injected current, i.e. emerging from the surface the semiconductor structure, and (2) transversal quantum heat converter, with the electromagnetic field propagating in a perpendicular direction to the injected current, i.e. emerging from a lateral surface of the semiconductor structure. We found operation conditions for the physical characteristics of the semiconductor structure. We studied the dependence of the dissipative rates, coupling coefficients, and threshold currents as functions of the i-region thickness, which enables the control of these quantities in a large field of values. We found that the operation conditions do not depend on the i-layer thickness. When this thickness is decreased, the injected current and the corresponding super radiant power increase. However, these quantities of interest can not be indefinitely increased, especially due to the temperature variation induced by the heat propagation throughout the structure, which tends to produce an atomic detuning of the quantum dot layers. We highlighted the super radiant dissipative dynamics under a step current injection, and thermal fluctuations of the conduction electrons and holes.

8. References

[1] Kent D. Choquette and John F. Klem, Long wavelength vertical cavity surface emitting laser, US 6,931,042 B2 (US Patent Office, Aug. 16, 2005).

[2] Ashkan A. Arianpour, James P. McCanna, Joshua R. Windmiller, Semeon Y. Litvin, Photovoltaic device employing a resonator cavity, US 2008/0128023 A1 (US Patent Office, Jun. 05, 2008).

[3] E. Stefanescu, Master equation and conversion of environmental heat into coherent electromagnetic energy, Prog. Quantum Electron. 34 (2010) 349-408.

[4] Eliade Stefanescu, Lucien Eugene Cornescu, Longitudinal quantum heat converter, US 20090007950 (US Patent Office, 01-08-2009), http://www.freepatentsonline.com/y2009/0007950.html.

[5] Eliade Stefanescu, Lucien Eugene Cornescu, Transversal quantum heat converter, US 20100019618 (US Patent Office, 01-28-2010), http://www.freepatentsonline.com/y2010/0019618.html.

[6] Eliade Stefanescu, Lucien Eugene Cornescu, Quantum injection system, US 20090007951 (US Patent Office, 01-08-2009), http://www.freepatentsonline.com/y2009/0007951.html.

[7] E. Stefanescu, W. Scheid, and A. Sandulescu, Non-Markovian master equation for a system of Fermions interacting with an electromagnetic field, Ann. Phys. 323 (2008) 1168-1190.

[8] T. Fließbach, Statistische Physik, Lehrbuch zur Theoretischen Physik IV (Elsevier, München 2007).

[9] G. W. Ford, J. T. Lewis, and R. F. O'Connell, Master Equation for an Oscillator Coupled to the Electromagnetic Field, Ann. Phys. 252 (1996) 362-385.

[10] V. M. Axt and S. Mukamel, Nonlinear Optics of semiconductor and moleclar nanostructures, Rev.Mod..Phys. 70 (1998) 145-287.

[11] Eliade Stefanescu, Dynamics of a Fermi system with resonant dissipation and dynamical detailed balance, Physica A 350 (2005) 227-244.

[12] Haward Carmichael, an Open Quantum System Approach to Quantum Optics, in Lecture notes in Physics (Springer Verlag, Berlin 1993).

[13] Hartmut Haug and Antti-Pekka Jauho, Quantum Kinetics in Transport and Optics of Semiconductors (Springer-Verlag, Berlin, Heidelberg, New York, 1998).

[14] Günter Mahler and Volker A. Weberruß, Quantum Networks - Dynamics of Open Nanostructures (Springer-Verlag, Berlin, Heidelberg, New York, 1995).

14

Quantum Mechanics of Semiconductor Quantum Dots and Rings

I. Filikhin, S.G. Matinyan and B. Vlahovic
North Carolina Central University
USA

1. Introduction

The progress of semiconductor physics in the decade 1970-1980 is connected with gradual deviation from the electronic band structure of ideal crystal of Bloch picture (Bloch, 1928) where, unlike atomic world with its discrete and precisely defined, in the limits of uncertainty relation, energy levels, energy of bound electron is a multivalued function of momentum in the energy band and density of states are continuous (For the earlier short but comprehensive survey see (Alferov, 1998)).

In principle, Bloch theory deals with infinite extension of lattice, with the understandable (and important) surface effects. The decreasing of the size of the object to a few micrometers principally does not change the picture of the extended crystal qualitatively. It takes a place until one reaches the scale where the size quantization essentially enters the game and we can speak about microscopic limit of matter. What generally divides macroscopic limit of the solid state from the microscopic one? It is defined by some correlation length (or, more generally, all such relevant lengths)): for carriers it is mean free path length l or Broglie length $l_B = h / p$ (p-momentum), which is smaller. One may say that the quantum mechanical properties of matter clearly reveal if $l / a \geq 1$, where a is the size of the lattice constant. In the opposite limit $l / a < 1$, matter is considered macroscopically.

In this light, it is worthy to remind that as long as 1962, L. V. Keldysh (Keldysh, 1962 as cited in Bimberg et al., 1999) considered electron motion in a crystal with periodic potential with the period that is much larger than the lattice constant. In this limit he discovered so called minizones and negative resistance. Just in this limit $l / a \geq 1$ we expect the size quantization with its discrete levels and coherence in the sense that electron can propagate across the whole system without scattering, its wave function maintains a definite phase. In this limit, mesoscopic (term coined by van Kampen (1981) relates to the intermediate scale dividing the macro and micro limits of matter) and nanoscopic objects (Quantum Wells (QW), Wires and Dots (QD)) shown very interesting quantum mechanical effects. In this limit many usual rules of macroscopic physics may not hold. For only one example, rules of addition of resistance both in series and parallel are quite different and more complicated (Landauer, 1970; Anderson et al., 1980; Gefen, et al., 1984).

Closing this brief introduction concerning some aspects of genuine quantum objects (QW, Q Wires, QD) we would like to emphasize the conditional sense of the notion of dimensions in this world: in the limit $l/a \geq 1$ dimensions are defined as difference between real spatial dimension (in our world D =3) and numbers of the confined directions: Quantum Well: D =2, Quantum Wire: D =1, Quantum Dot: D =0. However, for example, QD which will be one of our subject for study, has very rich structure with many discrete levels, their structure define the presence or absence of Chaos, as we will see below, inside QD. Minimal size of QD is defined by the condition to have at least one energy level of electron (hole) or both: $a_{\min} = \pi\hbar / \sqrt{2m^* \Delta E} \approx 4$ nm, where ΔE is average distance between neighboring energy levels. Maximal size of QD is defined by the conditions that all three dimensions are still confined. It depends, of course, on temperature: at room temperature it is 12 nm (GaAs), 20 nm (InAs) ($\Delta E \approx 3kT$). The lower temperature, the wider QD is left as quantum object with D =0 and the number of energy levels will be higher.

2. Effective model for semiconductor quantum dots

The effective potential method has been developed (Filikhin et al., 2006) to calculate the properties of realistic semiconductor quantum dot/ring (QD/QR) nanostructures with the explicit consideration of quantum dot size, shape, and material composition. The method is based on the single sub-band approach with the energy dependent electron effective mass. In this approach, the confined states of carriers are formed by the band gap offset potential. Additional effective potential is introduced to account for cumulative band gap deformations due to strain and piezoelectric effects inside the quantum dot nanostructure. The magnitude of the effective potential is selected in such a way as to reproduce experimental data for a given nanomaterial. Additionally, an analog of the Kane formula (Kane, 1957) is implemented in the model to take into account the non-parabolicity of the conduction/valence band. The resulting nonlinear eigenvalue problem for the Schrödinger equation is solved by means of the iterative procedure with the adjusted effective electron mass and non-parabolicity parameter, where in each iteration step the Schrödinger equation is numerically linearized and solved by the finite element method.

At present, simulations based on this approach are performed for the InGaAs/GaAs quantum dots and quantum rings of different sizes and configurations under different external conditions. The obtained results show that the residual strain and conduction band non-parabolicity effects greatly affect the device related properties of semiconductor quantum dots. The results are in good agreement with available experimental data, closely matching energy level and effective mass data extracted from capacitance–voltage experiments. The method also allows one to accurately simulate spin-orbital coupling effects for the electrons in excited states, as well as the presence of admixtures, such as Ga. Our calculations of the Coulomb shifts of the exciton complexes (positively and negatively charged trions, biexcitons) in the InGaAs/GaAs quantum dots with 22%-25% Ga fraction match very well both capacitance-voltage and photoluminescence measurements. To best reproduce the experimental data, Ga fraction in the InGaAs/GaAs quantum dots should not exceed 25%.

Commonly used numerical approaches, such as the 8-band ***kp***-theory, density functional theory, or atomistic pseudo-potential technique, take into account inter-band interactions, strain and piezoelectric effects in quantum dots in an *ab initio* manner. Such methods are very computationally intensive and time-consuming. The important advantage of the effective model is that the high accuracy of calculations is obtained at a very low computational cost - calculations can typically be completed using a 3 GHz PC with 1 GB of memory in less than 20 minutes. The effective potential method satisfactorily reproduces the results of the realistic simulations, thus offering an independent evaluation of the electronic confinement effects calculated within others models.

2.1 Formalism

2.1.1 Schrödinger equation and effective mass approximation

In the present review a semiconductor 3D heterostructure (QD or QR) is modeled utilizing a ***kp***-perturbation single sub-band approach with quasi-particle effective mass (Harrison, 2005; Manasreh, 2005; Yu & Cardona, 2005). The energies and wave functions of a single carrier in a semiconductor structure are solutions the Schrödinger equation:

$$(H_{kp} + V_c(\vec{r}))\Psi(\vec{r}) = E\Psi(\vec{r}) \tag{1}$$

Here H_{kp} is the single band ***kp***-Hamiltonian operator, $H_{kp} = -\nabla \frac{\hbar^2}{2m^*(\vec{r})}\nabla$, m^* is the electron/hole effective mass for the bulk, which may depend on coordinate, and $V_c(\vec{r})$ is the confinement potential. The confinement of the single carrier is formed by the energy misalignment of the conduction (valence) band edges of the QD material (index 1) and the substrate material (index 2) in the bulk. $V_c(\vec{r})$ is so called "band gap potential". The magnitude of the potential is proportional to the energy misalignment. The band structure of the single band approximation can be found in many textbooks (see, for example, (Harrison, 2005; Manasreh, 2005; Yu & Cadona, 2005). * (see the input below) $\Psi(\vec{r})$ and its derivative $1/m^*(\vec{n},\nabla)\Psi(\vec{r})$ on interface of QD and the substrate are continues.

2.1.2 The non-parabolicity of the conduction band. The Kane formula

Traditionally applied in the macroscopic scale studies parabolic electron spectrum needs to be replaced by the non-parabolic approach, which is more appropriate to nano-sized quantum objects (Wetzel et al., 1996; Fu et al., 1998). The Kane formula (Kane, 1957) is implemented in the model to take into account the non-parabolicity of the conduction band. The energy dependence of the electron effective mass is defined by the following formula:

$$\frac{m_0}{m^*} = \frac{2m_0P^2}{3\hbar^2}\left(\frac{2}{E_g + E} + \frac{1}{E_g + \Delta + E}\right). \tag{2}$$

Here m_0 is free electron mass, P is Kane's momentum matrix element, E_g is the band gap, and Δ is the spin-orbit splitting of the valence band.

Taking into account the relation (2) the Schrödinger equation (1) is expressed as follows

$$(H_{kp}(E) + V_c(\vec{r}))\Psi(\vec{r}) = E\Psi(\vec{r}). \tag{3}$$

Here $H_{kp}(E)$ is the single band ***kp***-Hamiltonian operator $H_{kp}(E) = -\nabla \frac{\hbar^2}{2m^*(E,\vec{r})}\nabla$,

$m^*(E,\vec{r})$ is the electron/hole effective mass and $V_c(\vec{r})$ is the band gap potential. As a result, we obtain a non-linear eigenvalue problem.

Solution of the problem (3)-(2) results that the electron/hole effective mass in QD (or QR) varies between the bulk values for effective mass of the QD and substrate materials. The same it is given for the effective mass of carriers in the substrate. The energy of confinement states of carries is rearranged by the magnitude of the band gap potential V_c.

The Schrödinger equation (1) with the energy dependence of effective mass can be solved by the iteration procedure (Li et al., 2002; Voss, 2005; Filikhin et al., 2004, 2005).

$$H_{kp}(m^{*k-1}_i)\Psi^k(\vec{r}) = E^k\Psi^k(\vec{r}),$$
$$m^{*k}_i = f_i(E^k), \tag{4}$$

where k is the iteration number, i refers to the subdomain of the system; $i=1$ for the QD, $i=2$ for the substrate. $H_{kp}(m^{*k}_i)$ is the Hamiltonian in which the effective mass does not depend on energy and is equal to the value of m^{*k}_i, f_i is the function defined by the relation (2). For each step of the iterations the equation (1) is reduced to Schrödinger equation with the effective mass of the current step which does not depend on energy. At the beginning of iterations the bulk value of the effective mass is employed. Obtained eigenvalue problem can be solved numerically (by the finite element method, for example). After that, a new value for effective mass is taken by using Eq. (2) and procedure is repeated. The convergence of the effective mass during the procedure has a place after 3-5 steps. As an example, the typical convergences for election effective mass and confinement energy of single electron are displayed in Fig. 1 for the InAs/GaAs QR (Filikhin et al., 2005). Description of other methods for the solution of the problem (3)-(2) can be found in (Betcke & Voss, 2011).

Remarks: at the first, in the present review the consideration was restricted by the electron and heavy hole carriers, and, the second, the Coulomb interaction was excluded. Often the linear approximation for the function $m^*_i/m_0 = f(E,r)$ is used. We also will apply the linear fit in the present chapter.

2.1.3 Effective approach for strained InAs/GaAs quantum structures: Effective potential

Here we propose the effective potential method to calculate the properties of realistic semiconductor quantum dot/ring nanostructures with the explicit consideration of quantum dot size, shape, and material composition. The method is based on the single sub-band approach with the energy dependent electron effective mass (Eq. (3)). In this approach, the confined states of carriers are formed by the band gap offset potential. Additional effective potential is introduced to simulate the cumulative band gap deformations due to strain and piezoelectric effects inside the quantum dot nanostructure. The magnitude of the effective potential is selected in such a way that it reproduces experimental data for a given nanomaterial.

We rewrite the Schrödinger equation (3) in the following form:

$$(H_{kp}(E) + V_c(\vec{r}) + V_s(\vec{r}))\Psi(\vec{r}) = E\Psi(\vec{r})\,. \tag{5}$$

Here $H_{kp}(E)$, as before, is the single band ***kp***-Hamiltonian operator $H_{kp}(E) = -\nabla \frac{\hbar^2}{2m^*(E,\vec{r})}\nabla$.

As previously, $m^*(E,\vec{r})$ is the electron (or hole) effective mass, and $V_c(\vec{r})$ is the band gap potential, $V_s(\vec{r})$ is the effective potential. $V_c(\vec{r})$ is equal zero inside the QD and is equal to V_c outside the QD, where V_c is defined by the conduction band offset for the bulk (see Section 1.22). The effective potential $V_c(\vec{r})$ has an attractive character and acts inside the volume of the QD. This definition for the effective potential is schematically illustrated by Fig. 2 for the conduction band structure of InAs/GaAs QD. In the figure, the confinement potential of the simulation model with effective potential V_s is denoted as "strained". The band gap potential for the conduction band (valence band) can be determinate as V_c =0.594 eV (V_c =0.506 eV). The magnitude of the effective potential can be chosen to reproduce experimental data. For example, the magnitude of V_s for the conduction (valence) band chosen in (Filikhin et al., 2009) is 0.21 eV (0.28 eV). This value was obtained to reproduce results of the 8-th band ***kp***-calculations of (Schliwa et al., 2007) for InAs/GaAs QD. To reproduce the experimental data from (Lorke et al., 2000), the V_s value of 0.31 eV was used in (Filikhin, et al. 2006) for the conduction band.

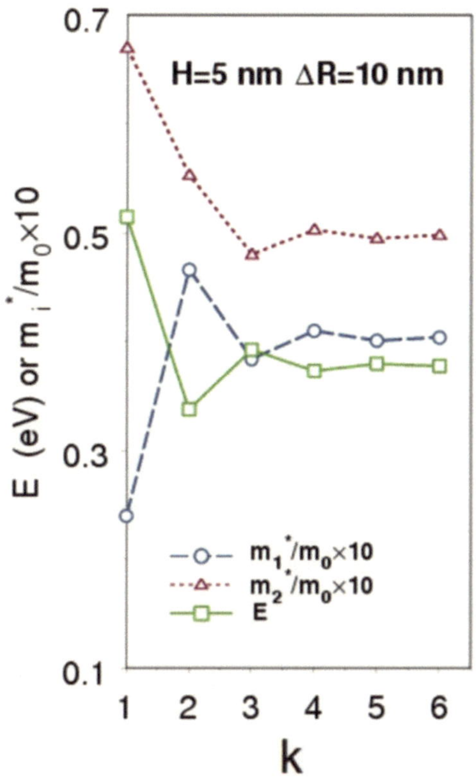

Fig. 1. Convergence of the iterative procedure (4) for the confinement energy E (solid line) and electron effective mass m^*_i/m_0 calculated for InAs/GaAs QR (dashed line) and GaAs substrate (dotted line). Here the height of QR is H, radial width is ΔR and inner radius is R_1 (R_1 =17 nm), V_c =0.77 eV.

Possibility for the substitution of the function describing the strain distribution in QD and the substrate was firstly proposed in (Califano & Harrison, 2000). Recent works (Zhao & Mei, 2011; Li, Bin & Peeters, 2011) in which the strain effect taken into account rigorously applying the analytical method of continuum mechanics allow us to say that the approximation of the effective potential is appropriate.

In the next sub-section of the section 2 we will review the results obtained in both these approximations as the non-parabolic one as well as the effective potential method.

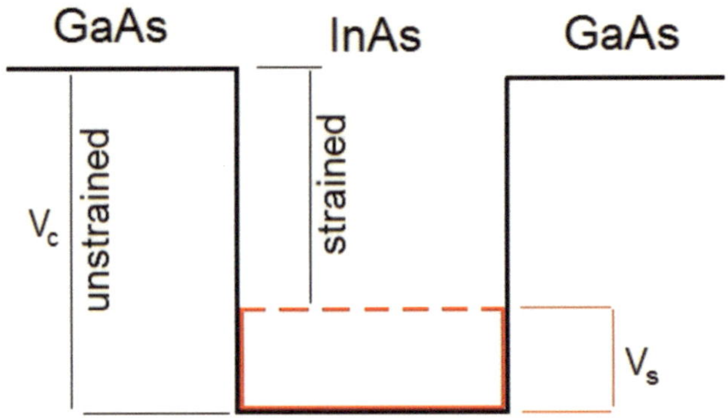

Fig. 2. Effective potential V_s and band gap structure of the conductive band of InAs/GaAs QD.

2.2 Electron energy in quantum rings with varieties of geometry: Effect of non-parabolicity

In this section a model of the InAs/GaAs quantum ring with the energy dispersion defined by the Kane formula (2) (non-parabolic approximation) based on single sub-band approach is considered. This model leads to the confinement energy problem with three-dimensional Schrödinger equation in which electron effective mass depend on the electron energy. This problem can be solved using the iterative procedure (4). The ground state energy of confined electron was calculated in (Filikhin et al., 2004, 2005, 2007a) where the effect of geometry on the electron confinement states of QR was studied and the non-parabolic contribution to the electron energy was estimated. The size dependence of the electron energy of QR and QD was subject of several theoretical studies (Li & Xia, 2001; Li et al., 2002). We present here, unlike the previous papers, a general relation for the size dependence of the QR energy.

Consider is semiconductor quantum ring located on the substrate. Geometrical parameters of the semi-ellipsoidal shaped QR are the height H, radial width ΔR and inner radius R_1. It is assumed that $H / \Delta R$ << 1 which is appropriate technologically. QR cross section is schematically shown in Fig. 3. The discontinuity of conduction band edge of the QR and the substrate forms a band gap potential, which leads to the confinement of electron.

The band gap potential $V_c(\vec{r})$ is equal to zero inside the QR ($V_c(\vec{r})$ =0) and it is equal to the confinement potential E_c outside of the QR: The spatial dependence of the electron effective mass is given as $m^*(E,\vec{r}) = m_i^*(E)$, i =1,2,3, where m_1^* is the effective mass in the material of QR ($\vec{r}$ $\in$ E1), and $m_2^*(E)$, m_3^* are the effective mass of the substrate material ($\vec{r}$ $\in$ E2 and

E3). Within each of the regions E1, E2 and E3 m_i^* does not depend on the coordinates. The effective mass m_3^* is equal to a constant bulk value. The energy dependence of the electron effective mass from the E1 and E2 subdomains is defined by the formula (2) (Kane, 1957). The equation (1) satisfies the asymptotical boundary conditions: $\Psi(\vec{r})|_{|\vec{r}|\to\infty}\to 0$, $\vec{r}\in$ substrate and $\Psi(\vec{r})|_{|\vec{r}|\in S}=0$, where S is free surface of QR. On the surface of boundaries with different materials the wave function and the first order derivative $(n,\nabla\Psi)/m_i^*$ are continuous (the surface normal $\vec{n}$).

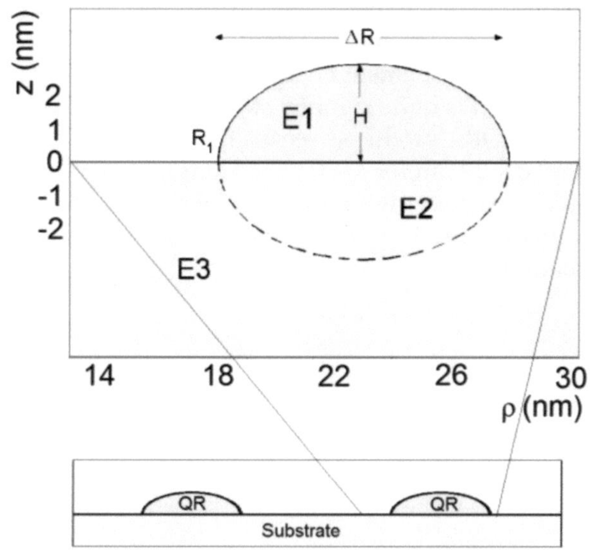

Fig. 3. Profile of cross section of quantum ring (E1) and substrate (E2 and E3). Cylindrical coordinates ρ and z shown on axis.

The Schrödinger equation (3) was numerically solved by the finite element method and iterative procedure (4). The following typical QR/substrate structures with experimental parameters were chosen: InAs/GaAs and CdTe/CdS. The parameters of the model are given in Tabl. 1 for the each hetero-structure.

QR/Substrate	m^*_1/m^*_2	m^*_1/m^*_2 (eV)	$\frac{2m_0P_1^2}{\hbar^2}/\frac{2m_0P_2^2}{\hbar^2}$	Δ_1/Δ_2
InAs/GaAs	0.024/0.067	0.77	22.4/24.6	0.34/0.49
CdTe/CdS	0.11/0.20	0.66	15.8/12.0	0.80/0.07

Table 1. Parameters of the QR and substrate materials

It has to be noted that the effective mass substrate calculated for the InAs/GaAs and CdTe/CdS QRs is slightly differ from the bulk values within area E2. One can consider a simpler model when the properties of the area E2 and E3 are similar. It means that the wave function of electron does not penetrated by surface of QR (area E1) essentially. The simple model does not change qualitative results of these calculations.

Analysis of the results of numerical calculations shows that the ground state energy of QR can be best approximated as a power function of the inverse values of the height and the radial width:

$$E \approx a(\Delta R)^{-\gamma} + bH^{-\beta}, \tag{6}$$

where the coefficients γ =3/2 and β=1 were obtained numerically by the least square method. An example of this relation is illustrated in Fig. 4 for InAs/GaAs QR. Parameters a and b remain constant except for extremely low values of H and ΔR. Our analysis also reveals a significant numerical difference between the energy of QR electron ground states, calculated in non-parabolic and parabolic approximations. The results of the calculation with parabolic approximation are represented in the Fig. 4 by the dashed lines. Computation of the electron confinement energy of QRs for different materials shows that the non-parabolic contribution is quite significant when chosen QR geometrical parameters are close to those of the QRs produced experimentally: H < 7 nm, R < 30 nm for InAs/GaAs, H < 5 nm, R < 20 nm for CdTe/CdS. Magnitude of this effect for InAs/GaAs can be greater than 30%. According with this fact the coefficients a and b in Eq. (6) also depend on the approximation used: a/b =3.4/1.9 for the non-parabolic and a/b =6.2/3.0 for parabolic approximation.

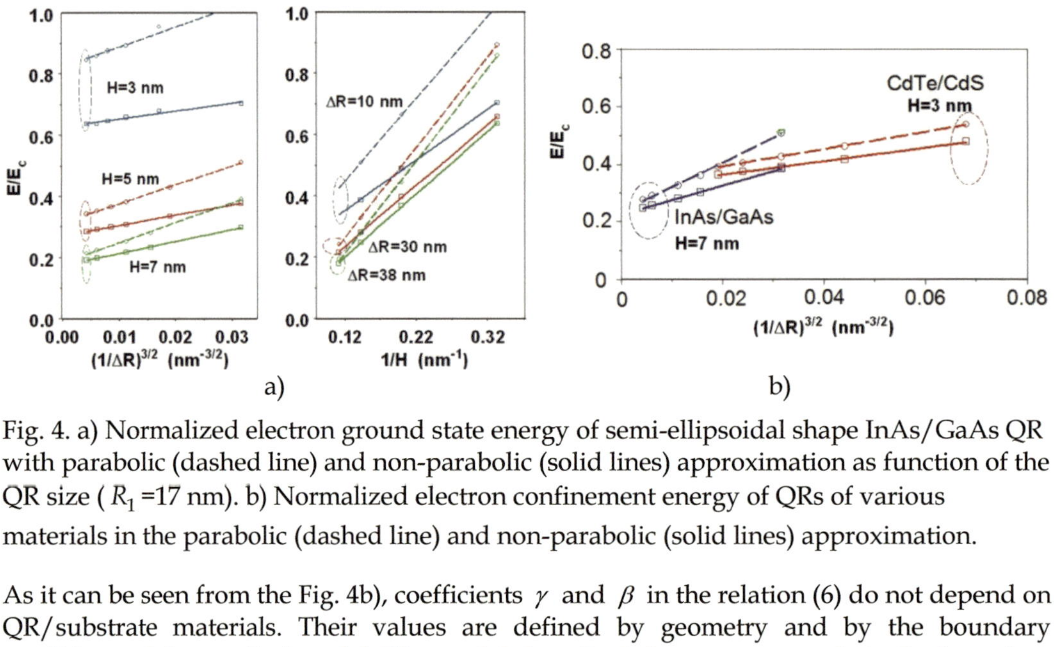

Fig. 4. a) Normalized electron ground state energy of semi-ellipsoidal shape InAs/GaAs QR with parabolic (dashed line) and non-parabolic (solid lines) approximation as function of the QR size (R_1 =17 nm). b) Normalized electron confinement energy of QRs of various materials in the parabolic (dashed line) and non-parabolic (solid lines) approximation.

As it can be seen from the Fig. 4b), coefficients γ and β in the relation (6) do not depend on QR/substrate materials. Their values are defined by geometry and by the boundary conditions of the applied model. The model described above corresponds to the boundary condition as "hard wall at one side" (top side of the QR). For the model without the walls when the QR embedded into the substrate one can obtain γ =1, and β=1/3. In contrast with it, the coefficients a and b depend on the QR/substrate material set essentially.

Concluding, we have shown that for wide QR sizes the non-parabolicity effect does considerably alter the energy of the electron states, especially when the height or width of QR is relatively small.

2.3 The C-V measurements and the effective model: Choosing the parameters

The well-established process of QDs formation by epitaxial growth and consecutive transformation of QDs into InAs/GaAs quantum rings (QR) (Lorke et al., 2000) allows the production of 3D structures with a lateral size of about 40-60 nm and a height of 2-8 nm. In produced QDs and QRs it is possible directly to observe discrete energy spectra by applying capacitance-gate-voltage (CV) and far-infrared spectroscopy (FIR). In this section we will show how the effective model works using as an example the CV data. We use results of the CV experiment from (Lorke et al., 2000; Emperador et al., 2000; Lei et al., 2010) for QD and QR.

The effective mass of an electron in QD and QR is changing from the initial bulk value to the value corresponding to the energy given by the Kane formula (2). Results of the effective model calculations for the InAs/GaAs QR are shown in Fig. 5. The effective mass of an electron in the InAs QR is close to that of the bulk value for the GaAs substrate. Since the effective mass in the QD is relatively smaller, as it is clear from Fig. 5, for QD the electron confinement is stronger; the s-shell peak of the CV trace is lower relative upper edge of conduction band of GaAs. The lower s-shell peak corresponds to the tunneling single electron into the QD. The pictures is a starting point for the choosing the parameters of the effective potential model. In this section we follow the paper (Filikhin et al., 2006a) where the semi-ellipsoidal InAs/GaAs QD has been considered. The average sizes of InAs/GaAs QD reported in (Lorke et al., 2000) were: H =7 nm (the height) and R=10 nm (the radius). A cross section of the quantum dot is shown in Fig. 6a). The quantum dot has rotation symmetry. Thus the cylindrical coordinate was chosen in Eq. (5) which defines the effective model. For each step of iterative procedure (4) the problem (3-2) is reduced to a solution of the linear eigenvalue problem for the Schrödinger equation.

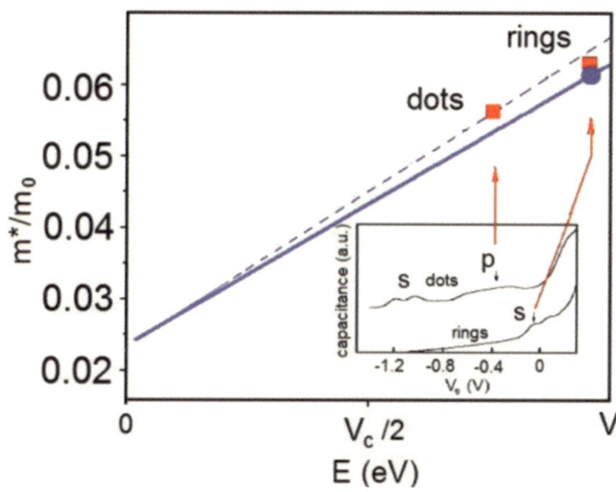

Fig. 5. Calculated (circle) and experimentally obtained by (Lorke et al., 2000; Emperador et al., 2000) (squares) values for the electron effective mass and the confinement energies of the electron s- and p-levels of QD and QR. The solid line is obtained by the Kane formula (2), and the dashed line connects the bulk values of the effective mass. The insert: the capacitance-gate voltage traces (Lorke et al., 2000).

Taking into account the axial symmetry of the quantum dot (ring) considered, this equation may be written in the cylindrical coordinates (ρ, z, ϕ) as follows:

$$(-\frac{\hbar^2}{2m^*}(\frac{\partial^2}{\partial\rho^2}+\frac{1}{\rho}\frac{\partial}{\partial\rho}-\frac{l^2}{\rho^2}+\frac{\partial^2}{\partial z^2})+V_c(\rho,z)+V_s(\rho,z)-E)\Phi(\rho,z)=0. \quad (7)$$

The wave function is of the form: $\Psi(r)=\Phi(\rho,z)exp(il\phi)$, where l=0, ± 1, ± 2... is the electron orbital quantum number. For each value of the orbital quantum number l, the radial quantum numbers $n=0,1,2,...$ are defined corresponding to the numbers of the eigenvalues of (4) which are ordered in increasing. The effective mass m^* must be the mass of electron for QD or for the substrate depending on the domain of the Eq. (3) is considered.

The wave function $\Phi(\rho,z)$, and its first derivative in the form $\frac{\hbar^2}{2m^*}(\vec{n},\nabla)\Phi$, have to be continuous throughout the QD/substrate interface, where $\vec{n}$ is the normal vector to the interface curve. The Neumann boundary condition $\frac{\partial}{\partial\rho}\Phi(\rho,z)=0$ is established for $\rho=0$.

The asymptotical boundary conditions is $\Phi(\rho,z)\to 0$, when $\rho\to\infty$, $|z|\to\infty$ (QD is located near the origin of z-axes).

When quantum dots are in an external perpendicular magnetic field, as it will be considered below, the magnetic potential term must be added to the potentials of Eq. (7) (Voskoboinikov et al., 2000) in the form $V_m(\rho)=\frac{1}{2m^*}(\beta\hbar l+\frac{\beta^2}{4}\rho^2)$, where $\beta=eB$, B is the magnetic field strength, and e is the electron charge. We consider the case of a magnetic field normal to the plane of the QD and do not take into account the spin of electron because the observed Zeeman spin-splitting is small.The confinement potential in Eq. (7) was defined as follows: V_c = 0.7($E_g^S - E_g^{QD}$); $V_c=0.77$ eV. The parameters of the QD and substrate materials were $m^*_{bulk,1}/m^*_{bulk,2}$ =0.024/0.067, E_g^{QD}/E_g^S =0.42/1.52, $\frac{2m_0P_1^2}{\hbar^2}/\frac{2m_0P_2^2}{\hbar^2}$ =20.5/24.6, Δ_1/Δ_2=0.34/0.49. The magnitude of the effective potential V_s was chosen as 0.482 eV. There are three electron confinement states: the s, p, and d, as shown in the Fig. 6b). The energy of the s single electron level measured from the top of the GaAs conduction band can be obtained from CV experimental data. To explain it, in Fig. 6c) the capacitance-gate-voltage trace from (Miller et. al., 1997) is shown. The peaks correspond to the occupation of the s and p energy shells by tunneled electrons. The Coulomb interaction between electrons results to the s-shell splits into two levels and the p-shell splits into four levels taking into account the spin of electron and the Pauli blocking for fermions. The gate voltage-to-energy conversion coefficient $f=7$ ($\Delta E=e\Delta V_g/f$) was applied to recalculate the gate voltage to the electron energy. The value of the effective potential V_s was chosen in order to accurately reproduce the observed s-wave level localization with respect to the bottom of GaAs conduction band. The approximate size of this energy region is 180 meV.

The non-parabolic effect causes a change in the electron effective mass of QD with respect to the bulk value. According to the relation Eq. (2), the effective electron mass for InAs is sufficiently increased from the initial value of 0.024 m_0 to 0.054 m_0, whereas for GaAs substrate it is slightly decreased from $0.067\,m_0$ to $0.065\,m_0$ within the region where the wave function is out of the quantum dot. The obtained value of the electron effective mass of InAs in QD is close to the one ($0.057\,m_0 \pm 0.007$) extracted in (Miller et. al., 1997) from the CV measurements of orbital Zeeman splitting of the p level.

Appling the obtained effective model, one can take into account the effect the Coulomb interaction between electrons (the Coulomb blockade). The goal is to reproduce the C-V data presented in Fig. 6 for the InAs QD. The calculations (Filikhin et al., 2006a) have been carried out using the perturbation procedure, proposed in (Warburton et al., 1998). The Coulomb energy matrix elements were calculated by applying single electron wave functions obtained from the numerical solution of Eq. (7). Both the direct terms of E_{ij}^{c} and the exchange terms E_{ij}^{x} of the Coulomb energy between electron orbitals with angular momentum projection of $\pm i$ and $\pm j$ were calculated (notation is given in (Warburton et al., 1998)). The results of calculations of the electron energies of the s, p and d levels are shown in Fig. 7 (*Cal.* 2). The s shell Coulomb energy was found to be close to the experimental value which is about 20 meV.

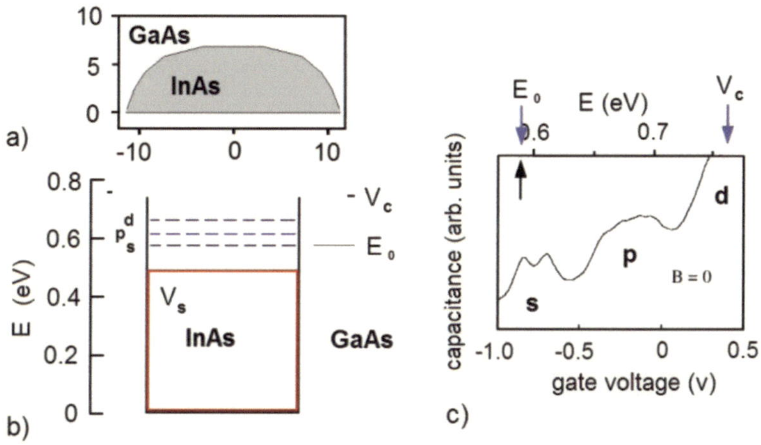

Fig. 6. a) A cross section of the quantum dot. The dimensions are given in nm. b) Localization of the s, p and d single electron levels relatively to the bottom of the GaAs conduction band. V_c is the band-gap potential, V_s is the effective potential simulating the sum of the band-gap deformation potential, the strain-induced potential and the piezoelectric potential. c) The capacitance-gate-voltage trace (Miller et. al., 1997). The peaks correspond to the occupation of the s and p energy shells by tunneled electrons. The arrows denote the s level (E_0) and the bottom of the GaAs conduction band.

Returning to the Fig. 5 we have to note that the effective potential obtained for InAs/GaAs QD has to be corrected for the case of the InAs/GaAs quantum rings. The reason is the topological, geometrical dependence of the depth of the effective potential. This dependence is weak for the considered QD and QR. The corresponding V_s potentials have the magnitude of 0.482 eV and 0.55 eV for QD and QR, respectively. Accordingly to the experimental data the electron effective mass in quantum dots and rings is changing from $0.024\,m_0$ to $(0.057 \pm 0.007)\,m_0$ (Miller et. al., 1997) and $0.063\,m_0$ (Lorke et al., 2000), respectively. The Kane's formula describes these variations well as it is shown in Fig. 5. The calculated values for the effective masses for quantum dots and rings are 0.0543 m_0 and 0.0615 m_0, respectively (Filikhin et al., 2006).

Correct choice of the average QD profile is important for an analysis of the C-V data. It was shown in (Filikhin et al., 2008), where the calculation of the energy shifts due to the Coulomb interaction between electrons tunneling into the QD was performed for comparison with the C-V experiments.

One can see in Fig. 7 that the agreement between our results and the experimental data is satisfactory well. Slight disagreement can be explained by uncertainty in the QD geometry which has not been excluded by available experimental data. In (Filikhin et al., 2008) it was shown that small variations of the QD cross section lead to significant changes in the levels presented in Fig. 7. The variations of the QD profile we considered are shown in Fig. 8a, and the results of calculations for the electron energies are presented in Fig. 8b) for s, p and d -shell levels. The results of the calculations shown in Fig. 8 reveal rather high sensitivity to these variations of the QD profile. In particular, the spectral levels shift is noticeable due to a small deformation of the QD profile. Thus, we have seen that the average QD profile is important when we are comparing the result of the calculations and the experimental data.

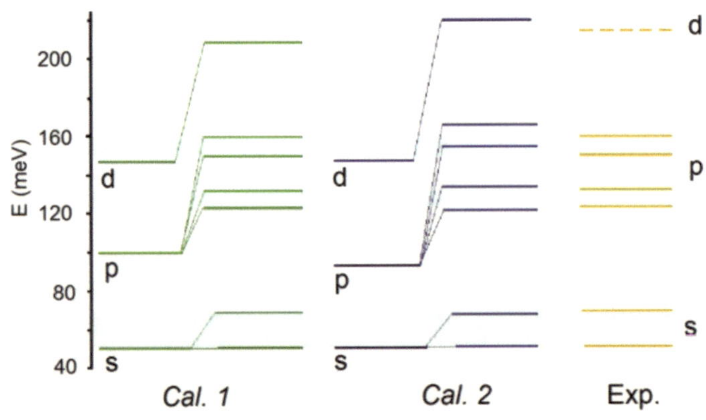

Fig. 7. Energies of the electrons occupying a few first levels of the quantum dot at zero magnetic field. The calculations *Cal.* 1 are that of parabolic model (Warburton et al., 1998). Our calculations are denoted by *Cal.* 2. The splitting of the single electron levels of a corresponding energy shell is presented. CV experimental data are taken from (Warburton et al., 1998).

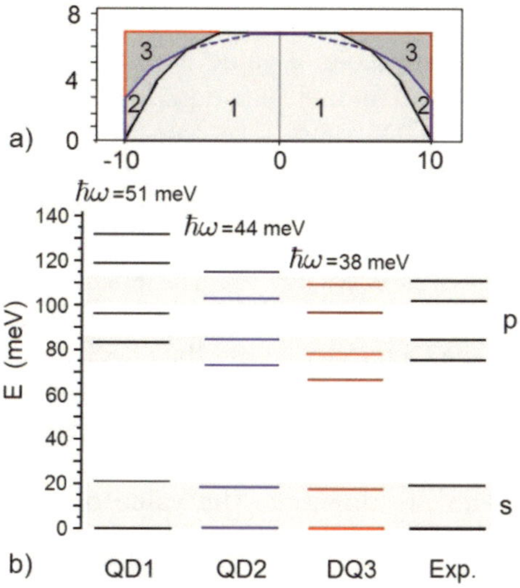

Fig. 8. a) Cross sections of the QD. The dimensions are given in nm. b) Excitation energies of the electrons occupying s and p -energy shells of the InAs/GaAs quantum dot for various QD profiles are shown in Figure 7a). CV experimental data are taken from (Warburton et al., 1998). Here $\hbar\omega$ is the excitation energy $\hbar\omega = E_{(0,0)} - E_{(0,1)}$, where $E_{(n,l)}$ is a single electron energy of the (n,l) state.

Finally, we may conclude that the effective model of QD/substrate semiconductor structure with the energy dependent effective mass and realistic 3D geometry taken into account, can quantitatively well interpret the CV spectroscopy measurements.

2.4 Electron effective mass in the InAs/GaAs QD

In this section we present the effective model based on another version of the band structure model for InAs/GaAs QDs proposed in (Filikhin et al., 2008). The cross section of the semi-ellipsoidal shaped InAs QD embedded in a GaAs substrate is shown in Fig. 6a). Band gap structure model was defined by choosing for the conduction band κ^{CB} =0.54, and for the valence band κ^{VB} =0.46 (Duque et al., 2005). Using experimental values $E_{g,1}$ =0.42 eV, $E_{g,2}$ =1.52 eV we obtain V_c =0.594 eV for the conduction band and. V_c .=0.506 eV for the valence band. The band structure model for InAs/GaAs QDs is shown in Fig. 9.

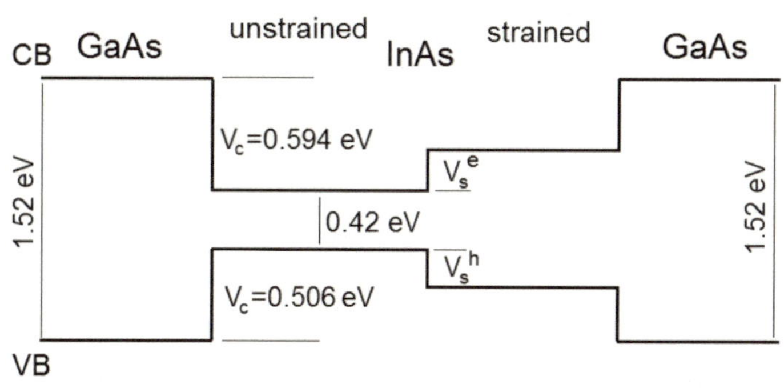

Fig. 9. Band structure model for InAs/GaAs QDs. CB (VB) is conduction (valence) band.

Bulk effective masses of InAs and GaAs are $m^*_{0,1}$ =0.024 m_0 and $m^*_{0,2}$ =0.067 m_0, respectively. For the effective mass of the heavy hole, a value of m^* =0.4 m_0 for both the QD and the substrate was used. The band gap model just described is for "unstrained" InAs/GaAs structures. Realistic models for QDs must take into account the band-gap deformation potential, the strain-induced potential, and the piezoelectric potential, in addition to the band-gap potential. These effects can be included by introducing an effective potential V_s. The magnitude of the potential has been chosen (Filikhin et al., 2006) to reproduce experimental data and the value of 0.31 eV was used for V_s. The effect of non-parabolicity, taken into account in the effective model, leads to a change of the effective electron mass in the QD relative to its bulk value. For the QD under study, the effective mass for InAs increases from the initial bulk value of 0.024 m_0 to 0.057 m_0 which coincides with the experimental value $0.057m_0 \pm 0.007m_0$, obtained in CV measurements through the Zeeman splitting of p-shell levels. This result is shown in Fig. 10. In accordance with Eqs. (2)-(3), the effective electron masses in the s, p and d states are different. The value of the effective mass, mentioned

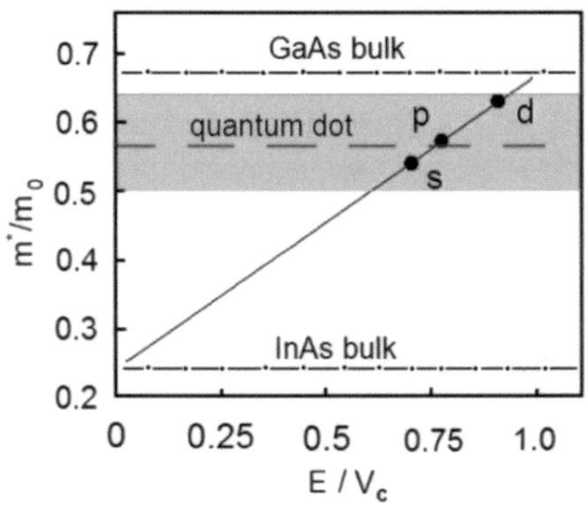

Fig. 10. Effective mass of electron and single electron energy of s, p, d-levels in InAs/GaAs QD. Dashed line corresponds to the experimental value. The grey color stripe shows the experimental uncertainty.

above, corresponds to the one for the p-state. The effective mass for s-shell is slightly less and is equal to 0.054 m_0. The differences of the effective masses are small and cannot be extracted from this experiment due to the large experimental uncertainties (Miller et al., 1998).

2.5 Experimental data for InAs/GaAs QR and the effective model

In this section we continue the description of the effective model use on the example of InAs/GaAs quantum ring. The geometry of the self-assembled QRs, reported in (Lorke et al., 2000), is shown in Fig. 11 (Geometry 1). The InGaAs QRs have a height of about 2 nm, an outer diameter of about 49 nm, and an inner diameter of about 20 nm. Also, three-dimensional QR geometry (Geometry 2), which follows from the oscillator model (Lei et al., 2010) is used. The confinement of this model is given by the parabolic potential:

$U(r) = \frac{1}{2} m^* \omega (r - r_0)^2$, where ω, r_0 are parameters (Chakraborty & Pietiläinen, 1994). The QR geometry is dictated by the relation between the adopted oscillator energy and a length l as follows (Szafran & Peeters, 2005):

$$l = \sqrt{2\hbar / m^* \omega}\ . \tag{8}$$

Here the width d for the considered rings is defined by $d = 2l$. The obtained geometry with the parameters m^* and ω from (Lei et al., 2010) is shown in Fig. 11 (Geometry 2); m^*=0.067 m_0 and ω =15 meV. The center radius of QR is 20 nm.

Results of the effective model calculations for the ground state energy of electron in a magnetic field are shown in Fig. 12. (Filikhin et al. 2011a) The picture of the change of the orbital quantum number of the ground state is similar to that obtained in (Lei et al., 2010) with the oscillator model. The change occurred at 2.2 T and 6.7 T. The obtained energy fits the experimental data rather well.

It has to be noted here that one cannot reproduce this result using the geometry proposed in (Lei et al., 2010) (Geometry 1) for this QR. The correspondence between the confinement potential parameters of the oscillator model and the real sizes of quantum objects has to be established by Eq. (8). Only using the geometry followed from Eq. (8) we reproduce result of (Lei et al., 2010), as is shown in Fig. 12. The strength parameter of the effective potential, in the case of the Geometry 2, was chosen to be 0.382 eV, which is close to that for QD from (Filikhin et al., 2008), where V_s =0.31 eV. The difference is explained by the topology dependence of the effective potentials (see section above and also (Filikhin et al., 2006)).

Note that the considered QRs are the plane quantum rings with the condition H <<D (for height and diameter of QR), which enhances the role of the lateral size confinement effect. To qualitatively represent the situation shown in Fig. 12, one can used an approximation for the 3D QR based on the formalism of one dimensional ideal quantum ring. Additional electron energy, due to the magnetic field, can be calculated by the relation: $E = \hbar^2 / (2m^* R^2)(l + \Phi / \Phi_0)^2$ (see for instance (Emperador et al., 2000)), where fluxes are $\Phi = \pi R^2 B$, $\Phi_0 = h / e$.. ($\Phi_0 = 4135.7$ T nm²); R is radius of the ideal ring. The Aharonov-Bohm (AB) (Aharonov & Bohm, 1959) period ΔB (Aronov & Sharvin, 1987) is given by the relation: $\Delta B = \Phi_0 / \pi / R^2$. Using the root mean square (rms) radius for R (R=20.5 nm), one can obtain $\Delta B / 2$ =1.56 T and $\Delta B / 2 + \Delta B$ =4.68 T for the ideal ring. This result is far from the result of 3D calculations shown in Fig. 12 where $\Delta B / 2 \approx 2.2$ T and $\Delta B / 2 + \Delta B \approx 6.7$ T are determined. Note here that the electron root mean square radius $R_{n,l}$ is defined by the relation $R^2_{n,l} = \int |\Phi^N_{n,l}(\rho,z)|^2 \rho^3 d\rho dz$, where $\Phi^N_{n,l}(\rho,z)$ is the normalized wave function of electron state described by the quantum numbers (n,l).

One can obtain better agreement by using the radius for the most probable localization of the electron $R_{loc.}$, defined at the maximum of the square of the wave function. The electron is mostly localized near 17.1 nm, for B=0. With this value, the ideal ring estimation leads to the values for $\Delta B / 2$ and $\Delta B / 2 + \Delta B$ as 2.25 T and 6.75 T, respectively. That agrees with the result of the 3D calculations (see Fig. 12). Obviously, the reason for this agreement is the condition $H << D$, for the considered QR geometry as it was mentioned above. The mostly localized position of the electron in QR depends weakly on the magnetic field. We present .. as a function of the magnetic field B in Fig. 13. $R_{loc.}(B)$ is changed in an interval of ±1 nm around the mean value $R_{loc.}(0)$ of 17 nm. It is interesting to note that the magnetization of a single electron QR demonstrates the same behavior as it does for $R_{loc.}(B)$ if the one dimensional ring is used (see (Voskoboynikov et al., 2002) for details).

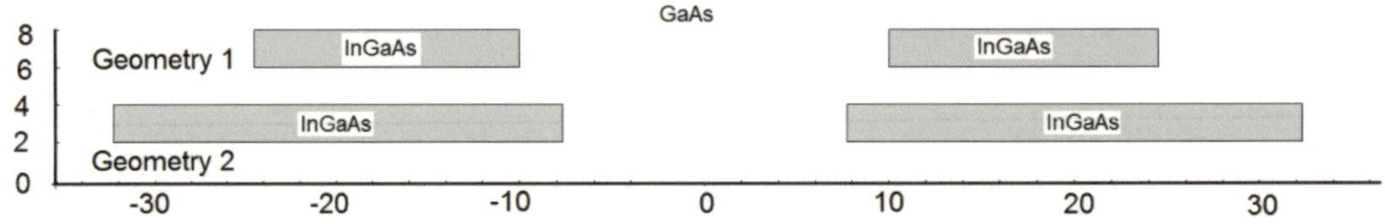

Fig. 11. QR cross section profile corresponding to Geometry 1 and Geometry 2; sizes are in nm.

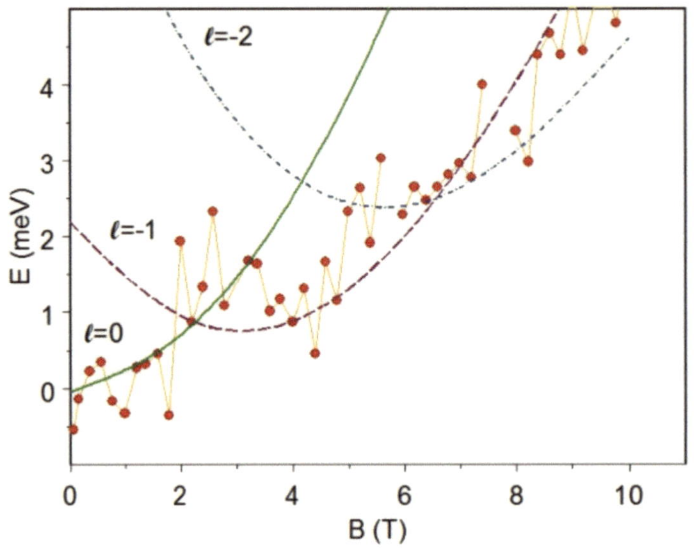

Fig. 12. Additional energy of an electron in QR in a magnetic field B. The C-V experimental energies (circles) were obtained in (Lei, et al. 2010) by using a linear approximation $\Delta E = e\Delta V_g / f$, with the lever arm $f = 7.84$. The curves $l = 0, -1, -2$ are the results of our calculations multiplied by a factor of 1.18 (Lei, et al. 2010).

Additionally we compare the results of calculations for the QR geometry parameters corresponding to Geometry 1 and Geometry 2 in Fig. 11 with the far-infrared (FIR) data, reported in (Emperador, et al. 2000). The results are presented in Fig. 14. One can see that the QR geometry proposed in (Lei et al., 2010) leads to a significant difference between the FIR data and the effective model calculations (see Fig. 14a), whereas the results obtained with Geometry 2 are in satisfactory agreement with the data (Fig. 14b). Again we conclude that the QR geometry of (Lei et al., 2010) does not provide an adequate description of electron properties of the InAs/GaAs QRs measured in (Lorke et al., 2000; Lei et al., 2010).

To summarize, we wish to point out that the problem of reliable theoretical interpretation of the C-V (and FIR) data for InAs/GaAs quantum rings is far from resolved. Obtained geometry can be considered as a possible version of geometry for experimentally fabricated QR.

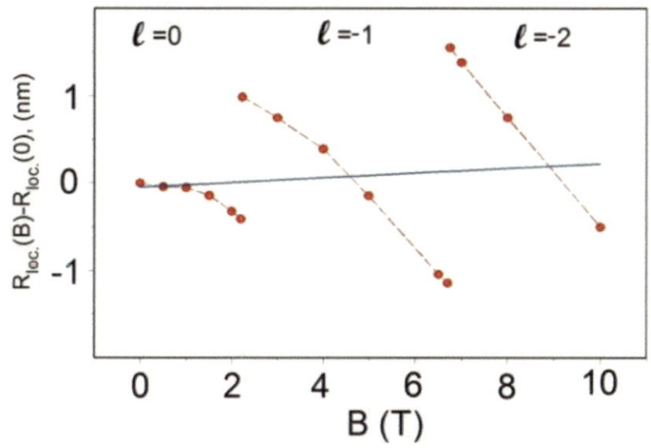

Fig. 13. The radius ($R_{loc.}$) of the most localized position of an electron as a function of a magnetic field B. The electron of the ground state is considered. The circles indicate the calculated values and the solid line indicates the result of the least squares fitting of the calculated values. The orbital quantum number of the ground state is shown.

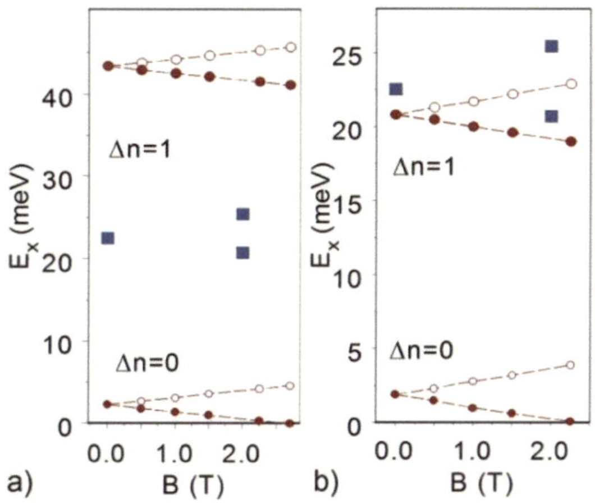

Fig. 14. Solid squares represent the observed resonance positions (Emperador, et al. 2000) of the FIR transmissions at various magnetic field B. Calculated energies of the excited states with $|\ \Delta l\ |=1$ are marked by the circles. a) QR with shape given by Geometry 1, b) QR with shape given by Geometry 2. The orbital quantum number of the ground state is $l = 0$. The quantum number n is changed as shown.

2.6 Material mixing in InGaAs/GaAs quantum dots

The fabrication process of nano-sized self-assembled InAs/GaAs quantum dots and quantum rings may be accompanied by the material mixing in the initially pure InAs QDs due to interdiffusion of the QD/substrate materials. This mixing cannot be precisely controlled, resulting in QDs with spatially inhomogeneous Ga fractions that are not well specified. In this section we show an application of the effective model to study $In_xGa_{1-x}As$ QDs with significant Ga fractions.

InAs QDs having a semi-ellipsoidal shape embedded into the GaAs substrate are considered (see Fig. 6a)). The effective potential V_s =0.31 eV, which was found in (Filikhin et al., 2008) for pure InAs QDs, reproduces the capacitance-gate-voltage experiments satisfactory well (in Table 2 these results described in the column "0% Ga (V_s =0.31eV)"). It was assumed that a realistic approach must therefore take into account material mixing. The results of the effective model calculations for Ga fractions of 10%, 20% and 25% are listed in Table 2 (Filikhin et al., 2009). The calculations was performed varying the Ga fraction in QDs for strength parameters V_s^e =0.21 eV and V_s^h =0.28 eV of the potential. The effective electron mass, the band gap and the effective potential for $In_xGa_{1-x}As$ changed linearly with respect to the value of the Ga fraction, assuming a homogeneous distribution of Ga in the QD volume. The experimental value of the transition energy for recombination of an exciton pair (E_{ex}) in the ground state is matched by calculations corresponding to a Ga fraction of approximately 22% in the QDs. Thus we conclude that the data obtained in CV and PLexperiments to this QD may be related with mixing in QD of 22%. It has to note that calculations with the 22% in the QDs (V_s =0.21eV) and pure InAs QDs (used V_s =0.31eV) demonstrate some uncertainties in the QD geometry and the Ga fraction and may lead to non-unique descriptions of the same experimental data.

Ga fraction	10%	20%	25%	0% (V_s =0.31eV)	Exp.
m^*/m_0	0.050	0.056	0.057	0.057	0.057 ± 0.007
$\Delta E(e)$ $\Delta E(h)$	238 245	205 217	188 151	185 206	204
$e_1 - e_0$ $e_2 - e_1$	50 55	48 53	46 52	46 52	44 49
$h_0 - h_1$ $h_1 - h_2$	10 12	10 11	9 11	10 11	
E^c_{e0e0}	21.0	20.9	20.8	20.8	21.5 (or 18.9)
E^c_{e0e1}	18.1	18.0	17.9	18.0	24 (or 13.0)
E^c_{e1e1}	17.0	17.0	16.9	17.0	~18.0
E^c_{h0h0}	25.1	24.9	24.7	25.1	24
E^c_{e0h0}	22.8	22.6	22.5	22.7	33.3
E_{ex}	1014	1075	1160	1106	1098
d_{00}	0.08	0.08	0.08	0.08	0.4 ± 0.1

Table 2. Calculated single electron (hole) energy-level spacing $e(h)$, electron (hole) binding energy $\Delta E(e)$ ($\Delta E(h)$), electron-electron, electron-hole and hole-hole Coulomb energies $E^c_{\alpha\beta}$ ($\alpha,\beta = e,h$), excitonic band gap E_{ex} (in meV), exciton dipole moment d_{00} (in nm) and effective mass of the QD material for semi-ellipsoidally shaped InGaAs QDs (Ga fraction in %) embedded in GaAs. Electron (hole) energy of the ground state is measured from the GaAs conduction (valence) band. The value of the effective mass is given for the p -wave electron level.

In (Filikhin et al. 2009) it was brought argument for existence of the essential mixing of the Ga-fraction in QD. The effective model with the material mixing was tested by comparison with available experimental data for the Coulomb shifts of the transition energies for positive (X^+) and negative (X^-) charged trions and biexcitons (XX) as a function of the neutral exciton (X) recombination energy. Results of these calculations for various base size parameters of QDs are depicted in Fig. 15, along with experimental data. The root mean square fit of the experimental data from (Rodt et al., 2005) shown by the dashed lines in Fig. 15. The vertical line shows the transition energy that corresponds to the limit of the QD sizes for which there are only two electron and two heavy hole levels. In this case the Coulomb shifts are calculated by combinations of the Coulomb energies of electron-electron, electron-hole and hole-hole pairs:

$$E(XX)-(X)=E^c_{ee}-E^c_{eh}+E^c_{hh}-E^c_{eh},\ E(X^-)-(X)=E^c_{ee}-E^c_{eh},\ E(X^+)-(X)=E^c_{hh}-E^c_{eh}.$$

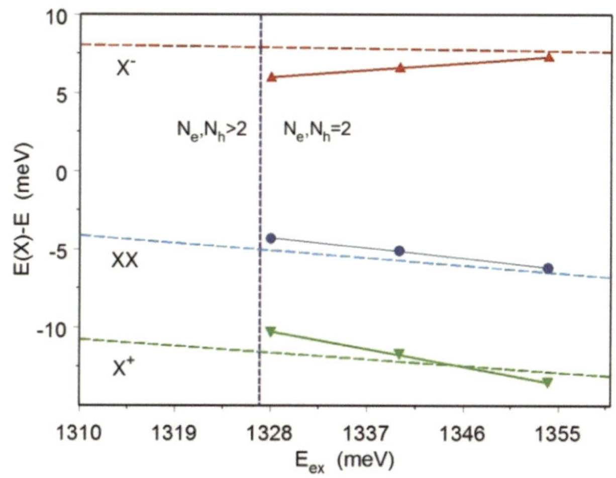

Fig. 15. Coulomb shifts of transition energies for positively (X^+) and negatively (X^-) charged trions and biexcitons (XX) as a function of neutral exciton (X) recombination energy. Results of the calculations for various base size parameters of QDs are marked by solid triangles (X^-, X^+) and dots (XX). The dashed lines correspond to root mean square fits to experimental data from (Rodt et al., 2005). The solid lines correspond to root mean square fits to the calculated results. The vertical line shows the transition energy, which corresponds to the limit of the QD sizes for which there are only two electrons and two heavy holes levels. The amount of the Ga fraction in our calculations is equal to 25%.

When there are several interacting carrier pairs, the calculations must be performed with more intricate scheme using perturbation theory. The value of the Ga fraction in our calculations was 25%. Calculations were performed for three QD geometries. A lens-shaped geometry with a height of 3.5 nm and base sizes of 9 nm, 10 nm and 11 nm were used. The effective model results in Fig. 15 demonstrate qualitative agreement with the experimental data for the aforementioned confinement region. The calculated results are very sensitive to the value of the Ga fraction. In particular, increasing the fraction shifts the X^+ and X^- energies to the region of large exciton energies (X). At the same time, the Coulomb shifts decrease in absolute value within the region of the X-energies with N_h=2. Decreasing the Ga fraction gives the opposite results.

We can conclude that in the framework of an effective model one can reproduce the CV and PL experimental data for InGaAs/GaAs QDs. In these calculations the amount of the Ga-fraction was taken to be about 22%. Taking into account this value for Ga-fraction we also reproduce the measured transition energies and Coulomb shifts for excitonic complexes (X^- , X^+ , XX) in the limit of two interacting pairs of carriers in the QDs.

3. Quantum chaos in single quantum dots

3.1 Quantum chaos

Quantum Chaos concerns with the behavior of quantum systems whose classical counterpart displays chaos. It is quantum manifestation of chaos of classical mechanics.

The problem of quantum chaos in meso - and nano-structures has a relatively long history just since these structures entered science and technology. The importance of this problem is related to wide spectrum of the transport phenomena and it was actively studied in the last two decades (Beenakker & van Houten, 1991; Baranger & Stone, 1989; Baranger et al., 1991). One of the main results of these studies, based mainly on the classical and semi-classical approaches, is that these phenomena sensitively depend on the geometry of these quantum objects and, first of all, on their symmetry: Right - Left (RL) mirror symmetry, up-down symmetry and preserving the loop orientation inversion symmetry important in the presence of the magnetic field (Whitney et al., 2009; Whitney et al., 2009a).

These results are well -known and discussed widely. There is another, actively studied in numerous fields of physics, aspect which ,in essence, is complimentary to the above mentioned semi classical investigations: Quantum Chaos with its inalienable quantum character , including, first of all, Nearest Neighbor level Statistics (NNS) which is one of the standard quantum-chaos test.

Mathematical basis of the Quantum Chaos is a Random Matrix Theory (RMT) developed by Wigner, Dyson, Mehta and Goudin (for comprehensive review see book (Beenakker & van Houten, 1991)). RMT shows that the level repulsion of quantum systems (expressed by one of the Wigner-Dyson -like distributions of RMT) corresponds to the chaotic behavior and, contrary, level attraction described by Poisson distribution tells about the absence of chaos in the classical counterpart of the quantum system. This theorem-like statement checked by numerous studies in many fields of science. For the completeness, we add that there are other tests of Quantum Chaos based on the properties of the level statistics: Δ_3 statistics (spectral rigidity $\Delta_3(L)$), Number variance $\Sigma_2(L)$), spectral form-factor, two- and multipoint correlation functions, two level cluster function $Y_2(E)$ etc. They play an important subsidiary role to enhance and refine the conclusions emerging from the NNS.

The present review surveys the study of the NNS of nanosize quantum objects - quantum dots (QD) which demonstrate atom-like electronic structure under the regime of the size confinement. To use effectively NNS, we have to consider so called weak confinement regime where the number of levels can be of the order of several hundred. QD of various shape embedded into substrate are considered here under the effective model (Filikhin et al., 2010). We use the sets of QD/substrate materials (Si/SiO_2, $GaAs/Al_{0.7}Ga_{0.25}As$, $GaAs/InAs$).

3.2 The nearest neighbor spacing statistics

For the weak confinement regime (for the Si/SiO_2 QD, the diameter $D \geq 10$ nm), when the number of confinement levels is of the order of several hundred (Filikhin et al. 2010), we studied NNS statistics of the electron spectrum. The low-lying single electron levels are marked by E_i, $i = 0,1,2,...N$. One can obtain the set $\Delta E_i = E_i - E_{i-1}$, $i = 1,2,3,...N$ of energy differences between neighboring levels. An example of the energy spectrum and set of the neighbor spacings for Si/SiO_2 QD are in Fig. 16. We need to evaluate the distribution function $R(\Delta E)$, distribution of the differences of the neighboring levels. The function is normalized by $\int R(\Delta E)d\Delta E = 1$. For numerical calculation, a finite-difference analog of the distribution function is defined by following relation:

$$R_j = N_j / H_{\Delta E} / N \, , \; j = 1,...M \, ,$$

where $\sum N_j = N$ represents total number of levels considered, $H_{\Delta E} = ((\Delta E)_1 - (\Delta E)_N) / M$ is the energy interval which we obtained by dividing the total region of energy differences by M bins. N_j ($j = 1,2,...M$) is the number of energy differences which are located in the j -th bin.

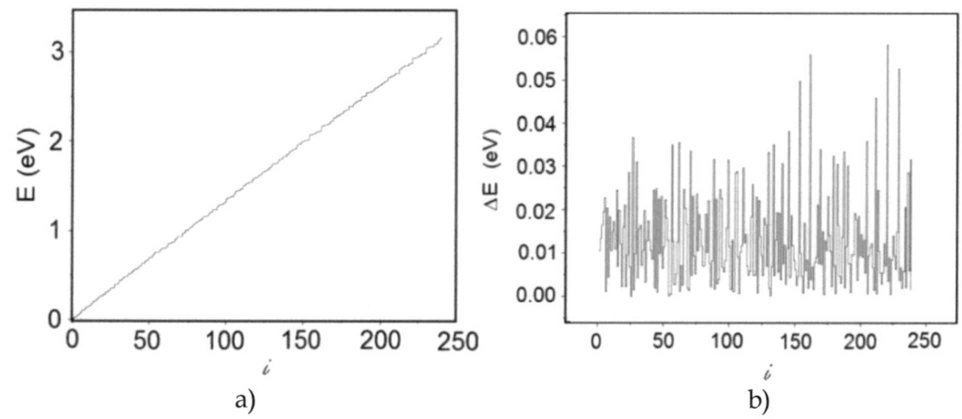

Fig. 16. a) The energy levels and b) the neighbor spacings $\Delta E_i = E_i - E_{i-1}$, $i = 1,2,..N$,of the spherical Si/SiO_2 QD with diameter D=17 nm.

The distribution functions $R(\Delta E)$ is constracted using the smoothing spline method. If R_j, $j = 1,2,..M$, are calculated values of the distribution functions corresponding to ΔE_j, the smoothing spline is constructed by giving the minimum of the form $\sum_{j=1}^{M}(R_j - R(\Delta E_j))^2 + \int R''(\Delta E)^2 d(\Delta E) / \lambda$. The parameter $\lambda > 0$ is controlling the concurrence between fidelity to the data and roughness of the function sought for. For $\lambda \to \infty$ one obtains an interpolating spline. For $\lambda \to 0$ one has a linear least squares approximation.

We studied neighbouring level statistics of the electron/hole spectrum treated by way considered above. The Si quantum dots having strong difference of electron effective mass in two directions is considered as appropriate example for the study of role of the effective mass asymmetry. In this study we do not include the Coulomb potential between electrons and holes. The shape geometry role is studied for two and three dimensions.

3.3 Violation of symmetry of the QD shape and nearest neighbor spacing statistics

Distribution functions for the nearest neighboring levels are calculated for various QD shapes (Filikhin et al., 2010). Our goal here to investigate the role of violation of the symmetries of QD shape on the chaos. The two and three dimensional models are considered. Existing of any above mentioned discrete symmetry of QD shape leads to the Poisson distribution of the electron levels.

In Fig. 17 the numerical results for the distribution functions of Si/SiO_2 QD are presented. The QD has three dimensional spherically shape. We considered the two versions of the shape. The first is fully symmetrical sphere, and the second shape is a sphere with the cavity damaged the QD shape. The cavity is represented by semispherical form; the axis of symmetry for this form does not coincide with the axis of symmetry of the QD. In the first case, the distribution function is the Poisson-like distribution. The violation symmetry in the second case leads to non-Poisson distribution.

We fit the non-Poissonian distribution function $R(\Delta E)$ using the Brody distribution (Brody et al., 1999):

$$R(s) = (1+\beta)bs^{\beta}\exp(-bs^{1+\beta}), \tag{9}$$

with the parameter β =1.0 and $b = (\Gamma[(2+\beta)/(1+\beta)]/D)^{1+\beta}$, D is the average level spacing. Note that for the Poisson distribution the Brody parameter is equal zero.

If the QD shape represents a figure of rotation (cylindrical, ellipsoidal and others) then the 3D Schrödinger equation is separable. In cylindrical coordinates the wave function is written by the following form $\psi(\vec{r}) = \Phi(\rho,z)\exp(il\varphi)$, where $l = 0,\pm1,\pm3,...$ is the electron orbital quantum number. The function $\Phi(\rho,z)$ is a solution of the two dimensional equation for cylindrical coordinates ρ and z.

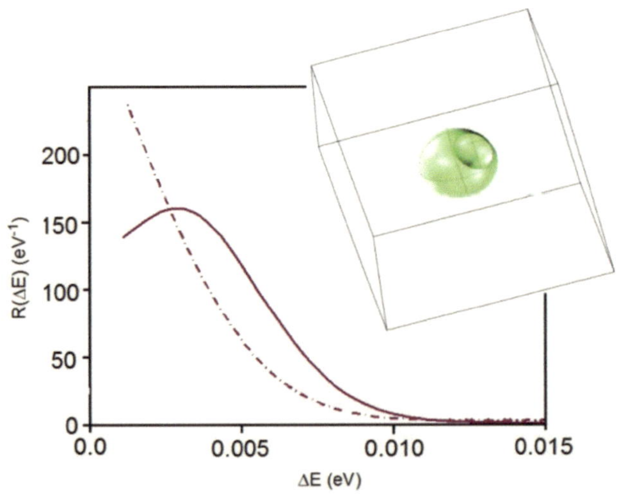

Fig. 17. Distribution functions for electron neighboring levels in Si/SiO_2 QD for spherical-like shape with cut. The Brody parameter β =1.0. The geometry of this QD is shown in 3D. The QD diameter is 17 nm (inset).

Our results for the distribution function for the ellipsoidal shaped Si/SiO_2 QD are presented in Fig. 18a) (left). In the inset we show the cross section of the QD. The fitting of the calculated values for $R(\Delta E)$ gives the Poisson-like distribution. For the case of QD shape with the break of the ellipsoidal symmetry (Fig. 18b) (left)) by the cut below the major axis we obtained a non-Poisson distribution.

Fig. 18 (right) shows the that slightly deformed rhombus-like shape leads to the NNS with Brody parameter β =1 (10). It is obvious why systems with different discrete symmetries reveal Poisson statistics: the different levels of the mixed symmetry classes of the spectrum of the quantum system are uncorrelated.

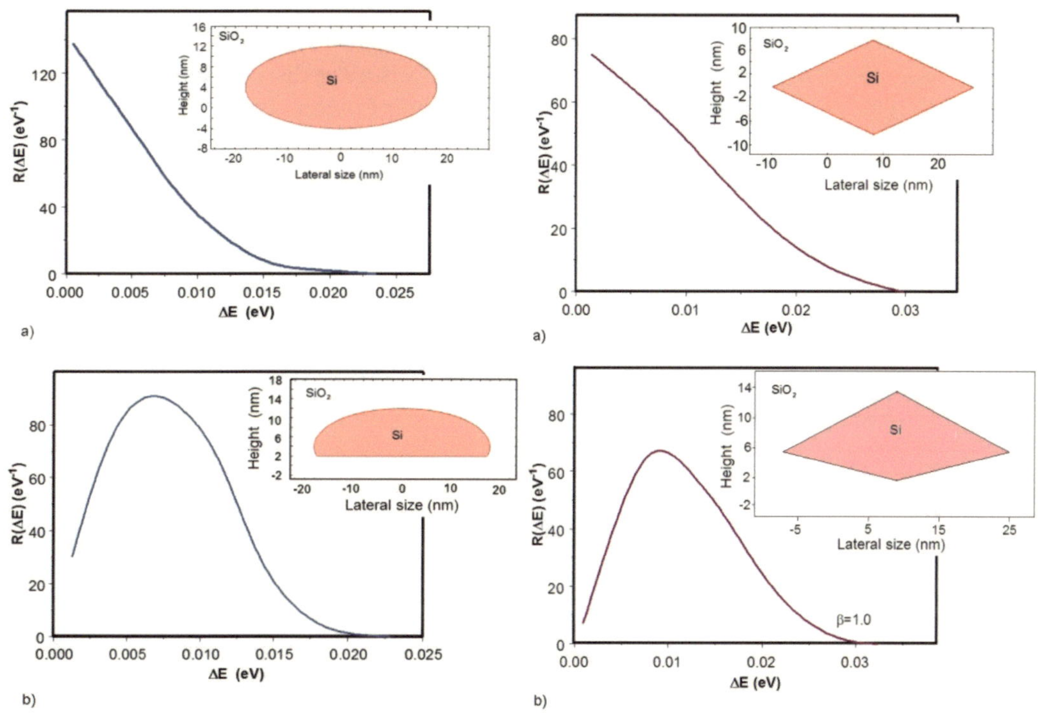

Fig. 18. (Left) Distribution functions for electron neighboring levels in Si/SiO_2 QD for different shapes: a) ellipsoidal shape, b) ellipsoidal like shape with cut. Brody parameter β is defined to be equal 1.02 for the fitting of this distribution. The 3D QD shape has rotation symmetry. Cross section of the shapes is shown in the inset.
(Right). Violation of the shape Up-Down symmetry for Si/SiO_2 QD. Distribution functions for electron neighboring levels in Si/SiO_2 QD for different shapes: a) with rhombus cross section, b) with slightly deformed rhombus cross section. The 3D QD shape has rotation symmetry. The Brody parameter β for the curve fitting this distribution is shown. Cross section of the shapes is shown in the inset.

In Schrödinger equation (7) in the asymptotical region of ρ one can neglect the two terms $\frac{1}{\rho}\frac{\partial}{\partial\rho}$ and $-\frac{l^2}{\rho^2}$ of this equation. The solution of Eq. (7) can demonstrate the same properties of the solution of the Schrödinger equation for 2D planar problem in Cartesian coordinates with the same geometry of QD shape in the asymptotical region. We illustrate this fact by Fig. 19. In this figure the violation of the shape Up-Down symmetry for 2D Si/SiO_2 QD is clarified. We compare the distribution functions for QD with "regular" semi-ellipsoidal shape (dashed curve in Fig. 19 a) and for QD with the semi-ellipsoidal shape having the cut (solid curve) as it are shown in Fig. 19 b). In the first case there is Up-Down symmetry of the QD shape. Corresponding distribution functions is Poissonian type. In second case the symmetry is broken by cut. The level statistics become non Poissonian. We have qualitative the same situation as for QD having rotation symmetry in 3D, presented in Fig. 18 (left) for the QD shape with rotation symmetry in cylindrical coordinates. The relation between the symmetry of QD shape and NNS is presented by Fig. 20 where we show the results of calculation of NNS for the 2D InAs/GaAs quantum well (QW). The two types of the statistics are presented in Fig. 20(left). The Poisonian distribution corresponds to shapes shown in Fig. 20 (b)-(d)(left) with different type of symmetry. The non-Poissonian distribution has been obtained for the QW shape with cut (a) which violated symmetry of initial shape (b), which is square having left-right symmetry, up-down symmetry, and diagonal reflection symmetry. The shape of the Fig. 20c) has only diagonal reflection symmetry. In Fig. 20d) the left-right symmetry of the shape exists only. The electron wave function of the high excited state, which contour plot is shown with the shape contour in Fig. 20(left), reflects the symmetry properties of the shapes.

Concluding, we can note that, obviously, the topological equivalent transformations of QD shape (keeping at least one discrete symmetry) do not lead to the non Poissonian distribution of the levels.

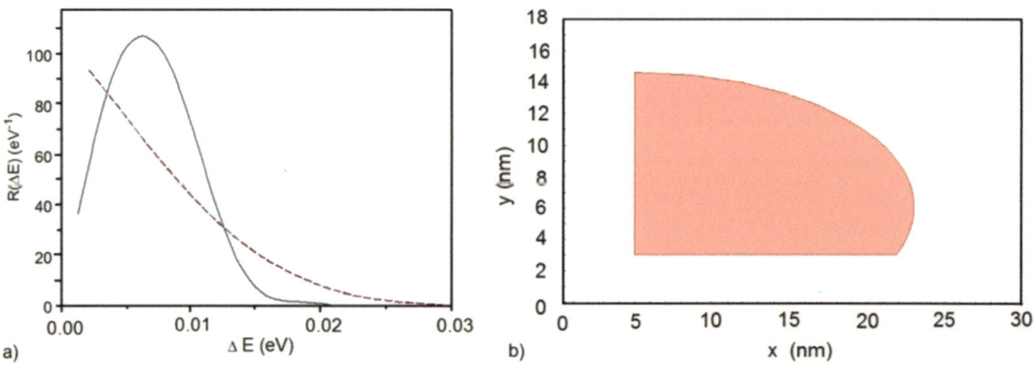

Fig. 19. Violation of the shape Up-Down symmetry for two dimensional Si/SiO_2 QD. a) Distribution functions for electron neighboring levels for the "regular" semi-ellipsoidal shape (dashed curve), for the semi-ellipsoidal shape with the cut (solid curve). b) The shape of the QD with cut (in Cartesian coordinates).

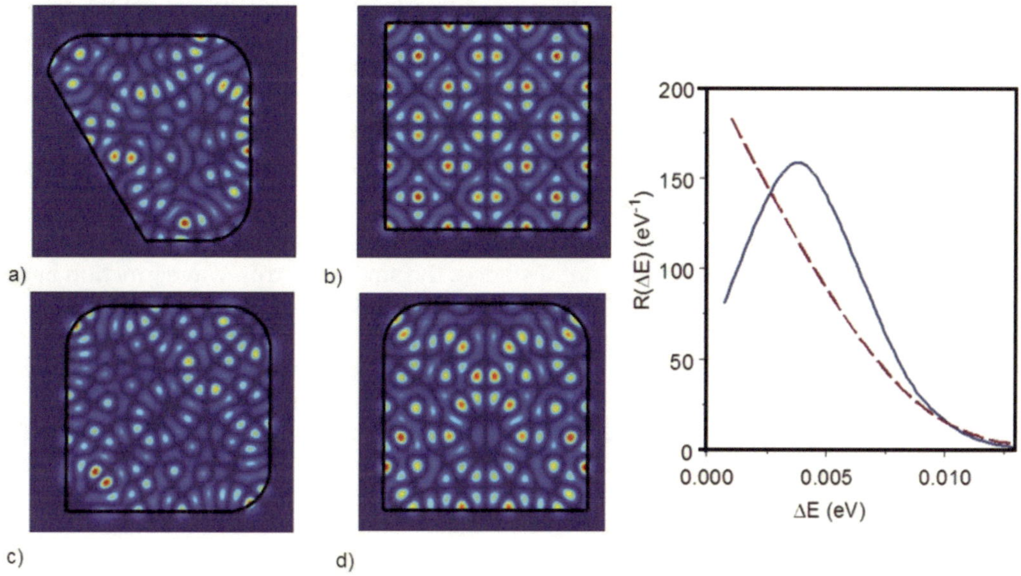

Fig. 20. Shape of the 2D InAs/GaAs quantum dots (Left). The black curves mean the perimeters. The electron wave function contour plots of the excited state (with energy about 0.5 eV are shown). The corresponding types of the level statistics are shown (Right). The shape a) leads to non-Poissonian statistics (solid curve). The shapes b)-d) result to the Poissonian statistics (dashed curve).

4. Double quantum dots and rings: New features

4.1 Disappearance of quantum chaos in coupled chaotic quantum dots

In the previous section, we investigated the NNS for various shape of the single quantum dots (SQD) in the regime of the weak confinement when the number of the levels allows to use quite sufficient statistics. Referring for details to (Filikhin et al., 2010), we briefly sum up the main conclusions of previous section: SQDs with at least one mirror (or rotation) symmetry have a Poisson type NNS whereas a violation of this symmetry leads to the Quantum Chaos type NNS.

In this section we study quantum chaotic properties of the double QD (DQD). By QD here we mean the three dimensional (3D) confined quantum object, as well its 2D analogue - quantum well (QW). In three dimensional case we use an assumption of the rotational symmetry of QD shape. The presented effective approach is in good agreement with the experimental data and previous calculations in the strong confinement regime (Filikhin et al., 2010). Here, in the regime of weak confinement, as in (Filikhin et al., 2010), we also do not consider Coulomb interaction between electron and hole: Coulomb effects are weak when the barrier between dots is thin leading to the strong interdot tunneling and dot sizes are large enough. In these circumstances, studied in detail in (Bryant, 1993) (see also for short review a monograph (Bimberg et al., 1999), one may justify disregard of the Coulomb effects. The physical effect, we are looking for, has place just for thin barriers; to have sufficient level statistics, we need large enough QDs ($\geq$100 nm for InAs/GaAs QW).

Thus, we consider tunnel coupled two QDs with substrate between, which serves as barrier with electronic properties distinct from QD. Boundary conditions for the single electron Schrödinger equation are standard. We take into account the mass asymmetry inside as well outside of QDs (Filikhin et al., 2010). To avoid the complications connected with spin-orbit coupling, *s*-levels of electron are only considered in the following. We would like to remind that the selection of levels with the same quantum numbers is requisite for study of NNS and other types of level statistics.

Whereas at the large distances between dots each dot is independent and electron levels are twofold degenerate, expressing the fact that electron can be found either in one or in the other isolated dot, at the smaller inter-dot distances the single electron wave function begins to delocalize and extends to the whole DQD system. Each twofold degenerated level of the SQD splits by two, difference of energies is determined by the overlap, shift and transfer integrals (Bastard, 1990). Actually, due to the electron spin, there is fourfold degeneracy, however that does not change our results and below we consider electron as spinless. Note that the distance of removing degeneracy is different for different electron levels. This distance is larger for levels with higher energy measured relative to the bottom quantum well (see Fig. 23 below). By the proper choice of materials of dots and substrate one can amplify the "penetration" effects of the wave function.

Below we display some of our results for semiconductor DQDs. The band gap models are given in (Filikhin et al., 2010). Fig. 21 shows distribution function for two Si/SiO_2 QDs of the shape of the 3D ellipsoids with a cut below the major axis. Isolated QD of this shape, as we saw in the previous section, is strongly chaotic. It means that distribution function of this QD can be well fitted by Brody formula with the parameter which is close to unity (Filikhin et al, 2010). We see that the corresponding up-down mirror symmetric DQD shows Poisson-like NNS. Note that these statistics data involved 300 confined electron levels, which filled the quantum well from bottom to upper edges. We considered the electron levels with the orbital momentum *l*=0, as was mentioned above. The orbital momentum of electron can be defined due to rotational symmetry of the QD shape.

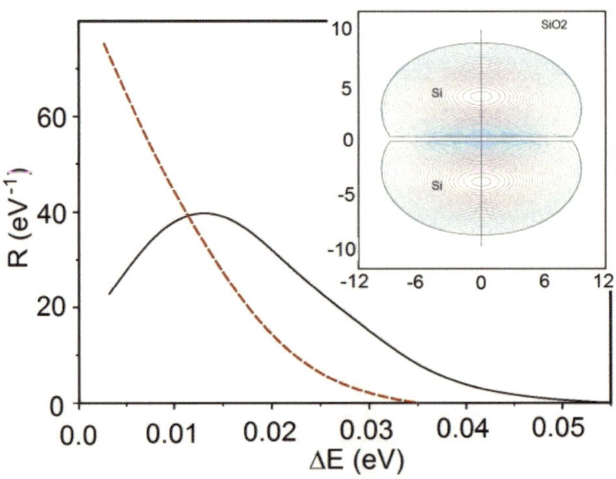

Fig. 21. The electron wave function of the ground state is shown by the contour plot. (The lower figure) Distribution functions for energy differences of the electron neighbori between QDs in InAs/GaAs DQD.

In Fig. 22, SQD (2D quantum well) without both type of symmetry reveals level repulsion, two tunnel coupled dots show the level attraction. From the mirror symmetry point of view, the chaotic character of such single object is due to the lack of the R-L and up-down mirror symmetries. The symmetry requirements in this case, for the coupled dots are less restrictive: presence of one of the mirror symmetry types is sufficient for the absence of quantum chaos.

Dependence NNS on the interdot distance shows a gradual transition to the regular behaviour with intermediate situation when Poisson-like behavior coexists with chaotic one: they combine but the level attraction is not precisely Poisson-like. Further decreasing distance restores usual Poisson character (see Fig. 22). Fig. 23 shows how the degeneracy gradually disappears with the distance b between QDs in InAs/GaAs DQD.

Finally, we would like to show the disappearance of the Quantum Chaos when chaotic QW is involved in the "butterfly double dot" (Whitney, 2009) giving huge conductance peak in the semi-classical approach. Fig. 24 shows the NNS for chaotic single QW of (Whitney, 2009) by dashed line. Mirror (up-down and L-R) symmetry is violated. The NNS for an L-R mirror symmetric DQW is displayed by solid line in Fig. 24. It is clear that Quantum Chaos disappears.

We conjecture that the above mentioned peak in conductance of (Whitney, 2009) and observed here a disappearance of Quantum Chaos in the same array are the expression of the two faces of the Quantum Mechanics with its semi-classics and genuine quantum problem of the energy levels of the confined objects, despite the different scales (what seems quite natural) in these two phenomena (several micrometers and 10–100 nm, wide barrier in the first case and narrow one in the second). We have to emphasize here that the transport properties are mainly the problem of the wave function whereas the NNS is mainly the problem of eigenvalues. Similar phenomena are expected for the several properly arranged coupled multiple QDs and QD superlattices. In the last case, having in mind, for simplicity, a linear array, arranging the tunnel coupling between QDs strong enough, we will have wide mini-bands containing sufficient amount of energy levels and the gap between successive mini-bands will be narrow. Since the levels in the different mini-bands are uncorrelated, the overall NNS will be Poissonian independently of the chaotic properties of single QD. We would like to remark also that our results have place for 3D as well as for 2D quantum objects. It is important to notice that the effect of reduction of the chaos in a system of DQD could appear for interdot distances larger than considered, for instance in figure 22, if an external electrical field is applied. By properly designed bias the electric field will amplify wave function "penetration" effectively reducing a barrier between QDs.

Thus, we have shown that the tunnel coupled chaotic QDs in the mirror symmetric arrangement have no quantum chaotic properties, NNS shows energy level attraction as should to be for regular, non-chaotic systems. These results are confronted with the huge conductance peak found by the semi-classical method in (Whitney, 2009). We think that our results have more general applicability for other confined quantum objects, not only for the quantum nanostructures, and may be technologically interesting. Concerning the last issue, problem is what easer: try to achieve regular, symmetric shape of SQDs, or, not paying attention to their irregular, chaotic shape arrange more or less symmetric mutual location (Ponomarenko et al., 2008).

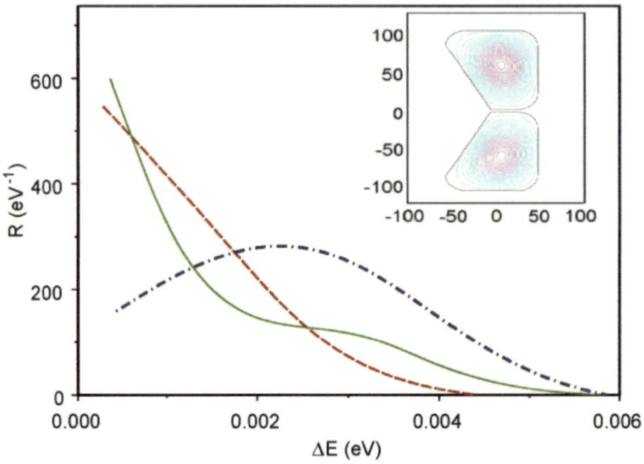

Fig. 22. Distribution functions for energy differences of the electron neighboring levels in the 2D InAs/GaAs DQW calculated for various distances b between QWs. Dashed (solid) line corresponds to b =4 nm (b =2 nm). Distribution functions of single QW is also shown by the dot-dashed line. The DQW shape is shown in inset (sizes are in nm).

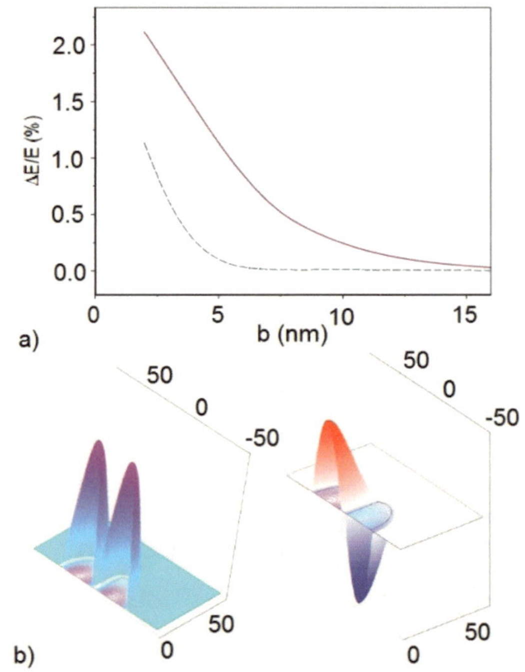

Fig. 23. (The upper figure) Doublet splitting ΔE of single electron levels dependence on the distance b between QDs in InAs/GaAs DQD. The ground state (E =0.23 eV) level splitting is ΔE expressed by dashed line. The solid line corresponds to doublet splitting of a level which is close to upper edge of the quantum well (E =0.56 eV). The shape of DQD is the same as in Fig. 21 (The lower figure). The electron wave functions of the doublet state: the ground state (left) and first excited state (right), are shown.

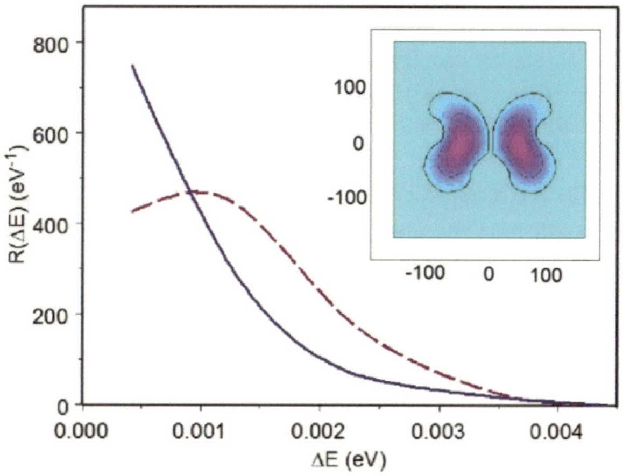

Fig. 24. Distribution functions for electron neighboring levels in InAs/GaAs single QW (dashed line) and DQW (solid line). Shape of DQW is shown in the inset. The electron wave function of the ground state is shown by the contour plot in the inset. Data of the statistics include 200 first electron levels.

4.2 Electron transfer between pair of concentric quantum rings in magnetic field

Quantum rings are remarkable meso- and nanostructures due to their non-simply connected topology and attracted much attention last decade. This interest supported essentially by the progress in the fabrication of the structures with wide range of geometries including single and double rings. This interest rose tremendously in the connection to the problem of the persistent current in mesoscopic rings (Buttiker et al., 1983) Transition from meso - to nano - scale makes more favorable the coherence conditions and permits to reduce the problem to the few or even to single electron.

Application of the transverse magnetic field B leads to the novel effects: Whereas the quantum dots (QDs) of the corresponding shape (circular for two dimensional (2D), cylindrical or spherical for 3D) has degeneracy in the radial n and orbital l quantum numbers, QR due to the double connectedness in the absence of the magnetic field B has degeneracy only in l, and the nonzero B lifts the degeneracy in l, thus making possible the energy level crossing at some value of B, potentially providing the single electron transition from one state to the another.

Use the configurations with double concentric QR (DCQR) reveals a new pattern: one can observe the transition between different rings in the analogy with atomic phenomena. For the DCQR, the 3D treatment is especially important when one includes the inter-ring coupling due to the tunneling. The dependence on the geometries of the rings (size, shape and etc.) becomes essential.

We investigate the electron wave function localization in double concentric quantum rings (DCQRs) when a perpendicular magnetic field is applied. In weakly coupled DCQRs can be arisen the situation, when the single electron energy levels associated with different rings may be crossed. To avoid degeneracy, the anti-crossing of these levels has a place. In this DCQR the electron spatial transition between the rings occurs due to the electron level anti-crossing. The anti-crossing of the levels with different radial quantum numbers (and equal orbital quantum numbers) provides the conditions when the electron tunneling between rings becomes possible. To study electronic structure of the semiconductor DCQR, the single sub-band effective mass approach with energy dependence was used (see section 2 of this Chapter). Realistic 3D geometry relevant to the experimental DCQR fabrication was employed taken from (Kuroda et al., 2005; Mano et al., 2005). The GaAs QRs and DQRs rings, embedded into the $Al_{0.3}Ga_{0.7}As$ substrate, are considered (Filikhin, et al., 2011). The strain effect between the QR and the substrate materials was ignored here because the lattice mismatch between the rings and the substrate is small. Due to the non-parabolic effect taken into account by energy dependence effective mass of electron in QR, the effective mass of the electron ground state is calculated to be the value of 0.074 m_0 that is larger than the bulk value of 0.067 m_0. For the excited states, the effective mass will increase to the bulk value of the $Al_{0.3}Ga_{0.7}As$ substrate. Details of this calculation one can find in (Filikhin, et al. 2011).

Electron transfer in the DCQR considered is induced by external factor like a magnetic or electric fields. Probability of this transfer strongly depends on the geometry of DCQR. The geometry has to allow the existing the weakly coupled electron states. To explain it, we note that DCQR can be described as a system of double quantum well. It means that there is duplication of two sub-bands of energy spectrum (see (Manasreh, 2005) for instance) relative the one for single quantum object. In the case of non-interacting wells (no electron tunneling between wells) the each sub-band is related with left or right quantum well. The wave function of the electron is localized in the left or right quantum well. When the tunneling is possible (strong coupling state of the system), the wave function is spread out over whole volume of the system. In a magnetic field, it is allowed an intermediate situation (weak coupled states) when the tunneling is possible due to anti-crossing of the levels. Anti-crossing, of course, is consequence of the impossibility to cross of levels with the same space symmetry (von Neumann & Wigner, 1929; Landau & Lifshitz, 1977).

There is a problem of notation for states for DCQR. If we consider single QR (SQR) then for each value of the orbital quantum number $|l|=0,1,2...$ in Eq. (7) we can definite radial quantum number n =1,2,3,... corresponding to the numbers of the eigenvalues of the problem (7) in order of increasing. One can organize the spectrum by sub-bands defined by different n. When we consider the weakly coupled DCQR, in contrast of SQR, the number of these sub-bands is doubled due to the splitting the spectrum of double quantum object (Bastard, 1990). Electron in the weakly coupled DCQR can be localized in the inner or outer ring. In principle, in this two ring problem one should introduce a pair of separate sets of quantum numbers (n_i,l) where index i =1,2 denoted the rings where electron is localized. However, it is more convenient, due to the symmetry of the problem, to have one pair (n,l) numbers ascribed to both rings (inner or outer), in other words, we use a set of quantum numbers $(n,l),p$ where p is dichotomic parameter attributed to the electron localization ("inner" or "outer").

Since we are interested here in the electron transition between rings and, as we will see below, this transition can occur due to the electron levels anti-crossing followed a tunneling, we concentrate on the changing of the quantum numbers n. The orbital quantum numbers must be equal providing the anti-crossing of the levels with the same symmetry (see Landau & Lifshitz, 1977). Thus, the anti-crossing is accompanied by changing the quantum numbers n and p of the $(n,l),p$ set.

Strongly localized states exist in the DCQR with the geometry motivated by the fabricated DCQR in (Kuroda et al., 2005; Mano et al., 2005). The wave functions of the two s-states of the single electron with n =1,2 are shown in Fig. 25, where the electron state n =1 is localized in outer ring, and the electron state n =2 is localized in inner ring. Moreover all states of the sub-bands with n =1,2, and $|l|$=1,2,3... are well localized in the DCQR. The electron localization is outer ring for n =1, $|l|$=0,1,2,..., and inner ring for n =2, $|l|$=0,1,2....

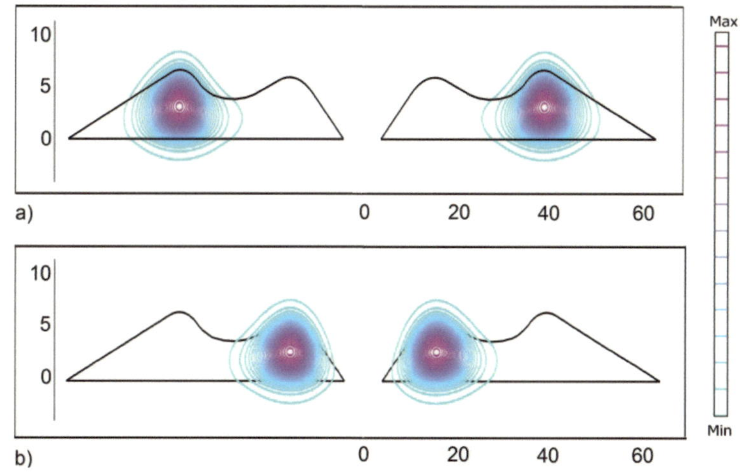

Fig. 25. The squares of wave functions for the a) (1,0) ,outer ($E = 0.072$ eV) and b) (2,0) ,inner ($E = 0.080$ eV) states are shown by contour plots. The contour of the DCQR cross-section is given. The sizes are in nm.

The difference of properties of the two sub-bands can be explained by competition of two terms of the Hamiltonian of Eq. (7) and geometry factor. The first term includes first derivative of wave function over ρ in kinetic energy; the second is the centrifugal term. For $|l|\neq 0$ the centrifugal force pushes the electron into outer ring. One can see that the density of the levels is higher in the outer ring. Obviously, the geometry plays a role also. In particular, one can regulate density of levels of the rings by changing a ratio of the lateral sizes of the rings.

Summarizing, one can say that for B =0 the well separated states are only the states $(1,l),p$ and $(2,l),p$. Thus, used notation is proper only for these states. The wave functions of the rest states $(n > 2,l)$ are distributed between inner and outer rings. These states are strongly coupled states.

Crossing of electron levels in the magnetic field B are presented in Fig. 26 There are crossings of the levels without electron transfer between the rings. This situation is like when we have crossing levels of two independent rings. There are two crossings when the orbital quantum number of the lower state is changed due to the Aharonov-Bohm effect. It occurs at about 0.42 T and 2.5 T. There are two anti-crossings: the first is at 4.8 T, another is at 5.2 T. These anti-crossings are for the states with different n; the first are states (1,0) and (2,0) and the second are states (1,-1) and (2,-1). In these anti-crossings the possibility for electron tunneling between rings are realized. In Fig. 27 we show how the root mean square (rms) of the electron radius is changed due to the tunneling at anti-crossing. One can see from Fig. 26 that the electron transition between rings is only possible when the anti-crossed levels have different radial quantum numbers and equal orbital quantum numbers, in accordance of (von Neuman & Vigner, 1929).

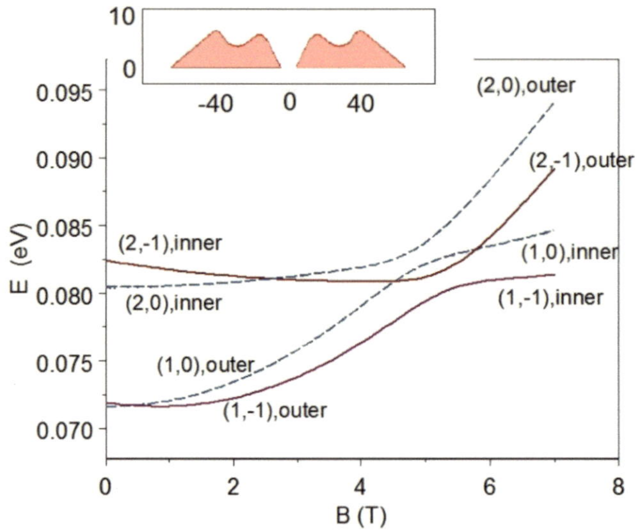

Fig. 26. Single electron energies of DCQR as a function of magnetic field magnitude B. Notation for the curves: the double dashed (solid) lines mean states with l=0 (l=-1) with n =1,2. The quantum numbers of the states and positions of the electron in DCQR are shown. The cross section of the DCQR is given in the inset.

Transformation of the profile of the electron wave function during the process of anti-crossing with increasing B is given in Fig. 28. The electron state (1,-1), outer is considered as "initial" state of an electron (B=0). The electron is localized in outer ring. Rms radius is calculated to be R=39.6 nm. For B=5.2 T the second state is the tunneling state corresponding to the anti-crossing with the state (0,-1). The wave function is spreaded out in both rings with R=32.7 nm. The parameter p has no definite value for this state. The "final" state is considered at B=7 T. In this state the electron was localized in inner ring with R=17.6 nm. Consequently connecting these three states of the electron, we come to an electron trapping, when the electron of outer ring ("initial" state) is transferred to the inner ring ("final" state). The transfer process is governed by the magnetic field.

Note that the energy gap between anti-crossed levels which one can see in Fig. 26 can be explained by the general theory for double interacting quantum well (Bastard, 1990). The value of the gap depends on separation distance between the rings, governed by the overlapping wave functions corresponding to the each ring, and their spatial spread which mainly depends on radial quantum number of the states (Filikhin et al., 2011).

Other interesting quantum system is one representing QR with QD located in center of QR. The cross section of such heterostructure ($GaAs/Al_{0.3}Ga_{0.7}As$) is shown in Fig. 29a. In Fig. 29b we present the results of calculations for electron energies of the (1,0) and (3,0) states in the magnetic field B (Filikhin et al., 2011). Once more we can the level anti-crossing (for about of 12.5 T). This anti-crossing is accompanied by exchange of electron localization between the QD and the QR. In other words if initial state (for B <12.5 T) of electron was the state (1,0),outer, then the "final" state (for B >12.5 T) will be (1,0),inner. It can be considered as one of possibilities for trapping of electron in QD.

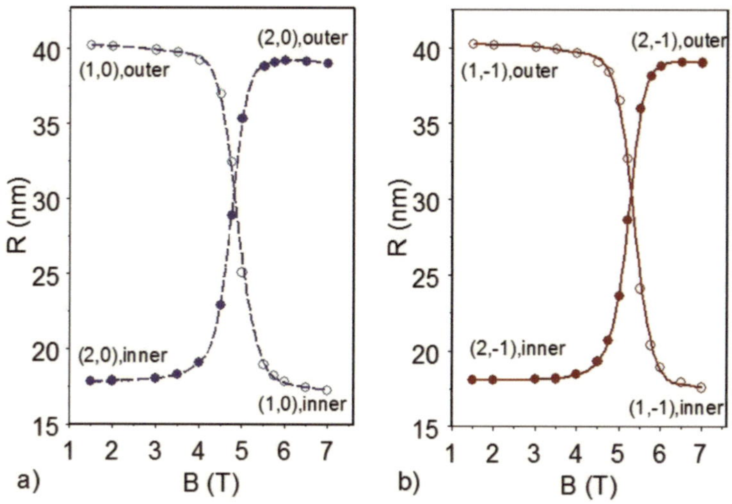

Fig. 27. Rms radius of an electron in DCQR as a function of magnetic field for the states a) $((n=1,2),l=0)$ and b) $((n=1,2),l=-1)$ near point of the anti-crossing. The calculated values are shown by solid and open circles. The dashed (solid) line, associated with states of $l=0$ ($l=-1$), fits the calculated points.

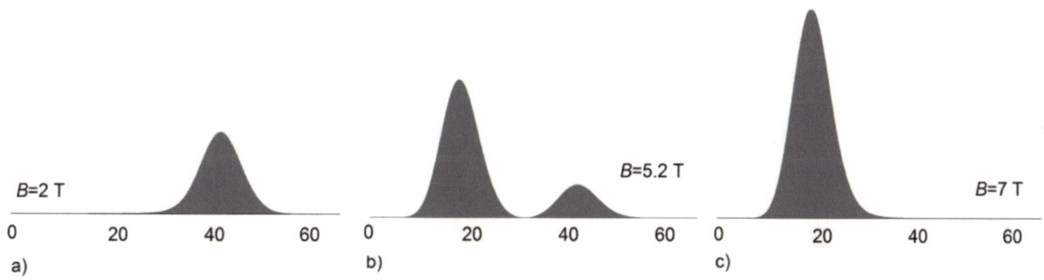

Fig. 28. Profiles of the normalized square wave function of electron in the states a) $(1,-1)$,outer; b) $(1,-1)$,n/a and c) $(1,-1)$,inner for different magnetic field B. The a) is the "initial" state (B =0) with R =39.6 nm, the b) is the state of electron transfer (B =5.2 T) with R =32.7 nm, the c) is the "final" state (B =7 T) with R =17.6 nm. The radial coordinate ρ is given in nm (see Fig. 26 for the DCQR cross section).

One can see from Fig. 29b that the energy of the dot-localized state grows more slowly than the envelope ring-localized state. At the enough large B the dot-localized state becomes the ground state (Szafran et al., 2004). In other words, when the Landau orbit of electron becomes smaller then dot size, electron can enter the dot without an extra increase of kinetic energy.

Concluding, we made visible main properties of this weakly coupled DCQD established by several level anti-crossings that occurred for the states with different radial quantum number n (n =1,2) and equal orbital quantum number l. One may conclude that the fate of the single electron in DCQRs is governed by the structure of the energy levels with their crossing and anti-crossing and is changing with magnetic field. The above described behavior is the result of the nontrivial excitation characteristic of the DCQRs. Effect of the trapping of electron in inner QR (or QD) of DCQR may be interesting from the point of view of quantum computing.

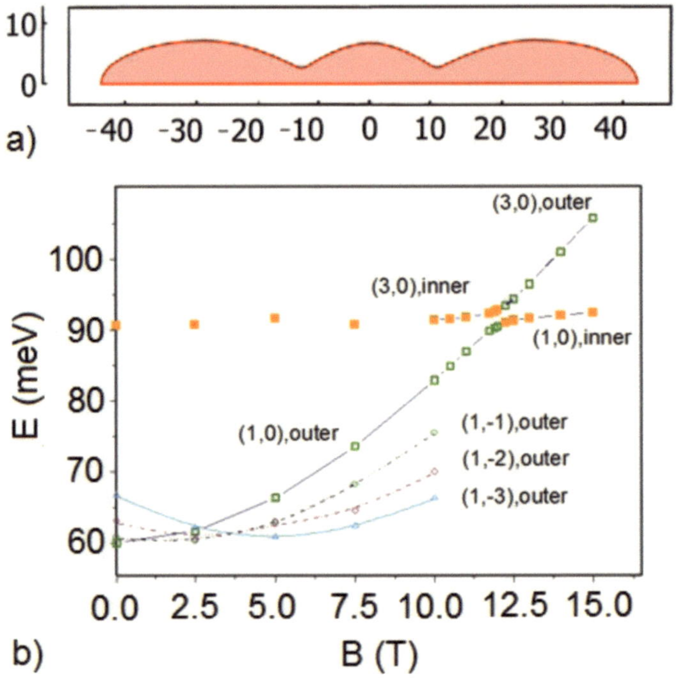

Fig. 29. a) Cross section of the QR with QD system. Sizes are given in nm. b) Energies of the (1,0) and (3,0) states in the magnetic field B for the QR with QD system. The open symbols show that the electron is localized in the ring. The solid squares show that the electron localized in QD.

5. Acknowledgment

This work is supported by NSF CREST award; HRD-0833184 and NASA award NNX09AV07A

6 References

Aharonov Y. & Bohm, D. (1959). Significance of Electromagnetic Potentials in the Quantum Theory, *Physical Review*, Vol. 115, Issue 3, (1959), pp. 485–491

Alferov, Zh.I. (1998).The history and future of semiconductor heterostructures, *Semiconductors*, Vol. 32, No. 1 (January 1998), pp. 1-14, ISSN 1063-7826 (Print); 1090-6479 (Online)

Anderson P.W. ; Thouless D.J.; Abrahams, E.; Fisher, D. S. (1980). New method for a scaling theory of localization. *Physical Review B*, Vol. 22, Issue 8, (1980), pp. 3519–3526, ISSN 1550-235x (online), 1098-0121 (print)

Aronov A.G. & Sharvin, Yu.V. (1987). Magnetic flux effects in disordered conductors, *Review Modern Physics*. Vol. 59, (1987), pp. 755–779, ISSN 1539-0756 (online), 0034-6861 (print)

Baranger, H.U. & Stone, A.D. (1989). Quenching of the Hall resistance in ballistic microstructures: A collimation effect, *Physical Review Letters*, Vol. 63, (1989), pp. 414–417, ISSN 1079-7114 (online) 0031-9007 (print)

Baranger, H.U.; DiVincenzo, D.P.; Jalabert R.A. & Stone, A.D. (1991). Classical and quantum ballistic-transport anomalies in micro-junctions, *Physical Review B*, Vol. 44, Issue 19, (1991), pp. 10637–10675, ISSN 1550-235x (online), 1098-0121 (print)

Bastard, G. (1990).*Wave Mechanics applied to Semiconductor Heterostructures*, John Wiley and Sons Inc. (Jan 01, 1990), p. 357, ISBN 0-471-21708-1, New York, NY (USA)

Beenakker, C.W.J. & van Houten, H. (1991). *Solid State Physics: Advances in Research and Applications*, Ed. H. Ehrenreich and D. Turnbull, Academic Press (April 2, 1991), Vol. 44, (1991), p.1-454, ISBN-10: 0126060444, N-Y, USA

BenDaniel D. J. & Duke C. B. (1966). Space-Charge Effects on Electron Tunneling, *Physical Review*, Vol. 152, Issue 2, (1966), pp. 683–692,

Betcke, M.M. & Voss, H. (2011). Analysis and efficient solution of stationary Schrödinger equation governing electronic states of quantum dots and rings in magnetic field, https://www.mat.tu-harburg.de/ins/forschung/rep/rep143.pdf; to appear in CiCT (2011)

Bimberg, D.; Grundmann, M.; Ledentsov, N.N. (1999). *Quantum Dot Heterostructure*, John Wiley and Sons Inc. (1999), ISBN 0-471-97388-2, Chichester, England

Bloch, F. (1928). Über die Quantenmechanik der Elektronen in Kristallgittern, *Zeitschrift fur Physik*, Vol. 52, Issue 7-8, (1928), pp. 555–600, ISSN 0939-7922 (Print), 1431-5831 (Online)

Brody, T. A.; Flores, J.; French, J. B.; Mello, P. A.; Pandey, A.; Wong, S. S. M. (1981). Random-matrix physics: spectrum and strength fluctuations, *Reviews of Modern Physics*, Vol. 53, Issue 3, (1981) pp. 385–479, ISSN 1539-0756 (online), 0034-6861 (print).

Bryant, G.W (1993). Electrons in coupled vertical quantum dots: Interdot tunneling and Coulomb correlation, *Physical Review B*, Vol. 48, Issue 11, (1993), pp. 8024–8034, ISSN 1550-235x (online), 1098-0121 (print)

Bryant, G.W. (1992). Indirect-to-direct crossover of laterally confined excitons in coupled quantum wells. *Physical Review B*, Vol. 46, Issue 3, (1992), pp. 1893–1896, ISSN 1550-235x (online), 1098-0121 (print)

Buttiker, M. , Imry Y. & Landauer, R. (1983). Josephson behavior in small normal one-dimensional rings, *Physics Letters A*, Vol. 96, Issue 7, 1(8 July 1983), pp. 365-367, ISSN 0375-9601

Califano, M. & Harrison, P. (2000). Presentation and experimental validation of a single-band, constant-potential model for self-assembled InAs/GaAs quantum dots, *Physical Review B*, Vol. 61, (2000), pp. 10959–10965, ISSN 1550-235x (online), 1098-0121 (print)

Chakraborty, T. & Pietiläinen, P. (1994). Electron-electron interaction and the persistent current in a quantum ring. *Physical Review B*, Vol. 50, pp. 8460–8468, ISSN 1550-235x (online), 1098-0121 (print)

Duque, C.A.; Porras-Montenegro, N.; Barticevic, Z.; Pachecoc, M.; Oliveira, L.E. (2005). Electron-hole transitions in self-assembled InAs/GaAs quantum dots: Effects of applied magnetic fields and hydrostatic pressure, *Microelectronics Journal* 36, (2005), pp. 231-233, ISSN 0026-2692

Emperador, A.; Pi, M.; Barranco, M.; Lorke, A. (2000). Far-infrared spectroscopy of nanoscopic InAs rings, *Physical Review B*, Vol. 62, (2000), pp. 4573. 4573–4577, ISSN 1550-235x (online), 1098-0121 (print)

Filikhin I, E. Deyneka & Vlahovic, B (2007a) Numerical modeling of experimentally fabricated InAs/GaAs quantum rings, *Molecular Simulation*, Vol. 33, Issue 7, (2007), pp. 589-592, ISSN 0892-7022 (Print), 1029-0435 (Online)

Filikhin, I.; Deyneka, E. & Vlahovic, B. (2006a). Non parabolic model for InAs/GaAs quantum dot capacitance spectroscopy, *Solid State Communications* 140, (2006), pp. 483–486, ISSN 0038-1098

Filikhin, I.; Deyneka, E.; Melikyan, H. & Vlahovic, B. (2005). Electron States of Semiconductor Quantum Ring Electron States of Semiconductor Quantum Ring with Geometry and Size Variations, *Journal Molecular Simulation*, Vol. 31, (2005), pp. 779-782, Print ISSN 0892-7022, Online ISSN 1029-0435

Filikhin, I.; Matinyan, S.; Nimmo, J. & Vlahovic, B. (2011). Electron Transfer between Weakly Coupled Concentric Quantum Rings, *Physica E: Low-dimensional Systems and Nanostructures*, Vol. 43, No. 9, (July 2011), pp. 1669-1676, ISSN: 1386-9477

Filikhin, I.; S.G. Matinyan, S.G.; Schmid, B.K. & Vlahovic, B. (2010). Electronic and level statistics properties of Si/SiO2 quantum dots, *Physica E: Low-dimensional Systems and Nanostructures*, Vol. 42, Issue 7, (May 2010), pp. 1979-1983, ISSN: 1386-9477.

Filikhin, I.; Suslov, V. M.; Wu, M. & Vlahovic, B. (2008). Effective approach for strained InAs/GaAs quantum structures. *Physica E: Low-dimensional Systems and Nanostructures*, Vol. 40, (2008), pp. 715–723, ISSN: 1386-9477

Filikhin, I.; Suslov, V.M. & Vlahovic, B. (2006). Modeling of InAs/GaAs quantum ring capacitance spectroscopy in the nonparabolic approximation, *Physical Review B*, Vol. 73, (2006), pp. 205332-205336, ISSN 1550-235x (online), 1098-0121 (print)

Filikhin, I.; Suslov, V.M. & Vlahovic, B. (2009) InGaAs/GaAs quantum dots within an effective approach, *Physica E: Low-dimensional Systems and Nanostructures*, Vol. 41, (2009), pp. 1358–1363, ISSN: 1386-9477

Filikhin, I.N., Deyneka, E. & Vlahovic, B. (2004). Energy dependent effective mass model of InAs/GaAs quantum ring, *Modelling and Simulation in Materials Science and Engineering*, Vol. 12, (2004), pp. 1121-1126, ISSN 0965-0393 (Print); ISSN 1361-651X (Online)

Filikhin, I; Suslov, V.M. & Vlahovic, B. (2011a). C-V Data and Geometry Parameters of Self-Assembled InAs/GaAs Quantum Rings, accepted for publication in *Journal of Computational and Theoretical Nanoscience* (2011), ISSN:1546-1955

Fu, H.; Wang, L.-W.; Zunger, A. (1998). Applicability of the k.p method to the electronic structure of quantum dots, *Physical Review B*, Vol. 57, (1998), pp. 9971, ISSN 1550-235x (online), 1098-0121 (print)

Gefen Y.; Imry, Y.; Azbel, M. Ya. (1984). Quantum Oscillations and the Aharonov-Bohm Effect for Parallel Resistors. *Physical Review Letters*, Vol. 52, Issue 2, (1984), pp. 129–132, ISSN 1079-7114 (online); 0031-9007 (print)

Harrison, P. (2005). *Quantum Wells, Wires And Dots: Theoretical And Computational Physics of Semiconductor Nanostructures*, Wiley-Interscience, (2005), pp. 1-497, ISBN:9780470010808

Kane, E.O. (1957), Band Structure of Indium Antimonide, *Journal of Physics and Chemistry of Solids*, Pergamon Press, Vol. 1, (1957), pp. 249-261, ISSN 0022-3697

Keldysh, L. V. (1962). *Soviet Physics Solid State,* Vol. 4, (1962), pp. 2265, ISSN: 0038-5654

Landau, L.D. & Lifschitz, E. M. *Quantum Mechanics (Non-Relativistic Theory),* 3rd ed. Pergamon Press, (1977), ISBN 0080209408, Oxford, England

Landauer, R. (1970). Electrical resistance of disordered one-dimensional lattices, *Philosophical Magazine,* Vol. 21, Issue 172, (1970) pp. 863-867, ISSN 1478-6443

Lei, W.; Notthoff, C.; Lorke, A.; Reuter, D. & Wieck, A. D. (2010). Electronic structure of self-assembled InGaAs/GaAs quantum rings studied by capacitance-voltage spectroscopy, Applied Physics Letters, Vol. 96, (2010), pp. 033111-033114, Print: ISSN 0003-6951, Online: ISSN 1077-3118

Li, Bin & Peeters F. M. (2011). Tunable optical Aharonov-Bohm effect in a semiconductor quantum ring. *Physical Review B,* Vol. 83, (28 March 2011) pp. 115448-13, ISSN 1550-235x (online), 1098-0121 (print)

Li, S.S. & Xia, J.B. (2001). Electronic states of InAs/GaAs quantum ring, *Journal Applied Physics,* Vol. 89, (2001), pp. 3434-3438, Print: ISSN 0021-8979, Online: ISSN 1089-7550

Li, Y.; Voskoboynikov, O.; Lee, C.P. (2002). Computer simulation of electron energy states for three-dimensional InAs/GaAs semiconductor quantum ring, *Proceedings of the International Conference on Modeling and Simulating of Microsystems* (San Juan, Puerto Rico, USA), p. 543, Computational Publication, (2002), Cambridge

Lorke, A.; Luyken, R.J.; Govorov, A.O. & Kotthaus, J.P.; Garcia, J.M. & Petroff, P. M. (2000). Spectroscopy of Nanoscopic Semiconductor Rings, *Physical Review Letters,* Vol. 84, Issue 10, (March, 2000), pp. 2223–2226, ISSN 1079-7114 (online); 0031-9007 (print)

Manasreh, O. (2005). *Semiconductor Heterojunctions and Nanostructures (Nanoscience & Technology),* McGraw-Hill, (2005), p. 554, ISBN 0-07-145228-1, New York

Miller, B.T.; Hansen, W.;Manus, S.; Luyken, R.J.; Lorke, A; Kotthaus, J.P.; Huant, S.; Medeiros-Ribeiro, G; Petroff, P.M. (1997). Few-electron ground states of charge-tunable self-assembled quantum dots, *Physical Review B,* Vol. 56, (1997), pp. 6764–6769, ISSN 1550-235x (online), 1098-0121 (print)

Ponomarenko, L.A.; Schedin, F.; Katsnelson, M.I.; Yang, R.; Hill, E. W.; Novoselov K. S.; Geim A. K. (2008). Chaotic Dirac Billiard in Graphene Quantum Dots, *Science,* Vol. 320, No. 5874, (18 April 2008), pp. 356-358, ISSN 0036-8075 (print), 1095-9203 (online).

Rodt, S.; Schliwa, A.; Pötschke, K.; Guffarth, F.& Bimberg, D. (2005). Correlation of structural and few-particle properties of self-organized InAs⁄GaAs quantum dots, *Physical Review B,* Vol. 71, Issue 15, (2005), pp. 155325-155232, ISSN 1550-235x (online), 1098-0121 (print)

Schliwa, A.; Winkelnkemper, M. & Bimberg, D. Impact of size, shape, and composition on piezoelectric effects and electronic properties of In(Ga)As⁄GaAs quantum dots, *Physical Review B,* Vol. 76, (2007) pp. 205324-205341, ISSN 1550-235x (online), 1098-0121 (print)

Szafran, B. & Peeters, F.M. (2005). Few-electron eigenstates of concentric double quantum rings, *Physical Review B,* Vol. 72, (2005), pp. 155316-1555325, ISSN 1550-235x (online), 1098-0121 (print)

Szafran, B.; Peeters, F.M. & Bednarek, S. (2004). Exchange energy tuned by asymmetry in artificial molecules, *Physical Review B,* Vol. 70, Issue 20, (2004), pp. 205318-205323, ISSN 1550-235x (online), 1098-0121 (print).

T. Kuroda, T. Mano, T. Ochiai, S. Sanguinetti, K. Sakoda, G. Kido and N. Koguchi, (2005). Optical transitions in quantum ring complexes, *Physical Review B*, Vol. 72, Issue 20, (2005), pp. 205301–205309, ISSN 1550-235x (online), 1098-0121 (print)

T. Mano, T. Kuroda, S. Sanguinetti, T. Ochiai, T. Tateno, J. Kim, T. Noda, M. Kawabe, K. Sakoda, G. Kido, and N. Koguchi, (2005). Self-Assembly of Concentric Quantum Double Rings, *Nano Letters* , Vol. 5, No. 3, (2005), pp. 425–428, Print Edition ISSN 1530-6984, Web Edition ISSN 1530-6992

V. Neumann, J. & Wigner, E. (1929). Über das Verhalten von Eigenwerten bei adiabatischen Prozessen, *Physikalische Zeitschrift*, Vol. 30, (1929), pp. 467-468, OCLC Number 1762351

Voskoboynikov, O.; Li, Yiming; Lu, Hsiao-Mei; Shih, Cheng-Feng & Lee C.P. (2002). Energy states and magnetization in nanoscale quantum rings, *Physical Review B*, Vol. 66, (2002), pp. 155306-155312, ISSN 1550-235x (online), 1098-0121 (print).

Voss, H. (2005), Electron energy level calculation for a three dimensional quantum dot, Advances in Computational Methods in Sciences and Engineering 2005, selected papers from the *International Conference of Computational Methods in Sciences and Engineering 2005* (ICCMSE 2005) pp. 586 - 589, Leiden, The Netherlands; Editors: Theodore Simos and George Maroulis. ISBN 10 9067644412, 9067644439, 9067644447

Warburton R. J.; Miller, B. T.; Durr, C. S.; Bodefeld, C.; Karrai, K.; Kotthaus, J.P.; Medeiros-Ribeiro, G.; Petroff, P. M.; Huant, S. (1998). Coulomb interactions in small charge-tunable quantum dots: A simple model, *Physical Review B*, Vol. 58, Issue 24, (1998), pp. 16221–16231, ISSN 1550-235x (online), 1098-0121 (print)

Wetzel, C.; Winkler, R.; Drechsler M. & Meyer, B. K.;Rössler, U.; Scriba J. & Kotthaus, J. P.; Härle, V. & Scholz, F. (1996). Electron effective mass and nonparabolicity in Ga0.47In0.53As/InP quantum wells, *Physical Review B*, Vol. 53, (1996), pp. 1038–1041, ISSN 1550-235x (online), 1098-0121 (print)

Whitney, R.S.; Marconcini P. & Macucci, M. (2009a). Huge Conductance Peak Caused by Symmetry in Double Quantum Dots, *Physical Review Letters*, Vol. 102, (2009), pp. 186802-186806, ISSN 1079-7114 (online) 0031-9007 (print)

Whitney, R.S.; Schomerus, H. & Kopp, M. (2009). Semiclassical transport in nearly symmetric quantum dots. I. Symmetry breaking in the dot, *Physical Review E*, Vol. 80, Issue 5, (2009), 056209-056225, ISSN 1550-2376 (online), 1539-3755 (print)

Yu, P. & Cardona, M. (2005). *Fundamentals of Semiconductors: Physics and Materials Properties* (3rd ed.). Springer. Section 2.6, (2005), pp. 68, ISBN 3-540-25470-6

Zhao Q. & Mei, T. (2011). Analysis of electronic structures of quantum dots using meshless Fourier transform k·p method, *Journal of Applied Physics*, Vol. 109, Issue 6, (2011), pp. 063101-13, Print: ISSN 0021-8979, Online: ISSN 1089-7550

15

Non-Equilibrium Green Functions of Electrons in Single-Level Quantum Dots at Finite Temperature

Nguyen Bich Ha
Institute of Materials Science, Vietnam Academy of Science and Technology, Cau Giay, Hanoi
Vietnam

1. Introduction

In the quantum field theory with the vacuum being the ground state the Green functions are the vacuum expectation values of the chronological, retarded or advanced products of the field operators (Bjoken & Drell, 1964; Itzykson & Zuber, 1985; Peskin & Schroeder, 1995). They are the generalized functions of the real time variables t_i (and also other spatial coordinates). For the application of the Green function technique to the study of the time-independent phenomena in equilibrium many-body systems at a finite temperature, the Matsubara imaginary time Green functions were introduced and widely used (Abrikosov et al., 1975; Bruuns & Flensberg, 2004; Haken, 1976). They are the mean values over a statistical ensemble at a finite temperature of the chronological products of the imaginary time-dependent operators. Both these types of Green functions are inadequate for the application to the study of the time-dependent phenomena in the many-body systems with a finite density and at a finite temperature, in particular the non-equilibrium systems. For the application to the study of the time-dependent dynamical processes in non-equilibrium many-body systems Keldysh (Keldysh, 1965) has introduced a more general class of time-dependent Green functions at finite temperature and density. They are the mean values of the time-ordered products of quantum operators in the Heisenberg picture over statistical ensembles of many-body systems with finite densities and at finite temperatures (which may be non-vanishing). The simplest example is the two-point Green function

$$G_{ab}(t) = -i\langle T[a(t)b(0)]\rangle = -i\frac{Tr\{e^{-\beta H}T[a(t)b(0)]\}}{Tr\{e^{-\beta H}\}} , \tag{1.1}$$

where $a(t)$ and $b(t)$ are two quantum operators in the Heisenberg picture, H is the total Hamiltonian of the system, β and T are the Boltzmann constant and the temperature.

Having shown that the Green functions at finite density and temperature of the form (1.1) can be analytically continued with respect to the time variable t to become the functions of a complex variable z analytical in the stripe $-\beta < \text{Im } z < 0$ parallel to the real axis, Keldysh (Keldysh, 1965) has proposed to consider these functions as the quantum statistical averages of the linear combinations of the products of ordered operators depending on complex variables as complex times. For the definition of the ordering of the complex variables z, z' it was proposed to use some contour C in above-mentioned stripe with some initial point t_0 on the real axis and the final point $t_0 - i\beta$ such that all the complex numbers z, z'... belong to this contour. Then the "chronological" ordering T_C of the complex times z, z' ... is defined as the ordering along the contour C. The complex time-dependent operators $a(z)$, $b(z)$ and $\overline{a}(z), \overline{b}(z)$, for example, are defined in the analogy with the operators in the Heisenberg picture

$$\begin{aligned} a(z) &= e^{iHz} a(0) e^{-iHz}, \\ b(z) &= e^{iHz} b(0) e^{-iHz}, \\ \overline{a}(z) &= e^{iHz} a^{+}(0) e^{-iHz}, \\ \overline{b}(z) &= e^{iHz} b^{+}(0) e^{-iHz}. \end{aligned} \tag{1.2}$$

As the generalization of formula (1.1) one defines the two-point Green function of two operators $a(z)$ and $b(z')$, for example, depending on two complex times $z, z' \in C$, as follows:

$$G_{ab}(z - z')_C = -i\langle T_C[a(z)b(z')]\rangle = -i\frac{Tr\{e^{-\beta H}T_C[a(z)b(z')]\}}{Tr\{e^{-\beta H}\}}, \tag{1.3}$$

T_C denoting the "chronological" ordering along the contour C. The Green functions of the form (1.3), usually called the Keldysh complex time-dependent Green functions at finite density and temperature, some time also simply called non-equilibrium Green functions, are widely used in quantum statistical physics and many-body theories (Chou et al., 1985; Kapusta, 1989; Le Bellac, 1996).

In practice we need to know the Green functions at the real values of the time variables. For the convenience we chose the contour C to consists of four parts $C = C_1 \cup C_2 \cup C_3 \cup C_\infty$, C_1 being the part of the straight line over and infinitely close to the real axis from some point $t_0 + io$ to infinity $+\infty + io$, C_2 being the part of the straight line under and infinitely close to the real axis from infinity $+\infty - io$ to the point $t_0 - io$, C_3 and C_∞ being the segments $[t_0, t_0 - i\beta]$ and $[+\infty + io, +\infty - io]$ parallel to the axis Oy (figure 1).

The contributions of the segment $[+\infty + io, +\infty - io]$ to all physical observables are negligibly small, because of its vanishing length. Therefore this segment plays no role, and the contour C can be considered to consist of only three parts C_1, C_2 and C_3. Then the function $G_{ab}(z - z')_C$ with the complex time variables z and z' on the contour C effectively becomes a set of nine functions of two variables, each of which has the values on one among three segments C_1, C_2 and C_3. When both variables z and z' belong to the line C_1, the function (1.3) is the quantum statistical average of the usual chronological product of two quantum operators $a(t)$ and $b(t')$ in the Heisenberg picture over a statistical ensemble of a many-body system at finite density and temperature, and can be denoted by $G_{ab}(t - t')_{11}$. When both variables z and z' belong to the line C_3, the function (1.3) is reduced to the Matsubara imaginary time Green function and can be denoted $G_{ab}(-i\tau + i\tau')_{33}$.

In the study of stationary physical processes one often uses the complex time Green functions of the form (1.3) in the limit $t_0 \to -\infty$. Because the interaction must satisfy the "adiabatic hypothesis" and therefore vanishes at this limit, the segment C_3 also gives no contribution to the stationary physical processes. In this case the contour C can be considered to consist of only two segments C_1 and C_2, and the complex time Green function (1.3) effectively becomes a set of four functions of real variables $G_{ab}(t-t')_{11}$, $G_{ab}(t-t')_{12}$, $G_{ab}(t-t')_{21}$, $G_{ab}(t-t')_{22}$.

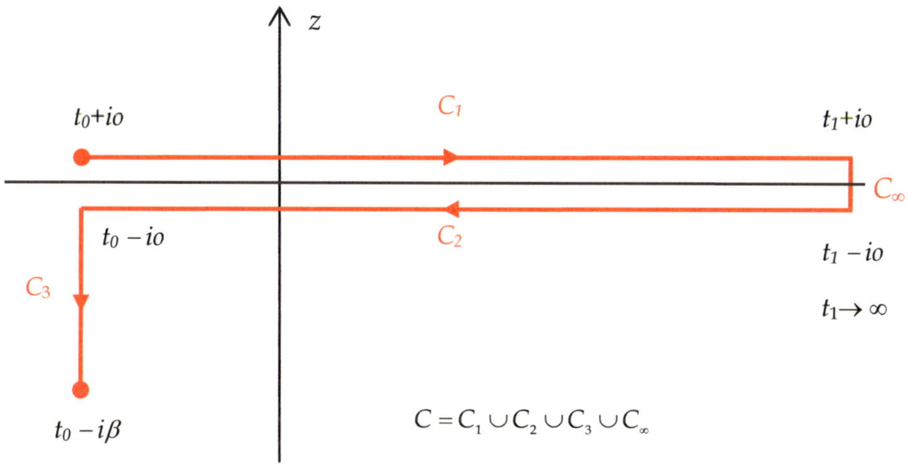

Fig. 1. Contour C consists of four parts $C=C_1\cup C_2\cup C_3\cup C_\infty$.

The electrons transport through a single-level quantum dot (QD) connected with two conducting leads has been the subject for theoretical and experimental studies in many works since the early days of nanophysics (Choi et al., 2004; Costi et al., 1994; Craco & Kang, 1999; Fujii & Ueda, 2003; Hershfield et al., 1991; Inoshita et al., 1993; Izumida et al., 1997, 1998, 2001; Konig & Gefen, 2005; Meir et al., 1991, 1993; Ng, 1993; Nguyen Van Hieu & Nguyen Bich Ha, 2005, 2006; Nguyen Van Hieu et al., 2006a, 2006b; Pustilnik & Glasman, 2004; Sakai et al., 1999; Swirkowicz et al., 2003, 2006; Takagi & Saso, 1999a, 1999b; Torio et al., 2002; Wingreen & Meir, 1994; Yeyati et al., 1993). Two observable physical quantities, which can be measured in experiments on electron transport, are the electron current through the QD and the time-averaged value of the electron number in the QD. Both can be expressed in terms of the single-electron Green functions. In the pioneering theoretical works (Meir et al., 1991, 1993) on the electron transport through a single-level QD, the differential equations for the non-equilibrium Green functions were derived with the use of the Heisenberg equations of motion for the electron destruction and creation operators. Due to the presence of the strong Coulomb interaction between electrons in the QD, the differential equations for the single-electron Green functions contain multi-electron Green functions, and all the coupled equations for these Green functions form an infinite system of differential equations. In order to have a finite closed system of equations, one can assume some approximation to decouple the infinite system of equations. Moreover, since the electron transport is a non-equilibrium process, one should work with the Keldysh formalism of non-equilibrium complex time Green functions.

As the simplest explanation of the calculation methods for establishing the differential equations of non-equilibrium Green functions and deriving their exact solutions, in Section 2 we present the theory of non-equilibrium Green functions of free electron in a single-level quantum system. In Section 3 we study non-equilibrium Green functions of interacting electron in an isolated single-level QD. The elaborated calculation methods are then applied in Section 4 to the study of non-equilibrium Green functions of electrons in a single-level QD connected with two conducting leads. Due to the electron tunneling between QD and conducting leads there does not exist a closed finite system of differential equations for some finite number of Green functions. In order to truncate the infinite system of differential equations for the infinite number of Green functions we can apply some suitable approximation. In Section 5 the mean-field approximation was used to truncate the infinite system of differential equations for the Green functions. As the result we establish a closed system of Dyson equations for a finite number of Green functions. This system of differential equations can be exactly solved. The asymptotic analytical expressions of these Green functions at the resonances, Kondo and Fano resonances, are derived in Section 6. Section 7 is the Conclusion.

2. Non-equilibrium Green function of free electrons in a single-level quantum system

For the demonstration of the calculation methods to derive the differential equations and the expressions of the non-equilibrium Green functions let us consider a simplest quantum system - that of free electrons at a single energy level E. Denote by c_σ and c_σ^+ the destruction and creation operators of the electron with the spin projection $\sigma=\uparrow,\downarrow$ in the Schrödinger picture and by H_0 the Hamiltonian of this system. We have

$$H_0 = E\sum_\sigma c_\sigma^+ c_\sigma \ . \tag{2.1}$$

The non-equilibrium Green function of electron system with Hamiltonian (2.1) is defined as follows:

$$S_{\sigma\sigma'}^E(z-z')_C = -i\langle T_C[c_\sigma(z)\overline{c}_{\sigma'}(z')]\rangle = -i\frac{Tr\{e^{-\beta H_0}T_C[c_\sigma(z)\overline{c}_{\sigma'}(z')]\}}{Tr\{e^{-\beta H_0}\}}, \tag{2.2}$$

where

$$\begin{aligned} c_\sigma(z) &= e^{iH_0 z}c_\sigma e^{-iH_0 z}, \\ \overline{c}_\sigma(z') &= e^{iH_0 z'}c_\sigma^+ e^{-iH_0 z'}. \end{aligned} \tag{2.3}$$

Note that at the real values t of the time variable we have $\overline{c}_\sigma(t) = c_\sigma^+(t) = c_\sigma(t)^+$. Complex time-dependent operators $c_\sigma(z)$ and $\overline{c}_\sigma(z)$ satisfy Heisenberg quantum equation of motion

$$\begin{aligned} i\frac{dc_\sigma(z)}{dz} &= -[H_0, c_\sigma(z)], \\ i\frac{d\overline{c}_\sigma(z)}{dz} &= -[H_0, \overline{c}_\sigma(z)]. \end{aligned} \tag{2.4}$$

From the canonical anti-commutation relations

$$\begin{aligned} \{c_\sigma, c_{\sigma'}^+\} &= \delta_{\sigma\sigma'}, \\ \{c_\sigma, c_{\sigma'}\} &= \{c_\sigma^+, c_{\sigma'}^+\} = 0 \end{aligned} \tag{2.5}$$

it follows that

$$i\frac{dc_\sigma(z)}{dz} = Ec_\sigma(z)\,,$$
$$i\frac{d\overline{c}_\sigma(z)}{dz} = -E\overline{c}_\sigma(z)\,. \qquad (2.6)$$

Green function $S^E_{\sigma\sigma'}(z-z')_C$ with both variables z and z' ranging over contour C is a set of nine functions $S^E_{\sigma\sigma'}(z-z')_{ij}$ with the variable $z(z')$ ranging over the segment $C_i(C_j)$. All they have the form

$$S^E_{\sigma\sigma'}(z-z')_C = \delta_{\sigma\sigma'}S^E(z-z')_C\,,$$
$$S^E_{\sigma\sigma'}(z-z')_{ij} = \delta_{\sigma\sigma'}S^E(z-z')_{ij}\,. \qquad (2.7)$$

First consider three cases when both variables z and z' belong to one and the same segment C_i, i = 1, 2, 3. For $z,z' \in C_{1,}\,, z = t + io,\,, z' = t' + io,$ T_C is the usual chronological ordering T of the real times t and t':

$$T[c_\sigma(t)c^+_{\sigma'}(t')] = \theta(t-t')c_\sigma(t)c^+_{\sigma'}(t') - \theta(t'-t)c^+_{\sigma'}(t')c_\sigma(t).$$

Using one of equations (2.6) and one of anti-commutation relations (2.5), we derive the differential equation for $S^E_{\sigma\sigma'}(z-z')_{11} = S^E_{\sigma\sigma'}(t-t')_{11}$ and obtain

$$\left[i\frac{d}{dt} - E\right]S^E(t-t')_{11} = \delta(t-t'). \qquad (2.8.1)$$

For $z,z' \in C_{2,}\,, z = t - io,\,, z' = t' - io,$ T_C is the anti-chronological ordering T^{-1} reverse to the usual chronological ordering T of the real times t and t':

$$T^{-1}[c_\sigma(t)c^+_{\sigma'}(t')] = \theta(t'-t)c_\sigma(t)c^+_{\sigma'}(t') - \theta(t-t')c^+_{\sigma'}(t')c_\sigma(t).$$

In this case we have the differential equation

$$\left[i\frac{d}{dt} - E\right]S^E(t-t')_{22} = -\delta(t-t'). \qquad (2.8.2)$$

For $z,z' \in C_{3,}\,, z = t_0 - i\tau,\,, z' = t_0 - i\tau',$ T_C becomes the usual chronological ordering T_τ of the real values τ and τ' in the imaginary times $i\tau$ and $i\tau'$, $0 \le \tau, \tau' \le \beta$, and we have

$$S^E_{\sigma\sigma'}(z-z')_{33} = S^E_{\sigma\sigma'}(-i\tau + i\tau')_{33} = -i\mathbb{S}^E_{\sigma\sigma'}(\tau-\tau'), \qquad (2.9)$$

where $\mathbb{S}^E_{\sigma\sigma'}(\tau-\tau')$ is the Matsubara imaginary time-dependent two-point Green function in statistical physics

$$\mathbb{S}^E_{\sigma\sigma'}(\tau-\tau') = \delta_{\sigma\sigma'}\mathbb{S}^E(\tau-\tau') = \left\langle T_\tau\left[\gamma_\sigma(\tau)\overline{\gamma}_{\sigma'}(\tau')\right]\right\rangle, \qquad (2.10)$$

where

$$\gamma_\sigma(\tau) = e^{H_0\tau}c_\sigma e^{-H_0\tau},$$
$$\overline{\gamma}_\sigma(\tau) = e^{H_0\tau}c^+_\sigma e^{-H_0\tau} \qquad (2.11)$$

and

$$T_\tau\left[\gamma_\sigma(\tau)\bar{\gamma}_{\sigma'}(\tau')\right] = \theta(\tau-\tau')\gamma_\sigma(\tau)\bar{\gamma}_{\sigma'}(\tau') - \theta(\tau'-\tau)\bar{\gamma}_{\sigma'}(\tau')\gamma_\sigma(\tau). \tag{2.12}$$

The Heisenberg quantum equation of motion for imaginary time-dependent operators (2.11) has the form

$$\begin{aligned} \frac{d\gamma_\sigma(\tau)}{d\tau} &= [H_0, \gamma_\sigma(\tau)], \\ \frac{d\bar{\gamma}_\sigma(\tau)}{d\tau} &= [H_0, \bar{\gamma}_\sigma(\tau)]. \end{aligned} \tag{2.13}$$

From the anti-commutation relations (2.5) it follows that

$$\begin{aligned} \frac{d\gamma_\sigma(\tau)}{d\tau} &= -E\gamma_\sigma(\tau), \\ \frac{d\bar{\gamma}_\sigma(\tau)}{d\tau} &= E\bar{\gamma}_\sigma(\tau), \end{aligned} \tag{2.14}$$

and therefore

$$\left[\frac{d}{d\tau} + E\right]\mathbb{S}^E(\tau-\tau') = \delta(\tau-\tau'). \tag{2.15}$$

In the analogy with relations (2.9) we set

$$S^E(-i\tau+i\tau')_{33} = -i\mathbb{S}^E(\tau-\tau') \tag{2.16}$$

and rewrite equation (2.15) in the form similar to equations (2.8.1) and (2.8.2):

$$\left[i\frac{d}{d(-i\tau)} - E\right]S^E(-i\tau+i\tau')_{33} = i\delta(\tau-\tau'). \tag{2.17}$$

Now consider six other cases when two variables z and z' belong to different segments C_i and C_j with $i \neq j$. For $z = t + io \in C_1$ and $z' = t' - io \in C_2$ the values of z always precede those of z' with respect to the ordering along the contour C and therefore

$$T_C[c_\sigma(t+io)\bar{c}_{\sigma'}(t'-io)] = -\bar{c}_{\sigma'}(t'-io)c_\sigma(t+io).$$

Similarly, for $z = t + io \in C_1$ or $z = t - io \in C_2$ and $z' = t_0 - i\tau \in C_3$ we have

$$T_C[c_\sigma(t \pm io)\bar{c}_{\sigma'}(t_0 - i\tau)] = -\bar{c}_{\sigma'}(t_0 - i\tau)c_\sigma(t \pm io).$$

On the contrary, for $z = t - io \in C_2$ and $z' = t' + io \in C_1$ the values of z' always precede those of z with respect to the ordering along the contour C and therefore

$$T_C[c_\sigma(t-io)\bar{c}_{\sigma'}(t'+io)] = c_\sigma(t-io)\bar{c}_{\sigma'}(t'+io).$$

Similarly, for $z = t_0 - i\tau \in C_3$ and $z' = t' + io \in C_1$ or $z' = t' - io \in C_2$ we have

$$T_C[c_\sigma(t_0 - i\tau)\overline{c}_{\sigma'}(t' \pm io)] = c_\sigma(t_0 - i\tau)\overline{c}_{\sigma'}(t' \pm io).$$

In all six later cases the differential equations for corresponding functions $S^E(z-z')_{ij}$, $i \neq j$, are six homogeneous ones:

$$\begin{aligned}
&\left[i\frac{d}{dt} - E\right]S^E(t-t')_{12} = \left[i\frac{d}{dt} - E\right]S^E(t-t')_{21} \\
&= \left[i\frac{d}{dt} - E\right]S^E(t-t_0+i\tau)_{13} = \left[i\frac{d}{dt} - E\right]S^E(t-t_0+i\tau)_{23} \\
&= \left[i\frac{d}{d(-i\tau)} - E\right]S^E(t_0-i\tau-t)_{31} = \left[i\frac{d}{d(-i\tau)} - E\right]S^E(t_0-i\tau-t)_{32} = 0.
\end{aligned} \tag{2.18}$$

By introducing a new notation

$$\delta(z-z')_C = \begin{cases} \delta(t-t') & \text{for } z,z' \in C_1, \\ -\delta(t-t') & \text{for } z,z' \in C_2, \\ i\delta(\tau-\tau') & \text{for } z,z' \in C_3, \\ 0 & \text{for } z \in C_i, z' \in C_j, i \neq j \end{cases} \tag{2.19}$$

we rewrite equations (2.8.1), (2.8.2), (2.17) and (2.18) in the unified form

$$\left[i\frac{d}{dz} - E\right]S^E(z-z')_C = \delta(z-z')_C\,. \tag{2.20}$$

From above presented reasonnings and relations determining nine functions $S^E(z-z')_{ij}$, and formula (2.1) for total Hamiltonian, it is straightforward to derive explicit expressions of these functions. They depend on the average electron number with a definite spin projection

$$n = \langle n_\sigma \rangle = \langle c_\sigma^+ c_\sigma \rangle = \frac{e^{-\beta H_0}}{1+e^{-\beta H_0}}\,. \tag{2.21}$$

We obtain following results:

$$S^E(z-z')_{11} = -i[\theta(t-t') - n]e^{-iE(t-t')}, \tag{2.22.1}$$

$$S^E(z-z')_{22} = -i[\theta(t'-t) - n]e^{-iE(t-t')}, \tag{2.22.2}$$

$$S^E(z-z')_{33} = -i[\theta(\tau-\tau') - n]e^{-E(\tau-\tau')}, \tag{2.22.3}$$

$$S^E(z-z')_{12} = ine^{-iE(t-t')}, \tag{2.22.4}$$

$$S^E(z-z')_{21} = -i(1-n)e^{-iE(t-t')}, \tag{2.22.5}$$

$$S^E(z-z')_{13} = S^E(z-z')_{23} = ine^{E\tau}e^{-iE(t-t_0)}, \tag{2.22.6}$$

$$S^E(z-z')_{31} = S^E(z-z')_{32} = -i(1-n)e^{-E\tau}e^{-iE(t_0-t)}. \tag{2.22.7}$$

They satisfy above presented differential equations (2.8.1), (2.8.2), (2.17) and (2.18), respectively.

For concluding this Section we consider the Fourier transformation of the functions $S^E(z-z')_{ij}$:

$$S^E(z-z')_{ij} = \frac{1}{2\pi}\int d\omega\, e^{-i\omega(t-t')}\tilde{S}^E(\omega)_{ij}, \qquad i,j=1,2, \tag{2.23.1}$$

$$S^E(z-z')_{i3} = \frac{1}{2\pi}\int d\omega\, e^{-i\omega(t-t_0+i\tau)}\tilde{S}^E(\omega)_{i3}, \qquad i=1,2, \tag{2.23.2}$$

$$S^E(z-z')_{3i} = \frac{1}{2\pi}\int d\omega\, e^{-i\omega(t_0-t-i\tau)}\tilde{S}^E(\omega)_{3i}, \qquad i=1,2, \tag{2.23.3}$$

$$S^E(z-z')_{33} = -i\frac{1}{\beta}\sum_{\nu} e^{i\varepsilon_\nu(\tau-\tau')}\tilde{S}^E_\nu, \tag{2.23.4}$$

$$\varepsilon_\nu = (2\nu+1)\frac{\pi}{\beta}, \qquad \nu = 0, \pm 1, \pm 2, \ldots$$

From the expressions (2.22.1)-(2.22.7) of the functions $S^E(z-z')_{ij}$ it follows that

$$\begin{aligned}\tilde{S}^E(\omega)_{11} &= \frac{1}{\omega - E + io} + i2\pi n\delta(\omega - E)\\ &= P\frac{1}{\omega - E} - i\pi\frac{1-e^{-\beta E}}{1+e^{-\beta E}}\delta(\omega - E),\end{aligned} \tag{2.24.1}$$

$$\begin{aligned}\tilde{S}^E(\omega)_{22} &= -\frac{1}{\omega - E - io} + i2\pi n\delta(\omega - E)\\ &= -P\frac{1}{\omega - E} - i\pi\frac{1-e^{-\beta E}}{1+e^{-\beta E}}\delta(\omega - E),\end{aligned} \tag{2.24.2}$$

where P means the principal value,

$$\tilde{S}^E(\omega)_{12} = i2\pi n\delta(\omega - E), \tag{2.24.3}$$

$$\tilde{S}^E(\omega)_{21} = -i2\pi(1-n)\delta(\omega - E), \tag{2.24.4}$$

$$\tilde{S}^E(\omega)_{i3} = i2\pi n\delta(\omega - E), \qquad i=1,2, \tag{2.24.5}$$

$$\tilde{S}^E(\omega)_{3i} = -i2\pi(1-n)\delta(\omega - E), \qquad i=1,2, \tag{2.24.6}$$

$$\tilde{S}^E_\nu = \frac{1}{i\varepsilon_\nu + E}. \tag{2.24.7}$$

The explicit expressions of Green functions of free electrons presented in this Section are often used in the theoretical studies of non-equilibrium processes by means of the perturbation theory.

3. Non-equilibrium Green functions of electrons in isolated single-level quantum dot

The calculation methods and reasonnings presented in the preceding Section are now applied to the study of the Keldysh non-equilibrium Green functions of interacting electrons in the simplest nanosystem - the isolated single-level quantum dot (QD) with total Hamiltonian

$$H = E\sum_{\sigma} c_{\sigma}^{+}c_{\sigma} + UN_{\uparrow}N_{\downarrow}\,, \tag{3.1}$$

where U is the value of a potential energy, $\sigma = \uparrow,\downarrow$ denotes the spin projection (if $\sigma = \uparrow$ then $-\sigma = \downarrow$ and vice versa) and

$$N_{\sigma} = c_{\sigma}^{+}c_{\sigma} \tag{3.2}$$

is the number of electrons with the spin projection σ. The second term in Hamiltonian (3.1) is the potential energy of the Coulomb electron-electron interaction (two electrons with different spin projections in one and the same energy level). The interacting nanosystem with total Hamiltonian (3.1) is an exactly solvable model. There are four exactly determined eigenstates and eigenvalues of H: the vacuum with vanishing energy, two degenerate single-electron states with two different spin projections and the same energy E, and a two-electron state with total energy $2E+U$. The Keldysh complex time-dependent two-point Green function of two operators $c(z)$ and $\overline{c}(z')$ is defined as follows

$$G_{\sigma\sigma'}(z-z')_C = -i\langle T_C[c_{\sigma}(z)\overline{c}_{\sigma'}(z')]\rangle = -i\frac{Tr\{e^{-\beta H}T_C[c_{\sigma}(z)\overline{c}_{\sigma'}(z')]\}}{Tr\{e^{-\beta H}\}} \tag{3.3}$$

with total Hamiltonian (3.1). They have the form

$$G_{\sigma\sigma'}(z-z')_C = \delta_{\sigma\sigma'}G(z-z')_C\,. \tag{3.4}$$

As in the preceding Section, we choose the contour C to consist of three segments C_1, C_2 and C_3. Then $G(z-z')_C$ becomes the set of nine functions $G(z-z')_{ij}$, $i,j = 1,2,3$. The calculations of these functions are straightforward, as they have been done in the preceding Section for free electrons at a single energy level. We obtain following results:

$$\begin{aligned} G(z-z')_{11} = \frac{i}{Z}\{&-\theta(t-t')[e^{-iE(t-t')} + e^{-\beta E}e^{-i(E+U)(t-t')}] \\ &+\theta(t'-t)[e^{-\beta E}e^{-iE(t-t')} + e^{-\beta(2E+U)}e^{-i(E+U)(t-t')}]\}\,, \end{aligned} \tag{3.5.1}$$

$$\begin{aligned} G(z-z')_{22} = \frac{i}{Z}\{&-\theta(t'-t)[e^{-iE(t-t')} + e^{-\beta E}e^{-i(E+U)(t-t')}] \\ &+\theta(t-t')[e^{-\beta E}e^{-iE(t-t')} + e^{-\beta(2E+U)}e^{-i(E+U)(t-t')}]\}\,, \end{aligned} \tag{3.5.2}$$

$$G(z-z')_{33} = i\mathcal{G}(\tau-\tau'),$$
$$\mathcal{G}(\tau) = \frac{1}{Z}\{ [\theta(\tau)e^{-\tau E} - \theta(-\tau)e^{-(\tau+\beta)E}] \tag{3.5.3}$$
$$+e^{-\beta E}[\theta(\tau)e^{-\tau(E+U)} - \theta(-\tau)e^{-(\tau+\beta)(E+U)}]\},$$

$$G(z-z')_{12} = \frac{i}{Z}\left\{e^{-\beta E}e^{-iE(t-t')} + e^{-\beta(2E+U)}e^{-i(E+U)(t-t')}\right\}, \tag{3.5.4}$$

$$G(z-z')_{21} = \frac{i}{Z}\left\{-e^{-iE(t-t')} - e^{-\beta E}e^{-i(E+U)(t-t')}\right\}, \tag{3.5.5}$$

$$G(z-z')_{13} = G(z-z')_{23}$$
$$= \frac{i}{Z}\left\{e^{-\beta E}e^{\tau E}e^{-iE(t-t_0)} + e^{-\beta(2E+U)}e^{\tau(E+U)}e^{-i(E+U)(t-t_0)}\right\}, \tag{3.5.6}$$

and

$$G(z-z')_{31} = G(z-z')_{32}$$
$$= \frac{i}{Z}\left\{-e^{-\tau E}e^{-iE(t_0-t')} + e^{-\beta E}e^{-\tau(E+U)}e^{-i(E+U)(t_0-t')}\right\}. \tag{3.5.7}$$

In the study of non-equlibrium dynamical processes by means of the perturbation theory one often needs to use the Fourier transformation of four functions $G(z-z')_{ij}$ with i,j =1,2:

$$G(z-z')_{ij} = G(t-t')_{ij} = \frac{1}{2\pi}\int d\omega\, e^{-i\omega(t-t')}\tilde{G}(\omega)_{ij}. \tag{3.6}$$

We have following exact expressions of their Fourier transforms:

$$\tilde{G}(\omega)_{11} = \frac{1}{Z}\left\{\frac{1}{\omega-E+io} + \frac{e^{-\beta E}}{\omega-E-io} + \frac{e^{-\beta E}}{\omega-E-U+io} + \frac{e^{-\beta(2E+U)}}{\omega-E-U-io}\right\}$$
$$= \frac{1}{Z}\left\{\left[1+e^{-\beta E}\right]P\frac{1}{\omega-E} - i\pi\left[1-e^{-\beta E}\right]\delta(\omega-E)\right. \tag{3.7.1}$$
$$\left.+e^{-\beta E}\left[1+e^{-\beta(E+U)}\right]P\frac{1}{\omega-E-U} - i\pi e^{-\beta E}\left[1-e^{-\beta(E+U)}\right]\delta(\omega-E-U)\right\},$$

$$\tilde{G}(\omega)_{22} = \frac{1}{Z}\left\{-\frac{1}{\omega-E+io} - \frac{e^{-\beta E}}{\omega-E-io} - \frac{e^{-\beta E}}{\omega-E-U-io} - \frac{e^{-\beta(2E+U)}}{\omega-E-U+io}\right\}$$
$$= -\frac{1}{Z}\left\{\left[1+e^{-\beta E}\right]P\frac{1}{\omega-E} + i\pi\left[1-e^{-\beta E}\right]\delta(\omega-E)\right. \tag{3.7.2}$$
$$\left.+e^{-\beta E}\left[1+e^{-\beta(E+U)}\right]P\frac{1}{\omega-E-U} + i\pi e^{-\beta E}\left[1-e^{-\beta(E+U)}\right]\delta(\omega-E-U)\right\},$$

$$\tilde{G}(\omega)_{12} = \frac{2\pi i}{Z}\{\delta(\omega-E) + e^{-\beta E}\delta(\omega-E-U)\}, \tag{3.7.3}$$

$$\tilde{G}(\omega)_{21} = -\frac{2\pi i}{Z}\{ e^{-\beta E}\delta(\omega - E) + e^{-\beta(2E+U)}\delta(\omega - E - U)\} \tag{3.7.4}$$

with

$$Z = 1 + 2e^{-\beta E} + e^{-\beta(2E+U)}.$$

Now we derive the system of differential equations for two-point Green functions $G_{\sigma\sigma'}(t-t')_{ij}$. Consider first the function with $i = j = 1$:

$$G_{\sigma\sigma'}(t-t')_{11} = -i\langle T[c_\sigma(t)\overline{c}_{\sigma'}(t')]\rangle . \tag{3.8}$$

We have

$$i\frac{dG_{\sigma\sigma'}(t-t')_{11}}{dt} = \delta_{\sigma\sigma'}\delta(t-t') - i\left\langle T\left[i\frac{dc_\sigma(t)}{dt}\overline{c}_{\sigma'}(t')\right]\right\rangle . \tag{3.9}$$

From the Heisenberg quantum equation of motion

$$i\frac{dc_\sigma(t)}{dt} = -[H, c_\sigma(t)] \tag{3.10}$$

with total Hamiltonian (3.1) it follows that

$$i\frac{dc_\sigma(t)}{dt} = Ec_\sigma(t) + UN_{-\sigma}c_\sigma(t). \tag{3.11}$$

Substituting this expression of $i\frac{dc_\sigma(t)}{dt}$ into the r.h.s. of equation (3.9), we obtain

$$\left[i\frac{d}{dt} - E\right]G_{\sigma\sigma'}(t-t') = \delta_{\sigma\sigma'}\delta(t-t') + UH_{\sigma\sigma'}(t-t')_{11}, \tag{3.12}$$

where

$$\begin{gathered} H_{\sigma\sigma'}(t-t')_{11} = -i\langle T[N_{-\sigma}c_\sigma(t)\overline{c}_{\sigma'}(t')]\rangle = \\ -i\theta(t-t')\langle [N_{-\sigma}c_\sigma(t)\overline{c}_{\sigma'}(t')]\rangle + i\theta(t'-t)\langle [\overline{c}_{\sigma'}(t')N_{-\sigma}c_\sigma(t)]\rangle . \end{gathered} \tag{3.13}$$

Thus the differential equation for $G_{\sigma\sigma'}(t-t')_{11}$ contains a new Green function $H_{\sigma\sigma'}(t-t')_{11}$. In order to derive the differential equation for this new Green function it is necessary to calculate the time derivatives of both sides of equation (3.13). Note that $N_{-\sigma}$ commutes with H and therefore does not depend on t. Moreover, it has following property

$$N_{-\sigma}^2 = N_{-\sigma} .$$

Multiplying both sides of relation (3.11) with $N_{-\sigma}$ and using these two above-mentioned properties of $N_{-\sigma}$, we obtain

$$i\frac{d}{dt}[N_{-\sigma}c_\sigma(t)] = (E+U)[N_{-\sigma}c_\sigma(t)] . \tag{3.14}$$

Differentiating both sides of equation (3.13) and using relation (3.14), we derive following differential equation for the new Green function $H_{\sigma\sigma'}(t-t')_{11}$:

$$\left[i\frac{d}{dt}-(E+U)\right]H_{\sigma\sigma'}(t-t')=n\delta_{\sigma\sigma'}\delta(t-t'). \tag{3.15}$$

Thus both $G_{\sigma\sigma'}(t-t')_{11}$ and $H_{\sigma\sigma'}(t-t')_{11}$ have the common form

$$\begin{aligned} G_{\sigma\sigma'}(t-t')_{11} &= \delta_{\sigma\sigma'}G(t-t')_{11}, \\ H_{\sigma\sigma'}(t-t')_{11} &= \delta_{\sigma\sigma'}H(t-t')_{11}, \end{aligned} \tag{3.16}$$

where $G(t-t')_{11}$ and $H(t-t')_{11}$ must satisfy differential equations

$$\left[i\frac{d}{dt}-E\right]G(t-t')_{11}=\delta(t-t')+UH(t-t')_{11}, \tag{3.17}$$

$$\left[i\frac{d}{dt}-(E+U)\right]H(t-t')_{11}=n\delta(t-t'). \tag{3.18}$$

In preceding Section we have shown that

$$\left[i\frac{d}{dt}-E\right]S^{E}(t-t')_{11}=\delta(t-t')$$

(equation (2.8.1)). Therefore

$$\left[i\frac{d}{dt}-(E+U)\right]S^{E+U}(t-t')_{11}=\delta(t-t'). \tag{3.19}$$

Equations (3.18) and (3.19) show that $\frac{1}{n}H(t-t')_{11}$ satisfies the same inhomogeneous differential equation as $S^{E+U}(t-t')$ does. It follows that

$$H(t-t')_{11}=nS^{E+U}(t-t')_{11}, \tag{3.20}$$

and the differential equation for $G(t-t')_{11}$ becomes

$$\left[i\frac{d}{dt}-E\right]G(t-t')_{11}=\delta(t-t')+nS^{E+U}(t-t')_{11}. \tag{3.21}$$

Similarly, it can be shown that the Green function

$$G_{\sigma\sigma'}(t-t')_{22}=-i\left\{\theta(t'-t)\langle c_{\sigma}(t)\overline{c}_{\sigma'}(t')\rangle-\theta(t-t')\langle\overline{c}_{\sigma'}(t')c_{\sigma}(t)\rangle\right\} \tag{3.22}$$

has the form

$$G_{\sigma\sigma'}(t-t')_{22}=\delta_{\sigma\sigma'}G(t-t')_{22}, \tag{3.23}$$

and $G(t-t')_{22}$ satisfies differential equation

$$\left[i\frac{d}{dt}-E\right]G(t-t')_{22}=-\delta(t-t')+nUS^{E+U}(t-t')_{22} \tag{3.24}$$

etc. In general, Keldysh complex time Green function

$$G_{\sigma\sigma'}(z-z')_C=-i\langle T_C[c_\sigma(z)\overline{c}_\sigma(z')\rangle \tag{3.25}$$

has the form

$$G_{\sigma\sigma'}(z-z')_C=\delta_{\sigma\sigma'}G(z-z')_C, \tag{3.26}$$

and $G(z-z')_C$ satisfies differential equation

$$\left[i\frac{d}{dt}-E\right]G(z-z')_C=\delta(z-z')_C+nUS^{E+U}(z-z')_C. \tag{3.27}$$

4. Non-equilibrium Green functions of electrons in single-level quantum dot connected with two conducting leads

Consider the single-electron transistor (SET) consisting of a single-level quantum dot (QD) connected with two conducting leads through two potential barriers. The electron transport through this SET was investigated experimentally and studied theoretically in many works (Choi et al., 2004; Costi et al., 1994; Craco & Kang, 1999; Fujii & Ueda, 2003; Hershfield et al., 1991; Inoshita et al., 1993; Izumida et al., 1997, 1998, 2001; Meir et al., 1991, 1993; Ng, 1993; Pustilnik & Glasman, 2004; Sakai et al., 1999; Swirkowicz et al., 2003, 2006; Takagi & Saso, 1999a, 1999b; Torio et al., 2002; Wingreen & Meir, 1994; Yeyati et al., 1993). It was assumed that the electron system in this SET has following total Hamiltonian

$$\begin{aligned}H=&E\sum_\sigma c_\sigma^+c_\sigma+UN_\uparrow N_\downarrow+\sum_k\sum_\sigma\{E_a(k)a_\sigma^+(k)a_\sigma(k)+E_b(k)b_\sigma^+(k)b_\sigma(k)\}\\&+\sum_k\sum_\sigma\{V_a(k)a_\sigma^+(k)c_\sigma+V_a(k)^*c_\sigma^+a_\sigma(k)+V_b(k)b_\sigma^+(k)c_\sigma+V_b(k)^*c_\sigma^+b_\sigma(k)\}.\end{aligned} \tag{4.1}$$

In order to define the complex time-dependent Green functions we introduce the complex time-dependent quantum operators

$$\begin{aligned}&c_\sigma(z)=e^{iHz}c_\sigma e^{-iHz}, \quad \overline{c}_\sigma(z)=e^{iHz}c_\sigma^+e^{-iHz},\\&a_\sigma(k,z)=e^{iHz}a_\sigma(k)e^{-iHz}, \quad \overline{a}_\sigma(k,z)=e^{iHz}a_\sigma^+(k)e^{-iHz},\\&b_\sigma(k,z)=e^{iHz}b_\sigma(k)e^{-iHz}, \quad \overline{b}_\sigma(k,z)=e^{iHz}b_\sigma^+(k)e^{-iHz}.\end{aligned} \tag{4.2}$$

The Keldysh non-equilibrium Green functions of electrons are defined as follows:

$$G_{\sigma\sigma'}^{c\overline{c}}(z-z')_C=\delta_{\sigma\sigma'}G^{c\overline{c}}(z-z')_C=-i\langle T_C[c_\sigma(z)\overline{c}_{\sigma'}(z')\rangle, \tag{4.3}$$

$$H_{\sigma\sigma'}^{c\overline{c}}(z-z')_C=\delta_{\sigma\sigma'}H^{c\overline{c}}(z-z')_C=-i\langle T_C[N_{-\sigma}(z)c_\sigma(z)\overline{c}_{\sigma'}(z')\rangle, \tag{4.4}$$

$$G^{a\bar{c}}_{\sigma\sigma'}(\mathbf{k};z-z')_C = \delta_{\sigma\sigma'}G^{a\bar{c}}(\mathbf{k};z-z')_C = -i\langle T_C[a_\sigma(\mathbf{k};z)\bar{c}_{\sigma'}(z')]\rangle, \tag{4.5}$$

$$H^{a\bar{c}}_{\sigma\sigma'}(\mathbf{k};z-z')_C = \delta_{\sigma\sigma'}H^{a\bar{c}}(\mathbf{k};z-z')_C = -i\langle T_C[N_{-\sigma}(z)a_\sigma(\mathbf{k};z)\bar{c}_{\sigma'}(z')]\rangle, \tag{4.6}$$

$$\begin{aligned} G^{ac\bar{c}\bar{c}}_{\sigma\sigma'}(\mathbf{k};z-z')_C &= \delta_{\sigma\sigma'}G^{ac\bar{c}\bar{c}}(\mathbf{k};z-z')_C \\ &= -i\langle T_C[a_{-\sigma}(\mathbf{k};z)c_\sigma(z)\bar{c}_{-\sigma}(z)\bar{c}_{\sigma'}(z')]\rangle, \end{aligned} \tag{4.7}$$

$$\begin{aligned} G^{cc\bar{a}\bar{c}}_{\sigma\sigma'}(\mathbf{k};z-z')_C &= \delta_{\sigma\sigma'}G^{cc\bar{a}\bar{c}}(\mathbf{k};z-z')_C \\ &= -i\langle T_C[c_{-\sigma}(z)c_\sigma(z)\bar{a}_{-\sigma}(\mathbf{k};z)\bar{c}_{\sigma'}(z')]\rangle, \end{aligned} \tag{4.8}$$

$$\begin{aligned} G^{aa\bar{c}\bar{c}}_{\sigma\sigma'}(\mathbf{k},\mathbf{l};z-z')_C &= \delta_{\sigma\sigma'}G^{aa\bar{c}\bar{c}}(\mathbf{k},\mathbf{l};z-z')_C \\ &= -i\langle T_C[a_{-\sigma}(\mathbf{k};z)a_\sigma(\mathbf{l};z)\bar{c}_{-\sigma}(z)\bar{c}_{\sigma'}(z')]\rangle, \end{aligned} \tag{4.9}$$

$$\begin{aligned} G^{ac\bar{a}\bar{c}}_{\sigma\sigma'}(\mathbf{k},\mathbf{l};z-z')_C &= \delta_{\sigma\sigma'}G^{ac\bar{a}\bar{c}}(\mathbf{k},\mathbf{l};z-z')_C \\ &= -i\langle T_C[a_{-\sigma}(\mathbf{k};z)c_\sigma(z)\bar{a}_{-\sigma}(\mathbf{l};z)\bar{c}_{\sigma'}(z')]\rangle, \end{aligned} \tag{4.10}$$

and similarly for the others $G^{b\bar{c}}_{\sigma\sigma'}(\mathbf{k};z-z')_C$, $H^{b\bar{c}}_{\sigma\sigma'}(\mathbf{k};z-z')_C$, $G^{bc\bar{c}\bar{c}}_{\sigma\sigma'}(\mathbf{k};z-z')_C$, $G^{cc\bar{b}\bar{c}}_{\sigma\sigma'}(\mathbf{k};z-z')_C$, $G^{ab\bar{c}\bar{c}}_{\sigma\sigma'}(\mathbf{k},\mathbf{l};z-z')_C$, $G^{ac\bar{b}\bar{c}}_{\sigma\sigma'}(\mathbf{k},\mathbf{l};z-z')_C$ etc.

Because there is no magnetic interaction, all Green functions (4.3)-(4.10) and other ones are proportional to $\delta_{\sigma\sigma'}$. From Heisenberg quantum equations of motion and equal-time canonical anti-commutation relations for the electron destruction and creation operators it follows the differential equations for these operators:

$$i\frac{dc_\sigma(z)}{dz} = Ec_\sigma(z) + UN_{-\sigma}(z)c_\sigma(z) + \sum_{\mathbf{k}}\left[V_a^*(\mathbf{k})a_\sigma(\mathbf{k};z) + V_b^*(\mathbf{k})b_\sigma(\mathbf{k};z)\right], \tag{4.11}$$

$$i\frac{d\bar{c}_\sigma(z)}{dz} = -Ec_\sigma(z) - UN_{-\sigma}(z)\bar{c}_\sigma(z) - \sum_{\mathbf{k}}\left[V_a(\mathbf{k})\bar{a}_\sigma(\mathbf{k};z) + V_b(\mathbf{k})\bar{b}_\sigma(\mathbf{k};z)\right], \tag{4.12}$$

$$i\frac{da_\sigma(\mathbf{k};z)}{dz} = E_a(\mathbf{k})a_\sigma(\mathbf{k};z) + V_a(\mathbf{k})c_\sigma(z), \tag{4.13}$$

$$i\frac{d\bar{a}_\sigma(\mathbf{k};z)}{dz} = -E_a(\mathbf{k})\bar{a}_\sigma(\mathbf{k};z) - V_a^*(\mathbf{k})\bar{c}_\sigma(z) \tag{4.14}$$

and similarly for $b_\sigma(\mathbf{k};z)$ and $\bar{b}_\sigma(\mathbf{k};z)$.

By using differential equation (4.11) and the equal-time canonical anti-commutation relation between $c_\sigma(z)$ and $\bar{c}_{\sigma'}(z)$, it is easy to derive the differential equation for the Green function $G^{c\bar{c}}_{\sigma\sigma'}(z-z')_C$

$$\begin{aligned} \left[i\frac{d}{dz} - E\right]G^{c\bar{c}}_{\sigma\sigma'}(z-z')_C &= \delta_{\sigma\sigma'}\delta(z-z')_C + UH^{c\bar{c}}_{\sigma\sigma'}(z-z') \\ &+ \sum_{\mathbf{k}}\left[V_a^*(\mathbf{k})G^{a\bar{c}}_{\sigma\sigma'}(\mathbf{k};z-z')_C + V_b^*(\mathbf{k})G^{b\bar{c}}_{\sigma\sigma'}(\mathbf{k};z-z')_C\right], \end{aligned} \tag{4.15}$$

which contains Green functions $H^{c\bar{c}}_{\sigma\sigma'}(z-z')_C$, $G^{a\bar{c}}_{\sigma\sigma'}(\mathbf{k};z-z')_C$ and $G^{b\bar{c}}_{\sigma\sigma'}(\mathbf{k};z-z')_C$. These new functions must satisfy following differential equations which can be also derived by using differential equations (4.11)-(4.14):

$$\left[i\frac{d}{dz}-(E+U)\right]H^{c\bar{c}}_{\sigma\sigma'}(z-z')_C = n\delta_{\sigma\sigma'}\delta(z-z')_C$$
$$+\sum_{\mathbf{k}}\left[V^*_a(\mathbf{k})H^{a\bar{c}}_{\sigma\sigma'}(\mathbf{k};z-z')_C + V^*_b(\mathbf{k})H^{b\bar{c}}_{\sigma\sigma'}(\mathbf{k};z-z')_C \right.$$
$$\left. -V_a(\mathbf{k})G^{cca\bar{c}}_{\sigma\sigma'}(\mathbf{k};z-z')_C - V_b(\mathbf{k})G^{ccb\bar{c}}_{\sigma\sigma'}(\mathbf{k};z-z')_C\right], \tag{4.16}$$

where

$$n = \langle c^+_\uparrow c_\uparrow\rangle = \langle c^+_\downarrow c_\downarrow\rangle, \tag{4.17}$$

$$\left[i\frac{d}{dz}-E_a(\mathbf{k})\right]G^{a\bar{c}}_{\sigma\sigma'}(\mathbf{k};z-z')_C = V_a(\mathbf{k})G^{c\bar{c}}_{\sigma\sigma'}(z-z')_C \tag{4.18}$$

and similarly for $G^{b\bar{c}}_{\sigma\sigma'}(\mathbf{k};z-z')_C$.

In Section 2 we have established the differential equation (2.20) for the Keldysh non-equilibrium Green function of a free electron. If the free electron has energy $E_a(\mathbf{k})$, then it is denoted by $S^{Ea(\mathbf{k})}(z-z')_C$ and must satisfy differential equation

$$\left[i\frac{d}{dz}-E_a(\mathbf{k})\right]S^{Ea(\mathbf{k})}(z-z')_C = \delta(z-z')_C. \tag{4.19}$$

Using this function, we obtain following expression of the solution of equation (4.18)

$$G^{a\bar{c}}_{\sigma\sigma'}(\mathbf{k};z-z')_C = V_a(\mathbf{k})\int dz'' S^{Ea(\mathbf{k})}(z-z'')_C G^{c\bar{c}}_{\sigma\sigma'}(z''-z')_C \tag{4.20}$$

and similarly for $G^{b\bar{c}}_{\sigma\sigma'}(\mathbf{k};z-z')_C$. Substituting the expression of the form (4.20) for $G^{a\bar{c}}_{\sigma\sigma'}(\mathbf{k};z-z')_C$ and $G^{b\bar{c}}_{\sigma\sigma'}(\mathbf{k};z-z')_C$ into the r.h.s. of differential equation (4.15) for $G^{c\bar{c}}_{\sigma\sigma'}(z-z')_C$, we rewrite this equation in a new form

$$\left[i\frac{d}{dz}-E\right]G^{c\bar{c}}_{\sigma\sigma'}(z-z')_C = \delta_{\sigma\sigma'}\delta(z-z')_C + UH^{c\bar{c}}_{\sigma\sigma'}(z-z')_C$$
$$+\int_C dz''\Sigma^{(1)}(z-z'')_C G^{c\bar{c}}_{\sigma\sigma'}(z''-z')_C, \tag{4.21}$$

where $\Sigma^{(1)}(z-z')_C$ is following self-energy part

$$\Sigma^{(1)}(z-z')_C = \sum_{\mathbf{k}}\left[|V_a(\mathbf{k})|^2 S^{Ea(\mathbf{k})}(z-z')_C + |V_b(\mathbf{k})|^2 S^{Eb(\mathbf{k})}(z-z')_C\right]. \tag{4.22}$$

The differential equation for $H^{c\bar{c}}_{\sigma\sigma'}(z-z')_C$ contains new functions $H^{a\bar{c}}_{\sigma\sigma'}(\mathbf{k};z-z')_C$, $H^{b\bar{c}}_{\sigma\sigma'}(\mathbf{k};z-z')_C$, $G^{acc\bar{c}}_{\sigma\sigma'}(\mathbf{k};z-z')_C$, $G^{bcc\bar{c}}_{\sigma\sigma'}(\mathbf{k};z-z')_C$, $G^{cca\bar{c}}_{\sigma\sigma'}(\mathbf{k};z-z')_C$ and $G^{ccb\bar{c}}_{\sigma\sigma'}(\mathbf{k};z-z')_C$, which must satisfy following differential equations

$$\left[i\frac{d}{dz}-E_a(\mathbf{k})\right]H^{a\bar{c}}_{\sigma\sigma'}(\mathbf{k};z-z')_C=V_a(\mathbf{k})H^{c\bar{c}}_{\sigma\sigma'}(z-z')_C \tag{4.23}$$

and similarly for $H^{b\bar{c}}_{\sigma\sigma'}(\mathbf{k};z-z')_C$,

$$\begin{aligned}\left[i\frac{d}{dt}-E_a(\mathbf{k})\right]G^{ac\bar{c}\bar{c}}_{\sigma\sigma'}(\mathbf{k};z-z')_C&=\left\langle\left\{a_{\sigma}(\mathbf{k})c_{\sigma}c^{+}_{-\sigma},c^{+}_{\sigma'}\right\}\right\rangle\delta(z-z')_C\\&+V_a(\mathbf{k})\left[H^{c\bar{c}}_{\sigma\sigma'}(z-z')_C-G^{c\bar{c}}_{\sigma\sigma'}(z-z')_C\right]\\&+\sum_{\mathbf{l}}\left[V^{*}_a(\mathbf{l})G^{aa\bar{c}\bar{c}}_{\sigma\sigma'}(\mathbf{k},\mathbf{l};z-z')_C+V^{*}_b(\mathbf{l})G^{ab\bar{c}\bar{c}}_{\sigma\sigma'}(\mathbf{k},\mathbf{l};z-z')_C\right]\\&-\sum_{\mathbf{l}}\left[V_a(\mathbf{l})G^{ac\bar{a}\bar{c}}_{\sigma\sigma'}(\mathbf{k},\mathbf{l};z-z')_C+V_b(\mathbf{l})G^{ac\bar{b}\bar{c}}_{\sigma\sigma'}(\mathbf{k},\mathbf{l};z-z')_C\right],\end{aligned} \tag{4.24}$$

and similarly for $G^{bc\bar{c}\bar{c}}_{\sigma\sigma'}(\mathbf{k};z-z')$,

$$\begin{aligned}\left\{i\frac{d}{dt}-\left[2E-E_a(\mathbf{k})+U\right]\right\}G^{cc\bar{a}\bar{c}}_{\sigma\sigma'}(\mathbf{k};z-z')_C&=\left\langle\left\{c_{-\sigma}c_{\sigma}a^{+}_{-\sigma}(\mathbf{k}),c^{+}_{\sigma'}\right\}\right\rangle\delta(z-z')_C\\&-V^{*}_a(\mathbf{k})\left[H^{c\bar{c}}_{\sigma\sigma'}(z-z')_C-G^{c\bar{c}}_{\sigma\sigma'}(z-z')_C\right]+\\&+\sum_{\mathbf{l}}\left\{V^{*}_a(\mathbf{l})\left[G^{ac\bar{a}\bar{c}}_{\sigma\sigma'}(\mathbf{l},\mathbf{k};z-z')_C+G^{ca\bar{a}\bar{c}}_{\sigma\sigma'}(\mathbf{l},\mathbf{k};z-z')_C\right]\right.\\&\left.+V^{*}_b(\mathbf{l})\left[G^{bc\bar{a}\bar{c}}_{\sigma\sigma'}(\mathbf{l},\mathbf{k};z-z')_C+G^{cb\bar{a}\bar{c}}_{\sigma\sigma'}(\mathbf{l},\mathbf{k};z-z')_C\right]\right\}\end{aligned} \tag{4.25}$$

and similarly for $G^{cc\bar{b}\bar{c}}_{\sigma\sigma'}(\mathbf{k};z-z')_C$.

The presented calculations for deriving differential equations of Green function showed that there does not exist a closed system of a finite number of differential equations for a finite number of Green functions. Some approximation should be used for truncating the infinite system of all differential equations at some step. The mean-field approximation is the most appropriate one. In order to apply this approximation we rewrite equations (4.23)-(4.25) in the form of integral equations:

$$H^{a\bar{c}}_{\sigma\sigma'}(\mathbf{k};z-z')_C=V_a(\mathbf{k})\int_C dz''S^{Ea(\mathbf{k})}(z-z'')_C H^{c\bar{c}}_{\sigma\sigma'}(z-z'')_C \tag{4.26}$$

and similarly for $H^{b\bar{c}}_{\sigma\sigma'}(\mathbf{k};z-z')_C$,

$$\begin{aligned}G^{ac\bar{c}\bar{c}}_{\sigma\sigma'}(\mathbf{k};z-z')_C&=\left\langle\left\{a_{-\sigma}(\mathbf{k})c_{\sigma}c^{+}_{-\sigma},c^{+}_{\sigma'}\right\}\right\rangle S^{Ea(\mathbf{k})}(z-z')_C\\&+V_a(\mathbf{k})\int_C dz''S^{Ea(\mathbf{k})}(z-z'')_C\left[H^{c\bar{c}}_{\sigma\sigma'}(z''-z')_C-G^{c\bar{c}}_{\sigma\sigma'}(z''-z')_C\right]\\&+\int_C dz''S^{Ea(\mathbf{k})}(z-z'')_C\sum_{\mathbf{l}}\left[V^{*}_a(\mathbf{l})G^{aa\bar{c}\bar{c}}_{\sigma\sigma'}(\mathbf{k},\mathbf{l};z''-z')_C+V^{*}_b(\mathbf{l})G^{ab\bar{c}\bar{c}}_{\sigma\sigma'}(\mathbf{k},\mathbf{l};z''-z')_C\right]\\&-\int_C dz''S^{Ea(\mathbf{k})}(z-z'')_C\sum_{\mathbf{l}}\left[V_a(\mathbf{l})G^{ac\bar{a}\bar{c}}_{\sigma\sigma'}(\mathbf{k},\mathbf{l};z''-z')_C+V_b(\mathbf{l})G^{ac\bar{b}\bar{c}}_{\sigma\sigma'}(\mathbf{k},\mathbf{l};z''-z')_C\right],\end{aligned} \tag{4.27}$$

and similarly for $G^{bc\bar{c}\bar{c}}_{\sigma\sigma'}(\mathbf{k};z-z')_C$,

$$\begin{aligned}
&G_{\sigma\sigma'}^{cc\bar{a}\bar{c}}(\mathbf{k};z-z')_C = \left\langle \left\{ c_{-\sigma}c_{\sigma}a_{-\sigma}^{+}(\mathbf{k}), c_{\sigma'}^{+} \right\} \right\rangle S^{2E+U-Ea(\mathbf{k})}(z-z')_C \\
&-V_a^{*}(\mathbf{k})\int_C dz'' S^{2E+U-Ea(\mathbf{k})}(z-z')_C \left[H_{\sigma\sigma'}^{c\bar{c}}(z''-z')_C - G_{\sigma\sigma'}^{c\bar{c}}(z''-z')_C \right] \\
&+\int_C dz'' S^{2E+U-Ea(\mathbf{k})}(z''-z')_C \sum_{\mathbf{l}} \left\{ V_a^{*}(\mathbf{l}) \left[G_{\sigma\sigma'}^{ac\bar{a}\bar{c}}(\mathbf{l},\mathbf{k};z''-z')_C + G_{\sigma\sigma'}^{ca\bar{a}\bar{c}}(\mathbf{l},\mathbf{k};z''-z')_C \right] \right. \\
&\qquad\qquad \left. + V_b^{*}(\mathbf{l}) \left[G_{\sigma\sigma'}^{bc\bar{a}\bar{c}}(\mathbf{l},\mathbf{k};z''-z')_C + G_{\sigma\sigma'}^{cb\bar{a}\bar{c}}(\mathbf{l},\mathbf{k};z''-z')_C \right] \right\},
\end{aligned} \tag{4.28}$$

and similarly for $G_{\sigma\sigma'}^{cc\bar{b}\bar{c}}(\mathbf{k};z-z')_C$. Substituting these solutions into the r.h.s. of the differential equation (4.16) for $H_{\sigma\sigma'}^{c\bar{c}}(z-z')_C$, we rewrite this equation in the new form

$$\begin{aligned}
&\left[i\frac{d}{dz} - (E+U) \right] H_{\sigma\sigma'}^{c\bar{c}}(z-z')_C = n\delta_{\sigma\sigma'}\delta(z-z')_C + \int_C dz'' \Sigma^{(1)}(z-z'')_C H_{\sigma\sigma'}^{c\bar{c}}(z''-z')_C \\
&+\sum_{\mathbf{k}} \left\{ V_a^{*}(\mathbf{k}) \left\langle \left\{ a_{-\sigma}(\mathbf{k}) c_{\sigma} c_{-\sigma}^{+}, c_{\sigma'}^{+} \right\} \right\rangle S^{Ea(\mathbf{k})}(z-z')_C \right. \\
&\qquad \left. + \left| V_a(\mathbf{k}) \right|^2 \int_C dz'' S^{Ea(\mathbf{k})}(z-z'')_C \left[H_{\sigma\sigma'}^{c\bar{c}}(z''-z')_C - G_{\sigma\sigma'}^{c\bar{c}}(z''-z')_C \right] \right\} \\
&+\int_C dz'' \sum_{\mathbf{k}} V_a^{*}(\mathbf{k}) S^{Ea(\mathbf{k})}(z-z'')_C \sum_{\mathbf{l}} \left[V_a^{*}(\mathbf{l}) G_{\sigma\sigma'}^{aa\bar{c}\bar{c}}(\mathbf{k},\mathbf{l};z''-z')_C + V_b^{*}(\mathbf{l}) G_{\sigma\sigma'}^{ab\bar{c}\bar{c}}(\mathbf{k},\mathbf{l};z''-z')_C \right] \\
&-\int_C dz'' \sum_{\mathbf{k}} V_a^{*}(\mathbf{k}) S^{Ea(\mathbf{k})}(z-z'')_C \sum_{\mathbf{l}} \left[V_a(\mathbf{l}) G_{\sigma\sigma'}^{ac\bar{a}\bar{c}}(\mathbf{k},\mathbf{l};z''-z')_C + V_b(\mathbf{l}) G_{\sigma\sigma'}^{ac\bar{b}\bar{c}}(\mathbf{k},\mathbf{l};z''-z')_C \right] \\
&-\sum_{\mathbf{k}} \left\{ V_a(\mathbf{k}) \left\langle \left\{ c_{-\sigma}c_{\sigma}a_{-\sigma}^{+}(\mathbf{k}), c_{\sigma'}^{+} \right\} \right\rangle S^{2E+U-Ea(\mathbf{k})}(z-z'')_C \right. \\
&\qquad \left. + \left| V_a(\mathbf{k}) \right|^2 \int_C dz'' S^{2E+U-Ea(\mathbf{k})}(z-z'')_C \left[H_{\sigma\sigma'}^{c\bar{c}}(z''-z')_C - G_{\sigma\sigma'}^{c\bar{c}}(t''-t) \right] \right\} \\
&-\int_C dz'' \sum_{\mathbf{k}} V_a(\mathbf{k}) S^{2E+U-Ea(\mathbf{k})}(z-z'')_C \sum_{\mathbf{l}} \left\{ V_a^{*}(\mathbf{l}) \left[G_{\sigma\sigma'}^{ac\bar{a}\bar{c}}(\mathbf{l},\mathbf{k};z''-z')_C + G_{\sigma\sigma'}^{ca\bar{a}\bar{c}}(\mathbf{l},\mathbf{k};z''-z')_C \right] \right\} \\
&-\int_C dz'' \sum_{\mathbf{k}} V_a(\mathbf{k}) S^{2E+U-Ea(\mathbf{k})}(z-z'')_C \sum_{\mathbf{l}} V_b^{*}(\mathbf{l}) \left[G_{\sigma\sigma'}^{bc\bar{a}\bar{c}}(\mathbf{l},\mathbf{k};z''-z')_C + G_{\sigma\sigma'}^{cb\bar{a}\bar{c}}(\mathbf{l},\mathbf{k};z''-z')_C \right]
\end{aligned} \tag{4.29}$$

+ similar terms with suitable interchange $(a \leftrightarrow b)$.

5. Dyson equations for non-equilibrium Green functions of electrons in single-level quantum dot connected with two conducting leads and their solutions

The r.h.s. of equation (4.29) for Green function $H_{\sigma\sigma'}^{c\bar{c}}(z-z')_C$ contains multi-electron Green functions $G_{\sigma\sigma'}^{aa\bar{c}\bar{c}}(\mathbf{k},\mathbf{l};z-z')_C$, $G_{\sigma\sigma'}^{ab\bar{c}\bar{c}}(\mathbf{k},\mathbf{l};z-z')_C$, $G_{\sigma\sigma'}^{ac\bar{a}\bar{c}}(\mathbf{k},\mathbf{l};z-z')_C$, $G_{\sigma\sigma'}^{ac\bar{b}\bar{c}}(\mathbf{k},\mathbf{l};z-z')_C$, $G_{\sigma\sigma'}^{ca\bar{a}\bar{c}}(\mathbf{k},\mathbf{l};z-z')_C$, $G_{\sigma\sigma'}^{bc\bar{a}\bar{c}}(\mathbf{k},\mathbf{l};z-z')_C$, $G_{\sigma\sigma'}^{bc\bar{a}\bar{c}}(\mathbf{k},\mathbf{l};z-z')_C$ and similar ones with suitable interchange (a ↔ b). In order to decouple this equation from those for other multi-electron Green functions we apply the mean-field approximation to the products of four operators. For example

$$\left\langle T_C\left[\bar{a}_{-\sigma}(\mathbf{l};z) a_{-\sigma}(\mathbf{k};z) c_{\sigma}(z) \bar{c}_{\sigma'}(z') \right] \right\rangle \approx \left\langle \bar{a}_{-\sigma}(\mathbf{l};z) a_{-\sigma}(\mathbf{k};z) \right\rangle \left\langle T_C\left[c_{\sigma}(z) \bar{c}_{\sigma'}(z') \right] \right\rangle \tag{5.1}$$

with

$$\left\langle \bar{a}_{-\sigma}(\mathbf{l};z)a_{-\sigma}(\mathbf{k};z)\right\rangle = \delta_{\mathbf{kl}}\left\langle \bar{a}_{-\sigma}(\mathbf{k})a_{-\sigma}(\mathbf{k})\right\rangle = \delta_{\mathbf{kl}}n_a(\mathbf{k}), \tag{5.2}$$

where $n_a(\mathbf{k})$ is the density of electrons with momentum **k** and spin projection σ or $-\sigma$ in the lead "a" at the given temperature

$$n_a(\mathbf{k}) = \frac{e^{-\beta Ea(\mathbf{k})}}{1+e^{-\beta Ea(\mathbf{k})}}. \tag{5.3}$$

Note that

$$\left\langle T_C\left[c_\sigma(z)\bar{c}_{\sigma'}(z')\right]\right\rangle = iG^{c\bar{c}}_{\sigma\sigma'}(z-z')_C.$$

As the result we have

$$G^{ac\bar{a}\bar{c}}_{\sigma\sigma'}(\mathbf{k,l};z-z')_C \approx -\delta_{\mathbf{kl}}\left[1-n_a(\mathbf{k})\right]G^{c\bar{c}}_{\sigma\sigma'}(z-z')_C \tag{5.4}$$

and similarly for $G^{bc\bar{b}\bar{c}}_{\sigma\sigma'}(\mathbf{k,l};z-z')_C$. Applying the mean-field approximation to each of others above-mentioned multi-electron Green functions in any manner, we always obtain the vanishing mean value in the lowest order the perturbation theory with respect to the effective tunnelling coupling constants $V_{a,b}(\mathbf{k})$. Note that these functions enter the r. h. s. of the equation (29) with the coefficients of the second order with respect to the effective tunnelling coupling constants. This means that in this second order they do not give contributions. Thus in the second order approximation the equation (4.29) is simplified and becomes

$$\begin{aligned}\left[i\frac{d}{dz}-(E+U)\right]H^{c\bar{c}}_{\sigma\sigma'}(z-z')_C &= n\delta_{\sigma\sigma'}\delta(z-z')_C+\Delta(z-z')_C\\ &+\int_C dz''\Sigma^{(2)}(z-z')_C H^{c\bar{c}}_{\sigma\sigma'}(z''-z')_C-\int_C dz''\Sigma^{(3)}(z-z'')_C G^{c\bar{c}}_{\sigma\sigma'}(z''-z')_C,\end{aligned} \tag{5.5}$$

where

$$\begin{aligned}\Delta(z-z')_C = \sum_{\mathbf{k}}\Big[&V_a^*(\mathbf{k})\left\langle\left\{a_{-\sigma}(\mathbf{k})c_\sigma c^+_{-\sigma},c^+_{\sigma'}\right\}\right\rangle S^{Ea(\mathbf{k})}(z-z')_C\\ &-V_a(\mathbf{k})\left\langle\left\{c_{-\sigma}c_\sigma a^+_{-\sigma}(\mathbf{k}),c^+_{\sigma'}\right\}\right\rangle S^{2E+U-Ea(\mathbf{k})}(z-z')_C+(a\to b)\Big],\end{aligned} \tag{5.6}$$

$$\Sigma^{(2)}(z-z')_C = \sum_{\mathbf{k}}\left\{\left|V_a(\mathbf{k})\right|^2\left[2S^{Ea(\mathbf{k})}(z-z')_C+S^{2E+U-Ea(\mathbf{k})}(z-z')_C\right]+(a\to b)\right\}, \tag{5.7}$$

$$\Sigma^{(3)}(z-z')_C = \sum_{\mathbf{k}}\left\{n_a(\mathbf{k})\left|V_a(\mathbf{k})\right|^2\left[S^{Ea(\mathbf{k})}(z-z')_C+S^{2E+U-Ea(\mathbf{k})}(z-z')_C\right]+(a\to b)\right\}. \tag{5.8}$$

Note that in the r.h.s. of equations (5.6)-(5.8), there appear the crossing terms containing $S^{2E+U-Ea(\mathbf{k})}(z-z')_C$. They must disappear in the non-crossing approximation. Two equations (4.21) and (5.5) form the closed system of Dyson equations for two Green functions $G^{c\bar{c}}_{\sigma\sigma'}(z-z')_C$ and $H^{c\bar{c}}_{\sigma\sigma'}(z-z')_C$.

To proceed further we note that

$$\left\langle \left\{ a_{-\sigma}(\mathbf{k}) c_{\sigma} c_{-\sigma}^{+}, c_{\sigma'}^{+} \right\} \right\rangle = -\delta_{\sigma\sigma'} \left\langle a_{-\sigma}(\mathbf{k}) c_{-\sigma}^{+} \right\rangle,$$
$$\left\langle \left\{ c_{-\sigma} c_{\sigma} a_{-\sigma}^{+}(\mathbf{k}), c_{\sigma'}^{+} \right\} \right\rangle = -\delta_{\sigma\sigma'} \left\langle a_{-\sigma}(\mathbf{k}) c_{-\sigma}^{+} \right\rangle^{*}, \tag{5.9}$$

where $\left\langle a_{-\sigma}(\mathbf{k}) c_{-\sigma}^{+} \right\rangle$ is a limiting value of the Green function $G_{-\sigma-\sigma}^{ac}(\mathbf{k};t)_{11}$:

$$\left\langle a_{-\sigma}(\mathbf{k}) c_{-\sigma}^{+} \right\rangle = iG_{-\sigma-\sigma}^{ac}(\mathbf{k};+0)_{11}. \tag{5.10}$$

For evaluating the vertex (5.6) in the second order with respect to the tunnelling coupling constants $V_{a,b}(\mathbf{k})$ we calculate the limiting value (5.10) in the first order. Introduce the Fourier transformations of the Green functions, for example

$$G^{a\bar{c}}(\mathbf{k};t)_{ij} = \frac{1}{2\pi} \int_{-\infty}^{+\infty} d\omega\, e^{-i\omega t} \tilde{G}^{a\bar{c}}(\mathbf{k};\omega)_{ij},$$

$$S^{Ea(\mathbf{k})}(t)_{ij} = \frac{1}{2\pi} \int_{-\infty}^{+\infty} d\omega\, e^{-i\omega t} \tilde{S}^{Ea(\mathbf{k})}(\omega)_{ij}, \tag{5.11}$$

$$G^{c\bar{c}}(t)_{ij} = \frac{1}{2\pi} \int_{-\infty}^{+\infty} d\omega\, e^{-i\omega t} \tilde{G}^{c\bar{c}}(\omega)_{ij} \quad \text{for } i, j = 1, 2.$$

From the equation (4.20) it follows that

$$\tilde{G}^{a\bar{c}}(\mathbf{k};\omega)_{11} = V_{a}(\mathbf{k}) \tilde{S}^{Ea(\mathbf{k})}(\omega)_{11} \tilde{G}^{c\bar{c}}(\omega)_{11}. \tag{5.12}$$

For deriving $\tilde{G}^{a\bar{c}}(\mathbf{k};\omega)_{11}$ in the first order with respect to the constant $V_{a}(\mathbf{k})$ it is enough to use the expression of $\tilde{G}^{c\bar{c}}(\omega)_{11}$ in the case of the vanishing tunnelling coupling constant and have

$$\begin{aligned} \tilde{G}^{a\bar{c}}(\mathbf{k};\omega)_{11} = \frac{1}{Z} V_{a}(\mathbf{k}) & \left[\frac{n_a(\mathbf{k})}{\omega - io - E_a(\mathbf{k})} + \frac{1 - n_a(\mathbf{k})}{\omega + io - E_a(\mathbf{k})} \right] \\ & \times \left[\frac{e^{-\beta E}}{\omega - io - E} + \frac{e^{-\beta(2E+U)}}{\omega - io - E - U} + \frac{1}{\omega + io - E} + \frac{e^{-\beta E}}{\omega + io - E - U} \right], \end{aligned} \tag{5.13}$$

$$Z = 1 + 2e^{-\beta E} + e^{-\beta(2E+U)}.$$

It is easy to calculate the limit

$$G^{a\bar{c}}(\mathbf{k};+0)_{11} = \lim_{\varepsilon \to +0} \frac{1}{2\pi} \int_{-\infty}^{+\infty} d\omega\, e^{-i\omega\varepsilon}\, \tilde{G}^{a\bar{c}}(\mathbf{k};\omega)_{11} \tag{5.14}$$

by using the residue theorem and obtain

$$\left\langle a_{-\sigma}(\mathbf{k}) c_{-\sigma}^{+} \right\rangle = -\nu_{a}(\mathbf{k}) V_{a}(\mathbf{k}), \tag{5.15}$$

where

$$\nu_a(\mathbf{k})=\frac{1}{Z}\left\{\frac{e^{-\beta E}-\left[1+e^{-\beta E}\right]n_a(\mathbf{k})}{E-E_a(\mathbf{k})}+e^{-\beta E}\frac{e^{-\beta(E+U)}-\left[1+e^{-\beta(E+U)}\right]n_a(\mathbf{k})}{E+U-E_a(\mathbf{k})}\right\}. \tag{5.16}$$

The formula (5.6) becomes

$$\Delta(z-z')_C=\sum_{\mathbf{k}}\left\{\left|V_a(\mathbf{k})\right|^2\nu_a(\mathbf{k})\left[S^{Ea(\mathbf{k})}(z-z')_C-S^{2E+U-Ea(\mathbf{k})}(z-z')_C\right]+(a\to b)\right\}. \tag{5.17}$$

The system of Dyson equations (4.21) and (5.5) is the mathematical tool for the study of the electron transport through a single-level QD. Since this is a stationary process one can apply the Keldysh non-equilibrium Green function formalism in the limit $t_0\to-\infty$. Because the interaction vanishes at this limit, the contour C can be considered to consist of only two segment $C_1=[-\infty+io,+\infty+io]$ and $C_2=[\infty-io,-\infty+io]$. In this case each complex time-dependent Green function $S^E(z-z')_C$, $G^{c\bar{c}}(z-z')_C$, $H^{c\bar{c}}(z-z')_C$, $\Sigma^{(\alpha)}(z-z')_C$, $\alpha=1,2,3$ or $\Delta(z-z')_C$ becomes a set of four real time-dependent functions $S^E(t-t')_{ij}$, $G^{c\bar{c}}(t-t')_{ij}$, $H^{c\bar{c}}(t-t')_{ij}$, $\Sigma^{(\alpha)}(t-t')_{ij}$, $\alpha=1,2,3$ or $\Delta(t-t')_{ij}$ with their Fourier transforms $\tilde{S}^E(\omega)_{ij}$, $\tilde{G}^{c\bar{c}}(\omega)_{ij}$, $\tilde{H}^{c\bar{c}}(\omega)_{ij}$, $\tilde{\Sigma}^{(\alpha)}(\omega)_{ij}$, $\alpha=1,2,3$ or $\tilde{\Delta}(\omega)_{ij}$ with $i,j=1,2$. Considering them as the elements of corresponding 2×2 matrices $\hat{S}^E(\omega)$, $\hat{G}^{c\bar{c}}(\omega)$, $\hat{H}^{c\bar{c}}(\omega)$, $\hat{\Sigma}^{(\alpha)}(\omega)$, $\alpha=1,2,3$ or $\hat{\Delta}(\omega)$, and setting

$$\hat{\eta}=\begin{pmatrix}1 & 0\\ 0 & -1\end{pmatrix}, \tag{5.18}$$

we rewrite the system of Dyson equations (4.21) and (5.5) in the matrix form

$$\hat{G}(\omega)=\hat{S}^E(\omega)+U\hat{S}^E(\omega)\hat{\eta}\hat{H}(\omega)+\hat{S}^E(\omega)\hat{\eta}\hat{\Sigma}^{(1)}(\omega)\hat{\eta}\hat{G}(\omega), \tag{5.19}$$

$$\begin{aligned}\hat{H}(\omega)=n\hat{S}^{E+U}(\omega)+\hat{S}^{E+U}(\omega)\hat{\eta}\hat{\Delta}(\omega)+\hat{S}^{E+U}(\omega)\hat{\eta}\hat{\Sigma}^{(2)}(\omega)\hat{\eta}\hat{H}(\omega)\\ -\hat{S}^{E+U}(\omega)\hat{\eta}\hat{\Sigma}^{(3)}(\omega)\hat{\eta}\hat{G}(\omega).\end{aligned} \tag{5.20}$$

From these matrix equations we derive two systems of algebraic equations, each of which consists of four equations for four functions $\tilde{G}(\omega)_{i1}$ and $\tilde{H}(\omega)_{i1}$ or $\tilde{G}(\omega)_{i2}$ and $\tilde{H}(\omega)_{i2}$, $i=1,2$. The observable physical quantities are expressed in terms of these functions.

For the application let us calculate the Green function $\tilde{G}(\omega)_{11}$. By solving the system of equation (5.19) and (5.20) we obtain following result:

$$\tilde{G}(\omega)_{11}=\frac{Z(\omega)}{Y(\omega)}, \tag{5.21}$$

$$\begin{aligned}Z(\omega)=&\left\{B(\omega)[1+\Omega_{22}^{(1)}(\omega)]-UD_{22}(\omega)\right\}\left\{B(\omega)\tilde{S}^E(\omega)_{11}+UC_1(\omega)\right\}\\ &-\left\{B(\omega)\Omega_{12}^{(1)}(\omega)-UD_{12}(\omega)\right\}\left\{B(\omega)\tilde{S}^E(\omega)_{21}+UC_2(\omega)\right\},\end{aligned} \tag{5.22}$$

$$\begin{aligned}Y(\omega)=&\left\{B(\omega)[1-\Omega_{11}^{(1)}(\omega)]+UD_{11}(\omega)\right\}\left\{B(\omega)[1+\Omega_{22}^{(1)}(\omega)]-UD_{22}(\omega)\right\}\\ &+\left\{B(\omega)\Omega_{12}^{(1)}(\omega)-UD_{12}(\omega)\right\}\left\{B(\omega)\Omega_{21}^{(1)}(\omega)-UD_{21}(\omega)\right\},\end{aligned} \tag{5.23}$$

$$B(\omega)=[1+\Omega_{22}^{(2)}(\omega)][1-\Omega_{11}^{(2)}(\omega)]+\Omega_{12}^{(2)}(\omega)\Omega_{21}^{(2)}(\omega), \tag{5.24}$$

$$\begin{aligned} C_i(\omega)= &\left\{\tilde{S}^E(\omega)_{i1}[1+\Omega_{22}^{(2)}(\omega)]-\tilde{S}^E(\omega)_{i2}\Omega_{21}^{(2)}(\omega)\right\}\lambda_{11}(\omega) \\ &-\left\{\tilde{S}^E(\omega)_{i1}\Omega_{12}^{(2)}(\omega)+\tilde{S}^E(\omega)_{i2}[1-\Omega_{11}^{(2)}(\omega)]\right\}\lambda_{21}(\omega), \end{aligned} \tag{5.25}$$

$$i=1,2$$

$$\begin{aligned} D_{ij}(\omega)= &\left\{\tilde{S}^E(\omega)_{i1}[1+\Omega_{22}^{(2)}(\omega)]-\tilde{S}^E(\omega)_{i2}\Omega_{21}^{(2)}(\omega)\right\}\Omega_{1j}^{(3)}(\omega)- \\ &-\left\{\tilde{S}^E(\omega)_{i1}\Omega_{12}^{(2)}(\omega)+\tilde{S}^E(\omega)_{i2}[1-\Omega_{11}^{(2)}(\omega)]\right\}\Omega_{2j}^{(3)}(\omega), \end{aligned} \tag{5.26}$$

$$\begin{aligned} \Omega_{ij}^{(1)}(\omega)&=\tilde{S}^E(\omega)_{i1}\tilde{\Sigma}^{(1)}(\omega)_{1j}-\tilde{S}^E(\omega)_{i2}\tilde{\Sigma}^{(1)}(\omega)_{2j}, \\ \Omega_{ij}^{(2)}(\omega)&=\tilde{S}^{E+U}(\omega)_{i1}\tilde{\Sigma}^{(2)}(\omega)_{1j}-\tilde{S}^{E+U}(\omega)_{i2}\tilde{\Sigma}^{(2)}(\omega)_{2j}, \\ \Omega_{ij}^{(3)}(\omega)&=\tilde{S}^{E+U}(\omega)_{i1}\tilde{\Sigma}^{(3)}(\omega)_{1j}-\tilde{S}^{E+U}(\omega)_{i2}\tilde{\Sigma}^{(3)}(\omega)_{2j}, \end{aligned} \tag{5.27}$$

$$\lambda_{ij}(\omega)=n\tilde{S}^{E+U}(\omega)_{ij}+\tilde{S}^{E+U}(\omega)_{i1}\tilde{\Delta}(\omega)_{1j}-\tilde{S}^{E+U}(\omega)_{i2}\tilde{\Delta}(\omega)_{2j}, \tag{5.28}$$

$$i=1,2, \quad j=1,2.$$

The expressions (5.21)-(5.28) of $\tilde{G}(\omega)_{11}$ and similar ones for the Fourier transforms of other Green functions contain the self-energies $\tilde{\Sigma}^{(\alpha)}(\omega)_{ij}, \alpha=1,2,3$. Because the tunnelling coupling constants $V_a(\mathbf{k})$ and $V_b(\mathbf{k})$ have small values, the contributions of these self-energies, in general, give small corrections to the Green functions. However, the self-energies may be divergent at some special values of ω. At some points near these special values the denominator $Y(\omega)$ may vanish and the Green functions have the resonances. The formulae (5.23)-(5.28) would be used for the rigorous study of the behaviour of $\tilde{G}(\omega)_{11}$ at the resonances. This will be done in the subsequent Section.

6. Kondo and Fano resonances in electron transport through single-level quantum dot

In this Section we study the appearance of the resonances in the expressions of the Fourier transforms of the Green functions when the denominator $Y(\omega)$ is vanishing. The expression of $Y(\omega)$ consists of the terms of two types: the finite terms which do not depend on the Fourier transforms $\tilde{\Sigma}^{(\alpha)}(\omega)_{ij}$ of the self-energies and those proportional to $\tilde{\Sigma}^{(\alpha)}(\omega)_{ij}, \alpha=1,2,3$. The functions $\tilde{\Sigma}^{(\alpha)}(\omega)_{ij}$ contain the small tunnelling coupling constants $V_a(\mathbf{k})$ and $V_b(\mathbf{k})$. They are determined by following formulae:

$$\tilde{\Sigma}^{(1)}(\omega)_{ij}=\sum_{\mathbf{k}}\left\{|V_a(\mathbf{k})|^2\tilde{S}^{Ea(\mathbf{k})}(\omega)_{ij}+(a\to b)\right\}, \tag{6.1}$$

$$\tilde{\Sigma}^{(2)}(\omega)_{ij}=\sum_{\mathbf{k}}\left\{|V_a(\mathbf{k})|^2[2\tilde{S}^{Ea(\mathbf{k})}(\omega)_{ij}+\tilde{S}^{2E+U-Ea(\mathbf{k})}(\omega)_{ij}]+(a\to b)\right\}, \tag{6.2}$$

$$\tilde{\Sigma}^{(3)}(\omega)=\sum_{\mathbf{k}}\left\{n_a(\mathbf{k})|V_a(\mathbf{k})|^2[\tilde{S}^{Ea(\mathbf{k})}(\omega)_{ij}+\tilde{S}^{2E+U-Ea(\mathbf{k})}(\omega)_{ij}]+(a\to b)\right\}, \tag{6.3}$$

Introducing the spectral functions

$$\Gamma^{(\alpha)}_{a,b}(\omega)=\pi\sum_{\mathbf{k}}\left[\frac{e^{-\beta E a,b(\mathbf{k})}}{1+e^{-\beta E a,b(\mathbf{k})}}\right]^{\alpha}\left|V_{a,b}(\mathbf{k})\right|^{2}\delta[\omega-E_{a,b}(\mathbf{k})]\,, \tag{6.4}$$

$$a=0,1,2,$$

we rewrite them in the new form convenient for the study of their divergence:

$$\begin{aligned}\tilde{\Sigma}^{(1)}(\omega)_{11}&=\frac{1}{\pi}P\int d\omega'\frac{1}{\omega-\omega'}\left[\Gamma^{(0)}_{a}(\omega')+\Gamma^{(0)}_{b}(\omega')\right]\\&\quad-i\left[\Gamma^{(0)}_{a}(\omega)+\Gamma^{(0)}_{b}(\omega)-2\Gamma^{(1)}_{a}(\omega)-2\Gamma^{(1)}_{b}(\omega)\right],\end{aligned} \tag{6.5}$$

$$\tilde{\Sigma}^{(1)}(\omega)_{12}=2i\left[\Gamma^{(1)}_{a}(\omega)+\Gamma^{(1)}_{b}(\omega)\right],$$

$$\tilde{\Sigma}^{(1)}(\omega)_{21}=-2i\left[\Gamma^{(0)}_{a}(\omega)+\Gamma^{(0)}_{b}(\omega)-\Gamma^{(1)}_{a}(\omega)-\Gamma^{(1)}_{b}(\omega)\right],$$

$$\begin{aligned}\tilde{\Sigma}^{(1)}(\omega)_{22}&=-\frac{1}{\pi}P\int d\omega'\frac{1}{\omega-\omega'}\left[\Gamma^{(0)}_{a}(\omega')+\Gamma^{(0)}_{b}(\omega')\right]\\&\quad-i\left[\Gamma^{(0)}_{a}(\omega)+\Gamma^{(0)}_{b}(\omega)-2\Gamma^{(1)}_{a}(\omega)-2\Gamma^{(1)}_{b}(\omega)\right].\end{aligned}$$

$$\begin{aligned}\tilde{\Sigma}^{(2)}(\omega)_{11}&=\frac{1}{\pi}P\int d\omega'\left[\frac{2}{\omega-\omega'}+\frac{1}{\omega-2E-U+\omega'}\right]\left[\Gamma^{(0)}_{a}(\omega')+\Gamma^{(0)}_{b}(\omega')\right]-\\&\quad-2i\left[\Gamma^{(0)}_{a}(\omega)+\Gamma^{(0)}_{b}(\omega)-2\Gamma^{(1)}_{a}(\omega)-2\Gamma^{(1)}_{b}(\omega)\right]-\\&\quad-i\left[\Gamma^{(0)}_{a}(2E+U-\omega)+\Gamma^{(0)}_{b}(2E+U-\omega)\right.\\&\quad\left.-2\Gamma^{(1)}_{a}(2E+U-\omega)-2\Gamma^{(1)}_{b}(2E+U-\omega)\right],\end{aligned} \tag{6.6}$$

$$\tilde{\Sigma}^{(2)}(\omega)_{12}=4i\left[\Gamma^{(1)}_{a}(\omega)+\Gamma^{(1)}_{b}(\omega)\right]+2i\left[\Gamma^{(1)}_{a}(2E+U-\omega)+\Gamma^{(1)}_{b}(2E+U-\omega)\right],$$

$$\begin{aligned}\tilde{\Sigma}^{(2)}(\omega)_{21}&=-4i\left[\Gamma^{(0)}_{a}(\omega)+\Gamma^{(0)}_{b}(\omega)-\Gamma^{(1)}_{a}(\omega)-\Gamma^{(1)}_{b}(\omega)\right]-\\&\quad-2i\left[\Gamma^{(0)}_{a}(2E+U-\omega)+\Gamma^{(0)}_{b}(2E+U-\omega)\right.\\&\quad\left.-\Gamma^{(1)}_{a}(2E+U-\omega)-\Gamma^{(1)}_{b}(2E+U-\omega)\right],\end{aligned}$$

$$\begin{aligned}\tilde{\Sigma}^{(2)}(\omega)_{22}&=-\frac{1}{\pi}P\int d\omega'\left[\frac{2}{\omega-\omega'}+\frac{1}{\omega-2E-U+\omega'}\right]\left[\Gamma^{(0)}_{a}(\omega')+\Gamma^{(0)}_{b}(\omega')\right]-\\&\quad-2i\left[\Gamma^{(0)}_{a}(\omega)+\Gamma^{(0)}_{b}(\omega)-2\Gamma^{(1)}_{a}(\omega)-2\Gamma^{(1)}_{b}(\omega)\right]-\\&\quad-i\left[\Gamma^{(0)}_{a}(2E+U-\omega)+\Gamma^{(0)}_{b}(2E+U-\omega)\right.\\&\quad\left.-2\Gamma^{(1)}_{a}(2E+U-\omega)-2\Gamma^{(1)}_{b}(2E+U-\omega)\right].\end{aligned}$$

$$\tilde{\Sigma}^{(3)}(\omega)_{11} = \frac{1}{\pi} P \int d\omega' \left[\frac{1}{\omega-\omega'} + \frac{1}{\omega-2E-U+\omega'} \right] \left[\Gamma_a^{(1)}(\omega') + \Gamma_b^{(1)}(\omega') \right] -$$
$$- i\left[\Gamma_a^{(1)}(\omega) + \Gamma_b^{(1)}(\omega) - 2\Gamma_a^{(2)}(\omega) - 2\Gamma_b^{(2)}(\omega) + \Gamma_a^{(1)}(2E+U-\omega) \right.$$
$$\left. + \Gamma_b^{(1)}(2E+U-\omega) - 2\Gamma_a^{(2)}(2E+U-\omega) - 2\Gamma_b^{(2)}(2E+U-\omega) \right], \tag{6.7}$$

$$\tilde{\Sigma}^{(3)}(\omega)_{12} = 2i\left[\Gamma_a^{(2)}(\omega) + \Gamma_b^{(2)}(\omega) \right] + 2i\left[\Gamma_a^{(2)}(2E+U-\omega) + \Gamma_b^{(2)}(2E+U-\omega) \right],$$

$$\tilde{\Sigma}^{(3)}(\omega)_{21} = -2i\left[\Gamma_a^{(1)}(\omega) + \Gamma_b^{(1)}(\omega) - \Gamma_a^{(2)}(\omega) - \Gamma_b^{(2)}(\omega) + \Gamma_a^{(1)}(2E+U-\omega) \right.$$
$$\left. + \Gamma_b^{(1)}(2E+U-\omega) - \Gamma_a^{(2)}(2E+U-\omega) - \Gamma_b^{(2)}(2E+U-\omega) \right],$$

$$\tilde{\Sigma}^{(3)}(\omega)_{22} = -\frac{1}{\pi} P \int d\omega' \left[\frac{1}{\omega-\omega'} + \frac{1}{\omega-2E-U+\omega'} \right] \left[\Gamma_a^{(1)}(\omega') + \Gamma_b^{(1)}(\omega') \right] -$$
$$-i\left[\Gamma_a^{(1)}(\omega) + \Gamma_b^{(1)}(\omega) - 2\Gamma_a^{(2)}(\omega) - 2\Gamma_b^{(2)}(\omega) + \Gamma_a^{(1)}(2E+U-\omega) \right.$$
$$\left. + \Gamma_b^{(1)}(2E+U-\omega) - 2\Gamma_a^{(2)}(2E+U-\omega) - 2\Gamma_b^{(2)}(2E+U-\omega) \right].$$

The integrals in the r.h.s. of formulae (6.5)-(6.7) may be divergent at definite values of the frequency ω which will be called the divergence points. Although the functions $\tilde{\Sigma}^{(\alpha)}(\omega)_{ij}$, *i*, *j* = 1, 2, contain the small tunnelling coupling constants $V_a(\mathbf{k})$ and $V_b(\mathbf{k})$, near each divergence point some of them may become comparable with the finite terms in $Y(\omega)$. When $Y(\omega)$ vanishes due to the cancellation between the finite terms and those containing divergent integrals, there appear the resonances. Therefore in order to study the resonances it is necessary to investigate the divergence of the integrals in the r.h.s. of the formulae (6.5)-(6.7).

The functions $\tilde{\Sigma}^{(\alpha)}(\omega)_{11}$ and $\tilde{\Sigma}^{(\alpha)}(\omega)_{22}$ contain the dispersion integrals with the spectral functions $\Gamma_{a,b}^{(\alpha)}(\omega)$. Denote μ_a and μ_b the chemical potentials of the systems of conducting electrons in the leads "*a*" and "*b*". From the definition (6.4) with

$$E_{a,b}(\mathbf{k}) = E_{a,b}^{(0)}(\mathbf{k}) - \mu_{a,b},$$

where $E_{a,b}^{(0)}(\mathbf{k})$ are the kinetic energies of the conducting electrons in the leads, $E_{a,b}^{(0)}(\mathbf{k}) \geq 0$, it follows that $\Gamma_a^{(\alpha)}(\omega)$ vanishes at $\omega < -\mu_a$ and similarly for $\Gamma_b^{(n)}(\omega)$. Therefore the dispersion integrals in formulae (6.5)-(6.7) have the form

$$K_{a,b}^{(n)} = P \int_{-\mu_{a,b}}^{\Omega_{a,b}} d\omega' \frac{\Gamma_{a,b}^{(n)}(\omega')}{\omega'-\omega} \tag{6.8}$$

and

$$L_{a,b}^{(n)} = P \int_{-\mu_{a,b}}^{\Omega_{a,b}} d\omega' \frac{\Gamma_{a,b}^{(n)}(\omega')}{\omega'-2E-U+\omega}, \tag{6.9}$$

where Ω_a is the top of the energy band of the conducting electrons in the leads "a" and similarly for Ω_b. For the study of the divergence of the integrals we replace approximately the values of $\Gamma_a^{(\alpha)}(\omega)$ in the interval $-\mu < \omega < \Omega_a$ by a constant Γ_a and similarly for $\Gamma_b^{(n)}(\omega)$. Then at zero temperature

$$\begin{aligned} K_{a,b}^{(0)} &= \Gamma_{a,b} I_{a,b} , \\ K_{a,b}^{(1)} &= K_{a,b}^{(2)} = \Gamma_{a,b} I'_{a,b} , \\ L_{a,b}^{(0)} &= \Gamma_{a,b} J_{a,b} , \\ L_{a,b}^{(1)} &= L_{a,b}^{(2)} = \Gamma_{a,b} J'_{a,b} , \end{aligned} \tag{6.10}$$

with

$$I_{a,b} = P \int_{-\mu_{a,b}}^{\Omega_{a,b}} d\omega' \frac{1}{\omega - \omega'} , \tag{6.11}$$

$$I'_{a,b} = P \int_{-\mu_{a,b}}^{0} d\omega' \frac{1}{\omega - \omega'} , \tag{6.12}$$

$$J_{a,b} = P \int_{-\mu_{a,b}}^{\Omega_{a,b}} d\omega' \frac{1}{\omega - 2E - U + \omega'} , \tag{6.13}$$

$$J'_{a,b} = P \int_{-\mu_{a,b}}^{0} d\omega' \frac{1}{\omega - 2E - U + \omega'} . \tag{6.14}$$

Usually $\Omega_a(\Omega_b)$ is very large in comparison with $\mu_a(\mu_b)$ and ω. Therefore we have

$$I_{a,b} \approx \ln\left|\frac{\omega + \mu_{a,b}}{\Omega_{a,b} - \mu_{a,b}}\right| , \tag{6.15}$$

$$I'_{a,b} \approx \ln\left|\frac{\omega + \mu_{a,b}}{\omega}\right| , \tag{6.16}$$

$$J_{a,b} \approx \ln\left|\frac{\Omega_{a,b} - \mu_{a,b}}{\omega - 2E - U - \mu_{a,b}}\right| , \tag{6.17}$$

$$J'_{a,b} \approx \ln\left|\frac{\omega - 2E - U}{\omega - 2E - U - \mu_{a,b}}\right| . \tag{6.18}$$

It is obvious that I_a is divergent at $\omega \to -\mu_a$, I'_a is divergent at $\omega \to -\mu_a$ and $\omega \to 0$, J_a is divergent at $\omega \to 2E + U + \mu_a$ and J'_a is divergent at $\omega \to 2E + U + \mu_a$ and $\omega \to 2E + U$. For I_b, I'_b, J_b and J'_b we have similar results.

If the temperature T of the system is low enough, but does not vanish,

$$\mu_{a,b} >> kT > 0\,,$$

then instead of the divergence of $I'_{a,b}$ at the Fermi surface $\omega \to 0$ we have the limit

$$\lim_{\omega\to 0} I'_{a,b} = \ln\frac{2e\mu_{a,b}}{kT}\,, \tag{6.19}$$

and, similarly, instead of the divergence of $J'_{a,b}$ at $\omega \to 2E+U$ we have the limit

$$\lim_{\omega\to 2E+U} J'_{a,b} = -\ln\frac{2e\mu_{a,b}}{kT}\,. \tag{6.20}$$

For the simplicity we set $\mu_a = \mu_a = \mu$.

From the results of the study of resonances of Green function $\tilde{G}^{c\bar{c}}(\omega)_{11}$ and the explicit expressions (5.21)-(5.28) determining this function we obtain its asymptotic behavious at the divergence points of $\tilde{\Sigma}^{(\alpha)}(\omega)_{ij}$:

a. As $\omega \to -\mu$ and at low temperature $T \approx 0$, the Green function in equation (5.21) has asymptotic form:

$$\begin{aligned}\tilde{G}^{c\bar{c}}(\omega)_{11} \approx &-\frac{\left[1-n+\frac{1}{2}(E+\mu)\frac{\lambda}{\Gamma}\right]U}{E+2U+\mu}\,\frac{1}{\frac{1}{2}(E+\mu)+\frac{2\Gamma}{\pi}\ln\left|\frac{\omega+\mu}{\Omega}\right|+2i\Gamma} \\ &-\frac{E+\mu+\left[1+n-(E+U+\mu)\frac{\lambda}{\Gamma}\right]U}{E+2U+\mu}\,\frac{1}{(E+U+\mu)+\frac{2\Gamma}{\pi}\ln\left|\frac{\omega+\mu}{\Omega}\right|+2i\Gamma}\,.\end{aligned} \tag{6.21}$$

If $E+\mu > 0$, then $\tilde{G}^{c\bar{c}}(\omega)_{11}$ has two resonances at two points

$$\omega_1^{(\pm)} = -\mu \pm \Omega e^{-\frac{\pi}{4\Gamma}(E+\mu)} \tag{6.22}$$

and two resonances at two points

$$\omega_2^{(\pm)} = -\mu \pm \Omega e^{-\frac{\pi}{2\Gamma}(E+U+\mu)}\,. \tag{6.23}$$

Between these four resonances there are the dips. If $E+\mu < 0$ but $E+U+\mu > 0$, then $\tilde{G}^{c\bar{c}}(\omega)_{11}$ has only two resonances at the points $\omega_2^{(\pm)}$. If $E+U+\mu < 0$, then in the neighbourhood of the point $\omega = -\mu$, the Green function $\tilde{G}^{c\bar{c}}(\omega)_{11}$ has no resonance. All four points $\omega_1^{(\pm)}$ and $\omega_2^{(\pm)}$ are very close to the point $\omega = -\mu$ and the resonances at $\omega_1^{(\pm)}$ and $\omega_2^{(\pm)}$ look like a resonance at $\omega = -\mu$. The origin of these resonances is the presence of the Fano quasi-bound state at the lower edge of the energy band of the conducting electrons. If they exist, they would be called the Fano resonances.

b. As $\omega \to 0$ and at $T = 0$, the Green functions $\tilde{G}^{c\bar{c}}(\omega)_{11}$ has asymptotic form

$$\tilde{G}^{c\bar{c}}(\omega)_{11} \approx -\frac{E+(1-n)U}{E(E+U)+\frac{2\Gamma U}{\pi}\ln\left|\frac{\mu}{\omega}\right|+2i(3E+2U)\Gamma}. \tag{6.24}$$

If $E(E+U) < 0$, then $\tilde{G}^{c\bar{c}}(\omega)_{11}$ has two resonances at the points

$$\omega_3^{(\pm)} = \pm\mu\exp\left\{-\frac{\pi}{2\Gamma}\frac{|E(E+U)|}{U}\right\}, \tag{6.25}$$

which are very close to the point $\omega = 0$. At $\omega = 0$ and $0 < T < T_K$,

$$T_K = \frac{1}{k}\mu\exp\left\{-\frac{\pi}{2}\frac{|E(E+U)|}{\Gamma U}\right\}, \tag{6.26}$$

where k is the Boltzmann constant, instead of formula (6.24) we have

$$\tilde{G}^{c\bar{c}}(0)_{11} \approx \frac{\pi}{2}\frac{E+(1-n)U}{\Gamma U}\frac{1}{\ln|T/T_K| - i\pi(3E+2U)\Gamma}. \tag{6.27}$$

The resonances in the neighbourhood of the point $\omega = 0$ have the same physical origin as the Kondo effect due to the scattering of electrons by a magnetic impurity. They are the Kondo resonances.

c. As $\omega \to 2E+U$ and at $T = 0$, the Green function $\tilde{G}^{c\bar{c}}(\omega)_{11}$ has asymptotic form:

$$\tilde{G}^{c\bar{c}}(\omega)_{11} \approx \frac{E+nU}{E(E+U)-\frac{2\Gamma U}{\pi}\ln\left|\frac{\mu}{\omega-2E-U}\right|-2iE\Gamma}. \tag{6.28}$$

Therefore if $E(E+U) > 0$, then $\tilde{G}^{c\bar{c}}(\omega)_{11}$ has also two resonances at the points

$$\omega_4^{(\pm)} = 2E+U \pm \mu e^{-\frac{\pi}{2\Gamma}\frac{E(E+U)}{U}} \tag{6.29}$$

which are very close to the point $\omega = 2E+U$. At $\omega = 2E+U$ and $0 < T < T'_K$,

$$T'_K = \frac{1}{k}\mu\exp\left\{-\frac{\pi}{2}\frac{E(E+U)}{\Gamma U}\right\}, \tag{6.30}$$

instead of formula (6.28) we have

$$\tilde{G}^{c\bar{c}}(2E+U)_{11} \approx \frac{\pi}{2}\frac{E+nU}{\Gamma U}\frac{1}{\ln|T/T'_K| - i\pi E\Gamma}. \tag{6.31}$$

The resonances in the neighbourhood of the point $\omega = 2E+U$ are the Kondo resonances of the crossing terms.

d. As $\omega \to 2E+U+\mu$ and at low temperature $T \approx 0$, the Green functions $\tilde{G}^{c\bar{c}}(\omega)_{11}$ has asymptotic form:

$$\tilde{G}^{c\bar{c}}(\omega)_{11} \approx \frac{1-n+(E+U+\mu)\frac{\lambda}{\Gamma}}{(E+U+\mu)+\frac{2\Gamma}{\pi}\ln\left|\frac{\omega-2E-U-\mu}{\Omega}\right|-2i\Gamma}\frac{U}{E+\mu}. \tag{6.32}$$

If $E+U+\mu>0$, then $\tilde{G}^{c\bar{c}}(\omega)_{11}$ has two resonances at the points

$$\omega_5^{(\pm)} = 2E+U+\mu \pm \Omega e^{-\frac{\pi}{2}\frac{E+U+\mu}{\Gamma}}, \tag{6.33}$$

which are very close to the point $\omega = 2E+U+\mu$. They are the Fano resonances of the crossing terms.

7. Conclusion

The present Chapter is an introductory review of the Keldysh non-equilibrium Green functions of electrons in simplest nanosystems: isolated single-level QD and single-level QD connected with two conducting leads. In the case of an isolated single-level QD the closed system of a finite number of differential equations for a finite number of Green functions was established by using the Heisenberg quantum equations of motion for the electron destruction and creation operators. The exact expressions of the Green functions were derived. In the case of the nanosystem consisting of a single-level QD connected with two conducting leads there does not exist a finite closed system of differential equations for some finite number of Green functions. In the differential equations for n-point Green functions there appear the contributions from (n+2)-point Green functions. Therefore, the exact system of differential equations contains an infinite number of equations for an infinite number of Green functions. In order to truncate this infinite system of differential equations we have applied the mean-field approximation to the products of four electron quantum operators and limited at the terms of the second order with respect to the effective tunnelling coupling constants. As the result we have derived a closed system of Dyson equations for two types of 2-point Green functions. All the crossing terms are included into the equations. The exact solution of the system of Dyson equations may have the resonances of four types in the dependence on the physical parameters of the system: the Kondo resonances at the Fermi surface, whose origin is similar to that of the Kondo effect in the scattering of electrons on magnetic impurities, the Fano resonances due to the presence of the electron quasi-bound state at the lower edge of the energy band of the conducting electrons, the Kondo resonances in the crossing terms and the Fano resonances in the crossing terms. The analytical asymptotic expressions of the single-electron Green function at these resonances were derived. These results agree well with the numerical calculations in references on the electron Green functions in QD (Yeyati et al., 1993; Costi et al., 1994; Izumida et al., 1997, 1998, 2001; Sakai et al., 1999; Torio et al., 2002).

The theoretical study of the non-equilibrium Green functions of electrons in QDs would signify the beginning of the development of the quantum dynamics of physical processes in QD-based nanodevices. The next step would be the elaboration of the theory of non-equilibrium Green functions of phonons in QDs as well as of electrons and phonons of interacting electron-phonon systems in QDs. The quantum dynamical theory of QD-based optoelectronic and photonic nanodevices necessitates also the study of non-equilibrium Green functions of electrons and phonons confined in QDs in the presence of the electron-phonon interactions as well as the interaction of photons with confined electron-phonon systems. The methods and reasonnings presented in this Chapter could be generalized for the application to the study of all above-mentioned non-equilibrium Green functions.

8. Acknowledgment

I would like to express the gratitude to Institute of Materials Science and Vietnam Academy of Science and Technology for the support to my work on the subject of this review during many years. I thank also Academician Nguyen Van Hieu for suggesting the main ideas of the series of publications on this subject.

9. References

Abrikosov, A. A; Gorkov, L. P. & Dzyaloshinski, I. E. (1975). *Methods of Quantum Field Theory in Statistical Physics,* Dover Publications, ISBN 0486632288, New York

Bjoken, J. D & Drell, S. D. (1964). *Relativistic Quantum Mechanics,* Mc Graw-Hill, ISBN 0070054932, New York

Bruus, H. & Flensberg, K. (2004). *Many-Body Quantum Theory in Condensed Matter Physics,* Oxford University Press, ISBN 0198566336, Oxford New York

Choi, M.-S.; Sánchez, D. & López, R. (2004). Kondo effect in a quantum dot coupled to ferromagnetic leads: A numerical renormalization group analysis, *Physical Review Letters,* Vol. 92, No. 5, p. 056601 (4 pages); ISSN 0031-9007 (print), 1079-7114 (online)

Chou, K. C.; Su, Z. B.; Lao, B. L. & Yu, L. (1985). Equilibrium and nonequilibrium formalisms made unified, *Physics Reports,* Vol. 118, Nos. 1&2, pp. 1-131, ISSN 0370-1573

Costi, T. A.; Hewson, A. C. & Zlatic, V. (1994). Transport coefficients of the Anderson model via the numerical renormalization group, *Journal of Physics: Condensed Matter,* Vol. 6, No. 13, p. 2519 (1994), ISSN 0953-8984 (print), 1361-648X (online)

Craco, L & Kang, K. (1999). Perturbation treatment for transport through a quantum dot, *Physical Review B,* Vol. 59, No. 19, pp. 12244–12247; ISSN 1098-0121 (print), 1550-235X (online)

Fujii, T. & Ueda, K. (2003). Perturbative approach to the nonequilibrium Kondo effect in a quantum dot, *Physical Review B,* Vol. 68, No. 15, p.155310 (5 pages); ISSN 1098-0121 (print), 1550-235X (online)

Haken, H. (1976). *Quantum Field Theory of Solids: An Introduction,* North-Holland Pub. Co., ISBN 0720405459, Amsterdam

Hershfield, S.; Davies, J. H. & Wilkins J. W. (1991). Probing the Kondo resonance by resonant tunneling through an Anderson impurity, *Physical Review Letters,* Vol. 67, No. 26 , pp. 3720-3723; ISSN 0031-9007 (print), 1079-7114 (online)

Inoshita, T.; Shimizu, A.; Kuramoto, Y & Sakaki, H. (1993). Correlated electron transport through a quantum dot: The multiple-level effect, *Physical Review B,* Vol. 48, No. 19, pp. 14725–14728; ISSN 1098-0121 (print), 1550-235X (online)

Itzykson, C. & Zuber, J. B. (1987). *Quantum Field Theory*, Mc Graw-Hill, ISBN 0070320713, New York

Izumida,W.; Sakai, O. & Shimizu, Y. (1997). Many body effects on electron tunneling through quantum dots in an Aharonov-Bohm circuit, many body effects on electron tunneling through quantum dots in an Aharonov-Bohm circuit, *Journal of the Physical Society of Japan,* Vol. 66, No. 3, pp. 717-726; ISSN 0031-9015 (print), 1347-4073 (online)

Izumida, W.; Sakai, O. & Shimizu, Y. (1998). Kondo effect in single quantum dot systems - Study with numerical renormalization group method, *Journal of the Physical Society of Japan,* Vol. 67, No. 7, pp. 2444-2454; ISSN 0031-9015 (print), 1347-4073 (online)

Izumida, W.; Sakai, O. & Suzuki, S. (2001). Kondo effect in tunneling through a quantum dot, *Journal of the Physical Society of Japan,* Vol. 70, No. 4, pp. 1045-1053; ISSN 0031-9015 (print), 1347-4073 (online)

Kapusta, J. I. (1989). *Finite Temperature Field Theory*, Cambridge University Press, ISBN 0-521-35155-3, Cambridge (UK)

Keldysh, L. V. (1965). Diagram technique for nonequilibrium processes, *Soviet Physics- JETP, Vol.* 20, No. 4, pp. 1018-1026, ISSN 0038-5646

Konig, J. & Gefen, Y. (2005). Nonmonotonic charge occupation in double dots, *Physical Review B,* Vol. 71, No. 20, p. 201308(R) (4 pages), ISSN 1098-0121 (print), ISSN 1550-235X (online)

Le Bellac, M. (1996). *Thermal Field Theory*, Cambridge University Press, ISBN 0-521-65477-7, Cambridge (UK)

Meir, Y.; Wingreen, N. S. & Lee, P. A. (1991). Transport through a strongly interacting electron system: Theory of periodic conductance oscillations, *Physical Review Letters,* Vol. 66, No. 23, pp. 3048-3051; ISSN 0031-9007 (print), 1079-7114 (online)

Meir, Y.; Wingreen, N. S. & Lee, P. A. (1993). Low-temperature transport through a quantum dot: The Anderson model out of equilibrium, *Physical Review Letters,* Vol. 70, No. 17, pp. 2601-2604; ISSN 0031-9007 (print), 1079-7114 (online)

Ng, T. K. (1993). Nonlinear resonant tunneling through an Anderson impurity at low temperature, *Physical Review Letters,* Vol. 70, No. 23, pp. 3635-3638; ISSN 0031-9007 (print), 1079-7114 (online)

Nguyen Van Hieu & Nguyen Bich Ha (2005). Quantum theory of electron transport through single-level quantum dot, *Advances in Natural Sciences,* Vol. 6, No. 1, pp. 1-18, ISSN 0866-708X

Nguyen Van Hieu & Nguyen Bich Ha (2006). Time-dependent Green functions of quantum dots at finite temperature, *Advances in Natural Sciences,* Vol. 7, No. 2, pp.153-165, ISSN 0866-708X

Nguyen Van Hieu; Nguyen Bich Ha & Nguyen Van Hop (2006a). Dyson equations for Green functions of electrons in open single-level quantum dot, *Advances in Natural Sciences,* Vol. 7, No. 1, pp. 1-12, ISSN 0866-708X

Nguyen Van Hieu; Nguyen Bich Ha; Gerdt, V. P.; Chuluunbaatar, O.; Gusev, A. A.; Pali, Yu. G. & Nguyen Van Hop (2006b). Analytical asymptotic expressions for the Green's function of the electron in a single-level quantum dot at the Kondo and the Fano resonances, *Journal of the Korean Physical Society,* Vol. 53, No. 96, pp. 3645-3649, ISSN 0374-4884 (print), 1976-8524 (online)

Peskin, M. E. & Schroeder, D. V. (1995). *An Introduction to Quantum Field Theory*, Addison-Wesley, ISBN 0201503972, New York

Pustilnik, M. & Glasman, L. (2004). Kondo effect in quantum dots, *Journal of Physics: Condensed Matter*, Vol. 16, No. 16, p. R513; ISSN 0953-8984 (print), 1361-648X (online)

Sakai, O.; Suzuki, S.; Izumida, W. & Oguri, A. (1999). Kondo effect in electron tunneling through quantum dots - Study with quantum Monte Carlo method, *Journal of the Physical Society of Japan*, Vol. 68, No. 5, pp. 1640-1650; ISSN 0031-9015 (print), 1347-4073 (online)

Swirkowicz, R.; Barnas, J. & Wilczynski, M. (2003). Nonequilibrium Kondo effect in quantum dots, *Physical Review B*, Vol. 68, No. 19, p. 195318 (10 pages); ISSN 1098-0121 (print), 1550-235X (online)

Swirkowicz, R.; Wilczynski, M. & Barnas, J. (2006). Spin-polarized transport through a single-level quantum dot in the Kondo regime, *Journal of Physics: Condensed Matter*, Vol. 18, No. 7, p. 2291; ISSN 0953-8984 (print), 1361-648X (online)

Takagi, O. & Saso, T. (1999a). Magnetic field effects on transport properties of a quantum dot studied by modified perturbation theory, *Journal of the Physical Society of Japan*, Vol. 68, No. 6, pp.1997-2005; ISSN 0031-9015 (print), 1347-4073 (online)

Takagi, O. & Saso, T. (1999b). Modified perturbation theory applied to the Kondo-type transport through a quantum dot under magnetic field, *Journal of the Physical Society of Japan*, 68, No. 9, pp. 2894-2897; ISSN 0031-9015 (print), 1347-4073 (online)

Torio, M. E.; Hallberg, K.; Ceccatto, A. H. & Proetto C. R. (2002). Kondo resonances and Fano antiresonances in transport through quantum dots, *Physical Review B*, Vol. 65, No. 8, p. 085302 (5 pages); ISSN 1098-0121 (print), 1550-235X (online)

Wingreen, N. S. & Meir, Y. (1994). Anderson model out of equilibrium: Noncrossing-approximation approach to transport through a quantum dot, *Physical Review B*, Vol. 49, No. 16, pp. 11040–11052; ISSN 1098-0121 (print), 1550-235X (online)

Yeyati, A. L.; Martin-Rodero, A. & Flores, F. (1993). Electron correlation resonances in the transport through a single quantum level, *Physical Review Letters*, Vol. 71, No. 18 , pp. 2991–2994; ISSN 0031-9007 (print), 1079-7114 (online)

16

Electron Scattering Through a Quantum Dot

Leonardo Kleber Castelano[1], Guo-Qiang Hai[2] and Mu-Tao Lee[3]
[1]*Departamento de Física, Universidade Federal de São Carlos*
[2]*Instituto de Física de São Carlos, Universidade de São Paulo*
[3]*Departamento de Química, Universidade Federal de São Carlos*
Brazil

1. Introduction

Electron scattering and transport through quantum dots (QDs) in a semiconductor nanostructure have been intensively studied (Engel & Loss, 2002; Fransson et al., 2003; Konig & Martinek, 2003; Koppens et al., 2006; Qu & Vasilopoulos, 2006; Zhang et al., 2002). The spin-dependent transport properties are of particular interest for its possible applications, *e.g.*, the QD spin valves (Konig & Martinek, 2003), the quantum logic gates using coupled QDs, as well as the spin-dependent transport in single-electron devices (Seneor et al., 2007), etc.. In such systems, the electron-electron exchange potential and the electron spin states have been utilized and manipulated (Burkard et al., 2000; Gundogdu et al., 2004; Sarma et al., 2001; Wolf et al., 2001). A thorough quantitative understanding of spin-dependent transport properties due to electron-electron interaction is therefore important for a successful construction of these devices. Theoretically the transport through QDs has been studied by different approaches such as transfer matrix, nonequilibrium Green's functions, random matrix theory, as well as those methods built on the Lippmann-Schwinger (L-S) equation (Castelano et al., 2007a;b).

In this chapter, we develop a theoretical method to study electron scattering through a quantum dot (QD) of N-electrons embedded in a semiconductor nanostructure. We construct the scattering equations including electron-electron interaction to represent the process of a free electron scattered by the QD confined in a two-dimensional (2D) or in a quasi-one-dimensional (Q1D) semiconductor system. The generalized multichannel Lippmann-Schwinger equations(Bransden & McDowell, 1977; Joachain, 1975) are solved for these systems by using the method of continued fractions (MCF). As an example, we apply this method to a one-electron QD case and obtain scattering cross-sections in 2D and conductances in Q1D systems resulting from both the singlet- and triplet-coupled continuum states of two electrons (incident and QD electron) during the electron transport.

This chapter is organized as follows. In Sec. 2 we present our general theoretical approach and numerical method. In Sec. 3, we describe the electron scattering through a quantum dot in a 2D system. The scattering for a quantum dot confined in a Q1D system is presented in Sec. 4. In Sections 5 and 6, we show our numerical results for the scattering through a one-electron QD within both the one-channel and the multichannel models. We conclude in Sec. 7.

2. Theoretical approach

2.1 The system: Incident electron + quantum dot

The system under investigation consists of an incident free electron and a quantum dot of N electrons as shown schematically in Fig. 1. The incident electron is scattered by both the QD potential and by the confined electrons inside the QD. The Schrödinger equation of the system is given by

$$(H - \mathcal{E}_i)\Psi_i(\tau; \mathbf{r}_{N+1}, \sigma_{N+1}) = 0\,, \tag{1}$$

where τ represents collectively the spatial and spin coordinates of the N electrons localized in the QD and $\mathbf{r}_{N+1} = (x_{N+1}, y_{N+1})$ and σ_{N+1} denote the spatial and spin coordinates of the incident electron. The total energy of the system is $\mathcal{E}_i$, where the subscript i represents a set of quantum numbers required to uniquely specify the initial quantum state of the system. Explicitly, the total Hamiltonian of the system can be written as

$$H = H_0(\mathbf{r}_{N+1}) + H_{\mathrm{QD}}(\tau) + V_{\mathrm{int}}(\mathbf{r}_1, \mathbf{r}_2, ..., \mathbf{r}_N, \mathbf{r}_{N+1})\,, \tag{2}$$

where $H_0(\mathbf{r}_{N+1}) = -\hbar^2\nabla^2_{N+1}/2m^* + V_{\mathrm{QD}}(\mathbf{r}_{N+1})$, $H_{\mathrm{QD}}(\tau)$ is the Hamiltonian of the QD of N electrons, and V_{int} is the interaction potential between the incident electron at $\mathbf{r}_{N+1}$ and the N electrons in the QD

$$V_{\mathrm{int}}(\mathbf{r}_1, \mathbf{r}_2, ..., \mathbf{r}_N, \mathbf{r}_{N+1}) = \frac{e^2}{\epsilon_0^*}\sum_{i=1}^{N}\frac{1}{|\mathbf{r}_{N+1} - \mathbf{r}_i|}, \tag{3}$$

where ϵ_0^* is the dielectric constant of the semiconductor material and m^* is the electron effective mass. The Hamiltonian for an unperturbed QD is given by

$$H_{\mathrm{QD}}(\tau) = \sum_{i=1}^{N}\left(-\frac{\hbar^2}{2m^*}\nabla_i^2 + V_{\mathrm{QD}}(\mathbf{r_i})\right) + \frac{e^2}{\epsilon_0^*}\sum_{i\neq j}^{N}\frac{1}{|\mathbf{r}_i - \mathbf{r}_j|}, \tag{4}$$

where the first term in the *rhs* of Eq. (4) describes N independent electrons in the QD of confinement potential $V_{\mathrm{QD}}(\mathbf{r})$ and the second term gives the Coulomb interactions among these electrons. The eigenenergy and eigenfunction of this N-electron QD are denoted by ε_n and Φ^n, respectively. They are determined by the following Schrödinger equation

$$H_{\mathrm{QD}}(\tau)\Phi^n = \varepsilon_n\Phi^n, \tag{5}$$

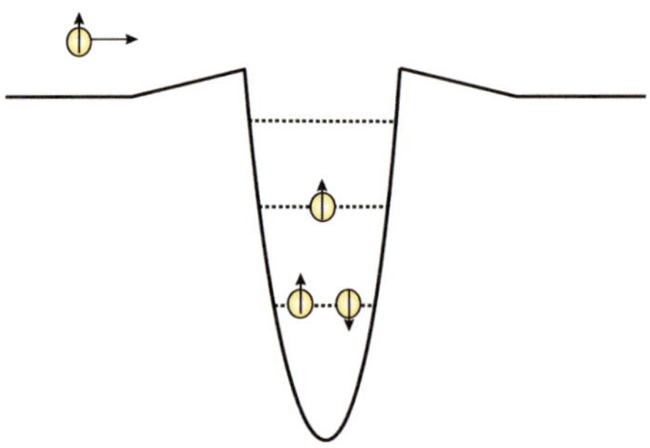

Fig. 1. Representation of the incident electron and the target, which in this case is a quantum dot containing 3 electrons.

with $n = 0, 1, 2, 3...$. The ground state of the QD is labeled by $n = 0$ and the excited states by $n \geq 1$. The eigenstates of the QD can be obtained using, *e.g.*, the restricted or unrestricted Hartree-Fock (HF) methods (Szabo & Ostlund, 1982).

2.2 Scattering equations including electron-electron interaction

In order to extract scattering properties of the system (QD + incident electron), we can write the total wave-function Ψ_i of the system as a superposition of the QD wave-function Φ^n and the incident electron wave-function,

$$|\Psi_i\rangle = \sum_{n=0}^{\infty} |\mathcal{A}(\Phi^n \psi_{ni})\rangle, \tag{6}$$

where ψ_{ni} describes the wave-functions of the incident (scattered) electron in the continuum states corresponding to a quantum transition from an initial state i to a final state n. The operator $\mathcal{A}$ warrants the antisymmetrization property between the QD electrons and the incident electron, defined by,

$$\mathcal{A} = \frac{1}{\sqrt{N+1}} \sum_{p=1}^{N+1} (-1)^{N+1-p} \mathcal{P}_{N+1,p} \tag{7}$$

where $\mathcal{P}_{N+1,p}$ is the permutation operator which exchanges the electrons at $\mathbf{r}_{N+1}$ and $\mathbf{r}_p$. From Eqs. (1), (2) and (6), we obtain

$$\sum_{n=0}^{\infty} \left(-\frac{\hbar^2}{2m^*} \nabla^2_{N+1} + V_{\text{QD}} + H_{\text{QD}} + V_{\text{int}} \right) |\mathcal{A}(\Phi^n \psi_{ni})\rangle = \mathcal{E}_i \sum_{n=0}^{\infty} |\mathcal{A}(\Phi^n \psi_{ni})\rangle. \tag{8}$$

The total energy of the system $\mathcal{E}_i$ is composed of two parts. The first part is the kinetic energy of the incident (scattering) electron and the second is the energy of the N-electron QD in a particular configuration, *i.e.*, $\mathcal{E}_i = \frac{\hbar^2 k_i^2}{2m^*} + \varepsilon_i = \frac{\hbar^2 k_n^2}{2m^*} + \varepsilon_n$, for different eigenstates of the QD $(i, n = 0, 1, 2, ...)$ or different scattering channels. These different channels appear because the incident electron can probably be scattered inelastically, leaving the QD in a different state from its initial. A projection of Eq. (8) onto a particular QD state $|\Phi^m\rangle$ leads to the following scattering equation for the incident electron,

$$\frac{\hbar^2}{2m^*} \left(\nabla^2 + k_m^2 \right) \psi_{mi}(\mathbf{r}) = \sum_{n=0}^{\infty} V_{mn}(\mathbf{r}) \psi_{ni}(\mathbf{r}) \tag{9}$$

for $i, m = 0, 1, 2, ...$, where $\mathbf{r} = \mathbf{r}_{N+1}$ and $V_{mn} = V^{\text{st}}_{mn} + V^{\text{ex}}_{mn}$ with V^{st}_{mn} the static potential and V^{ex}_{mn} the exchange potential due the nonlocal interaction, giving by

$$V^{\text{st}}_{mn}(\mathbf{r}) = V_{\text{QD}}(\mathbf{r}) \delta_{mn} + \frac{e^2}{\epsilon_0^*} \sum_{j=1}^{N} \langle \Phi^m | \frac{e^{-\lambda |\mathbf{r} - \mathbf{r}_j|}}{|\mathbf{r} - \mathbf{r}_j|} | \Phi^n \rangle, \tag{10}$$

and

$$V_{mn}^{\rm ex}(\mathbf{r})\psi_{ni}(\mathbf{r}) = (H_0(\mathbf{r}) - \frac{\hbar^2 k_m^2}{2m^*})\langle\Phi^m|\mathcal{A}'(\Phi^n\psi_{ni})\rangle + \frac{e^2}{\epsilon_0^*}\sum_{j=1}^{N}\langle\Phi^m|\frac{1}{|\mathbf{r}-\mathbf{r_j}|}|\mathcal{A}'(\Phi^n\psi_{ni})\rangle, \quad (11)$$

respectively, where $\mathcal{A}' = \sum_{p=1}^{N}(-1)^{N+1-p}\mathcal{P}_{N+1,p}$. In Eq. (10) we have introduced a screening $e^{-\lambda|\mathbf{r}-\mathbf{r}'|}$ on the direct Coulomb potential for two reasons: (i) the ionized impurities in the semiconductor nanostructure and/or the external electrodes screen the direct Coulomb potential and (ii) at $|\mathbf{r}| \to \infty$ limit the scattering potential should decay faster than $1/|\mathbf{r}|$. The screening length is given by λ^{-1}. Notice that we do not consider the screening on the exchange potential because this potential is non-zero inside the QD only. Inclusion of the screening on the exchange potential in Eq. (11) is possible but it will not affect much our results and considerably complicates the numerical calculation. The scattering equation is a system of coupled integro-differential equations. The corresponding generalized L-S equation for such a multichannel scattering problem is given by

$$\psi_{mi}(\mathbf{r}) = \varphi_i(\mathbf{r})\delta_{mi} + \sum_{n=0}^{\infty}\int d\mathbf{r}' G^{(0)}(\mathbf{k}_m,\mathbf{r},\mathbf{r}')V_{mn}(\mathbf{r}')\psi_{ni}(\mathbf{r}'), \text{ for } i,m = 0,1,2\ldots \quad (12)$$

with an incident plane wave $\varphi_i(\mathbf{r}) = e^{i\mathbf{k}_i.\mathbf{r}} = e^{ik_i x}$ in the x-direction.

2.3 Method of continued fractions

The method of continued fractions (MCF) (Horacek & Sasakawa, 1984) is an iterative method to solve the L-S equation. To apply this method for a multi-channel scattering we have firstly to rewrite Eq. (12) in a matrix form:

$$\widetilde{\Psi} = \widetilde{\varphi} + \widetilde{G}^{(0)}\widetilde{V}\widetilde{\Psi}. \quad (13)$$

In the first step to start the MCF, we use the scattering potential $\widetilde{V} = V^{(0)}$ and the free electron wave-function $\widetilde{\varphi} = |\varphi^{(0)}\rangle$ in Eq. (13). Afterwards, we define the nth-order weakened potential as

$$V^{(n)} = V^{(n-1)} - \frac{V^{(n-1)}|\varphi^{(n-1)}\rangle\langle\varphi^{(n-1)}|V^{(n-1)}}{\langle\varphi^{(n-1)}|V^{(n-1)}|\varphi^{(n-1)}\rangle}, \quad (14)$$

where

$$|\varphi^{(n)}\rangle = \widetilde{G}^{(0)}V^{(n-1)}|\varphi^{(n-1)}\rangle. \quad (15)$$

The nth-order correction of the **T** matrix can be obtained through

$$T^{(n)} = \langle\varphi^{(n-1)}|V^{(n-1)}|\varphi^{(n)}\rangle + \langle\varphi^{(n)}|V^{(n)}|\varphi^{(n)}\rangle$$

$$\times\left[\langle\varphi^{(n)}|V^{(n)}|\varphi^{(n)}\rangle - T^{(n+1)}\right]^{-1}\langle\varphi^{(n)}|V^{(n)}|\varphi^{(n)}\rangle. \quad (16)$$

Hence, we can stop the iteration when the potential $V^{(N)}$ becomes weaker enough. In the numerical calculation, we start with $T^{(N+1)} = 0$ and evaluate $T^{(N)}, T^{(N-1)}, ...,$ and $T^{(1)}$.

Therefore the **T** matrix is given by

$$\mathbf{T} = \langle \varphi^{(0)} | V^{(0)} | \varphi^{(0)} \rangle + T^{(1)} \frac{\langle \varphi^{(0)} | V^{(0)} | \varphi^{(0)} \rangle}{\langle \varphi^{(0)} | V^{(0)} | \varphi^{(0)} \rangle - T^{(1)}}. \tag{17}$$

3. Quantum dot embedded in a two-dimensional system

The Green's function $G^{(0)}(\mathbf{k}, \mathbf{r}, \mathbf{r}')$ in 2D is given by

$$G^{(0)}(\mathbf{k}, \mathbf{r}, \mathbf{r}') = -\frac{2m^*}{\hbar^2}(i/4)H_0^{(1)}(k|\mathbf{r} - \mathbf{r}'|), \tag{18}$$

where $H_0^{(1)}$ is the usual zero order Hankel's function (Morse & Feshbach, 1953).

At $|\mathbf{r}| \to \infty$ limit, the asymptotic form of Eq. (12) for the scattered wave-function in a 2D system is given by

$$\psi_{mi}(\mathbf{r}) \underset{|\mathbf{r}|\to\infty}{\longrightarrow} e^{ik_i x}\delta_{mi} + \frac{2m^*}{\hbar^2}\sqrt{\frac{i}{k_m}}\frac{e^{+ik_m r}}{\sqrt{r}} f_{k_m,k_i}(\theta), \tag{19}$$

where $f_{k_m,k_i}(\theta)$ is the scattering amplitude

$$f_{k_m,k_i}(\theta) = -\frac{1}{4}\sqrt{\frac{2}{\pi}} < \mathbf{k}_m | T(E) | \mathbf{k}_i > \tag{20}$$

with

$$< \mathbf{k}_m | T(E) | \mathbf{k}_i > = \sum_{n=0}^{\infty} \int d\mathbf{r}' e^{-i\mathbf{k}_m \cdot \mathbf{r}'} V_{mn}(\mathbf{r}') \psi_{ni}(\mathbf{r}').$$

The momenta of the initial and final states of the incident (scattered) electron are $\mathbf{k}_i$ and $\mathbf{k}_m$, respectively, and θ is the scattering angle between them. It is evident from Eq. (12) and its boundary condition Eq. (19) that the different scattering channels are coupled to each other through the interaction potential V_{mn}.

In the above procedure in dealing with the electron scattering through a QD, both the electron-electron exchange and correlation interactions are present in this system. However, a complete correlation effect is difficult to include in a practical calculation. In order to do so, besides an exact solution for the N-electron QD, a full sum over all the intermediate states n in the scattering equation [Eq. (9)] is needed, which is a formidable task in a self-consistent calculation. In an alternative way, the correlation effects can be considered by adding an effective correlation potential in the scattering equation (Joachain, 1975). In the present work, we focus on the exchange effects on the scattering process and limit the sum over n to a few lowest energy levels of the QD. For this reason, we prefer to call the nonlocal interaction potential V_{mn}^{ex} in Eq. (11) as exchange potential, though the correlation can be partially included.

The differential cross-section (DCS) for a scattering from initial state i (*i.e.* the incident electron of kinetic energy $E_i = \frac{\hbar^2 k_i^2}{2m^*}$ and the QD in the state ε_i) to final state m (*i.e.* $E_m = \frac{\hbar^2 k_m^2}{2m^*}$ and the QD in the state m) is given by

$$\sigma_{mi}(\theta) = \frac{k_m}{k_i^2}|f_{k_m,k_i}(\theta)|^2. \tag{21}$$

The integral cross-section (ICS) which is an energy dependent quantity can be found by

$$\Gamma_{mi}(E_i) = \int_0^{2\pi} \sigma_{mi}(\theta) d\theta. \tag{22}$$

When the incident electron is scattered to a state of the same energy and the QD keeps in the same state ($m = i$), the scattering is called elastic. Otherwise, the scattering is inelastic. A possible scattering is the so-called super-elastic scattering ($E_m > E_i$) where the incident electron is scattered out with a higher energy by an QD initially in an excited state. Because the different scattering channels are coupled to each other, we have to solve the multichannel L-S equation to obtain the scattering probabilities through different channels simultaneously for the same total energy of the system.

3.1 Partial wave expansion

In two dimensions the angular momentum basis is given by (Adhikari, 1986),

$$\Theta_l(\phi) = \sqrt{\frac{\kappa_l}{2\pi}} \cos(l\phi) \tag{23}$$

where $l = 0, 1, 2, ...$, $\kappa_l = 2$ for $l \neq 0$ and $\kappa_l = 1$ for $l = 0$. In applying the partial wave expansion in the multi-channel scattering problem Eq. (12), we expand all functions, *i.e.*, the incident free electron wavefunction $\varphi_i(\mathbf{r})$, the Green's function $G^{(0)}(\mathbf{k}_m, \mathbf{r}, \mathbf{r}')$, and the scattered electron wavefunction $\psi_{mi}(\mathbf{r})$, in the angular momentum basis as follows,

$$\varphi_i(\mathbf{r}) = \sum_{l,l'=0}^{\infty} \sqrt{\frac{\kappa_l}{2\pi}} i^l J_l(kr) \delta_{ll'} \Theta_l(\phi_r) \Theta_{l'}(\phi_k), \tag{24}$$

and

$$\psi_{mi}(\mathbf{r}) = \sum_{l,l'=0}^{\infty} \psi_{mi}^{l,l'}(k,r) \Theta_l(\phi_r) \Theta_{l'}(\phi_k), \tag{25}$$

where ϕ_r and ϕ_k are the variables due to expansion on the position $\mathbf{r}$ and momentum $\mathbf{k}$, respectively. The expansion on the Green's function yields the following expression,

$$G^{(0)}(\mathbf{k}_m, \mathbf{r}, \mathbf{r}') = -\frac{i\pi}{2} \sum_{l=0}^{\infty} \sqrt{\frac{\kappa_l}{2\pi}} J_l(k_m r_<) H_l^{(1)}(k_m r_>) \Theta_l(\phi_r) \Theta_l(\phi_{r'}), \tag{26}$$

where $k = k_i$, $r_< = \min(r, r')$, $r_> = \max(r, r')$, $J_l(k_m r)$ ($Y_l(k_m r)$) is the Bessel (Neumann) function and $H_l^{(1)}(k_m r) = J_l(k_m r) + iY_l(k_m r)$ is the Hankel function (Morse & Feshbach, 1953). Using the partial wave expansion the Lippmann-Schwinger equation can be reduced to a set of radial equations. The radial Lippmann-Schwinger equation corresponding to Eq. (12) is given by,

$$\psi_{mi}^{l,l'}(k,r) = \sqrt{\frac{\kappa_l}{2\pi}} i^l J_l(kr) \delta_{ll'} \delta_{mi} + \sum_{l''=0}^{\infty} \sum_{n=0}^{\infty} \int_0^{\infty} r' dr' g_0^l(k_m, r, r') V_{mn}^{l,l''}(r') \psi_{ni}^{l'',l'}(r'), \tag{27}$$

where

$$g_0^l(k_m, r, r') = \frac{-i\pi}{2} \sqrt{\frac{\kappa_l}{2\pi}} J_l(k_m r_<) H_l^{(1)}(k_m r_>) \tag{28}$$

and

$$V_{mn}^{l,l''}(r') = \int_0^{2\pi} d\phi_{r'}\Theta_l(\phi_{r'})V_{mn}(\mathbf{r}')\Theta_{l''}(\phi_{r'}). \tag{29}$$

We see that, when the partial wave method is used, there is a change in the continuum variable ϕ to a partial wave l. Consequently, the wave function $\psi_{mi}(\mathbf{r})$ becomes a matrix function with elements $\psi_{mi}^{l,l'}(k,r)$.

The partial wave expansion for the exchange potential is a little subtle due to its non-locality. Here, we show some details about how the partial wave expansion is applied in this case. We take as an example the exchange potential which couples the channels n and m for a single electron spin-orbital α [see Eq. (11)],

$$V_{mn}^{\text{ex}}(\mathbf{r})\psi_{ni}(\mathbf{r}) = -\frac{e^2}{\epsilon_0^*}\zeta_\alpha^n(\mathbf{r})\int d\mathbf{r}_1\zeta_\alpha^{m*}(\mathbf{r}_1)\frac{1}{|\mathbf{r}-\mathbf{r}_1|}\psi_{ni}(\mathbf{r}_1). \tag{30}$$

The partial wave expansion of the spin-orbital function is given by

$$\zeta_\alpha^n(\mathbf{r}) = \sum_{l=0}^{\infty}\zeta_{n\alpha}^l(r)\Theta_l(\phi_r). \tag{31}$$

The product of two different functions can also be expanded in the angular momentum basis as follows,

$$\psi_{ni}(\mathbf{r})\zeta_\alpha^{m*}(\mathbf{r}) = \sum_{l,l'}\Pi_{ni;m\alpha}^{l,l'}(r)\Theta_l(\phi_r)\Theta_{l'}(\phi_k), \tag{32}$$

where

$$\Pi_{ni;m\alpha}^{l,l'}(r) = \sum_{\lambda,\lambda'}\frac{\psi_{ni}^{\lambda,l'}(k,r)\zeta_{m\alpha}^{\lambda'*}(r)}{2\sqrt{2\pi}}\sqrt{\frac{\kappa_\lambda\kappa_{\lambda'}}{\kappa_l}}\left(\delta_{l,\lambda+\lambda'}+\delta_{l,|\lambda-\lambda'|}\right). \tag{33}$$

Using the above relation, we obtain Eq. (30) in the partial wave expansion form,

$$V_{mn}^{\text{ex}}(\mathbf{r})\psi_{ni}(\mathbf{r}) = -\frac{e^2}{\epsilon_0^*}\zeta_\alpha^n(\mathbf{r})\sum_{l,l'}\Theta_l(\phi_r)\Theta_{l'}(\phi_k) \tag{34}$$
$$\times\int_0^{\infty} r_1 dr_1\Pi_{ni;m\alpha}^{l,l'}(r_1)\int_0^{2\pi}\frac{\Theta_l(\theta)d\theta}{\sqrt{r^2+r_1^2-2rr_1\cos(\theta)}},$$

where $\theta = \phi_r - \phi_{r_1}$. To solve the angular integral we use the generating function of the Legendre Polynomials (Morse & Feshbach, 1953),

$$\frac{1}{\sqrt{r^2+r_1^2-2rr_1\cos(\theta)}} = \sum_{j=0}^{\infty}\frac{r_<^j}{r_>^{j+1}}P_j(\cos\theta), \tag{35}$$

where $r_< = \min(r,r_1)$, $r_> = \max(r,r_1)$ and $P_j(\cos\theta)$ are the Legendre Polynomials. Thus the angular integral that we need to solve is

$$c_{l,j} = \int_0^{2\pi} d\theta\Theta_l(\theta)P_j(\cos\theta). \tag{36}$$

Substituting the Eqs. (35) and (36) into Eq. (34) we finally obtain the exchange potential

$$V_{mn}^{\mathrm{ex}}(\mathbf{r})\psi_{ni}(\mathbf{r}) = -\frac{e^2}{\epsilon_0^*}\zeta_\alpha^n(\mathbf{r})\sum_{l,l'}\Theta_l(\phi_r)\Theta_{l'}(\phi_k)\sum_{j=0}^{\infty}\int_0^\infty r_1 dr_1 \Pi_{ni;m\alpha}^{l,l'}(r_1)c_{l,j}\frac{r_<^j}{r_>^{j+1}}. \tag{37}$$

In the numerical calculations, we firstly evaluate the coefficients $c_{l,j}$ given by Eq. (36). Then the integration on r_1 in Eq. (37) is performed for each iteration in the MCF. Finally we multiply the result by $-\frac{e^2}{\epsilon_0^*}\zeta_\alpha^n(\mathbf{r})$.

Within the one-channel approximation ($i = m = n = 0$), the calculations can be further simplified by using the concept of phase shift. Considering a central potential $V(r)$ ($l = l' = l''$), Eq. (27) becomes

$$\psi^l(k,r) = \sqrt{\frac{\kappa_l}{2\pi}}i^l J_l(kr) + \int_0^\infty r'dr' g_0^l(k,r,r')V(r')\psi^l(k,r') \tag{38}$$

where $\psi^l(k,r) = \psi_{00}^{l,l}(k,r)$. To define the phase-shift we write the asymptotic form of the above equation as

$$\psi^l(k,r) \underset{r\to\infty}{\longrightarrow} A_l\sqrt{\frac{1}{kr}}\cos(kr - \frac{l\pi}{2} - \frac{\pi}{4} - \Delta_l), \tag{39}$$

where Δ_l is the phase-shift. Comparing the coefficients of e^{ikr} and e^{-ikr} of Eq. (39) with the asymptotic form of Eq. (38) one can obtain the following relations

$$A_l = 2\sqrt{\frac{\kappa_l}{\pi}}i^l e^{i\Delta_l}, \tag{40}$$

and

$$e^{i\Delta_l}\sin\Delta_l = \frac{-\pi}{2i^l}\int_0^\infty r'dr' J_l(kr')V(r')\psi^l(r'). \tag{41}$$

On the other hand, from the definition of the scattering amplitude in Eq. (19), we can express the scattering amplitude f_{k_0,k_0} in terms of the phase-shift (Adhikari, 1986) Δ_l,

$$f_{k_0,k_0}(\theta) = 2\sum_{l=0}^{\infty}\sqrt{\frac{\kappa_l}{\pi}}e^{i\Delta_l}\sin\Delta_l\Theta_l(\theta). \tag{42}$$

The corresponding DCS is $\sigma_{00}(\theta) = \left|f_{k_0,k_0}(\theta)\right|^2/k$ and the ICS is given by

$$\Gamma_{00} = \frac{4}{k}\sum_{l=0}^{\infty}\kappa_l\sin^2\Delta_l. \tag{43}$$

4. Quantum dot confined in a quasi-one-dimensional structure

In this section, we study the electron scattering through a QD confined in a quasi-one-dimensional structure. The quantum dot is considered to be confined in the y-direction and the incident (scattered) electron moves in the x-direction. Far from the QD, the electron is free to propagate in the x-direction. In this limit, the Schrödinger equation in the y-direction is given by

$$\left[-\frac{\hbar^2}{2m^*}\frac{d^2}{dy^2} + V_c(y)\right]\chi_n(y) = \varepsilon_n\chi_n(y). \tag{44}$$

We choose the confining potential as being parabolic $V_c(y) = \frac{1}{2}m^*\omega_y^2 y^2$. The solution of equation (44) for this potential is given by (Merzbacher, 1970):

$$\chi_n(y) = \frac{e^{-\frac{y^2}{2l_y^2}}}{(\pi l_y^2)^{1/4}} \frac{H_n(\frac{y}{l_y})}{\sqrt{2^n!}} \tag{45}$$

where H_n are the Hermite's polinomials (Morse & Feshbach, 1953), $\varepsilon_n = \hbar\omega_y(n+1/2)$, and $l_y = \sqrt{\hbar/m^*\omega_y}$. The eigenfunctions $\chi_n(y)$ are also called transversal modes.

As the basis composed of $\chi_n(y)$ is complete, we are able to expand the wave-function in such a system on this basis,

$$\Psi_i(\mathbf{r}) = \sum_{n=0}^{\infty} \chi_n(y)\psi_{ni}(x), \tag{46}$$

where i refers to the incident wave-vector. By introducing this result into the Schrödinger equation, multiplying it by $\chi_m^*(y)$, and integrating in the y-direction, we find the following coupled equations:

$$\left(\frac{\hbar^2}{2m^*}\frac{d^2}{dx^2} + \frac{\hbar^2 k_n^2}{2m^*}\right)\psi_{ni}(x) = \sum_{m=0}^{\infty} V_{m,n}(x)\psi_{mi}(x), \tag{47}$$

where $\frac{\hbar^2 k^2}{2m^*} = \frac{\hbar^2 k_n^2}{2m^*} + \varepsilon_n$ and

$$V_{m,n}(x) = \int dy \chi_m^*(y) V(\mathbf{r}) \chi_n(y). \tag{48}$$

The Green's function is defined as being the solution of the equation:

$$\left(\frac{\hbar^2}{2m^*}\frac{d^2}{dx^2} + \frac{\hbar^2 k_n^2}{2m^*}\right) G_n(x,x') = \delta(x-x'), \tag{49}$$

which allows to rewrite the solution of Eq. (47) as a Lippmann-Schwinger equation in one-dimension,

$$\psi_{ni}(x) = \varphi_n(x) + \frac{2m^*}{\hbar^2}\sum_{m=0}^{\infty}\int dx' G_n(x,x') V_{m,n}(x')\psi_{mi}(x'), \tag{50}$$

where $\varphi_n(x) = exp(ik_n x)\delta_{n,i}/\sqrt{k_n}$. The Green's function to each sub-band in a Q1D system is equal to:

$$G_n(x,x') = \frac{-i}{2k_n} e^{ik_n|x-x'|}. \tag{51}$$

Because the energy of the incident electron is $\frac{\hbar^2 k^2}{2m^*} = \frac{\hbar^2 k_n^2}{2m^*} + \varepsilon_n$, there is the possibility of $\varepsilon_n > \frac{\hbar^2 k^2}{2m^*}$ and of k_n being a pure imaginary number. In such a situation, we must replace k_n by $i|k_n|$ in Eq. (50) and the eigenfunctions $\psi_{ni}(x)$ are not localized anymore.

Taking the limit $x \to \infty$ in Eq. (50), we obtain:

$$\psi_{ni}(x) \underset{x\to\infty}{\longrightarrow} \frac{e^{ik_n x}}{\sqrt{k_n}}\left[\delta_{ni} + \frac{m^*}{i\hbar^2}\sum_{m=0}^{\infty}\int dx' \frac{e^{-ik_n x'}}{\sqrt{k_n}} V_{m,n}(x')\psi_{mi}(x')\right]. \tag{52}$$

The scattering matrix T can be found through the following result

$$T_{ni} = \sum_{m=0}^{\infty} \int_{-\infty}^{\infty} dx' \frac{e^{-ik_n x'}}{\sqrt{k_n}} V_{m,n}(x')\psi_{mi}(x'). \tag{53}$$

Another quantity that we can obtain is transmission probability t_{ni}, which by definition satisfies the following equation (Vargiamidis et al., 2003):

$$\Psi_i(\mathbf{r}) \underset{x\to\infty}{\longrightarrow} \sum_{n=0}^{\infty} t_{ni} \frac{e^{ik_n x}}{\sqrt{k_n}} \chi_n(y). \tag{54}$$

Multiplying Eq. (52) by $\chi_n(y)$, then adding $n = 0$ to ∞, and comparing the resulting equation to Eq. (54) we obtain the following expression for t_{ni}:

$$t_{ni} = \delta_{ni} + \frac{m^*}{i\hbar^2} \sum_{m=0}^{\infty} \int_{-\infty}^{\infty} dx' \frac{e^{-ik_n x'}}{\sqrt{k_n}} V_{m,n}(x')\psi_m(x'). \tag{55}$$

So we can relate the matrix T with the scattering transmission probability t_{ni} by:

$$t_{ni} = \delta_{ni} + \frac{m^*}{i\hbar^2} T_{ni}. \tag{56}$$

These quantities are useful to determine the conductance (Fisher & Lee, 1981; Imry & Landauer, 1999). In the Q1D system with multiple scattering channels, the conductance can be obtained by using the Landauer-Büttiker equation (Buttiker et al., 1985; Landauer, 1957; 1970; 1975),

$$G = \frac{e^2}{\pi\hbar} Tr(tt^{\dagger}), \tag{57}$$

where t is the matrix whose elements are exactly given by Eq. (55).

5. Applications and numerical results

In the previous sections, we presented a theoretical model that describes the quantum scattering through a quantum dot with N-electrons confined. However, we apply this model to the case where only one electron is confined in the quantum dot. Although this is the simplest case, it reveals basic information for a more complicated system. In this section, we describe the details of this particular case considering the elastic and inelastic scattering in the 2D system in sub-sections 5.1 and 5.2, respectively. The scattering through a confined QD in the Q1D system will be discussed in sub-section 5.3.

5.1 Elastic scattering

Here we describe in details how the elastic scattering can be accounted for. To do so, we start by considering the electron in the ground state of energy ε_1. The total Hamiltonian for this system (incident electron + confined electron) is given by:

$$H(\mathbf{r}_1, \mathbf{r}_2) = \frac{-\hbar^2\nabla_2^2}{2m^*} + V_{QD}(\mathbf{r}_2) + H_{QD}(\mathbf{r}_1) + V(\mathbf{r}_1, \mathbf{r}_2) \tag{58}$$

where H_{QD} is the QD Hamiltonian and $V(\mathbf{r}_1, \mathbf{r}_2)$ is the Coulomb interaction potential between the pair of electrons. The total wave function should be written as linear combination of

Slater determinants, as shown in sub-section 2.2. There are four possible combinations for two electrons,

$$|\overline{\Phi^1}, \overline{\psi_{11}}> = \frac{1}{\sqrt{2}} \begin{vmatrix} \Phi^1(\mathbf{r}_1)\beta(1) & \psi_{11}(\mathbf{r}_1)\beta(1) \\ \Phi^1(\mathbf{r}_2)\beta(2) & \psi_{11}(\mathbf{r}_2)\beta(2) \end{vmatrix}, \tag{59}$$

$$|\Phi^1, \psi_{11}> = \frac{1}{\sqrt{2}} \begin{vmatrix} \Phi^1(\mathbf{r}_1)\alpha(1) & \psi_{11}(\mathbf{r}_1)\alpha(1) \\ \Phi^1(\mathbf{r}_2)\alpha(2) & \psi_{11}(\mathbf{r}_2)\alpha(2) \end{vmatrix}, \tag{60}$$

$$|\Phi^1, \overline{\psi_{11}}> = \frac{1}{\sqrt{2}} \begin{vmatrix} \Phi^1(\mathbf{r}_1)\alpha(1) & \psi_{11}(\mathbf{r}_1)\beta(1) \\ \Phi^1(\mathbf{r}_2)\alpha(2) & \psi_{11}(\mathbf{r}_2)\beta(2) \end{vmatrix}, \tag{61}$$

$$|\psi_{11}, \overline{\Phi^1}> = \frac{1}{\sqrt{2}} \begin{vmatrix} \psi_{11}(\mathbf{r}_1)\alpha(1) & \Phi^1(\mathbf{r}_1)\beta(1) \\ \psi_{11}(\mathbf{r}_2)\alpha(1) & \Phi^1(\mathbf{r}_2)\beta(2) \end{vmatrix}, \tag{62}$$

where $\Phi^1(\mathbf{r})$ is the wave function of the confined electron in the QD, $\alpha(i)$ and $\beta(i)$ correspond to spin-up ($\uparrow$) and spin-down ($\downarrow$), respectively. The index (i) denotes to which electron the spin refers to.

Because the total Hamiltonian (Eq. (58)) commutes with the total spin operator (S^2) and its component in the z-direction (S_z), the Hamiltonian eigenfunctions must be eigenfunctions of both S_z and S^2. The first two determinants of Slater in equations (59 and 60) are eigenfunctions of S_z and S^2, but the equations (61 and 62) are not eigenfunctions of S^2. Thus, we have to construct linear combinations between these Slater determinants (Eqs. (61 and 62)) in order to obtain eigenfunctions of S_z and S^2. These combinations can be written as follows:

$$\begin{aligned} |\Psi^s> &= \frac{1}{\sqrt{2}}\left[|\Phi^1, \overline{\psi_{11}}> + |\psi_{11}, \overline{\Phi^1}>\right] = \\ &= \frac{1}{\sqrt{2}}\left[\psi_{11}(\mathbf{r}_1)\Phi^1(\mathbf{r}_2) + \psi_{11}(\mathbf{r}_2)\Phi^1(\mathbf{r}_1)\right]\left(\frac{|\downarrow,\uparrow> - |\uparrow,\downarrow>}{\sqrt{2}}\right) \end{aligned} \tag{63}$$

and

$$\begin{aligned} |\Psi^t> &= \frac{1}{\sqrt{2}}\left[|\Phi^1, \overline{\psi_{11}}> - |\psi_{11}, \overline{\Phi^1}>\right] = \\ &= \frac{1}{\sqrt{2}}\left[\psi_{11}(\mathbf{r}_1)\Phi^1(\mathbf{r}_2) - \psi_{11}(\mathbf{r}_2)\Phi^1(\mathbf{r}_1)\right]\left(\frac{|\downarrow,\uparrow> + |\uparrow,\downarrow>}{\sqrt{2}}\right). \end{aligned} \tag{64}$$

Equation (63) corresponds to the wave function of the singlet state and Equation (59, 60 and 64) correspond to wave functions of the triplet states. Since the Hamiltonian (Eq. (58)) does not have a explicit spin-dependent potential, the state of total spin is conserved before and after the collision. In such a way, the total wave function of the system (incident electron + confined electron) can be written as:

$$\Psi(\mathbf{r}_1, \mathbf{r}_2) = \Phi^1(\mathbf{r}_1)\psi_{11}(\mathbf{r}_2) \pm \Phi^1(\mathbf{r}_2)\psi_{11}(\mathbf{r}_1), \tag{65}$$

where the positive (negative) sign refers to the spin singlet (triplet) state. In order to determine the potential for the scattered electron, we have to calculate the following equation:

$$< \Phi^1(\mathbf{r}_1)|H(\mathbf{r}_1, \mathbf{r}_2)|\Psi(\mathbf{r}_1, \mathbf{r}_2) >= E < \Phi^1(\mathbf{r}_1)|\Psi(\mathbf{r}_1, \mathbf{r}_2) >, \tag{66}$$

where

$$E = \varepsilon_1 + \frac{\hbar^2 k_1^2}{2m^*}. \tag{67}$$

The left hand side of Eq. (66) can be rewritten as:

$$< \Phi^1|H|\Psi > = < \Phi^1|H^1_{QD}|\Psi > + < \Phi^1|H^2_{QD}|\Psi > + < \Phi^1|V^{1,2}|\Psi >, \tag{68}$$

where the superscript is related to each electron the operator is operating on. The first term of Equation (68) is equal to:

$$< \Phi^1|H^1_{QD}|\Psi > = \varepsilon_1 \left[< \Phi^1|\Phi^1 > \psi_{11} \pm < \Phi^1|\psi_{11} > \Phi^1 \right]. \tag{69}$$

While the second term of Eq. (68) is given by

$$< \Phi^1|H^2_{QD}|\Psi > = < \Phi^1|\Phi^1 > H^2_{QD}\psi_{11} \pm \varepsilon_1 < \Phi^1|\Phi^1 > \psi_{11}. \tag{70}$$

The third term of Eq. (68) can be written as:

$$< \Phi^1|V^{1,2}|\Psi > = < \Phi^1|V^{1,2}|\Phi^1 > \psi_{11} \pm < \Phi^1|V^{1,2}|\psi_{11} > \Phi^1. \tag{71}$$

By substituting Eqs.(69, 70, and 71) into Eq. (68), we obtain the following result:

$$\begin{aligned} &(H^2_{QD} - \frac{\hbar^2 k_1^2}{2m^*})\psi_{11} + < \Phi^1|V^{1,2}|\Phi^1 > \psi_{11} \pm \\ &\pm \left(< \Phi^1|V^{1,2}|\psi_{11} > \Phi^1 + (\varepsilon_1 - \frac{\hbar^2 k_1^2}{2m^*}) < \Phi^1|\psi_{11} > \Phi^1 \right) = 0. \end{aligned} \tag{72}$$

The previous equation can be further simplified as

$$-\frac{\hbar^2}{2m^*}(\nabla^2 + k_1^2)\psi_{11}(\mathbf{r}) + \left[V^{\mathrm{st}}(\mathbf{r}) \pm V^{\mathrm{ex}}(\mathbf{r}) \right] \psi_{11}(\mathbf{r}) = 0, \tag{73}$$

where

$$V^{\mathrm{st}}(\mathbf{r}) = < \Phi^1|V^{1,2}|\Phi^1 > + V_{QD}(\mathbf{r}), \tag{74}$$

and

$$V^{\mathrm{ex}}_{11}(\mathbf{r})\psi_{11}(\mathbf{r}) = \Phi^1(\mathbf{r}) \left[< \Phi^1|V^{1,2}|\psi_{11} > + (\varepsilon_1 - \frac{\hbar^2 k_1^2}{2m^*}) < \Phi^1|\psi_{11} > \right]. \tag{75}$$

Finally, the Lippmann-Schwinger equation corresponding to Eq. (73) is given by

$$\psi_{11}(\mathbf{r}) = \varphi_1(\mathbf{r}) + \int d\mathbf{r}' G^{(0)}\mathbf{k}_1, \mathbf{r}, \mathbf{r}') \left[V^{st}(\mathbf{r}') \pm V^{ex}(\mathbf{r}') \right] \psi_{11}(\mathbf{r}'), \tag{76}$$

which can be numerically solved by the method of continuous fractions.

From Eq. (76), we observe that the exchange potential is different when the two electrons form a singlet spin state (plus sign) or triplet spin state (minus sign). In order to calculate the spin-dependent scattering, we have to calculate separately the scattering considering the singlet state and the triplet state. Moreover, the cross sections are given by

$$\sigma^s_{11}(\theta) = \frac{1}{k_1}|f^s_{11}(\theta)|^2, \tag{77}$$

for the singlet state, and

$$\sigma_{11}^{t}(\theta) = \frac{1}{k_1}|f_{11}^{t}(\theta)|^2, \tag{78}$$

for the triplet state, where

$$f_{11}^{s}(\theta) = -\frac{1}{4}\sqrt{\frac{2}{\pi}}\int d^2\mathbf{r}' e^{-i\mathbf{k}_1'.\mathbf{r}'}[V^{st}(\mathbf{r}') + V^{ex}(\mathbf{r}')]\psi_{11}(\mathbf{r}'), \tag{79}$$

$$f_{11}^{t}(\theta) = -\frac{1}{4}\sqrt{\frac{2}{\pi}}\int d^2\mathbf{r}' e^{-i\mathbf{k}_1'.\mathbf{r}'}[V^{st}(\mathbf{r}') - V^{ex}(\mathbf{r}')]\psi_{11}(\mathbf{r}'). \tag{80}$$

A quantity that we can obtain is the spin-unpolarized cross section (su-ICS), which is given by the statistical average of possible configurations, *i.e.*,

$$\sigma_{11}^{su}(\theta) = \frac{1}{4k_1}(|f_{11}^{s}(\theta)|^2 + 3|f_{11}^{t}(\theta)|^2) \tag{81}$$

where the factor three that multiplies the squared modulus of the scattering amplitude of the triplet state is due to the existence of three different triplet states, which are scattered with equal probability. Another quantity that we can extract from the calculation is the spin-flip (sf) cross-section (da Paixão et al., 1996; Hegemann et al., 1991), which measures the probability of the incident electron changes its spin after being scattered,

$$\sigma_{11}^{sf}(\theta) = \frac{1}{4k_1}|f_{11}^{t}(\theta) - f_{11}^{s}(\theta)|^2 \tag{82}$$

In the last expression the factor three does not appear because only one of the triplet states can change its spin $\frac{1}{\sqrt{2}}(|\uparrow\downarrow> + |\downarrow\uparrow>)$.

5.2 Multi-channel scattering

A very important process that we can study by using the previous formalism is the multi-channel scattering, which reveals the probability of an incident electron to promote an excitation or the decay of electrons within the quantum dot. A priori the number of channels of excitation and decay are infinite, but obviously when doing calculations, this number must be truncated. In the case of the parabolic potential of a 2D quantum dot, the first excited energy level is doubly degenerate with an angular momentum $l = \pm 1$. To consider the possible channels of scattering described by the ground state ϵ_1 and by the first excited state ϵ_2, we must consider three coupled channels because the degeneracy of the first excited energy level must be included in the calculation. As the probability of finding the electron in the ground state initially is higher and as the excitation to the first excited state is more likely, we consider only three coupled channels. The calculation details can be found in Ref.(Castelano, 2006). Here we will present some numerical results in Section 6.2.

5.3 Scattering through the QD confined in the Q1D structure

As already discussed in Section 4, the Lippmann-Schwinger equation for the confined QD includes several sub-bands. However, as a first example, we consider only the lowest transversal sub-band due to the confinement in the y-direction. Thus, the L-S equation for a single sub-band is given by:

$$\psi_1(x) = \frac{e^{ik_1 x}}{\sqrt{k_1}} + \frac{2m^*}{\hbar^2} \int dx' G_1(x,x') V_{1,1}(x') \psi_1(x'), \tag{83}$$

where $\frac{\hbar^2 k^2}{2m^*} = \frac{\hbar^2 k_1^2}{2m^*} + \varepsilon_1$. The potential and Green's function for the one sub-band case are respectively given by

$$V_{1,1}(x) = \int dy \chi_1^*(y) V(\mathbf{r}) \chi_1(y), \tag{84}$$

and

$$G_1(x,x') = \frac{-i}{2k_1} e^{ik_1|x-x'|}. \tag{85}$$

We also consider only one confined electron in the QD. The electron wave function of the ground state of the QD can be approximated as

$$\Phi_1(x,y) = \frac{1}{\sqrt{\pi l_x l_y}} \exp\left(-\frac{x^2}{2l_x^2} - \frac{y^2}{2l_y^2}\right), \tag{86}$$

where $l_x = \sqrt{\hbar/m^*\omega_x}$ and $l_y = \sqrt{\hbar/m^*\omega_y}$. Here we consider the QD confining potential in the x-direction as a finite parabolic one with confinement frequency ω_x. The calculation of the exchange potential is more complicated in the Q1D system. However, when we use the wave function of the harmonic oscillator, we can partially obtain analytical expressions for the exchange potential.

Just as in the elastic scattering in the 2D case without extra confinement, the exchange potential is different when the two electrons form a singlet or a triple spin state. So, we have to calculate separately the scattering for the different spin sates. The T matrices can be obtained by the following equations:

$$T_{11}^s = \int_{-\infty}^{\infty} dx' \frac{e^{-ik_1 x'}}{\sqrt{k_1}} \left[V_{1,1}^{st}(x') + V_{1,1}^{ex}(x')\right] \psi_{11}(x'), \tag{87}$$

and

$$T_{11}^t = \int_{-\infty}^{\infty} dx' \frac{e^{-ik_1 x'}}{\sqrt{k_1}} \left[V_{1,1}^{st}(x') - V_{1,1}^{ex}(x')\right] \psi_{11}(x'), \tag{88}$$

where

$$V_{1,1}^{st}(x) = \int dy \chi_1^*(y) V^{st}(\mathbf{r}) \chi_1(y), \tag{89}$$

and

$$V_{1,1}^{ex}(x)\psi_{11}(x) = \int dy \chi_1^*(y) V^{ex}(\mathbf{r}) \chi_1(y) \psi_{11}(x). \tag{90}$$

The static potential $V_{1,1}^{st}(x)$ and the exchange potential $V_{1,1}^{ex}(x)$ are analog to the Eqs. (10) and (11). The transmission probability for the electrons behaving as singlet and triplet states can also be obtained, see Section 4.

6. Results and analysis

In this section, we present the numerical results of scattering of the incident electron through a quantum dot containing just one confined electron. The first step we have to do is to calculate the eigenfunctions and the eigenenergies of the Hamiltonian Eq. (4) for N = 1, which has the following form:

$$\left[-\frac{\hbar^2}{2m^*}\nabla^2 + V_{QD}(r)\right]\Phi^n = H_{QD}\Phi^n = \varepsilon_n\Phi^n, \tag{91}$$

where

$$V_{QD}(r) = \begin{cases} \frac{1}{2}m^*\omega_0^2(r^2 - R_0^2), & r < R_0, \\ 0, & r > R_0. \end{cases} \tag{92}$$

Usually, the QD is modeled by an infinite parabolic potential $V_{QD}(r) = \frac{1}{2}m^*\omega_0^2 r^2$. However,

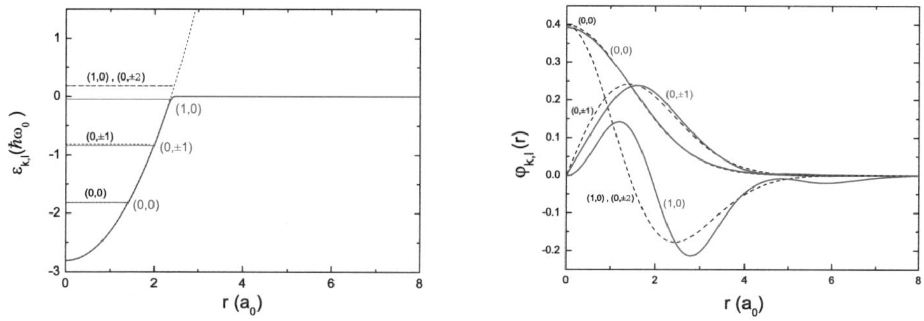

(a) Energy Levels (b) Wave Functions

Fig. 2. (a) Energy levels and (b) wave functions of the infinite (black dashed curves) and finite (blue solid curves) parabolic potential for R_0 =2.37 a_0. The indexes (k, l) denote the radial and angular quantum numbers, respectively.

we are dealing with the scattering processes of an incident electron through the QD, we must employ a potential that goes to zero at the infinity. Therefore, we use the finite parabolic potential (Eq. (92)) as the QD potential. We solve Eq. (91) by expanding the wave function Φ^n in the Fock-Darwin basis (Darwin, 1930; Fock, 1928). The eigenenergies and eigenfunctions are determined by diagonalizing the matrix within the Fock-Darwin basis. Figure 2 (a) compares the energy levels of the infinite and finite parabolic potential. From Figs. 2 (a) and (b) we can see that the ground state $(0, 0)$ and the first excited state $(0, \pm 1)$ of the finite parabolic potential are not very different from the infinite parabolic potential for R_0 =2.37 a_0, where $a_0 = \sqrt{\hbar/m^*\omega_0}$. However, the state $(1, 0)$ is quite different for the two potentials. We also see that the finite parabolic potential with $R_0 = 2.37a_0$ supports three discrete levels only.

6.1 Elastic scattering

The differential cross sections (DCS) for elastic scattering are shown in Figs. 3 (a), (b), and (c) for different incident electron energies $E_0 = \hbar^2 k_0^2/2m^*$= 0.6 meV, 1.7 meV, and 4.2 meV, respectively . In order to understand the role of the electron spin in the scattering, we compare the DCS due to the static potential (blue solid curve) with the spin unpolarized scattering

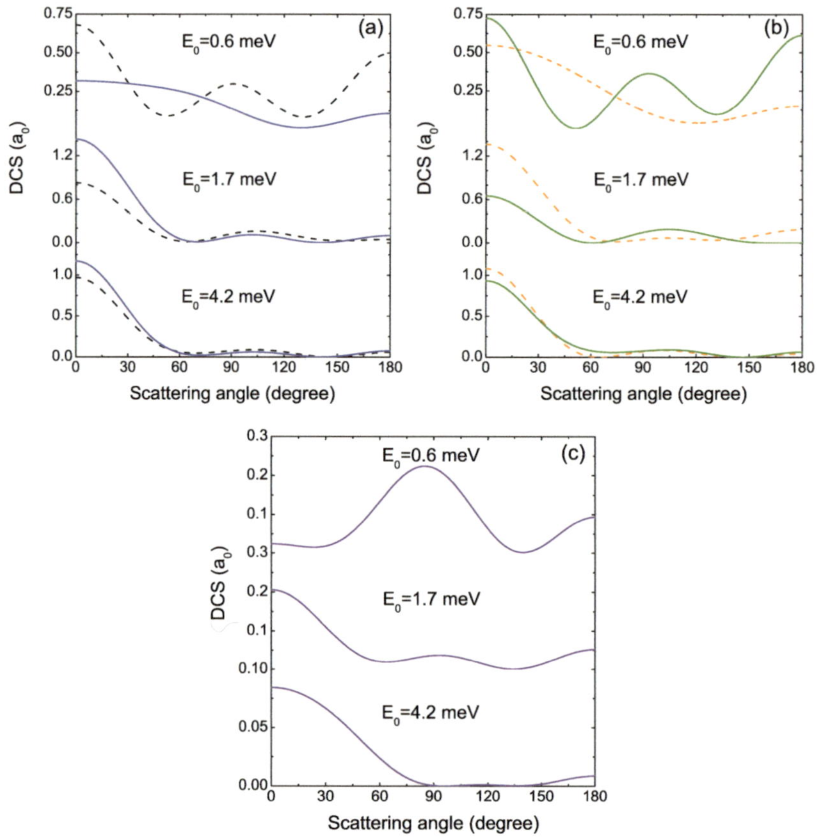

Fig. 3. The elastic DCS's obtained within the one-channel model for electron scattering by the one-electron QD of $\hbar\omega_0 = 5$ meV and $r_0 = 35$ nm. The incident electron energies are indicated in the figures. (a) The spin-unpolarized DCS with (the dashed curves) and without (the solid curves) the exchange potential; (b) The DCS due to the singlet state (the dashed curves) and the triplet state (the solid curves); and (c) The spin-flip DCS.

(black dashed curve) in Figure 3 (a). It is evident that the electron spin is of significant contribution to low energy (E_0=0.6 meV) and/or small scattering angles. The exchange effect on the scattering originates from the two different coupling states between the incident and the QD electron (*i.e.*, the singlet and the triplet states) during the collision. The difference due to the spin states for low-energy and small scattering angles is evident in Figure 3 (b), which compares the DCS of the singlet (orange dashed curve) to that of the triplet (green solid curve) state. For higher energies E_0=1.7 meV and E_0=4.2 meV, the DCS for singlet and for triplet are similar. We observe that the spin-flip DCS in Figure 3 (c) reaches to maximum for angles close to $\pi/2$ for E_0=0.6 meV, while for E_0=1.7 meV and E_0=4.2 meV its maximum appears at angles close to zero.

Figure 4 (a) shows the integral cross section (ICS) for the elastic scattering by static potential (blue solid curve) and spin unpolarized (black dashed curve). Once again, the importance of the dependence on the spin emerges at low energies. The ICS without including the exchange potential is very different from that considering the electron exchange effects for

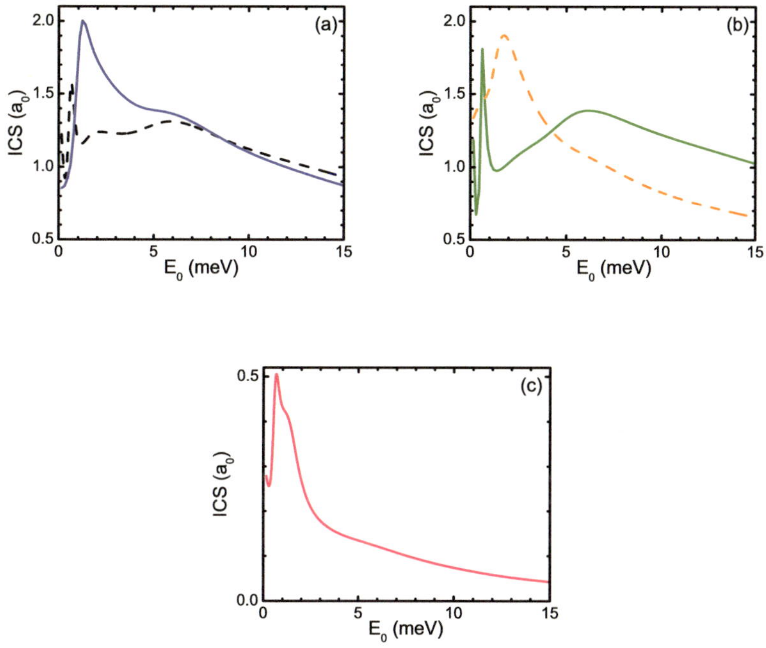

Fig. 4. The elastic ICS as a function of the incident electron energy. (a) The spin-unpolarized ICS with (the dashed curves) and without (the solid curves) the exchange potential; (b) The ICS due to the singlet state (the dashed curves) and the triplet state (the solid curves); and (c) The spin-flip ICS. The parameters for the QD are $\hbar\omega_0 = 5$ meV and $r_0 = 35$ nm

small incident electron energy. However, the ICS is dominated by the static potential at higher energies.

The integral cross section for the singlet (orange dashed curve) and for the triplet (green solid curve) are shown in Figure 4 (b). Note that in the both cases, as well as in the ICS of the static potential (Fig. 4 (a)), a resonant scattering occurs. These resonances can be explained by the analyzing the phase shifts as shown in Figs. 6 (a), (b), and (c). Generally, the phase shifts are functions that smoothly vary as a function of energy. However, under certain circumstances a sudden change of the phase shifts happens in a energy range and a dramatic change in the cross section takes places for these energies, as can be verified by Eq.(43). A physical explanation to this fact can be found when we consider the Schrödinger equation for a central potential, in the basis of angular momentum (equivalent to Eq. (38)),

$$\left[\frac{1}{r}\frac{d}{dr}\left(r\frac{d}{dr}\right)+\frac{l^2}{r^2}+\frac{2m^*}{\hbar^2}V(r)\right]\psi^{l,l}(k,r)+k^2\psi^{l,l}(k,r)=0. \tag{93}$$

We may identify in Eq. (93) an effective potential $V_{ef} = \frac{l^2}{r^2} + \frac{2m^*}{\hbar^2}V(r)$, for each different angular momentum l. Figure 5 shows this effective potential as a function of r for $V(r) = V_{QD}(r)$. By assuming that there is a metastable state with energy E_r, as sketched in Figure 5, one can prove that when the electron energy E_0 reaches E_r, a rapidly varying phase shift

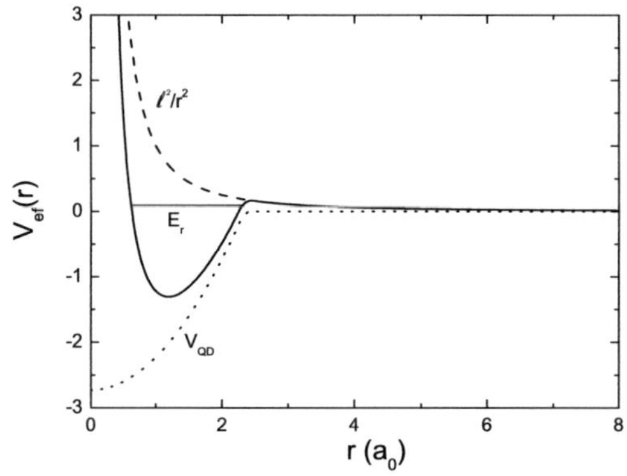

Fig. 5. The effective potential as a function of r is represented by the solid curve. Also, we plot the QD potential (dotted curve) and centrifugal barrier (dashed curve).

occurs and a resonance appears in the ICS (Joachain, 1975). Because this resonance depends on the potential's shape $V(r)$, it is usually called shape resonance.

So, the resonance that appears in the ICS for the static potential (Fig. 4 (a)) with energy $E_0 =$ 1.22 meV corresponds to the rapid fluctuation of the phase shift Δ_2 shown in Figure 6 (a). In this case $\Delta_2 \approx \pi/2$ for the value of the energy $E_0 = 1.22$ meV. We also note the appearance of another broader resonance at $E_0 = 6.14$ meV, which corresponds to the rapid increasing of Δ_3 seen in Figure 6 (a). The singlet resonance (E_0 =1.72 meV) and the triplet resonance (E_0 =0.57 meV) shown in Fig. 4 (b) are also resulting from the rapidly varying Δ_2 plotted in Figs.6 (b) and (c).

In Figure 4 (c) we present the integral elastic spin-flip cross section. The spin-flip cross section depends on the scattering amplitudes of the singlet and the triplet as shown by equation (82). As the ICS modulus (the square of the scattering amplitudes) of singlet and triplet states are very distinct at small energies, the spin-flip cross section is of maximum at the same energy range, as shown in Figure 4 (c).

Figure (7) shows the elastic ICS varying the size of the QD, for the scattering by the static potential (Coulomb without exchange). When the radius R_0 is increased, the potential of QD becomes more negative and the resonance energy E_r (Fig. 5) decreases. Thus, we see that the resonance peak in Fig. (7) shifts to lower energy values when R_0 increases. Therefore, the potential becomes deeper and the metastable state E_r decreases, consequently, the shape resonance shifts to lower energies. The shape resonance may disappear when the radius is further increased. In this situation, the state E_r becomes a real bound state instead of a metastable sate.

6.2 Multi-channel scattering

When we consider the three-channel scattering, we have nine possibilities of scattering, *i.e.*, the incident electron has initially energy $E_0 = \hbar^2 k_i^2/2m^*$ and can be scattered with the energy

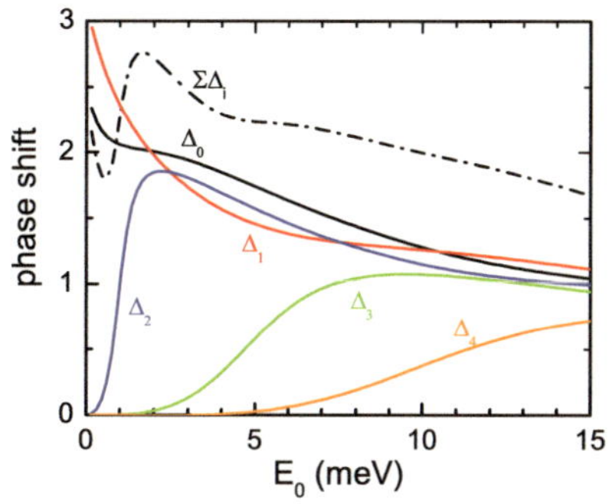

(a) Static potential

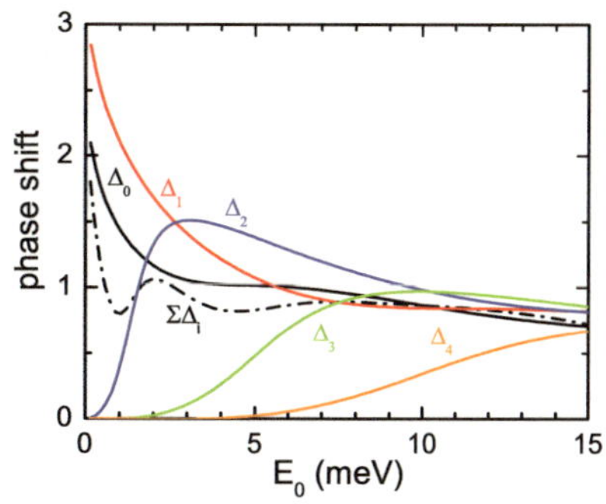

(b) Singlet state

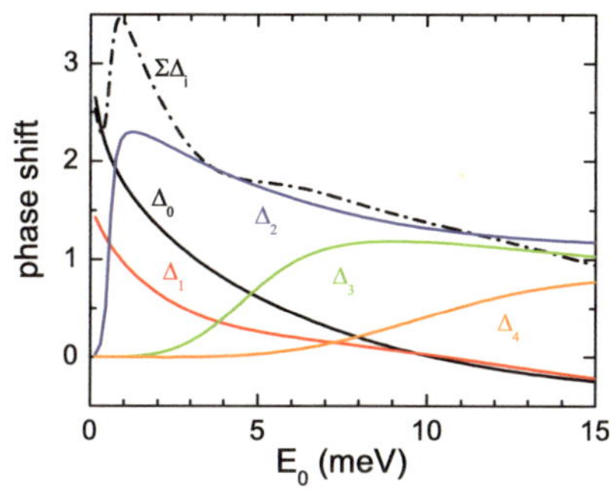

(c) Triplet state.

Fig. 6. The phase shift Δ_l as a function of the incident electron energy for the partial waves with angular momentum l =0, 1, 2, 3, and 4.

$\frac{\hbar^2 k_f^2}{2m} = \frac{\hbar^2 k_i^2}{2m} + \varepsilon_i - \varepsilon_f$, with i and f =1, 2, and 3. If $\varepsilon_f = \varepsilon_i$ the scattering is elastic and if $\varepsilon_f > \varepsilon_i$ there is a excitation. Finally, if $\varepsilon_f < \varepsilon_i$ there is a decay.

In our case, as $\varepsilon_2 = \varepsilon_3$ the probability of exciting or decaying for either of one of these two states is exactly the same. Thus, we calculate the cross section considering elastic and inelastic scattering $\varepsilon_1 \rightarrow \varepsilon_2$. Fig. 8 shows the ICS for the elastic channel (a) and for the excitation channel (b). The black (blue) solid curve represents the spin-unpolarized potential (static), while the dashed curves represent the respective ICS when only one channel is considered. For E_0 ≈7 meV, we notice that the ICS for the excitation channel in Fig. 8 (b) has a maximum, while the elastic channel in Fig. 8(a) exhibits a minimum, which is obvious from the probability current conservation. For E_0 >8 meV, the behavior of the ICS shown in Fig. 8 (a) is very similar to the results considering the elastic scattering (dashed curves).

In Figure 9, the ICS is shown for scattering by three-channels, where the green (orange) solid curve represents the scattering by the potential of the triplet (singlet) and the dashed curves represent their ICS when only one channel is considered. In the case of scattering for the triplet state, we verify that when E_0 >9 meV the elastic scattering (Figure 9 (a)) for three-channels is equal to one-channel and therefore, the probability of excitation is practically null (Figure

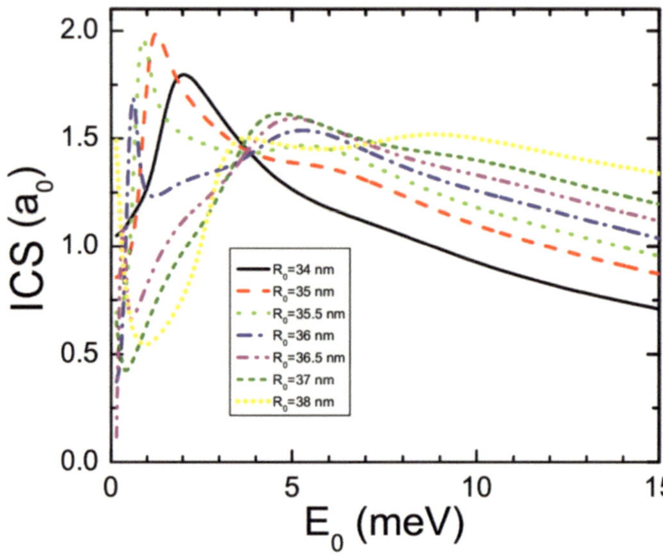

Fig. 7. The elastic ICS as a function of the incident electron energy, considering the static potential for different sizes (R_0) of the QD with $\hbar\omega_0 = 5$ meV.

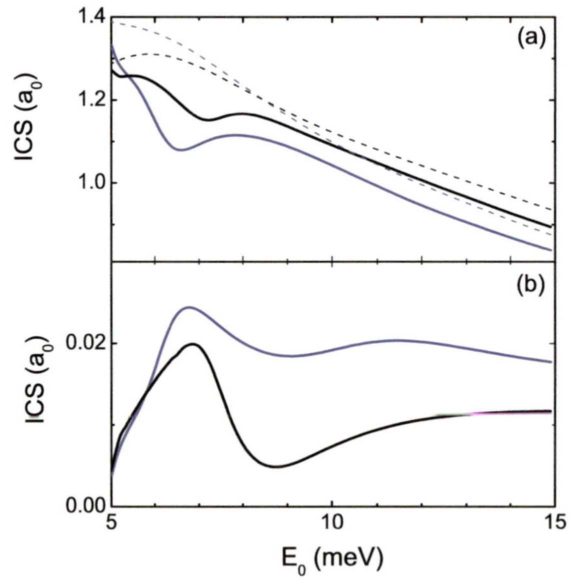

Fig. 8. The ICS for the three-channel scattering as a function of electron energy (R_0 =35 nm and $\hbar\omega_0 = 5$ meV). (a) Elastic channel and (b) excitation channel from the ground state to the first excited state ($l = \pm 1$). Black (blue) curve shows the ICS for the spin-unpolarized (static) scattering. Dashed curves are the respective ICS within the one-channel scattering approximation.

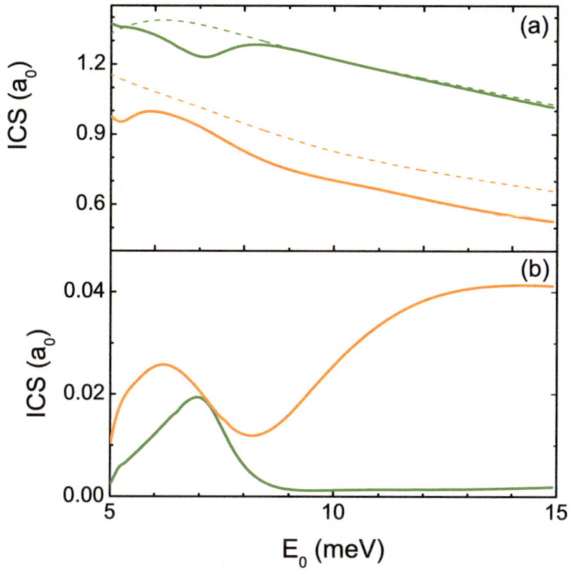

Fig. 9. The same as Fig.8 but now the orange (green) curve shows the ICS for the triplet (singlet) spin state. Dashed curves are the respective ICS within the one-channel scattering approximation.

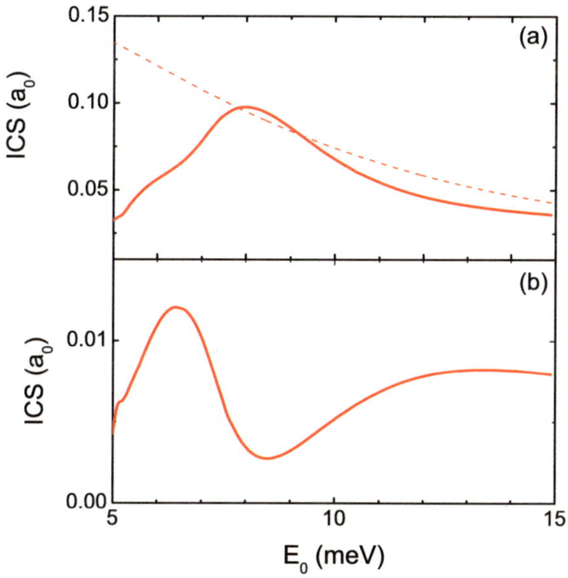

Fig. 10. The same as Fig.8 but now for the spin-flip ICS. The dashed curve is the respective spin-flip ICS within the one-channel approximation.

9 (b)). In the case of the singlet state, the behavior is contrary, *i.e.*, when E_0 >9 meV the excitation probability begins to increase (Fig 9 (b)), thus showing that the scattering may be completely different depending on the spin state of electrons.

In Figure 10, the spin-flip ICS is shown considering three-channels of scattering, where the solid curve represents the scattering from (a) elastic and (b) excitation channel. The dashed curve represents the spin-flip ICS considering only one channel of scattering. The spin-flip ICS for one-channel presents a maximum for $E_0 \approx 8$ meV and a minimum in the same energy range for the excitation channel. We also found that for $E_0 > 9$ meV, the spin-flip cross section of three-channels is similar to the elastic scattering (dashed curve).

6.3 Scattering in the quasi-one-dimensional system

In this section, we apply the MCF to solve the Lippmann-Schwinger equation for the electron scattering through the QD confined in the Q1D structure. The convergency of MCF is very accurate in this case and achieves a precision of 10^{-4} for the transmission probability in approximately 20 interactions. To probe our numerical method, we consider one electron confined in the QD with radius R_0=45 nm. Moreover, two different cases for confined potential in x-direction with $\hbar\omega_x$ =5 meV and $\hbar\omega_x$ =3 meV are tested. The obtained results are shown in Fig. 11. In both cases, the screening length is fixed as $\lambda^{-1} = l_x$ and the confinement frequency the y-direction is set different from that in the x-direction with $\omega_y = 1.7\omega_x$. In Fig. 11 (a) and (b), we plot the transmission probability as a function of the incident electron energy assuming different scattering situations: (i) the QD potential only (black dotted curve), (ii) the static potential (red dash-dotted curve), (ii) the singlet state (orange dashed curve), and (iv) the triplet state (green solid curve). From the results, we see that the confinement potential (or frequency ω_x) of the QD strongly affects the transmission probability through the QD. Furthermore, the Coulomb potential alters considerably the scattering. The electron-electron exchange potential splits the resonant peak into two due to different spin states of the system. It shows that, when the incident electron has anti-parallel (parallel) spin with the confined electron in the QD, the transmission probability is enhanced (suppressed) significantly at low energy. This is a kind of spin filter effect if we could control the spin state of the confined electron.

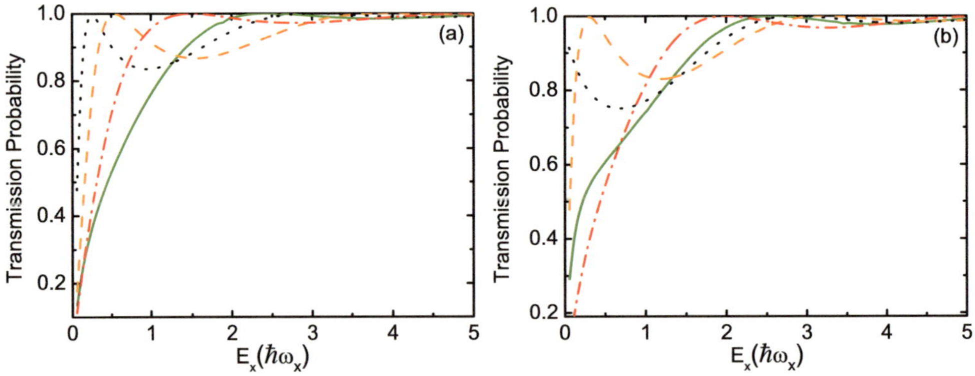

(a) $\hbar\omega_x = 3$ meV (b) $\hbar\omega_x =$ 5meV.

Fig. 11. Transmission probability as a function of the incident electron's energy assuming different scattering situations: only the QD potential (black dotted curve), the static potential (red dash-dotted curve), the singlet state (orange dashed curve), and the triplet state (green solid curve).

7. Conclusion remarks

We presented a theoretical approach to calculate the electron scattering and transport through an N-electron QD embedded in a 2D and a Q1D semiconductor structure. The multichannel L-S equations are solved numerically using the iterative method of continued fractions considering the electron-electron interactions. From this method, we can study the multichannel scattering including the excited states of the QD. The electron transport property due to elastic and inelastic scattering, as well as its dependence on the spin states of the system can be obtained in great precision.

We applied this method to the case where only one electron is confined in the QD. The results indicate a rapid convergency of the numerical method for the electron scattering in both 2D and 1D systems. We found that the electron-electron exchange effects are relevant when the kinetic energy of incident electron is small. For a QD of more electrons, we need firstly to find the eigenstates of the QD with electron-electron interactions. In principal, the scattering processes can be calculated according to the total wave-function defined by Eq. (6).

8. Acknowledgments

This work was supported by FAPESP and CNPq. (Brazil).

9. References

Adhikari, S. K. (1986). *Am. J. Phys.* 54: 362.

Bransden, B. H. & McDowell, M. R. C. (1977). *Phys. Rep.* 30: 207.

Burkard, G., Engel, H. A. & Loss, D. (2000). *Fortschr. Phys.* 48: 965.

Buttiker, M., Imry, Y., Landauer, R. & Pinhas, S. (1985). *Phys. Rev. B* 31: 6207.

Castelano, L. K. (2006). PhD thesis, University of Sao Paulo.

Castelano, L. K., Hai, G.-Q. & Lee, M.-T. (2007a). *Phys. Rev. B* 76: 165306.

Castelano, L. K., Hai, G.-Q. & Lee, M.-T. (2007b). *Phys. Stat. Sol. (c)* 4: 466.

da Paixão, F. J., Lima, M. & McKoy, V. (1996). *Phys. Rev. A* 53: 1400.

Darwin, C. (1930). *Proc. Cambridge Philos. Soc.* 27: 86.

Engel, H.-A. & Loss, D. (2002). *Phys. Rev. B* 65: 2002.

Fisher, D. S. & Lee, P. A. (1981). *Phys. Rev. B* 23: 6851.

Fock, V. (1928). *Z. Phys.* 47: 446.

Fransson, J., Holmstrom, E., Eriksson, O. & Sandalov, I. (2003). *Phys. Rev. B* 67: 205310.

Gundogdu, K., Hall, K. C., Boggess, T. F., Deppe, D. G. & Shchekin, O. B. (2004). *Appl. Phys. Lett.* 84: 2793.

Hegemann, T., Oberste-Vorth, M., Vogts, R. & Hanne, G. F. (1991). *Phys. Rev. Lett.* 66: 2968.

Horacek, J. & Sasakawa, T. (1984). *Phys. Rev. A* 28: 2151.

Imry, Y. & Landauer, R. (1999). *Rev. Mod. Phys.* 71: s306.

Joachain, C. J. (1975). *Quantum collision theory*, North-Holland, Amsterdam.

Konig, J. & Martinek, J. (2003). *Phys. Rev. Lett.* 90: 166602.

Koppens, F. H. L., Buizert, C., Tielrooij, K. J., Vink, I. T., Nowack, K. C., Meunier, T., Kowenhoven, L. P. & Vandersypen, L. M. K. (2006). *Nature* 442: 766.

Landauer, R. (1957). *IBM J. Res.* 1: 223.

Landauer, R. (1970). *Phil. Mag.* 21: 863.

Landauer, R. (1975). *Z. Phys.* B24: 247.

Merzbacher, E. (1970). *Quantum Mechanics*, John Wiley and Sons, Inc., New York.

Morse, P. M. & Feshbach, H. (1953). *Methods of Theoretical Physics*, McGraw-Hill, New York.
Qu, F. Y. & Vasilopoulos, P. (2006). *Phys. Rev. B* 74: 245308.
Sarma, S. D., Fabian, J., Hu, X., & Zutic, I. (2001). *Solid State Commun.* 119: 207.
Seneor, P., Bernand-Mantel, A. & Petroff, F. (2007). *J. Phys.: Condens. Matter* 19: 165222.
Szabo, A. & Ostlund, N. (1982). *Modern Quantum Chemistry*, Macmillan Publishing, New York.
Vargiamidis, V., Valassiades, O. & Kyriakos, D. S. (2003). *Phys. Stat. Sol. (b)* 236: 597.
Wolf, S. A., Awschalom, D. D., Buhrman, R. A., Daughton, J. M., von Molnár, S., Roukes, M. L., Chtchelkanova, A. Y. & Treger, D. M. (2001). *Science* 294: 1488.
Zhang, P., Xue1, Q.-K., Wang, Y. & Xie, X. C. (2002). *Phys. Rev. Lett.* 89: 286803.

17

Coherent Spin Dependent Transport in QD-DTJ Systems

Minjie Ma[1], Mansoor Bin Abdul Jalil[1,2] and Seng Ghee Tan[1,3]
[1]*Computational Nanoelectronics and Nano-Device Laboratory, Electrical and Computer Engineering Department, National University of Singapore*
[2]*Information Storage Materials Laboratory, Department of Electrical and Computer Engineering, National University of Singapore*
[3]*Data Storage Institute, A *STAR (Agency of Science, Technology and Research) Singapore*

1. Introduction

As the dimension of devices reduces to nano-scale regime, the spin-dependent transport (SDT) and spin effects in quantum dot (QD) based systems become significant. These QD based systems have attracted much interest, and can potentially be utilized for spintronic device applications. In this chapter, we consider nano-scale spintronic devices consisting of a QD with a double barrier tunnel junction (QD-DTJ)(schematically shown in Fig. 1). The DTJ couples the QD to two adjacent leads which can be ferromagnetic (FM) or non-magnetic (NM).

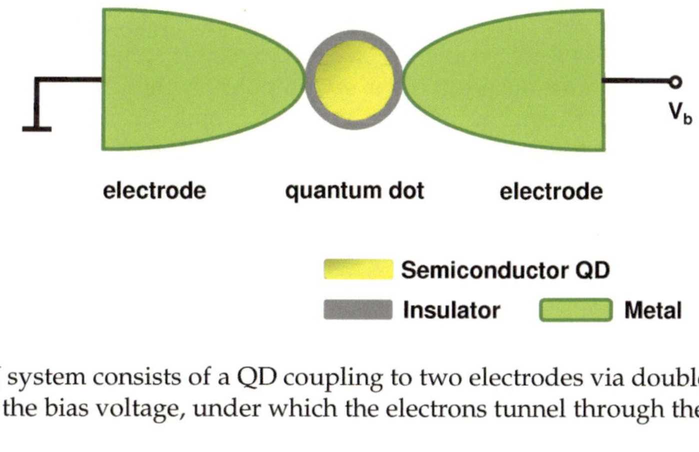

Fig. 1. QD-DTJ system consists of a QD coupling to two electrodes via double tunnel junctions. V_b is the bias voltage, under which the electrons tunnel through the QD one by one.

In a QD-DTJ system, the electron tunneling is affected by the quantized energy levels of the QD, and can thus be referred to as single electron tunneling. The single electron tunneling process becomes spin-dependent when the leads or the QD is a spin polarizer, where the density of states (DOS) for spin-up and spin-down electrons are different. The interplay of SDT with quantum and/or single electron charging effects makes the QD-DTJ systems interesting. In such QD-DTJ systems, it is possible to observe several quantum spin phenomena, such as spin blockade (Shaji et al. (2008)), Coulomb blockade (CB) (Bruus & Flensberg (2004)), cotunneling (Weymann & Barnaś (2007)), tunnel magnetoresistance (TMR) (Rudziński & Barnaś (2001)), spin transfer torque (Mu et al. (2006)) and Kondo effect (Katsura (2007); Lobo et al. (2006); Potok et al. (2007)). The complex spin and charge transport properties of QD-DTJ systems have attracted extensive theoretical (Bao et al. (2008); Braig & Brouwer (2005); Jauho et al. (1994); Kuo & Chang (2007); Ma et al. (2008); Meir & Wingreen (1992); Meir et al. (1991; 1993); Mu et al. (2006); Qi et al. (2008); Qu & Vasilopoulos (2006); Souza et al. (2004); Zhang et al. (2002); Zhu & Balatsky (2002)) and experimental ((Deshmukh & Ralph, 2002; Hamaya et al., 2007; Pasupathy et al., 2004; Potok et al., 2007)) investigations recently. These studies may ultimately lead to the utilization of such devices in diverse applications such as single spin detector (Wabnig & Lovett (2009)) and STM microscopy (Manassen et al. (2001)).

The theoretical study of the SDT through these DTJ systems are mainly based on two approaches, namely the master equation (ME) approach and the Keldysh nonequilibrium Green's function (NEGF) approach. For coherent transport across QD-DTJ devices, quantum transport methods are applied, such as the linear response (Kubo) method applicable for small bias voltage, and its generalization, the NEGF method for arbitrary bias voltage. Since the objective of the study in this Chapter is for device application over a wide voltage range, we focus on the latter. The NEGF method has been employed to analyze various transport properties of QD-DTJ systems, such as TMR, tunneling current (Weymann & Barnaś (2007)) and conductance. These analyses were conducted based on the Anderson model (Meir et al. (1993); Qi et al. (2008)), for collinear or noncollinear (Mu et al. (2006); Sergueev et al. (2002); Weymann & Barnaś (2007)) configurations of the magnetization of the two FM leads, or in the presence of spin-flip scattering in the QD (Lin & D.-S.Chuu (2005); Souza et al. (2004); Zhang et al. (2002)).

In this Chapter, based on the NEGF approach, we study the SDT through two QD-DTJ systems. In Section. 2, the electronic SDT through a single energy-level QD-DTJ is theoretically studied, where the two FM leads enable the electron transport spin-dependent. In the study, we systematically incorporate the effect of the spin-flip (SF) within the QD and the SF during tunneling the junction between the QD and each lead, and consider possible asymmetry between the coupling strengths of the two tunnel junctions. Based on the theoretical model, we first investigate the effects of both types of SF events on the tunneling current and TMR; subsequently, we analyze the effect of coupling asymmetry on the QD's electron occupancies and the charge and spin currents through the system (Ma et al. (2010)).

In Section. 3, we studied the SDT through a QD-DTJ system with finite Zeeman splitting (ZS) in the QD, where the two leads which sandwich the QD are NM. The spin-dependence of the electron transport is induced by the ZS caused by the FM gate attached to the QD. A fully polarized tunneling current is expected through this QD-DTJ system. The charge and spin currents are to be analyzed for the QD-DTJ systems with or without ZS.

2. Single energy level QD

The QD-DTJ device under consideration is shown in Fig. 2. It consists of two FM leads and a central QD in which a single energy level is involved in the electron tunneling process. The SDT through the QD-DTJ is to be theoretically modeled via the Keldysh NEGF approach (Caroli et al. (1971); Meir & Wingreen (1992)). In the transport model, the limit of small correlation energy is assumed, in the case where the energy due to electron-electron interaction in the QD is much smaller than the thermal energy or the separation between the discrete energy levels in the QD (Fransson & Zhu (2008)).

2.1 Theory

For the QD-DTJ device shown in Fig. 2, the full Hamiltonian consists of the lead Hamiltonian H_α, the QD Hamiltonian H_d, and the tunneling Hamiltonian H_t. The explicit form of the Hamiltonian is given by

$$H = \sum_\alpha H_\alpha + H_d + H_t, \tag{1}$$

where

$$H_\alpha = \sum_{k\sigma} \epsilon_{\alpha k\sigma} a^\dagger_{\alpha k\sigma} a_{\alpha k\sigma}, \tag{2}$$

$$H_d = \sum_{\sigma} \epsilon_{\sigma\sigma} a^\dagger_{\sigma} a_{\sigma} + \sum_{\sigma} \epsilon_{\sigma\bar{\sigma}} a^\dagger_{\bar{\sigma}} a_{\sigma}, \tag{3}$$

$$H_t = \sum_{\alpha k\sigma} \left(t_{\alpha k\sigma,\sigma} a^\dagger_{\sigma} a_{\alpha k\sigma} + t^*_{\alpha k\sigma,\sigma} a^\dagger_{\alpha k\sigma} a_{\sigma} \right) + \sum_{\alpha k\sigma} \left(t_{\alpha k\sigma,\bar{\sigma}} a^\dagger_{\bar{\sigma}} a_{\alpha k\sigma} + t^*_{\alpha k\sigma,\bar{\sigma}} a^\dagger_{\alpha k\sigma} a_{\bar{\sigma}} \right). \tag{4}$$

In the above, $\epsilon_{\sigma\sigma}$ is the single energy level in the QD, $\epsilon_{\sigma\bar{\sigma}}$ denotes the coupling energy of the spin-flip within quantum dot (SF-QD) from spin-σ to spin-$\bar{\sigma}$ state, $t_{\alpha k\sigma,\sigma}$ ($t_{\alpha k\sigma,\bar{\sigma}}$) is the coupling between electrons of the same (opposite) spin states in the lead and the QD. $\alpha = \{L, R\}$ is the lead index for the left and right leads, $\sigma = \{\uparrow, \downarrow\}$ stands for up- and down-spin, and k is the momentum, $\epsilon_{\alpha k\sigma}$ represents the energy in the leads. The operators $a^\dagger_\nu$ (a_ν) and $a^\dagger_\sigma$ (a_σ) are the creation (annihilation) operators for the electrons in the leads and the QD, respectively.

2.1.1 Tunneling current and tunnel magnetoresistance

The tunneling current through the QD-DTJ system can be expressed as the rate of change of the occupation number $N = \sum_\sigma a^\dagger_\sigma a_\sigma$ in the QD,

$$I = e\dot{N} = \frac{ie}{\hbar} \langle [H, N] \rangle. \tag{5}$$

Without loss of generality, we can calculate the tunneling current in Eq. (5) by considering the tunneling current I_L through the left junction between the left lead and the QD. Evaluating the commutator in Eq. (5) in terms of creation and annihilation operators gives

$$I = I_L = \frac{ie}{\hbar} \sum_{Lk\sigma,\sigma'} \left(t_{Lk\sigma,\sigma'} \langle a^\dagger_{Lk\sigma} a_{\sigma'} \rangle - t^*_{Lk\sigma,\sigma'} \langle a^\dagger_{\sigma'} a_{Lk\sigma} \rangle \right). \tag{6}$$

In Eq. (6), one may replace the creation and annihilation operators by the lesser Green's functions, which are defined as $G^<_{\sigma',Lk\sigma}(t) = i\langle a^\dagger_{Lk\sigma} a_{\sigma'}(t) \rangle$ and $G^<_{Lk\sigma,\sigma'}(t) = i\langle a^\dagger_{\sigma'} a_{Lk\sigma}(t) \rangle$ (Meir & Wingreen (1992)). Eq. (6) then takes the form of

$$I_L = \frac{e}{\hbar} \sum_{Lk\sigma,\sigma'} \left(t_{Lk\sigma,\sigma'} G^<_{\sigma',Lk\sigma}(t) - t^*_{Lk\sigma,\sigma'} G^<_{Lk\sigma,\sigma'}(t) \right). \tag{7}$$

After performing a Fourier transform on Eq. (7), $G^<_{Lk\sigma,\sigma'}(\epsilon)$ can be expressed in form of the left lead's and QD's Green's functions, under the assumption of non-interacting

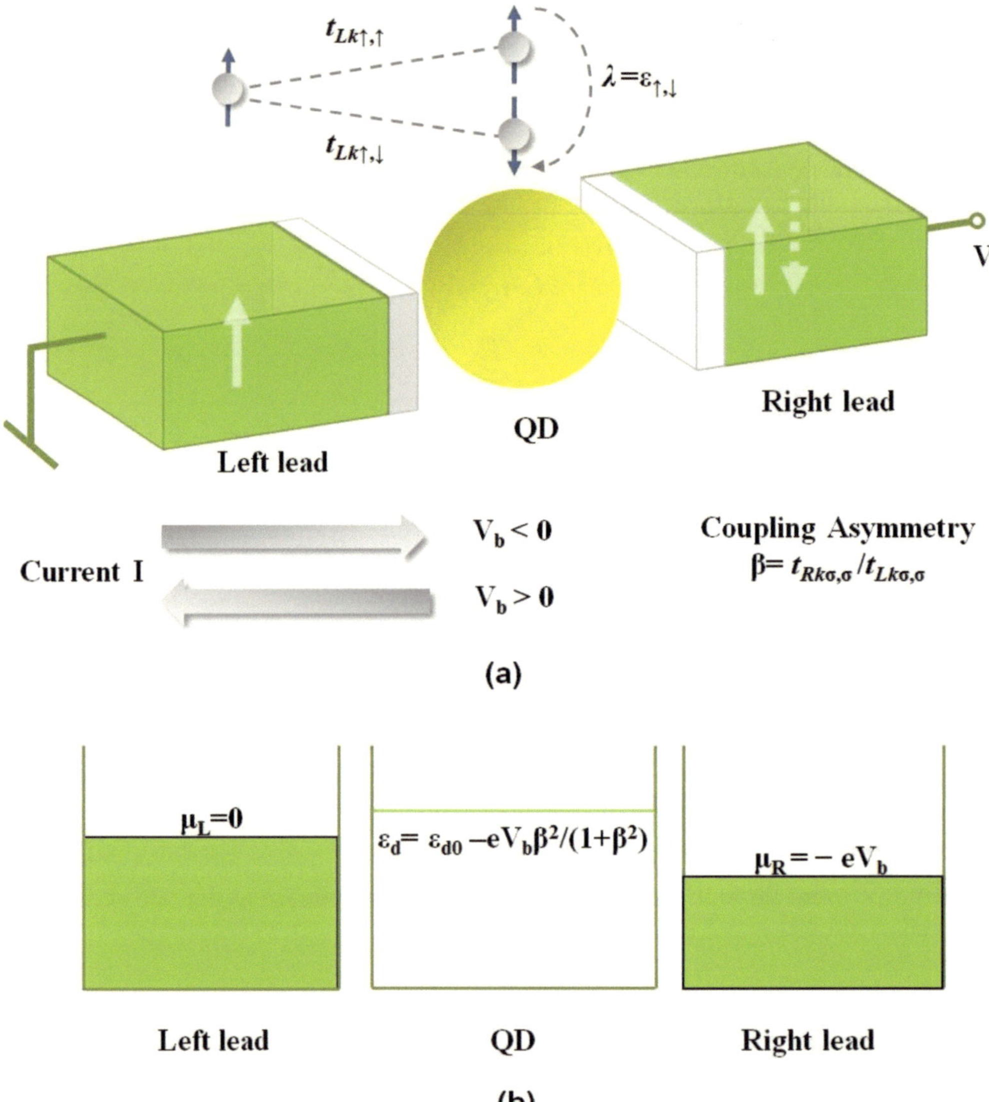

Fig. 2. (a) Schematic diagram of the QD-DTJ structure consisting of a QD sandwiched by two FM leads; (b) the schematic energy diagram for the system in (a). In (a), the arrows in the leads indicate magnetization directions, which can either be in parallel (solid) or antiparallel (dashed) configuration, V_b denotes the bias between the two leads, λ characterizes the strength of the SF-QD, $t_{Lk\uparrow,\downarrow}$ describes the SF-TJ from the up-spin state in left lead and the down-spin state in the QD, $t_{Lk\uparrow,\uparrow}$ shows the coupling between the same electron spin states in left lead and QD, and $\beta = t_{Rk\sigma,\sigma}/t_{Lk\sigma,\sigma}$ represents the coupling asymmetry between the left and right tunneling junctions. In (b), μ_L and μ_R are the chemical potentials of left and right leads respectively, and ϵ_d (ϵ_{d0}) denotes the single energy level of the QD with or without bias voltage.

leads. Taking into account the contour-ordered integration over the time loop, the corresponding Dyson's equations for $G^{<}_{Lk\sigma,\sigma'}(\epsilon)$ can then be obtained (Mahan (1990)), i.e., $G^{<}_{Lk\sigma,\sigma'}(\epsilon) = \sum_{\sigma''} t_{Lk\sigma,\sigma''}[g^{t}_{Lk\sigma,Lk\sigma}(\epsilon)G^{<}_{\sigma'',\sigma'}(\epsilon) - g^{<}_{Lk\sigma,Lk\sigma}(\epsilon)G^{\bar{t}}_{\sigma'',\sigma'}(\epsilon)]$ and $G^{<}_{\sigma',Lk\sigma}(\epsilon) = \sum_{\sigma''} t^{*}_{Lk\sigma,\sigma''}\ [g^{<}_{Lk\sigma,Lk\sigma'}(\epsilon)G^{t}_{\sigma'',\sigma'}(\epsilon) - g^{\bar{t}}_{Lk\sigma,Lk\sigma}(\epsilon)G^{<}_{\sigma''\sigma'}(\epsilon)]$, where $G^{t} = \theta(t)G^{>} + \theta(-t)G^{<}$ and $G^{\bar{t}} = \theta(-t)G^{>} + \theta(t)G^{<}$ are the time-ordered and anti-time-ordered Green's functions respectively, $G^{<}_{\sigma'',\sigma'}(t) = -i\langle a^{\dagger}_{\sigma'}a_{\sigma''}(t)\rangle$, and the g's are the corresponding unperturbed Green's functions of the leads, whose lesser Green's function and greater Green's function are in form of $g^{<}_{Lk\sigma,Lk\sigma}(\epsilon) = i2\pi f_{L\sigma}(\epsilon)\delta(\epsilon - \epsilon_{L\sigma})$ and $g^{>}_{Lk\sigma,Lk\sigma}(\epsilon) = -i2\pi[1 - f_{L\sigma}(\epsilon)]\delta(\epsilon - \epsilon_{L\sigma})$, where $f_{L\sigma}(\epsilon) = (1 + \exp(\frac{\epsilon - \mu_L}{k_B T}))^{-1}$ is the Fermi-Dirac function, μ_L is the chemical potential, $\epsilon_{L\sigma}$ is the energy for electrons with spin σ in the left lead, k_B is the Boltzmann constant and T is the temperature of the device. With this, the current in Eq. (7) can be expressed in terms of the Green's functions wholly of the leads and the QD, i.e.,

$$I_L = \frac{e}{\hbar}\sum_{Lk\sigma,\sigma'}\int_{-\infty}^{\infty}\frac{d\epsilon}{2\pi}\{t_{Lk\sigma,\sigma'}\sum_{\sigma''}t^{*}_{Lk\sigma,\sigma''}[g^{<}_{Lk\sigma,Lk\sigma'}(\epsilon)G^{t}_{\sigma'',\sigma'}(\epsilon) - g^{\bar{t}}_{Lk\sigma,Lk\sigma}(\epsilon)G^{<}_{\sigma''\sigma'}(\epsilon)]$$
$$-t^{*}_{Lk\sigma,\sigma'}\sum_{\sigma''}t_{Lk\sigma,\sigma''}[g^{t}_{Lk\sigma,Lk\sigma}(\epsilon)G^{<}_{\sigma'',\sigma'}(\epsilon) - g^{<}_{Lk\sigma,Lk\sigma}(\epsilon)G^{\bar{t}}_{\sigma'',\sigma'}(\epsilon)]\}. \quad (8)$$

By applying the identities $G^{t} + G^{\bar{t}} = G^{<} + G^{>}$ and $G^{>} - G^{<} = G^{r} - G^{a}$ to Eq. (8), we obtain after some algebra (Mahan (1990)):

$$I_L = \frac{ie}{\hbar}\sum_{Lk\sigma,\sigma'}\int_{-\infty}^{\infty}\frac{d\epsilon}{2\pi}t_{Lk\sigma,\sigma'}|_{\epsilon=\epsilon_v}t^{*}_{Lk\sigma,\sigma''}|_{\epsilon=\epsilon_v}\{f_v(\epsilon)[G^{r}_{\sigma',\sigma''}(\epsilon) - G^{a}_{\sigma',\sigma''}(\epsilon)] + G^{<}_{\sigma',\sigma''}(\epsilon)\}. \quad (9)$$

We now introduce the density of states for the electrons in the FM leads, denoted by $\rho_{\alpha\sigma}(\epsilon)$. For the electrons in the left FM lead, the density of states is $\rho_{L\sigma}(\epsilon) = [1 + (-1)^{\sigma} p_L]\,\rho_{L0}(\epsilon)$, while for the electrons in the right FM lead, it is $\rho_{R\sigma}(\epsilon) = \left[1 + (-1)^{a+\sigma} p_R\right]\rho_{R0}(\epsilon)$, where $\sigma = \{0, 1\}$ for spin-up/down electrons, $a = \{0, 1\}$ for parallel/antiparallel alignment of the two FM leads' magnetization, $\rho_{\alpha 0} = (\rho_{\alpha\uparrow} + \rho_{\alpha\downarrow})/2$, and p_α is the polarization of the lead α. For the summation over k in Eq.(9), one may apply the continuous limit approximation $\sum_{\{Lk\sigma\}} \rightarrow \sum_{\{L\sigma\}}\int d\epsilon\ \rho_{L\sigma}(\epsilon)$. The current then can be expressed as

$$I_L = \frac{ie}{\hbar}\sum_{v=\{L\sigma\}}\int d\epsilon\ \mathrm{tr}\left\{f_v(\epsilon)\mathbf{\Gamma}_v\left[\mathbf{G}^{r}(\epsilon) - \mathbf{G}^{a}(\epsilon)\right] + \mathbf{\Gamma}_v\mathbf{G}^{<}(\epsilon)\right\}, \quad (10)$$

where $\mathbf{\Gamma}_v$ and $\mathbf{G}^{(r,a,<)}(\epsilon)$ are (2×2) coupling and Green's function matrices, given by

$$\mathbf{\Gamma}_{L\sigma}(\epsilon) = 2\pi\rho_{L\sigma}(\epsilon)\begin{pmatrix} |t_{L\sigma,\sigma}(\epsilon)|^2 & |t^{*}_{L\sigma,\sigma}(\epsilon)t_{L\sigma,\bar{\sigma}}(\epsilon)| \\ |t^{*}_{L\sigma,\bar{\sigma}}(\epsilon)t_{L\sigma,\sigma}(\epsilon)| & |t^{*}_{L\sigma,\bar{\sigma}}(\epsilon)t_{L\sigma,\bar{\sigma}}(\epsilon)| \end{pmatrix}, \quad (11)$$

$$\mathbf{G}^{(r,a,<)}(\epsilon) = \begin{pmatrix} G^{(r,a,<)}_{\sigma,\sigma}(\epsilon) & G^{(r,a,<)}_{\bar{\sigma},\sigma}(\epsilon) \\ G^{(r,a,<)}_{\sigma,\bar{\sigma}}(\epsilon) & G^{(r,a,<)}_{\bar{\sigma},\bar{\sigma}}(\epsilon) \end{pmatrix}. \quad (12)$$

In Eq.(11), $t_{L\sigma,\sigma}(t_{L\sigma,\bar{\sigma}})$ applies for the case of spin-σ electron tunneling to the spin-σ ($\bar{\sigma}$) state with (without) spin-flip. In low-bias approximation, $\Gamma_{L\sigma}(\epsilon)$=$2\pi\rho_{L\sigma}(\epsilon)|t^*_{L\sigma,\sigma'}(\epsilon)$ $t_{L\sigma,\sigma''}(\epsilon)|$ is taken to be constant (zero) within (beyond) the energy range close to the lead's electrochemical potential where most of the transport occurs, i.e., $\epsilon \in [\mu_\alpha - D, \mu_\alpha + D]$, where D is constant (Bruus & Flensberg (2004)). Based on the kinetic equation (Meir & Wingreen (1992)), the lesser Green's function $G^<_{\sigma',\sigma''}(\epsilon)$ can be written as $G^<_{\sigma',\sigma''}(\epsilon) = iG^r_{\sigma',\sigma''}(\epsilon)G^a_{\sigma',\sigma''}(\epsilon)[\Gamma_{L\sigma}f_{L\sigma}(\epsilon) + \Gamma_{R\sigma}f_{R\sigma}(\epsilon)]$, where $G^r_{\sigma',\sigma''}(t) = -i\theta(t)\langle\{a_{\sigma'},^\dagger a_{\sigma''}(t)\}\rangle$ and the advanced Green's function $G^a_{\sigma,\sigma}(\epsilon) = [G^r_{\sigma,\sigma}(\epsilon)]^*$. $\Gamma_{\alpha\sigma}$ is the aforementioned coupling strength, and $f_{\alpha\sigma} = \left(1 + \exp(\frac{\epsilon-\mu_{\alpha\sigma}}{k_B T})\right)^{-1}$ is the Fermi-Dirac function of lead α, with $\mu_{\alpha\sigma}$ being the chemical potential of that lead. When a bias voltage of V_b is between the two leads, the leads' electrochemical potentials are, respectively, given by $\mu_{L\sigma} = 0$ and $\mu_{R\sigma} = -eV_b$.

Considering that the current from the left lead to the QD is equal to the current from the QD to the right lead, one may calculate the current in a symmetric form, i.e., $I = \frac{I_L + I_R}{2}$. The final form for the total current is then given by

$$I = \frac{e}{h}\sum_\sigma \int d\epsilon \, [f_{L\sigma}(\epsilon) - f_{R\sigma}(\epsilon)] \, \mathrm{tr}\{\mathbf{G}^a\mathbf{\Gamma}_{R\sigma}\mathbf{G}^r\mathbf{\Gamma}_{L\sigma}\}. \tag{13}$$

In this QD-DTJ system, there exists the tunnel magnetoresistance (TMR) effect, which is caused by the difference between the resistance in parallel and antiparallel configurations of the two FM leads' magnetization. The TMR is given by

$$\mathrm{TMR} = \frac{R^{AP} - R^P}{R^P} = \frac{I^P - I^{AP}}{I^{AP}}, \tag{14}$$

where I^P (I^{AP}) is the tunneling current in parallel (antiparallel) configuration of the two leads' magnetization.

During the course of analyses, we would also consider the state of the QD, which is characterized by its occupancy. The QD's occupancy with electrons of spin-σ can be obtained by considering the lesser Green's function of the QD, i.e.,

$$\langle n_\sigma \rangle = \frac{1}{2\pi}\mathrm{Im}\int d\epsilon G^<_{\sigma,\sigma}(\epsilon). \tag{15}$$

2.1.2 Retarded Green's function

To calculate the tunneling current in Eq. (13), one has to obtain the explicit expression for the retarded Green's functions $G^r_{\sigma\sigma'}(\epsilon)$ of the QD. This can be done by means of the (equation-of-motion) EOM method. By definition, the general form of a retarded Green's function is given by $G^r_{\sigma,\sigma'}(t) = -i\theta(t)\langle\{a_\sigma(t), a^\dagger_{\sigma'}\}\rangle$. In the EOM method, the analytical expression for $G^r_{\sigma,\sigma'}(t)$ is obtained by firstly differentiating $G^r_{\sigma,\sigma'}(t)$ with respect to time. This yields

$$\begin{aligned} i\partial_t G^r_{\sigma,\sigma'}(t) &= \delta(t-t')\delta_{\sigma\sigma'} - i\theta(t-t')\langle\{i\partial_t a_\sigma(t), a^\dagger_{\sigma'}\}\rangle \\ &= \delta(t-t')\delta_{\sigma\sigma'} - i\theta(t-t')\langle\{-[H, a_\sigma], a^\dagger_{\sigma'}\}\rangle. \end{aligned} \tag{16}$$

Based on Eq. (16), for the QD-DTJ system with Hamiltonian in Eq. (1), one may obtain a closed set of equations involving $G^r_{\sigma,\sigma'}(\epsilon)$ after Fourier transform,

$$1 = (\epsilon + i\eta - \epsilon_{\sigma\sigma})G^r_{\sigma,\sigma} - \sum_{\nu=\{\alpha,k,\sigma'\}} t_{\nu,\sigma}G^r_{\nu,\sigma} - \epsilon_{\bar{\sigma}\sigma}G^r_{\bar{\sigma},\sigma}, \tag{17}$$

$$0 = (\epsilon + i\eta - \epsilon_{\bar{\sigma}\bar{\sigma}})G^r_{\bar{\sigma},\sigma} - \sum_{\nu=\{\alpha,k,\sigma'\}} t_{\nu,\bar{\sigma}}G^r_{\nu,\sigma} - \epsilon_{\sigma\bar{\sigma}}G^r_{\sigma,\sigma}, \tag{18}$$

$$0 = (\epsilon + i\eta - \epsilon_{\nu})G^r_{\nu,\sigma} - \sum_{\sigma'=\{\sigma,\bar{\sigma}\}} t^*_{\nu,\sigma'}G^r_{\sigma',\sigma}, \text{ where } \nu = \{\alpha, k, \sigma\}, \tag{19}$$

$$0 = (\epsilon + i\eta - \epsilon_{\nu})G^r_{\nu,\sigma} - \sum_{\sigma'=\{\sigma,\bar{\sigma}\}} t^*_{\nu,\sigma'}G^r_{\sigma',\sigma}, \text{ where } \nu = \{\alpha, k, \bar{\sigma}\}. \tag{20}$$

By solving the equation array of Eqs. (17) to (20), one reaches the explicit expressions for the retarded Green's functions (those in Eq. 12) of the QD:

$$G^r_{\sigma,\sigma} = \frac{1}{\epsilon + i\eta - \epsilon_{\sigma\sigma} - \Sigma^{\sigma}_{\sigma\sigma}(\epsilon) - \Sigma^{\bar{\sigma}}_{\sigma\sigma}(\epsilon) - \dfrac{\prod\limits_{\sigma'=\{\sigma,\bar{\sigma}\}} \left(\epsilon_{\bar{\sigma}'\sigma'} + \Sigma^{\sigma'}_{\sigma'\bar{\sigma}'}(\epsilon) + \Sigma^{\bar{\sigma}'}_{\bar{\sigma}'\sigma'}(\epsilon)\right)}{\epsilon + i\eta - \epsilon_{\bar{\sigma}\bar{\sigma}} - \Sigma^{\sigma}_{\bar{\sigma}\bar{\sigma}}(\epsilon) - \Sigma^{\bar{\sigma}}_{\bar{\sigma}\bar{\sigma}}(\epsilon)}}, \tag{21}$$

$$G^r_{\bar{\sigma},\sigma} = \frac{1}{-\epsilon_{\bar{\sigma}\sigma} - \Sigma^{\sigma}_{\sigma\bar{\sigma}}(\epsilon) - \Sigma^{\bar{\sigma}*}_{\sigma\bar{\sigma}}(\epsilon) + \dfrac{\prod\limits_{\sigma'=\{\sigma,\bar{\sigma}\}} \left(\epsilon + i\eta - \epsilon_{\bar{\sigma}'\sigma'} - \Sigma^{\sigma'}_{\bar{\sigma}'\bar{\sigma}'}(\epsilon) - \Sigma^{\bar{\sigma}'}_{\bar{\sigma}'\bar{\sigma}'}(\epsilon)\right)}{\epsilon_{\sigma\bar{\sigma}} + \Sigma^{\sigma}_{\bar{\sigma}\sigma}(\epsilon) + \Sigma^{\bar{\sigma}*}_{\sigma\bar{\sigma}}(\omega)}}, \tag{22}$$

where the self energy $\Sigma^{\sigma}_{\sigma'\sigma''}(\epsilon) = \sum_{\{\alpha,k\}} \frac{t_{\alpha k\sigma,\sigma'} t^*_{\alpha k\sigma,\sigma''}}{\epsilon + i\eta - \epsilon_{\alpha k\sigma}}$, $\Sigma^{\sigma*}_{\sigma'\sigma''}(\epsilon) = \sum_{\{\alpha,k\}} \frac{t^*_{\alpha k\sigma,\sigma'} t_{\alpha k\sigma,\sigma''}}{\epsilon + i\eta - \epsilon_{\alpha k\sigma}}$, with $\sigma, \sigma', \sigma'' \in \{\uparrow, \downarrow\}$.

2.2 Results and discussion

Based on the electron transport model developed in Sec. 2.1, one may analyze the SDT properties, such as the spectral functions, the tunneling charge current, spin current, the TMR and the electron occupancies of the QD. The SDT model enables one to investigate the effects of the SF-QD and SF-TJ events and the effect of the coupling asymmetry (CA) on the SDT properties as well.

To focus on the above effects, one may assume that, i) proportional and spin independent lead-QD coupling across the two junctions, i.e., $t_{\alpha k\uparrow,\uparrow}$=$t_{\alpha k\downarrow,\downarrow}$=$t_{\alpha k\sigma,\sigma}$=$t_{\alpha}$, and t_R=βt_L=t; ii) junction and spin independent strength of SF-TJ, i.e., $t_{\alpha k\uparrow,\downarrow}$=$t_{\alpha k\downarrow,\uparrow}$=$t_{\alpha k\sigma,\bar{\sigma}}$=$v_{\alpha}$, and v_R=$\beta v_L = v$; and iii) spin independence of SF-QD, i.e., $\epsilon_{\uparrow\downarrow}$=$\epsilon_{\downarrow\uparrow}$=$\lambda$, iv) the chemical potential of the left and right leads are μ_L=0, $\mu_R = -eV_b$; and v) spin independence of the energy level of the QD, i.e., $\epsilon_{\sigma\sigma}$=$\epsilon_{\bar{\sigma}\bar{\sigma}}$=$\epsilon_d$= $\epsilon_{d0} - eV_b\frac{\beta^2}{1+\beta^2}$, where ϵ_{d0} is the QD's energy level without bias voltage. Based on the assumptions i)-v) and Eq. (22), one can readily deduce the spin symmetry of the off-diagonal Green's functions, i.e., $G^r_{\uparrow,\downarrow} = G^r_{\downarrow,\uparrow} = G^r_{\sigma,\bar{\sigma}}$. For simplicity, in the following discussion, the form of $G^r_{\sigma\sigma'}$ is used to replace the form of $G^r_{\sigma,\sigma'}$ for retarded Green's function.

2.2.1 Spin-flip effects

Firstly, one may evaluate the four elements of the retarded Green's function (GF) matrix [given in Eq. (12)], $G^r_{\uparrow\uparrow}$, $G^r_{\uparrow\downarrow}$, $G^r_{\downarrow\uparrow}$ and $G^r_{\downarrow\downarrow}$. Based on Eqs. (21) and (22), one may obtain the respective spectral functions, $-2\text{Im}G^r_{\uparrow\uparrow}$, $-2\text{Im}G^r_{\uparrow\downarrow}$, $-2\text{Im}G^r_{\downarrow\uparrow}$, and $-2\text{Im}G^r_{\downarrow\downarrow}$. Spectral functions provide information about the nature of the QD's electronic states which are involved in the tunneling process, regardless whether the states are occupied or not. The spectral functions can be considered as a generalized density of states.

If one neglects the SF-QD or SF-TJ events in the QD-DTJ system, there is no mixing of the spin-up and spin-down electron transport channels. In such QD-DTJ system, the two off-diagonal Green's functions, $G^r_{\sigma\bar{\sigma}}(\sigma = \{\uparrow,\downarrow\})$ become zero [this can be confirmed by considering Eq. (22)], and so are their respective spectral functions. Thus, we focus on the spectral functions corresponding to the diagonal components of the retarded GF matrix. Those spectral functions are analyzed as a function of energy under both parallel and antiparallel configuration of the two FM leads' magnetization, in Figs. 3(a) to (d). A broad peak is observed corresponding to the QD's energy level ($\epsilon = \epsilon_d$). The broad peak can be referred to as "QD resonance". The broadening of the QD resonance is caused by the finite coupling between the QD and the leads, since the QD resonance is a δ function for an isolated QD with no coupling to leads. The width of the QD resonance reflects the strength of coupling between QD and leads; the stronger the coupling is, the broader the energy spread is, hence, a wider peak.

Under zero-bias [shown in Figs. 3 (a) and (b)], one may note three distinct features of the spectral functions:

1. A second resonance peak which corresponds to the leads' potentials, $\mu_L = \mu_R = 0$ eV. The peak can be referred to as the "lead resonance".
2. The lead resonance for the spin-up spectral function ($-2\text{Im}G^r_{\uparrow\uparrow}$) has a broader and lower profile compared to that of the spin-down spectral function, when the QD-DTJ system is in the parallel configuration. This indicates that the excitation at the lead energy has a larger energy spread for spin-up carriers due to the polarization of the lead.
3. The spin-up and spin-down spectral functions are identical in the antiparallel alignment, due to the spin symmetry of the system in antiparallel configuration.

The spectral functions under an finite bias voltage ($V_b = 0.2$ eV) are shown in Figs. 3 (c) and (d). It is observed that,

1. the lead resonance splits into two peaks at the respective left lead and right lead potentials, $\epsilon = \mu_L = 0$ and $\epsilon = \mu_R = -eV_b = -0.2$ eV.
2. In the parallel configuration, the lead resonance of the spin-down electrons is higher (lower) than that of the spin-up electrons at μ_L (μ_R). This is due to the spin-dependence of the electron tunneling between leads and QD.
3. The antiparallel alignment of leads' magnetization gives rise to similar magnitude of the two lead resonances for both spin-up and down spectral functions, due to the spin symmetry of the two spin channels.

Next, one may investigate the SF-TJ effects on the electron transport through the QD-DTJ system, where the SF-TJ strength $v \neq 0$. Figure 4 shows the effect of SF-TJ on the spectral function of diagonal GFs. With the SF-TJ effects, both the QD resonance and the lead

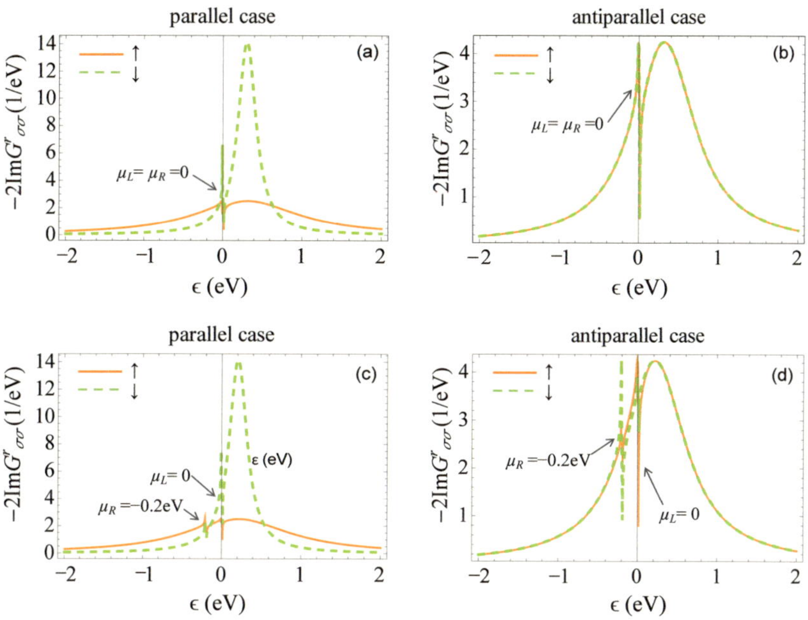

Fig. 3. Spectral functions for spin-up (solid line) and spin-down (dashed line) retarded Green's functions, as a function of electrons' energy, in parallel and antiparallel case. Other parameters are t=0.5 eV, v=0, $\epsilon_{d0} = 0.3$eV, $\rho_0 = 0.7(eV)^{-1}$, $p_L = p_R = 0.7, \lambda = 0, \beta = 1, T =$ 0.3 K.

resonance at $\epsilon = 0$ are enhanced while the lead resonance at $\epsilon = -eV_b$ is suppressed. This indicates that the increasing SF-TJ helps the tunneling to proceed primarily in the vicinity of the QD's energy level, resulting in an effective decrease in the coupling between the same spin states in leads and QD.

Based on the SDT model, one may analyze the effects of the SF-QD events (denoted by λ) on the spectral functions of the diagonal retarded GFs($G^r_{\uparrow\uparrow}$ and $G^r_{\downarrow\downarrow}$) of the QD-DTJ system, for both parallel and antiparallel alignments, as shown in Figure 5. At the QD energy level $\epsilon_d = 0.2$ eV, the presence of the SF-QD causes a symmetric split of the QD resonance, resulting in the suppression of tunneling via the lead resonances. The splitting of the QD resonance indicates that the two effective energy levels within the QD are involved in the tunneling process. This split translates to an additional step in the $I - V$ characteristics, which will be discussed later in Fig. 7.

Considering the off-diagonal GF's ($G^r_{\sigma\bar{\sigma}}$), the spectral functions are ploted in Figure 6, for both parallel and antiparallel alignment, under varying SF-TJ strengths (ν) and SF-QD strengths (λ). As shown in Figs. 6(a)-(d), without SF-TJ or SF-QD effects, the off-diagonal spectral functions vanish (the solid lines), i.e., the transport proceeds independently in the spin-up and spin-down channels. The presence of either the SF-TJ ($\nu > 0$) or the SF-QD ($\lambda \neq 0$) enhances the magnitudes of the off-diagonal spectral functions monotonically, indicating stronger mixing of the tunneling transport through the two spin channels.

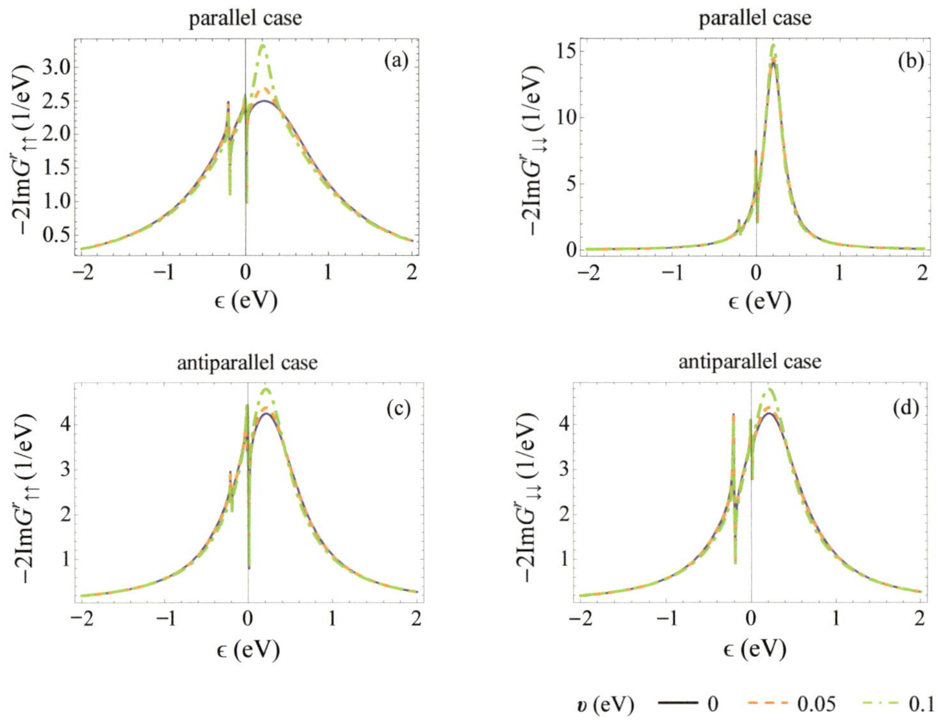

Fig. 4. Spectral functions for the diagonal retarded Green's functions, as a function of electrons' energy, with varied SF-TJ strength (v) between leads and QD, in (a)-(b) parallel and (c)-(d) antiparallel alignment of two leads' magnetization, where $\lambda = 0$ eV, V_b=0.2 V. Other parameters are the same with those in Fig. 3.

The individual effects from SF-TJ or SF-QD on the tunneling current and the TMR are then investigated, as shown in Figs. 7.

The $I - V_b$ characteristics in Figs. 7(a)-(b) and Figs. (d)-(e) show a step at the threshold voltage V_{th}, which is required to overcome the Coulomb blockade (CB). The threshold voltage is given by $V_{th} = 2\epsilon_{d0}$. Considering the bias voltage regions, one may find the following features of the $I - V_b$ characteristics,

1. Within the sub-threshold bias range ($V < V_{th}$), the current is still finite due to thermally assisted tunneling at finite temperature.
2. The sub-threshold current is particularly large in the parallel configuration, due to the stronger lead-QD coupling and hence a greater energy broadening of the QD's level.
3. Overall, the parallel current exceeds the antiparallel current for the entire voltage range considered, due to the nonzero spin polarization of the FM lead.
4. Beyond the threshold voltage (i.e. $V_b \gg V_{th}$), the tunneling current saturates since only a single QD level is assumed to participate in the tunneling transport.

In the presence of SF-TJ, the tunneling currents in the parallel and antiparallel configurations are found to be significantly enhanced for bias voltage exceeding the threshold ($V_b > V_{th}$), as shown in Figs. 7(a) and (b). The enhancement in current stems from the overall stronger coupling between the lead and the QD. Additionally, the degree of enhancement of the

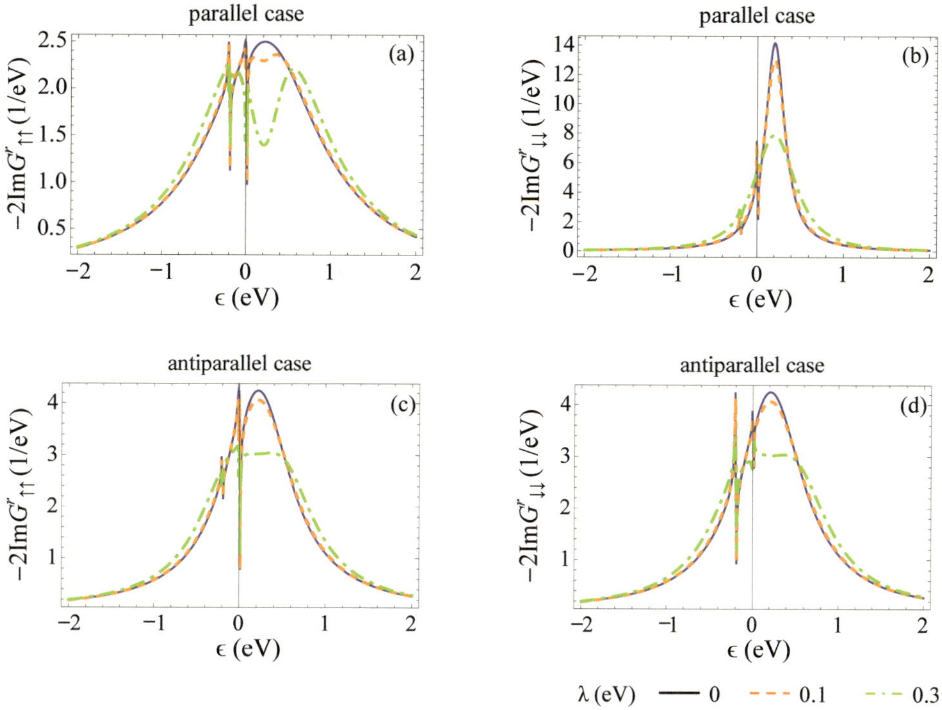

Fig. 5. Spectral function as a function of electrons' energy, with varied SF-QD strength (λ), in (a)-(b) parallel and (c)-(d) antiparallel cases, where $v = 0$ eV. Other parameters are the same with those in Fig. 3.

tunneling current is more pronounced for the parallel alignment of the FM leads. This results in an enhancement of the TMR for the voltage range above the threshold, as shown in Fig. 7(c).

When SF-QD events are present in the system, two new features show up in the $I - V_b$ characteristics, in Figs. 7 (d) and (e). First, the current step at the threshold bias V_{th} splits into two, at $V_b = V_{th} \pm \lambda$, respectively. The presence of the additional step is due to the splitting in the QD resonances observed in the spectral functions of Fig. 5. Secondly, the presence of SF-QD suppresses the current saturation value at large bias voltage (i.e., $V_b \gg V_{th} + \lambda$). The decrease is more pronounced in the antiparallel configuration, resulting in the enhancement of the TMR with the increasing SF-QD probability, as shown in Fig. 7(f).

When both SF processes (Fig. 8) exist in the QD-DTJ system, the two types of SF have competing effects on the tunneling current at large bias voltage exceeding the threshold. The SF-TJ (SF-QD) tends to enhance (suppress) the tunneling current within the bias voltage region exceeding the threshold voltage. This competitive effect is shown for the overall I-V_b characteristics in Figs. 8 (a)-(b). Evidently, the effect caused by one SF mechanism is mitigated by the other for both parallel and antiparallel alignments. However, both SF mechanism contribute to the asymmetry of tunneling current between the parallel and antiparallel cases, leading to an additive effect on the TMR for voltage bias region beyond the threshold voltage, as shown in Fig. 8 (c). The competitive effect on current and collaborative effect on TMR make it possible to attain simultaneously a high TMR and tunneling current density.

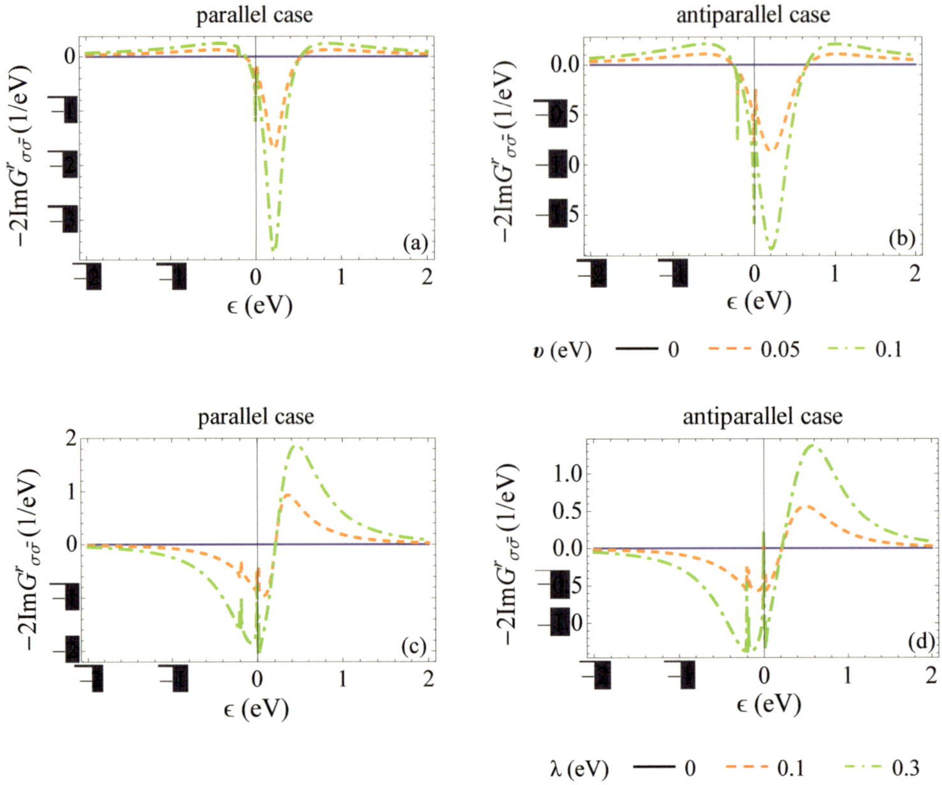

Fig. 6. (a),(b): Off-diagonal spectral functions as a function of energy, for varying SF-TJ strength (υ) between leads and QD, in the absence of SF-QD (i.e., $\lambda = 0$ eV).(c),(d):Off-diagonal Spectral functions as a function of electrons' energy, with varied SF-QD strength (λ), in parallel (left) and antiparallel (right) case, where υ=0 eV. Other parameters are the same with those in Fig. 3.

2.2.2 Coupling asymmetry effects

Recent experimental studies (Hamaya et al. (2009; 2007)) of QD-DTJ structures revealed that the SDT characteristics are strongly dependent on the coupling asymmetry (CA) between the two junctions. Such asymmetry is inherent in the sandwich structure, given the exponential dependence of the coupling strength on the tunnel barrier width.

One may study the effect of the junction CA on the overall spin and charge current characteristics of the QD-DTJ system. The degree of the CA is characterized by the ratio of the right and left junction coupling parameter. The CA is denoted by β and $\beta = t_{Rk\sigma,\sigma}/t_{Lk\sigma,\sigma}$. The spin-up (spin-down) components of the tunneling current can be represented as $I_\uparrow$ $(I_\downarrow)$, based on which the spin current is defined to be the difference between the two components, $I_s = I_\uparrow - I_\downarrow$. In the following, one may focus on the parallel alignment of the magnetization of the two leads of the QD-DTJ system, since the magnitude of the spin current is the greatest in this case (see Mu et al. (2006)).

For simplicity but without loss of generality, one may assume β to be spin-independent, i.e, $\beta = t_{Rk\uparrow,\uparrow}/t_{Lk\uparrow,\uparrow} = t_{Rk\downarrow,\downarrow}/t_{Lk\downarrow,\downarrow} = t_{Rk\sigma,\sigma}/t_{Lk\sigma,\sigma}$. In Sec. 2.1, the coupling strength is defined as

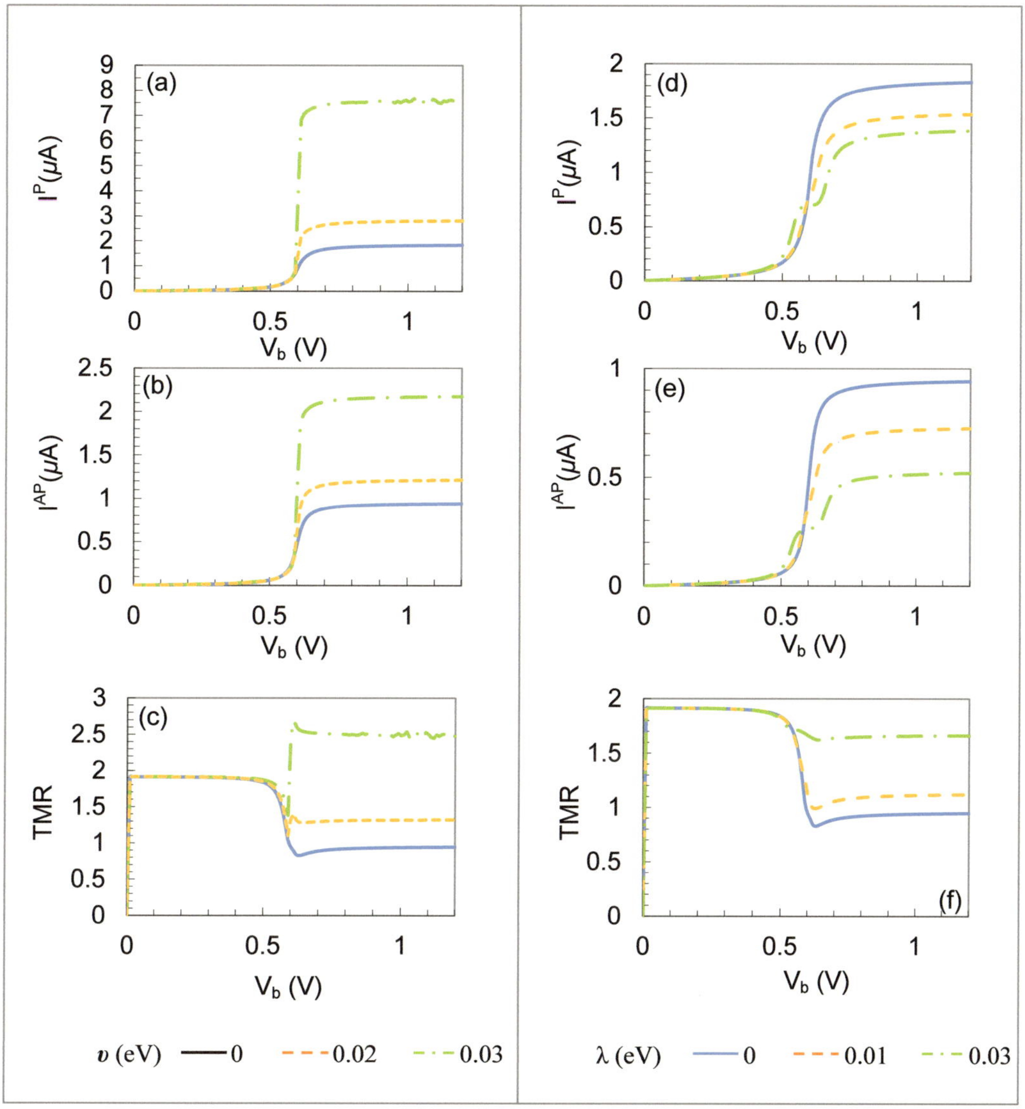

Fig. 7. Current as a function of bias voltage, with varying SF-TJ strength (υ) between the lead and the dot, in (a) parallel and (b) antiparallel cases, or with varying SF-QD strength (λ) in (d) parallel and (e) antiparallel cases. (c)/(f): Tunnel magnetoresistance (TMR) as a function of bias voltage under increasing SF-TJ/SF-QD strength, respectively. In plots (a)-(c), λ=0 eV, and in plots (d)-(f), υ=0, while other parameters are the same with those in Fig. 3.

$\Gamma_{\alpha\sigma} = 2\pi\rho_{\alpha\sigma}|t^*_{\alpha k\sigma,\sigma}\, t_{\alpha k\sigma,\sigma}| = [1 + (-1)^\sigma p_\alpha]\, 2\pi\rho_{\alpha 0}|t^*_{\alpha k\sigma,\sigma}\, t_{\alpha k\sigma,\sigma}| = [1 + (-1)^\sigma p_\alpha]\, \Gamma_{\alpha 0}$. If assuming identical intrinsic electron density of states and identical polarization of the two leads, i.e., $\rho_{\alpha 0} = \rho_0$, $p_\alpha = p$, one may obtain that $\beta = \sqrt{\Gamma_{R\sigma}/\Gamma_{L\sigma}} = \sqrt{\Gamma_{R0}/\Gamma_{L0}}$.

We consider the I-V characteristics for the charge current and spin current, shown in Fig. 9 for two different CA values. These two values were chosen so that $\beta_1 = 1/\beta_2$, meaning

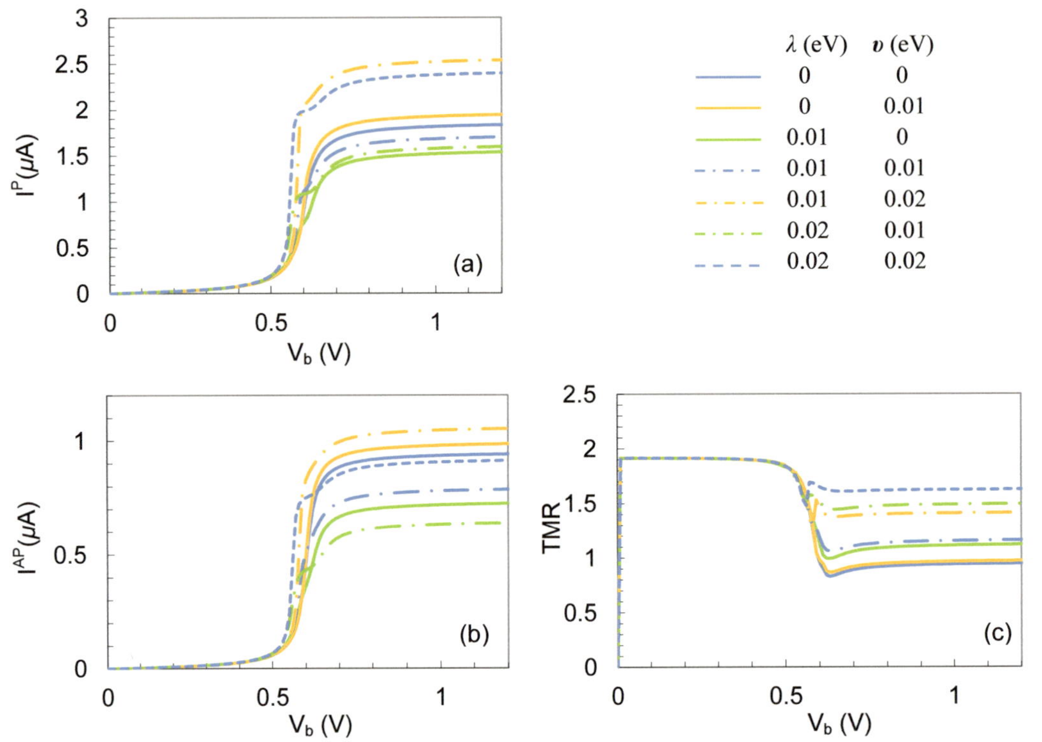

Fig. 8. (a),(b): Current as a function of bias voltage in parallel case and antiparallel case, and (c) TMR as a function of bias voltage, with varying SF-TJ strength (υ) and varying SF-QD (λ). Other parameters are the same as those in Fig. 3.

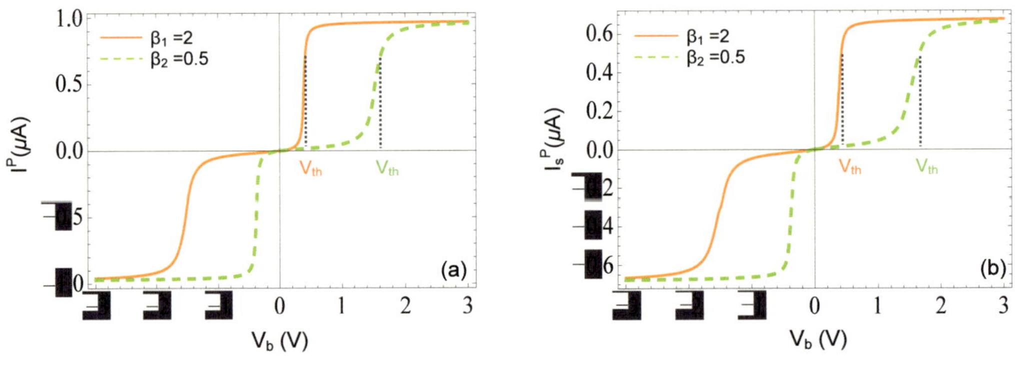

Fig. 9. (a) Charge current I as a function of bias voltage V_b and (b) spin current (I_s) as a function of bias voltage V_b, with two different coupling asymmetry β, in the parallel alignment of the leads' magnetization. The coupling asymmetry is denoted by $\beta = t_{Rk\sigma,\sigma} / t_{Lk\sigma,\sigma} = \sqrt{\Gamma_{R\sigma}/\Gamma_{L\sigma}}$, where $\Gamma_{L\sigma} = (1 \pm p_L)\Gamma_{L0}$ and $\Gamma_{R\sigma} = (1 \pm p_R)\Gamma_{R0}$. Other parameters are $\Gamma_{L0} = 0.012$ eV and $\Gamma_{R0} = \Gamma_{L0} \times {\beta_1}^2$ for β_1 case, $\Gamma_{L0} = 0.006$ eV and $\Gamma_{R0} = \Gamma_{L0} \times {\beta_1}^2$ for β_2 case, $\epsilon_{d0} = 0.3$eV, $p_L = p_R = 0.7$, $T = 100$ K, υ=0 eV, λ=0 eV.

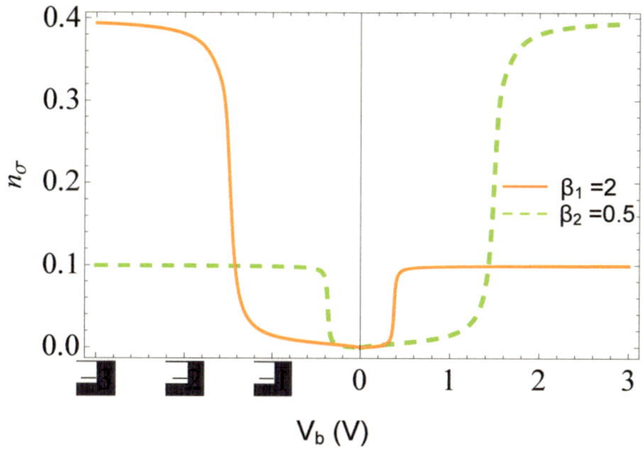

Fig. 10. The occupancy of the quantum dot as a function of bias voltage, with two different coupling asymmetry β, in the parallel alignment of the leadsąf́ magnetization. Other parameters are the same with those Fig. 9.

that the left (right) junction in β_1 system is the right (left) junction for β_2 system. It is found that when the coupling strength of the right junction is four times as strong as that of the left junction, i.e., $\beta = 2$, both the magnitude of the charge and spin currents beyond the threshold voltage are the same as those for the reverse case ($\beta = 0.5$). This is due to the fact that the total resistance of the two QD-DTJ system is maintained regardless of the coupling asymmetry reversal. However, the CA affects the threshold voltage V_{th}. This is due to the different shifts of the QD energy level under positive and negative bias voltage, i.e., $V_{th} = 2\epsilon_d$, where $\epsilon_d = \epsilon_{d0} - \frac{\beta^2}{1+\beta^2} eV_b$. The CA effect on the charge current I-V characteristics is consistent with the experimental results observed by K. Hamaya *et al.* for an asymmetric Co/InAs/Co QD-DTJ system (Fig. 2(a) of Ref. Hamaya et al. (2007)).

Next, one may investigate the CA effect on the QD occupancies, which are obtained by integrating the spectral function in Eq. (15). The QD occupancies for both spin-up and spin-down electrons are shown in Fig. 10. The occupancies for spin-up and spin-down electrons in the QD actually coincide since the QD-DTJ system is operated in the parallel configuration of the leads' magnetization. Moreover, as β is increased from 0.5 to 2, the QD occupancies of both spin orientations decrease. This decrease is reasonable since as Γ_L is decreased with respect to Γ_R, the coupling which allows the electron to tunnel to the QD from the source (left lead) is reduced, while the coupling which allows the electron to tunnel out of the QD to the drain (right lead) is enhanced. In this circumstance, electrons start to have a higher occupancy in the QD for $\beta < 1$ case, where $\Gamma_L > \Gamma_R$.

2.3 Summary

In summary, the SDT through a QD-DTJ system is theoretically studied. In the SDT model described in Sec.2.1, well-separated QD levels are assumed such that only a single energy level are involved in the SDT process, and the correlation between different energy levels is then neglected. The spectral functions, QD electron occupancies, tunneling charge current, spin current as well as TMR are evaluated based on the Keldysh NEGF formulism and EOM method, with consideration of the effects of the SF-TJ events, SF-QD events, and the CA between the two tunnel junctions on the SDT of the system.

3. QD with Zeeman splitting

In the last section, the SDT is studied for the QD-DTJ system where the spin dependence of the electron transport is caused by the spin polarization in the FM leads. In this section, one may analyze the SDT through the QD-DTJ system where the leads sandwiching the QD are non-magnetic (NM), and a FM gate is applied above the QD. The electron transport through this QD-DTJ system is spin-dependent due to the Zeeman splitting (ZS) generated in the QD. In this QD-DTJ system, one may expect a fully polarized current to tunnel through (Recher et al. (2000)). A fully spin-polarized current is important for detecting or generating single spin states (Prinz (1995; 1998)), and thus is of great importance in the realization of quantum computing (Hanson et al. (2007); Kroutvar et al. (2004); Loss & DiVincenzo (1998); Moodera et al. (2007); Petta et al. (2005); Wabnig & Lovett (2009)).

The QD-DTJ system is schematically shown in Fig. 11. The magnetic field generated by the FM gate is assumed to be spatially localized such that it gives rise to a ZS of the discrete energy levels of the QD, but negligible ZS in the energy levels of the NM electrodes. When the bias voltage V_b between the two NM electrodes, and the size of the ZS in the QD are appropriately tuned, a fully polarized spin current is observed in this QD-DTJ system. The polarization of the current depends on the magnetization direction of the FM gate. Here, the down (up)-spin electrons have spins which are aligned parallel (antiparallel) to the magnetization direction of the FM gate.

3.1 Theory

The Hamiltonian of the QD-DTJ system is in form of

$$H = \sum_{\alpha} H_{\alpha} + H_d + H_t, \tag{23}$$

where

$$H_{\alpha} = \sum_{k\sigma} \epsilon_{\alpha k\sigma} a^{\dagger}_{\alpha k\sigma} a_{\alpha k\sigma}, \tag{24}$$

$$H_d = \sum_{\sigma} \epsilon_{\sigma} a^{\dagger}_{\sigma} a_{\sigma} + U n_{\uparrow} n_{\downarrow}, \tag{25}$$

$$H_t = \sum_{\alpha k\sigma} \left(t_{\alpha k\sigma,\sigma} a^{\dagger}_{\sigma} a_{\alpha k\sigma} + t^{*}_{\alpha k\sigma,\sigma} a^{\dagger}_{\alpha k\sigma} a_{\sigma} \right), \tag{26}$$

where $\alpha = \{L, R\}$ is the lead index for the left and right leads, k is the momentum, $\sigma = \{\uparrow, \downarrow\}$ is the spin-up and spin-down index, $a^{\dagger}$ and a are the electron creation and annihilation operators, ϵ_{σ} is the energy level in the QD for electrons with spin-σ, U is the Coulomb blockade energy when the QD is doubly occupied by two electrons with opposite spins, and $t_{\alpha k\sigma,\sigma}$ describes the coupling between the electron states with spin-σ in the lead α and the QD.

In our model, we consider only the lowest unoccupied energy level of the QD ϵ_{σ} since most of the overall transport occurs via that level. With the presence of an applied magnetic field B, the lowest unoccupied energy level is given by $\epsilon_{\sigma} = \epsilon_d + \frac{(-1)^{\sigma}}{2} g\mu_B B$, where $\sigma = 0$ ($\sigma = 1$) for up-spin (down-spin) electrons, g is the electron spin g-factor, μ_B is the Bohr magneton, and $g\mu_B B$ is the ZS between the two spin states $\epsilon_{\downarrow}$ and $\epsilon_{\uparrow}$. Under an applied bias V_b between the

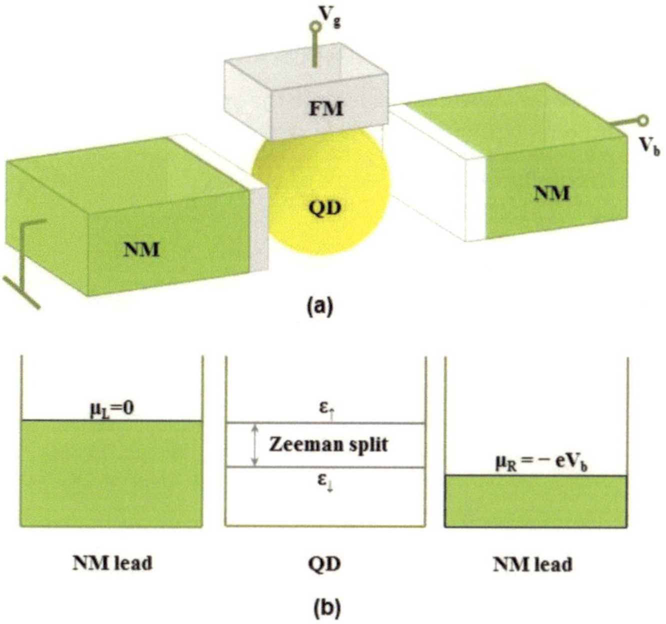

Fig. 11. (a) Schematic diagram of the QD-DTJ set up, which consists of two NM electrodes, one QD and one FM gate. (b) schematic energy diagram of the QD-DTJ system in (a), where V_b is the bias voltage, $\epsilon_\downarrow = \epsilon_d - g\mu_B B/2$ ($\epsilon_\uparrow = \epsilon_d + g\mu_B B/2$) is the energy level for spin-down(up) electrons, respectively, $g\mu_B B$ is the Zeeman splitting between $\epsilon_\downarrow$ and $\epsilon_\uparrow$, g is the electron spin g-factor, μ_B is the Bohr magneton, B is the applied magnetic field generated by the FM gate, and $\epsilon_d = \epsilon_{d0} - eV_b\beta^2/(1+\beta^2)$ is the single energy level of the QD without applied magnetic field, with ϵ_{d0} being the single energy level under zero bias voltage and β being the coupling asymmetry between the two tunnel junctions. We assume a symmetrical QD-DTJ system where $\beta = 1$.

leads and in the absence of B-field, the QD's energy level is modified as $\epsilon_d = \epsilon_{d0} - eV_b\beta^2/(1+\beta^2)$, where ϵ_{d0} is the energy level at zero bias voltage, and $\beta = t_{Rk\sigma,\sigma}/t_{Lk\sigma,\sigma}$ denotes the asymmetry of the coupling in the left and right tunnel junctions. In the following, a symmetric QD-DTJ system is assumed where $\beta = 1$.

Based on the Hamiltonian, the tunneling current is evaluated via the NEGF formalism introduced in Section. 2.1. The charge and spin current are defined as $I_c = I_\downarrow + I_\uparrow$ and $I_s = I_\downarrow - I_\uparrow$, respectively, where the tunneling current of spin-σ electron tunneling through the QD-DTJ system is given by

$$I_\sigma = \frac{e}{h}\int d\epsilon \left[f_{L\sigma}(\epsilon) - f_{R\sigma}(\epsilon)\right] G^a_{\sigma,\sigma}\Gamma_{R\sigma}G^r_{\sigma,\sigma}\Gamma_{L\sigma}. \tag{27}$$

Here, $\Gamma_{\alpha\sigma}(\epsilon) = 2\pi\rho_{\alpha\sigma}(\epsilon)t_{\alpha\sigma,\sigma}(\epsilon)t^*_{\alpha\sigma,\sigma}(\epsilon)$, $G^a_{\sigma,\sigma}(\epsilon) = [G^r_{\sigma,\sigma}(\epsilon)]^*$, $G^r_{\sigma,\sigma}(t) = -i\theta(t)\langle\{a_\sigma(t), a^\dagger_\sigma\}\rangle$.

The explicit form of $G^r_{\sigma,\sigma}(\epsilon)$ is given by

$$G^r_{\sigma,\sigma}(\epsilon) = \frac{1-n_{\bar\sigma}}{\epsilon + i\eta - \epsilon_\sigma - \Sigma(\epsilon)} + \frac{n_{\bar\sigma}}{\epsilon + i\eta - \epsilon_{\bar\sigma} - \Sigma(\epsilon)}, \tag{28}$$

where the self energy terms are $\Sigma(\epsilon) = \sum_{\{\alpha,k\}} \frac{t_{\alpha k\sigma,\sigma} t^*_{\alpha k\sigma,\sigma}}{\epsilon + i\eta - \epsilon_{\alpha k\sigma}}$. The coupling coefficients $t_{\alpha k\sigma,\sigma}$ and $t^*_{\alpha k\sigma,\sigma}$ are spin-independent since the two leads are NM.

Based on Eqs. (27) and (28), one can then calculate the spin-σ current $I_\uparrow$ and $I_\downarrow$, and hence the charge and spin current, which are defined as $I_c = I_\downarrow + I_\uparrow$ and $I_s = I_\downarrow - I_\uparrow$, respectively.

In the EOM method, the following Hartree-Fock decoupling decoupling approximation of (Lacroix (1981); Sergueev et al. (2002)) is applied,

$$\left\langle \left\{ a^\dagger_{k\alpha\sigma} a_\sigma a_{k\alpha\bar{\sigma}}, a^\dagger_{\bar{\sigma}} \right\} \right\rangle = \left\langle a^\dagger_{k\alpha\sigma} a_\sigma \right\rangle \left\langle \left\{ a_{k\alpha\bar{\sigma}} a^\dagger_{\bar{\sigma}} \right\} \right\rangle, \tag{29}$$

$$\left\langle \left\{ a^\dagger_{\sigma} a_{k\alpha\sigma} a_{k\alpha\bar{\sigma}}, a^\dagger_{\bar{\sigma}} \right\} \right\rangle = \left\langle a^\dagger_{\sigma} a_{k\alpha\sigma} \right\rangle \left\langle \left\{ a_{k\alpha\bar{\sigma}} a^\dagger_{\bar{\sigma}} \right\} \right\rangle, \tag{30}$$

$$\left\langle \left\{ a_\sigma a^\dagger_{k\alpha\sigma} a_{\bar{\sigma}}, a^\dagger_{\bar{\sigma}} \right\} \right\rangle \simeq 0, \tag{31}$$

$$\left\langle \left\{ a_{k\alpha\sigma} a^\dagger_{\sigma} a_{\bar{\sigma}}, a^\dagger_{\bar{\sigma}} \right\} \right\rangle \simeq 0, \tag{32}$$

$$\left\langle a^\dagger_{k\alpha\sigma} a_\sigma \right\rangle = \left\langle a^\dagger_{\sigma} a_{k\alpha\sigma} \right\rangle \simeq 0. \tag{33}$$

3.2 Spin polarized current

Based on the SDT model in Sec. 3.1, one may obtain the $I - V$ characteristics of the system for both spin current I_s and charge current I_c, as shown in Fig. 12.

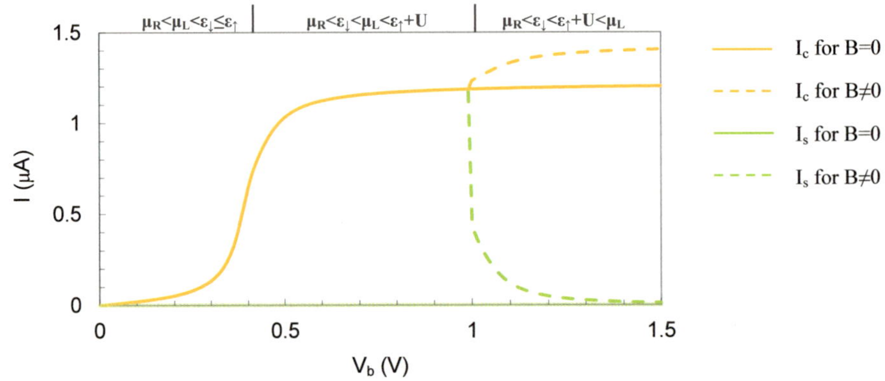

Fig. 12. Charge current (I_c) and spin current (I_s) as a function of the bias voltage, with ($B \neq 0$) or without ($B = 0$) ZS. $V_g = 0$. The following parameters are assumed: lowest unoccupied energy level in the QD under zero bias voltage ϵ_{d0}=0.2eV, the Coulomb blockade energy $U = 0.26$ eV, the Zeeman splitting due to the FM gate is $g\mu_B B = 0(0.36$ meV) for $B = 0$ ($B \neq 0$) case, the gate voltage $V_g = 0$, and temperature $T = 3$ K.

In the absence of a FM gate, i.e., with zero magnetic field ($B = 0$) applied to the QD, the magnitude of the charge current I_c is the same as that of the system with a FM gate, within the bias region $\mu_R < \epsilon_\downarrow < \mu_L < \epsilon_\uparrow + U$. In this region, the spin current I_s is zero for the system without a FM gate, where the device is spin-symmetric and the transport across it is spin-independent.

For the system with a FM gate, both the charge current I_c and spin current I_s of the system show the three distinct regions with respect to the bias voltage,

1. $\mu_R < \mu_L < \epsilon_\downarrow < \epsilon_\uparrow$, where both I_c and I_s are negligible due to the suppression of electron tunneling by Coulomb blockade;
2. $\mu_R < \epsilon_\downarrow < \mu_L < \epsilon_\uparrow + U$, where due to spin blockade, only the spin-down channel contributes to the transport across the system, resulting in a fully spin-down polarized current with $I_s = I_c$;
3. and $\mu_R < \epsilon_\downarrow < \epsilon_\uparrow + U < \mu_L$, where it is energetically favorable for both types of spins to tunnel across the device, leading to zero spin current.

The sign of the spin polarization of the tunneling current can be *electrically* modulated, i.e., by means of a gate voltage V_g. The gate voltage modulation of the QD energy level ϵ_d can result in the switching of the spin polarization of current, without requiring any corresponding change to the magnetization of the FM gate. If the energy diagram of the system satisfies $\epsilon_\downarrow - eV_g < \mu_R < \epsilon_\uparrow - eV_g < \mu_L$, a fully spin-up polarized current will thus flow continuously through the system.

3.3 Summary

In summary, the SDT through a QD-DTJ system is analyzed with NM leads and FM gate. Under the applied magnetic field from the FM gate, the energy level in the QD splits to two due to ZS effect. The two energy levels can be modulated by the gate voltage applied to the FM gate. Based on the SDT model developed by NEGF formulism and EOM method, the $I - V_b$ characteristics is analyzed, and a fully spin-down polarized current is obtained when the system is operated under a proper bias voltage between the two leads. Additionally, by utilizing the gate voltage modulation instead of switching the magnetization of the FM gate, the polarization of the current can be reversed from spin-down to spin-up by electrical means.

4. Conclusion

In conclusion, the SDT is theoretically studied for the QD-DTJ systems where a QD is sandwiched by two adjacent leads. The tunneling current through the systems is shown to be rigorously derived via the Keldysh NEGF approach and EOM method. The SF events, CA, ZS and FM gating are systematically incorporated in the SDT models. Considering these effects, one may analyze the SDT properties of QD-DTJ systems, including the tunneling current (charge current and spin current), the TMR, the spectral functions and the occupancies of the QD. The SF-TJ and the SF-QD events are found to have competitive effects on the tunneling current. The presence of CA effectively modifies the threshold voltage, and gives rise to additional bias voltage dependence of the QD's electron occupancy and the charge and spin currents. The FM gate attached to the QD can be utilized to generate a bipolar spin polarization of the current through QD-DTJ systems. The above investigations done have yielded a better understanding of the SDT in QD-DTJ systems.

5. Acknowledgement

We gratefully acknowledge the SERC Grant No. 092 101 0060 (R-398-000-061-305) for financially supporting this work.

6. List of Abbreviations

CA coupling asymmetry
CB Coulomb blockade
DOS density of states
DTJ double tunnel junction
EOM equation of motion
FM ferromagnetic
GF Green's function
ME master equation
NEGF nonequilibrium Green's function
NM non-magnetic
QD quantum dot
SDT spin dependent transport
SF spin-flip
TMR tunnel magnetoresistance
ZS Zeeman splitting

7. References

Bao, Y. J., Tong, N. H., Sun, Q. F. & Shen, S. Q. (2008). Conductance plateau in quantum spin transport through an interacting quantum dot, *Europhys. Lett.* 83: 37007:1–5.

Braig, S. & Brouwer, P. W. (2005). Rate equations for coulomb blockade with ferromagnetic leads, *Phys. Rev. B* 71: 195324:1–9.

Bruus, H. & Flensberg, K. (2004). *Many-body Quantum Theory in Condensed Matter Physics*, Oxford University Press, New York.

Caroli, C., Combescot, R., Nozieres, P. & Saint-James, D. (1971). Direct calculation of the tunneling current, *J. Phys. C* 4: 916.

Deshmukh, M. M. & Ralph, D. C. (2002). Using single quantum states as spin filters to study spin polarization in ferromagnets, *Phys. Rev. Lett.* 89: 266803:1–4.

Fransson, J. & Zhu, J.-X. (2008). Spin dynamics in a tunnel junction between ferromagnet, *New J. Phys.* 10: 013017:1–9.

Hamaya, K., Kitabatake, M., Shibata, K., Jung, M., Ishida, S., Taniyama, T., Hirakawa, K., Arakawa, Y. & Machida, T. (2009). Spin-related current suppression in a semiconductor quantum dot spin-diode structure, *Phys. Rev. Lett.* 102: 236806:1–4.

Hamaya, K., Masubuchi, S., Kawamura, M., Machida, T., Jung, M., Shibata, K., Hirakawa, K., Taniyama, T., Ishida, S. & Arakawa, Y. (2007). Spin transport through a single self-assembled inas quantum dot with ferromagnetic leads, *Appl. Phys. Lett.* 90: 053108.

Hanson, R., Kouwenhoven, L. P., Petta, J. R., Tarucha, S. & Vandersypen, L. M. K. (2007). *Rev. Mod. Phys.* 79: 1217.

Jauho, A. P., Wingreen, N. S. & Meir, Y. (1994). Time-dependent transport in interacting and noninteracting resonant-tunneling systems, *Phys. Rev. B* 50: 5528–5544.

Katsura, H. (2007). Nonequilibrium kondo problem with spin-dependent chemical potentials: Exact results, *J. Phys. Soc. Jpn* 76: 054710.

Kroutvar, M., Ducommun, Y., Heiss, D., Bichler, M., Schuh, D., Abstreiter, G. & Finley, J. J. (2004). Optically programmable electron spin memory using semiconductor quantum dots, *Nature* 432: 81–84.

Kuo, D. M. T. & Chang, Y. (2007). Tunneling current spectroscopy of a nanostructure junction involving multiple energy levels, *Phys. Rev. Lett.* 99: 086803:1–4.

Lacroix, C. (1981). Density of states for the anderson model, *J. Phys. F: Met. Phys.* 11: 2389.

Lin, K.-C. & D.-S.Chuu (2005). Anderson model with spin-flip-associated tunneling, *Phys. Rev. B* 72: 125314:1–11.

Lobo, T., Figueira, M. S. & Foglio, M. E. (2006). Kondo effect in a quantum dot-the atomic approach, *Nanotechnology* 17: 6016.

Loss, D. & DiVincenzo, D. P. (1998). Quantum computation with quantum dots, *Phys. Rev. A* 57: 120–126.

Ma, M. J., Jalil, M. B. A. & Tan, S. G. (2008). Sequential tunneling through a two-level semiconductor quantum dot system coupled to magnetic leads, *J. Appl. Phys.* 104: 053902:1–6.

Ma, M. J., Jalil, M. B. A. & Tan, S. G. (2010). Coupling asymmetry effect on the coherent spin current through a ferromagnetic lead/quantum dot/ferromagnetic lead system, *IEEE Trans. Magnetics* 46: 1495.

Mahan, G. D. (1990). *Many-Particle Physics*, 2nd edn, Plenum, New York.

Manassen, Y., Mukhopadhay, I. & Rao, N. R. (2001). Electron-spin-resonance stm on iron atoms in silicon, *Phys. Rev. B* 61: 16223–16228.

Meir, Y. & Wingreen, N. S. (1992). Landauer formula for the current through an interacting electron region, *Phys. Rev. Lett.* 68: 2512–2515.

Meir, Y., Wingreen, N. S. & Lee, P. A. (1991). Transport through a strongly interacting electron system: Theory of periodic conductance oscillations, *Phys. Rev. Lett.* 66: 3048–3051.

Meir, Y., Wingreen, N. S. & Lee, P. A. (1993). Low-temperature transport through a quantum dot: The anderson model out of equilibrium, *Phys. Rev. Lett.* 70: 2601–2604.

Moodera, J. S., Santos, T. S. & Nagahama, T. (2007). The phenomena of spin-filter tunnelling, *J. Phys.: Condens. Matter* 19: 165202.

Mu, H. F., Su, G. & Zheng, Q. R. (2006). Spin current and current-induced spin transfer torque in ferromagnet-quantum dot-ferromagnet coupled systems, *Phys. Rev. B* 73: 054414:1–9.

Pasupathy, A., Bialczak, R., Martinek, J., Grose, J., Donev, L., McEuen, P. & Ralph, D. C. (2004). The kondo effect in the presence of ferromagnetism, *Science* 306: 86–89.

Petta, J. R., Johnson, A. C., Taylor, J. M., Laird, E. A., Yacoby, A., Lukin, M. D., Marcus, C. M., Hanson, M. P. & Gossard, A. C. (2005). Coherent manipulation of coupled electron spins in semiconductor quantum dots, *Science* 309: 2180–2184.

Potok, R. M., Rau, I. G., Shtrikman, H., Oreg, Y. & Goldhaber-Gordon, D. (2007). Observation of the two-channel kondo effect, *Nature* 446: 167–171.

Prinz, G. (1995). Spin-polarized transport, *Phys. Today* 48: 58.

Prinz, G. (1998). Magnetoelectronics, *Science* 282: 1660–1663.

Qi, Y., Zhu, J.-X., Zhang, S. & S.Ting, C. (2008). Kondo resonance in the presence of spin-polarized currents, *Phys. Rev. B* 78: 045305:1–5.

Qu, F. & Vasilopoulos, P. (2006). Spin transport across a quantum dot doped with a magnetic ion, *Appl. Phys. Lett.* 89: 122512:1–3.

Recher, P., Sukhorukov, E. V. & Loss, D. (2000). Quantum dot as spin filter and spin memory, *Phys. Rev. Lett.* 85: 1962–1965.

Rudziński, W. & Barnaś, J. (2001). Tunnel magnetoresistance in ferromagnetic junctions: Tunneling through a single discrete level, *Phys. Rev. B* 64: 085318:1–10.

Sergueev, N., Sun, Q. F., Guo, H., Wang, B. G. & Wang, J. (2002). Spin-polarized transport through a quantum dot: Anderson model with on-site coulomb repulsion, *Phys. Rev. B* 65: 165303.

Shaji, N., Simmons, C. B., Thalakulam, M., Klein, L. J., Qin, H., Luo, H., Savage, D. E., Lagally, M. G., Rimberg, A. J., Joynt, R., Friesen, M., Blick, R. H., Coppersmith, S. N. & Eriksson, M. A. (2008). Spin blockade and lifetime-enhanced transport in a few-electron si/sige double quantum dot, *Nature Physics* 4: 540–544.

Souza, F. M., Egues, J. C. & Jauho, A. P. (2004). Tmr effect in a fm-qd-fm system, *Brazilian Journal of Physics* 34: 565–567.

Wabnig, J. & Lovett, B. W. (2009). A quantum dot single spin meter, *New J. Phys.* 11: 043031.

Weymann, I. & Barnaś, J. (2007). Cotunneling through quantum dots coupled to magnetic leads: Zero-bias anomaly for noncollinear magnetic configurations, *Phys. Rev. B* 75: 155308:1–12.

Zhang, P., Xue, Q. K., Wang, Y. P. & Xie, X. C. (2002). Spin-dependent transport through an interacting quantum dot, *Phys. Rev. Lett.* 89: 286803:1–4.

Zhu, J.-X. & Balatsky, A. V. (2002). Quantum electronic transport through a precessing spin, *Phys. Rev. Lett.* 89: 286802:1–4.

18

The Thermopower of a Quantum Dot Coupled to Luttinger Liquid System

Kai-Hua Yang[1], Yang Chen[1], Huai-Yu Wang[2] and Yan-Ju Wu[3]
[1]*College of Applied Sciences, Beijing University of Technology*
[2]*Department of Physics, Tsinghua University*
[3]*College of Applied Sciences, Beijing University of Technology*
China

1. Introduction

With the progress in nanofabrication technique and nanometer scale materials, research on electron transport properties of mesoscopic systems has become a very active field in condensed matter physics. Considerable researches have been mainly focused on charge transport in nanostructures and nanodevices. Besides the charge transport, a detailed understanding of heat transport through mesoscopic systems is of equally importance (Afonin, 1995; Small, 2003) because they can provide additional information on the kinetics of carriers not available in the measurement of current voltage characteristics (Heremans, 2004). For instance, thermoelectric properties are very sensitive to dimensionality, the electronic spectrum near the Fermi level, scattering processes (Koch, 2004), electron-phonon coupling strength (Yang1, 2010) and electron-hole symmetry (Small, 2003). There have been several theoretical studies on the thermopower S, which mainly focused on quantum dot (QD) coupled to the normal Fermi liquid (FL) leads (Boese, 2001; Dong, 2002; Kim, 2003; Krawiec, 2007; Yang1, 2010), denoted hereafter as FL-QD-FL. As for systems containing a quantum dot coupled to one-dimensional (1D) interacting electron leads, although their charge transport phenomena have been investigated (Yang2, 2001; Yang3, 2010), yet there have been much less efforts devoted to the thermoelectric properties of them (Kane, 1996; Krive, 2001; Romanovsky, 2002). It is well known that the 1D interacting electron systems can be described by the Luttinger liquied (LL) theory (Luttinger, 1963), which holds some unique features such as spin-charge separation, suppression of the electron tunneling density of states, power-law dependence of the electrical conductance on temperature and bias voltage, etc.. The LL behaviour has been experimentally reported in single- and multi-wall carbon nanotubes (Bockrath, 1999; Kanda, 2004; Yao, 1999) and fractional quantum Hall edge states (Chang, 1996). Recently, the use of carbon nanotubes as a thermoelectric material has gained great interest due to their 1D structure. The thermopower of single-walled carbon nanotubes have been measured in experiments (Bradley, 2000; Choi, 1999; Collins, 2000; Hone, 1998; Kong, 2005; Small, 2003). For example, Kong *et al.* have shown a linear temperature dependence of the thermopower at low temperature (Kong, 2005). Small *et al.* have observed strong modulations of thermopower as the function of gate voltage Vg in individual Carbon nanotubes (Small, 2003). Dresselhaus *et al.* have found that the low-dimensional thermoelectric materials performed better than bulk ones (Dresselhaus, 1999). Several theoretical works have been developed to predict the enhancement of thermopower in nanoscaled materials by the intralead electron interaction (Kane, 1996; Krive, 2001; Krive2, 2001; Romanovsky, 2002). Krive *et al.* (Kane, 1996; Krive, 2001) used a phenomenological approach to investigate the thermopower of a LL wire containing an impurity. In spite of the above work, an explicit thermopower formula in the LL leads was not given out. In the following, we use the notation S to denote the thermopower S of systems comprosing LL and S_0 to those comprising noninteracting FL. Theoretically, the thermopower S of a LL with an impurity can be represented by the thermopower S_0 multiplied by an interaction-dependent renormalization factor. Alternatively, one may intentionally introduce a QD into the LL, denoted as LL-QD-LL. Thus, he may attach a QD, instead of an impurity atom, to the end of a carbon nanotube. A quantum dot is experimentally more controllable than an impurity. For instance, the energy of a quantum dot can be tuned by the gate voltage V_g.

In this chapter, we will first give the stationary thermopower formula of a QD coupled to LL leads through tunneling junctions, a system of LL-QD-LL (see Fig.1) by applying the nonequilibrium Green function technique (Haug, 1996) instead of phenomenal theories. And then we later turn our attention to the time-dependent phenomena. The generalized thermopower formula is obtained under time-dependent gate voltage. Although there are many studies on time-dependent nonequilibrium transport, the research on the time-dependent thermopower and the formula under the ac field are still lack. Here we will fill the blanks for the low dimension system.

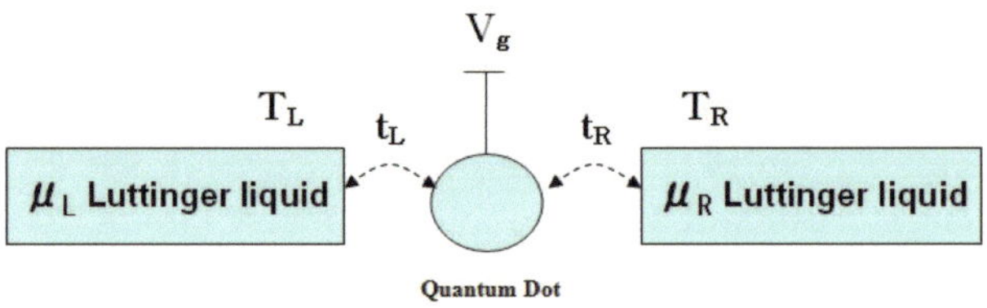

Fig. 1. The two-terminal electron transport through a single-level quantum dot weakly coupled to the Luttinger liquid leads with the chemical potentials μ_L and μ_R. Electrons tunnel from one lead to another by hopping on and off the dot level with the energy ε. The position of the dot levels with respect to the Fermi energy can be uniformly shifted by applying a voltage V_g to the gate electrode.

2. The Model

In the considered LL-QD-LL system, the QD is weakly connected with semi-infinite LL electrodes. The Hamiltonian of this system includes three parts:

$$H = H_{leads} + H_{dot} + H_T, \tag{1}$$

which represents the Hamiltonians of the left and right LL leads ($H_{leads} = H_L + H_R$), the central dot and the tunneling interactions between them H_T, respectively. Firstly, we present a detailed discussion of bosonization for a continuum model of length L with open boundary conditions (Eggert0, 1992; Fabrizio, 1995; Furusaki0, 1994) and electron-electron interaction.

The specify Hamiltonian of the LL leads can be easily written

$$H_{L/R} = H_{0\mathrm{L/R}} + H_{intL/R}, \tag{2}$$

where the first term represents the kinetic energy,

$$H_0 = \sum_{\sigma=\uparrow,\downarrow} \int_0^L dx \psi_s^\dagger(x)\varepsilon(-i\partial_x)\psi_s(x), \tag{3}$$

and the second one describes the electron-electron interaction,

$$H_{int} = \frac{1}{2}\sum_{\sigma\sigma'} \int dxdy \psi_\sigma^\dagger(x)\psi_{\sigma'}^\dagger(y)U_{\sigma\sigma'}(x-y)\psi_{\sigma'}(y)\psi_\sigma(x), \tag{4}$$

ε_k is the dispersion law of the 1D band, and $\psi_\sigma(x)$ is the spin σ electron annihilation operator subject to the open boundary conditions:

$$\psi_\sigma(0) = \psi_\sigma(L) = 0. \tag{5}$$

We apply the boundary conditions Eq.(5) to expand electron annihilation operator ψ which takes the form

$$\psi_\sigma(x) = \sqrt{\frac{2}{L}} \sum_k \sin(kx) c_{\sigma k}, \tag{6}$$

with $k = \pi n/L$, n being a positive integer. Usually a 1D system with periodic boundary conditions has two Fermi points $\pm k_F$. Here we only have single Fermi point given by $k = k_F$. The 1D fermion field ψ_σ can be expanded about the Fermi point k_F in terms of the left moving and right moving fields as

$$\psi_\sigma(x) = e^{ik_F x}\psi_{\sigma R}(x) + e^{-ik_F x}\psi_{\sigma L}(x). \tag{7}$$

In the case of periodic boundary conditions (Haldane, 1981), these left moving and right moving fields are not independent and satisfy

$$\psi_{\sigma L}(x) = -\psi_{\sigma R}(-x). \tag{8}$$

Then the fermion fields automatically satisfy the boundary conditions

$$\psi_\sigma(0) = 0, \tag{9}$$

whereas the condition

$$\psi_\sigma(L) = 0 \tag{10}$$

implies that the operator $\psi_{\sigma R}(x)$ should obey

$$\psi_{\sigma R}(-L) = \psi_{\sigma R}(L). \tag{11}$$

Therefore, we can actually work with the right moving field only, the left moving one is then defined by the above relation. Thus the field $\psi_{\sigma R}(x)$ can be defined for all x, and obeying the periodicity condition with the period $2L$:

$$\psi_{\sigma R}(x+2L)=\psi_{\sigma R}(x). \tag{12}$$

In terms of the right moving field, the kinetic energy terms in the Hamiltonian Eq.(1) takes the form

$$H_0=v_F\sum_{\sigma}\int_{-L}^{L}dx\psi_{\sigma R}^{\dagger}(x)(-i\partial_x)\psi_{\sigma R}(x) \tag{13}$$

where we have linearized the electron spectrum. The single fermion operators for right-moving electrons with spin σ on lead α can be bosonized in the position representation by applying the periodic boundary condition Eq.(12) as

$$\psi_{\sigma R}(x)\approx\frac{\eta_\sigma}{\sqrt{2L}}e^{ik_Fx}e^{i\phi_\sigma(x)}. \tag{14}$$

The operator η_σ is real fermion and satisfies the anti-commutation relations

$$\{\eta_\sigma,\eta_\sigma'\}=\delta_{\sigma\sigma'}, \tag{15}$$

with δ is Delta function. Eq.(15) assure the correct anti-commutation rules for electron operators with different σ. In order to calculate the correlation function, a method of dealing with this was suggested by Luther and Perchel (Luther, 1974). It used a limiting process, where the wave function contained s parameter α', and the limit $\alpha'\to 0$ is taken at the end of the calculation of the correlation function. Using the parameter α', we can represent the electron operator $\psi_{\sigma R}(x)$ as

$$\psi_{\sigma R}(x)\approx\lim_{\alpha'\to 0}\frac{1}{\sqrt{2\pi\alpha'}}e^{ik_Fx}e^{i\phi_\sigma(x)}. \tag{16}$$

Where α' is a short-distance cutoff of the order of the reciprocal of the Fermi wave number k_F. The phase field $\phi_\sigma(x)$ satisfies periodic boundary condition:

$$\phi_\sigma(x+2L)=\phi_\sigma(x) \tag{17}$$

and can be expressed as follows

$$\phi_\sigma(x)=\sum_{q>0}\sqrt{\frac{\pi}{qL}}e^{iqx-\alpha' q/2}a_q+H.c, \tag{18}$$

here, $a_q^{\dagger}$ and a_q are the creation and annihilation operators of bosons. These operators satisfy the canonical bosonic commutation relations $[a_q,a_{q'}^{\dagger}]=\delta(q-q')$. $q=\pi n/L$, n is an integer. The density of right moving electrons is given by

$$\rho_{\sigma R}(x)\approx\frac{\partial_x\phi_\sigma(x)}{2\pi} \tag{19}$$

Applying the boundary conditions Eq.(8), we have

$$\rho_{\sigma L}(-x)=\rho_{\sigma R}(x). \tag{20}$$

The bosonized form of the kinetic energy is

$$H_0 = \pi v_F \sum_{\sigma} \int_{-L}^{L} dx : \rho_{\sigma R}(x)\rho_{\sigma R}(x) := v_F \sum_{\sigma q>0} q b^{\dagger}_{\sigma q} b_{\sigma q}, \tag{21}$$

where colon represents the normal order form of the operators.

In order to deal with electron-electron interacting terms in Hamiltonian Eq.(1), we continue to make use of the above bosonization procedure expressing the electron interaction Hamiltonian in terms of the right moving Fermi field $\psi_{\sigma R}$ only.

Before we turn to the interaction effects, we introduce the bosonic variables corresponding to charge and spin excitations:

$$b_{\rho(\sigma)q} = \frac{1}{\sqrt{2}}(b_{\uparrow q} \pm b_{\downarrow q}) \tag{22}$$

and

$$\rho_{\rho(\sigma)} = \frac{1}{\sqrt{2}}(\rho_{\uparrow q} \pm \rho_{\downarrow q}). \tag{23}$$

The interaction part of the Hamiltonian contains several terms classified as: the diagonal terms in the electron densities and the mixing left and right densities term. Consequently, the Hamiltonian becomes

$$H = \sum_{\nu=\rho(\sigma)} \{\sum_{q>0} v^0_{\nu} q[b^{\dagger}_{\nu q} b_{\nu q} - \frac{g_{2\nu}}{4\pi}(b_{\nu q} b_{\nu q} + b^{\dagger}_{\nu q} b^{\dagger}_{\nu q})] \tag{24}$$

where

$$v_{\nu q} = v_F + \frac{g_{4\nu} + g_{2\nu}}{2\pi}, \tag{25}$$

where v_F is the Fermi velocity, g_4 and g_2 represent forward scatterings; in our work, we will not consider the backscattering interaction. In the absence of backscattering, the Luttinger Hamiltonian, H_{LL}, is exactly soluble using the technique of bosonization. In order to express the Hamiltonian in diagonal form, we introduce the canonically conjugate Boson operators, in a standard way by the Bogolubov rotation,

$$b_{\nu q} = \cosh(\varphi_\nu)\widetilde{b}_{\nu q} - \sinh(\varphi_\nu)\widetilde{b}^{\dagger}_{\nu q} \tag{26}$$

where

$$\tanh(2\varphi_\nu) = -\frac{\widetilde{g}_\nu}{2\pi v^0_\nu} \tag{27}$$

The Hamiltonian can be achieved by the canonical transformation in terms of $\widetilde{b}_{\nu q}$ and $\widetilde{b}^{\dagger}_{\nu q}$

$$\widetilde{H} = UHU^{\dagger} = \sum_{\nu q>0} v_\nu q b^{\dagger}_{\nu q} b_{\nu q}, \tag{28}$$

where

$$v_\nu = \frac{v^0_\nu}{\cosh(2\varphi_\nu)}. \tag{29}$$

The unitary operator U is defined by

$$U = exp\{\frac{1}{2}\sum_{\nu,q>0} \varphi_\nu (b^\dagger_{\nu q} b^\dagger_{\nu q} - b_{\nu q} b_{\nu q})\} \tag{30}$$

In the next step we find how the Fermi operators transform by applying U. Employing the method of Mattis and Lieb (Mattis, 1964), after lengthy but straightforward calculations, we arrive at the expression for the electron annihilation operator in terms of free bosons for the case of the interacting Fermi system with open boundaries:

$$U\psi_{\sigma R}(x,t)U^\dagger \approx \frac{\eta_\sigma}{\sqrt{2\pi\alpha'}} \exp\{i\sum_\nu \varepsilon_{\nu\sigma}[\frac{c_\nu}{\sqrt{2}}\phi_\nu(x - v_\nu t) - \frac{s_\nu}{\sqrt{2}}\phi_\nu(-x - v_\nu t)]\} \tag{31}$$

where $\varepsilon_\nu \sigma$ is $+1$ unless $\sigma =\downarrow$ and $\nu = \sigma$ when its value is -1. We have defined

$$c_\nu = \cosh(\varphi_\nu), s_\nu = \sinh(\varphi_\nu). \tag{32}$$

In the continuum limit, the Hamiltonian can be expressed

$$H_{L/R} = \hbar v_c \int_0^\infty k a_k^\dagger a_k dk. \tag{33}$$

This Hamiltonian describes the propagation of the charge density fluctuations in the leads with renormalized velocity v_c. From Krönig's relation (Krönig, 1935), the kinetic term has been written in a quadratic form of the density operators, because the bosons are defined as excitations above an N particle ground state, Hamiltonian must include terms that include the energy of the different bosonic ground states. These terms are not required for the calculations in this chapter, and are hence omitted.

The Hamiltonian of the single-level QD takes the form of

$$H_{dot} = \varepsilon d^\dagger d, \tag{34}$$

where ε is the energy of the electron on the dot, and $d^\dagger$ and d are fermionic creation and annihilation operators satisfying canonical commutation relation $\{d, d^\dagger\} = 1$.

The tunneling Hamiltonian is given by the standard expression:

$$H_T = \sum_\alpha (t_\alpha d^\dagger \psi_\alpha + h.c.), \tag{35}$$

where t_α is the electron tunneling constant and $\psi^\dagger_\alpha, \psi_\alpha$ $(\alpha = L/R)$ are the Fermi field operators at the end points of the left/right lead. The operator ψ_α could be written in a "bosonized" form (Furusaki, 1998)

$$\psi_\alpha = \sqrt{\frac{2}{\pi\alpha'}} \exp\left[\int_0^\infty dq \frac{e^{(-\alpha' q/2)}}{\sqrt{2K_{\rho\alpha} q}}(a_{q\alpha} - a^\dagger_{q\alpha}) + \sigma \int_0^\infty dq \frac{e^{(-\alpha' q/2)}}{\sqrt{2q}}(b_{q\alpha} - b^\dagger_{q\alpha})\right], \tag{36}$$

where and $K_{\rho\alpha} = e^{2\varphi_v}$ is the interaction parameter in the "fermionic" form of the LL Hamiltonian (33), which restricts the LL parameter g to vary between 0 and 1. The noninteracting case corresponds to $v_c = v_F$ and $K_{\rho\alpha} = 1$. For repulsive interactions, $K_{\rho\alpha} < 1$. Because of the SU(2) spin symmetry under no magnetic filed, $K_\sigma = 1$. Thus the correlation functions the end point of the left LL lead without the coupling to the quantum dot $\langle\psi_\sigma^\dagger(0,t)\psi_\sigma(0,0)\rangle$ can be obtained after long calculation

$$\langle\psi_\sigma^\dagger(0,t)\psi_\sigma(0,0)\rangle_L = \frac{c_A}{\alpha'}\{\frac{i\Lambda}{\pi T}\sinh[\frac{\pi T(t-i\delta)}{\hbar}]\}^{-1/g_L}. \tag{37}$$

Where c_A is a dimensionless constant of order 1, Λ is a high-energy cutoff or a band width, δ is positive infinitesimal, and $g_L^{-1} = \frac{1}{2}(1/K_{\rho L}+1)$. $\psi_\sigma(0,t) = e^{iH_L t/\hbar}\psi_\sigma(0,0)e^{-iH_L t/\hbar}$. Similarly, the correlation function at the end point of the the right lead is obtained as the above method. The electron-electron interaction parameters of the left and right LL leads are assumed equal $g_L = g_R = g$ for convenience.

3. The thermopower formula under no ac field

The charge current J_L flowing from the left lead L into the quantum dot can be evaluated as follows:

$$J_L(t) = -\frac{e}{\hbar}\left\langle\frac{d}{dt}N_L\right\rangle = \frac{ie}{\hbar}\left\langle t_L d^\dagger(t)\psi_L(t) - h.c.\right\rangle. \tag{38}$$

We introduce the time-diagonal parts of the correlation functions: $G_{dL}^<(t,t') = i\langle\psi_L^\dagger(t')d(t)\rangle$ and $G_{Ld}^<(t,t') = i\langle d^\dagger(t')\psi_L(t)\rangle$. With the help of the Langreth analytic continuation rules (Haug, 1996). By means of them, it is easy to express the current as $J_L = 2eRe(t_L^* G_{Ld}^<(t,t'))$. After applying Langreth theorem of analytic continuation, the average current can then be expressed as

$$J_L = \frac{e}{2\pi}|t_L|^2\int d\omega Re[G_d^r(\omega)g_L^<(\omega) + G_d^<(\omega)g_L^a(\omega)]. \tag{39}$$

In terms of a long derivation, we can easily establish an expression for the expectation value of the electric current

$$J_L = \frac{e}{2\pi}|t_L|^2|t_R|^2\int d\omega G_d^r G_d^a\left[g_L^<(\omega)g_R^>(\omega) - g_L^>(\omega)g_R^<(\omega)\right], \tag{40}$$

where $G_d^{r(a)}$ is retarded (advanced) Green function of the quantum dot and $\Gamma_{L/R}$, proportional to $|t_{L/R}|^2$, describes the effective level broadening of the dot. $g_\alpha^{<(>)}(\omega)$ is the Fourier transform of the lesser (greater) Green function at the end point of the left (right) LL lead without the coupling to the QD, which has been obtained by (Furusaki, 1998):

$$g_\alpha^{<,>}(\omega) = \pm i\frac{T_\alpha}{|t_\alpha|^2}\exp[\mp(\omega-\mu_\alpha)/2T_\alpha]\gamma_\alpha(\omega-\mu_\alpha), \tag{41}$$

Now we define the Luttinger liquid distribution functions $F_{L/R}^{<,>}$ as

$$F_\alpha^{<,>}(\omega) = \frac{1}{2\pi}e^{\frac{\mp(\omega-\mu_\alpha)}{2T_\alpha}}\left(\frac{\pi T_\alpha}{\Lambda}\right)^{1/g-1}\frac{|\Gamma[\frac{1}{2g}+i\frac{\omega-\mu_\alpha}{2\pi T_\alpha}]|^2}{\Gamma(1/g)} \quad (\alpha = L/R). \tag{42}$$

The function $F^{<}(\omega)$ is the electron occupation number for interacting electrons which is analogous to the Fermi distribution function $f(\omega)$ of noninteracting electrons and $F^{>}(\omega)$ is analogous to $1 - f(\omega)$ of FL leads. $T_{L/R}$ is temperature and $\mu_{L/R}$ the chemical potential of the left or right lead where $\mu_L = \mu + \eta V$ and $\mu_R = \mu + (\eta - 1)$. $\Gamma(z)$ is the Gamma function. Following the derivation in Ref. (Yang3, 2010), the current can be obtained as

$$J_L = \frac{e}{2\pi} \int d\omega T(\omega) \left[F_L^{<}(\omega) F_R^{>}(\omega) - F_L^{>}(\omega) F_R^{<}(\omega) \right], \tag{43}$$

with $T(\omega) = \Gamma_L \Gamma_R G_d^r(\omega) G_d^a(\omega)$ is the transmission probability. If $g = 1$, Eq. (43) will degrade to the usual well-known current expression for a FL-QD-FL system.

Our goal is to find the general thermopower formula of the model described by the Hamiltonian Eq. (62). The thermopower S is defined in terms of the voltage V generated across the quantum dot when temperature gradient $\Delta T = T_L - T_R$ is much less than T_L and T_R and when current J is zero (Cutler, 1969):

$$S \equiv - \lim_{\Delta T \to 0} \frac{V}{\Delta T}|_{J=0} = -\frac{1}{eT} \frac{L_{12}}{L_{11}}. \tag{44}$$

where L_{12} and L_{11} are linear response coefficients when the current J_L is presented by small bias voltages and small temperature gradients ΔT:

$$J_L = L_{11} \frac{\delta \mu}{T} + L_{12} \frac{\delta T}{T^2} = \frac{e}{2\pi} \int d\omega T(\omega) \left\{ \left[\frac{\partial F(\omega)}{\partial \mu} \right]_T \delta V + \left[\frac{\partial F(\omega)}{\partial T} \right]_\mu \delta(\Delta T) \right\}, \tag{45}$$

where $F(\omega) = F_L^{<}(\omega) F_R^{>}(\omega) - F_L^{>}(\omega) F_R^{<}(\omega)$. Comparing both sides of the Eqs.(78) (let $e = 1, \hbar = 1$), we obtain

$$L_{11} = \frac{T}{2\pi} \int d\omega T(\omega) \left[\frac{\partial F(\omega)}{\partial \mu} \right]_T, \tag{46}$$

$$L_{12} = \frac{T^2}{2\pi} \int d\omega T(\omega) \left[\frac{\partial F(\omega)}{\partial T} \right]_\mu. \tag{47}$$

The formulas Eqs.(46) and Eqs.(47) are independent of the approximation adopted in deriving the retarded (advanced) Green function. However, the partial derivatives $\frac{\partial F}{\partial \mu}$ and $\frac{\partial F}{\partial T}$ are not yet expressed evidently. In the following we will show the explicit expression for L_{11} and L_{12}. The linear expansion of the Luttinger liqiud distribution function becomes

$$F_\alpha(\omega) = F(\omega) + \frac{\partial F_\alpha(\omega)}{\partial \mu_\alpha}|_{\mu_\alpha = \mu, T_\alpha = T}(\mu_\alpha - \mu) + \frac{\partial F_\alpha(\omega)}{\partial T_\alpha}|_{\mu_\alpha = \mu, T_\alpha = T}(T_\alpha - T). \tag{48}$$

In order to achieve a compact expression, we define $F_1 = F_L^{<} F_R^{>}$ and $F_2 = F_R^{<} F_L^{>}$, and expand them to the first order derivatives:

$$F_1 = F_1(\mu, T) + \frac{\partial F_1}{\partial V}|_T \delta V + \frac{\partial F_1}{\partial(\Delta T)}|_\mu \delta(\Delta T) \tag{49}$$

and

$$F_2 = F_2(\mu, T) + \frac{\partial F_2}{\partial V}|_T \delta V + \frac{\partial F_2}{\partial(\Delta T)}|_\mu \delta(\Delta T), \tag{50}$$

where $F_1(\mu,T)$ and $F_2(\mu,T)$ are the equilibrium LL distribution functions, and $F_1(\mu,T) = F_2(\mu,T)$. Then

$$F_1 - F_2 = \frac{\partial(F_1-F_2)}{\partial V}|_T \delta V + \frac{\partial(F_1-F_2)}{\partial(\Delta T)}|_\mu \delta(\Delta T)$$
$$= \frac{\partial F}{\partial V}|_T \delta V + \frac{\partial F}{\partial(\Delta T)}|_\mu \delta(\Delta T). \tag{51}$$

Substituting of Eq.(51) into Eq.(43)enables one to obtain the expressions of $\frac{\partial F}{\partial V}$ and $\frac{\partial F}{\partial(\Delta T)}$ required in Eqs. (46) and (47). We arrive at that

$$\frac{\partial F_L^{<,>}}{\partial V} = \eta\left\{\pm\frac{1}{2T}F^{<,>} - \frac{i}{2\pi T}\Psi(\frac{1}{2g}+i\frac{\omega-\mu}{2\pi T})F^{<,>} + \frac{i}{2\pi T}\Psi(\frac{1}{2g}-i\frac{\omega-\mu}{2\pi T})F^{<,>}\right\}, \tag{52}$$

and

$$\frac{\partial F_R^{<,>}}{\partial V} = (\eta-1)\left\{\pm\frac{1}{2T}F^{<,>} - \frac{i}{2\pi T}\Psi(\frac{1}{2g}+i\frac{\omega-\mu}{2\pi T})F^{<,>} + \frac{i}{2\pi T}\Psi(\frac{1}{2g}-i\frac{\omega-\mu}{2\pi T})F^{<,>}\right\}, \tag{53}$$

In derivation we have used the relation $|\Gamma(x+iy)|^2 = \Gamma(x+iy)\Gamma(x-iy)$ and $\Gamma'(z) = \psi(z)\Gamma(z)$ with $\psi(z)$ is the Digamma function. Then substituting the Eqs.(52) and Eqs.(53) into the Eq. (51), we obtain

$$\frac{\partial F}{\partial V} = \frac{1}{T}F^{>}F^{<}. \tag{54}$$

With the same deriving process, we obtain the partial derivation with respect to temperature as

$$\frac{\partial F}{\partial(\Delta T)} = \frac{\omega-\mu}{T^2}F^{>}F^{<}. \tag{55}$$

It follows from Eqs. (54), (55), (46) and (47) that

$$L_{11} = \frac{T}{h}\int d\omega T(\omega)\frac{1}{T}F^{>}F^{<} \tag{56}$$

and

$$L_{12} = \frac{T^2}{h}\int d\omega T(\omega)\frac{\omega-\mu}{T^2}F^{>}F^{<}, \tag{57}$$

with $T(\omega)|_{\delta V=0,\delta(\Delta T)=0}$. And they become functions related to the QD density of states and LL distribution function. We stress that Eqs. (56) and (57) are the linear response coeffcients in a LL-QD-LL system. These equations will naturally degrade to those of a FL-QD-FL system if $g = 1$. The thermopower can be obtained by the equation Eq. (44) in which the current equals to zero. Substituting Eqs. (56) and (57) into Eq. (44), we have

$$S = \frac{1}{T}\frac{\int d\omega(\omega-\mu)T(\omega)F^{>}F^{<}}{\int d\omega T(\omega)F^{>}F^{<}}. \tag{58}$$

As shown in Eq. (78), when temperature difference between the leads is zero, conductance is then given by $G = e^2 L_{11}/T$. Comparison of the explicit expressions of the conductance G and thermopower S exhibits that the latter contains information different from the former.

In calculation, the Green functions of the QD are required as shown in Eq. (40). The retarded Green function is defined by $G^r(t) = -i\theta(t)\langle\{d(t), d^\dagger(0)\}\rangle$ and can be derived by means of the equation of motion method. Its analytical expression is

$$G_d^r(\omega) = \frac{1}{\omega - \varepsilon - \Sigma^r(\omega)}, \tag{59}$$

where the retarded self-energy is originated from the tunneling into the leads and is given by:

$$\Sigma^r(\omega) = -\frac{i}{2} \sum_{\alpha=L,R} \Gamma_\alpha [F_\alpha^<(\omega) + F_\alpha^>(\omega)]. \tag{60}$$

In the next section we will give our numerical results and discuss the thermoelectric properties.

4. Numerical results

The expressions (56) and (57)enable us to calculate numerically the conductance and thermopower as functions of the applied voltage and temperature. It is assumed that the system is of structural symmetric: $\Gamma_L = \Gamma_R = \Gamma$. In calculation we take the coupling strength Γ as the energy unit and set the Fermi level of the lead to be zero. Then the energy level ε of the QD represents the gate voltage Vg. No other bias is applied, *i.e.*, we always consider the zero bias case.

Figures 1(a) and (b) show the gate voltage dependence of the conductance and thermopower, respectively. The conductance varies smoothly, which is in agreement with the previous scanning gate microscopy experiments (Small2, 2003; Woodside, 2002). The thermopower S varies rapidly with the variation of the gate voltage and can reach a large absolute value at low temperature. Obviously, the gate voltage violates electron-hole asymmetry and its value tunes the thermopower. Experiments did show the features (Small, 2003; Small3, 2004).

From Fig. 1 it is seen that the conductance is an even function of ε , while the thermopower is an odd function: $S(\varepsilon) = -S(-\varepsilon)$, which is coincide to experiments (Staring, 1993). It is easily understood that the Hamiltonian in this paper possesses electron-hole symmetry: as V_g is changed to $-V_g$, the form of the Hamiltonian remains unchanged if the electrons are simultaneously converted to holes. This is the foundation of discussing the symmetry relations for the dependence of the G and S on V_g. When both bias voltage V and the current J_L change their signs, the sign of the conductance $G = dJ_L/dV$ remains unchanged. Subsequently, $G(-\varepsilon) = G(\varepsilon)$. On the other hand, the temperature difference is irrelevant to the change of the current carriers, *i.e.*, the kinetic coefficient $L_{12} = dJ_L/d(\delta T)$, changes sign: $L_{12}(\varepsilon) = -L_{12}(-\varepsilon)$ when V_g changes sign. Thus we conclude that $G(-\varepsilon) = G(\varepsilon)$. Note that G is proportional to L_{11}. Therefore, the thermopower is also an odd function $S = -\frac{1}{e^2 T}\frac{L_{12}}{L_{11}}$ of the gate voltage: $S(\varepsilon) = -S(-\varepsilon)$. The result is qualitative agreement with the experiments (Dzurak, 1997; Egger, 1997; Moller, 1998) where there was no evidence of other significant contributions of transport mechanisms, such as phonon drag, to the observed thermopower.

Comparing the curves G in Fig. 1(a) and S in Fig. 1(b), we find that the sign of S coincides with dG/dV_g. The reason of the lowering of the conductance with the energy level increasing of the quantum dot is ascribed to the change of the electron tunneling from the resonant tunneling to sequential behavior.

Now we turn to the effect of temperature on the conductance and thermopower. Figure 2 plots their curves. Figure 2(a) shows that at low temperature the conductance scales as power laws with respect to temperature, $G(T) \propto T^{\alpha}$ where $\alpha = 2/g - 2$ ($g < 1$). This functional form and the power index are in good agreement with experimental results (Bockrath, 1999; Kong, 2005; Yao, 1999). Some theoretical works with respect to impurity-contained systems gained the same conclusion (Dresselhaus, 1999; Krive, 2001; Krive2, 2001; Romanovsky, 2002). The temperature-dependent power-law scalings of conductance is associated with the suppression of tunneling to a LL in which the density of states vanishes as a power law in the energy of the tunneling electron, and the suppression becomes stronger with the decrease of g, which manifests a signature for electron-electron correlations (Harman, 2002; Kane, 1996). With increasing temperature, the mechanism of electron transport gradually turns from a resonant tunneling-like process to a sequential process. At higher temperature, the conductance is inverse to the temperature. This reflects that the effect of the electron-electron interaction on transport mechanism decays. In the temperature range between, there appears a conductance peak.

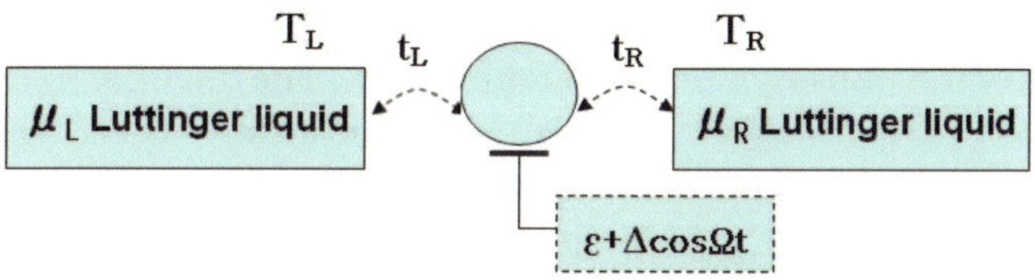

Fig. 2. The schematic picture of the two-terminal electron transport through a single-level quantum dot weakly coupled to the Luttinger liquid leads with the chemical potentials μ_L and μ_R . The position of the dot levels with respect to the Fermi energy can be uniformly shifted by applying a voltage V_g to the gate electrode.

In Fig. 2(b), the thermopower S shows linear behavior as temperature rises from zero. This is because in the low temperature regime electron tunneling transport mechanism is dominant. This behavior is the same as that of a LL containing an impurity (Krive2, 2001) and in agreement with experiments (Hone, 1998; Kong, 2005). With the electron-electron interaction enhancement, the thermopower is also increased, which has been proved in experiment (Lerner, 2008). We fit numerically the thermopower relation between the thermopower S of the LL and S_0 of Fermi liquid at low temperature with $S = (\frac{3}{2g} - \frac{1}{2})S_0(T)$ where $S_0(T) \propto T$. The electron-electron interaction in LL systems enhances and renormalizes the thermopower. In the limit of strong interaction $g << 1$, this thermopower S can be expressed as $S(T) \propto S_0(T)/g$. In this case, the thermopower of the LL is enhanced by a factor of order of magnitude of $3/(2g)$. Figure 2(b) reveals that at a fix temperature, a smaller g results in a lager S value. Hence, a larger slope of the $S - T$ curve at low temperature means a stronger interaction in LL leads. It is worth to note that the thermopower S of LL is much greater than the value S_0 of FL ($g = 1$), which reflects that the intralead electron interactions in the LL enhance the electron-hole asymmetry. With further increasing temperature a peak-like structure emerges. This is due to the mechanism at low temperature of electron transport switching from a tunnelling process to a diffusive process at high temperature. It is worth mentioning that the result is qualitatively agreement with the works (Romanovsky, 2002) with respect to an impurity in the LL lead connected to noninteracting electrons or a FL. At low temperature, a small potential barrier can strongly influence the transport properties of a LL system, so that the thermopower induced by electron backscattering dominates. This behavior is similar the case of an impurity (Romanovsky, 2002), where the impurity backscattering is considered to be a main origin of the thermopower. The impurity can also be modeled as tunnelling junctions between two decoupled semi-infinite LLs (Collins, 2000), and the tunneling junction between impurity and the LL is described by the tunneling Hamiltonian (Barnabe, 2005; Goldstein, 2010)[45-47].

At high temperature, the thermomotion of electrons become predominant and the interaction between them will be less important. Thus discrepancy between the LL and FL systems will disappear. As a result, in the high temperature limit, S becomes identical to S, as shown in Fig. 2(b). We recall that in a weak interaction system, the thermopower S_M is related to conductance G by Mott's formula (Kane, 1996):

$$S_M = -\frac{\pi^2}{3}\frac{k_B^2 T}{e}\frac{\partial lnG}{\partial \mu}, \tag{61}$$

which was originally derived for bulk systems. Note that this approximation is independent of the specific form of the transmission probability $T(\omega)$. The quantity S_M is different from S_0 of a noninteraction FL. Dependence of the zero bias conductance G on the chemical potential can be in practice measured under the variation of the gate voltage V_g. Since the gate voltage shifts the energy levels of the QD, one may assume that $\frac{\partial lnG}{\partial \mu}, \frac{\partial lnG}{\partial V_g}$. Then Eq. (26) becomes $S_M = -\frac{\pi^2}{3}\frac{k_B^2 T}{e}\frac{dlnG}{dV_g}|_{E_F}$. Figure 3 shows the variation of the thermopower S obtained from Eq. (23) and S_M from Mott relation Eq. (26) at $T = 1.0$ for four electron-electron interactions. It is seen that the relation between and G holds qualitatively for weak electron-electron interaction (Appleyard, 2000; Kane, 1996; Krive, 2001). However, even in the noninteraction case $g = 1$, there is some quantitative difference between S and S_M. The difference is obviously enhanced by the strong electron-electron interaction. Experiments (Bockrath2, 2001) evidenced the deviation from the Mott formula Eq. (26). We interpret it as a manifestation of many-body effects in the 1D electron gas. The intralead electron interactions affect the thermopower through the dependence of the transmission probability on electron-electron interaction.

From the above numerical results, we can observe that both the thermopower and conductance manifest linear temperature-dependent power-law scaling, a behavior the same as that of an impurity-contained LL system (Kane, 1996) at low temperature. The electron-electron interaction in the leads brings a significant improvement of the thermopower, a conclusion similar to that of a LL with an impurity (Krive, 2001; Krive2, 2001). As is well known, in a perfectly electron-hole symmetric system, the thermopower $S = 0$. The strong suppression of thermopower arises from the exact counteraction of the currents of electrons and holes induced by temperature gradient, which results in a zero net electric current. Only when the electron-hole symmetry is broken, the nonzero thermopower emerges. Our numerical fittings show that at low temperature the thermopower S can be expressed by the thermopower S_0 of noninteracting electrons multiplied by an

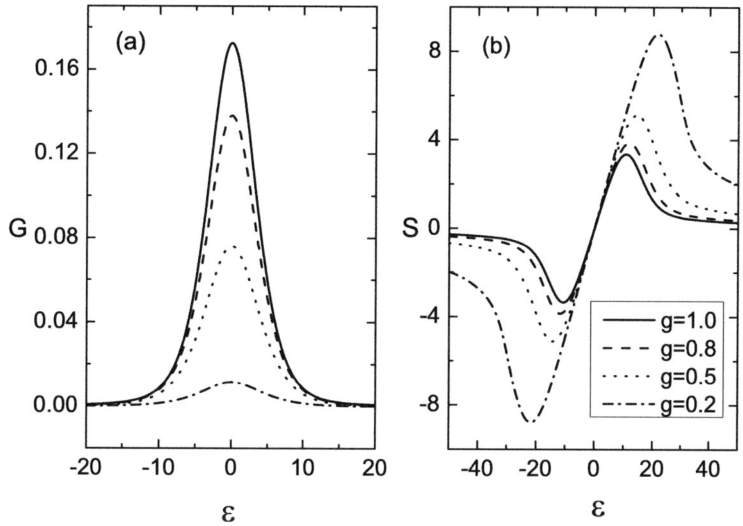

Fig. 3. The dependence of the conductance G and thermopower S on the gate voltage with $T = 1.0$ for g=0.2, 0.5, 0.8 and 1.0. The thermopower is strongly modulated by the gate voltage.

interaction factor as: $S = (3/2g - 1/2)S_0(T)$. In the limit of the strong intralead electron interaction, we have $S(T) \approx 3S_0(T)/2g$ which has an additional 3/2 factor compared to $S(T) \approx S_0(T)/g$ of the impurity-induced thermopower in 1D systems (Romanovsky, 2002). A slight deviation from the electron-hole symmetry will cause a considerable thermopower. In low-dimensional materials, the electron-hole asymmetry is usually strong and can be modulated experimentally by tuning external parameters, such as gate voltage and magnetic field. Our results reveal that how the thermopower of a LL system containing a QD is modulated by tuning the gate voltage.

5. The thermopower formula with a time-dependent gate voltage

The thermopower formula has been derived at the stationary system. Below we will derive an expression for the time-dependent thermopower for the Luttinger liquid leads connected to the central region. It is well known that more rich physics could be exploited if the QD device is subject to a microwave irradiation field. The perturbations of ac fields can give rise to some very interesting phenomena, such as photon-electron pumping effect, the turnstile effect, the sideband effect, and photon-assisted tunneling (Blick, 1995; Kouwenhoven et al., 1997; Tien, 1963). It has been reported that the microwave spectroscopy is a possible tool to probe the energy spectrum of small quantum systems (Wiel, 2002). So the photon-assisted tunneling could provide a new way of understanding the electron-electron influence on the transport properties of the dot. Indeed, the influence of the ac field on the current-voltage characteristics in the strongly correlated interaction model was discussed by some authors. The essential effect of photon-assisted tunneling on transport properties is that the electrons tunneling through the system can exchange energy with microwave fields, opening new inelastic tunneling channels and introducing many effects. The measurement of the thermopower at ac field frequency in the order of GHz regime may offer more value information on the electron interaction. However, the explicit thermopower formula under the time-dependent gate has been still lacking. Here we will fill the blanks. We start out by introducing a model for a QD coupled to the LL leads under a time-dependent gate voltage. The Hamiltonian of the system (see Fig.2) can be described as follows

$$H = H_{leads} + H_{\mathrm{D}} + H_{\mathrm{T}}. \tag{62}$$

where $H_{leads} = H_L + H_R$ represents the Hamiltonians of the left and right LL leads and its standard form is given as above, $H_{\mathrm{D}} = \varepsilon(t)d^\dagger d$ is the Hamiltonian of the QD, with $\{d^\dagger, d\}$ the creation/annihilation operators of the electron in the QD, $\varepsilon(t) = \varepsilon + \Delta cos\Omega t$, ε is the time-independent single electron energy of the QD without microwave fields, Δ and Ω are the amplitude and frequency of the ac gate voltage, respectively. It causes an alternating current through the dot. H_T is the tunneling Hamiltonian and can be written as

$$H_{\mathrm{T}} = \sum_\alpha (t_\alpha d^\dagger \psi_\alpha + h.c.)). \tag{63}$$

by applying a time-dependent canonical transformation (Bruder, 1994)(hereafter $\hbar = 1$) to Hamiltonian H_D

$$U_1(t) = \exp\left[i\int_{-\infty}^{t} dt'\Delta\cos(\Omega t')d^\dagger d\right]. \tag{64}$$

Under this transformation, we obtain $H_D'(t) = U_1(t)H(t)U_1^{-1}(t) - iU_1(t)\partial_t U_1^{-1}(t) = \varepsilon d^\dagger d$, the time-dependence of the gate voltage ε is removed. Instead, the electron tunnel coupling

$$t_\alpha(t) = t_\alpha \exp\left[-i\int_{-\infty}^{t} dt'\Delta\cos(\Omega t')\right]. \tag{65}$$

is now time-dependent. The current operator which describes tunneling from the L lead into the QD at time t is found to be: (in units of $\hbar = 1$))

$$J_L(t) = ie\left[t_L(t)\psi_L^\dagger d - t_L^*(t)d^\dagger\psi_L\right]. \tag{66}$$

Using nonequilibrium-Green-function technique and Langreth theorem of analytic continuation, the current can then be expressed as:

$$J_L(t) = -2eRe\int dt_1 t_L(t)[G^r(t,t_1)g_L^<(t_1,t) + G^<(t,t_1)g_L^a(t_1,t)]t_L^*(t_1). \tag{67}$$

where $G^r(t,t_1)$ and $G^<(t,t_1)$ are the Green's function of the QD. The retarded Green function $G^r(t,t_1)$ and lesser Green function $G^<(t,t_1)$ can be calculated from the following Dyson equation:

$$G^r(t,t_1) = g^r(t,t_1) + \int\int d\tau d\tau' g^r(t,\tau)\Sigma^r(\tau,\tau')G^r(\tau',t_1). \tag{68}$$

and the Keldysh equation

$$G^<(t,t_1) = \int\int d\tau d\tau' G^r(t,\tau)\Sigma^<(\tau,\tau')G^a(\tau',t_1). \tag{69}$$

where $g^r(t,t_1)$ is the free retarded Green function of isolated dot which depends only on the time difference $t - t_1$. $\Sigma^{r/a,<}(\tau,\tau') = \sum_{\alpha=L,R} t_\alpha^*(\tau)g_\alpha^{r/a,<}(\tau,\tau')t_\alpha(\tau')$ is the self-energy.

We now make Fourier transformation over the two times t and t' which switches from the time-domain into energy representation through a double-time Fourier-transform defined as (Wang, 1999; Xing, 2004)

$$F(\omega,\omega_1) = \int dt dt_1 F(t,t_1) e^{i\omega t} e^{-i\omega_1 t_1} \tag{70}$$

and

$$F(t,t_1) = \int \frac{d\omega}{2\pi}\frac{d\omega_1}{2\pi} F(\omega,\omega_1) e^{-i\omega t} e^{i\omega_1 t_1}. \tag{71}$$

And with the help of the above equation with respect to $\tau = (t+t_1)/2$ and let $t' = t - t_1$.

$$\langle F(t,t_1)\rangle = \lim_{T\to\infty} \frac{1}{2T}\int_{-T}^{T} F(\tau + t'/2, \tau - t'/2) d\tau, \tag{72}$$

we finally obtain following expression for Dyson equation

$$G^r(\omega) = g^r(\omega) + +g^r(\omega)\Sigma^r(\omega,\omega_1)G^r(\omega_1,\omega). \tag{73}$$

and the following expression for the lesser Green function from Eq. (69)

$$G^<(\omega) = G^r(\omega)\Sigma^<(\omega)G^a(\omega). \tag{74}$$

Where the time-average greater (lesser)self-energy $\Sigma^>(\omega)$ $(\Sigma^<(\omega))$ is which can be obtained by the time-average double-time self energy

$$\Sigma^{>,<}(\omega) = \sum_{\alpha=L/R,n} J_n^2(\frac{\Delta}{\Omega})|t_\alpha|^2 g_\alpha^{>,<}(\omega + n\Omega). \tag{75}$$

After using Langreth theorem of analytic continuation, and taking Fourier transformation over the current equation Eq. (67) and the time-averaged tunneling current can be expressed as,

$$I_L = \frac{e}{2\pi}\sum_{m,n} J_m^2 J_n^2 |t_L|^2 |t_R|^2 \int d\omega G_d^r G_d^a \left[g_{Ln}^<(\omega) g_{Rm}^>(\omega) - g_{Ln}^>(\omega) g_{Rm}^<(\omega)\right]. \tag{76}$$

with $g_{Lm/Rm}(\omega) = g_{L/R}(\omega + m\Omega)$.

As the above step, we also introduce the electron occupation number for interacting electrons $F(\omega)$, then we finally obtains the photon-assisted tunneling current

$$J = e\sum_{m,n} J_m^2 J_n^2 \int \frac{d\omega}{2\pi}\Gamma_L\Gamma_R G^r G^a (F_{Rm}^>(\omega)F_{Ln}^<(\omega) - F_{Rm}^<(\omega)F_{Ln}^>(\omega)), \tag{77}$$

where $J_m(z)$ is the mth-order Bessel function and $F_{Lm/Rm}(\omega) = F_{L/R}(\omega + m\Omega)$. The more detail derivation process of the time-dependent current formula Eq. (77) can be found in the work (Yang3, 2010).

In the next we only give the time-dependent thermopower formula using the above procedure. Under the small bias voltages and small temperature gradients and with the help of the linear expansion, we have

$$J_L = L_{11}\frac{\delta\mu}{T} + L_{12}\frac{\delta T}{T^2} = \frac{e}{2\pi}\sum_{m,n} J_m^2 J_n^2 \int d\omega T(\omega) \left\{ \left[\frac{\partial F(\omega)}{\partial\mu}\right]_T \delta V + \left[\frac{\partial F(\omega)}{\partial T}\right]_\mu \delta(\Delta T) \right\}, \quad (78)$$

where $F_{mn}(\omega) = F_{Ln}^{<}(\omega)F_{Rm}^{>}(\omega) - F_{Ln}^{>}(\omega)F_{Rm}^{<}(\omega)$. omparing both sides of the Eqs.(78) (let $e = 1, \hbar = 1$), we obtain

$$L_{11} = \frac{T}{2\pi}\sum_{m,n} J_m^2 J_n^2 \int d\omega T(\omega) \left[\frac{\partial F_{mn}(\omega)}{\partial\mu}\right]_T, \quad (79)$$

$$L_{12} = \frac{T^2}{2\pi}\sum_{m,n} J_m^2 J_n^2 \int d\omega T(\omega) \left[\frac{\partial F_{mn}(\omega)}{\partial T}\right]_\mu, \quad (80)$$

where the partial derivatives $\frac{\partial F_{mn}}{\partial\mu}$ and $\frac{\partial F_{mn}}{\partial T}$ are not yet expressed evidently. In the following we will show the explicit expression for L_{11} and L_{12}. In order to obtain L_{11} and L_{12}, we must arrive at $\frac{\partial F_{Lm}^{<,>}}{\partial V}$ and $\frac{\partial F_{Lm}^{<,>}}{\partial T}$. With the help of the linear expansion of the Luttinger liqiud distribution function and expand them to the first order derivatives, the Luttinger liqiud distribution function becomes

$$\frac{\partial F_{Lm}^{<,>}}{\partial V} = \eta \left\{ \pm\frac{1}{2T} - \frac{i}{2\pi T}\Psi_{+m} + \Psi_{-m} \right\} F_{Lm}^{<,>}, \quad (81)$$

and

$$\frac{\partial F_{Rm}^{<,>}}{\partial V} = (\eta - 1) \left\{ \pm\frac{1}{2T} - \frac{i}{2\pi T}\Psi_{+m} + \Psi_{-m} \right\} F_{Rm}^{<,>}, \quad (82)$$

where $\Psi_{\pm m} \equiv \Psi(\frac{1}{2g} \pm i\frac{\omega+m\Omega-\mu}{2\pi T})$. Substituting of Eq.(81) and Eq.(82) into Eq.(77) enables one to obtain the expressions of $\frac{\partial F_{mn}}{\partial V}$ and $\frac{\partial F_{mn}}{\partial(\Delta T)}$ required in Eqs. (79) and (80). After a long calculation using the same steps above we finally obtain

$$\begin{aligned}\frac{\partial F_{mn}}{\partial V} &= \frac{\partial[F_{Lm}^{<}F_{Rn}^{>} - F_{Lm}^{>}F_{Rn}^{<}]}{\partial V} \\ &= F_m^{<}F_n^{>}(\frac{1}{T} + \frac{i}{2\pi T}(\Psi_{-m} - \Psi_{+m} - \Psi_{-n} + \Psi_{+n})),\end{aligned} \quad (83)$$

With the same deriving process, we obtain the partial derivation with respect to temperature as

$$\begin{aligned}\frac{\partial F_{mn}}{\partial V} &= \frac{\partial[F_{Lm}^{<}F_{Rn}^{>} - F_{Lm}^{>}F_{Rn}^{<}]}{\partial T} \\ &= F_m^{<}F_n^{>}[\frac{(\omega_m + \omega_n)}{2T^2} - \frac{i\omega_m}{2\pi T^2}(\Psi_{+m} - \Psi_{-m}) + \frac{i\omega_n}{2\pi T^2}(\Psi_{+n} - \Psi_{-n})],\end{aligned} \quad (84)$$

with $\omega_m = \omega + m\Omega$ and $F_m = F(\omega + m\Omega)$. In terms of the linear expansion of the time-dependent current formula, The coefficients $L_{11/12}$ of the linear response theory can be determined from the corresponding correlation functions.

$$L_{11} = \frac{T}{h}\int d\omega T(\omega) \sum_{m,n=-\infty}^{\infty} J_m^2(\frac{\Delta}{\Omega})J_n^2(\frac{\Delta}{\Omega})F_m^< F_n^> \left[\frac{1}{T} + \frac{i}{2\pi T}(\Psi_{-m} - \Psi_{+m} - \Psi_{-n} + \Psi_{+n})\right] \tag{85}$$

and

$$L_{12} = \frac{T^2}{h}\int d\omega T(\omega) \sum_{m,n=-\infty}^{\infty} J_m^2(\frac{\Delta}{\Omega})J_n^2(\frac{\Delta}{\Omega})F_m^< F_n^> \left[\frac{(\omega_m + \omega_n)}{2T^2} - \frac{i\omega_m}{2\pi T^2}(\Psi_{+m} - \Psi_{-m}) + \frac{i\omega_n}{2\pi T^2}(\Psi_{+n} - \Psi_{-n})\right]. \tag{86}$$

From the expression of the coefficients $L_{11/12}$, we can see the coefficients $L_{11/12}$ containing a additional term caused by the time-dependent gate voltage.

The time-dependent zero bias conductance is then given by $G(0) = \frac{e^2}{T}L_{11}$, and the time-dependent thermopower can be obtained from the ratio between voltage gradient ΔV and and temperature gradient ΔT between the two reservoirs, when both left and right time-dependent electric currents cancel

$$S = -\frac{\Delta V}{\Delta T}|_{<I(t)=0>}. \tag{87}$$

Thus the conductance and thermopower take the form

$$G = \frac{1}{h}\int d\omega T(\omega) \sum_{m,n=-\infty}^{\infty} J_m^2(\frac{\Delta}{\Omega})J_n^2(\frac{\Delta}{\Omega})F_m^< F_n^> \left[\frac{1}{T} + \frac{i}{2\pi T}(\Psi_{-m} - \Psi_{+m} - \Psi_{-n} + \Psi_{+n})\right], \tag{88}$$

and

$$S = \frac{\int d\omega T(\omega) \sum_{m,n=-\infty}^{\infty} J_m^2(\frac{\Delta}{\Omega})J_n^2(\frac{\Delta}{\Omega})F_m^< F_n^> (\frac{1}{T} + \frac{i}{2\pi T}\psi_{mn})}{T^2 \int d\omega T(\omega) \sum_{m,n=-\infty}^{\infty} J_m^2(\frac{\Delta}{\Omega})J_n^2(\frac{\Delta}{\Omega})F_m^< F_n^> [\frac{(\omega_m+\omega_n)}{2T^2} - \frac{i\omega_m}{2\pi T^2}\psi_{\pm m} + \frac{i\omega_n}{2\pi T^2}\psi_{\pm n}]}, \tag{89}$$

where $\psi_{mn} = \Psi_{-m} - \Psi_{+m} - \Psi_{-n} + \Psi_{+n}$, $\psi_{\pm m} = \Psi_{+m} - \Psi_{-m}$ and $\psi_{\pm n} = \Psi_{+n} - \Psi_{-n}$. When no ac filed, the above formula return to the equation (58). This formula describes the time-averaged thermopower through the LL-QD-LL system in the presence of ac fields which contains more information than the equation (58). The numerical results of the time-dependent thermopower will be published in the future.

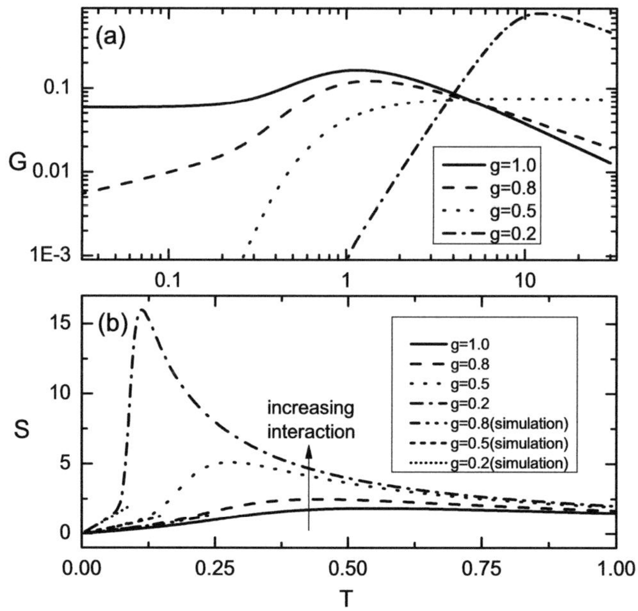

Fig. 4. The conductance (a) and thermopower S (b) as a function of temperature with $\varepsilon = 2.0$ for g=0.2, 0.5, 0.8 and 1.0. At low temperature, the conductance exhibits a power-law dependence of the temperature and the thermopower manifests the linear and positive temperature dependence, respectively. The interaction factor g can be inferred from the slopes of thermopower. With the enhancement of the electron-electron, the thermopower is increased.

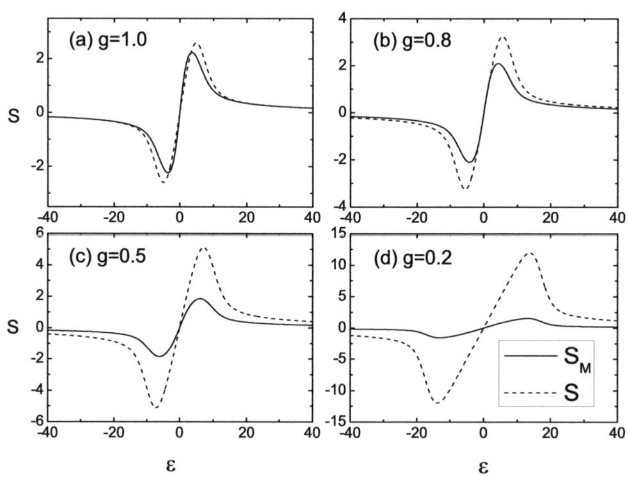

Fig. 5. Thermopowers S calculated by Eq. (23) (dash lines) and SM calculated by Mott relation Eq. (26) (solid lines) at T=1.0 for (a) g=1.0, (b) g=0.8, (c) g=0.5 and (d) g=0.2.

6. Acknowledgment

This work is supported by the Beijing Novel Program 2005B11, by Beijing Natural Science Foundation (1112003) and Fund of Beijing Municipal Education Commission.

7. References

Afonin V.V. & Rudin A.M. ,(1995).Thermoelectric resonant transport through the Anderson impurities. *Phys. Rev. B*, Vol. 51, No.24, June 1995, 18025–18028.

Appleyard N. J., Nicholls J. T., Pepper M., Tribe W. R., Simmons M. Y.& Ritchie D. A., (2000).Direction-resolved transport and possible many-body effects in one-dimensional thermopower. *Phys. Rev. B*, vol. 62, December 2000, 16275(R)–16278(R).

Barnabé-Théiault X. , Sedeki A., Meden V. & Schóhammer K.,(2005).Junctions of one-dimensional quantum wires: Correlation effects in transport.*Phys. Rev. B*, vol. 71,May 2005, 205327-1–205327-13.

Blick R.H., et al.,(1995).Photon-assisted tunneling through a quantum dot at high microwave frequencies. *Appl. Phys. Lett.* vol. 67,October 1995,3924-3926.

Bockrath M.,Cobden D. H.,Lu J.,Rinzler A. G.,Smalley R. E.,Balents L. & McEuen P. L., (1999).Luttinger-liquid behaviour in carbon nanotubes.*Nature* vol. 397, 598-601.

Bockrath M.,Liang W.,Bozovic D., Hafner J. H., Lieber C.M., Tinkham M.& Park H., (2001). Resonant Electron Scattering by Defects in Single-Walled Carbon Nanotubes.*Science* vol. 291, 283-285.

Boese D. & Fazio R.,(2001). Thermoelectric effects in Kondo-correlated quantum dots. *Europhys. Lett.* vol. 56, November 2001, 576–582.

Bradley K., Jhi S.-H., Collins P.G., Hone J., Cohen M.L., Louie S.G.& Zettl A.,(2000).Is the Intrinsic Thermoelectric Power of Carbon Nanotubes Positive? *Phys. Rev. Lett.* vol. 85, November 2000, 4361-4364.

Bruder C. & Schoeller H.,(1994).Charging effects in ultrasmall quantum dots in the presence of time-varing fields. *Phys. Rev. Lett.* vol. 72,February 1994,1076-1079.

Chang A. M., Pfeiffer L. N. & West K. W. (1996).Observation of Chiral Luttinger Behavior in Electron Tunneling into Fractional Quantum Hall Edges.*Phys. Rev. Lett.* vol. 77, September 1996, 2538–2541.

Choi E.S., Suh D.S., Kim G.T., Kim D.C., Park Y.W., Liu K., Duesberg G.& Roth S.,(1999)Magnetothermopower of singlewall carbon nanotubenewtwork . *Synth. Met.* vol. 103, 2504-2505.

Collins P.G., Bradley K., Ishigami M. & Zettl A., (2000).Extreme Oxygen Sensitivity of Electronic Properties of Carbon Nanotubes *Science* vol. 287, 1801-1804.

Cutler M. & Mott N. F.,(1969). *Phys. Rev.*, vol. 181, 1336.

Dong B. & Lei X.L. ,(2002). Effect of the Kondo correlation on the thermopower in a quantum dot. *J. Phys.: Condens. Matter* vol. 14,November 2002, 11747-11756.

Dresselhaus MS, Dresselhaus G, Sun X, Zhang Z, Cronin SB& Koga T., (1999).Low-dimensional thermoelectric materials. *Phys Solid State* vol. 41,May 1999,679-682.

Dzurak A. S., Smith C. G., Barnes C. H. W., Pepper M., Martin-Moreno L., Liang C. T., Ritchie D. A. & Jones G. A. C., (1997).Thermoelectric signature of the excitation spectrum of a quantum dot.*Phys. Rev. B* vol. 55, April 1997, 10197(R)-10200(R).

Egger R., Gogolin A.,(1997). Effective Low-Energy Theory for Correlated Carbon Nanotubes. *Phys. Rev. Lett.* vol. 79,December 1997, 5082-5085.

Eggert S. & Affleck I.,(1992).Magnetic impurities in half-integer-spin Heisenberg antiferromagnetic chains. *Phys. Rev. B*,vol. 46,May 1992, 10866-10883; Eggert S., Johannesson H., & Mattsson A.,(1996).Boundary Effects on Spectral Properties of Interacting Electrons in One Dimension. *Phys. Rev. Lett.*, vol. 76,February 1996, 1505-1508.

Fabrizio M. & Gogolin A. O.,(1995).Interacting one-dimensional electron gas with open boundaries. *Phys. Rev. B* vol. 51,June 1995, 17827-17841.

Furusaki A., (1998). Resonant tunneling through a quantum dot weakly coupled to quantum wires or quantum Hall edge states.*Phys. Rev. B* vol. 57,March 1998, 7141-7148.

Furusaki A. & Nagaosa N.,(1994),Kondo effect in a Tomonaga-Luttinger liquid. *Phys. Rev. Lett.* 72,February 1994, 892-895.

Goldstein M. & Berkovits, (2010).Density of states of a dissipative quantum dot coupled to a quantum wire. *Phys. Rev. B* vol. 82, December 2010,235315-1–235315-9; Goldstein M., Gefen Y., & Berkovits R.,(2011). Entanglement entropy and quantum phase transitions in quantum dots coupled to Luttinger liquid wires. *Phys. Rev. B*, vol. 83, June 2011, 245112-1–245112-11.

Haldane F.D.,(1981).'Luttinger liquid theory' of one-dimensional quantum fluids. I. Properties of the Luttinger model and their extension to the general 1D interacting spinless Fermi gas. *J. Phys. C* vol. 14,July 1981, 2585-2610.

Harman T.C., Taylor P. J., Walsh M. P.& LaForge B. E.,(2002).Quantum Dot Superlattice Thermoelectric Materials and Devices. *Science* vol. 297, 2229–2232 .

Haug H., Jauho A.-P., (1996)*Quantum Kinetics in Transport and Optics of Semiconductors*, (Springer-Verlag).

Heremans J.P. ,Thrush C.M. & D.T. Morelli,(2004).Thermopower enhancement in lead telluride nanostructures . *Phys. Rev. B* ,Vol. 70, September 2004, 115334-1–115334-5, and references cited therein.

Hone J., Ellwood I.,Muno M., AriMizel,Cohen M.L., Zettl A., Rinzler A.G. & Smalley R.E. , Thermoelectric Power of Single-Walled Carbon Nanotubes.(1998). *Phys. Rev. Lett.* vol. 80, February 1998, 1042-1045.

Kanda A, Tsukagoshi K, Aoyagi Y & Ootuka Y (2004). Gate-Voltage Dependence of Zero-Bias Anomalies in Multiwall Carbon Nanotubes. *Phys. Rev. Lett.* vol. 92,January 2004. 036801-1–036801-4.

Kane C.L. & Fisher M.P.A.,(1996).Thermal Transport in a Luttinger Liquid . *Phys. Rev. Lett.* vol. 76, September 1995, 3192–3195.

Kim T.-S. & Hershfield S., (2003). Thermoelectric effects of an Aharonov-Bohm interferometer with an embedded quantum dot in the Kondo regime . *Phys. Rev. B* vol. 67,April 2003, 165313-1–165313-15.

Koch J., Oppen F. von ,Oreg Y., & Eran Sela,(2004). Thermopower of single-molecule devices.*Phys. Rev. B* ,Vol.70, November 2004, 195107-1–195107-12.

Kong W.J., Lu L., Zhu H.W., Wei B.Q.& Wu D.H.,(2005). *J. Phys.: Condens. Matter* vol. 17,March 2005, 1923-1928.

Kouwenhoven L.P., Schön G. & Soun L.L.,(1997). Mesoscopic Electron Transport. Kluwer Academic Publisher,ISBN 0-7923-4737-4, The Netherlands.

Krawiec M. & Wysokinski K.I.,(2007). Thermoelectric phenomena in a quantum dot asymmetrically coupled to external leads. *Phys. Rev. B* vol.75, April 2007, 155330-1–155330-6;Krawiec M. & Wysokinski K.I.,(2006) Thermoelectric effects in strongly interacting quantum dot coupled to ferromagnetic leads. *Phys. Rev. B* vol. 73, Fermber 2006, 075307-1–075307-7; *Physica B* vol. 378, 2006, 933–934.

Krive I.V. ,Bogachek E.N.,Scherbakov A.G. & Landman Uzi,(2001). Interaction enhanced thermopower in a Luttinger liquid. *Phys. Rev. B* vol. 63, February 2001, 113101-1–113101-4čż Romanovsky I.A. ,Krive I.V. ,Bogachek E.N. & Landman U.,(2002).Thermopower of an infinite Luttinger liquid.*Phys. Rev. B* vol. 65, Fermber 2002, 075115-1–075115-9.

Krive I.V. , Romanovsky I.A., Bogachek E.N., Scherbakov A.G. & Landman U.,(2001). Thermoelectric effects in a Luttinger liquid. *Low Temp. Phys.* vol. 27, Octember 2001, 821-830.

Krönig R. de L.,(1935).Zur neutrinotheorie des lichtes III. *Physica* 2,August 1935, 968-980.

Lerner I. V., Yudson V. I.,& Yurkevich I. V.,(2008).Quantum Wire Hybridized With a Single-Level Impurity. *Phys. Rev.Lett.* vol. 100,June 2008, 256805-1–256805-4 .

Luther A.& Peschel I.,(1974).Single-particle states, Kohn anomaly, and pairing fluctuations in one dimension. *Phys. Rev. B*, vol. 9, April 1974, 2911-2919.

Luttinger J. M.,(1963).An Exactly Soluble Model of a Many-Fermion System. *J. Math. Phys.*, vol. 4, September 1963, 1154-1162.

Mattis D.C. & Lieb E.H.,(1965).Exact Solution of a Many-Fermion System and Its Associated Boson Field. *J. Math. Phys.* vol. 6,September 1964,February,304-312.

Möller S., Buhmann H., Godijn S. F.& Molenkamp L. W.,(1998). Charging Energy of a Chaotic Quantum Dot. *Phys. Rev. Lett.* vol. 81,June 1998, 5197-5200.

Romanovsky I.A. ,Krive I.V.,Bogachek E.N. & Landman U. ,(2002).Thermopower of an infinite Luttinger liquid.*Phys. Rev. B* vol. 65, Fermber 2002, 075115-1–075115-9.

Small J. P. , Perez K. M. & Kim P.,(2003).Modulation of Thermoelectric Power of Individual Carbon Nanotubes. *Phys. Rev. Lett.*,Vol. 91,No.25,December 2003, 256801-1–256801-4.

Small J.P., Shi L.& Kim P., (2003).Mesoscopic thermal and thermoelectric measurements of individual carbon nanotubes. *Solid State Commun* vol. 127,March 2003, 181-186.

Small J. P.& Kim P.,(2004). THERMOPOWER MEASUREMENT OF INDIVIDUAL SINGLE WALLED CARBON NANOTUBES. *Microscale Thermophysical Engineering* .vol.8,July 2003, 1–5.

Staring A.A.M., Molenkamp L.W., Alphenhaar B.W., Houten H. van, Buyk O.J.A., Mabesoone M.A.A., Beenakker C.W.J. & Foxon C.T.,(1993).Coulomb-Blockade Oscillations in the Thermopower of a Quantum Dot. *Europhys. Lett.* vol. 22,April 1993, 57-62.

Tien P.K. & Gordon J.P.,(1963).Multiphoton Process Observed in the Interaction of Microwave Fields with the Tunneling between Superconductor Films. *Phys. Rev.* vol. 129, January 1963, 647-677.

Wang B., Wang J.& Guo H.,(1999).Current Partition: A Nonequilibrium Green's Function Approach. *Phys. Rev. Lett.* vol. 82, January 1999, 398-401.

Wiel W.G. van der, Franceschi S. De, Elzerman J.M., Fujisawa T., Tarucha S. & Kouwenhoven L.P.,(2003).Electron transport through double quantum dots. *Rev. Mod. Phys.*, vol. 75,December 2002,1-22.

Woodside M.T.& McEuen P. L., (2002).Scanned Probe Imaging of Single-Electron Charge States in Nanotube Quantum Dots. *Science* vol. 296, 1098–1101.

Xing Y.X., Wang B., Wei Y.D., Wang B.G. & Wang J.,(2004).Spin pump in the presence of a superconducting lead. *Phys. Rev. B* vol. 70,December 2004, 245324-1–245324-8.

Yang K.-H. ,Wu Y.-P. & Zhao Y.L. ,(2010). The shot noise in a vibrating molecular dot in the Kondo regime. *Europhys. Lett.*, vol. 89, February 2010, 37008-1–37008-6.

Yang Yi-feng & Lin Tsung-han ,(2001)Nonequilibrium transport through a quantum dot weakly coupled to Luttinger liquids. *Phys. Rev. B* vol. 64, Nomber 2001,233314-1–233314-4.

Yang K.-H.,Wu Y.-P. & Zhao Y.-L. ,(2010).The shot noise of the quantum dot weakly coupled to Luttinger liquid. *Phys. Lett. A* vol. 374, December,2009, 917–922; Yang K.-H.,Wu Y.-J.,Wu Y.-P.& Zhao Y.-L.,(2011).Photon-assisted shot noise through a quantum dot coupled to Luttinger liquid . *Phys. Lett. A* vol.375,December 2010, 747–755; Yang K.-H. & Zhao Y.-L., (2010).Phonon-assisted shot noise through a quantum dot weakly coupled to Luttinger liquid. *Physica E* vol. 42,May 2010, 2324–2330.

Yao Z.,Postma H. W. Ch. ,Balents L., & Dekker C. ,(1999). Carbon nanotube intramolecular junctions.*Nature* vol. 402, 273-276.